Contents

Evolware

Vision

Speech/Hearing

Signal/Image Processing and Recognition

Parallel GAs

Combinatorial Optimisation

Scheduling/Timetabling

Telecommunications — General

Telecommunications — FAP

Applications — General Heuristics

Evolutionary ANNs I — RBFs

Evolutionary ANNs II

Evolutionary ANNs III

Reinforcement Learning

Genetic Programming

George D. Smith
Nigel C. Steele
Rudolf F. Albrecht

Artificial Neural Nets and Genetic Algorithms

Proceedings of the International Conference in Norwich, U.K., 1997

Springer-Verlag Wien GmbH

Dr. George D. Smith
School of Information Systems
University of East Anglia, Norwich, U.K.

Dr. Nigel C. Steele
Division of Mathematics
School of Mathematical and Information Sciences
Coventry University, Coventry, U.K.

Dr. Rudolf F. Albrecht
Institut für Informatik
Universität Innsbruck, Innsbruck, Austria

Originally published by Springer-Verlag Wien 1998

Camera-ready copies provided by authors and editors

Graphic design: Ecke Bonk
Printed on acid-free and chlorine-free bleached paper
SPIN 10635776

With 384 Figures

ISBN 978-3-211-83087-1 ISBN 978-3-7091-6492-1 (eBook)
DOI 10.1007/978-3-7091-6492-1

Preface

This is the third in a series of conferences devoted primarily to the theory and applications of artificial neural networks and genetic algorithms. The first such event was held in Innsbruck, Austria, in April 1993, the second in Ales, France, in April 1995. We are pleased to host the 1997 event in the mediaeval city of Norwich, England, and to carry on the fine tradition set by its predecessors of providing a relaxed and stimulating environment for both established and emerging researchers working in these and other, related fields.

This series of conferences is unique in recognising the relation between the two main themes of artificial neural networks and genetic algorithms, each having its origin in a natural process fundamental to life on earth, and each now well established as a paradigm fundamental to continuing technological development through the solution of complex, industrial, commercial and financial problems. This is well illustrated in this volume by the numerous applications of both paradigms to new and challenging problems.

The third key theme of the series, therefore, is the integration of both technologies, either through the use of the genetic algorithm to construct the most effective network architecture for the problem in hand, or, more recently, the use of neural networks as approximate fitness functions for a genetic algorithm searching for good solutions in an 'incomplete' solution space, i.e. one for which the fitness is not easily established for every possible solution instance.

Turning to the contributions, of particular interest is the number of contributions devoted to the development of 'modular' neural networks, where a divide and conquer approach is adopted and each module is trained to solve a part of the problem. Contributions also abound in the field of robotics and, in particular, evolutionary robotics, in which the controllers are adapted through the use of some evolutionary process. This latter field also provided a forum for contributions using other related technologies, such as fuzzy logic and reinforcement learning.

Furthermore, we note the relatively large number of contributions in telecommunications related research, confirming the rapid growth in this industry and the associated emergence of difficult optimisation problems. The increasing complexity of problems in this and other areas has prompted researchers to harness the power of other heuristic techniques, such as simulated annealing and tabu search, either in their 'pure' form or as hybrids. The contributions in this volume reflect this trend. Finally, we are also pleased to continue to provide a forum for contributions in the burgeoning and exciting field of evolutionary hardware.

We would like to take this opportunity to express our gratitude to everyone who contributed in any way to the completion of this volume. In particular, we thank the members of the Programme Committee for reviewing the submissions and making the final decisions on the acceptance of papers, Romek Szczesniak (University of East Anglia) for his unenvious task of preparing the LaTeX source file, Silvia Shilgerius (Springer-Verlag) for the final stages of the publication process and, not least, to all researchers for their submissions to ICANNGA97.

We hope that you enjoy and are inspired by the papers contained in this volume.

George D. Smith	Nigel C. Steele	Rudolf F. Albrecht
Norwich	Coventry	Innsbruck

Classification

Intelligent Data Analysis/Evolution Strategies

Coevolution and Control

Process Control/Modelling

LCS/Prisoner's Dilemma

ICANNGA 97
International Conference on Artificial Neural Networks and Genetic Algorithms
Norwich, UK, April 2 – 4, 1997

International Advisory Committee

Programme Committee

Obstacle Identification by an Ultrasound Sensor Using Neural Networks

D. Diep[1], A. Johannet[1], P. Bonnefoy[2] and F. Harroy[2]
[1] LGI2P - EMA/EERIE, Parc Scientifique G. Besse, 30000 Nîmes, FRANCE.
[2] IMRA Europe, 220 rue Albert Caquot, 06904 Sophia Antipolis, FRANCE
Email: diep@eerie.fr

Abstract

This paper presents a method for obstacle recognition to be used by a mobile robot. Data are made of range measurements issued from a phased array ultrasonic sensor, characterized by a narrow beam width and an electronically controlled scan. Different methods are proposed: a simulation study using a neural network, and a signal analysis using an image representation. Finally, a solution combining both approaches has been validated.

1 Introduction

The development of an autonomous mobile robot is still a difficult task. Generally three types of problems are studied: the first deals with locomotion (stability, efficiency) the second deals with reflex actions (obstacle avoidance) and the third with navigation in order to reach a goal. The major difficulties encountered in such a task is the extreme variability of the environment with which the robot interacts, and the noise inherent in the real world. Obviously nobody tries to develop a robot able to evolve in all types of environment but the variability intrinsic to even a specific type of environment is sufficient to lead to a relative failure of the traditional methods of modelling [1]. In this context, the neural networks approach appears to be an alternative solution in which the robot learns to adapt to the environment rather than learns all the reactions to each possible event. Within the wide field of research dealing with the development of mobile robots, starting from works centred on obstacle avoidance [9], this study focuses on the neural identification of obstacles using an original ultrasound sensor.

2 The Ultrasonic Sensor

Ultrasound sensors are usually used as proximity sensors, but they lack bearing directivity which generally prevents us from obtaining any accurate information. In order to reduce this drawback we have proposed an original sensor including several individual ultrasound emitter-receivers [3,4]. The ultrasonic sensor concerned consists of an array of 7 transmitters simultaneously emitting acoustic waves at the frequency of 40 kHz (Figure 1).

The phase of each emitter can be adjusted individually, so that the beam width of the resultant wave will have a restricted size, and its bearing direction may be fixed (Figure 2).

Echoes coming from reflectors are detected by two receivers, and the reflectors' range and orientation can be determined by measuring the time of flight, i.e. the time duration between the transmission and the reception of a signal. The sensor is thus analogous to a sonar system, upon whose main principles the ultrasound system was developed.

3 Simulation Study

The first part of the work consists of modelling the sensor and the echoes in order to find out by simula-

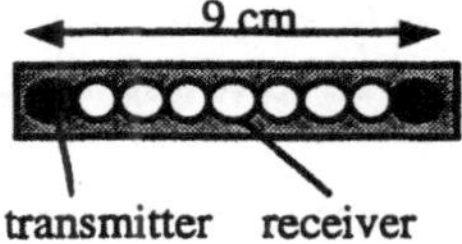

Figure 1: Configuration of the transducers.

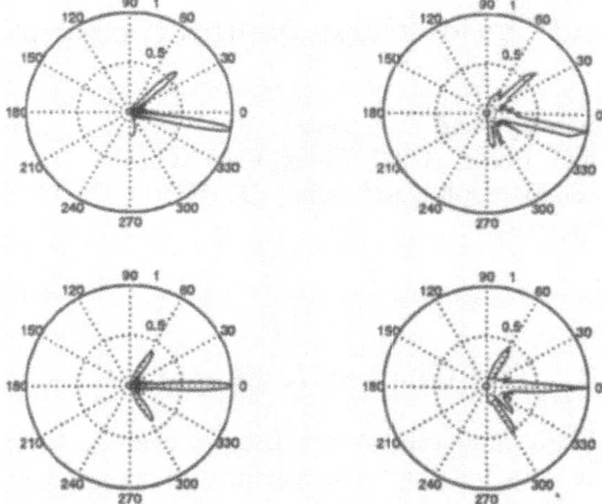

Figure 2: Directivity diagram for a transmission at $-10°$ and $0°$: (a) theoretical, (b) experimental.

Figure 3: Simulated situations for a mobile robot.

tion the best way to identify simple obstacles such as walls, doors and pillars. Assuming that the distance between the obstacle and the sensor can be computed from the time of flight, a multilayer network was used in order to classify the obstacles used. The inputs which seem to be relevant are the distance between the obstacle and the sensor for 9 emission directions in front of the robot, stepping from $-32°$ to $+32°$.

Data collected were issued from a software program simulating the dynamical behavior of a mobile robot equipped with the ultrasonic sensor [7]. Figure 3 shows different situations encountered by the robot when moving along in a room.

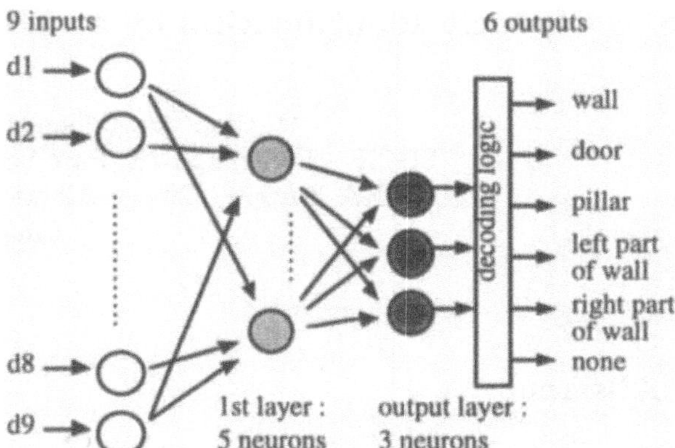

Figure 4: Architecture of the network.

The learning was performed with a hundred examples by standard backpropagation in order to classify 6 types of obstacles including the particular scene where there is no obstacle. Inputs called *d*1 to *d*9 on Figure 4 were the distances measured along each direction of transmission.

The results obtained were quite good with 92% well classified and 3% of error evaluated on a test set [2]. Nevertheless, this simulation allowed us to demonstrate one principal limitation: the problem of the apparent size of the obstacle, which increases when the obstacle is nearer to the sensor. This problem cannot be solved by the neural net and has to be treated beforehand. Secondly when we tried to compare the results obtained with the true signals, it appeared that it was not possible to compute the distance between the obstacle and the sensor in the case of a large angle of bearing, without additional information on the amplitude of signals. In conclusion, in spite of the good results, the modelling approach of this first treatment was not sufficiently realistic to be applied to a real concrete case.

4 Signal Analysis

The second study we carried out took into account the problems listed above and had two goals: firstly to estimate the distance to the obstacle and secondly to find a way to characterise a type of obstacle independently of its distance. For this, a method based on a signal modeling method used in [6] was

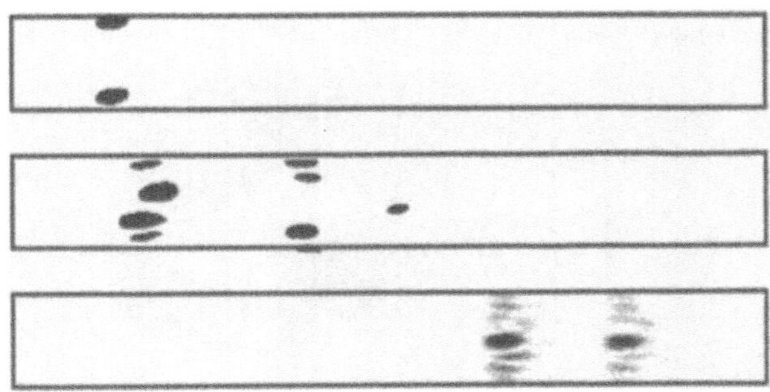

Figure 5: Images from walls, corners and edges.

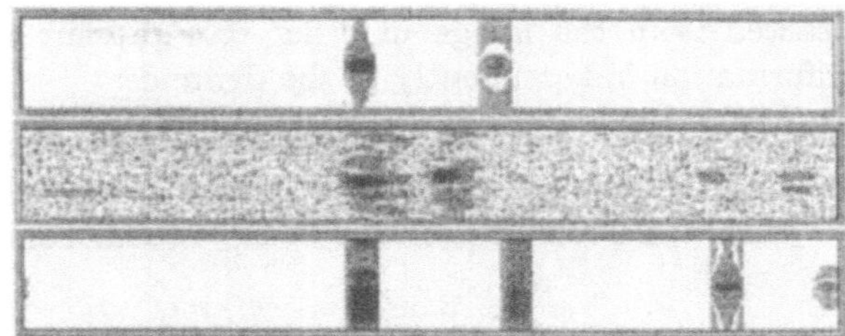

Figure 6: Simulated image of a corner, original image, simulated image of a wall /edge.

employed: first the distance is estimated including all the angular reflections, afterwards the signal is compared to a simulated reference signal computed from the previously estimated range [8]. In practice, the array of transmitters was programmed to make an acquisition at each degree between $-30°$ and $+30°$ for 512 samples (the acquisition for each direction was done at 50 kHz, so 512 samples gave a visibility window of 1.8 m). All the values collected were gathered together to form an image of 61×512 pixels. (Figure 5).

According to the nature, the orientation, and the distance of the obstacle, the images are very different, be it for the number of echoes or for their position. Furthermore, each type of obstacle studied does not always give the same response, depending on its orientation and its distance. Then, these 'images' were analysed in order to extract some kind of constant pattern for each obstacle. Then, for a few simple obstacles (wall, corner, edge as classified in [6]) the reflection pattern could be easily explained depending on the height of the sensor and the distance between the sensor and the obstacle. Based on this analysis, a simulation generates an artificial reflection image for each type of obstacle, which is then compared to the real image (Figure 6).

Operating on the real image, the mean amplitude of each of the 512 vectors is computed (mean amplitude versus distance). Hence, the darkest echo on an image corresponds to the minimum of this mean amplitude, which gives the distance between the obstacle and the sensor. A similar operation is performed for the angle to obtain the direction of the obstacle. Once the distance and the angle have been found, the recognition is performed by making a comparison between the real image and the simulated image for the three types of obstacles considered. A series of 26 measurements was performed in a room, the sensor being located at various distances and orientation angles from the obstacles. In all cases, the distance to the obstacle was accurately estimated by the sensor with a margin of error less than 1 cm. Among the different kinds of obstacles, 21 shapes (i.e. 81% of the total number) were correctly recognised. The estimation of the angle was correct for 18 obstacles (69%). In some cases, the values found by this method were incorrect, so two ways were used to empirically improve the performances: the first was based on the comparison of the values found for the two channels (one for a left sensor, the other for the right sensor), and the second calculates the disparity in the distance for the two channels to find the angle.

5 Recognition with Neural Network

The logical follow-up to the previous study was to integrate neural networks in order to: first implement the computation of various thresholds intervening during the recognition process, and second to enable adaptations to various wall coverings. The problem was the following: starting from the previously described images (61×512), we want to classify the scene viewed by a robot in three categories: wall, edge or corner. Using the estimation of the distance D between the sensor and the obstacle described above, and assuming in the case of a corner that the sensor is located roughly at the same distance from both walls, several features were extracted from the image in order to represent the information independently of the distance:

- energy (i.e. the integral value) of the first peek (i.e. the first echo received) located at the distance D, which is in any case issued from a wall,
- energy at the distance $\sqrt{2}D$ (location of a possible corner).
- energy at the distance $\sqrt{D^2 + H^2}$, where H is the height of the sensor above the floor level (echo reflecting from the ground at the foot of a wall).
- energy at the distance $\sqrt{2}\sqrt{D^2 + H^2}$ (echo reflecting from the ground at the foot of a corner).

These characteristics, called E_1, E_2, E_3, E_4, plus the estimated distance D for each ultrasonic receiver (right and left) led to a total amount of 10 inputs for the network (Figure 8).

A first study showed that, with the chosen coding, the classes (walls, corners, edges) were not linearly separable, so a multilayer neural network was necessary. Nevertheless, because of the well known problems of convergence inherent in the use of the backpropagation learning rule, we begin with a simpler network where the learning operates only on the first layer, whereas the second layer computes logical combinations. This type of network had been used for the recognition of zip code [5] and gave in this case very surprising and satisfactory results.

The principle of the method is the following: we consider that the classes to separate are non linearly separable one from all the others, but their representation is good, and the classes are linearly separable one class from another one. Then it is possible to compute the separation with several straight lines rather than one more complicated curve. This type of configuration can be illustrated in a smaller dimension with only two inputs in Figure 7.

The learning is performed on the first layer of the network: each neuron defines a straight line which separates one class from another using a simple learning rule (such as perceptron learning rule). For example the line S_1 in Figure 7 separates the class of 'Corners' from the class of 'Edges'. The final interpretation is computed by a logical function:

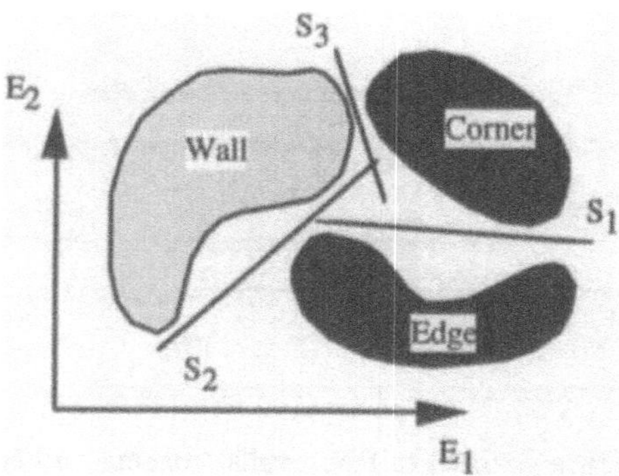

Figure 7: Example of classification with combination of straight lines. The classes are separated one from another because the separation of one class from all classes is not possible using straight lines.

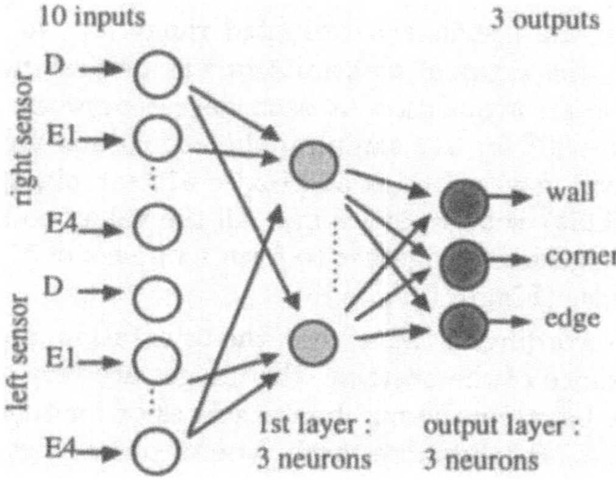

Figure 8: Architecture of the network.

for example in the Figure 7, the class of 'corners' is identified in the upper part of the line S_1, AND in the right part of the line S_2. This logical combination operating on the responses of the neurons of the first layer can be implemented using a neural formalism and leads to a multilayer neural network (Figure 8).

Real tests were performed on the same measurements as previously and the neural network behaves very satisfactorily, because 100% of the learning examples, which were the same 26 measurements as in section 5, were well classified. During the test phase the network worked well on straightforward obstacles. Nevertheless the main problem encountered was, for several measurements, the interpretation of what the obstacle was: for instance the extremity of a wall was perhaps considered as an edge, and, depending on the angle, a part of a corner might be considered as a wall. During the generalisation phase such ambivalence has to be tolerated.

6 Conclusion

In conclusion, for the identification of obstacles by ultrasound sensors no direct method can work well because of the complexity of the problem and the presence of noise. Therefore we proposed a method which takes into account the behaviour of reflected ultrasound waves in order to extract some features from the signals, and then to take a decision using a neural network. This method had proved efficient for a small set of data. Further work will have to be done in order to generalize this result to more complex environments.

7 Acknowledgements

The authors would like to thank M. Denis Roux and M. Gérard Cauvy, students from the University of Montpellier for their enthusiasm and their work on this difficult problem, including hardware and software difficulties.

References

[1] R.A. Brooks. Intelligence without representation. *Artificial Intelligence*, 47:139, 1991.

[2] G. Cauvy. Etude par réseau de neurones d'un sonar pour robot mobile. Technical report, DEA-USTL, Montpellier, 1995.

[3] D. Diep and K. El Kherdali. Un radar ultra-sons pour la localisation d'un robot mobile. In *Journées SEE Capteurs en Robotique*, 1993.

[4] K. El Kherdali. *Etude, conception et réalisation d'un radar ultra-sonore.* PhD thesis, USTL, Montpellier, 1992.

[5] S. Knerr, L. Personnaz, and G. Dreyfus. Handwritten digit recognition by neural networks with single-layer training. *IEEE Trans. Neural Networks*, 1992.

[6] R. Kuc and M.W. Siegel. Physically based simulation model for acoustic sensor robot navigation. *IEEE Trans.*, PAMI 9(6), November 1987.

[7] C. Moschetti. Neural network — a connectionist way for artificial intelligence & application to acoustic recognition of shapes. Technical report, IMRA-ESSI DESS, Sophia-Antipolis, 1994.

[8] D. Roux, D. Diep, P. Bonnefoy, and F. Harroy. Reconnaissance d'obstacles avec un capteur ultra-sonore. In *4ieme Congrès Français d'Acoustique*, Marseille, 1997.

[9] I. Sarda and A. Johannet. Behaviour learning by ARP: From Gait learning to obstacle avoidance by neural networks. In D. W. Pearson, N. C. Steele, R. F. Albrecht (editors), *Artificial Neural Networks and Genetic Algorithms*, pages 464–467. Springer-Verlag, Wien New York, 1995.

A Modular Reinforcement Learning Architecture for Mobile Robot Control

R. M. Rylatt, C. A. Czarnecki and T. W. Routen
Department of Computer Science, De Montfort University,
Leicester, LE1 9BH, UK
Email: {rylatt, cc, twr}@dmu.ac.uk

Abstract

The paper presents a way of extending complementary reinforcement backpropagation learning (CRBP) to modular architectures using a new version of the gating network approach in the context of reactive navigation tasks for a simulated mobile robot. The gating network has partially recurrent connections to enable the co-ordination of reinforcement learning across both modules [illegible] successive time steps. The experiments reported explore the possibility that architectures based on this approach can support concurrent acquisition of different reactive navigation related competences while the robot pursues light-seeking goals.

1 Introduction

Schemes for the control of mobile robots based on a stimulus-response view of behaviour offer an alternative to traditional AI approaches that relied on much more computationally demanding representational structures. The aim is to achieve effective autonomous real-time performance in unstructured and uncertain domains. As a representative example, Brooks' subsumption architecture [2] relies on the idea of multiple behavioural layers concurrently active and competing for control of the robot or agent, mediated by some kind of arbitration scheme that is often based on simple prioritisation. However, the problems of co-ordinating behaviours, or action selection, is a central concern for this branch of adaptive autonomous agent research. It can be argued that schemes like subsumption offer *ad hoc* engineering solutions conceived too prescriptively in observer space. For example, Rylatt *et al.* [9], and Molland *et al.* [7] have discussed respectively the role of learning and of short-term memory in achieving run-time adaptivity. Rylatt *et al.* [8] also survey approaches based on neural networks to explore the argument that this kind of substrate is an inherently more promising basis for achieving the necessary flexibility of behaviour. Another issue is whether this alternative substrate also implies architectural modularity. Ziemke [13] argues that a monolithic neural network can acquire modular features (learn its own control structure) during the process of adapting to an environment at run-time. However, a contra-indication is provided by our knowledge of brain structure, where there is good evidence for predetermined functional modularity. Obviously this kind of modularity is the result of phylogenetic adaptation, or evolution, rather than the kind of ontogenetic changes that could be compared to the run- time adaptation of an artificial autonomous agent. Taking broad inspiration from the biological existence proof, our initial approach was to define modules in relation to distinct sensory modalities of the agent. More details of the architecture are given in Section 2. Section 3 discusses some experimental results. Section 4 concludes with a summary of the achievements to date, some reflections on their implications and an outline of further work.

2 Reinforcement Learning in Modular Architectures

Different forms of reinforcement learning in neural networks have been described. The general approach in this paper is of the kind discussed by Williams [12], known as associative reinforcement learning: a neural network architecture reacts to the environment by emitting a time varying vector of effector outputs in response to a time-varying vec-

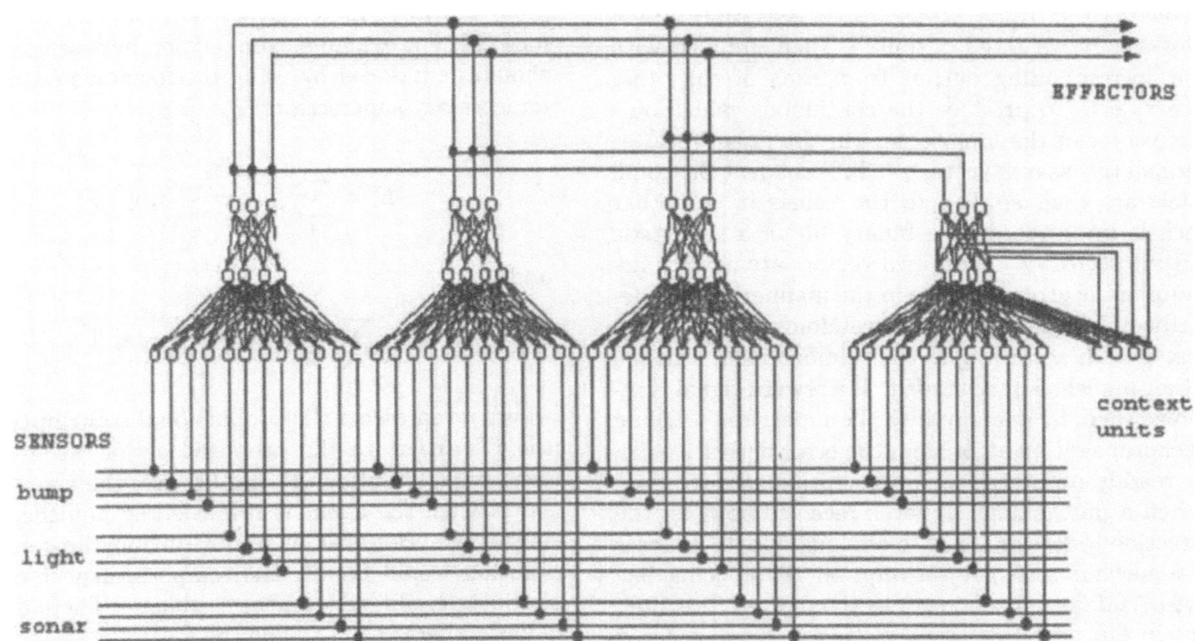

Figure 1: Modular neural network architecture.

tor of sensor inputs and learns to maximise a time-varying scalar reinforcement signal that is some task-dependent function of the input and output patterns unknown to the controller. Meeden *et al.* [6] applied complementary reinforcement back-propagation (CRBP), a form of associative reinforcement learning originally described by Ackley and Littman [1]) to a simple monolithic neural network controller for a car-like mobile robot; we have adapted it for use in modular neural network architectures, which presented a particular set of problems. In broad outline, the architectures are inspired by the Addam architecture [11] but as we use trial and error rather than supervised learning the principle of control is different. Our early work used an explicitly algorithmic (if we regard neural networks capable of simulation on Turing machines as implicitly algorithmic) approach to the temporally extended credit and blame assignment problems [10]. In the work reported in this paper we have been able to replace the arbitration algorithm with a gating network [5], originally devised for static, or time-implicit, problems, to which we have added a partially recurrent connections [4] as a way of solving problems of credit assignment arising from both the temporally extended nature of the domain and the architectural structure. An example of the architecture is shown in Figure 1 — in this version, although the modularity reflects the number of sensory modalities, each module has access to the whole input space; another version assigns a different sensor group to each module. In each net, competence in one of three modality-related tasks is expected to develop through trial and error:

- light-seeking using light sensor data;
- wall avoidance using active-sonar range data;
- avoiding low obstacles ('invisible' to the sonars) using bump detector data;

In each inchoate expert net a vector of sensor inputs i is propagated forward through the hidden layer to reach the vector of sigmoid output units, o, each of which takes on a value in the range (0,1). Each of the outputs for each net is multiplied by the corresponding output from the gating network, normalised as $\frac{\exp(x_j)}{\sum_i exp(x_i)}$. Each of the resultant probabilitstically weighted outputs is then summed with the corresponding output from each of the other expert nets to produce the continuous-valued output vector of the architecture in the range (0,1) — termed the 'search vector', s. Independent Bernoulli trials are then applied to the values in s so that each is interpreted as a binary bit in a stochastic output vector O. These two vectors are used to determine the error measure in the manner shortly described. In this way, initially random moves are suggested and, according to the reinforcement scheme, either punished or rewarded. If a reward signal is received then, by analogy with the supervised learning backpropagation algorithm, the error derivative can be readily obtained, so we backpropagate $(O - s)$. When a punishment signal is received however the direction to force s is not so obvious. CRBP chooses a somewhat stronger assumption than 'being like not-o,' taking $((1 - O) - s)$ as the desired direction, but in our case this assumption can be considered stronger still as we can use a little domain knowledge to ensure the encoding of our steering vectors makes the binary complements equate to opposite directions — reversing the direction of motion when punished may often be a reasonable one to adopt. Although this scheme may appear flawed (in the sense that the agent is 'learning to run before it can walk'), initially, the principle of a rich interaction between control levels and sensory modalities needs to be investigated in a search for flexible behaviour patterns that are not excessively constrained in the design time decision space. We also suggest that there is biological evidence for this kind of learning in that imperfectly mastered neuro-motor skills are gradually improved whilst the organism seeks higher level goals — an animal does not wait until it can walk perfectly before it moves to feed or flees from danger.

The aim of our reinforcement learning scheme can be rephrased as the intention that each module should become an expert at mapping a particular subset of the input domain onto the output range. In static, or time-implicit domains, gating networks of the kind described by Jacobs *et al.* [5] have proved capable of selecting effective mixtures of 'experts'. Reinterpretation of the gating network error measures in terms of CRBP is relatively straightforward. For example, competition between experts should be induced by using the formulae (omitting unnecessary superscripts):

$$E = \sum_i p_i \parallel O - s_i \parallel^2 \tag{1}$$

and

$$E = \sum_i p_i \parallel (1 - O) - s_i \parallel^2 \tag{2}$$

where p represents the proportional contribution of the ith expert to the proposed action on a given time step. Equation (1) and (2) give the error measures when the agent is rewarded or punished, respectively. However, it is not obvious how such a solution would handle the temporal aspect of the modular credit assignment problem. The solution adopted here is to provide the gating network with Elman style [4] recurrent connections. It is well known that such nets can solve context dependent problems — in the temporal domain this can be interpreted as the ability to decide what happens next on the basis of what has gone before.

3 Experiments

Referring to Figure 2, the mobile agent extinguishes a light by coming into contact with it and this remotely switches on another light some distance away. The first light is positioned so that the agent has to navigate around an obstacle to reach the light source, thus overcoming the tendency of the first level module to be repelled. The next three lights are located in situations that are relatively straightforward or entail skirting obstacles and navigating through gaps between obstacles and walls. The most difficult light seeking task entails navigation down a narrow corridor. The position of the final light source goal requires the agent to return from the far end of the corridor back into open space. Thus each level of competence is likely to be exercised as the agent proceeds. To test the validity of using recurrent connections in the gating network, a control experiment was run in which no such connections were employed. Our observation

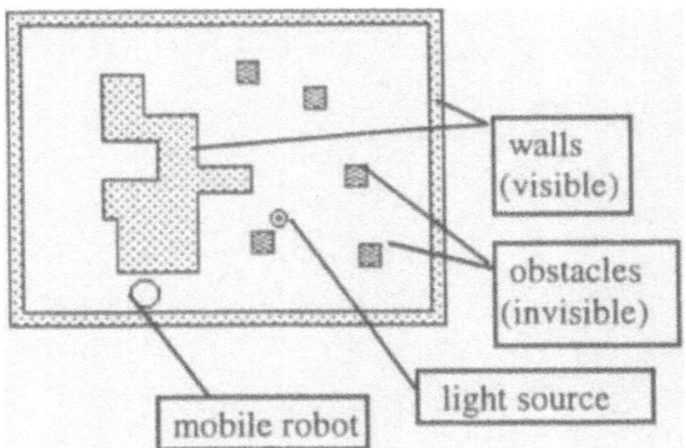

Figure 2: Experimental environment.

is that the presence of recurrent connections in the gating network appears to be decisive in determining the gating network's ability to select inchoate experts so as to assign credit and blame correctly across time steps — without recurrent connections the agent was unable to complete all the tasks and usually failed at tasks requiring relatively complicated manoeuvring.

4 Discussion

The specific contributions we have reported here are the extension of CRBP learning to a modular architecture, and the introduction of partially recurrent connections to a gating network in order to show that this approach has potential for mediating the actions of individual networks in a temporally extended domain. Our experiments show that architectures based on these principles are able to accomplish a series of tasks similar in type and arrangement to those reported in [11] and at a level of performance comparable to that achieved by our earlier explicit algorithmic control scheme [10]. It remains to be shown that the approach will scale well. The divide and conquer approach to problem solving is a universally accepted strategy in conventional software engineering but in the field of adaptive autonomous agents the questions of whether and how it should be applied are still open to debate. Underlying these concerns is the need for our agents to perform more complex and articulate tasks in uncertain and unstructured domains. Apart from its inherent lack of flexibility, the subsumption approach to building individual agents leads to ad hoc engineering solutions to highly specific tasks — a useful analogy might be that of a food processor with various task-oriented attachments — far from the emergent human-like intelligence promised at one time by Brooks [3]. A lesson for neural net based approaches is therefore to avoid predetermined modularization at the task level. In our work, a flexible approach to modularity that starts at the low level of the agent's own sensory modalities has shown some promise but, admittedly, the tasks we have devised are each closely associated with a particular sensory modality. Further investigation of possible architectural and task variations and analysis of the learning taking place in each module is now being undertaken. The development of genuinely autonomous agents entails extension of flexible control principles to higher cognitive levels; we hope that our approach can support progress in this direction.

References

[1] D. H. Ackley and M. L. Littman. Generalisation and scaling in reinforcement learning. In D. S. Touretsky, editor, *Advances in Neural Information Processing Systems*, pages 550–557. Morgan Kaufmann, San Mateo, CA, 1990.

[2] R. A. Brooks. A robust layered control system for a mobile robot. *IEEE Journal of Robotics and Automation*, RA-2:14–23, 1986.

[3] R. A. Brooks. Intelligence without representation. *Artificial Intelligence*, 47:131–159, 1991.

[4] J. Elman. Finding structure in time. *Cognitive Science*, 14:179–192, 1990.

[5] R. A. Jacobs, M. I. Jordan, S. J. Nowlan, and G. E. Hinton. Adaptive mixtures of local experts. *Neural Computation*, 3:337–345, 1991.

[6] L. Meeden, G. McGraw, and D. Blank. Emergent control and planning in an autonomous vehicle. In *Proceedings of the Fifteenth Annual Conference of the Cognitive Science Society*, 1994.

[7] R. Molland, T. Scutt, and P. Green. Extending low-level reactive behaviours using primitive behavioural memory. In *Proceedings of the International Conference on Recent Advances in Mechatronics*, pages 510–516, 1995.

[8] R. M. Rylatt, C. A. Czarnecki, and T. W. Routen. Connectionist learning in behaviour-based mobile robots: A survey. In *Artificial Intelligence Review*. Kluwer Academic Publishers. (to appear).

[9] R. M. Rylatt, C. A. Czarnecki, and T. W. Routen. A perspective on the future of behaviour-based robotics. In *Mobile Robotics Workshop Notes — Tenth Biennial Conference on Artificial Intelligence and Simulated Behaviour*, 1995.

[10] R. M. Rylatt, C. A. Czarnecki, and T. W. Routen. Learning behaviours in a modular neural net architecture for a mobile autonomous agent. In *Proceedings of the First Euromicro Workshop on Advanced Mobile Robots*, pages 82–86, 1996.

[11] G. M. Saunders, J. F. Kolen, and J. B. Pollack. The importance of leaky levels for behaviour based A.I. In *From Animals to Animats 3: Proceedings of the Third International Conference on Simulation of Adaptive Behaviour*, pages 275–281. MIT Press, 1994.

[12] R. J. Williams. On the use of backpropagation in associative reinforcement learning. In *Proceedings of the IEEE International Conference on Neural Networks*, pages 263–270, 1988.

[13] T. Ziemke. Towards adaptive perception in autonomous robots using second-order recurrent networks. In *Proceedings of the First Euromicro Workshop on Advanced Mobile Robots*, pages 89–98, 1996.

Timing without Time — An Experiment in Evolutionary Robotics

H. H. Lund
Department of Artificial Intelligence, University of Edinburgh,
5 Forrest Hill, Edinburgh EH1 2QL, Scotland, UK
Email: henrikl@aifh.ed.ac.uk

Abstract

Hybrids of genetic algorithms and artificial neural networks can be used successfully in many robotics applications. The approach to this is known as *evolutionary robotics*. Evolutionary robotics is advantageous because it gives a semi-automatic procedure to the development of a task-fulfilling control system for real robots. It is disadvantageous to some extent because of its great time consumption. Here, I will show how the time consumption can be reduced dramatically by using a simulator before transferring the evolved neural network control systems to the real robot. Secondly, the time consumption is reduced by realizing what are the sufficient neural network controllers for specific tasks. It is shown in an evolutionary robotics experiment with the Khepera robot, that a simple 2 layer feedforward neural network is sufficient to solve a robotics task that seemingly would demand encoding of time, for example in the form of recurrent connections or time input. The evolved neural network controllers are sufficient for exploration and homing behaviour with a very exact timing, even though the robot (controller) has no knowledge about time itself.

1 Introduction

When putting emphasis on developing adaptive robots, one can either choose to develop single robots with traditional learning techniques, or one can develop a whole population of robots with a simulated evolution process. The population based approach named *evolutionary robotics* has the advantage of requiring only a specification of a task-dependent fitness formula as opposed to traditional neural network learning techniques that demand a learning set so that each single action of a robot can be evaluated. The disadvantage of the evolutionary robotics approach is the time that it uses to reach a solution. This is because each single robot has to be evaluated for a number of time steps (e.g. 1500 steps of 100 ms each). If the population is large and the evolution has to run for many generation, then the time consumption when running online with real robots will be huge. Here, I describe how to overcome this problem in specific robotics tasks. This is done by designing an accurate simulator, in which the evolution of neural network control systems takes place before these evolved neural network control systems are transferred to the real robot in the real environment. The performances of the simulated and real robots are almost equal. This is due to the technique used to build the simulator. Sensory responses are simulated by using the sensory inputs from the robot itself rather than using a mathematical or symbolic description of the robot and its environment. Similarly, the possible motor responses of the robot are recorded and used in the simulator to determine the movement of the simulated robot in the simulated environment.

Another way to decrease the time consumption in evolutionary robotics is to determine the sufficient complexity of a controller for a given task. Many researchers try to evolve complex structures in order to have an open-ended evolutionary robotics, where it is possible to evolve any kind of task-fulfilling behaviour. Yet this might mislead us to think that the complex structures are necessary for the robot to achieve the tasks. In many cases, a much simpler structure can account for the behaviour, and the time used to search for a solution can therefore be reduced a lot by reducing the search space, when allowing only evolution of simpler structures that are known to be sufficient for accounting for the desired behaviour. In a biological context, this gives a tool to show how some behaviours, that are normally described as more complex by biologists, can be achieved with much simpler control systems. For example, tasks that seemingly demand an internal world map or an internal clock can be solved with simple neural network control systems, that do not have any memory units, recurrent connections or time inputs. This can be shown by evolving simple two layer feedforward neural networks (i.e. perceptrons with linear output) that connect the robot's sensory input (infra red sensors or ambient light sensors) with its motors.

It must be noted, that a robot with a specific physical structure is not the best robot to solve all tasks. Different tasks demands different robot body plans. For instance, a box pushing behaviour might demand a bigger body size than an obstacle avoidance behaviour, while quick turning could be obtained with a small wheel base and slow turning with a large wheel base. An evolutionary algorithm can be used to co-evolve robot controllers and robot body plans (the body plan of a robot includes the positions and number of sensors, the body size, the wheel base, the wheel radius, the motor time constant, etc.) for specific tasks, so that robot body plans that are adapted to each specific task are obtained [1, 3]. Here, however, I will concentrate on a robot with a pre-defined structure.

2 Experimental Setup and Method

In this experiment, I will show how a simple neural network controller with no recurrent connections or time input can solve an exploration and homing task with exact timing by evolving such simple controllers for the Khepera miniature mobile robot [7] (see Figure 1). The robot is supported by two wheels and two small Teflon balls. The wheels are controlled by two DC motors with incremental encoders (12 pulses per mm advance of the robot), and can move in both directions. The robot is provided with eight infra-red proximity and ambient light sensors. Six sensors are positioned on the front of the robot, the remaining two on the back.

As shown in [2], the time consumption when evolving neural network controllers on-line with the

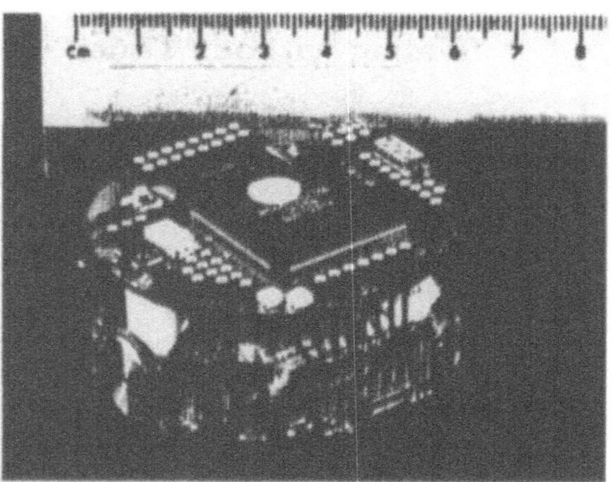

Figure 1: The Khepera miniature mobile robot that is used in the experiments.

Khepera robot is extremely high (in the order of weeks or months), so I chose to build a simulator for the Khepera robot and its environment. The neural network controllers were then evolved in the simulator and the best neural network controllers were afterwards transferred to the real robot in the real environment. In this way, the time consumption is reduced to less than one hour. The approach demands an accurate simulator from where the controllers can be transferred to the real robot in the real environment with no decrease in performance. For simple tasks, such a simulator can be obtained by using the *look-up table approach*, as shown in [2, 4, 5, 6].

In the look-up table approach, the robot itself is used to build the simulator. The sensor and motor responses, that are used in the simulator are not symbolic or mathematical description that an external observer believes characterise the robot and the environment, but rather samples taken with the robot itself. Therefore, the simulator becomes an accurate description of how the robot senses the environment and how the robot moves in the environment. In the present experiment, the environment was simply a 25 Watt light-bulb covered with white paper around the sides. The light-bulb was hanging 11 cm above a table, on which the Khepera robot could move. The exploration and homing task was defined as exploring as much of the ta-

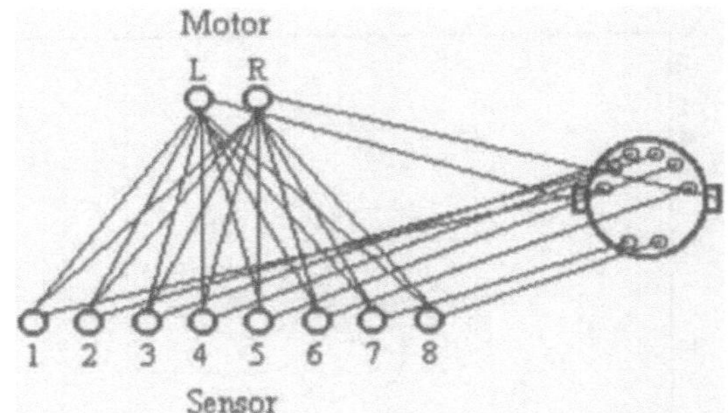

Figure 2: Connection between the Khepera robot and neural network controller. Additionally, there are two bias units.

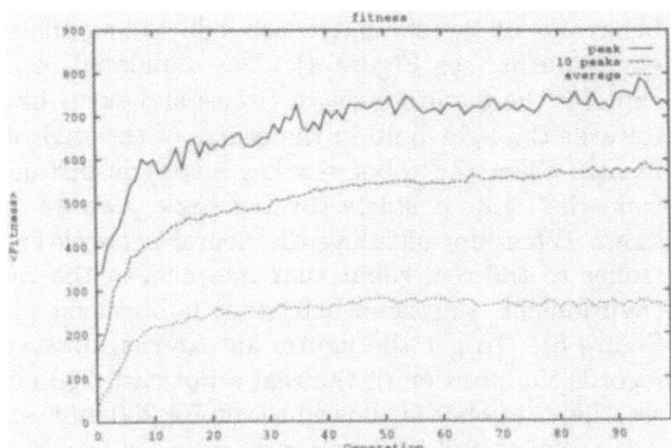

Figure 3: The peak, average of 10 peaks, and average fitness of each generation over the 10 runs with populations of 100 control systems.

ble as possible, but returning under the light-bulb within each 10 seconds — in this way the light-bulb worked as a 're-charging station'. In order to model the environment, the Khepera robot was placed under the light-bulb and allowed to turn 360 degrees while the activations of the 8 ambient light sensors were recorded at each 2 degrees. Then the robot was moved 2 cm backward and the sampling procedure was repeated. This was done for 20 distances. In this way, a (20,180,8) look-up table of the robot's sensory activation around the light-bulb was obtained. In constructing the look-up table for motor responses, I used a similar procedure. The motors were given all possible activations (which was set to 21 for each motor) one by one, and the displacement (angle and distance) of the robot was recorded. These look-up tables describe how the simulated robot senses and moves in the simulated environment.

A neural network control system for the Khepera robot can be a simple feedforward neural network that connects the robot's sensors with its motors (see Figure 2). In these experiments, I therefore used a feedforward neural network with 8 input units totally connected to 2 output units plus 2 bias units connected to the 2 output units. The sensory activation is normalised and fed directly to the 8 input units, while the activation of the 2 output units is used to set the motor activation of the robot.

A simple genetic algorithm was used to evolve the connection weights of the neural network control systems with the fixed, simple topology. Initially, a population of 100 networks with randomly chosen weights (in the interval -1.0 to 1.0) was constructed. Each of these neural network controllers was tested on the simulated robot in the simulated environment for 3 epochs of 500 actions. Then, the 20 most fit were selected to reproduce 5 times each in a reproduction procedure that included copying and mutation of 10 % of the weights (in the interval -0.1 to 0.1). The fitness formula that was used for selecting the best performing controllers was constructed by dividing the table into cells of 2×2 cm. The fitness of a controller was increased by one unit when it allowed the robot to move to a previously untouched cell, but only as long as it had been under the light-bulb within the last 10 seconds. This can be interpreted as the robot had energy to run for 10 seconds after being re-charged under the light bulb. In order to get high fitness, a neural network controller should therefore allow the robot to explore but always return to the light-bulb within 10 seconds of the last visit.

3 Results

The genetic algorithm was used in 10 runs with different initial random seed. In all 10 runs, the fitness increased quickly over the first 10 generations and then steadily with small increases over the last 90 generations. The average of the 10 runs is shown in Figure 3. It is very interesting to look at the behaviour of the simulated robot in the simulated environment (see Figure 4). The simulated robot explores the environment in circles and turns back towards the light-bulb in the centre of the environment. When the robot reaches a specific distance from the light, it starts turning back towards the light. When downloading the neural network controller to the real robot that interacts in the real environment, the same behaviour is obtained (see Figure 5). To get the figure, an external observer records the position of the real robot each 200 mill sec., i.e. the robot is allowed to run for 200 mill sec., it is stopped and its position is recorded (down to an accuracy of 1×1 cm) and the robot is again allowed to run for 200 mill sec. In total, the real robot in Figure 5 ran for 70 sec., which is approximately half the time of the corresponding one in simulation shown on Figure 4. The reason for a shorter run in reality is the very time consuming recording process. We are now constructing a video-tracking system to avoid this. The timing of the real robot is amazing. The robot moves out from the light-bulb, explores the environment, and returns to the position exactly under the light-bulb with the following timing: 8.0 sec, 8.2 sec, 7.4 sec, 8.8 sec, 8.0 sec, 7.4 sec, 7.6 sec, 7.2 sec. Other controllers result in another timing even closer to 10 seconds. Without having any knowledge whatsoever about the time, the neural network controller navigates the robot towards the light when time is running low. This amazing and surprising behaviour is due to the nature of the evolutionary algorithm that selects the controllers that allow the robot to return to light within 10 seconds. It is interesting to note, that the solutions found with the evolutionary algorithm allow the robot to move 'backwards'. By doing so, the robot obtains more knowledge about when to return towards the light source, since the six sensors placed on the 'back' of the robot sense the light emitted from the light source behind the robot. The evolutionary process puts pressure on controllers that allow the robot to turn back towards the light source when the input that the robot senses is of the kind that is exactly at the distance from which the robot can return to the light source before losing all its energy.

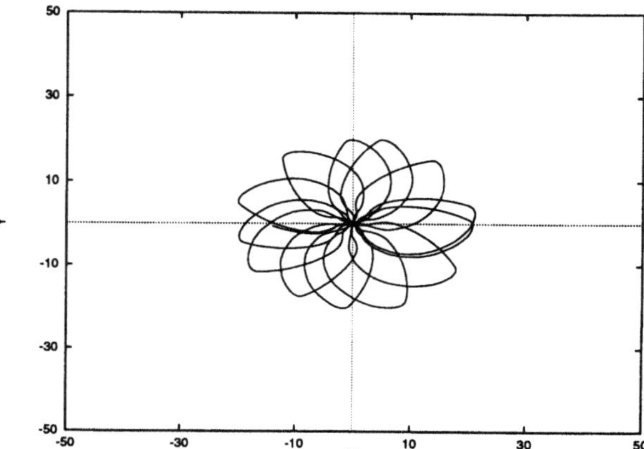

Figure 4: The behaviour of the simulated Khepera robot with the evolved perceptron from one of the 10 runs.

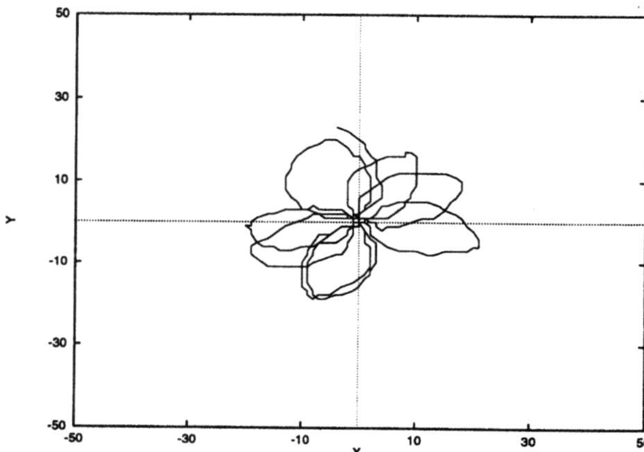

Figure 5: The behaviour of the real Khepera robot for 70 seconds. The position of the robot is recorded by an external observer each 200 ms with a resolution of 1×1 cm, i.e. the position is rounded to the nearest centimeter. Therefore, the path does not seem as smooth as in Figure 4. The actual path of the real Khepera robot is much smoother.

On the other hand, this goal could be achieved by returning to the light source before the distance-limit is reached, but the factor in the fitness formula for exploring the environment puts pressure on the robot to go as far away from the light as possible. Therefore, it becomes a specific input at a specific distance from the light source that is used to allow the robot to turn and start to navigate back towards the light source — distinguishing the inputs and making the returning response at a specific input is easier when the robot moves with the six sensors on the back side, so the evolution process has found this solution, which might not be the one that we would immediately imagine as the best.

4 Conclusion

The evolved robot uses its perception of the geometrical shape of the environment to navigate around the environment with very exact timing. Again, it should be emphasized, that this timing is done without any explicit knowledge about time. I therefore conclude that, for this kind of task, a knowledge about time (as for instance represented by providing the robot with additional input for time or by adding recurrent connections to the neurocontroller) is not necessary to solve the tasks efficiently, since this can be done with very simple neural networks that use the robot's perception of the geometrical shape of the environment.

5 Acknowledgement

H. H. Lund is supported by EPSRC grant nr. GR/K 78942 and the Danish Science Research Council.

References

[1] W.-P. Lee, J. Hallam, and H. H. Lund. A Hybrid GP/GA Approach for Co-evolving Controllers and Robot Bodies to Achieve Fitness-Specified Tasks. In *Proceedings of IEEE Third International Conference on Evolutionary Computation*, NJ, 1996. IEEE Press.

[2] H. H. Lund and J. Hallam. Sufficient Neurocontrollers can be Surprisingly Simple. Research Paper 824, Department of Artificial Intelligence, University of Edinburgh, 1996.

[3] H. H. Lund, J. Hallam, and W.-P. Lee. Evolving Robot Morphology. In *Proceedings of IEEE Fourth International Conference on Evolutionary Computation*, NJ, 1997. IEEE Press. Invited paper.

[4] H. H. Lund and O. Miglino. From Simulated to Real Robots. In *Proceedings of IEEE Third International Conference on Evolutionary Computation*, NJ, 1996. IEEE Press.

[5] O. Miglino, H. H. Lund, and S. Nolfi. Evolving Mobile Robots in Simulated and Real Environments. *Artificial Life*, 2(4):417–434, 1996.

[6] O. Miglino, K. Nafasi, and C. Taylor. Selection for Wandering Behavior in a Small Robot. *Artificial Life*, 2(1), 1995.

[7] F. Mondada, E. Franzi, and P. Ienne. Mobile robot miniaturisation: A tool for investigation in control algorithms. In *Experimental Robotics III. Lecture Notes in Control and Information Sciences 200*, pages 501–513, Heidelberg, 1994. Springer-Verlag.

Incremental Acquisition of Complex Behaviour by Structured Evolution

S. Perkins and G. Hayes
Department of Artificial Intelligence,
University of Edinburgh, 5 Forrest Hill, Edinburgh, Scotland
Email: s.perkins@ed.ac.uk, gmh@dai.ed.ac.uk

Abstract

In practice, general-purpose learning algorithms are not sufficient by themselves to allow robots to acquire complex skills — domain knowledge from a human designer is needed to bias the learning in order to achieve success. In this paper we argue that there are good ways and bad ways of supplying this bias and we present a novel evolutionary architecture that supports our particular approach. Results from preliminary experiments are presented in which we attempt to evolve a simple tracking behaviour in simulation.

1 Engineering vs. Evolution

In recent years search/optimization based methods have been widely touted as a way to design robot controllers without all that tedious mucking about with analysing the complex interaction between a robot and its environment.

Unfortunately, despite success in automatically designing controllers for a few simple tasks, pure learning methods do not scale to the complex tasks we would like our robots to perform, e.g. tasks involving visual sensing.

Increasingly, people are suggesting that what we need is a hybrid of the two approaches (e.g. [3]). Specifically: how can we use domain knowledge supplied by humans to speed up or bootstrap search-based methods? We explicitly recognize the tradeoff between engineering and search, and ask the questions: 'What bits of robot design are suited to human engineering?', 'What bits are suited to automated search?' and 'How do we combine them?'. Our proposed solution to this problem is called *structured evolution.*

2 Structured Evolution

The major principles of structured evolution are:

1. The job of the human designer is to determine the high-level structure of a task, and to devise appropriate environmental constraints to make training tractable.

2. The job of the evolutionary algorithm is to find low-level solutions to simple problems within this structure.

3. The designer shouldn't have to fiddle with the internal details of the evolutionary algorithm.

The first two points recognize that, while humans are generally quite good at decomposing complex tasks into simpler ones at a coarse scale, they are usually rather bad at imagining what a real robot is going to have to do in detail. Similarly, while learning/evolutionary algorithms can develop successful controllers for simple tasks, they are bad at determining the coarse scale structure of complex tasks. These complementary qualities suggest the above division of labour.

The last point makes the claim that, given this division, it is unnecessary for the designer to know about whatever internal representations, connections, weights, sub-symbolic rules etc. the low level learning algorithm is using. This frees the learning algorithm from the constraint of having to produce humanly intelligible solutions, and frees the designer from having to worry about what is going on at the lowest level. Instead, the designer is forced to think and analyze the task at hand at a level more suited to the human imagination.

We feel that these criteria can be met by a robot that acquires complex behaviour in an *incremental* fashion. The role of the designer is to specify a path of increasing competence from simple behaviour to complex, and a suitable training scheme to go with it. The role of the learning algorithm is to actually move the controller from one specified point on this path to the next. Note that the human trainer is only concerned with the external behaviour of the robot, and not with the internal workings of the learning mechanism.

3 Methods

There are many methods that a designer can use to provide external constraints to make incremental learning of complex tasks possible, including:
Task Decomposition: The robot is initially trained on sub-tasks of the complex task. Hopefully once it has these then the full task can be learned much more easily. Task decomposition has been used by a number of researchers to make learning of complex tasks tractable (e.g. [1, 2, 8]). However, we do *not* specify a specific controller hierarchy — that is left to emerge in response to the hierarchical training.
Training Explicit Representations: The robot is encouraged to develop internal representations for particular situations that are deemed likely to be useful in later learning. Since we are not 'allowed' to examine or specify internal values, this is done by training external responses e.g. 'stick up your right hand if there's a light in front of you' (or equivalently 'flash LED 1 ... ').
Good Reinforcement Policy Design: The evolutionary algorithm's job is made much easier if the robot receives frequent evaluations on its performance. [9] provides some discussion of these terms in terms of 'progress estimators' and 'heterogeneous reinforcement'. We are also looking at ways of using non-scalar reinforcement where it is available.
Enriched Learning Environments: The rate at which a robot receives rewards can be increased by simplifying or enriching the environment in the early stages of training so that the robot is more likely to achieve goals by accident. This should help speed up learning.
Simulator Training: We are keen to produce controllers for real robots. However, several researchers have shown that it is possible to train controllers in simulation and transfer them to a robot later e.g. [11]. Simulators also allow more informed evaluation functions.
Prohibiting Irrelevant Sensors/Actuators: Sometimes we can say for sure that a robot will not need a particular sensor/actuator for a task, so we can discourage the use of it.

4 An Evolutionary Architecture for Structured Evolution

The incremental approach to robot design required by structured evolution puts constraints on the learning architecture we can use. In particular we would like to be able to separately learn different sub-skills without forgetting old ones, and we would like new sub-skills to be able to take advantage of existing skills and internal representations.

Most evolutionary robotic systems attempt to evolve the whole controller in one go. They work with populations of monolithic controllers and attempt to evolve a single all-knowing controller. We feel this approach to be incompatible with the above aims — it is difficult to retain previously learned skills that are not immediately useful, and difficult to learn a single coordination system that can switch between them all.

We prefer a *multiple expert* approach where the total behaviour of the robot is a result of the interaction between a collection of experts. Ideally each expert is a specialist which is only active in a sub-portion of the total state space of the robot. Experts correspond quite naturally to our sub-tasks in structured evolution.

In our architecture, each expert is itself a population of individuals (called *agents*) and each population is (co-)evolved separately using a fairly conventional evolutionary architecture. The hope is that each population evolves a different specialization so that once converged, each population can be represented by its 'best' agent i.e. competition within populations, co-operation between populations.

Each agent takes input from sensors and calculates both an output action and a *validity.* This value says how confident the agent is that it is in a part of the state space where it is saying the right thing. If more than one agent ends up trying to influence the same actuator, then the one with the highest validity is chosen. In this way the state space is divided up into overlapping regions where different agents have priority.

Agents themselves are simple tree-structured genetic programming-like programs. Their function set is inspired by typical artificial neural network transfer functions (in earlier implementations agents were perceptrons) and terminal nodes are constants or sensory inputs.

5 Preliminary Experiments

We are interested in evolving complex visual behaviours such as tracking moving targets using a real 'pan/tilt head'. This task is particularly suitable for investigating structured evolution since the huge number of raw sensory inputs and difficulty of the task make it very difficult for pure reinforcement learning, while we do have a good idea how to design suitable controllers by hand.

Experiments are still at an early stage and currently we are working with the simpler task of trying to track a simulated bright target against a dark background. The target is initially positioned randomly at the edge of the robot's visual field. Evaluation is then given each cycle proportional to how much the centre of the image moved closer to the target. If the robot 'hits' the target or it goes out of sight, the target is randomly repositioned.

Visual sense inputs for agents sample varying sized regions of the input image in a natural neuron-inspired fashion. Separate actuator outputs are provided for the pan and tilt head velocity.

Evaluating agents within individual populations in our architecture poses problems. The only evaluation that can be given is to the robot as a whole. How do we evaluate the contribution of individual agents? Moreover, how do we cope with the fact that an agent may be evaluated in a part of the state space where it is not valid (and hence can't be blamed for things going wrong)? Our solution is as follows: every second or so, an agent is picked from each population to represent that population. The evaluation given to the robot over the next second is then given equally to all the chosen agents, with an extra weighting if they were particularly confident. We evaluate every agent in each population 20 times in different random parts of the state space, and with different random other agents active, and the total fitness is simply the weighted average of all these evaluations. Breeding then occurs by selecting single parents using rank-based selection and applying point mutation or random sub-tree replacement to generate new agents. The top 20% of the population is retained each generation.

In theory, it is possible to co-evolve several populations at the same time, but in practice interference between non-converged populations seems to make this difficult. Therefore we currently only evolve one new population each time we try to learn some new sub-skill or increase in task complexity, and freeze the others.

6 Results and Further Work

We present here some preliminary results using a simplified version of the above experiment where the robot is deemed to have 'hit' the target if the horizontal offset between the camera direction and the target reaches zero. Thus the population only has to worry about the pan velocity and can merely set the tilt velocity to zero. Figure 1 shows a typical run of 200 generations with a single population containing 50 agents.

These results demonstrate that the basic learning mechanism can at least learn something. The next stage is to learn the full task and we hope to be able to do this by freezing the converged 'pan' population above and then training an additional population to take care of the tilt component. We also wish to compare this with evolving the whole task in one go using one or two populations.

Eventually we hope to be able to use various structured evolution techniques to tackle the considerably harder problem of tracking moving targets (as opposed to bright ones).

Another thing we want to be able to do is to allow agents to influence each other by both allowing agents to take input from other populations, and allowing them to send inhibition or excitation to other populations. This will make the credit assignment even more complex and we hope to use a bucket-brigade technique [5] to allow fitness to flow between populations.

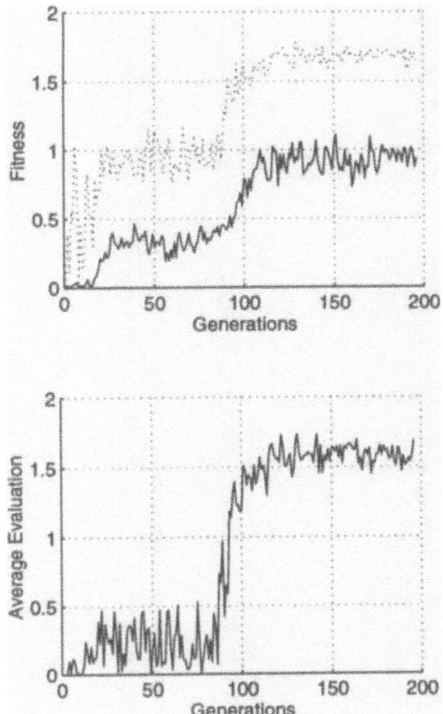

Figure 1: Results: The top graph shows the maximum and mean fitness of the population during a typical run. The bottom graph shows the average evaluation received during the test phase at the end of each generation during which an agent is picked randomly from the top 10% of the population each robot cycle for 100 cycles. The optimum average evaluation is about 1.7.

7 Related Work

There has been quite a lot of work on hierarchical training of skills in order to speed up learning in robots, e.g. [1, 2, 8]. Almost all of it involves an explicit decomposition of the controller itself however, which is something we hope to avoid.

The idea of emergently co-evolving different specialists within a controller population is quite old. Much of the work attempts to evolve such cooperation within a single population, e.g. classifier systems [6] and 'symbiotic neuro-evolution' [10]. Potter *et al.* [12] presents quite a similar architecture to our own in which multiple populations are used to evolve different specializations in a simple simulated robot foraging task, although they see the potential more for ***avoiding*** human design input rather than supporting it.

The idea of using a 'validity' to decompose the state space into areas of different priority for different agents is similar to Mark Humphry's 'W-Learning' [7].

One of our main aims is to train robots to perform complex visual behaviours. The evolutionary robotics group at Sussex University has similar aims e.g. [4] and has had some success evolving neural network controllers, although again they use a population of monolithic controllers and don't attempt to learn incrementally.

References

[1] J. H. Connell and S. Mahadevan. Rapid task learning for real robots. In Jonathan H. Connell and Sridhar Mahadevan, editors, *Robot Learning*, chapter 5, pages 105–139. Kluwer Academic Press, 1993.

[2] M. Dorigo and M. Colombetti. Robot shaping: Developing situated agents through learning. Technical Report TR-92-040, International Computer Science Institute, Berkley, CA 94704, April 1993.

[3] J. J. Grefenstette and A. C. Schultz. An evolutionary approach to learning in robots. In *Proc. Machine Learning Workshop on Robot Learning*, New Brunswick, NJ, 1994.

[4] I. Harvey, P. Husbands, and D. Cliff. Seeing the light: Artificial evolution, real vision. In D. Cliff, J.-A. Meyer, and S. Wilson, editors, *From Animals to Animats 3: Proc. 3rd Int. Conf. Simulation of Adaptive Behavior*. MIT Press, 1994.

[5] J. H. Holland. Adaptive algorithms for discovering and using general patterns in growing knowledge bases. *Int. Journal of Policy Analysis and Information*, 4(2):217–240, 1980.

[6] J. H. Holland and J. S. Reitman. Cognitive systems based on adaptive algorithms. In D.A. Waterman and F Hayes-Roth, (Eds.), *Pattern Directed Inference Systems*, pages 313–329. Academic Press, New York, 1978.

[7] M. Humphrys. Action selection methods using reinforcement learning. In *From Animals to Animats*

4: Proc. 4th Int. Conf. Simulation of Adaptive Behavior, Cape Cod, Massachusetts, USA, September 1996.

[8] L-J. Lin. Hierarchical learning of robot skills by reinforcement. In *International Conference on Neural Networks*, 1993.

[9] M. J. Mataric. Reward functions for accelerated learning. In William W. Cohen and Haym Hirsh, editors, *Machine Learning: Proceedings of the Eleventh International Conference*, San Fransisco CA, 1994. Morgan Kaufmann Publishers.

[10] D. E. Moriarty and R. Mikkulainen. Efficient reinforcement learning through symbiotic evolution. *Machine Learning*, (22), 1996.

[11] S. Nolfi and D. Parisi. Evolving non-trivial behaviours on real robots: An autonomous robot that picks up objects. In M. Gori and G. Soda, editors, *Proceedings of the fourth congress of the Italian Association of Artificial Intelligence*, pages 243–254, 15 Viale Marx, 00137 - Rome - Italy, 1995. Springer-Verlag.

[12] M. A. Potter, K. A. De Jong, and J. J. Grefenstette. A coevolutionary approach to learning in sequential decision rules. In *Proc. 6th Int. Conf. on Genetic Algorithms*, Pitsburgh, July 1995. Morgan Kaufmann.

Evolving Neural Controllers for Robot Manipulators

R. Salama and R. Owens,
Robotics and Vision Research Group, Department of Computer Science,
University of Western Australia, Nedlands, Australia
email: {rameri, robyn}@cs.uwa.edu.au

Abstract

We examine here the feasibility of using evolutionary techniques to produce controllers for a standard robot arm. The main advantage of our technique of solving path planning problems is that the neural network (once trained) can be used for the same robot, with a variety of start and target positions. The genetic algorithm learns, and encodes implicitly, the calibration parameters of both the robot and the overhead camera, as well as the inverse kinematics of the robot. The results show that the evolved neural network controllers are reusable and allow multiple start and target positions.

1 Introduction

Generally speaking, it is easier to recognise a good solution to a problem than it is to design one. This is the principle that underlies several approaches to problem solving. Examples include evolutionary programming, genetic algorithms, and simulated annealing.

Recent work has investigated the use of intelligent search over a large space of potential designs as an alternative to deliberate design. In a previous paper [5], we used a genetic algorithm to search a space of neural network based controllers for a hexapod robot.

In this paper, we extrapolate the results obtained from the previous work on hexapod robots to work on a UMI RTX robot manipulator arm working under visual guidance. Given the trajectory of the end effector of the robot, we wish to find the series of joint angle trajectories (*inverse kinematics*).

The vision system that provides guidance for the robot has to be calibrated with real world coordinates. Calibrating a camera generally involves determining the mapping between world points and image points for the camera. In this method the camera matrix is computed directly from the image positions corresponding to known world points [1].

For this paper, we evolve neural networks to control a robot arm. The neural network produces joint angle trajectories that move the robot from a start position to a final position using visual input. Husbands *et al.* [3] have done similar experiments with a wheeled robot.

2 The Problem

Figure 1 shows the layout of the RTX's workspace, with the camera situated above a planar board. The genetic algorithm must produce a neural network controller that moves the end effector of the robot to the target position from its start position in 2 dimensions. This is a kinematically redundant mechanism in that the robot has 3 degrees of freedom but is required to achieve only 2 positional coordinates. This means that there are many possible solutions to the problem. In the case of a robot controller, the inputs from the environment come from the robot's camera, and the outputs of the network control the robot's actuators; in this case they are angle values for each of the joints.

Traditionally, genetic algorithms encode a solution as a string which is then split to perform the operation of crossover. Sometimes mutation takes place, producing changes in certain bits of the string. We have chosen to use a matrix representation for the neural networks that will occupy our solution space. Each column represents connections from a particular node to every other node. The value of the element a_{ij} specifies the value of the weight of the connection from node i to node j.

2.1 Image Processing

The state of the robot, its position with respect to the target, and the position of the target have to be determined from video images grabbed by a CCD camera placed above the workspace. Due to camera noise and environmental effects some image processing has to take place for meaningful information to be extracted from the images. To facilitate the extraction of this information, there are retro-reflective markers around the workspace, which define the boundaries of the workspace. The target also has a large piece of retro-reflective tape placed over it, as does the end effector of the robot.

2.2 The Genetic Algorithm

A genetic algorithm attempts to modify a randomly generated population so that the characteristics of population members which define fitter traits are preserved. To do this it *generates* a population (which may or may not be totally random). It then *selects* pairs of members of the population to breed with each other in a process called *crossover*. The offspring may then be *mutated* in the hope that the occasional result of mutation is a fitter individual. Fitter individuals are given a greater chance to be selected for breeding. In this way the overall fitness of the population increases.

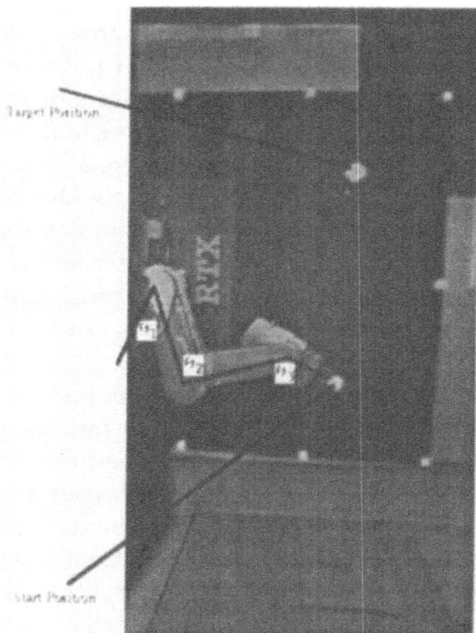

Figure 1: Typical board layout.

Each network is represented as a matrix of randomly generated weights. Each connection has a certain probability of existing. If a connection exists then the actual weight of the connection a_{ij} is determined randomly:

$$a_{ij} = \begin{cases} 0 & \text{if } a < v \\ w & \text{if } a \geq v \end{cases}$$

where $0 \leq a \leq 1$ is randomly generated from a uniform distribution, v is a threshold, and $|w| \leq m$ where m is the maximum possible weight. The term w is also randomly generated from a uniform distribution.

The neural networks are randomly arranged in a grid so that each network occupies an element of the grid. Each neural network can then be close to, or far from, other neural networks in the population using the metric induced from the grid. This is used to maintain diversity in the population. Allowing the population to have spatial behaviour is a strategy used by Ngo and Marks [4] in their genetic algorithm.

The mate selection process decides which neural networks breed with which others. Each neural network is assigned a maximum number of steps that it can take per generation. The neural networks are then allowed to wander over the population grid searching for mates. When a network has exhausted its search, it mates with the fittest individual that it has visited.

The crossover operator involves copying portions of the genetic strings of the parents so that a new organism with some of the characteristics of both parents is produced. The simplest involves the splitting of both parents at some random point. The offspring is the concatenation of the first half of one string with the second half of the other.

We need a crossover operator that can amalgamate entries from two connection matrices. The process which we use can be seen in Figure 2. This is column crossover and is analogous to crossover of connections *from* one node. It is possible to produce a crossover operator that is for connections *to*

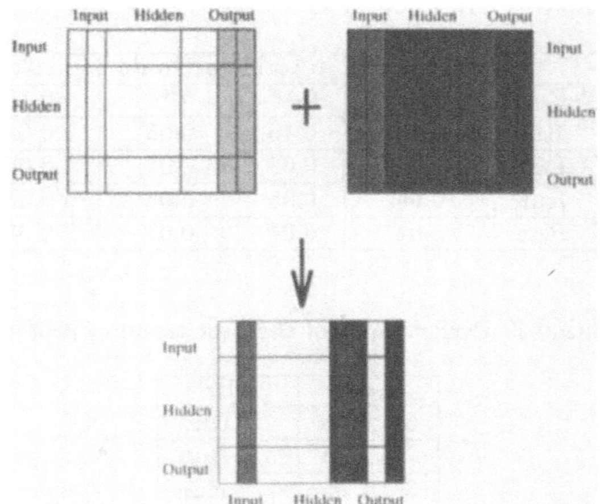

Figure 2: Crossover operation in matrices.

a node; this would be row crossover.

Mutation is the process where elements of the genetic makeup of an organism are changed. The change may increase the fitness of the organism. In general, mutation will introduce new genetic material into the population.

In the matrix representation of neural networks, it is possible to mutate the network by changing the weights of the connections in the connection matrix, or by adding or deleting connections in the connection matrix. The mutation ratio variable is represented by the quadruple $\{pRemove, pAdd, pRNode, pANode\}$, where *pRemove* is the probability that a connection is removed, *pAdd* is the probability that a connection is added, *pRNode* is the probability that all connections from a node are removed, and *pANode* is the probability that all connections from a node have random values assigned to them.

3 The Task

The task is for the robot to go from a start position to a target position. The neural networks that are evolved are three-layer, feedforward networks.

The inputs to the neural network show the current value of the joint angles of the arm $(\Theta_1, \Theta_2, \Theta_3)$, the distance from the target (Δ_1), and the number of times (η_τ) that this neural network has generated an invalid value of joint angles for the robot, as well as the current target that the robot is attempting to reach (x, y). The variable η_τ is especially significant since it allows us to provide more information about how well the neural network controller is performing. Since we are providing the neural network with the coordinates of the target we will be able to evolve controllers which can go to various targets and generalise for new targets. The inputs are not used directly by the neural network, but they are processed by a simple layer which attenuates the inputs for either the simulation or the real robot. The attenuation mechanism that we use is simply a filter that normalizes the inputs to the neural network.

The outputs of the neural network are the changes in each of the joint angles $(\delta_1, \delta_2, \delta_3)$.

4 Fitness Functions

The only part of the genetic algorithm which we have not examined is the module that evaluates the fitness of the neural network controller. The value assigned to the neural network controller by this module defines the type of mate that it selects. This module drives the search of the genetic algorithm in a particular direction.

The fitness function for this task is a function of the number of angle configurations that the neural network provides until the robot reaches the target, and the number of invalid angle configurations that the neural network generates. Since the robot can move anywhere in the workspace, the neural network which finds a solution in the least number of steps, without providing any invalid arm configurations has the best fitness.

We define the fitness function as

$$\begin{aligned} f_k &= 1/(\omega\eta_\tau(\delta + 0.1)), \\ F_{ij} &= \min\{f_1, f_2, ..., f_k\}, \end{aligned}$$

where f_k is the fitness of each individual test of the network, ω is the number of valid arm configurations, η_τ is the number of invalid arm configurations, δ is the final distance from the target, and F_{ij} is the fitness of the network in the (i, j) position.

The value of f_k is maximised when ω is 1, η_τ is 1, and δ is 0. This means that the robot has arrived at the target (δ is 0) in one step (ω is 1), and the number of invalid joint configurations that the controller made while moving the robot to the target is 0 (η_T is 1).

This fitness function returns the minimum fitness of all the tests that are run. This ensures that no network which is exceptionally good on only one test and bad on the others attains a high fitness. The best solution is one that performs at least as well as the worst solution. This approach has been adopted by Harvey [2].

5 Results

In these experiments we examine the evolution of neural controllers for the control of an RTX robot arm. The controller is evolved in a simulation, and then tested both in simulation and on the real arm.

The variables on each execution of the genetic algorithm are the number of generations over which the genetic algorithm is run, the size of the population, the number of hidden nodes for the neural network, the size of the grid for selection of partners, the number of steps that each organism takes while it is searching for a mate, the probability that a connection is removed or added and the probability that a node is removed or added, and the amount of noise that is added to the system.

For the following experiment the population size is 100 on a 10×10 grid, and each organism can wander for 5 steps in search of a mate. The genetic algorithm is run for 500 generations. The number of input nodes is 7, the number of hidden nodes is 9 and the number of output nodes is 3. The only variable is for mutation, which we change throughout the following experiment, as shown in Table 1.

After examining Table 2 we can see that when we have the lowest value of mutation rates the average of the average fitness (AAF) of the population is optimised. However, we also notice that the average best fitness (ABF) values occur when the mutation levels are the highest. Note that the second best overall fitnesses for the profiles shown is exhibited by the lowest mutation rates. Also the behaviour of the fitness for the organisms that are evolved using the lowest rates of mutation is more stable (as we would expect). Thus, we choose the lowest (non-zero) mutation rate of {0.04,0.04,0.03,0.03}.

Table 1: Mutation rates for the genetic algorithm.

	pRemove	pAdd	pRNode	pANode
1st Run	0.15	0.15	0.10	0.10
2nd Run	0.10	0.10	0.05	0.05
3rd Run	0.05	0.05	0.05	0.05
4th Run	0.05	0.05	0.03	0.03
5th Run	0.04	0.04	0.03	0.03

Table 2: Performance of the best training profiles.

AAF	ABF
5th Run	1st Run
1st Run	5th Run
3rd Run	2nd Run
4th Run	4th Run
2nd Run	3rd Run

The fittest neural network from the run of the genetic algorithm with that mutation rate had a fitness of 0.01423. A fitness of this value shows that if the robot took 7 steps ($\omega = 7$) to get to the target, and on the way to the target it made no errors ($\eta_T = 1$), then the final distance of the end effector from the target is 9.87mm.

In simulation for points that the robot was trained on the behaviour of the arm is excellent. As we can see in Figure 3 the performance of the network in conditions that it has been trained on is very accurate. It takes large steps till it gets close to the target and then slows down to get on top of the target.

In simulation for points that the robot was not trained on the behaviour of the arm is reasonably accurate. As we can see in Figure 4 the robot comes close to the target but does not reach it exactly. The networks are trained on points within the workspace of the robot, so we expect them to be more accurate for points within the workspace. If points outside of the workspace are given, the network will attempt to move the robot as close to the target as possible.

6 Conclusion and Discussion

An important aspect of this work is that we are trying to do in one step (with a neural network and genetic algorithm) that which normally requires

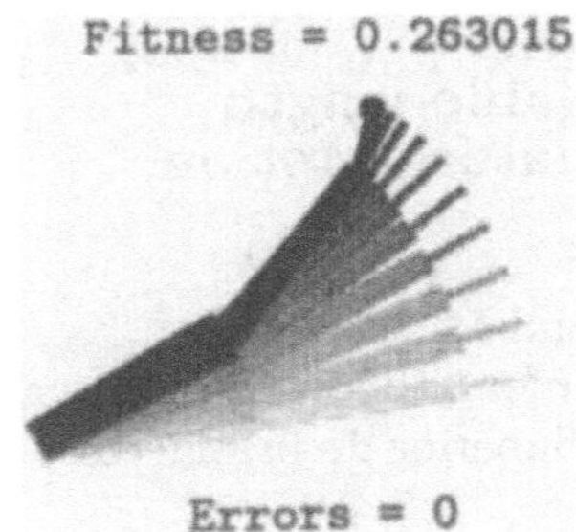

Figure 3: Training target: (400,400).

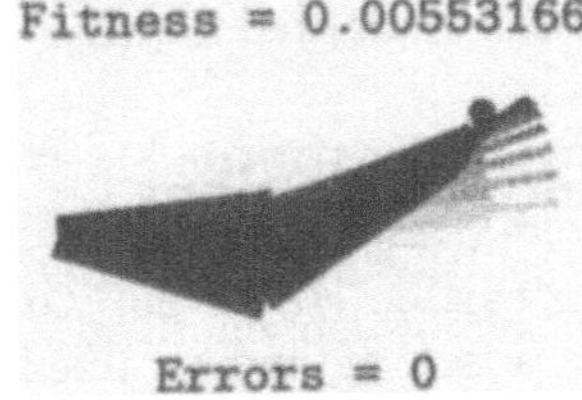

Figure 4: Testing target: (500,150).

the use of camera calibration, robot calibration and then solving inverse kinematics.

The inverse kinematics is especially troublesome since we operate in two dimensions only, where there is no explicit solution. We have shown that standard feedforward neural networks can be evolved to guide a robot manipulator from one position to another.

References

[1] D. H. Ballard and C. M. Brown. *Computer Vision.* Prentice Hall, Englewood Cliffs, NJ, 1982.

[2] I. Harvey. *The Artificial Evolution of Adaptive Behaviour.* PhD thesis, University of Sussex, April 1995.

[3] P. Husbands, I. Harvey, and D. Cliff. Analysing recurrent dynamical networks evolved for robot control. In *Proceedings of the Third IEE International Conference on Artificial Neural Networks (ANN93).* IEE Press, 1993.

[4] J. T. Ngo and J. Marks. Spacetime constraints revisited. In *Computer Graphics Proceedings*, pages 343–350. SIGGRAPH, 1993.

[5] R. Salama and P. Hingston. Evolving neural network controllers. In *Proceedings of the IEEE International Conference on Evolutionary Computation.* IEEE, Dec 1995.

Using Genetic Algorithms with Variable-length Individuals for Planning Two-Manipulators Motion

J. Riquelme[1], M.A. Ridao[2], E.F. Camacho[2] and M. Toro[1]
[1] Dpto. Lenguajes y Sistemas Informáticos. Facultad de Informática y Estadística.
[2] Dpto. Ingeniería de Sistemas y Automática. Escuela Superior de Ingenieros.
Universidad de Sevilla, Spain.

Abstract

A method based on genetic algorithms for obtaining coordinated motion plans of manipulator robots is presented. A decoupled planning approach has been used; that is, the problem has been decomposed into two subproblems: path planning and trajectory planning. This paper focuses on the second problem. The generated plans minimize the total motion time of the robots along their paths. The optimization problem is solved by evolutionary algorithms using a variable-length individuals codification and specific genetic operators.

1 Introduction

The problem is to plan a collision-free motion (obstacles and other robots), from an initial configuration to a goal configuration. The most extended approach to this problem is to decompose it into two subproblems: path planning and trajectory planning. Many algorithms to solve this problem can be found in the literature [1, 2, 3, 4].

The solution obtained by most of these algorithms is a robot trajectory. These trajectories are very difficult to implement in most industrial robots, because they require the internal controller of each articulation to be fully available to the user.

A method is presented in [5, 6] to minimize the total motion time of the robots along their paths. This method is used in this paper to find a collision-free motion plan for two robots, and evolutionary algorithms with three different chromosome codification are presented to solve the optimization problem. In this paper a genetic algorithm where the length of the individuals is variable [7] is proposed. Also, new genetic operators adapted to this codification are presented.

2 Problem Statement

The problem can be stated as: Given two robots R_1 and R_2, a set of known fixed obstacles and the initial and final configurations of R_1 and R_2; find a coordinated motion plan for the robots from their initial configuration to their final configuration avoiding collisions with environments obstacles and themselves. The use of a decoupled planning approach needs a fixed obstacle collision-free path to be previously obtained for each of the robots. The paths which the robots are expected to follow are assumed to be given as a parametrized curve in the joint space, where λ is the distance along the path. The *coordination space (CS)* is defined as the R^2 region.

$$CS = \{(\lambda^1, \lambda^2)/0 \leq \lambda^j \leq \lambda^j_{\max} \; with \; 1 \leq j \leq 2\}$$

Any path from (0,0) to $(\lambda^1_{\max}, \lambda^2_{\max})$ determines a coordinated execution of the two paths, and is called a *coordination path (CP)*. The *collision region (CR)* is defined as the set of points in CS where a collision between two manipulators is produced. In order to reduce the search space in CS, a discretization of each path has to be made, so the path is divided into several equal intervals. Let us number the intervals of each path from 1 to max_j and the ordered set of intervals is called Ω_j. A cell is defined as the subspace formed by one interval of the paths of each of the robots and is represented as the pair (n_1, n_2). With these discretized paths, CS is transformed into an array of cells, the *coordination diagram (CD)*. Let us notate $C_0 = (0, 0)$ and

$C_{max} = (max_1, max_2)$. A cell (n_1, n_2) is considered collision FREE if every point inside the cell does not belong to the collision region. Otherwise, it is considered an OBSTACLE cell.

Robots can be synchronized using ***synchronization points (SP)***, that is, a point in CD, which any CP will necessarily pass through. When the robot arrives at that place on its path, it will stop until the other robot arrives at its respective points. To avoid a collision it is possible to alter the CP defining the number and position of the SP, determining the total motion time.

Let us consider a rectangle formed by free cells in CD and let us consider the motion of the robots from the lower left corner cell to the upper right corner cell. Any trajectory defined for each robot between these two points in CD will always be a collision-free CP. This class of rectangles is called *free rectangles.*

Let us consider a set of free rectangles, connected in such a way that the upper right corner of one rectangle is the lower left corner of the next. Furthermore, the lower left corner of the first one is the lower left corner of the whole CD, and the upper right corner of the last rectangle is the upper right corner of CD. This set of rectangles is a *free rectangle sequence*, and the intersection points between two rectangles will be the SP. This constraint is very easy to implement using any robot programming language.

The problem can be stated as that of finding a Free Rectangle Sequence that minimizes the total execution time necessary for the robots to complete their whole path. The main variables used to find this sequence are the number of SP and the position of these points in. A complete description of the method can be found in [5, 6].

3 The Proposed Evolutionary Algorithms

Three different evolutionary algorithms to solve the optimization problem are presented in this paper. The main differences among them are the codification of the individuals and the genetic operators. While the first one uses the classical binary codification of genetic algorithm, the others use a specific integer codification with new genetic operators. One of them uses fixed-length chromosomes and the other variable-length individuals.

3.1 Fixed-Length Integer Codification

In the solution proposed in [5], each individual is formed by a fixed number of points represented by their absolute coordinates in Ω. Different crossover, mutations and specific genetic operators were studied in the same paper, obtaining good results. The main drawback of this codification is the use of fixed chromosome length. In order to avoid this disadvantage, a new variable length codification is proposed in this paper.

3.2 Variable-Length Integer Codification

Chromosome Representation of the Individuals

Each individual is represented by a increasing SP sequence. The chromosomes that represent each solution have variable lengths. If n is the length of a chromosome, then a SP sequence will be determined by $n+2$ points, where $(x_0, y_0) = (0, 0)$ and $(x_{n+1}, y_{n+1}) = (max_1, max_2)$. An individual is valid if it forms an increasing sequence of free rectangles. Finally, when a genetic operator is applied, it is possible for a non-increasing SP Sequence to be obtained (non-acceptable individual).

Generation of the Initial Population

The initial population is selected randomly. A uniform distribution of the number of points between 1 and $NMAX$ has shown to be inappropriate in the tests to represent the lenght of the individuals. The reason is that non-optimal solutions, with few points, can constitute local minima. For this, an increasing probability distribution from 1 (minimum) to $NMAX$ (maximum) was selected.

Once the number of points of an individual n is selected, its values are created as follows: two sets of n random values in [0,1] are generated, then they are ordered in an increasing way and projected on $[0, max_1]$ and $[0, max_2]$ respectively.

Fitness Measure

The evaluation of the fitness measure will consider valid and non-valid individuals. For valid ones, the fitness function gives the total execution time needed by the robots to complete their paths, when the SP are placed in the positions defined by the individual specifications (See [5] for a more detailed description).

The fitness function for non-valid individuals is completely different. The function must measure how far it is from a valid individual. The function considered is $f(N) = K + nco$, where K is an offset, i.e. a high value with respect to the value associated with the valid individuals, and nco is the number of obstacle cells inside the rectangle sequence.

Genetic Operators

Crossover Operator: Given two individuals S^1 and S^2 formed by sequences of n and m SP respectively, the idea is to obtain another SP sequence through genetic information exchange from parents S^1 and S^2. The following method is proposed: A SP of S^1 is randomly selected, called P. Then, Q, a SP of S^2 is selected. Q is the first SP that has both coordinates x and y greater than the respective coordinates of P, so the resulting child always will be an increasing SP sequence. It no point of S^2 verifies this limitation, then the child is returned as a copy of S^1. If point Q exists, a new individual is formed by the first points of S^1 (P inclusive), followed by the points of S^2 from Q (inclusive) to the end of S^2 (see Figure 1).

Mutation Operator: Two groups of mutation operators are proposed in this paper:

- *Slight Operators*, which vary the genetic information sightly. These operators accomplish a local search, that is, in the neighbourhood of a solution.

- *Strong Operators*, which modify the structure of a solution. The objective is to extend the search to new regions, permitting the algorithm to escape from local minima.

Experimental results show that the most important improvement are obtained with the slight mutation. Nevertheless, the strong mutations play a very important role in the process, avoiding the algorithm to remain trapped in a local minima,

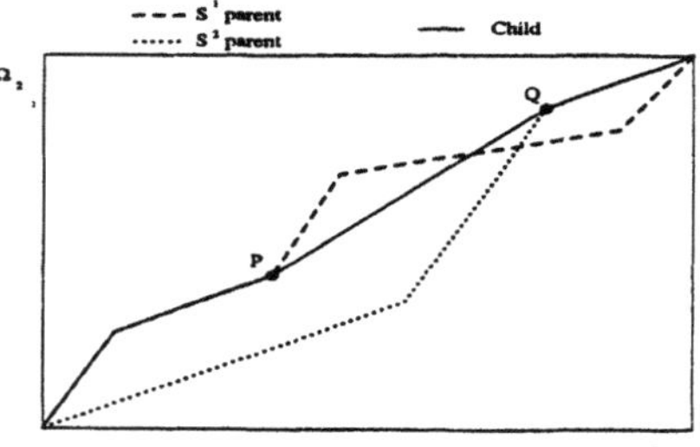

Figure 1: Crossover operator.

even when these mutations rarely improve the individuals. The Slight mutations are defined as follows: A SP (x_k, y_k) of individual S is selected randomly and also an integer m between $-MUTMAX$ and $MUTMAX$ is chosen too, where $MUTMAX$ is the maximum permitted mutation. The mutation consists of substituting the point (x_k, y_k) by (x_{k+m}, y_{k+m}) as can be seen in Figure 2, where the new point will be placed in the grey rectangle.

On the other hand, the strong mutations can be defined, for example, in the following way: A SP (x_k, y_k) from sequence S is selected randomly. The mutation consists of substituting that point by another one chosen in the rectangle defined by the previous SP as the lower left corner and the following one as the upper right corner (if they do not exist, (0,0) or (max_1, max_2) are considered respectively).

Restrictions to the Individuals: When these operators are applied, the resulting individual may not verify the condition of being an increasing sequence. To solve this problem, the conflicting coordinates are made equal to the same coordinates of the following point. For example, given two consecutive points $P = (x_k, y_k)$ and $P' = (x_{k+1}, y_{k+1})$, if $y_k > y_{k+1}$ then $y_k = y_{k+1}$ is made.

3.3 Evolutionary Algorithm Characteristics

An elitist evolutionary algorithm has been used, where the best individual of each generation is replicated in the following one. A percentage of the offspring is obtained through parents mutations selected with a probability proportional to its fitness. The rest of the offspring is obtained through parents crossover (also selected as a function of its fitness),

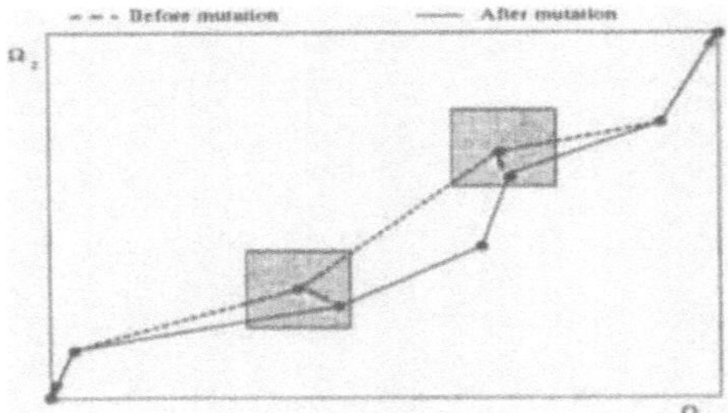

Figure 2: Segment mutation.

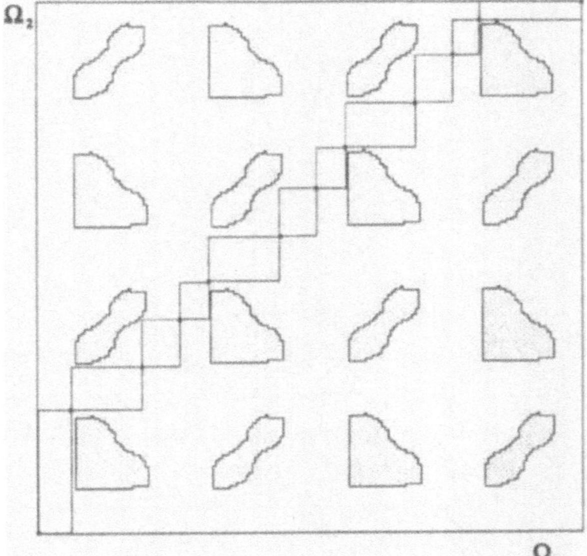

Figure 3: Coordination diagram and the best SP sequence.

and after, one of the above defined mutation operators is applied with a random probability.

4 Application Examples

The proposed algorithm has been implemented and applied to several examples in order to study its efficiency. Only one example is presented in this paper. The example corresponds to the motion of two SCORBOT and sixteen collision regions and 180180 cells (Figure 3). It is an iterative motion represented in Figure 4. The motion from Figure 4(1) to 4(2) and again to the initial configuration 4(3), is repeated twice.

Table 1: Results (time in seconds).

Codification	Avg.	σ	Min.
Binary	54.78	0.56	53.25
Integer fixed	42.23	2.78	37.63
Integer var.	41.19	1.16	38.63

To compare the different evolutionary algorithms, 40 simulations have been executed with a 100 individuals population and 300 generations. Individuals in binary codification and fixed-integer are formed by 20 SP. Also, the maximum number of SP of the initial population is 20 in integer-variable codification, in order to use similar initial information.

The results are shown in Table 1. The best results are obtained with integer variable codifications. Also, fixed codification reached a lower minimum value, but standard deviation is three times greater.

Finally, notice two other advantages in using variable codification. First, its computational cost of crossover and mutation operations may be even 60% of the fixed codification cost, because normally, the number of SP of the individuals to be processed is smaller in variable codification.. As a second advantage, in variable codification it is not necessary to previously know an estimation of the number of SP of the optimal solution.

5 Conclusions

This paper describes a method to generate collision free coordinated motion plans in multirobots systems. The method tries to find a SP sequence that

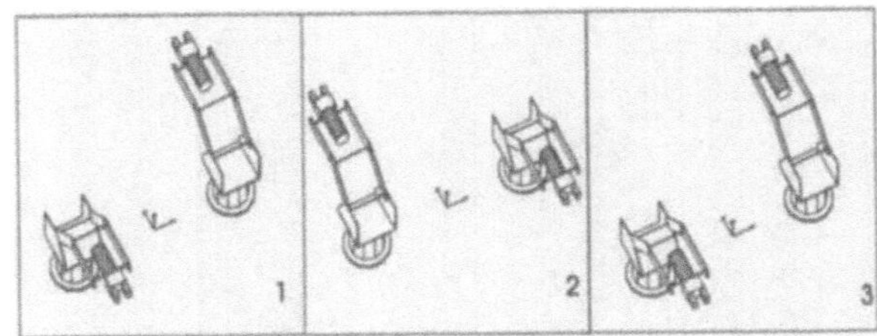

Figure 4: Initial position (1), intermediate (2) and goal position (3).

minimizes the total execution motion time. This optimization problem has been solved using different evolutionary algorithms. Three chromosome codifications and different genetic operators have been implemented. Better results are obtained in the application examples when using an integer codification. Nevertheless, in variable codification it is not necessary to previously know an estimation of the number of SP of the optimal solution.

References

[1] K. Kant and S. W. Zucker. Toward efficient trajectory planning: The path-velocity decomposition. *The Int. J. of Robotics Research*, 5(3):72–89, 1986.

[2] J. C. Latombe. *Robot Motion Planning*. Kluwer Academic Publishers, 1991.

[3] B. H. Lee and C. S. G. Lee. Collision-free motion planning of two robots. *IEEE Trans. on SMC*, 17(1):21–32, Jan–Feb 1987.

[4] P.A. O'Donnell and T. Lozano-Perez. Deadlock-free and collision-free coordination of two robots manipulators. In *Proc. of the IEEE Int. Conf. on Robotics and Automation*, pages 484–489, 1989.

[5] M. A. Ridao. *Generacion Automtica de Trayectorias Libres de Colisiones para Multiples Robots Manipuladores*. PhD thesis, Universidad de Sevilla, 1995.

[6] M. A. Ridao, J. Riquelme, E.Camacho, and M. Toro. Coordinated motion planning of manipulators by evolution strategies. In *Proc. of the 10th Int. Conf. on Applications of AI in Engineering*. Udine, 1995.

[7] S. F. Smith. *A Learning System Based on Genetic Adaptive Algorithms*. PhD thesis, University of Pittsburgh, 1980.

Ensembles of Neural Networks for Digital Problems

D. Philpot and T. Hendtlass
Centre for Intelligent Systems, School of Biophysical Sciences and Electrical Engineering,
Swinburne University of Technology, P.O. Box 218 Hawthorn 3122. Australia.
Email: dnp@stan.xx.swin.edu.au, tim@bsee.swin.edu.au

Abstract

Ensembles of neural networks is a technique that uses several different networks working together to find a relationship that exactly matches a many-to-one training set for digital classification problems. Each network learns a different region of the training space and all these regions fit together, like pieces of a jigsaw puzzle, to cover the entire training space. The individual networks are 'grown' as they are needed to form either cascades or branches of networks. The networks can be of any type such as backpropagation, cascade etc. However, virtually any other technique can be used in place of the networks: GA, EA, DRS, tabu search, nearest neighbour, and so on. Methods are discussed to improve the generalisation.

1 Introduction

This paper discusses a technique that involves using several neural networks working together to solve one digital problem. This ensemble of neural networks is 'grown', and typically takes the form of a cascade or branch of individual networks. Each network learns a separate region of the training space, and all the regions from these networks fit together, like pieces of a jigsaw puzzle, to cover the entire training space.

All training sets are learned with 100% accuracy, providing the set has a many-to-one relationship between inputs and outputs. The stability and robustness of this technique has meant that much of the guesswork that has traditionally been used to determine the various parameters such as architecture, learning coefficients, and so on, has been removed. Better parameters will often lead to a faster solution with fewer networks, but often they can be widely varied and a solution still found.

While test points are classified primarily on the learned (hyper)region they fall into, a technique for enhancing the generalisation capability of the ensemble in inter-region space, using relearning, is also described.

The individual networks do not have to be the same type. This means that the technique can take advantage of the many different architectures and learning rules that are available, choosing the best network for each of the different parts of the training space accordingly. An extension is also described that allows the networks to work together with other techniques such as evolutionary and genetic algorithms, DRS, tabu search, nearest neighbour, to use the extra advantages that these can also offer.

2 Training the Networks

The training process begins conventionally in that one network is trained on all of the training data until learning has stabilised. Once this has been achieved, only the points that were not learned by the first network are passed onto the second network for training. The second network then trains on these points until learning has stabilised, and only the unlearned points from the second network are passed onto a third network for training. This process continues until eventually all the training data has been learned. Each network learns a particular region of the training space and also records the range over which its outputs are valid. This range is divided into 'unlearned', 'learned ON' or 'learned OFF' regions (Figure 1). The learned ranges consist of actual output values for which there is only one desired output. The unlearned range(s)

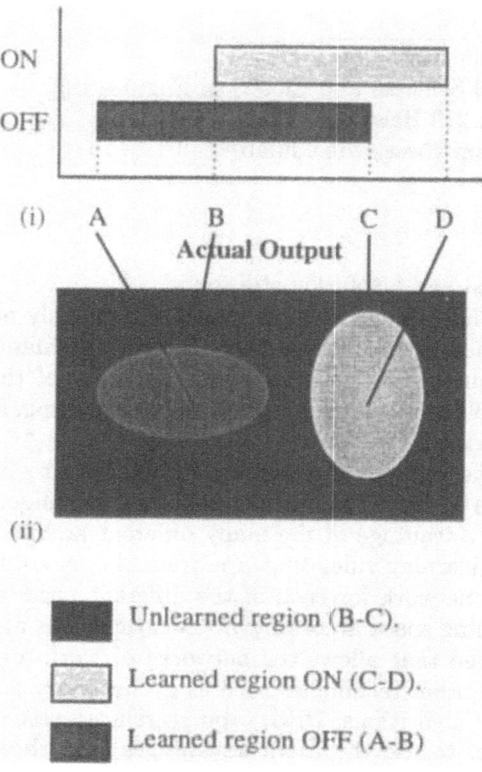

Figure 1: (i)A typical graph showing the relationship between desired output (ON or OFF) against the actual output from one network. The regions obtained from the actual output are: A-D: Full output range; A-C: Desired output OFF range. B-D: Desired output ON range; A-B: Learned desired OFF; B-C: Unlearned range; C-D: Learned desired ON. (ii) Indicates how the ranges (i) typically relate to a 2D training space.

consist of actual output values for which the desired output could be either of the desired outputs. The end result is a 'cascade' of networks whose combined relationship, using all the regions, matches the training data exactly (Figure 2). Often the regions learned by different networks do not abut. This can lead to testing points between these regions being incorrectly classified. These errors can often be reduced by relearning some of the points already learned by previous networks in addition to the unlearned points. The already learned points that are still passed onto the next network for relearning are those just inside the learned region that are also near to an unlearned region. Increasing the training set in this way helps ensure that all the problem space is learned and prevents testing points falling into unknown regions.

The training technique can also be modified to produce a 'branch' of networks rather than a 'cascade' of networks (Figure 3). A 'branch' of networks is produced by splitting an unlearned region into parts and passing the training points from each part to separate networks. This may be advantageous when a training set contains several separable features. A 'branch' of networks will often consist of more networks than a 'cascade' of networks. Despite this, training may be faster because networks are often trained on fewer points, and the relationships in these points are simpler. This can lead to better network generalisation. A 'branch' of networks will often lead to faster testing than a 'cascade' of networks.

3 Testing the Networks

When testing on a 'cascade' of networks, the inputs for the test point are presented to the first network that was trained. If the output for this network indicates that the test point falls in a 'learned' region, then it is considered classified. If, however, the test point falls in an 'unlearned' region, it is passed onto the second network in the 'cascade' and the same procedure is repeated. The test point keeps moving down the 'cascade' of networks until one of the networks classifies it. If training included relearning, a test input may be able to be classified by more than one network as the relearned region essentially consists of an overlap of these networks' learned regions. If these networks differ in their classification, a decision has to be made as to which classification to accept. Often the test point will lie further into the learned region of one network than any other. The further a test point is into a learned region, the more likely it is to be classified correctly. Accepting the classification associated with this network usually increases testing accuracy. Testing on a 'branch' of networks is similar to a cascade.

Cascade of Neural Nets

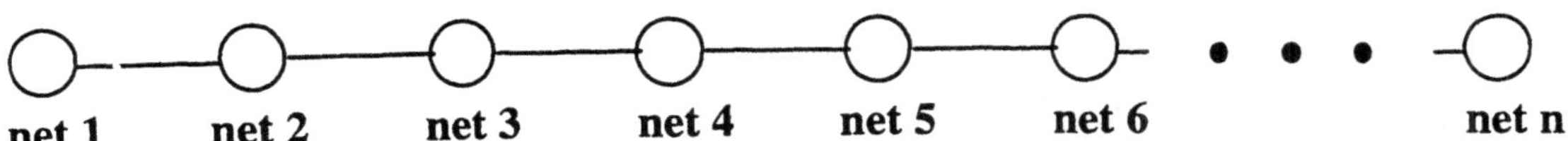

Figure 2: Unlearned points from the first net get passed onto the second network to learn. This process continues until all the training points have been learned. The end result is a 'cascade' of neural nets.

The main difference is that it depends on which unlearned region the outputs for the test point falls into as to which network the test point is passed onto. The test point continues branching from network to network until it is classified. This often leads to faster testing because each branch path is often shorter than the full 'cascade' of networks. As a result, a test point does not need to be tested on as many networks before being classified.

4 Different Network Types

The individual networks can be of various types. If an ensemble with the fewest networks is sought, any given training set is learned by several different network types, and the network that has learned the most points included in the ensemble. The disadvantage of this is that it is more time consuming. However, testing times can be dramatically reduced.

5 Extensions to Other Techniques

The technique can be easily extended to include other approaches such as evolved polynomials, nearest neighbour and so on. Each training set, made up of all the unlearned points from the previous network, is essentially treated as a completely new problem, even though it is a subset of the original training set. The result is that each of these training sets can be learned using any appropriate technique. Different types of problems will often require different techniques. Techniques, which could never solve the entire problem, may still be used to solve parts of the problem. For example, using an ensemble consisting of a number of networks to learn the general relationships in a data set, together with nearest neighbour to learn exceptions, may yield the higher accuracy of nearest neighbour while largely preserving the substantially reduced computational cost associated with neural networks.

Branch of Neural Networks

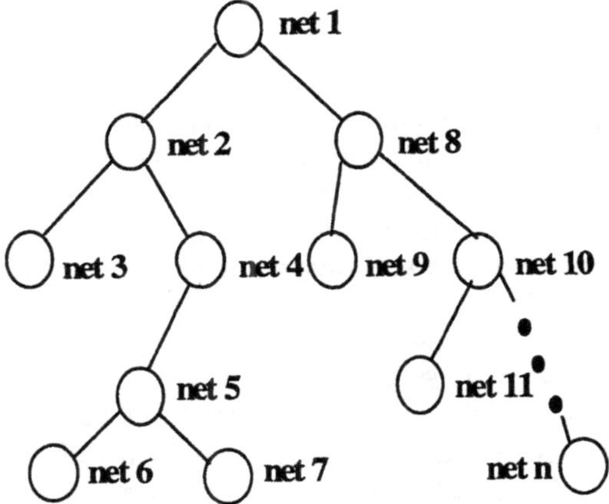

Figure 3: Unlearned regions can be split up and learned by different networks. The end result is a 'branch' of neural networks.

6 Results

Every many to one training set is learned perfectly, provided a large enough ensemble is used. Figure 4 above shows the results from a training set describing Mickey Mouse. The desired relationship (i) is the training data that was used to train the networks. The trained relationship without relearning (ii) shows that the training data was learned per-

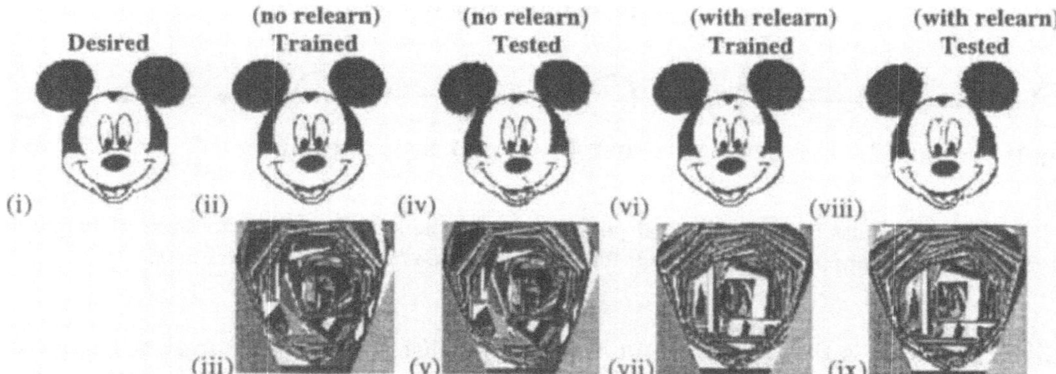

Figure 4: (i) Desired relationship. (ii) Trained relationship without relearn (matches Desired relationship exactly). (iii) Network map for trained data in (ii) indicates how good the generalisation is (93.80%). (v) Network map from (iv) above. (vi) Trained relationship with relearn (matches desired relationship exactly). (vii) Network map for trained data in (vi) (551 networks). (viii) Testing data using trained relationship in (vi) (93.88%). (ix) Network map from (viii) above.

fectly. The results shown in (iii) show which regions were learned by different networks. Each colour represents the learned region of a network (366 networks). The regions in (iii) all have straight boundaries because the networks in this case consisted only of an input and output layer, which are much faster to train. Other network types can give curved boundaries. The tested relationship (iv) gave an accuracy of 93.80%. The errors in the testing set show up as imperfections on Mickey's face. Figure 4 (v) shows which networks classified different regions of the testing data to give the relationship in (iv). Figure 4 (vi)-(ix) are the same as (ii)-(v) but using relearning.

Consider Frey and Slates' letter recognition problem [2] as an example of a complex many to one data set. This set consists of 16000 training points, 4000 test points, each point consists of 16 inputs and 26 outputs. Without using relearning and with an ensemble consisting exclusively of three layer backpropagation networks, the training set is learnt exactly and the test set is 88.55% correctly identified. When relearning was introduced (using the 30% of learned points closest to the boundary) and different types were used (chosen from basic backpropagation, cascade, GA and simple linear networks) the testing accuracy was 92.5%. This should be compared with Frey and Slates' best result of 82.7% using Holland style adaptive classifiers with exemplar-based induction, or 88.2% using one evolved neural network per letter [3]. The nearest neighbour approach alone can correctly identify 95.4% [1]. Work is currently being done to hybridise the trained networks with the nearest neighbour technique. This is expected to produce results that are comparable in accuracy with nearest neighbour, but with the much smaller computational cost from the trained networks.

References

[1] T.C. Fogarty. First nearest neighbour classification on Frey and Slates' letter recognition problem. *Machine Learning*, 9:387–388, 1992.

[2] P.W. Frey and D.J. Slate. Letter recognition using Holland-style adaptive classifiers. *Machine Learning*, 6:161–182.

[3] J.R. Podlena and T. Hendtlass. Evolving complex neural networks that age. In *Proc. ICEC95*. IEEE Press, 1995.

A Modular Neural Network Architecture with Additional Generalization Abilities for Large Input Vectors

A. Schmidt and Z. Bandar
The Intelligent Systems Group, Department of Computing, The Manchester Metropolitan University
Email: aschmidt@hydra.informatik.uni-ulm.de, Z.Bandar@doc.mmu.ac.uk

Abstract

This paper proposes a two layer modular neural system. The basic building blocks of the architecture are multilayer perceptrons trained with the backpropagation algorithm. Due to the proposed modular architecture the number of weight connections is less than in a fully connected multilayer perceptron. The modular network is designed to combine two different approaches of generalization known from connectionist and logical neural networks; this enhances the generalization abilities of the network. The architecture introduced here is especially useful in solving problems with a large number of input attributes.

1 Introduction

The multilayer perceptron (MLP) trained by the backpropagation (BP) algorithm has been used to solve real world problems in prediction, recognition, and optimization.

If the input dimension is small the network can be trained quickly. However for large input spaces the performance of the BP algorithm decreases [3]. In many cases it becomes difficult to find a parameter set which leads to convergence towards an acceptable minimum. This often makes it very difficult to find a useful solution, especially in recognition where large input spaces are common. Research is being done to overcome these problems; many of the ideas include modularity as a basic concept.

In [4] a locally connected adaptive modular neural network is described. This model employs a combination of BP training and a winner-take-all layer.

A modular neural system using a self organizing map and a multilayer perceptron is presented in [2]. It is applied to a cosmic ray space experiment. In this paper a modular neural network is proposed to enhance the generalization ability of neural networks for high dimensional inputs. The network consists of several MLPs. Each of the modules is trained by the BP algorithm. The number of weight-connections in the proposed architecture is significantly smaller than in a comparable monolithic network. The modular architecture is introduced, a training algorithm is given, the operation of the network is described, and experiments are presented.

2 The Network Architecture

The proposed network system consists of a layer of input modules and an additional decision module. All sub-networks are MLPs. Each input variable is connected to only one of the input modules. These connections are chosen at random. The outputs of all input modules are connected to the decision network. The structure is depicted in Figure 1.

The following parameters are assumed: the dimension of the input vector is l and the number of classes is k. One of the design issues is to select the number of inputs per module in the first layer (n); this decision determines the number of input modules $m = \lceil \frac{l}{n} \rceil$. (It is assumed that $l = m * n$; if this is not the case the spare inputs may be connected to constant inputs or the size of one of the networks may be altered.) Each network in the first layer has $\lceil \log_2 k \rceil$ outputs. This is the required number to represent all the classes in a binary code. The decision network has $m * \lceil \log_2 k \rceil$ inputs. The number of outputs is k, one neuron for each class. The number of weights is much less than in a fully connected monolithic MLP with the same number of hidden neurons.

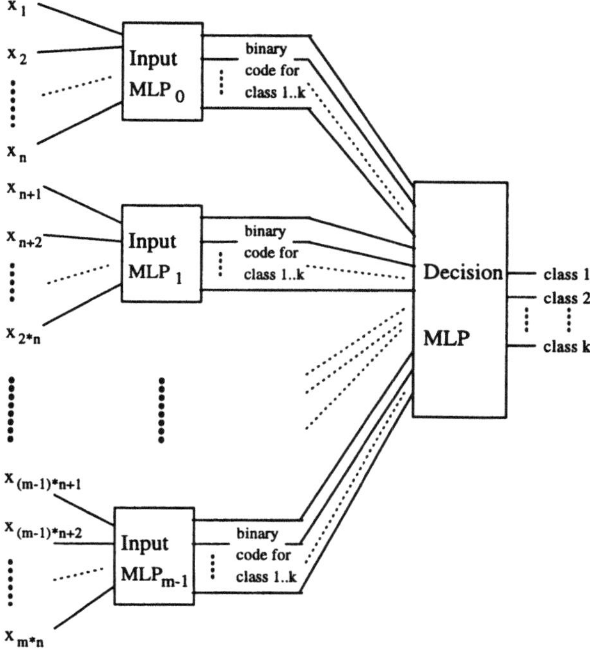

Figure 1: The multiple neural network architecture.

3 Training the System

The training occurs in two stages. All the modules are trained using the backpropagation algorithm [6].

In the first phase all sub-networks in the input layer are trained. The training set for each sub-network is selected from the original training set. The training pair for a single module consists of the components of the original vector which are connected to this particular network (as input vector) together with the desired output class represented in binary coding. All input modules can be trained in parallel very easily because they are all mutually independent.

In the second stage the decision network is trained. The training set for the decision module is built from the output of the input layer together with the original class number. To calculate the set each original input pattern is applied to the input layer; the resulting vector together with the desired output class (represented in a 1 out of k coding) form the training pair for the decision module.

The original training set is: $(x_1^j, x_2^j, \dots, x_l^j; d^j)$ for all $j = 1, \dots, t$. Here, $x_i^j \in R$ is the ith component of the jth input vector, d^j is the class number, and t is the number of training instances.

The module MLP_i is connected to:

$$x_{i\cdot n+1}, x_{i\cdot n+2}, \dots, x_{(i+1)\cdot n}.$$

The training set for the network MLP_i:

$$(x_{i\cdot n+1}^j, x_{i\cdot n+2}^j, \dots, x_{(i+1)\cdot n}^j; d_{BIN}^j), \forall j = 1, \dots, t.$$

The mapping performed by the input layer:

$$\Phi : R^{n*m} \mapsto R^{m*\lceil \log_2 k \rceil}.$$

The training set for the decision network:

$$(\Phi(x_1^j, x_2^j, \dots, x_l^j); d_{BIT}^j) \text{ and } j = 1, \dots, t.$$

The mapping of the decision network:

$$\Psi : R^{m*\lceil \log_2 k \rceil} \mapsto R^k.$$

4 Calculation of the Output

The mapping of the whole network is:

$$\Phi \circ \Psi : R^l \mapsto R^k.$$

The response r for a given test input $(a_1, a_2, \dots, a_l)$ is determined by the following function:

$$r = \Psi(\Phi(a_1, a_2, \dots, a_l)).$$

The k-dimensional output of the decision module is used to determine the class number for the given input. In the experiments the output neuron with the highest response was chosen as the calculated class. The differences between the winning neuron and the runner-up may be taken as a measure of accuracy.

5 On Generalization

The ability to generalize is the main property of neural networks. This is how neural networks can handle inputs which have not been learned but which are *similar* to inputs seen during the training phase. Generalization can be seen as a way of reasoning from a number of examples to the general case. This kind of reasoning is not valid in a logical context but can be observed in human behaviour.

The proposed architecture combines two methods of generalization. One way of generalizing is built-in to the MLP. Each of the networks has the ability to generalize on its input space. This type of generalization is common to connectionist systems.

The other method of generalization is due to the architecture of the proposed network. It is a way of generalizing according to the similarity of input patterns. This method of generalization is found in logical neural networks [1].

To explain the behaviour more concretely the following simplified example of a recognition system is given.

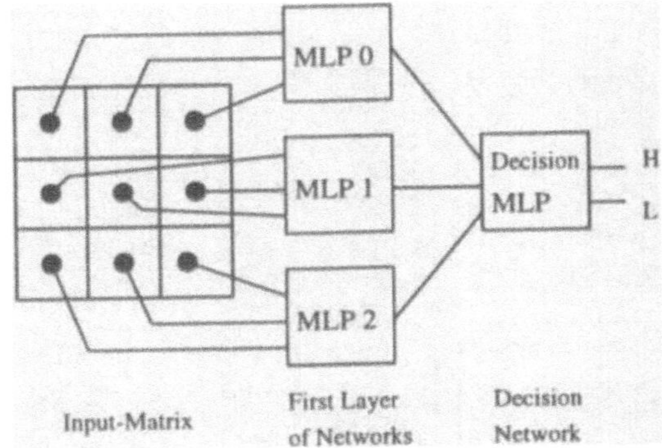

Figure 2: The example architecture.

A 3×3 input retina with the architecture shown in Figure 2 is assumed. Each of the nine inputs reads a continuous value between zero and one, according to the recorded gray level (black=1; white=0).

The network should be trained to recognize the simplified letters 'H' and 'L'. The training set is shown in Figure 3. The desired output of the input networks is '0' for the letter 'H' and '1' for the letter 'L'.

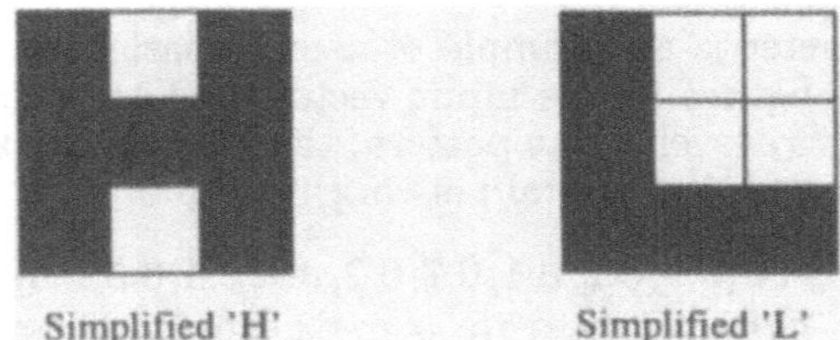

Figure 3: The training set.

The training subsets for the networks MLP_0, MLP_1, and MLP_2 are:

MLP_0	MLP_1	MLP_2
(1,0,1;0)	(1,1,1;0)	(1,0,1;0)
(1,0,0;1)	(1,0,0;1)	(1,1,1;1)

After completing the training of the first layer of networks it is assumed that the calculated output is equivalent to the desired output.The resulting training set for the decision network is:

$$(\Phi(1,0,1,1,1,1,1,0,1);1,0) = (0,0,0;1,0)$$

$$(\Phi(1,0,0,1,0,0,1,1,1);0,1) = (1,1,1;0,1)$$

After the training of the decision network the assumed response of the system to the training set is:

$$r_H = \Psi(\Phi(1,0,1,1,1,1,1,0,1)) = \Psi(0,0,0) = (1,0)$$

$$r_L = \Psi(\Phi(1,0,0,1,0,0,1,1,1)) = \Psi(1,1,1) = (0,1)$$

To show the different effects of generalization three distorted characters, shown in Figure 4 are used as the test set:

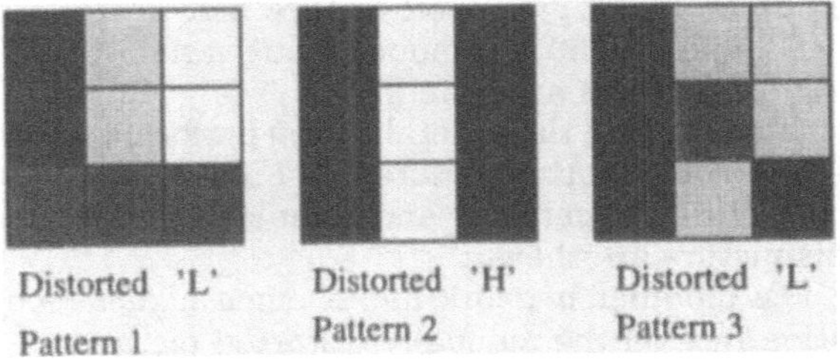

Figure 4: The test set.

The first character tests generalization within the input modules, the second shows the generalization on the number of correct sub-patterns, and the third character is an example of a combination of both. (The figures in the input vectors are according to the gray-level in the pattern; the outputs are taken from a typical neural network).

$$r_1 = \Psi(\Phi(0.9, 0.2, 0.1, 0.7, 0.2, 0.1, 0.5, 0.5, 0.5)) = \Psi(0.95, 0.86, 0.70) = (0.04, 0.96) \Rightarrow 'L'$$

$$r_2 = \Psi(\Phi(1.0, 0.0, 1.0, 1.0, 0.0, 1.0, 1.0, 0.0, 1.0)) = \Psi(0, 0.49, 0) = (0.91, 0.09) \Rightarrow 'H'$$

$$r_3 = \Psi(\Phi(0.9, 0.2, 0.2, 0.9, 0.5, 0.2, 0.9, 0.2, 0.9)) = \Psi(0.92, 0.65, 0.09) = (0.15, 0.89) \Rightarrow 'L'$$

6 Experiments

The proposed architecture was tested with different real world data sets. The number of input attributes was between 8 and 12000.

Throughout the experiment it appeared that the modular network converged for a large range of network parameters. Particularly for huge input spaces it was often very difficult to find an appropriate learning coefficient for a monolithic network, whereas convergence was no problem for the modular structure.

The time needed to train the modular network was much shorter than that for a monolithic network. In most cases it took less than half the time to train the network to a similar performance. For larger input spaces the training was up to ten times quicker (without parallel training).

For small input spaces (up to 60 attributes) the memorization and generalization performance of the modular network and a monolithic MLP were very similar on the real world data sets.

One task was to memorize five pictures of different faces. Each gray-level picture had a size of 75 by 90 pixels (6750 continuous input variables). The original pictures are from [5].

After training the generalization performance was tested with distorted pictures. In Figure 5 one training picture (upper left) and some degenerations of this picture are shown.

The modular network had a much higher recognition rate on the manually distorted pictures.

Another comparison was made on the ability to recognize noisy inputs. The noise on the pictures was generated randomly. In Figure 6 pictures with different noise-levels are shown. The modular network could recognize pictures with a significant higher noise-level than the single MLP; the results are shown in Figure 7.

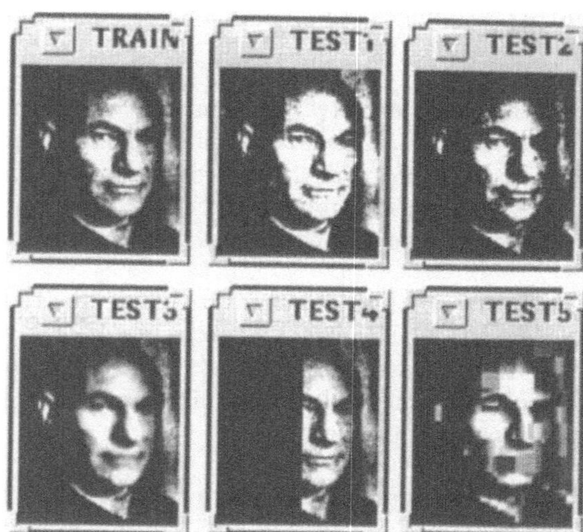

Figure 5: Original and distorted pictures.

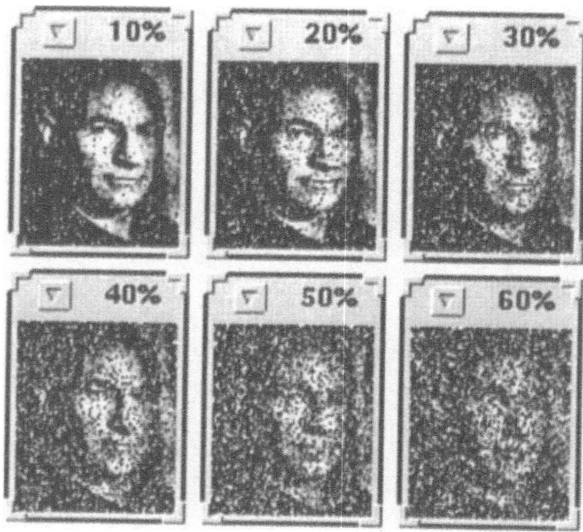

Figure 6: Examples of noisy test pictures.

From the above experiments it can be seen that the modular network has superior generalization abilities on high dimensional input vectors.

7 Limitations

The network is less useful for problems with very small input dimensions. The network has the ability to solve problems which are not linear separable. The proposed architecture has certain theoret-

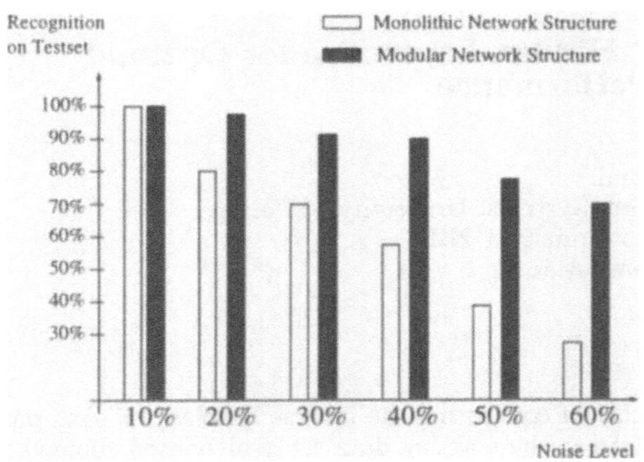

Figure 7: The performance on noisy inputs.

ical limitations; statistically neutral problems (like the XOR-problem) can not be learned. Monolithic MLPs are able to learn such problems but the generalization performance is very poor [7].

8 Conclusion

The usage of a modular architecture consisting of small MLPs to solve real world problems is demonstrated in this paper.

It is shown that for different real world data sets the training is much easier and faster with a modular architecture.

Two different approaches of generalization are combined in this model. It is demonstrated that this results in a generalization advantage on high dimensional input vectors.

Due to the independence of the modules in the input layer parallel training is readily feasible.

References

[1] I. Aleksander and H. Morton. *An Introduction to Neural Computing.* Chapman & Hall, second edition, 1995.

[2] R. Bellotti, M. Castellano, C. De Marzo, and G. Satalino. Signal/background classification in a cosmic ray space experiment by a modular neural system. In *Proc. of the SPIE - The International Society for Optical Engineering,* volume 2492, pages 1153–61. Springer-Verlag, 1995.

[3] T. Kohonen, G. Barna, and R. Chrisley. Statistical pattern recognition with neural networks: benchmarking studies. In *Proc. IEEE International Conference on Neural Networks,* pages 61–67. San Diego, 1988.

[4] L. Mui, A. Agarwal, A. Gupta, and P. Shen-Pei Wang. An adaptive modular neural network with application to unconstrained character recognition. *International Journal of Pattern Recognition and Artificial Intelligence,* 8(5):1189–1204, October 1994.

[5] University of Stuttgart. Picture directory. FTP from ftp.uni-stuttgart.de, /pub/graphics/pictures and tv_film/startrek/next_gen/portrait.

[6] D. E. Rumelhart, G. E. Hinton, and R. J. Williams. *Learning internal representations by error propagation,* volume I: Foundations. MIT Press, Cambridge, MA, 1986.

[7] J. V. Stone and C. J. Thorton. *Can Artificial Neural Networks Discover Useful Regularities?,* pages 201–205. 1995. Conference Publication No. 409 IEE. 26-28 June 1995.

Principal Components Identify MLP Hidden Layer Size for Optimal Generalisation Performance

M. Girolami
Department of Computing and Information Systems, University of Paisley,
High Street, Paisley, Scotland, PA1 2BE
Email: giro_ci0@paisley.ac.uk

Abstract

One of the major concerns when implementing a supervised artificial neural network solution to a classification or prediction problem, is the network's performance on unseen data. The phenomenon of the network overfitting the training data, is understood and reported in the literature. Most researchers recommend a 'trial and error' approach to selecting the optimal number of weights for the network, which is time consuming, or start with a large network and prune to an optimal size. Current pruning techniques based on approximations of the Hessian matrix of the error surface are computationally intensive and prone to severe approximation errors if a suitable minimal training error has not been achieved. We propose a novel and simple design heuristic for a three layer multi-layer perceptron (MLP) based on an eigenvalue decomposition of the covariance matrix of the middle layer output. This technique identifies the neurons which are contributing to the redundancy of data through the network and as such are additional effective network parameters which have a deleterious effect on the classifier surface smoothness. This technique identifies redundancy in the network data and so is not dependant on the network training having reached a minimal error value making the Levenberg-Marquardt approximation valid. We report on simulations using the double-convex benchmark which show the utility of the proposed method.

1 Introduction

The literature widely reports on second order based pruning methods which, it is generally accepted, have superseded magnitude based techniques. Hassibi *et al.* [3], dispensed with the assumption of a diagonal Hessian matrix of the error surface in developing the optimal brain surgeon (OBS) algorithm, this assumption was intrinsic to the optimal brain damage (OBD) algorithm, Le Cun *et al.* [6]. The cost of computing the inverse Hessian for each pattern in the training data set is alleviated somewhat by employing the Levenberg-Marquardt assumption [1], we have found that this makes the accuracy of computing the Hessian highly dependant on reaching the global minimum. For image or speech applications with high dimensional data, the practical requirements of manipulating the Hessian are highly taxing. For a typical raw digit recognition application with a 16×16 pixel input image and a $256 \times 10 \times 10$ network, the storage requirements for the Hessian are high (27 Mbytes). Hassibi *et al.* report dramatic reductions in the network size after the application of OBS, there is however no corresponding decrease in test error. We discuss this shortcoming of the OBS algorithm in the next section and raise question as to the practical utility of the algorithm implementation.

Moody *et al.* [8], have proposed the PCP (Principal Components Pruning) algorithm which gives a measure of the saliency of each weight based on the associated eigen value of the input data (to the layer) covariance matrix. Removal of the least significant eigennodes and projection of the weights onto the remaining eigenvectors and their transpose, causes each of the layer weights to act as eigenfilters on the input data. A PCA is being performed on the data input to the layer, we have found that for applications with continuous value outputs this technique successfully avoids overfitting of data in certain cases.

Motivated by Moody's PCP algorithm we propose that for a three layer MLP a simple eigenvalue analysis of the middle layer output will identify the excessive neurons in the layer, and so identifies the optimal size of the hidden layer. This simple design tool has been verified using the double convex benchmark, and has been successfully used in MLP design for satellite image cloud pattern classification.

2 Parameters Affecting Network Generalising Ability

Bishop [2] draws the analogy between the complexity of a network and the order of polynomials used in a regression, Shiavi [9] also details the problem of complexity in regression. This analogy is applied to ANN applications where the number of free parameters in the network is likened to the order of polynomial. A simple network with few weights will be unable to map the regression, whereas a large network with a great number of weights will be able to fit to the data points very well but will then oscillate between each point. If, as in the case of ANN's, the data is noisy an exact fit to the data points is to be discouraged. This problem has also been referred to as the bias-variance dilemma [1].

When considering an architecture for a feedforward network, in our case an MLP , there are certain constraints imposed on the problem. These are the dimensions of the input and output vectors. Although the input and output vector dimensions are fixed largely due to the specific problem domain, there is some latitude in setting these dimensions by the representation of the data. Prior processing of data may highlight redundant variables which may be discarded from the input. The output, for example, may be a binary 1 or 0, one output neuron with values 0 or 1, or two outputs with values 10, 01 may be chosen. This is largely dependant on the designers view of the problem, however, for our purposes we can consider the size of the input and output layers to be fixed. This then leaves only the number of nodes in the middle layer as the variable in the architecture selection, we are restricting ourselves to a three layer network.

3 Network Pruning and Destructive Pruning Algorithms

The simplest technique of pruning a network of excess parameters is to consider weights which have a small 'saliency' that is, those whose deletion will have least effect on the training error. It is plausible that weights with small magnitude will contribute least to the weighted sums of the neurons being fed and so their deletion will have limited effect on the overall error. It has been found that these magnitude-only based techniques are limited as it is not so obvious that a weight with small magnitude may indeed carry significant information and its deletion may have serious effects on the network output. Le Cun *et al.* [6] argue that there is a need to move from magnitude based saliency measures to more robust saliency measures for network pruning. Le Cun *et al.* devise a weight saliency based on the minimisation of the change in an objective function at the deletion of the particular weight. They start by taking a Taylor series expansion of the error criterion, this will then give the change in error experienced for a particular delta in the weights vector, or simply the deletion of a weight.

$$dE = \frac{\delta E^T}{\delta w} dw + \frac{1}{2} dw^T H dw + O(\| dw \|^3) \quad (1)$$

The important term is the Hessian matrix denoted in (1) as H, which is the matrix of second derivatives of the error surface with respect to the network weights. The goal is then to find a set of parameters dw which minimises the increase in error, that is dE. This then is the second order based saliency measure for the weights.

The major problem is the computational difficulty in computing the Hessian matrix, by making the assumption that the Hessian matrix is diagonal a simple saliency measure is developed. However, Le Cun reports that off diagonal terms start to affect OBD after a 30% reduction in the number of parameters. We have found in simulations that the diagonal assumption is rarely valid. OBD has been largely superseded by Hassibi and Storks optimal brain surgeon (OBS) [3], which dispenses with the diagonal assumption. A saliency measure and weight removal function are developed, this however now requires the computation of the inverse Hessian. Although a mathematically elegant method for the inverse Hessian computation is developed, there is a dependency on the validity of the Levenberg-Marquardt assumption being valid [3]. We have found that this sensitivity is high and that *a priori* knowledge of the region of validity is required before confident computation of a valid inverse Hessian can take place. Hassibi and Stork report on a network which is pruned from 58 weights down to 14. This is impressive, however, further examination of the paper shows that the test error is not reduced, that is generalisation is not enhanced. This is of course the case, as OBS will eliminate a weight with low saliency and automatically adjust the remaining weights, to retain the global minimum position of error. What this is doing is ensuring that as weights are removed the effective number of parameters [7] are kept the same, so if no retraining occurs then the effective complexity of the network is unchanged, and the generalisation performance will not be improved. Training will then effectively be training a simpler topology network and so should yield better test performance. The other point to note is that the reported network whose weights are reduced from 58 to 14 has an architecture of $17 \times 3 \times 1$, with one bias weight. The majority of pruning occurs, as it would, in the first layer. Eleven of the input nodes are effectively removed from the network, it is conjectured that the input data vectors have a large degree of redundancy within them, and the remaining six dimensions hold the essential information for the vector. The use of principal components analysis would identify these data correlations at minimal computational cost.

4 Principal Component Analysis

A full analysis of principal component analysis (PCA) can be found in Bishop [1], Haykin [4], we shall only present the details that are required to fully understand our proposed heuristic. Essentially PCA forms a projection from the original vector basis onto a basis which has a lower dimension; what requires to be identified is the optimal subspace and the associated basis vectors. It is shown that the optimal basis vectors are the eigenvectors of the covariance matrix of the input data set. So if a vector of dimensionality M is represented by a vector of dimension N where $N \ll M$, and the basis vector is the N most significant eigenvectors of the data covariance matrix, then the representation error is minimal.

So if we have a particular data set which has a high dimensionality, forming the covariance matrix and performing an eigenvalue decomposition, it will allow us to use the eigenvalues to assess which dimensions carry small amounts of information. We can assess the mean square error incurred by removing these eigenvectors. Using the significant principal components (eigenvalues) we can then represent the information from the original data set with an insignificant loss of information, by a data set of lower dimension. This then suggests a neural network design metric to assess the optimal number of nodes in the middle layer.

1. Train the network with an excessively large middle layer.
2. Pass the training data through the network in the forward pass and save the middle layer response.
3. Compute the covariance matrix of the middle layer output.
4. Perform an EVD on the covariance matrix.
5. Assess the number of significant neurons in the middle layer from the eigenvalue relative magnitudes.
6. Train a new network with a middle layer of size indicated by 5.

5 Simulations

In performing a classification task, the more complex the boundary between classes the more difficult is the problem for the MLP classifier. The double convex region benchmark [4] is a two dimensional problem where two interpenetrating convex shaped regions split the $X - Y$ plane into two distinct areas. All points will fall into either one or the other region, so the problem for the ANN classifier is to correctly classify a given series of x/y co-ordinates, assigning each to either region one or two. It can be seen that the problem is a highly nonlinearly separable classification task, the complexity of the task can be likened to the order of polynomial required in generating the boundary curve in a regression problem. The benefit of this benchmark is that the results can be visually checked, as can the training data sets ability to fully define the required regions. Figure 1 shows a diagram of the problem showing the two distinct regions. Also in Figure 2 there is a plot of the misclassification of a three layer MLP against the size of the hidden layer. It is clear that overfitting commences after more than 45 to 50 neurons appear in the middle layer.

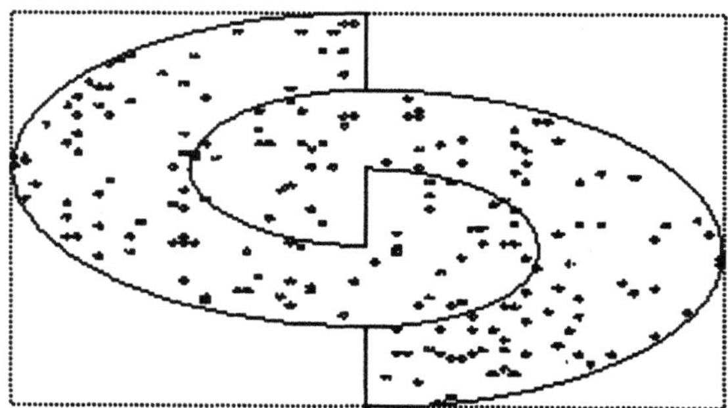

Figure 1: The double convex regions simulation.

Hidden Neurons vs. Percentage Mis-Classification

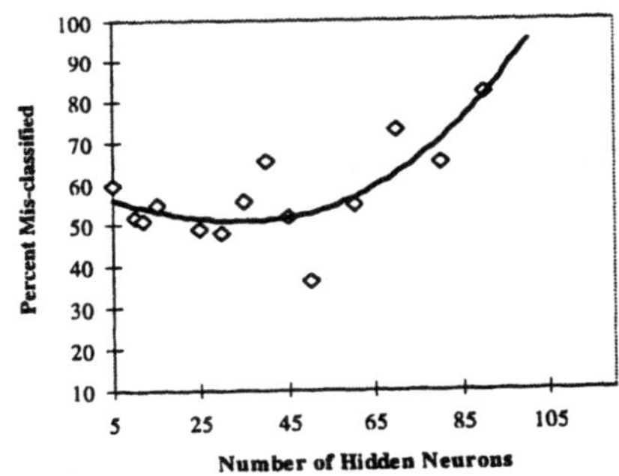

Figure 2: Classification performance.

Weight Saliency vs. Eigenvalues

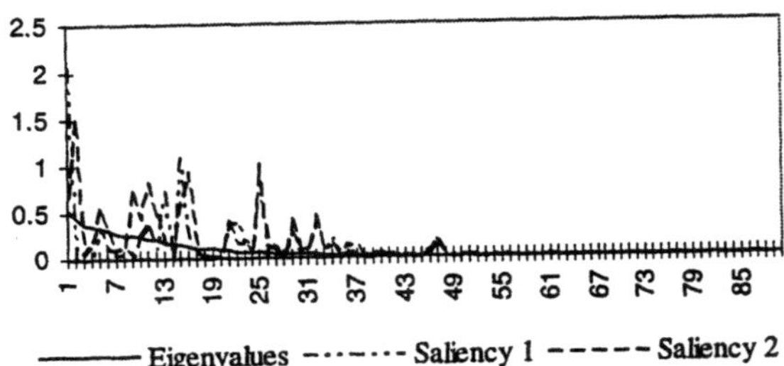

Figure 3: $2 \times 90 \times 2$ Network Weight Saliency.

We now take the middle layer output of a network which has a layer size of 90 neurons (which is severely overfitting the data) and perform the proposed design heuristic. An EVD gives a vector with all the eigenvalues of the covariance matrix, this is shown in Figure 3. It is clear from this that the trailing forty eigenvalues are insignificant and indeed the corresponding weight saliency is insignificant considering the values of saliency ranges from 1.9 down to 10^{-7}. This finding is in agreement with the results in Figure 2 which suggests that optimal generalisation occurs with 45 to 50 neurons in the middle layer. Figure 3 shows that the first 50 neurons have a significant contribution to the layer output and as such the trailing 40 neurons should be removed and the optimal hidden layer size is 50 neurons.

6 Conclusions

We have proposed a simple design heuristic for identifying the optimal number of nodes for the middle layer of a three layer MLP , this has been verified by simulation using the double-convex benchmark. In terms of computational loading, approximation validity and design utility this technique outperforms OBS and OBD in identifying the optimal number of neurons for generalisation performance. Further work includes the use of robust PCA networks as proposed by Karhunen *et al.* [5], to perform EVD on data with impulsive noise and outliers.

References

[1] C. Bishop. Neural networks for pattern recognition. *Oxford University Press*, 1995.

[2] C. Bishop. Regularization and complexity control in feedforward networks. In *International Conference on Artificial Neural Networks*, volume 1, pages 141–148, 1995.

[3] B. Hassibi, D.G. Stork, and G. Wolff. Optimal brain surgeon and general network pruning. In *IEEE International Conference on Neural Networks*, volume 1, pages 293–299, 1992.

[4] S. Haykin. *Neural Networks: A Comprehensive Foundation*. MacMillan Publishing, 1995.

[5] J. Karhunen and J. Joutsensalo. Generalisations of principal component analysis, optimisation problems and neural networks. *Neural Networks*, 8(4):549–562, 1995.

[6] Y. Le Cun, J.S. Denker, and S.A. Solla. Optimal brain damage. *Advances in Neural Information Processing Systems*, 2:598–605, 1990.

[7] J.E. Moody. The effective number of parameters: An analysis of generalisation and regularisation in nonlinear learning systems. In *Advances in Neural Informations Processing Systems*, pages 847–854. Morgan Kauffmann, 1992.

[8] J.E. Moody, A.U. Leen, and T.K. Leen. Fast pruning using principal components. In *Advances In Neural Information Processing*, volume 6. Morgan Kauffmann, 1994.

[9] R. Shiavi. *Introduction to Applied Statistical Signal Analysis*. Aksen Associates Incorporated Publishers, Irwin, 1991.

Bernoulli Mixture Model of Experts for Supervised Pattern Classification

N. Elhor, R. Bertrand and D. Hamad,
Centre d'Automatique de Lille, Bâtiment P2,
Université des Sciences et Technologies de Lille, F-59655
Villeneuve d'Ascq Cedex, France.
Email: en@cal.univ-lille1.fr

1 Introduction

Artificial neural networks have been applied to solve hard problems in different engineering domains, thanks to their capability of universal function approximators [4]. However, when these networks are used in their standard forms, 'black-box models', their performances are inferior to dedicated statistical solutions. Performances can be largely improved if we introduce prior knowledge in network architectures. If the real problem has an obvious decomposition, then it may be possible to design a network architecture by hand. Unfortunately, this is not always possible.

An alternative approach is to use a set of networks and to assign each one, thanks to an appropriate learning algorithm, to solve a part of the real hard problem. Network outputs are then combined together to form the solution to this problem. This is the principle of the mixture model of experts also called combined networks [5, 6]. In fact, performances of combined networks can be better than that obtained by the single best one [1]. The modular network architecture is defined as a set of K modules called expert networks and an integrating module called gating network. Each expert is constituted of a feedforward neural network while the gating network is composed of a linear network followed by a softmax non linearity. The gating network performs the function of a mediator among the expert networks.

Training networks requires that users define an error measure in order to adapt the network weights to achieve certain performance criteria. Jacobs *et al.* trained the modular network to model the probability distribution of the desired responses of a database. The likelihood function, or probablity distribution of targets, is viewed as an associative Gaussian mixture model and maximized by means of stochastic gradient ascent algorithm. Although this approach has been applied to pattern classification it is essentially adapted for regression problems [5, 6].

This paper presents a Bernoulli mixture model of experts for supervised pattern classification problem. Since the target variable is of binary type, the likelihood function must be considered as an associative Bernoulli mixture model. In fact, the idea of using Bernoulli error or cross-entropy for training feedforward networks has been suggested in [1]. Our purpose is to show that the proposed likelihood function is more suitable for pattern classification and accelerate the learning procedure.

2 Bernoulli Mixture Model of Experts

The idea of mixture model of experts has been first proposed by Jacobs *et al.* and successfully applied to the non linear control of a robot arm [5]. In this paper we are concerned with the supervised classification problem.

The network architecture is defined by a set of K modules called expert networks and an integrating module called gating network. Each expert is constituted of a simple perceptron while the gating network is composed of a linear network followed by a softmax non linearity. The gating network performs the function of a mediator among the K experts.

Let y_k denotes the output of the k-th expert network, π_k the output of the k-th neuron of the gating network and y the output of the modular network. The output y is given by:

$$y = \sum_{k=1}^{K} (\pi_k . y_k). \qquad (1)$$

Expert network equations Since each expert k is a simple perceptron, its output is defined by:

$$y_k = f(v_k). \tag{2}$$

where the transfert function $f(.)$ is of sigmoidal form:

$$f(v_k) = \frac{1}{1 + e^{-v_k}} \tag{3}$$

and,

$$v_k = \mathbf{X^T}.\mathbf{w_k} \tag{4}$$

with $\mathbf{w}_k = (w_{k0}, w_{k1}, \ldots, w_{kn})$ is the weight vector of connections between input neurons and the output neuron of the k-th expert and $\mathbf{X} = (x_0, x_1, \ldots, x_n)^T$ is the input vector.

Gating network equations The values π_k of the output neurons of the gating network are constrained to satisfy two requirements [5]:

$$0 \leq \pi_k \leq 1 \tag{5}$$

and,

$$\sum_{k=1}^{K} \pi_k = 1. \tag{6}$$

These two constraints are necessary if the outputs π_k are to be interpreted as *prior* probabilities. Given a set of unconstrained variables, $\{u_k, k = 1, 2, \ldots, K\}$, we may satisfy the contraints of Equations (5) and (6) by defining the output neurons of the gating network as following [2]:

$$\pi_k = \frac{e^{u_k}}{\sum_{j=1}^{K} e^{u_j}} \tag{7}$$

where u_k is given by:

$$u_k = \mathbf{X^T}.\mathbf{a_k} \tag{8}$$

with $\mathbf{a}_k = (a_{k0}, a_{k1}, \ldots, a_{kn})$ is the weight vector of connections between input neurons and the k-th output neuron of the gating network.

Probability density of the mixture model of experts Since desired responses are binary variables, the probability density of the desired response d, given the input vector $\mathbf{X}$ and that the k-th expert network is chosen, can be expressed as a Bernoulli law:

$$P(d/\mathbf{X}, k) = y_k^d.(1 - y_k)^{(1-d)}. \tag{9}$$

Chow *et al.* have shown that learning neural network by Bernoulli error measure is much more suitable for pattern classification problems [3]. Their appoach has been used for a multilayer perceptron. In this paper, we apply Bernoulli distribution for training mixture model of experts.

The probability density of the desired response given the input vector $\mathbf{X}$ is a linear combination of K different densities of Bernoulli, Bernoulli mixture model of experts:

$$P(d/\mathbf{X}) = \sum_{k=1}^{K} \pi_k.P(d/\mathbf{X}, k). \tag{10}$$

If we introduce the expression (9) in (10), we obtain:

$$P(d/\mathbf{X}) = \sum_{k=1}^{K} \pi_k.y_k^d.(1 - y_k)^{(1-d)}. \tag{11}$$

To assist in the formulation of the learning algorithm, we define the *posterior* probability associated with the output of the k-th expert network as:

$$h_k = \frac{\pi_k.y_k^d.(1 - y_k)^{(1-d)}}{\sum_{j=1}^{K} \pi_j.y_j^d.(1 - y_j)^{(1-d)}}. \tag{12}$$

This probability is conditional on both the input vector $\mathbf{X}$ and the desired response d. From this definition, we also note that, like the ***prior*** probabilities represented by the activations π_k, the ***posterior*** probabilities h_i satisfy two necessary conditions :

for all k

$$0 \leq h_k \leq 1 \tag{13}$$

and,

$$\sum_{k=1}^{K} h_k = 1. \tag{14}$$

3 Learning Algorithm

The learning data base is assumed to form a finite set of P paired observations $\{(\mathbf{X}_1, d_1), (\mathbf{X}_2, d_2), \ldots, (\mathbf{X}_p, d_p), \ldots, (\mathbf{X}_P, d_P)\}$ where $\mathbf{X}_p$ is the p-th vector of data and d_p is the index label: $d_p = 1$ if $\mathbf{X}_p$ belongs to the class C_1 and 0 if $\mathbf{X}_p$ belongs to the class C_2 (not C_1). For the rest of this section we shall eliminate the index p for simplicity of expressions.

The goal of the learning algorithm is to train the network to model the probability distribution of the desired responses. It attempts to find the synaptic weights of the network which maximize the log-likelihood function defined by:

$$l = ln[P(d/\mathbf{X})]. \tag{15}$$

By replacing (11) in (15) we obtain:

$$l = ln[\sum_{k=1}^{K} \pi_k.P(d/\mathbf{X}, k)]. \tag{16}$$

Learning procedure consists of the application of input vectors $\mathbf{X}$ to the expert networks and the gating network simultaneously and the calculation of the sensitivity of the expression (16) with respect to:

- the weights of the different expert networks $\mathbf{w}_k$, $k = 1, \ldots ,K$,
- the weights of the gating network $\mathbf{a}_k$, $k = 1, \ldots , K$.

The sensitivity of the log-likelihood with respect to $\mathbf{w}_k$ is given by:

$$\frac{\partial l}{\partial \mathbf{w}_k} = \frac{\sum_{k=1}^{K} \pi_k . \frac{\partial P(d/\mathbf{X}, k)}{\partial \mathbf{w}_k}}{\sum_{k=1}^{K} \pi_k.P(d/\mathbf{X}, k)} \tag{17}$$

where:

$$\frac{\partial P(d/\mathbf{X}, k)}{\partial \mathbf{w}_k} = h_k.(d - y_k).\mathbf{X} \tag{18}$$

The sensitivity of the log-likelihood with respect to $\mathbf{a}_k$ is given by:

$$\frac{\partial l}{\partial \mathbf{a}_k} = (h_k - \pi_k).\mathbf{X} \tag{19}$$

h_k is the posterior probability defined by Equation (12).

The adaptation rules, weights of different expert networks and the gating network, are given by:

$$\mathbf{w}_k(t + 1) = \mathbf{w}_k(t) + \alpha.h_k.(d - y_k).\mathbf{X} \tag{20}$$

$$\mathbf{a}_k(t + 1) = \mathbf{a}_k(t) + \alpha.(h_k - \pi_k).\mathbf{X}. \tag{21}$$

Note that, since we are using gradient ascent to maximize l, the learning rate α in the right-hand side of Equations (20) and (21) is $+\alpha$ and not $-\alpha$ with $0 \leq \alpha \leq 1$.

4 Experimental Results

Many experiments have been carried out in order to evaluate the performance of the Bernoulli mixture model of experts compared with the Gaussian mixture model of experts proposed in [5, 6]. However for the sake of clarity, only two examples are presented in the next section. For these experiments, synaptic weights of Bernoulli and Gaussian mixture models have been randomly initialized between [-0.5, 0.5]. The learning rate α has been fixed to 0.5.

4.1 The Two Spirals Problem

In this experiment we train Bernoulli and Gaussian mixture models to discriminate two spirals in 2-D plane. Each spiral has 30 points as it is depicted in Figure 1(a) and (b). Although the decision boundaries are similar as it is shown in Figure 1(a) and Figure 1(b), the Bernoulli mixture model converge after 1000 epoch while the Gaussian mixture model necessitates 10000 epochs.

4.2 The XOR problem

The data set used here is shown in Figure 2(a) and (b). There are four classes of samples and each one has 200 samples. The classes are drawn from four Gaussian distributions centered at (-2, -2), (2, 2), (-2, 2) and (2, -2). Their covariance matrices are all equal to 0.7.Id, where Id is the identity matrix. The two classes centered on (-2, -2) and (2, 2) have the same label, 'black square' in the Figure 2. The two remaining classes centered on (-2, 2) and (2, -2) have also the same label which is a 'white square' in Figure 2. Therefore the data set represents two clusters corresponding to the XOR problem.

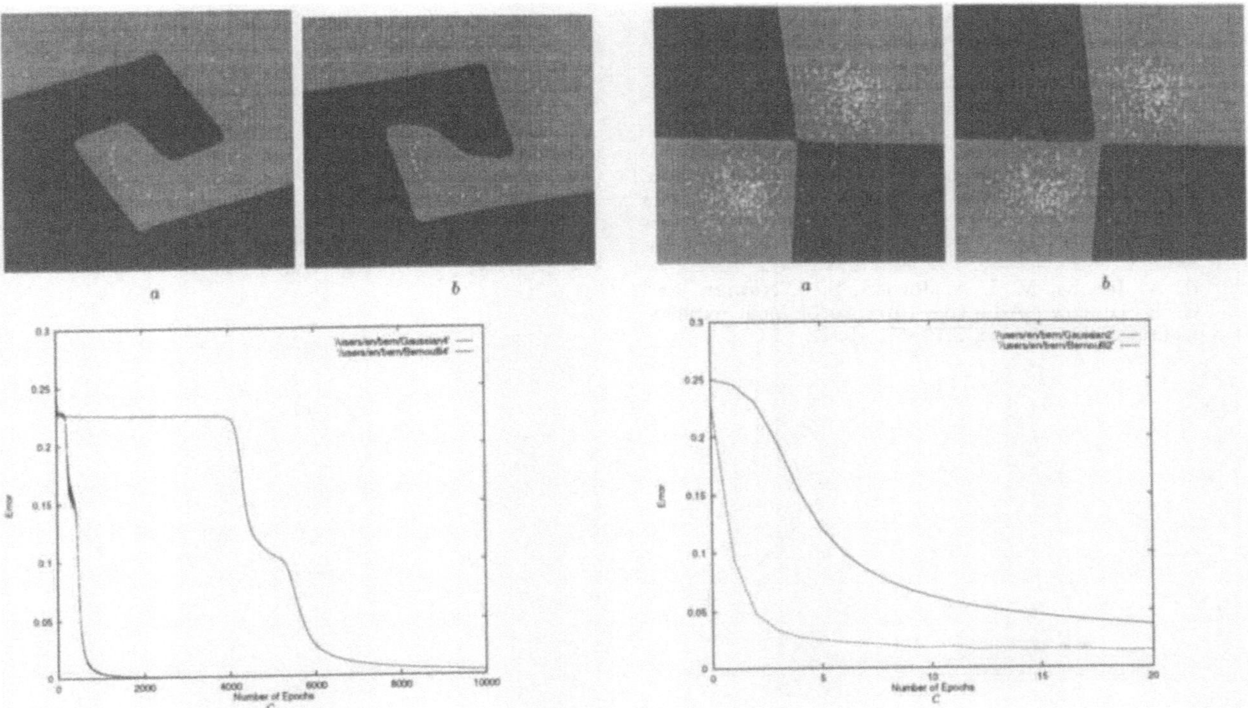

Figure 1. Two spirals training data set. Decision boundaries obtained by ten experts for Bernoulli (a) and ten experts for Gaussian mixture models (b), training error performances vs number of epochs (c).

Figure 2. XOR training data set. Decision boundaries given by two experts for Bernoulli (a) and two experts for Gaussian mixture models (b), training error performances vs number of epochs (c).

Bernoulli and Gaussian mixture models have been trained to discriminate the two clusters. Figure 2(a) and Figure 2(b) represent the decision boundaries between clusters obtained by each mixture model. Although decision boundaries seem to be similar for both models, the Figure 2(c) shows clearly that the training error performance of Bernoulli mixture model is better than the Gaussian one.

5 Conclusion

We have presented a modular network, called Bernoulli mixture model of experts, for supervised pattern classification. This network is constituted of a set of expert networks and a gating network which ponders the contribution of these experts to provide the ouput of the network. experts networks cooperate between them to decompose a complex classification task into simpler undertasks easily achieved by simple perceptron networks. Learning procedure has been given by maximization of a likelihood function which is a linear combination of Bernoulli distributions. The proposed model is much more suitable for pattern classification and accelerates learning times. This approach can be easily generalised and integrated in hierarchical network architectures.

References

[1] C. M. Bishop. *Neural Networks for Pattern Recognition.* Clarendon Press, Oxford, 1995.

[2] J. S. Bridle. Probabilistic interpretation of feedforward classification network outputs, with relationships to statistical pattern recognition. In F. Fogelman-Soulié and J. Hérault, editors, *Neuro-Computing: Algorithms, Architectures and Applications*, NATO Series, Vol. F68, pages 227–236, 1990.

[3] M. Y. Chow, A. Menozzi, J. Teeter, and J. P.Thrower. Bernoulli error measure approach to train feedforward artificial neural networks for classification problems. In *IEEE Inter. Conf. Neural*

Networks, volume 1, pages 44–49. Orlando, Florida, USA, 1994.

[4] S. Haykin. *Neural Networks a Comprehensive Foundation.* IEEE Computer Society Press, 1994.

[5] R. A. Jacobs and M. I. Jordan. A competitive modular connectionist architecture. In R. P. Lippmann, J. E. Moody, and D. J. Touretzky, editors, *Advances in Neural Information Processing Systems 3*, pages 767–773, San Mateo, CA, 1991. Morgan Kauffmann.

[6] R. A. Jacobs, M. I. A. Jordan, S. J. Nowlan, and G. E. Hinton. Adaptive mixture of local experts. *Neural Computation*, 3:79–87, 1991.

Electric Load Forecasting with Genetic Neural Networks

F.J. Marín[1] and F. Sandoval[2]
[1] Dpto. Electrónica, Universidad de Málaga, Campus Teatinos, 29071- Málaga, Spain
[2] Dpto. Tecnología Electrónica, Universidad de Málaga, Campus Teatinos, 29071- Málaga, Spain
Email: marin@ctima.uma.es

Abstract

This paper presents an evolution algorithm for optimizing a neural network architecture. The procedure establishes the structure and the training algorithm, as well as searching the minimal topology of the network, eliminating neurons and interconnection weights. The model network, this is, feedforward or feedback, can be selected by the user. This methodology is applied to the real problem of the forecasting in power system load in the city of Málaga (Spain) between the years 1992 and 1993. The results produced by the evolution algorithm are tested with a statistical regression analysis and with other training algorithms of paradigms of neural networks.

1 Introduction

Electric load forecasting is an integral part of the decision-making activities in power system planning. Load forecasting consists of finding a relationship between past input-output observed pairs $[u^{t-1}, y^{t-1}]$ and future outputs $y(t)$ given by the recurrent equation:

$$y(t) = g(u^{t-1}, y^{t-1}) + v(t) \tag{1}$$

where $v(t)$ is the noise and means that the next output is not an exact function of past data. The goal of forecasting is to find the most exact function $g()$, so that $v(t)$ shall be as small as possible. The function $g()$ is usually modeled with statistical techniques [5] or artificial neural networks (ANNs) [3].

This paper presents an evolution algorithm to optimize a neural network architecture, i.e. an evolutionary neural network. The procedure establishes the structure and the training algorithm, and searches the minimal topology of the network, eliminating unnecessary neurons and interconnection weights. The model network, that is, feedforward or feedback, can be selected by the user. This approach is applied to the real problem of forecasting in power system load and the results produced are compared with a statistical regression analysis and with other training algorithms of the ANNs paradigm.

2 Genetic Neural Networks

The representation scheme we propose is completely open to any kind of topology. The selected paradigm to solve a specific problem is totally transparent to the designer, his only task being to adapt the fitness function to reward or penalise the selected class of connectivity. Each individual's genotype is represented by a binary string with two independent chromosomes.

The first chromosome defines the underlying neural network. It includes the following fields: input units (I), output neurons (O), number of hidden layers (L), hidden neurons for each layer (H_l), type of connectivity (μ), and number of weight coding bits (W). The fields I, O, μ and W must be initially chosen by the user. The fields L and H_l are given by the user as maximum values. The field μ is coded with one bit only, where $\mu = 0$ means feedforward connectivity and μ=1 total connectivity. The connection weights are dynamically allocated in the second chromosome depending on the values of the first chromosome.

The genetic algorithm (GA) initially generates a random population of individuals that code specific ANNs. It should be noted that each individual has a completely different ANN topology, so that the search for the optimum topology can be achieved in a massively parallel way. Just one change in the first chromosome implies lots of changes, allowing or disabling connections, in the second chromosome. With this, the computational power is widely improved because many more zones of the search space are swept in shorter computational time.

The sigmoid activation function is used with output between 0.2 and 0.8 to avoid oversaturation on the network weights, with the resulting increase of learning speed. The fitness function carries out both the learning and the optimization of the ANN. To do that, we propose a fitness function which is composed of three terms, although there could be more: A first term to penalise the error between the target output and obtained output, $\theta(E) = \alpha.E$; a second term used to penalise a high number of hidden layers, $\zeta(H) = \beta.\sum H_l$; and the third one to penalise the number of weights and bias if $\mu = 0$, $\eta(W) = \delta.(\sum weights + \sum bias)$:

$$\text{fitness} = \theta(E) + \zeta(H) + \eta(W). \quad (2)$$

The constants α, β and δ are chosen according to the penalty degree we want for each term.

We have chosen the proportional selection operator (roulette wheel) [1], the two point crossover operator, and the per individual mutation operator, which act on the second chromosome, for individuals of the same size. The GA model belongs to the cooperative-coevolutive kind [4], initially conceived for function optimization. We have adapted it for ANN optimization, keeping the following features: the global population is divided into several independently coevolving populations (species), each one representing a subcomponent of the global solution, which is obtained by joining representative individuals of every species (co-operation) [2].

The available historical data have been provided by Compaia Sevillana de Electricidad (Spain), corresponding with the electric peak loads consumed in the distribution network of the province of Málaga during years 1992 and 1993. The input variables used for our simulations have been: $Pot(-6,-5,-1,0)$: consumed power six, five, and one days before, and present-day load, and $Tx(0,+1)$: present and predicted day maximum temperatures, and Pots: output power or consumed power in the predicted day. For all our simulations, the learning patterns are the historical data of 1992, validating each model results with data from January to September of 1993. To test the benefits of the proposed approach, it has been compared with the classical regression method and other ANN paradigms.

3 Results

Table 1 shows the results obtained in terms of the sum square error (SSE) and mean absolute percentage error (MAPE).

Table 1: Results obtained for a linear ANN, ARX model, radial basis function networks, Elman recurrent networks, and genetic feedforward and feedback networks, respectively. All the simulations have been trained with data from 1992 and have been validated with data from 1993.

	SSE92	SSE93	RE92 (%)	RE93 (%)
Lineal ANN (normaliz.)	2,6299	0,9800	4,0555	3,0417
Lineal ANN (no normal.)	195430	72822	4,0376	3,0287
ARX-Model	195433	72825	4,0484	3,0363
Radial Basis ANN (6x8x1)	2,5726	0,9426	4,0352	2,9595
Levenberg-Marquardt	1,86838	0,9230	3,4077	2,9354
Elman 6x5x1 (bipolar)	8,5854	3,6850	3,6191	2,8758
Genetic Feedforward	2,6347	0,9618	4,1676	2,9360
Genetic Feedback	2,5318	0,8947	4,1305	2,8986

Lineal ANN: This is the simplest ANN model. There are no hidden layers, the input signal is directly propagated to the output layer and the activation function is linear. The learning rule used is the least mean squares (LMS) with learning rate (delta rule). The results are presented in the first line, for data normalized in the interval [0,1]. The results are practically equal to those obtained with ARX model.

Autoregressive model with exogenous variables (ARX): This model searches for a linear dependence between the variables, in a way similar to a regression equation, supposing that, in Equation (1), $v(t)$ is white noise (gaussian normal distribution). The order is $AR(1\ 2\ 6\ 7)$ with the autoregressive parameters and order $X(1)$ with maximum temperature as the exogenous variable. Results are shown in the third row of Table 1 and, as expected, they are equal to those obtained with the linear ANN model.

Radial basis NNs: These are two-layered ANNs. The first (hidden) layer is self organizing with a nonlinear activation function that computes the analogy between the input patterns and the patterns that express the weights associated to each neuron. The second layer acts in a linear way. The output of the hidden neurons is a gaussian function centered around the distance between each input vector, x, and every weight vector of that hidden neuron, with an adaptable factor s_c called the influence range that acts as a normalizing factor and determines the width of the gaussian bell. For near outputs (small bell widths) s_c must be big enough, thus assuring a good generalization. The best results obtained corresponding to 8 neurons with $s_c = 4$ are shown in the fourth row of Table 1. Beginning with 9 neurons the network becomes overtrained. The weight values of the first layer are calculated by means of a competitive learning algorithm. This model's results are better than those obtained by linear networks and autoregressive models.

Backpropagation Neural Network: These are feedforward nets with one or more hidden layers. In our simulation, there is a hidden layer with 5 neurons and an output layer with one neuron, all with a sigmoidal activation function. The learning rule is a generalization of the Widrow-Hoff rule called generalized delta rule which uses the descending gradient approach (or Newton's method). During these simulations it frequently fell in local minima, even when a momentum was added with an adaptive learning rate. Because of that, the net was trained with the Levenberg-Marquardt method and adaptive learning rate. We obtained the best results for generalization from feedforward nets (Table 1) and, besides, in a short time (50 epochs).

Elman Recurrent Network: This is a backpropagation network with feedback connections in the hidden layer. The output layer is linear, working with unnormalized data. Among all the neural paradigms we have simulated, this is the one that achieves the best results in generalization (Table 1), with a relatively small number of iterations (about 1000 epochs). For greater numbers of epochs the network is overtrained.

Genetic Feedforward Network: The GA performs the design and learning of a feedforward ANN with a maximum number of five hidden neurons. The activation function is the sigmoid with slope $p = 1.0$, and the weights are limited to the interval [-10,10] with an 8-bit encoding. We took the proportionality constants of the fitness function (2) in order to reward the obtaining of a minimum MAPE better than the searching of a minimal network. The population consists of 100 individuals that evolve for 100 generations. Figure 1 shows a plot of the real historical data from year 1993 (dotted line) and the data obtained by the network (continuous line). The results are comparable to the best obtained with traditional ANN paradigms for feedforward networks,

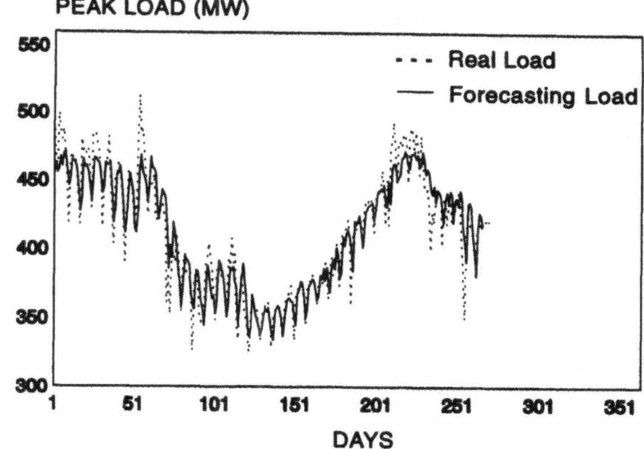

Figure 1: Real peak load of year 1993 (dotted line) and forecasted peak load (continue line) obtained with genetic feedforward networks, which has been trained with the data of year 1992.

but with the advantage that the genetic learning performs the search for the optimal topology, too, and it is able to avoid local minima. From a total of 41 connections, we found that 7 were eliminated.

Genetic Feedback Network: The net has 5 hidden units with autoconnections and lateral connections among themselves, and an output unit with bias weight. The GA parameters were kept the same as the feedforward case. The results were better than the ones obtained with feedforward networks and were comparable to the values output by the Elman feedback nets.

4 Conclusions

The goal of this paper is to use genetic algorithms to optimize the design of ANNs applied to electrical load forecasting. We prove that our model is capable of obtaining the optimal topology of the ANN, the connection weights among neurons, and uses the GA as a learning rule. The system allows both feedforward and feedback ANN paradigms, and any kind of connections among neurons.

The best experimental results are achieved with feedback ANNs. The genetic feedforward and feedback networks are practically comparable to their neural counterparts, with the added advantage that for feedforward networks the falling into local minima is avoided and optimal topologies are found.

5 Acknowledgments

This work has been partially supported by the Spanish Comisi6n Interministerial de Ciencia y Tecnologa (CICYT), Project N1. TIC95-0589.

References

[1] D.E. Goldberg. *Genetic Algorthims in Search Optimization and Machine Learning.* Addison Wesley, 1989.

[2] F. J. Marín, J. Ruano, and F. Sandoval. Optimization of artificial neural networks using a cooperative-coevolutive genetic algorithm. Technical report, University of Málaga, 1996.

[3] T. M. Peng, N. F. Hubele, and G. G. Karady. Advancement in the application of neural networks for short term load forecasting. *IEEE Trans. PWRS*, 7(1), 1992.

[4] M. A. Potter and K. A. DeJong. A cooperative coevolutionary approach to function optimization. In Y. Davidor, H-P. Schwefel, and R. Manner, editors, *Parallel Problem Solving from Nature III.* Springer-Verlag, 1994.

[5] H. K. Temraz and V. H. Quintana. Applications of the decomposition technique for forecasting the load of a large electric power network. *IEE Proceedings Hener. Transm. Distrib.*, 143(1), 1996.

Multiobjective Pressurised Water Reactor Reload Core Design using a Genetic Algorithm

G. T. Parks
Cambridge University Engineering Department,
Trumpington Street, Cambridge, CB2 1PZ, U.K.
Email: gtp@eng.cam.ac.uk

Abstract

The design of pressurised water reactor (PWR) reload cores is not only a formidable optimization problem but also in many instances a multiobjective problem. This paper describes a genetic algorithm (GA) designed to perform true multiobjective optimization on such problems.

1 Introduction

The PWR reload core designer's task is to identify the configuration of fresh and partially burnt fuel and burnable poisons (BPs) (control material) which optimizes the performance of the reactor over the ensuing cycle, while ensuring that various operational constraints are always satisfied.

A typical PWR core contains 193 fuel assemblies arranged with quarter core symmetry. At each refuelling one third or one quarter of these may be replaced. It is common practice for fresh fuel assemblies to carry a number of BP pins. It is also usual to rearrange old fuel in order to improve the characteristics of the new core. This shuffling can entail the exchange of corresponding assemblies between core quadrants, which is equivalent to changing the assembly orientations, or the exchange of different assemblies, which changes their locations and possibly their orientations also. Thus, a candidate core loading pattern (LP) of predetermined symmetry must specify:

- the fuel assembly to be loaded in each core location,
- the BP loading with each fresh fuel assembly, and
- the orientation of each assembly.

It is readily apparent that the search for the best LP is a formidable combinatorial optimization problem. The PWR reload core design problem has been tackled in many different ways [3], and one interesting point to emerge from a review of past work is the diversity in objective functions chosen. It is clear that the PWR reload core design problem is in reality a multiobjective optimization problem, where an improvement in one objective is often only gained at the cost of deteriorations in other objectives, hence trade-offs are necessary. There are two standard methods for treating multiobjective problems, if a traditional optimization algorithm which minimizes a single objective is to be employed. One is to construct a composite objective through a weighted summation of the individual objectives. The other is to place constraints on all but one of the objectives and to optimize the remaining one. Whichever method is used, the solution of the resulting single objective problem leads to the identification of just one point on the trade-off surface, the position of which depends on the designers preconceptions. To explore the trade-off surface further a large number of different optimization runs must be executed each with different weightings or constraints a potentially time-consuming and computationally expensive exercise if even attempted. This paper describes a GA based search method which is designed to perform true multiobjective optimization on such problems. This allows the trade-off surface between competing objectives to be identified in a single run and offers the designer a family of LPs lying on this surface, from which those worthy of further consideration can be chosen. An example illustrating the methods effectiveness is presented and analysed.

2 Genetic Algorithms

The use of GAs as an optimization tool for PWR reload core design has been the subject of some recent interest [2,7,9]. GAs differ from most optimization techniques by searching from one group (or population) of solutions to another, rather than from one solution to another, and it is this fact that makes them uniquely suited to multiobjective optimization.

2.1 Selection Procedure

The selection procedure determines the probability of each member of the current population being chosen to parent new offspring. When tackling a multiobjective optimization problem the concepts of Pareto optimality and dominance enable the current population to be ranked (allocated fitnesses) without imposing any preconceptions as to the relative importance of individual objectives [4]. A solution X is said to be dominated by solution Y if Y is better on all counts (objectives).

Using this definition, the entire population can be ranked by sorting through to identify all nondominated solutions, ranking these appropriately, then removing them from consideration, and repeating the procedure until all solutions have been ranked.

A selection probability can then be assigned to each solution based on its ranking in a manner similar to Baker's single criterion ranking selection procedure [1]. By biasing selection towards the better (on a multiobjective basis) solutions, the population can be expected, assuming that suitable crossover and mutation operators are used, to converge on the problem's trade-off surface.

It is well established that GAs perform better if an 'elitist' selection procedure is used, i.e., if the best solution found so far is always chosen as a parent, even if it is not a member of the current population. Of course, in multiobjective optimization there is probably more than one 'best' (nondominated) solution, so a slightly more elaborate procedure must be adopted.

While the algorithm is running an archive of nondominated solutions is maintained. After each trial solution has been evaluated it is compared with existing members of the archive. If it dominates any members of the archive, those are removed and the new solution is added. If the new solution is dominated by any members of the archive, it is not archived. If it neither dominates nor is dominated by any members of the archive, it is archived if it is sufficiently 'dissimilar' to existing archive members. (For these purposes the degree of dissimilarity between two LPs is defined in terms of their beginning-of-cycle (BOC) reactivity distributions [5], reactivity being a parameter related to the amount of fissile material in the fuel and thus a convenient indicator of other attributes of interest.) This dissimilarity requirement maintains diversity in the archive.

When the archive is full, any new nondominated solution replaces the most similar one in the archive. This means that nondominated solutions are discarded from the archive as the search progresses, but such discarding is unavoidable unless the archive size is set so large that all nondominated solutions encountered can be stored. Up to 25% of the parents for each new generation are chosen from the archive, thus introducing multiobjective elitism to the selection process. The population size is determined based on a measure of the number of changes that could be made to an individual LP and on the number of objectives to be optimized simultaneously, so, the larger the search space is, the larger the population.

2.2 Coding

Although traditional GAs map problems to strings of binary bits and manipulate these encodings, it is more natural, and therefore preferable [4], to represent each solution by three two-dimensional arrays, corresponding to the physical layout of the fuel assemblies, their BP loadings and their orientations respectively.

2.3 Fitness Evaluation

In the example which follows the Generalized Perturbation Theory based reactor model employed in FORMOSA-P [6] was used to evaluate LPs, but, in principle, the evaluation of objectives and constraints can be performed using any appropriate reactor physics code.

2.4 Constraints

Constraints can be either treated as hard, meaning that any LP violating them is ranked last in the population and does not survive, or as additional objectives to be minimized. The choice of treatment for any constraint is left to the users discretion.

2.5 Crossover Operator

The role of the crossover operator is to combine information from parent solutions to create offspring solutions with, it is hoped, better objective function values. For combinatorial problems application-specific crossover operators are required to guarantee that valid offspring are produced in this case to ensure that the fuel assembly inventory is maintained. For this application Poon's heuristic tie-breaking crossover (HTBX) operator [7] is used. HTBX maps the parent fuel arrays to reactivity-ranked arrays based on the assemblies' BOC reactivities. It then combines randomly selected complementary parts of these arrays through a cut and paste operation, and uses a simple tie-breaking algorithm to produce valid offspring reactivity-ranked arrays. Finally the assembly-ranking mapping is reversed to produce the offspring assembly LPs. The BP loadings and assembly orientations are all inherited from one or other parent. Thus, the BOC reactivity distribution (and it is hoped, in consequence, other attributes) of an offspring LP resembles, but is not necessarily identical to, parts of both parents.

2.6 Mutation Operator

The mutation operator makes small changes to trial LPs, and is necessary to ensure that the entire search space is accessible. For this application the mutation operator performs one, two or three fuel assembly shuffles, randomly allocating allowed BP loadings and orientations to the assemblies affected. It is used as an alternative to crossover, i.e. offspring are produced using either mutation or crossover but not both, the choice between operators being made randomly. The relative frequencies with which these operators are chosen are approximately 25% and 75% respectively, this ratio having been determined (by extensive testing) to give good performance on PWR reload design problems.

2.7 Initial Population

The initial population is generated by repeated application of the mutation operator to the initial LP supplied by the user/designer.

2.8 Search Termination

The search is terminated either after a user-specified number of generations, or when the search ceases to make progress. The latter is defined as occurring when no new solutions which dominate existing members of the archive are found for two successive generations.

3 Algorithm Performance

Figures 1 and 2 illustrate the performance of this multiobjective GA (MOGA) by plotting projections

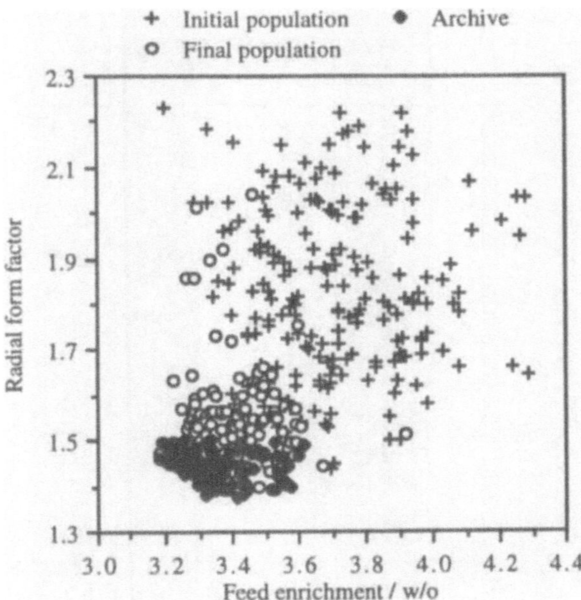

Figure 1: MOGA performance: radial form factor vs feed enrichment.

in two-objective space of the initial and final populations and the final archive contents for an optimization run on a PWR core in which there were three objectives:

- feed enrichment minimization (minimization of the enrichment of the fresh fuel),
- discharge burn-up maximization (maximization of the burn-up of the fuel to be discharged), and
- radial form factor (RFF) minimization (minimization of the ratio of the peak to average assembly power throughout the cycle).

In this run a population of 204 was evolved for 51 generations, so that 10 404 LPs were examined in all. It can be seen that the final population is on average much better than the initial population with respect to all three objectives. During the run 1186 LPs were archived, which is a good indication of the effectiveness of the search method. (Note that only solutions with RFF values below 1.50 were archived.) The archive of 100 LPs, and indeed the final population, gives the designer a clear picture of the trade-off surface between the competing objectives. The (expected) strong trade-off between feed

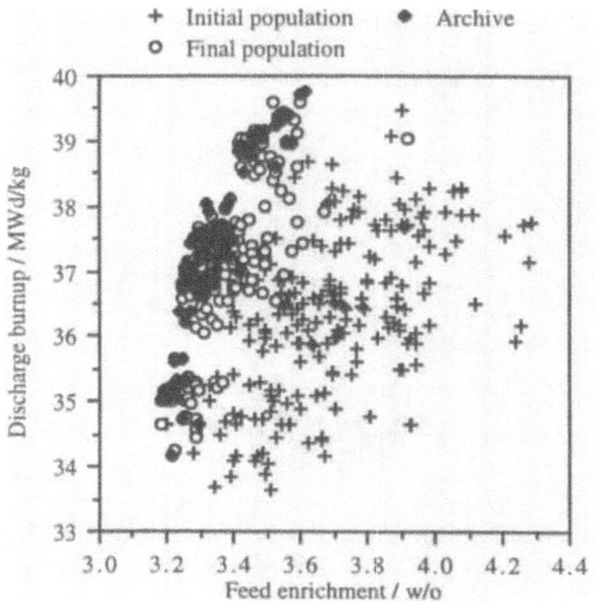

Figure 2: MOGA performance: discharge burn-up vs feed enrichment.

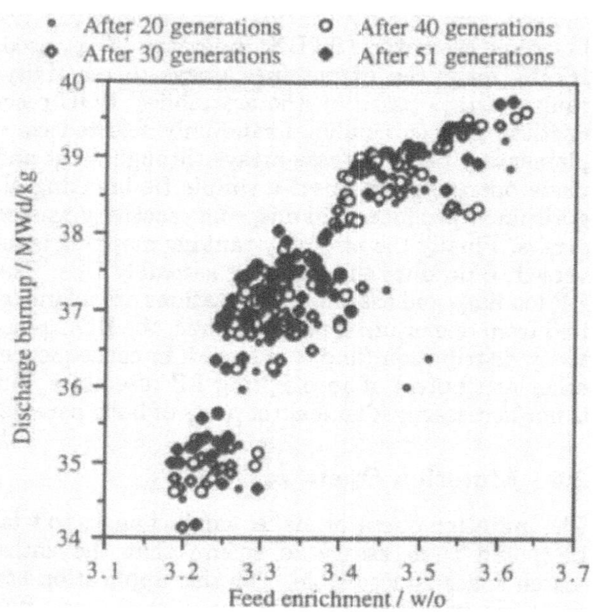

Figure 3: MOGA archive evolution: discharge burn-up vs feed enrichment.

enrichment and discharge burn-up is clearly shown. A trade-off between feed enrichment and RFF is also indicated. Figure 3 shows the evolution in two-objective space of the contents of the archive. It is apparent that the general shape of the trade-off surface is established quite early, and, therefore, if the purpose of performing multiobjective optimization is just to identify a part of the search space worthy of closer examination, the length of the MOGA run could be made significantly shorter, with an associated reduction in execution time. The number of LPs evaluations typically performed by the MOGA (10 404 in the example presented here) is similar in magnitude to those required by other state-of-the-art fuel management codes which employ stochastic search methods [2, 6, 8]. The MOGA's execution time is dominated by the evaluation of each LP by the core simulator, the overhead of running the GA being very small, and the choice of physics code is up to the user. Thus, the computational cost of a single run of the MOGA is similar to that of other current fuel management codes. However, it should be emphasized that the advantage of its use is not so much computational but the fact that it solves multiobjective optimization problems as true multiobjective optimization problems and does so in a single run.

4 Conclusions

The results show that a single MOGA run can present the reload core designer with a wealth of information about the range of achievable values of different (conflicting) objectives and the necessary trade-offs between them. Using this information the designer can make a much better informed decision about the areas of the search space worthy of closer examination. The MOGA is capable of working with any number of objectives, although obviously the more objectives there are the longer the run length required. As the selection procedure is only based on the ranking of solutions, the scaling of individual objectives is unimportant and thus running the code with different combinations of objectives is very easy. The ability of the MOGA to tackle a wide variety of problems and to interface with a wide variety of different physics codes makes it potentially a very powerful aid to the reload core designer.

5 Acknowledgements

The author gratefully acknowledges the financial support of Nuclear Electric plc and the collaborative co-operation and support of the North Carolina State University (NCSU) Electric Power Research Center.

References

[1] J. E. Baker. Adaptive selection methods for genetic algorithms. In J. J. Grefenstette, editor, *Proc. Int. Conf. on Genetic Algorithms and their Applications*, Hillsdale, NJ, 1985. Lawrence Erlbaum Associates.

[2] M. D. DeChaine and M. A. Feltus. Nuclear fuel management optimization using genetic algorithms. *Nucl. Technol.*, 111:109, 1995.

[3] T.J. Downar and A. Sesonske. Light water reactor fuel cycle optimization. *Adv. Nucl. Sci. Tech.*, 20:71, 1988.

[4] D.E. Goldberg. *Genetic Algorthims in Search Optimization and Machine Learning.* Addison Wesley, 1989.

[5] D. J. Kropaczek, G. T. Parks, G. I. Maldonado, and P. J. Turinsky. Application of simulated annealing to in-core nuclear fuel management optimization. In *Proc. 1991 Int. Top. Mtg. Advances in Mathematics, Computations and Reactor Physics*, 1991.

[6] D. J. Kropaczek, G. T. Parks, G. I. Maldonado, and P. J. Turinsky. The efficiency and fidelity of the in-core nuclear fuel management code FORMOSA-P. In Y. Ronen and E. Elias, editors, *Reactor Physics and Reactor Computations.* Ben Gurion University of the Negev Press, 1994.

[7] P. W. Poon and G. T. Parks. Application of genetic algorithms to in-core nuclear fuel management optimization. In *Proc. Joint Int. Conf. Mathematical Methods and Supercomputing in Nuclear Applications*, page 777, 1993.

[8] J. G. Stephens, K. S. Smith, K. R. Rempe, and T. J. Downar. Optimization of pressurized water reactor shuffling by simulated annealing with heuristics. *Nucl. Sci. Eng.*, 121:67, 1995.

[9] E. Tanker and A. Z. Tanker. Application of a genetic algorithm to core reload pattern optimization. In Y. Ronen and E. Elias, editors, *Reactor Physics and Reactor Computations.* Ben Gurion University of the Negev Press, 1994.

Using Artificial Neural Networks to Model Non-Linearity in a Complex System

P. Weller[1,2], A. Thompson[2] and R. Summers[1]
[1] Department of System Science, City University,
Northampton Square, London, EC1V 0HB, UK.
[2] Department of Nuclear Science & Technology,
Royal Naval College, Greenwich, London, UK.
Email: p.r.weller@city.ac.uk

Abstract

This paper describes an investigation into using artificial neural networks (ANNs) to model the non-linearities in a complex system, a nuclear reactor. A simple one compartment finite difference model of the plant is developed and an exact ANN equivalent formed directly without training. Conventional training using standard transfer functions available in ANN packages is compared with this. A novel method is used to produce the ANN training and test sets. A twenty five compartment model is built from directly formed ANNs and the results compared to a simulator model.

1 Introduction

Research in artificial neural network (ANN) applications for the nuclear industry has progressed steadily over a number of years. One reported application concerns the prediction of nuclear plant condition. Plant parameters are used as inputs to an ANN which has been trained on a series of operating conditions. The output of the network provides a guide to the present state of the reactor or can give a predictive value of a key parameter [1].

A second method of determining future plant condition is to predict the values for a set of parameters for the entire period of a single fault condition [2]. The initial inputs to the network are the pressure values for the first two time steps of a coolant leak, the output comprises the pressures for the next time period. These values are then fed back into the ANN as the input for the most recent time step. This recursive technique is continued for the entire leak period.

This work began to explore the possibilities of transient (leak) modelling using ANNs. In order to predict a wide range of reactor fault transients an ANN has to accurately describe the behaviour of a complex non-linear system. The system equations for this system are known and the ANN structure to exactly model this system is developed. The paper shows how this is achieved using certain structures in an ANN. These are identified as a unit to multiply two inputs and a repeated use of this unit with suitable connections. These features are not applicable to just nuclear reactors but also to many other non-linear systems.

The remainder of this paper presents a simple one compartment model of the system. This concept is then expanded to include the primary circuit of the reactor. Sample results are reported for the testing of each stage. Finally further developments of the system are discussed.

2 One Compartment Model

Consider a simple one compartmental model of a reactor as shown in Figure 1. The main body of the model represents the reactor pressure vessel, containing the core, while the loop is a steam generator and piping.

The conservation of energy equation to calculate the next pressure vessel temperature for this system is:

$$T_1^{(k+1)} = T_1^{(k)} + \frac{Qdt}{C_pM} + \frac{dt}{M}(M_{in}(T_{in}^{(k)} - T_1^{(k)})) \tag{1}$$

where: C_p = specific heat of liquid in reactor pressure vessel and dt = Time step between $(k+1)$ and k.

The intention is to construct an ANN to model this system and calculate the next temperature of the vessel. The network will be an exact equivalent of the system. The first two terms of Equation (1) are simple additions with the constant terms being modelled by the weightings between the ANN

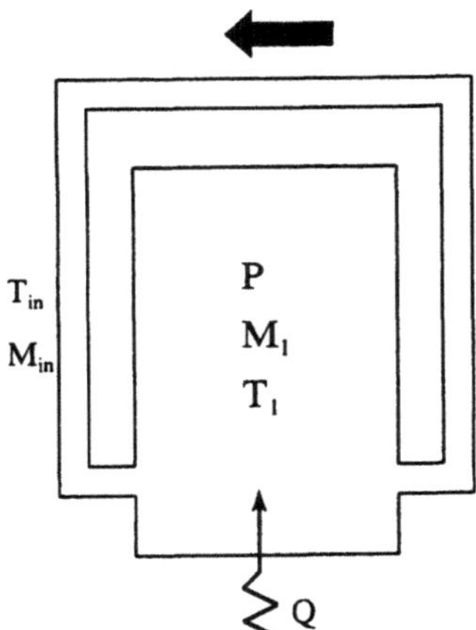

Figure 1: One compartment model of a nuclear reactor, where: P = Pressure in vessel, M_l = Mass of liquid in vessel, T_l = Temperature of liquid in vessel, T_{in} = Temperature of incoming liquid, M_{in} = Rate of change of liquid mass, Q = Heat into the system.

nodes. The third term contains a non-linearity, a multiplication, which cannot be modelled simply by the ANN. The energy equation is modelled by the network shown in Figure 2. The central part of the network models the multiplication while the addition of the first two terms is included in the weights and summations into the output node.

This network is a direct equivalent of Equation 1 but uses the following transfer function.

$$O_j = 0.5(\sum_{i=1}^{3}(I_i \cdot W_{ij}))^2 \qquad (2)$$

where: O_j = Output at node j, I_i = Input from node i and W_{ij} = Weighting between Nodes i and j.

While this feature is a useful theoretical tool it is not a standard function. It has no saturation level so the node output has no limit. A series of ANNs were developed to replace this transfer function with standard transfer functions. Firstly, the square element of the function was replaced by an ANN trained to square a single input. A second ANN was trained to multiply two inputs. Lastly, an ANN was trained to perform the entire function of the equation and calculate the change in temperature. In all cases the result from one calculation,

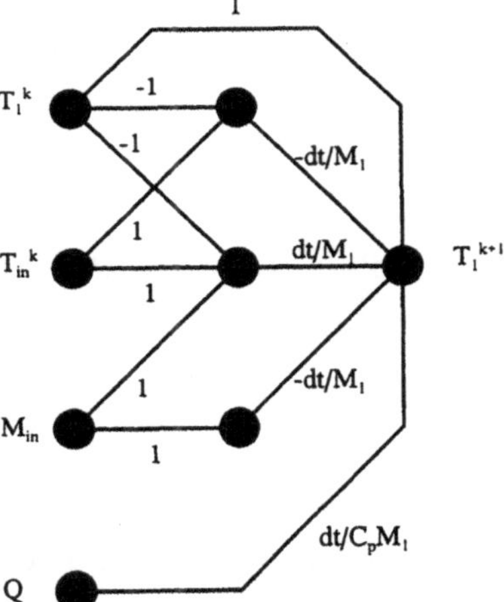

Figure 2: ANN representation of temperature equation.

T_{k+1} was fed back into the model as the input to determine the next temperature, T_{k+2}.

The first two ANNs mentioned above are still direct equivalents of the system equations. The training and test sets for these ANNs are simple lists of squares and multiples respectively. The third network, however, requires data from operating conditions to be able to develop relationships between the variables. The set of combinations of possible cases, even for a simple system, is large and an ANN training set designed to reflect these would also be sizeable. Instead three scenarios were produced that, although not necessarily a true reflection of the actual operating situations, give a guide to the relationships between the variables. An ANN was trained on this data set and used in a feedback program. A series of tests were performed, using values typical of a civil nuclear reactor. Two examples of the results obtained are shown below.

The results show that the ANN based predictions compare well to the simulator output. The lack of accuracy in the predictions from the squaring ANN model are due to the scaling factor applied to the network inputs.

It was found that one of the better ANNs for the limited range of flows and masses considered to date used a linear threshold function. The backpropagation training algorithm had developed a linear combination of inputs and weights to determine the future system temperatures of a sub-set of a non-linear system. Expanding the ANN gives the rela-

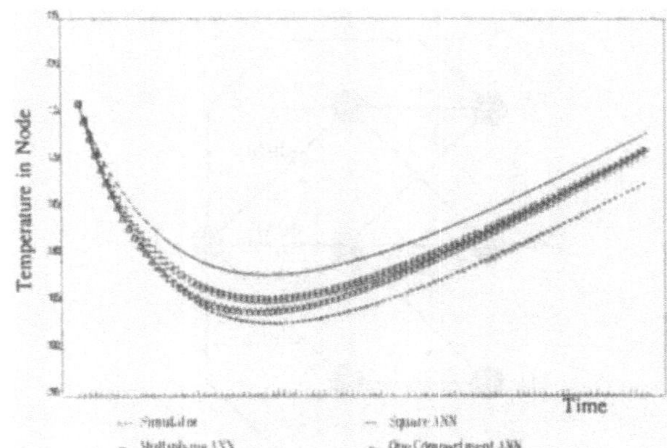

Figure 3: Results of transient, decreased flow and heat.

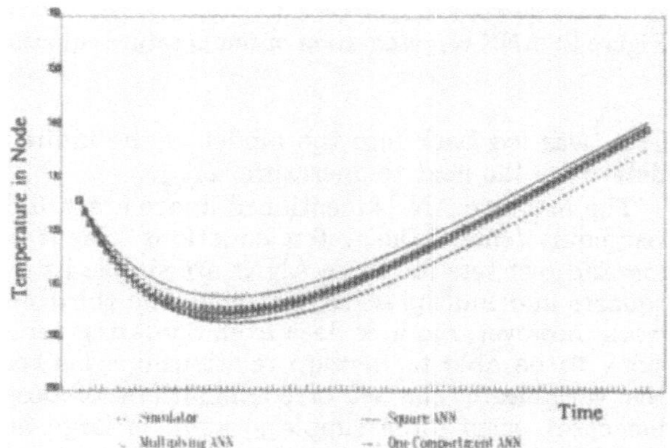

Figure 4: Results of transient, increased flow and heat.

tionship in Equation (3),

$$\begin{aligned}\text{Temp Diff} &= 3.99 - 0.05(T_1) + 0.05(T_{in}) \\ &\quad - 1.61e^{-4}(M_{in}) + 6.37e^{-10}(Q).\end{aligned} \tag{3}$$

3 Full Reactor Model

While the above model produced good results in modelling the non-linearity, more detail is required for a useful plant analyzer. A system, using one compartment models, was constructed to model the primary circuit of a nuclear reactor. This part of the reactor was divided into twenty-five key regions and each one was modelled by a one compartment model. The flow rate into each region was the total of flows from the compartments up-stream while the temperatures were the average of the corresponding up- stream temperatures. The ANN of the full compartment model was used for the calculation of change in temperatures. This ANN had to be modified to reflect different masses of liquid in the various compartments. The entire system was updated at the end of each time step, set at 0.1 seconds.

It was also possible to use this multi-compartmental model to include information on valve positions and settings. All the links associated with valves were included at the configuration stage of the system and set to an initial state. During the transient the valve states were changed either at defined times or when a set of criteria were satisfied, for example, the emergency systems. These changes in valve states were reflected in the corresponding system flows.

A large number of transients can be modelled by this system as each compartment predicts its own next temperature independently. The variables from upstream compartments are only considered in future calculation and then in terms of the four inputs to the ANN.

The above system was compared, for a number of transients, to a full simulator code. Typical results for three key regions are given in Figures 5 and 6.

The results confirm that the direct equivalent agrees with the results from the simulation program.

4 Discussion

This investigation arose from considering the number of transients that could be modelled by an ANN. Originally a number of different ANNs were envisaged for the prediction of reactor transients [4]. Each ANN was developed to predict a set of transients, grouped by common features or similar behaviour. However, it was realised that the grouping was very difficult. A better approach is to develop an ANN system capable of modelling many transients irrespective of their features. The compartmental system described here may be a method of achieving that goal. Some extreme scenarios may still require a bespoke ANN, but the majority of situations could be modelled by a version using the concepts developed in this paper.

The compartmental ANN approach has a wider appeal outside the nuclear industry. It is easily modified for any multi-compartment model providing that the links between compartments can be defined. A method of training such a system is presently under investigation. If successful it would

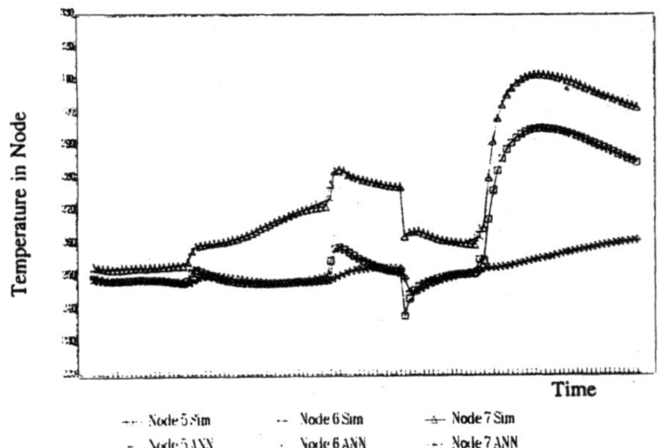

Figure 5: Temperature changes for steam generator loop.

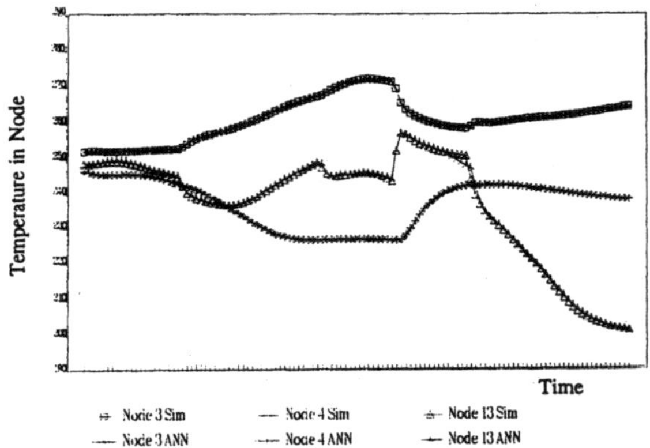

Figure 6: Temperature changes for reactor pressure vessel.

allow modelling of compartments whose variables are problematic to measure.

5 Conclusions

Several ideas have been presented in this paper. ANNs structures which model the commonly found non-linearity of multiplied inputs have been produced. The one compartment models have been connected together to produce a predictive model of a full nuclear reactor system. A network can be defined which is a direct equivalent of the equations in a simulation. This will reproduce any transient experienced by the system.

References

[1] A. G. Parlos, K. T. Chong, and A. F. Atiya. Empirical model development and validation with dynamic learning in the recurrent multilayer perception. *Nuclear Technology*, 106:271–290, 1994.

[2] A. C. Thompson, P. R. Weller, and R. Summers. The use of neural networks in a system transient code. In *Proceedings of ASME & JSME Fluid Engineering Symposium on Validation of System Transient Codes*, pages 168–173, 1995.

[3] V. R. Vemuri and R. D. Rogers, editors. *Artificial Neural Networks: Forecasting Time Series.* IEEE Computer Society Press, 1994.

[4] P. R. Weller, R. Summers, and A. C. Thompson. Using hierarchical neural networks for diagnosing and predicting the condition of a nuclear reactor. In *Proceedings of ICANN'95.* Paris, 1995.

Transit Time Estimation by Artificial Neural Networks

T. Tambouratzis[1], M. Antonopoulos-Domis[1], M. Marseguerra[2] and E. Padovani[2],
[1] Institute of Nuclear Technology - Radiation Protection, NCSR 'Demokritos', Aghia Paraskevi, Athens 153 10, Greece.
[2] Department of Nuclear Engineering, Polytechnic of Milan, Via Ponzio 34/3, 20133 Milano, Italy.
Email: tatiana@zeus.int-rpnet.ariadne-t.gr

Abstract

The use of interactive activation and competition (IAC) and backpropagation (BP) artificial neural networks (ANNs) for transit time estimation has been investigated in this piece of research. Owing to its competitive nature, the IAC ANN has been found able to correctly estimate the current transit time from short records of signals as well as to quickly follow changes in transit time and to detect when the transit time falls outside a predefined expected range. On the other hand, the interactive nature of the IAC ANN allows it to be robust to significant levels of noise and of the global component. A BP ANN has been appended to the IAC ANN, further allowing for the accurate estimation of decimated transit times.

1 Problem Formulation

Transit time estimation using noise signals is widely employed in industrial systems (e.g., chemical and oil industry). In boiling water nuclear reactors (BWRs), where it is important to estimate the in-core coolant velocity as a means of establishing normal operation, a widely used technique is to correlate the neutron noise signals of two axially separated neutron detectors (shown in Figure 1 as D_1 and D_2). The transit time (i.e. the time that it takes for the steam bubbles to travel from D_1 to D_2) is evaluated either from the maximum of the cross-correlation function (CCF) or from the slope of the phase of the cross-spectrum (CPSD) of the two signals. Subsequently — since the distance between the two detectors is known — the coolant velocity can be directly calculated.

The signals $s_1[i]$ and $s_2[i]$ ($i \in Z$) are described as:

$$\begin{aligned} s_1[i] &= l[i] + g[i] + n_1[i] \\ s_2[i] &= Al[i-d] + Bg[i] + n_2[i] \end{aligned} \tag{1}$$

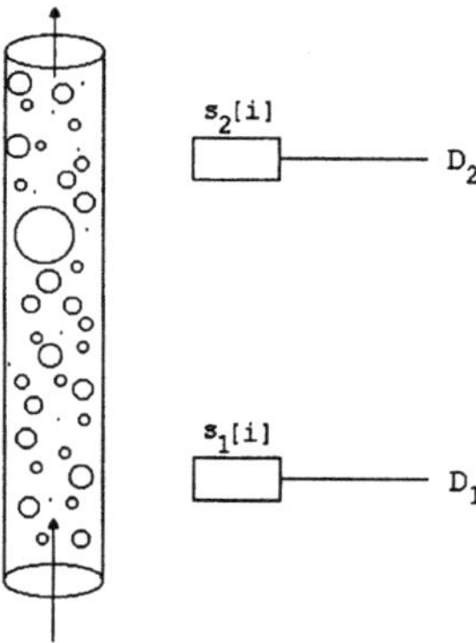

Figure 1: The rising steam bubbles and the pair of detectors.

where $l[i]$ denotes the local component that appears in $s_1[i]$ and in $s_2[i]$ with transit time d, $g[i]$ the global component, $n_1[i]$ and $n_2[i]$ uncorrelated random noise signals and A and B amplitude ratios. Techniques such as the CCF and the CPSD are sensitive to the ratio of global to local components of the neutron noise signals. Although the global component (which is usually contained at low frequencies) may be eliminated by high-pass filtering, it causes serious errors in transit time estimation when also present at higher frequencies. On the other hand, there is a growing interest in fast estimation of the transit time for monitoring purposes; as fast estimation involves short records of the noise signals the existing techniques are not adequately robust to the effects of the global component and uncorrelated noise. To this end, the use of artificial neural networks (ANNs) for transit time estimation is investigated. An ANN which is based on interactive activation and competition (IAC) [2, 4] has been employed and found able to correctly estimate the current transit time from short records of signals, to quickly follow changes in transit time and to detect when the coolant velocity falls outside a predefined expected range. A backpropagation (BP) ANN [3] has been appended to the IAC ANN, further allowing for the estimation of decimated transit times.

2 Signal Generation

In this piece of research, the neutron noise signals (neutron density random fluctuations) have been obtained by simulation using a one-dimensional homogeneous, two neutron-group model [1]. The excitation for the local component is bubbles with random size (random numbers) propagating upwards with constant velocity, while the excitation for the global component (coincident neutron density fluctuations over all space) is induced by simulating a randomly vibrating control rod (neutron absorber).

3 The IAC ANN

A purely competitive IAC ANN has been constructed for determining the transit time — and hence the velocity — in the in-core coolant velocity problem. Assuming that the expected range of transit times is $[d_1, d_N]$ with:

$$d_j = h_j \Delta t, \quad j = 1, 2, \ldots, N, \tag{2}$$

($h_j \in Z$ and $h_{j+1} = h_j + 1$, $j = 1, 2, \ldots, N-1$), the IAC ANN (shown in Figure 2) comprises a single cluster of N mutually inhibitory nodes where the jth node supports transit time d_j. Each node is fully connected (with inhibitory connections) to the remaining nodes of the IAC ANN and receives an environmental input which constitutes a function F of the match between the two sequences of signals $s_1[i]$ and $s_2[i]$. Instead of using isolated values of the two sequences in (1), the averaged values $as_1[k]$ ($k \in Z$) over M consecutive signal values have been employed for s_1,

$$as_1[k] = \frac{1}{M} \sum_{i=k}^{k+M-1} s_1[i] \tag{3}$$

and the N averaged values $as_2[k][j]$ ($j = 1, 2, \ldots, N$ and $k \in Z$) over M consecutive signal values have been employed for s_2 ,

$$as_2[k] = \frac{1}{M} \sum_{i=k+j}^{k+j+M-1} s_2[i] \tag{4}$$

Averaging over M consecutive values of the signals in (3)-(4) attenuates the effect of the uncorrelated noise; M has been kept small ($M = 10$) so as not to affect the speed of the IAC ANN decisions.

Beginning from 'resting' activation values for all the nodes of the IAC ANN, the nodes are updated in parallel at each presentation of the environmental inputs (for each value of k). As a result of the different values of the environmental input (3)-(4) which are accumulated over time, the node corresponding to the transit time becomes progressively more active and suppresses the activation values of the other nodes. If the transit time changes at some point, the activation values of the nodes in the ANN are reconfigured so that:

- the node corresponding to the new transit time becomes gradually more active,
- the activation value of the previously active node is lowered, and
- the rest of the activation values are not affected.

Table 1 lists the values of the most characteristic parameters for the IAC ANN with nodes working on the signals described in Section 1; *decay* denotes how quickly the nodes forget, while the combination of the threshold θ and the resting activation determines the competitive nature of the IAC ANN. This choice of parameter values ensures quick estimation of the transit time (only 2-8 updates are required when the IAC ANN begins the updating procedure or when it is reset) and rapid response (10- 17 updates) to changes in the transit time.

To further improve the accuracy of transit time estimation and the stability of the IAC ANN, averaged activation values $Activ_j$ ($j = 1, 2, \ldots, N$) over R consecutive IAC ANN updates are employed:

$$Activ_j = \frac{1}{R} \sum_{k=k_0}^{k_0+R-1} activ_j^k \tag{5}$$

where $activ_j^k$ denotes the activation value of the jth node at time k; $R = 20$, whereas transit time estimation is not considerably delayed. The following logical decisions have been appended to the IAC ANN and provide its decision after R updates:

Table 1: Parameter values of the IAC ANN.

PARAMETERS	VALUES/INTERVALS
decay	0.150
θ	0.050
activation	[0.000, 1.000]
resting activation	0.150

Table 2: Classification of the IAC ANN decisions.

TRANSIT TIME / IAC ANN DECISION	d_y		outside $[d_1, d_N]$
transit time d_x	correct $x = y$	error $x \neq y$	error
outside $[d_1, d_N]$	error		correct
no decision	no decision		no decision

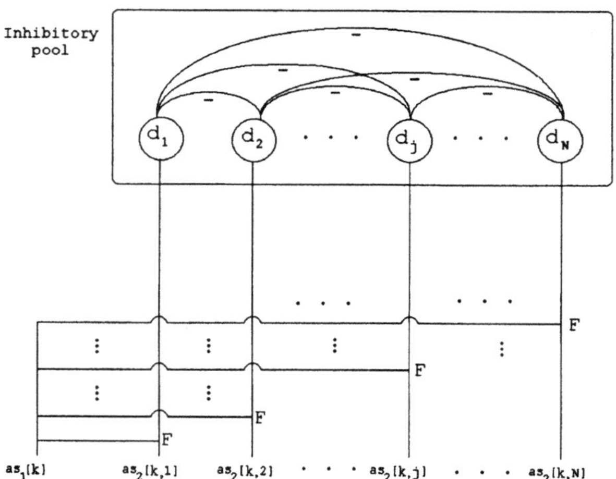

Figure 2: The IAC ANN for the estimation of the transit time.

1. Find x, i.e. the index of

$$Activ_x = \max_{j=1,\dots,N}\{Activ_j\}.$$

2. IF (($Activ_x > 0.1$) AND ($Activ_{x^*} \leq 0.1$, $\forall x^* \neq x$)), the IAC ANN supports transit time d_x.
3. IF (($Activ_x > 0.1$) AND ($\exists x^* \neq x : Activ_{x^*} > 0.1$), the IAC ANN does not make a decision.
4. IF ($Activ_x \leq 0.1$) the IAC ANN supports that the transit time is outside the expected range $[d_1, d_N]$.

By comparing the IAC ANN decision with the transit time, the responses of the IAC ANN can be classified into 'correct', 'error' and 'no decision', as explained in Table 2.

4 Transit Time Estimation

The following discussion describes the performance of the IAC ANN of Figure 2 for $N = 10$.

4.1 Integer Transit Time Estimation

If the transit time d assumes integer values, i.e. is an integer multiple of the sampling time $\Delta t (d = h\Delta t, h \in Z)$ the IAC ANN:

- Accurately and quickly establishes the transit time when this is kept constant and falls in the supported range $[d_1, d_{10}]$. The effect of the global component on correct decisions and 'no decisions' is evident; for high levels of the global component it becomes more probable for the IAC ANN to avoid making a decision than to make an erroneous one.
- Consistently detects when the transit time clearly falls outside the supported range. Prediction is most satisfactory when the transit time differs by more than $2\Delta t$ from the supported range of transit times, acceptable when it differs by $2\Delta t$ and impaired when it differs by Δt.
- Monitors the transit time, i.e. precisely follows its variations both inside and outside the supported interval. It is interesting to note that at the point of change 'no decisions' dramatically increase, while the IAC ANN correctly identifies the new transit time immediately afterwards.

4.2 Decimated Transit Time Estimation

It is also possible to estimate decimated transit times $d(d = h\Delta t, h \in \mathbb{R} - Z$ in (2), i.e. not an integer multiple of Δt) within the range supported by the IAC ANN.

Owing to the competitive nature of the IAC ANN, the winner node x is the one that supports transit time $d_x = h_x \Delta t (h_x \in Z)$, where x is the index of h_x such that $\|h_x - h\| = \min_{j=1,\dots,10}\{\|h_j - h\|\}$; however, no information concerning a more exact value

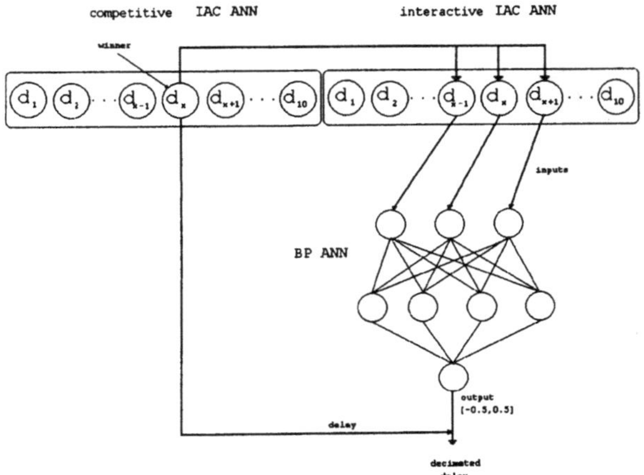

Figure 3: The combination of two IAC ANNs and a BP ANN for decimated transit time estimation.

of d can be gleaned from the activation value of the winner node. By adding a more interactive IAC ANN to the original IAC ANN and by appending a BP ANN (shown in Figure 3), decimated transit time estimation, within the range supported by the IAC ANN, can be performed:

- The interactive IAC ANN supports integer transit times $[d_1, d_{10}]$ and its values for the most characteristic parameters are given in Table 3; it is reset after every estimation of *Activ* (after every R updates). Although reliable decision making cannot be implemented, the activation values for the winner and its two neighbouring nodes (i.e. for nodes $x-1$, x, $x+1$ in Figure 3) correspond to the proximity of the transit time d to d_x and, subsequently, to d_{x-1} and d_{x+1} .
- Two BP ANNs have been used for approximating the mapping that appears between the activation values of the three nodes $x-1$, x, $x+1$ of the interactive IAC ANN and $d_x - d$, a 3-8-1 BP ANN when the winner of the competitive IAC ANN is one of the nodes $x = 2, 3, \ldots, 9$ and a 2-6-1 when the winner is one of the nodes $x = 1, 10$ (only one neighbour exists). Training of both BP ANNs has been performed with global-component-free patterns using $\alpha = 0.999$ and $\eta = 0.001$ and has been terminated when an error of 0.002 is reached.

During transit time estimation, both IAC ANNs

Table 3: Parameter values of the interactive IAC ANN.

PARAMETERS	VALUES/INTERVALS
decay	0.750
θ	0.050
activation	$[-0.500, 0.500]$
resting activation	0.100

receive the same signals and update the activation values of their nodes. Once the *Activ* values are calculated, the original IAC ANN performs decision making. If transit time d_x is supported and node x is also mostly active in the interactive IAC ANN, the values $Activ_{x-1}$, $Activ_x$ and $Activ_{x+1}$ of the interactive IAC ANN are input in the appropriate BP ANN; the output constitutes $d_x - d$ and is subtracted from to provide an estimate of d. Alternatively, if two or more neighbouring nodes demonstrate $Activ > 0.1$, the one with the highest value is chosen as the winner and the previous process is repeated. Finally, if the nodes with $Activ > 0.1$ are not adjacent or the transit time falls outside the supported range, no further action is taken.

5 Conclusions

An ANN estimator of the in-core coolant velocity in BWRs has been presented. The employed IAC ANN is able to correctly and quickly estimate the current transit time, to follow its changes and to detect when the velocity falls outside a predefined expected range. By using two IAC ANNs and a BP ANN, decimated transit time estimation is also accomplished.

References

[1] M. Antonopoulos, M. Marseguerra, and E. Padovani. On the fast estimation of transit times – application to bwr simulated data. In *Proceedings of the SMORN VII (Volume 2)*, 1995.

[2] S. Grossberg. Adaptive pattern classification and universal recoding: Parallel development and coding of neural feature detectors. *Biological Cybernetics*, 23:121–134, 1976.

[3] D. E. Rumelhart, G. E. Hinton, and R. J. Williams. Learning representations by back-propogating errors. *Nature*, 323:533–536, 1986.

[4] D. E. Rumelhart and D. Zipser. Feature discovery by competitive learning. *Cognitive Science*, 9:75–112, 1985.

Evolving Asynchronous and Scalable Non-uniform Cellular Automata

M. Sipper, M. Tomassini and M. S. Capcarrere
Logic Systems Laboratory, Swiss Federal Institute of Technology, IN-Ecublens,
CH-1015 Lausanne, Switzerland.
E-mail: {Moshe.Sipper, Marco Tomassinim, Mathieu Capcarrere}@di.epfl.ch.

Abstract

We have previously shown that non-uniform cellular automata (CA) can be *evolved* to perform computational tasks, using the *cellular programming* algorithm. In this paper we focus on two novel issues, namely, the evolution of asynchronous CAs, and the scalability of evolved synchronous systems. We find that asynchrony presents a more difficult case for evolution though good CAs can still be attained. We describe an empirically derived scaling procedure by which successful CAs of any size may be obtained from a particular evolved system. Our motivation for this study stems in part from our desire to attain realistic systems that are more amenable to implementation as "evolving ware," *evolware.*

1 Introduction

Cellular automata (CA) are dynamical systems in which space and time are discrete. A cellular automaton consists of an array of cells, each of which can be in one of a finite number of possible states, updated synchronously in discrete time steps according to a local, *identical* interaction rule. The state of a cell at the next time step is determined by the previous states of a surrounding neighborhood of cells; this transition is usually specified in the form of a *rule table*, delineating the cell's next state for each possible neighborhood configuration. The cellular array (grid) is n-dimensional, where $n = 1, 2, 3$ is used in practice (in this work we shall concentrate on $n = 1$).

CAs exhibit three notable features, namely, massive parallelism, locality of cellular interactions, and simplicity of basic components (cells). A major impediment preventing ubiquitous computing with CAs stems from the difficulty of utilizing their complex behavior to perform useful computations. Designing CAs to exhibit a specific behavior or perform a particular task is highly complicated, thus severely limiting their applications; automating the design (programming) process would greatly enhance the viability of CAs [5].

One possible approach, taken in this paper, is to employ artificial evolution. The model investigated by us is an extension of the original CA model, termed *non-uniform cellular automata.* Such automata function in the same way as uniform ones, the only difference being in the local cellular rules that need not be identical for all cells. Our approach involves the *cellular programming* algorithm, by which non-uniform CAs evolve to perform non-trivial, global computational tasks [7, 8, 9, 10, 11, 12] (for a description of the algorithm the reader is referred to these references).

The tasks for which non-uniform CAs were evolved via cellular programming include, among others, density and synchronization. Both involve two-state CAs, i.e., each cell can be in one of two states, 0 or 1, with connectivity radius $r = 1$, meaning that each cell is connected to one neighbor on either side (thus, each cell has $2r + 1$ neighbors, including itself). The one-dimensional density task is to decide whether or not the initial configuration contains more than 50% 1s, relaxing to a fixed-point pattern of all 1s if the initial density of 1s exceeds 0.5, and all 0s otherwise (Figure 1 [5, 7]). The term 'configuration' refers to an assignment of states to grid cells. In the one-dimensional synchronization task the CA, given any initial configuration, must reach a final configuration, within M time steps, that oscillates between all 0s and all 1s on successive time steps (Figure 2) [3, 7]. Spatially periodic boundary conditions are applied, resulting in a circular grid (for an $r = 1$ CA this means that the leftmost and rightmost cells are connected). It should be emphasized that both tasks comprise non-trivial computational problems for a small radius CA ($r \ll N$, where N is the grid size) [3, 5, 7].

Our previous studies involving cellular programming consisted of evolving *parallel cellular machines* to perform computational tasks. Our machine model was attained by considering a generalization of the original CA model, namely, non-uniform CAs, where cellular rules need not necessarily be identical. In this paper we study an additional generalization, namely, asynchronous CAs, as well as the scalability of evolved synchronous systems. Our motivation stems in part from our desire to attain more realistic systems that are amenable to implementation as "evolving ware," *evolware*. [7, 8].

2 Evolving Asynchronous CAs

One of the prominent features of the CA model is its synchronous mode of operation, meaning that all cells are updated simultaneously. A preliminary study of asynchronous CAs, where one cell is updated at each time step, was carried out in [4], where the different dynamical behavior of synchronous and asynchronous CAs was compared; the authors argued that some of the apparent self-organization of CAs is an artifact of the synchronization of the clocks. [13] noted that asynchronous updating makes it more difficult for information to propagate through the CA and that, furthermore, such CAs may be harder to analyze. Asynchronous CAs have also been discussed in [1, 6, 7], though it seems clear that they have received a limited amount of attention to date.

The issue investigated in this section is that of evolving asynchronous CAs to perform the density and synchronization tasks. The grid is partitioned into *blocks* in which synchronous updating takes place (i.e., all cells within a block are updated simultaneously), while the blocks themselves are updated asynchronously (rather than have all blocks updated at once); thus, inter-block updating is synchronous while intra-block updating is asynchronous. The number of blocks per grid, $\#_b$, is a tunable parameter, entailing a *scale* of asynchrony, ranging from complete synchrony ($\#_b = 1$) to complete asynchrony ($\#_b = N$). There are two main differences between our investigation and previous ones: (1) rather than consider only complete asynchrony ($\#_b = N$), we have introduced the above scale, and (2) asynchronous CAs were previously studied from a more abstract point of view, whereas we are interested in *evolving* them to perform a veritable *computation*.

Three models of asynchrony are considered, which differ in the scheduling of intra-block updating (inter-block updating is always synchronous):

Model 1 At every time step each block is updated *independently* of the others with probability p_{update}, chosen so as to insure that at least one block is updated per time step with probability ≥ 0.99.

Model 2 At each time step a different block is chosen at random without replacement, such that every $\#_b$ steps, *all* blocks are updated exactly once. We denote by *logical step* the succession of $\#_b$ time steps necessary for one full update cycle, in which all cells are updated (thus, one logical step is equivalent to one time step in the synchronous model, with respect to cell updating).

Model 3 All blocks are updated in a fixed, random order every logical step. This is similar to the second model, in that each cell is guaranteed to have updated its state every logical step, however, the (random) update order is fixed (rather than selected anew each logical step). Note that though the update order is deterministic, this model is interesting in that cells are not updated in a regular manner; neighboring cells may be updated at different points in time, which renders the computation more difficult.

Cyclic behavior cannot arise in the first model, since the notion of a logical step, i.e., a fixed number of time steps after which all cells will have been updated, does not exist; however, a fixed point, such as that desired for the density problem, can be attained. Models 2 and 3 can be applied to the synchronization problem since cyclic behavior may be attained, if one considers the CA's configuration every logical step, i.e., the alternation between all 0s and all 1s takes place every $\#_b$ time steps.

Our results for the density task show that model-1 asynchronous CAs can be evolved whose performance is comparable to the synchronous case (e.g. [7, 10]), provided the number of blocks does not exceed three ($\#_b \leq 3$); for $\#_b > 3$, successful asynchronous CAs did not evolve. Figure 1 demonstrates the operation of an evolved, non-uniform, model-1 asynchronous CA on the density task. For the synchronization task, successful model-3 CAs with $\#_b \leq 8$ were evolved (grid sizes considered were in the range $N \in [100, 150]$); applying model 2, no successful CA had emerged from the evolutionary process.

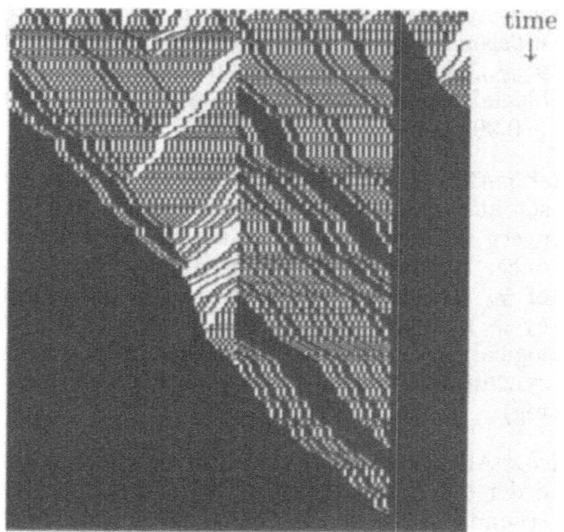

Figure 1: One-dimensional density task. Operation of a coevolved, non-uniform CA with connectivity radius $r = 1$. The CA is asynchronous, model 1. Grid size is $N = 150$, with two 75-cell blocks ($\#_b = 2$). White squares represent cells in state 0, black squares represent cells in state 1. The pattern of configurations is shown for the first 665 time steps, with time increasing down the page. The randomly generated initial configuration has a density of 1s greater than 0.5, and the CA relaxes to a fixed pattern of all 1s, which is the correct output.

The deterministic updating schedule of model 3 renders it easier for evolution to cope with, as compared with model 2. For both, however, an obstacle that hinders the evolutionary algorithm is the need to adapt to block boundaries. A 'good' rule in cell i may be of no use, or even detrimental, in cell $i + 1$, if a block boundary occurs between these two cells. Two strategies were observed to emerge through the evolutionary process in order to cope with this problem: either specialized rules are evolved at block boundaries (different than the rules present in the rest of the block), or a rule is evolved that is essentially insensitive to the presence or absence of a boundary.

3 Scaling Evolved CAs

In this section we return to *synchronous*, non-uniform CAs, our interest lying in the scalability issue. Essentially, this involves two separate matters: the evolutionary algorithm and the evolved solutions. As to the former, we note that as our cellular programming algorithm is local it scales better in terms of hardware resources than the standard (global) genetic algorithm; adding grid cells requires only local connections in our case whereas the standard genetic algorithm includes global operators such as fitness ranking and crossover [7]. In this section we concentrate on the second issue, namely, how can the grid size be modified given an evolved grid of a particular length, i.e., how can evolved solutions be scaled? This has been purported as an advantage of uniform CAs, since one can directly use the evolved rule in a grid of any desired size. However, this form of *simple* scaling does not bring about *task* scaling; as demonstrated, e.g., in [2] for the density task, performance decreases as grid size increases. Previously, we had attained successful systems for a random number generation task using a simple scaling scheme involving the duplication of the rules grid [12]. Below we report on a more sophisticated, empirically-obtained scheme that has proven successful.

Given an evolved non-uniform CA of size N, our goal is to obtain a grid of size N', where N' is given but arbitrary (N' may be $> N$ or $< N$), such that the original performance level is maintained. This requires an algorithm for determining which rule should be placed in each cell of the size N' grid, so as to preserve the original grid's "essence," i.e., its emergent global behavior. Thus, we must determine what characterizes this latter behavior. We first note that there are two basic rule structures of importance in the original grid (shown for $r = 1$):

- The *local structure* with respect to cell i, $i \in \{0, \ldots, N-1\}$, is the set of three rules in cells $i-1$, i, and $i+1$ (indices are computed modulus N since the grid is circular).
- The *global structure* is derived by observing the *zones* of identical rules present in the grid. For example, for the following evolved $N = 15$ grid:

$R_1R_1R_1R_1$	R_2R_2	R_3	$R_4R_4R_4R_4$	R_1	$R_5R_5R_5$

where R_j, $j \in \{1, \ldots, 5\}$, denotes a distinct rule, the number of zones is 6, and the global structure is given by the list $\{R_1, R_2, R_3, R_4, R_1, R_5\}$.

We have found that if these structures are preserved, the scaled CA's behavior is identical to that of the original one. A heuristic principle is to expand (or reduce) a zone of identical rules which spans at least four cells, while keeping intact zones of length three or less. It is straightforward to observe that a zone of length one or two should be left untouched, so as to maintain the local structure. As for a zone of length three, there is no a priori reason why it should be left unperturbed, rather, this has been found to hold empirically. A possible explanation may be that in such a three-cell zone the local structure $R_j R_j R_j$ appears only once, thereby comprising a "primitive" unit that must be maintained. As an example of this procedure, consider the above $N = 15$ CA— scaling this grid to size $N' = 19$ results in:

$R_1 \dots R_1$	$R_2 R_2$	R_3	$R_4 \dots R_4$	R_1	$R_5 R_5 R_5$

Note that both the local and global structures are preserved. We tested our scaling procedure on several CAs that were evolved to solve the synchronization task. The original grid sizes were $N = 100, 150$, which were then scaled to grids of sizes $N' = 200, 300, 350, 450, 500, 750$. In all cases the scaled grids exhibited the same performance level as that of the original ones. An example of a scaled system is shown in Figure 2.

Figure 2: One-dimensional synchronization task — example of a scaled CA. An evolved, size $N = 149$ CA was scaled to a size $N' = 350$ CA, shown above.

4 Conclusions

We studied the evolution of non-uniform CAs via cellular programming, concentrating on two novel issues, namely, asynchrony and scalability. We introduced three models of asynchrony, previously unstudied in this context, finding that asynchronous CAs can be evolved to perform the computational tasks in question. Though it seems that asynchrony presents a more difficult case for evolution, it is premature to draw any definitive conclusions at this point, since we have only considered two problems, using relatively small-size grids. We feel that successful asynchronous CAs can be evolved, though this will probably entail larger grids (coupled with larger blocks). We next described a scaling procedure for synchronous CAs, by which an evolved system of given size may be used to obtain augmented or reduced grids. Our tests suggest that this procedure yields scaled systems whose performance level is identical to the original one. Though preliminary, we hope that further studies along these lines will help deepen our knowledge of evolving cellular systems, as well as propel us toward the attainment of more realistic adaptive systems, that can ultimately be implemented as evolving ware, evolware.

References

[1] H. Bersini and V. Detour. Asynchrony induces stability in cellular automata based models. In R. A. Brooks and P. Maes, editors, *Artificial Life IV*, pages 382–387, Cambridge, Massachusetts, 1994. The MIT Press.

[2] J. P. Crutchfield and M. Mitchell. The evolution of emergent computation. *Proceedings of the National Academy of Sciences USA*, 92(23):10742–10746, 1995.

[3] R. Das, J. P. Crutchfield, M. Mitchell, and J. E. Hanson. Evolving globally synchronized cellular automata. In L. J. Eshelman, editor, *Proceedings of the Sixth International Conference on Genetic Algorithms*, pages 336–343, San Francisco, CA, 1995. Morgan Kaufmann.

[4] T. E. Ingerson and R. L. Buvel. Structure in asynchronous cellular automata. *Physica D*, 10:59–68, 1984.

[5] M. Mitchell, J. P. Crutchfield, and P. T. Hraber. Evolving cellular automata to perform computations: Mechanisms and impediments. *Physica D*, 75:361–391, 1994.

[6] M. A. Nowak, S. Bonhoeffer, and R. M. May. Spatial games and the maintenance of cooperation.

Proceedings of the National Academy of Sciences USA, 91:4877–4881, May 1994.

[7] M. Sipper. *Evolution of Parallel Cellular Machines: The Cellular Programming Approach*. Springer-Verlag, Heidelberg, 1997.

[8] M. Sipper. The evolution of parallel cellular machines: Toward evolware. *BioSystems*, 1996. (to appear).

[9] M. Sipper. Evolving uniform and non-uniform cellular automata networks. In D. Stauffer, editor, *Annual Reviews of Computational Physics*, volume V. World Scientific, Singapore, 1997. (to appear).

[10] M. Sipper. Co-evolving non-uniform cellular automata to perform computations. *Physica D*, 92:193–208, 1996.

[11] M. Sipper and E. Ruppin. Co-evolving architectures for cellular machines. *Physica D*, 1996. (to appear).

[12] M. Sipper and M. Tomassini. Generating parallel random number generators by cellular programming. *International Journal of Modern Physics C*, 7(2):181–190, 1996.

[13] S. Wolfram. Approaches to complexity engineering. *Physica D*, 22:385–399, October 1986.

One-Chip Evolvable Hardware: 1C-EHW

H. de Garis
Brain Builder Group, Evolutionary Systems Department,
ATR Human Information Processing Research Laboratories,
2-2 Hikaridai, Seika-cho, Soraku-gun, Kansai Science City, Kyoto, 619-02, Japan.
Email: degaris@hip.atr.co.jp

Abstract

This paper aims at stimulating discussion and research into a new concept called 'one-chip evolvable hardware (1C-EHW)' by proposing a generic architecture and methodology to generate the ultimate in evolvable hardware speed, where all the evolvable hardware components are placed on a single chip.

1 Introduction

My primary research interest is not in evolvable hardware, (even though I invented the concept [5]) but in building artificial brains [6, 8, 10, 11]. Recently, my colleague Felix Gers and I have managed to grow/evolve a system with 10 million artificial neurons [10], based on cellular automata [1], inside MIT's cellular automata machine 'CAM-8' [15, 16] (which updates 200 million cellular automata cells per second). By partitioning these 10 million neurons into 100-neuron 'modules' and evolving each module to perform some user defined function, it will be possible to build artificial brains. I have already published some 30 papers (e.g. [2, 3, 4, 7]) on the evolution [12] of neural network *dynamics* (another concept I invented [2]), so I am confident that building artificial brains will be a reality within less than a year. However, Gers and I are left with an immense practical problem, namely, 'How do you evolve 100,000 neural net modules in a reasonable time?'

To get a better feel for the above problem, imagine that these 100,000 modules are to be evolved within a 3 year time period. How fast would Gers and I have to evolve each module to achieve this goal? Assuming a 50 week year, and a 40 hour week, that will be 6000 hours for 100,000 modules, i.e. one module per 3.6 minutes. To evolve one module, let us assume that the genetic algorithm [12] measures the fitness of 2,160 chromosomes (a convenient but ballpark figure). So one chromosome fitness measurement takes 0.1 second. Assuming a 2D CA neural net model (we already have a 3D simulation model on a workstation, but we are not yet sure how to put a 3D version onto a 2D chip), a space of 10,000 CA cells in which to evolve each module, and a total of 1000 clock cycles (250 to grow the CA based neural net, and 750 to make the net transmit neural signals), i.e. 1000*10,000 CA cell updates in 0.1 second, i.e. 1 CA cell update in 10 nanoseconds. This is *on-chip electronic speed.* If it were necessary to transfer signals from chip to chip, then this can take milliseconds, especially if the transfer is bit serial. A multi-chip solution would be too slow. We would not be able to build the 100,000 modules in the 3 year target period. Actually the problem will be worse, because by 2001, Gers and I will have a billion artificial neurons, hence all the more need for 1C-EHW. (One alternative solution to obtain real speed is to go massively parallel, an option discussed in section 6 on future work).

Based on the above estimates, I convinced myself that what I needed was some kind of electronic system in which all the components for the evolution of the neural modules would be on-chip and hence as fast as possible. The concept of 'one-chip evolvable hardware (1C-EHW)' began to crystallize in my mind.

This 1C-EHW idea is, of course, not limited to growing neural net modules for artificial brains. It is quite general. I believe that 1C-EHW has a certain historical inevitability about it, because a one-chip approach offers the highest possible speed. Current, or near future, field programmable gate arrays (FPGAs) are probably big enough for 1C-EHW to become feasible. This paper hopes to attract attention to this idea, so that people more expert than I am in chip design and electronics in general, will be attracted to the topic and help to make it a reality.

This paper does not give a concrete architecture for a given application in 1C-EHW. Such a task is for future work, and may possibly be beyond the state-of-the-art. What it does hope to do, is stimulate the creation of a new sub-field within the broader field of evolvable hardware, namely 'EHW on a single chip', or 1C-EHW. This paper sets out to consider what the issues are in creating such a sub-branch.

2 Initial Thoughts

I will begin this section with a little bit of history. This paper on 1C-EHW is not the first time I have tried to stimulate electronics people to think about EHW. In fact, the rise of the field itself started in the summer of 1992, when I was visiting George Mason in Virginia University to discuss some ideas with an electronic engineering colleague. He told me about S-RAM based FPGAs and the general idea that hardware can be programmed, i.e. a software bit string can be used to configure or wire up an electronic circuit. Since I had already spent several years evolving neural net dynamics with a genetic algorithm (GA), having invented the technique [2], I immediately thought of the idea that one could consider the configuring bit string as a GA chromosome, and hence one could evolve hardware. I starting speaking about my idea and to write papers [5, 8]. A mere 4 years later the popularity of the idea has grown to the point, that there is now an international conference on the topic [13] and even a book [8].

Returning now to the present, I would like to give an overview of the proposal. The initial idea is to use the largest possible FPGA as the vehicle for implementing the idea of 1C-EHW. Maximum size in state-of-the-art FPGAs is of the order of 100,000 logic gates (1996). This may not be enough, but since the size of 'Moore-doublings' is now so large, we need only wait a few years before it should be possible. If there are too few gates in 1996, then we can still think about the issues in anticipation. It is also possible that FPGA designers might be inspired by the ideas in this paper to design new types of FPGAs more suitable for 1C-EHW.

The very large FPGA can be partitioned flexibly into functional regions, depending upon the individual application. One broadly applicable partitioning that I have been considering, consists of 3 regions. The first region contains a cache for the chromosome population used to govern the evolution, which can be down-loaded in nano-seconds to wherever it is needed within the chip. If several chromosomes are present in the population, this region can also implement certain aspects of the genetic algorithm. The second region is where the circuit itself can be evolved, using the chromosome instructions held in the first region. The third region is where the fitness of the circuit evolved in the second region is measured. The fitness results obtained in the third region are reported back to the first region. All of these functions are hardware programmable.

This approach should have a wide range of possible applications, and might be (partially) implementable on a single large reconfigurable FPGA. Applications suitable for such a chip are those for which the time for fitness measurement (in region 3) is *less* than the time needed for the evolving circuit to be configured and to execute its function. The outputs of the circuit are sent to the fitness measuring region for measurement. Of course, the application itself has to be suitable for such a one chip approach. Despite such restrictions, there should still be plenty of interesting applications. In this sense the architecture is generic.

3 Earlier Work

This section discusses earlier work by other people in the field of FPGAs and its relevance to the ideas in this paper. I consider some of this earlier work to be attractive enough to want to incorporate these ideas into my proposed methodology.

When I first had the idea that one could actually evolve hardware, the type of S-RAM based FPGAs available at the time (1992) were quite unsuitable for the task. I remember timing how long the whole process of changing and rerouting a hardware design took. It was of the order of 30 minutes, with most of the time taken up in the software controlled (re)routing process [5]. FPGAs were a revolution in the early days, because they enabled a quick turn around time when an engineer had to make a correction to a faulty circuit design. Prior to FPGAs, changes to factory based electronic chip designs took weeks to months. But for the purposes of EHW, 30 minutes was hopeless. There were other problems as well. Paraphrasing my original paper [5]:

1. The configuration bitstring (i.e. the software instruction which is used to configure or wire up the generic device) is inputted serially, i.e. one bit at a time, for (often) many thousands of bits. Even if one "mutates" only one bit, the whole bitstring has to be re-inputted (serially).

This downloading can take up to 30 seconds.

2. The routing, i.e. the choice of connections between the logic blocks and the I/O blocks of typical FPGAs is done by software and usually takes several to many minutes, depending upon the complexity of the circuit.

3. Generating the list of gating instructions, i.e. mapping all the routes between logic and I/O blocks, takes about a minute.

4. Most of the details as to how the circuit is wired up are company secrets, so one would be unable to know which bits in the bitstring correspond to which logic gates, etc.

5. For many FPGAs, it is not possible to send in just any old bit string of the appropriate length. Many such bitstrings would cause the FPGA to malfunction.

In view of these weaknesses I strongly suggested that the FPGA companies try to make an 'evolvable chip' which did not have the above weaknesses. Up to that point, FPGAs were not designed with evolvable hardware (EHW) in mind. The next step occurred with the recent arrival of Xilinx's XC6216 chip.

3.1 The Xilinx XC6216 Reconfigurable Co-Processor

At the time of writing (July 1996), I do not have such a chip in my possession, so I do not have first hand working experience with it (although this situation will change shortly). However Adrian Thompson of Sussex University, is doing pioneering work using this chip. (Adrian does ***intrinsic*** EHW, i.e. my term for 'on-chip', 'on-line', 'in-situ' evolution, where the circuit is reconfigured for each chromosome. *Extrinsic* EHW I defined to be 'off-line', 'simulated', 'off-chip' evolution, where only the elite simulated bitstring is actually downloaded to configure the circuit. In extrinsic EHW, most of the evolution occurs off-line, i.e. external to the circuit, hence extrinsic [5]). Thompson has already worked with the XC6216 and is the first person in the world (as far as I know) to succeed in doing real intrinsic EHW. According to him [14], the XC6216 chip is quite suitable for EHW. Thompson claims that the above weaknesses have been remedied by the Xilinx XC6216. For example, one does not have to download the whole bitstring, once a bit has been mutated, only the local region of the bitstring including the mutation is downloaded and only the local circuitry is modified. The XC6216 chip can accept an arbitrary bitstring, and it will not crash the chip. The XC6216 circuitry and bit string representation is public, it is not a company secret. One wonders whether Xilinx conceived the XC6216 with EHW in mind or if it is only a great coincidence that this chip is so suited to EHW.

3.2 MIT's Dynamically Programmable Gate Arrays (DPGAs)

The next innovation discussed in this section on features which will be included in the one-chip architecture, is de Hon's 'DPGA' [9]. This revolutionary concept allows the architecture of an FPGA to be changed at each tick of the clock, i.e. *a new computer every clock tick*. Prior to de Hon's work, to change the configuration of an FPGA meant down loading the configuration bitstring onto the chip externally, i.e. from outside the chip, an operation taking at least milliseconds. de Hon put 4 configuration bitstrings into an on-chip cache, and was thus able to download in nano-seconds, so he could change his on-chip architecture (i.e. he could reconfigure it) in nanoseconds. I consider this idea to be extremely useful and revolutionary in its implications, because its allows run-time reconfigurability. You can change the architecture of your system 'on the fly'.

As will be seen shortly, this idea of runtime reconfigurability will play an important role in the 1C-EHW architecture.

4 The Proposal

Before beginning to discuss details of the proposed architecture and methodology, several general remarks need to be made. Firstly, the architecture proposed here is supposed to be general, i.e. in the sense of flexible. Each application will have its own specific needs. What is being proposed is a general methodology rather than a concrete procedure. It is more a guide to the idea of a one-chip philosophy than a specific circuit. Specific circuit descriptions will be presented in later papers.

What is not clear at the time of writing is whether the ideas proposed in this paper are implementable this year (1996). If so, fine, but if not, then as mentioned earlier, one need only wait for a few more 'Moore-doublings' to achieve the necessary FPGA logic gate capacity. For example, Xilinx's XC4000 series has up to 125,000 logic gates (i.e. about 20,000 logic cells/blocks). By the turn of the century, this should be around the million gate mark. Of course, one can begin to investigate the one-chip approach with small examples, to see what the strengths and weaknesses of the approach are. This paper is aimed at raising the issue, and getting people thinking about the possibility of very fast intrinsic hardware evolution.

Figure 1 shows the top level functional layout of the design. The one-chip architecture consists of 3 main regions or areas. Note that the 'boundaries' of these regions are fluid, and will vary from application to application, depending on the users requirements. The actual chip will very likely be manufactured by a commercial company, and hence be extremely general. It will be the users task to specify (i.e. to program) the regions.

Region 1 contains a cache of the chromosomes of the application. The size of the population will depend upon the application and upon the logic-gate-number limit imposed by the chip size. The smallest population size will probably be two, namely the 'elite' chromosome (i.e. the 'best-so-far' chromosome, which gives the highest fitness score so far) and the 'current chromosome', which is currently being used to configure the application's circuit. If the current chromosome ends up with a higher fitness than the elite chromosome, it becomes the new elite chromosome. If several chromosomes can fit into the chip, then Region 1 also contains circuitry needed to handle the GA aspects of the evolution, e.g. the probability that a given chromosome passes into the next generation, crossover, etc. To generate a new 'current' chromosome, a copy is made of the best-so-far chromosome, and some of its bits are randomly flipped (i.e. mutated). Region 1 therefore needs to contain circuitry to perform this bit flipping.

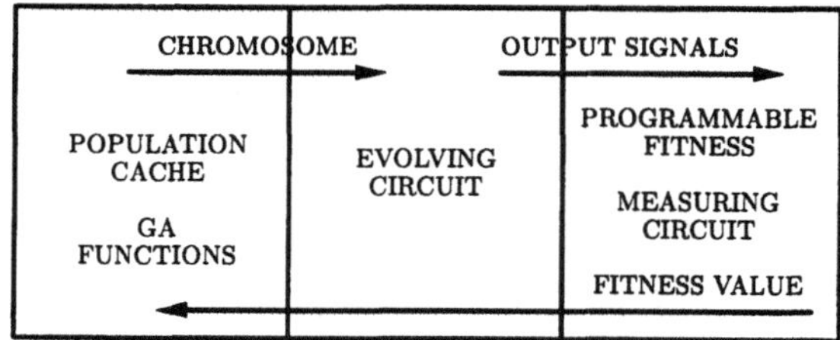

Figure 1: Generic One-Chip Evolvable Hardware (1C-EHW) Architecture.

The second region, Region 2, contains the circuit which is configured by the current chromosome in Region 1. Region 2 will need to contain two sub-regions, namely the circuitry necessary to interpret the instructions contained in the chromosome (called the 'interpreter'), and the sub-region in which the circuit itself is grown/configured. Perhaps these two sub-regions can be merged. For example, I want to use 1C-EHW to evolve neural net modules very quickly for 100,000-module artificial brains [10]. The instructions in the chromosome used to tell growing axons/dendrites to turn left or right etc can either be fed into a growing trail, or be distributed initially over the available space. When a growing trail hits a local growth instruction, it turns appropriately. To know how to turn appropriately, each local space needs to use some kind of look-up-table (LUT) which actually executes the instruction. This interpreter can either be located in some form of centralized memory or, in the limit, be distributed throughout the available space, and in many copies, one per turning point. This latter option will become obligatory when switching elements reach pico-second switching times in order to avoid signal delay problems, but is very expensive in terms of silicon real estate. So in this proposal, only centralized memory is proposed. The 'interpreter' sub-region can be programmed just once, and can hence be downloaded in milliseconds from outside the circuit. It remains intact throughout the entire evolution. If one wishes to change the interpreter on-the-fly, one could download the interpreter instructions from Region 1. However this initial proposal is not so ambitious.

The third region, Region 3, contains the circuitry used to measure the fitness of the circuit evolved in Region 2. The outputs from Region 2 are fed into Region 3, where they are used to obtain a fitness measurement of the evolved circuit in Region 2. If possible, it might be more efficient if this fitness value is measured incrementally, i.e. as each output value (e.g. a signal value at a given instant) is generated, it is fed into the fitness measurer which in turn incrementally updates its fitness value. The new fitness value $F(t+1)$ is a function of the previous fitness value $F(t)$ and the new signal value $SV(t+1)$.

$$F(t+1) = f[F(t), SV(t+1)]$$

If the application does not lend itself to such incrementalism, then the fitness evaluation can only begin after all the signal values have been stored.

This may be a much slower process and is to be avoided if possible. Another idea is to use a kind of 'pipeline processing'. For example, so long as the fitness evaluation time of Region 3 is less than the setup and execution time of Region 2 (otherwise Region 2 has to wait until Region 3 is finished), then while Region 2 is setting up and executing chromosome '$N+1$', Region 3 can be measuring the fitness of chromosome 'N'. This procedure may complicate the 'best-so-far' chromosome detection algorithm in Region 1.

All these regions are hardware programmable. Region 1, the interpreter sub-region of Region 2, and Region 3 can be programmed at the start, and once only, so it is not very important how long it takes. These initial hardware programming instructions can come from off-chip. After that, all the essential elements for fast evolution, i.e. the specification and construction of the circuit, its fitness measurement, and the creation of the next generation of chromosomes are all done on-chip at hardware speeds.

5 Constraints and Limitations

Common sense says that the usefulness of this one-chip approach depends upon the relative times needed for processing in the 3 regions. For example, if the time taken for processing in Region 2 plus the time taken to measure fitness in Region 3 is far longer than the time taken to load (from off-chip) the configuring chromosomes into Region 1, then there is no point of having a chromosome cache in Region 1 (i.e. no Region 1 is needed). So the 1C-EHW approach depends on the type of problem. If the set up and execution time in Region 2 is far longer than the time taken to measure the fitness of the chromosome in Region 3 (especially when one has to store all the outputs before performing the fitness measurement), then there is no point in putting Region 3 on-chip. Instead one could use a pipe-lining technique, and measure the fitness off-chip during the time the next chromosome executes. There are probably other types of restrictions of a similar nature.

The above objections are legitimate. Whether the 1C-EHW approach is worthwhile or not, will depend upon the individual application being considered. Potential users of 1C-EHW will need to make initial estimates of the relative execution times needed in the 3 Regions. For some cases, 1C-EHW will be quite useless. For others, it might be possible, with a bit of creativity and adjustment, to modify the application so that it fits the 1C-EHW criteria for suitability (e.g. by measuring the fitness incrementally, and keeping execution runs short). In other cases again, the 1C-EHW approach may be very suitable and hence allow considerable speed up in the evolution of hardware. As I said earlier, with my brain building work, I need to evolve 100,000 neural net modules, so I must have speed, speed and more speed. The whole brain building effort depends on getting enough speed to make the CAM-Brain Project [6, 10] successful.

It is possible that I may be overestimating the importance of the above constraints, because eventually, FPGAs will get so huge, that concerns about the relative execution times in the various Regions will become irrelevant. In time, it will simply be easier to put *everything* on-chip.

6 Future Work

One idea which I would like to see taken up in the future is that of wafer scale integration (WSI) using a possible immunity of evolved neural nets to chip fabrication faults. Using the full surface area of a silicon wafer allows one to position a huge number of artificial neurons, but such a large surface area will inevitably contain fabrication faults. However, since we are talking about EHW, it may be possible that *the evolution can take the faults into account*, and still be functional. Whether this idea is fully practical may depend on whether critical components of the wafer need to be fault-free. Perhaps these critical regions can be made rather small, so that the probability of error is small, so that the wafer is likely not to be trashed. A *potential marriage of EHW and WSI could be extremely interesting.* It would make the whole idea of 1C-EHW much more feasible, because the surface area would be hundreds to thousands of times larger than today's chip. (In fact, today's chips are small, precisely because of the fabrication faults. The smaller the chip, the lower the probability that it contains a fault, and hence the higher the 'yield', i.e. the percentage of perfectly functioning chips cut from a wafer).

A second fundamental idea for accelerating the speed of EHW, especially for my brain building work, is to update the CA cells in parallel, i.e. massively parallel. NTT is developing for me a 'content-addressable-memory' based device, for updating cellular automata, which is massively parallel. This device is expected to be able to update 10 billion to 100 billion CA cells per second, and should be completed by the end of 1997. If successful, this device could be incorporated into the 1C-EHW design, becoming an essential ingredient of the chip. Such an incorporation would make the chip more 'special-purpose' than a general FPGA. Alternatively, one could presumably implement the NTT design in an FPGA, but it would probably be slower than an ASIC ('application specific integrated circuit'). Generally speaking, an ASIC 1C-EHW design would be faster than a large commercial generic FPGA. Since for my brain work, I want speed, I would probably prefer an ASIC design.

A third idea worth considering for future development is that of 'field programmable multi-chip modules' (FPMCMs). Combining FPGAs and MCMs is the pioneering work of Professor Dai at UC Santa Cruz [13]. Quoting from their paper [13], 'A multichip module (MCM) is a device in which several integrated circuits are attached to a single substrate and then packaged as a unit.' MCMs allow an order of magnitude increase in the density of gates compared to a single chip module. For example, Dai's team has already produced a 200K gate FPMCM prototype (i.e. double the number of gates compared to state-of-the-art FPGAs). Using FPMCMs could eliminate several Moore-doublings while waiting for FPGAs to deliver enough gates to do non-trivial 1C-EHW. Perhaps FPMCM techniques may allow 1C-EHW now, but only if they offer speeds comparable to those of single chips. This in fact may be possible, because the multichips are packed very closely together inside the module and are closely integrated electronically, thus acting like one large chip. Dai's team talks about using FPMCMs for reconfigurable co-procesors, so this implies very high speed.

Finally, I will be buying an FPGA development system shortly, (possibly an FPMCM as well) and will attempt to implement some of the ideas in this paper in the near future. I hope others will want to do the same.

7 Summary

My work with artificial brains, i.e. the need to evolve 100,000 neural net modules in a reasonable time, necessitates performing this evolution as fast as possible. The fastest way to evolve hardware is to do it on-chip, hence the need for 'one-chip evolvable hardware' (1C-EHW). A generic 1C-EHW architecture and methodology was proposed in this paper. The vehicle proposed to support 1C-EHW is a large FPGA which consists of 3 main regions. The first region contains a cache of the chromosomes used to specify the circuit to be evolved, plus circuitry needed to perform the genetic algorithmic aspects. The second region contains the circuit to be evolved. The third region contains the fitness measuring circuit. All these regions are hardware programmable. The first and third regions (and the interpreter subregion of the second region) are programmed initially and once only. Since it is rather unimportant how long this takes, the instructions can come from off-chip. All the essential elements for fast hardware evolution, i.e. the growth/specification of the circuit, its fitness measurement, and the creation of the next generation of chromosomes are all performed on-chip at hardware speeds. The relative execution times in each of these regions will determine whether the 1C-EHW approach is sensible or not.

References

[1] E. F. Codd. *Cellular Automata.* Academic Press, NY, 1968.

[2] H. de Garis. Genetic programming: Building artificial nervous systems using genetically programmed neural network modules. In B. W. Porter and R. J. Mooney, editors, *Proc. 7th. Int. Conf. on Machine Learning*, pages 132–139. Morgan Kaufmann, 1990.

[3] H. de Garis. Genetic programming: Modular evolution for Darwin machines. In *Proc. Int. Joint Conf. on Neural Networks*, 1990.

[4] H. de Garis. *Genetic Programming*, chapter 8. John Wiley, NY, 1991.

[5] H. de Garis. Evolvable hardware: Genetic programming of a Darwin machine. In N. C. Steele, R. F. Abrecht. C. R. Reeves, editors, *Artificial Neural Networks and Genetic Algorithms*, pages 441–449. Springer-Verlag, Wien New York, 1993.

[6] H. de Garis. *An Artificial Brain - ATR's CAM-Brain Project Aims to Build/Evolve an Artificial Brain with a Million Neural Net Modules Inside a Trillion Cell Cellular Automata Machine*, volume 12, pages 215–221. Ohmsha Ltd. and Springer-Verlag, 1994.

[7] H. de Garis. *Genetic Programming: Evolutionary Approaches to Multistrategy Learning*, chapter 21. Morgan Kauffman, 1994.

[8] H. de Garis. *CAM-BRAIN: The Evolutionary Engineering of a Billion Neuron Artificial Brain by 2001 Which Grows/Evolves at Electronic Speeds Inside a Cellular Automata Machine (CAM).* In D.W. Pearson, N.C. Steele, R.F. Albrecht, editors, *Artificial Neural Networks and Genetic Algorithms*, pages 84–87. Springer-Verlag, Wien New York, 1995.

[9] A. de Hon. DPGA utilization and application. In *Proc. of the 1996 Int. Symp. on Field Programmable Gate Arrays, ACM/SIGDA*, February 1996.

[10] F. Gers and H. de Garis. Cam-brain: A new model for ATR's cellular automata based artificial brain project. In *Proc. Int. Conf. on Evolvable Systems ICES96*, 1996.

[11] F. Gers and H. de Garis. Porting a cellular automata based artificial brain to MIT's cellular automata machine CAM-8. In *Proc. Int. Conf. on Simulated Evolution & Learning, SEAL96*, 1996.

[12] D.E. Goldberg. *Genetic Algorthims in Search Optimization and Machine Learning.* Addison Wesley, 1989.

[13] T. Higuchi, D. Mange, H. Kitano, and H. Iba, editors. 1996. http://www.etl.go.jp:8080/etl/kikou/ICES96.

[14] A. Thompson. Silicon evolution. In *Proc. 1st Int. Conf. on Genetic Programming*, Cambridge, MA, 1996. MIT Press.

[15] T. Toffoli and N. Margolus. *Cellular Automata Machines.* MIT Press, Cambridge, MA, 1987.

[16] T. Toffoli and N. Margolus. *Cellular Automata Machines.* Addison-Wesley, 1990.

Evolving Low-Level Vision Capabilities with the GENCODER Genetic Programming Environment

P. Ziemeck and H. Ritter
Department of Computer Science, Bielefeld University
P.O.Box 100131, D-33501 Bielefeld, Germany
Email: {patrick,helge}@techfak.uni-bielefeld.de

Abstract

A new approach for the application of genetic programming to vision problems is presented. Sets of atomic subprograms are genetically combined to solve more advanced problems within low-level vision or image preprocessing. We present the main ideas and give a brief sketch of their implementation in the distributed simulation environment GENCODER. This system forms the basis for some introductory experiments obtained. Finally, some aspects of the gained results together with interesting possibilities for future research are portrayed.

1 A New Approach to Early Vision

The initial stages of the processing of visual information (or briefly: early vision), especially within the human visual system, perform a substantial part of the entire perceptual process. From a more technical point of view one could forsee - due to this strong contribution to the hard overall task - a considerable complexity of the underlying principles and functional subsystems. In accordance with that, (early) vision science is still characterized by an almost confusing variety of techniques.

Despite significant theoretical progress in many respects, practically useful computer vision systems still rely on numerous heuristics and solutions strictly tailored to special requirements of the particular task. Also the development of rigorous theories, especially in the presence of image discontinuities, still remains an open problem [1, 3].

This state of affairs certainly reflects the difficulty of the overall task. Many aspects of building a vision system have the format of a complex optimization problem. This, along with the diversity of available techniques for solving early vision subproblems strongly suggests the use of genetic search strategies for (i) automating the process of constructing task-specialized 'vision-frontends' from libraries of early vision modules and, (ii), facilitating the task specification by using an optimization criterion that is based on a set of examples of the desired image processing operation (e.g., in the form of typical source images paired with their optimal results).

To cope with the very high dimensional search space efficiently, we use the approach of *genetic programming* since it appears better suited to operate at the required high level of abstraction that is involved in the selection of visual modules, their combination into new functional subunits and their proper parametrization so that hierarchical structures can be built efficiently.

For successful application of GP to this problem domain, several prerequisites are required. First, we must chose a set of visual modules as primitives. Although this is an important step, the precise choices should not matter too much as long as the chosen set is sufficiently 'rich' since the genetic programming algorithm can then evolve subprograms with necessary processing abilities, if required. However, great care must be taken for an accurate definition of the signatures (parameters and return values) of the primitives, since the interface structure determines the compatibility of the modules to each other and thereby restricts the set of possible combinations for the genetic search process. Furthermore, the genetic programming approach apparently requires a suitable fitness measure, to evaluate the difference between a given optimal program result and the actual output. Below we will suggest a suitable gray-level image difference measurement based on local pixel histograms to avoid the severe drawbacks of pixel-by-pixel comparisons.

2 The GENCODER Environment

Technically, an important prerequisite is an implementation that supports concurrency in order to cope with the rather high computational demands of the genetic search and to make the approach scalable as more and cheaper computational resources become available. This aspect is addressed by the genetic code developer software. It is a complete environment to explore genetic programming techniques within the computer vision domain and is implemented in CM Modula-3 using object oriented software construction techniques together with language capabilities such as multithreading, easy graphical user interfacing or distributed systems construction support.

GENCODER consists of three independently running programs. The kernel part of the environment is the 'GENCODER workbench' ('GBench' for short). It maintains all state variables of the system, genetically constructs new visual frontends and passes them over the network for concurrent fitness evaluation by multiple instances of 'GTool'-interpreters.

These GTools understand 'image processing expressions' (IPEs), which are very much like lisp's symbolic expressions, but provide an additional range of simple image processing primitives such as thresholding or filter procedures as well as built in data structures for efficient image handling.

Each IPE represents a particular 'visual frontend' and is applied to a set of source images from the data examples. These images are compared to the specified optimal result examples by means of a image difference measurement. Each comparison yields a raw fitness value for the corresponding data example. Finally, an overall fitness is computed by means of averaging over the single raw fitness values and is returned by GTool. Both the image difference measurement and the evaluation function of the overall fitness can be controlled by the user or reprogrammed, if necessary. It is also possible to start multiple GTool interpreters on *one* host to take full advantage of the local network and symmetrical multiprocessing (SMP) architectures at the same time.

The effective usage of all distributed GTool interpreters for a given GENCODER task is achieved automatically by GBench. All task-specific data such as genetic search parameters [2] or the test set used for fitness evaluations on the one hand and more implementation related options like the communication setup to the interpreters on the other hand are maintained centrally within this program instance. In general, GBench could be called an integration platform for the application of given standard computing resources to a genetic search problem in computer vision.

Finally, the 'GScreen' graphical user interface provides access to all features of GBench. It is possible to set up new tasks, stop contemporary running ones, change their genetic search parameters, or retrieve intermediate results of a previously started GENCODER task. GScreen can also be shut down while GBench and the accompanied GTool interpreters continue to perform the desired task in background. Later the user can power up GScreen again to look at intermediate results.

3 Introductory Examples

3.1 Experimental Setup:

To show the basic ideas of the approach together with the capabilities of the software environment, we now describe three introductory simulation examples. For each experiment, the genetic search mechanism had to evolve image processing functions from *the same set* of pre-programmed elementary primitives shown in Table 1.

Table 1: Image processing primitives.

Primitive	LISP Syntax
Rescale [0,127] to [0,255]	(ResLow Img)
Rescale [128..255] to [0,255]	(ResHigh Img)
Threshold [0,127]	(ThrLow Img)
Threshold [128, 255]	(ThrHigh Img)
Add two images	(Add Img Img)
Subtract two images	(Sub Img Img)
Multiply two images	(Mul Img Img)
Divide two images	(Div Img Img)
Mask 2nd with 1st image	(Mask Img Img)
Laplace image	(Lap Img)
Gaussian blur image	(GBlur Img)

The tasks themselves were specified by way of examples, each consisting of a source image and an associated desired result image. Both were given as 256-graylevel 240 × 135 pixels images. In order to evaluate a particular program's fitness, a special fitness measurement was chosen. Since pixel by pixel comparisons cannot serve as a good criterion for the similarity of images (think of two identical line drawings slightly shifted), we used 'local' pixel histograms. Tesselating the entire image plane into a grid of subrectangles and averaging the difference $\sum_{i=0}^{255} |\mathcal{A}(i) - \mathcal{B}(i)|$ of two local histograms

$\mathcal{A}, \mathcal{B}$ for all subrectangles then yields our measurement for the images' similarity. If the subrectangles are sized down to one single pixel, the exact pixel by pixel comparison arises as a special case. During all experiments described below a subrectangle size of 5×5 pixels was chosen and the major genetic search parameters for all examples were adjusted to:

- 1000 individual programs per generation,
- maximally 60 generations,
- crossover and reproduction probabilities are 90 % and 10%, respectively,
- maximum depth for generated programs is 25.

3.2 Experiments

For all three experiments, we chose the 'xv-image' depicted in Figure 1 as the common source image. Within the first experiment the algorithm had to label the principal bright/dark image subregions, without having appropriate threshold operations as processing primitives, since 'ThrHigh' and 'ThrLow' (cf. Table 1) were not suitable to this task due to their wrong threshold values. The desired optimum result for this particular task is a picture not shown that is uniformly black below the 'horizon' and uniformly white above.

After 59 generations the best program was:

```
(Sub (ThrHigh (Sub Img (Mul (Lap Img)(Mul (ResLow
Img)(Mask Img (Lap Img))))))(ResLow (Lap (ResHigh
(Add(Sub(Lap Img)(Mul Img (Mask Img Img)))(ResHigh
(Add Img (ResHigh(Add(ResLow Img)(ResHigh(Lap Img)..
```

Applied to the source image this program yielded the resulting image of Figure 2.

Within the next experiment, the GENCODER environment had to genetically construct the built-in 'emboss' operator (a complex convolution) of John Bradley's XV image tool. All other settings and primitives were exactly the same as described above. The desired optimum result is shown in Figure 3.

After 38 generations the best program was:

```
(Mul (Sub (Add (GBlurr Img) (Sub (Add (Add
(Add Img (GBlurr Img))(Add Img (GBlurr Img)))
(Add (Add Img (Add Img (GBlurr Img))) Img))
(Sub (Add Img Img) Img)))(Sub (Add Img
(Add Img Img)) Img)) (GBlurr (Sub(Add Img
(Add Img Img)) Img)))
```

yielding the output image of Figure 4.

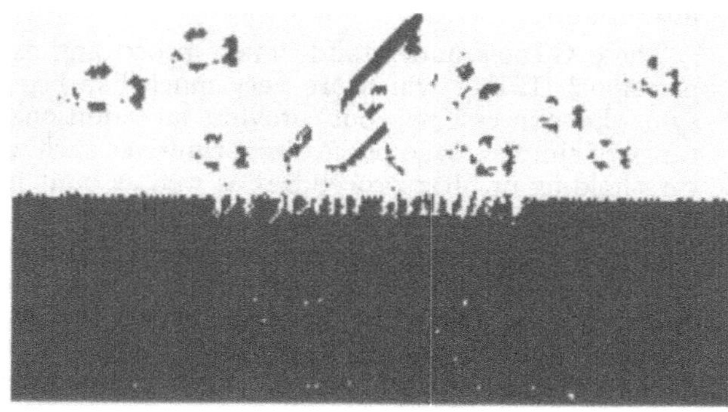

Figure 2: Best program's result image for task 1.

Figure 1: Source image for all examples.

Figure 3: Task 2: Optimal embossed image.

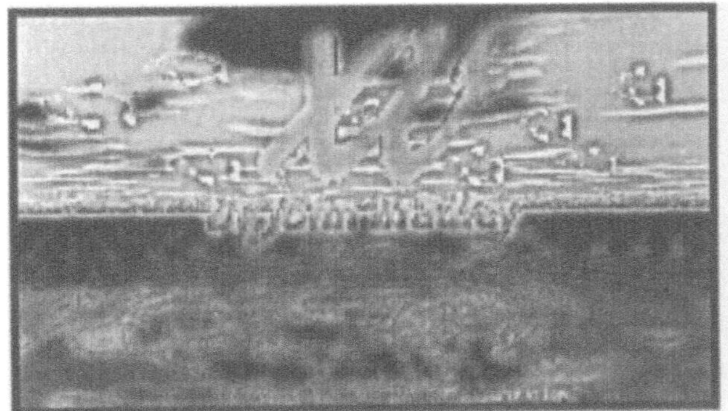

Figure 4: Best program's result image for task 2.

As a final task, the XV built-in 'edge detect' was approximated. This operator is implemented by two consecutive convolutions, detecting edges in X and Y directions separately and superposing the results. Again, nothing was changed concerning algorithm settings or function primitives available. The output of the original edge detection operator is shown in Figure 5.

After a total of only 18 (!) generations the best performing program was:

```
(Mul (Lap (Mul (Div (Mul (Div Img Img) Img)
(ResLow(Mul(Div Img (ResLow(Mul(Div Img Img)
Img))) Img))) Img))(Div(Mul(Div Img Img) Img)
(ResLow(Mul(Div(Mul(Div Img (Add Img (Sub(Sub
Img (Div Img Img)) Img))) Img) (ResLow(Mul(Div
Img Img)(Mul(Div Img Img) Img)))) Img))))
```

and the result is shown in Figure 6.

Figure 5: Optimal result for edge detection.

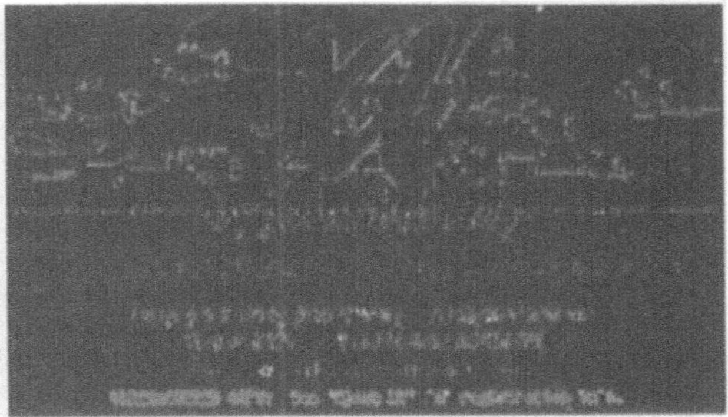

Figure 6: Best program's result image for task 3.

4 Discussion

These experiments demonstrate the principal feasibility of constructing non-trivial image processing capabilities from a library of simpler primitives, relying only on a task specification in the form of one or several examples, together with genetic search in the space of programs that use the initially chosen primitives.

We would like to emphasize several points. First, the initial library primitives (Table 1) were purposely chosen simple and incomplete; each primitive individually was by no means able to approximate the desired task even to a very moderate degree of accuracy. Second, despite the rather different tasks the system was able to generate good programs *starting from the same set of primitives and using the same search parameters* in all three cases. Furthermore, the search parameters could be chosen heuristically, with no further parameter tuning or other lengthy optimizations of design choices required.

In view of this high degree of parsimony in invested task specific knowledge, we find the performance of the genetically generated programs remarkable and very encouraging. We envisage that with a rather modest additional effort the example-driven synthesis of a number of interesting and already practically useful low-level vision operations, such as image enhancement, should be well within reach.

On a longer perspective, we shall attempt the 'evolution' of more complex visual capabilities. The investigation of systems, which cope with different kinds of image distortions simultaneously, appears rewarding and significant for real world applications. Also the automatic construction of systems capable to perform advanced classification tasks should be feasible within the framework presented.

If a set of early vision modules has been successfully constructed this way, automatic generation of higher level visual modules up to object recognition or image understanding might become possible. The investigation and isolation of information processing modules at different stages of vision could yield considerable insights into adequate overall structures for practically useful vision systems and may help to introduce a more taxonomic view of the variety of existing techniques.

Such an approach might also help to shed light on the importance of certain visual primitives and allow a comparison of the 'computational reach' of different module sets for solving tasks within a certain vision domain.

References

[1] J. Aloimonos and D. Shulman. 'Integration of Visual Modules - An Extension of the Marr Paradigm'. Academic Press, 1989.

[2] J. Koza. 'Genetic Programming - On the Programming of Computers by Means of Natural Selection'. MIT Press, 1992.

[3] T. Poggio, V. Torre, and C. Koch. 'Computational vision and regularization theory'. Nature 317, 1985.

NLRFLA: A Supervised Learning Algorithm for the Development of Non-Linear Receptive Fields

S. L. Funk[1], I. Kumazawa[1] and J. M. Kennedy[2]
[1] Kumazawa Laboratory, Department of Computer Science, Tokyo Institute of Technology, Ookayama 2-12-1, Tokyo 152, Japan.
[2] Department of Psychology, University of Toronto, Scarborough College.
Email: steve@cs.titech.ac.jp

Abstract

The non-linear receptive field (NLRF) neural network consists of a homogeneous, uniformly distributed series of locally connected non-linear receptive fields. Each receptive field exploits a set of local connections, with weights which are symmetrical around the center of the receptive field. The nonlinear behaviour is the result of three properties of the network. First, the activation is accumulated in the output layer units. Second, the recurrent feedback of activation from the output layer back onto itself. Third, the overlap of receptive fields. The nonlinear nature of the network allows it to perform relatively complex tasks, in spite of its simple architecture. The non-linear receptive field learning algorithm (NLRFLA) provides a way of finding the optimal set of connection weights for a given problem. The NLRFLA learning algorithm is essentially a recurrent backpropagation learning algorithm, with some special conditions.

1 Introduction

The processing of pattern information within the human perceptual system involves the extraction of features which occur consistently under the same conditions. The consistent relationship between the features in the pattern information, and the objects in the world allow the pattern information to convey meaning. In vision, there are a number of features which are of particular importance in the understanding of the perceived environment [7, 8]. However, there is evidence that much of this feature extraction is done by local processing structures [6]. These structures are known as receptive fields. While, in the past, there has been some work done on the computational capabilities of receptive fields, these have been primarily limited to linear filters [4], of one kind or another. The non-linear receptive field (NLRF) neural network differs from these systems to a rather large degree.

2 The Non-Linear Receptive Field Model

The NLRF network consists of a number of homogeneous, uniformly distributed, and overlapping, NLRFs. Each NLRF is a circular receptive field, 5 units wide (although this number is arbitrary). There are a set of afferent connections from the input layer to the output layer, and a set of lateral connections within the output layer (Figure 1).

Together, these connections form a receptive field, with a specific function. Under conditions where the receptive field is uniformly activated, the accumulation of activation only serves to exaggerate the receptive fields function. However, when the receptive field is not uniformly activated, the accumulation of activation allows the receptive field function to change in a way that is dependent as much upon the input pattern, as it is the original function (Figure 2).

Furthermore, the overlapping of these receptive fields, allows the network to operate as a coherent whole. The interaction is global because each receptive field is interacting with its neighbours, and

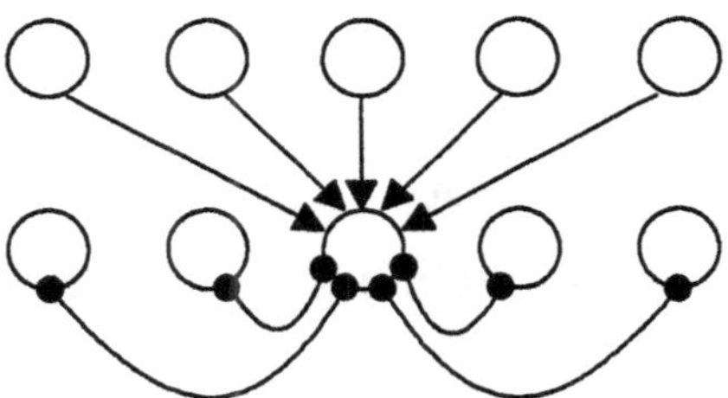

Figure 1: A cross-section view of a single nonlinear receptive field. In this case the NLRF is 5 units across, see [2].

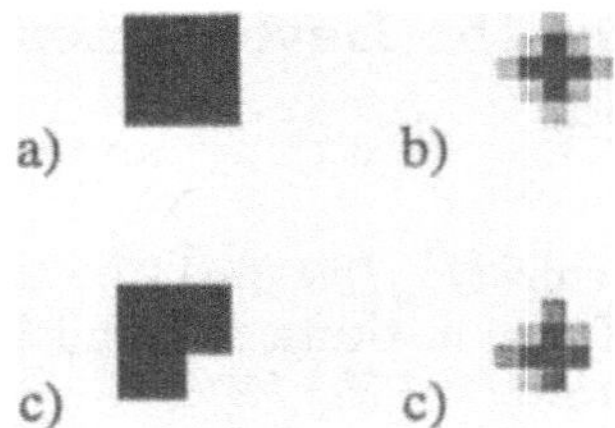

Figure 2: Input patterns to an NLRF. a) represents a uniform input patter, and b) its output pattern at 9 cycles. c) represents an incomplete input patter, d) its output at 9 cycles[3].

through these neighbours, with more distant receptive fields, as in Figure 3.

The adaptive nature of the receptive field's function means that each input pattern changes the terrain of the state space. The result is a simple network architecture, with a complex behaviour. It has been demonstrated that the NLRF architecture is useful for extracting a number of different features from a printed character [3]. However, the connection weights for each of these receptive fields were chosen by hand. In order to optimize the function of these receptive fields, and apply the NLRF architecture to other tasks, it is necessary to find an efficient means of developing each receptive field.

The problem of finding an appropriate learning algorithm is simplified in a number of ways. The uniform and consistent application of the receptive field means that only a single template need be developed. Furthermore, the architecture of the NLRF network is essentially the same as that of a simple recurrent network. So, the learning algorithm need only be concerned with the recurrent backpropagation learning of a single receptive field. Beyond this there is a single restriction which is placed upon the learning process. This restriction is the constraint of symmetry. All of the connections must be symmetrical about the center of the receptive field. To this end the connection weights concerned are averaged.

3 Recurrent Backpropagation Learning

The backpropagation learning algorithm is primarily concerned with solving the credit assignment problem [10]. When a pattern is presented to a three (or more) layer perceptron, the activation

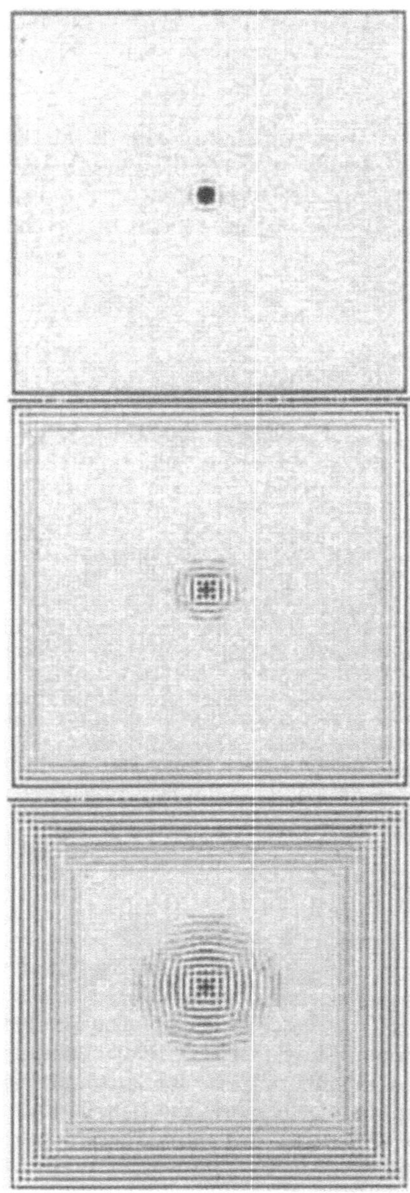

Figure 3: The long range influence of a receptive field in the accumulation of activation a) 5 cycles, b) 25 cycles, c) 50 cycles. Note that the input consists of an active receptive field against a background of uniform activity. The uneven accumulation of activation in the featureless field must be the result of the active receptive fields influence (as well as image border)[3].

flows from the units in the first layer through the first set of connections to the second layer. Then the activation from the second layer flows through the second set of connections to the third layer. If there is an error in the output pattern, it is difficult to assign credit for the error to a weight from the first set of connections or the second. This is because an error which is in the first set of connections will be propagated throughout the network until it reaches the third, output layer. The solution is to propagate the errors back through the network, and use this information to make small corrections to the connection weights.

The recurrent network involves a pattern of connectivity whereby activation is fed from one part of the network back into a part of the network that has already processed the information. Backpropagation learning can be applied to this network architecture also [1, 9]. The only critical difference is that the errors must be propagated back in time as well as across the ordinal space of the network. In applying the recurrent backpropagation learning algorithm to NLRFs it is important to take into account the constraint of symmetry. This condition, which is particular to the NLRFLA, is important in maintaining the uniform network structure. It is this uniform structure which allows the system to extract features in a shift-invariant manner. Symmetry is preserved by the averaging of weight values for connections from the center unit of the receptive field to all units of a given radial distance. The learning procedure itself can be described in three phases. The first is the feedforward activation phase (Figure 4). In this phase the network operates as it normally would, following the definition in [5]. First

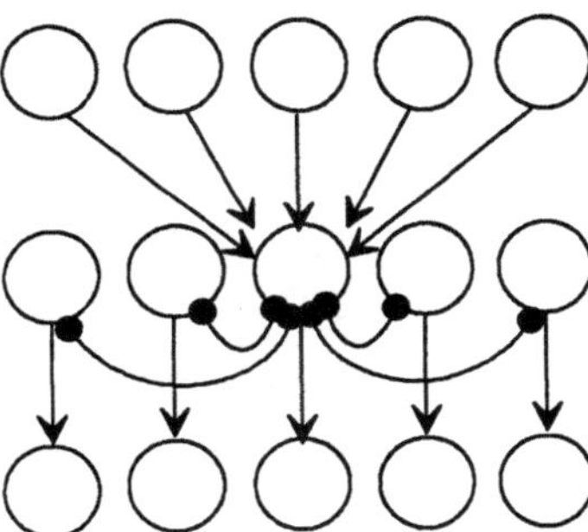

Figure 4: The feedforward spread of activation. The network is actually a 2 layer system, the 3rd layer simply represents the output, and is a copy of the second layer.

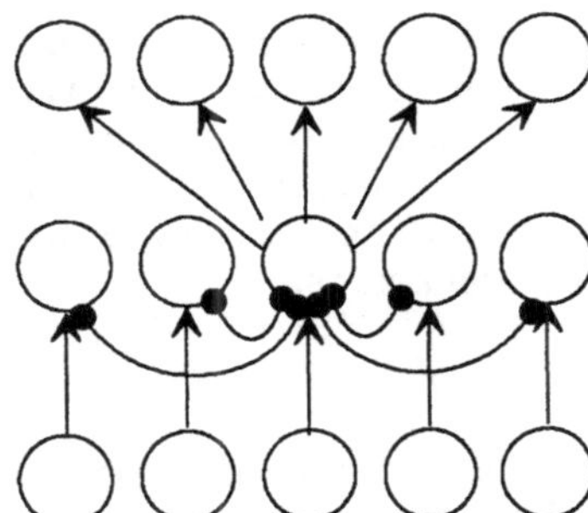

Figure 5: The backpropagation phase. In this case the third layer represents the error difference between the computed output, and the desired output.

the net activation is accumulated:

$$h_p = \sum_q w_{pq} V_q + \xi_p.$$

In h_p, the value ξ accounts for external inputs to the network. V_q is the activation of unit q, and w_{pq} is the connection weight from q to p. Each unit is then updated according to the following rule:

$$\tau \frac{dS_i}{dt} = -S_i + h_p$$

where:

$$V_i = g(S_i).$$

The function, $g(x)$, is a sigmoid function which takes an input from $-\infty$ to $+\infty$, and produces an output between 0.0 to 1.0.

The backpropagation phase is slightly more complex. The output resulting from the normal activation phase is compared with the desired output, and an error value is computed. This error value is then propagated backwards through the network until it reaches equilibrium (Figure 5).

The error value is constantly fed into the network from outside the network. Again, this is done for the entire network, as the 'error receptive field' is applied throughout the system.

The update procedure for the backpropagation of the errors is:

$$\tau \frac{dY_i}{dt} = -Y_i + g'(h_p) \sum_p w_{pi} Y_p + E_i$$

In this equation, the value Y is the activation value of a unit in the error network. The value $g'(h_p)$ is the derivative of the sigmoid function applied to the net input of unit p. The value E_i is the error output which is fed into the network from an external source. This is defined as:

$$E_i = T_i - V_i$$

where T_i is the target output. If the output is undefined, the E_i is set to 0.0.

In the weight modification phase, the NLRFLA deviates from the standard recurrent backpropagation algorithm, in several ways. The initial weight modifications are made to the receptive field, based only on the parameter values found at the center of the network. This is because the Learning Network is structured in such a way that the center represents the prototypical nonlinear receptive field. These modifications are based on the following equation:

$$\Delta w_{pq} = \eta V_p Y_q$$

where w_{pq} is the connection weight to be modified, V_p is the feedforward network activation of unit p, and Y_q is the error network activation for unit q. η is a small positive number.

4 Modifications to Recurrent Backpropagation

The departure from the standard learning algorithm involves the way that the connection weights are modified. One standard feature of NLRFs, as they are currently modelled, is the circular symmetry. This means that all connection weights at the same radial distance from the center of the receptive field, are the same. The recurrent backpropagation learning algorithm does not currently contain a method for doing this. So, in the case of a convex corner extracting NLRF, the resulting weights will clearly not be symmetrical (Figure 6).

In order to accommodate this, the connection weights for a given radial distance are averaged.

The network on which this procedure is carried out must also be considered. This is because of the way that the NLRF allows global computation through local connections. The activation seems to spread at a rate of approximately 1 unit per cycle. So, the network on which the learning takes place must reflect this fact. This means that the network size should be approximately receptive field size + 2(expected number of update cycles). The learning procedure is carried out for the central, prototypical NLRF, and its surrounding area. This prototype area determines the parameters for the learning procedure (Figure 7).

The next step in the further development of the NLRFLA, is to begin to adapt the algorithm to fit the specialized nature of the network. For example, the original system made use of a decay rate which is absent in the present algorithm.

5 Conclusions

In addition to the optimization of old NLRF templates, and the generation of new NLRF templates, the NLRFLA has the potential for much broader application. This broader application might involve the development of a heterogeneous NLRF network. Such a hetero-NLRF network might be able to provide enhanced performance in exchange for the specialization of network architecture. In addition the NLRFLA might eventually be extended to a self organizing map (SOM) type of system. A SOM has the advantage if being an unsupervised learning algorithm. However, before a SOM type of learning algorithm can be considered, it is important to understand 2 things. First, there should be some parametric way of controlling the development of the

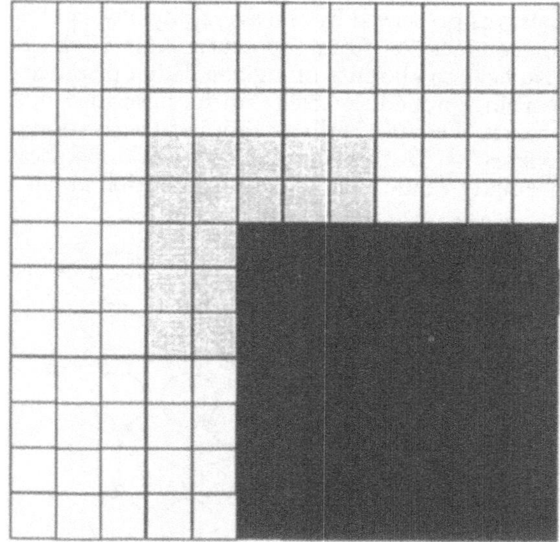

Figure 6: An input pattern, and the prototypical receptive field. Connection weights based on this input pattern will produce an asymmetrical receptive field. So, the connection weights are averaged for each unit of the same radial distance.

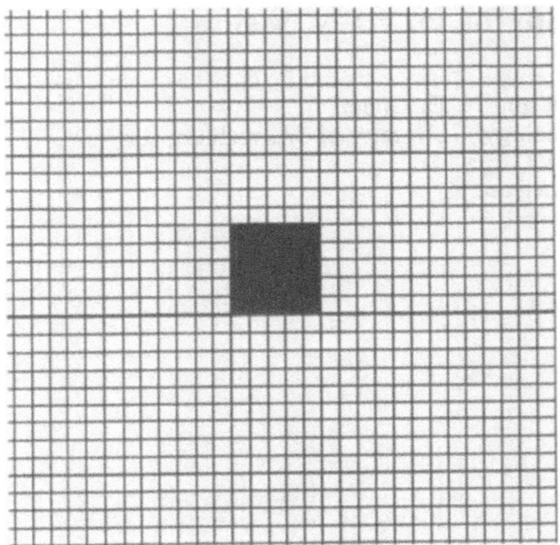

Figure 7: The central, prototypical receptive field. The entire network undergoes the normal feedforward, and backpropagation phases, but weight modification is based on the prototypical NLRF.

NLRF function. Second, there must be some way of mapping the input pattern to the optimal NLRF function. To these ends, the NLRFLA plays a critical role. The supervised nature of the NLRFLA allows us to study the way that an optimized NLRF is developed. In this way a SOM-NLRFLA might eventually be evolved.

References

[1] L. B. Almeida. Backpropogation in perceptrons with feedback. *Neural Computers*, pages 199–208, 1988.

[2] S. L. Funk, J. M. Kennedy, and I. Kumazawa. The nonlinear receptive field as a mechanism for the extraction of axes from outline drawings. In *Proc. 3rd International Workshop on Visual Form*, 1997. to appear.

[3] S. L. Funk, I. Kumazawa, and J. M. Kennedy. The role of non-linear receptive fields in shift-invariant feature extraction. In *Proceedings of the Meeting on Image Recognition and Understanding 96 (MIRU 96)*, 1996.

[4] S. Grossberg. A solution of the figure-ground problem for biological vision. *Neural Networks*, 6:463–483, 1993.

[5] J. J. Hopfield. Neurons with graded response have collective computational properties like those of two-state neurons. *Proceedings of the National Academy of Sciences*, 81:3088–3092, 1984.

[6] D. H. Hubel and T. N. Wiesel. Receptive fields of single neurones in the cat's straite cortex. *Journal of Physiology (London)*, 148, 1959.

[7] J. M. Kennedy. *Drawing and the Blind.* Yale, New Haven, 1993.

[8] J. M. Kennedy and S. L. Funk. Outline perception: Three theories of axis. In *Proceedings of the International Conference on Visual Coding*, 1995.

[9] F. J. Pineda. Generalization of back-propagation to recurrent neural networks. *Physical Review Letters*, 59:2229–2232, 1987.

[10] D. Rumelhart, G. Hinton, and R. Williams. *Learning Internal Representations by Error Propogation.* MIT Press, Cambridge, MA, 1986.

Fuzzy-tuned Stochastic Scanpaths for AGV Vision

I. J. Griffiths, Q. H. Mehdi and N. E. Gough
School of Computing and Information Technology, University of Wolverhampton
Wolverhampton, WV1 1SB, UK
Email: EX1131@wlv.ac.uk

Abstract

This paper details work on the development of an adaptive active vision system for an automated guided vehicle. An initial solution to the task of providing intelligent control of the saccades with which the AGV examines its environment is presented. A simple fuzzy logic technique suitable for implementation on a microcontroller is developed by using stochastic transition matrices. The results presented show the success of the technique in maintaining interest in objects previously located within the environment, locating new objects in an environment and making a compromise between the two.

1 Introduction

In order for an autonomous guided vehicle (AGV) to plan automatically a collision free path through an environment and to plan trajectory control to traverse the environment, information must be constantly available at local (AGV) and global (environment) levels. However the environment is subject to change as the AGV encounters the removal and introduction of static objects, other AGVs or personnel. This introduces uncertainty into the environment and a crucial problem arises: Can a compromise be found between monitoring a scene for the arrival of new objects and tracking known objects through the environment?

The University of Wolverhampton's AGV Research Group is researching the design and implementation of AGVs that can be used in a variety of environments. Previous work on vision systems [1, 4], navigation, object tracking and path planning [8, 9] have highlighted the need for a comprehensive strategy for the extraction of information from an environment that may alter with time. The present work is the initial formulation of an adaptive camera strategy to constantly monitor an environment for previously known and unknown objects based upon stochastic camera *scanpaths* that are tuned using fuzzy logic.

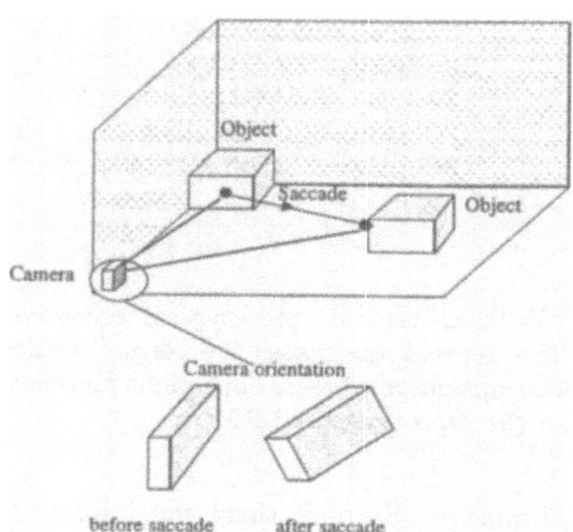

Figure 1: A saccade in an artificial vision system.

2 Saccades and Scanpaths

Eye movements are necessary while viewing scenes as situations arise where a target spans more than several degrees within the field of view. Figure 1 illustrates a single saccade in an active vision system. As detailed visual information is available at only a small central area of the eye, the fovea, the eye must be moved so that the area of interest is foveated [2, 3]. In order to do this a series of rapid eye movements are enacted and this motion is called a saccade [10]. Saccades typically can take place within 20 ms and during this time vision is suppressed.

It has been found that humans have repetitive patterns of saccades that are used to inspect scenes and these have been labelled scanpaths or *eye contour maps* [6, 7]. Efforts have been made to create stochastic models of eye movement fixation sequences [5]. These sequences have been modelled as strings and a quantitative way of measuring string similarity has been developed. The essential concepts of ***stochastic transition matrices*** are shown below.

$$\begin{array}{c} \begin{array}{c|cccc} & A & B & C & D \\ \hline A & 0 & 1 & 0 & 0 \\ B & 0 & 0 & 1 & 0 \\ C & 0 & 0 & 0 & 1 \\ D & 1 & 0 & 0 & 0 \end{array} \\ \text{(a)} \end{array} \qquad \begin{array}{c} \begin{array}{c|cccc} & A & B & C & D \\ \hline A & 0.01 & 0.93 & 0.06 & 0.03 \\ B & 0.1 & 0.1 & 0.8 & 0 \\ C & 0.1 & 0.1 & 0.7 & 0.1 \\ D & 1 & 0 & 0 & 0 \end{array} \\ \text{(b)} \end{array}$$

Each row of the matrices refers to the current position of the eye or camera and the individual elements of the row indicate the next position to be taken through a saccade. The first scanpath (a) can be described by the string ..BCDABCDA... as the transition between points is fixed at either 1, make transition, or 0, do not make transition. As only one possible saccade is available in each row the saccades will cycle in fixed patterns. The second matrix (b), could generate many different string sequences as the transitions from one state to the next are made on a basis of the probabilities. The pattern of saccades can now only be described statistically.

3 Fuzzy Logic Saccade Maps

We define the stochastic process as follows. The stochastic transition state matrix F which describes the state transition has the form:

$$F = \begin{array}{c} \\ q_1 \\ q_2 \\ q_3 \\ q_4 \end{array} \begin{array}{c} \begin{array}{cccc} q_1 & q_2 & \cdots & q_r \end{array} \\ \begin{bmatrix} p_{11} & p_{12} & \cdots & p_{1r} \\ p_{21} & p_{22} & \cdots & p_{2r} \\ \vdots & \vdots & & \vdots \\ p_{r1} & p_{r2} & \cdots & p_{rr} \end{bmatrix} \end{array}$$

where $0 \leq p_{ij} \leq 1$ denotes the probability that a transition will take place from state q_i to state q_j. Matrix F is non-stationary, that is to say the values $\{p_{ij}\}$ alter with time and these values are modified using the fuzzy methods detailed later.

The method of deciding a transition is made by means of a *weighted roulette method.* Suppose that the current state is q_1. The sum of all probabilities of state transitions, S_p, is calculated.

$$S_p = \sum_{j=1}^{r} p_{ij}.$$

A value D, the determined transition, is generated using random means.

$$D = S_p \cdot n$$

where

$$n \in [0,1].$$

The determined state q_j is derived as follows:

$$\sum_{x=1}^{j} p_{ix} \leq D \leq \sum_{x=1}^{j+1} p_{ix}.$$

Consequently states associated with higher probabilities of transitions will be taken more times than states with lower ones.

We can make use of the stochastic transition matrix by associating the states q_i to q_j with the saccade points that are available to the system. The matrix F is now the scanpath for the system and the probabilities $\{p_{ij}\}$ will define a fuzzy variable i, the ***interest*** of the scene at the particular saccade point.

Fuzzy logic is used to alter the probabilities $\{p_{ij}\}$ and provide the adaptive nature of the system. This is possible by using fuzzy logic to derive new values of $\{p_{ij}\}$ based upon simple fuzzy inputs. The fuzzy membership classes used are ***distance*** f_d and ***attention*** f_a, defined as follows:

$$f_d(d) = \begin{cases} 1.000 & d = 0 \\ 0.100 & 0 < d < 2 \\ 0.010 & 2 \leq d < 3 \\ 0.001 & d \geq 3 \end{cases}$$

$$f_a(t_i) = (T - t_i)/100T$$

where d is the minimum distance between a position and all points where an object is assumed to be located. T is the number of saccades since the system began and t_i is the value of T when the last saccade to the state q_i was made. The output of the fuzzy system p_{ij} is calculated for each element of F according to the following rule:

$$p_{ij} = Max[f_a(t_i), f_d(d)].$$

4 Simulation/Results

To illustrate the method, the scene of interest was modelled as a 10×10 grid of saccade points and the model was programmed in C. The initial value

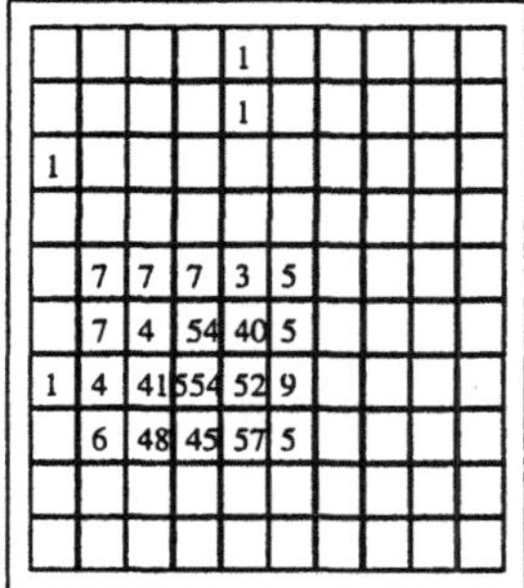

Figure 2: Saccade intensity in a scene of interest.

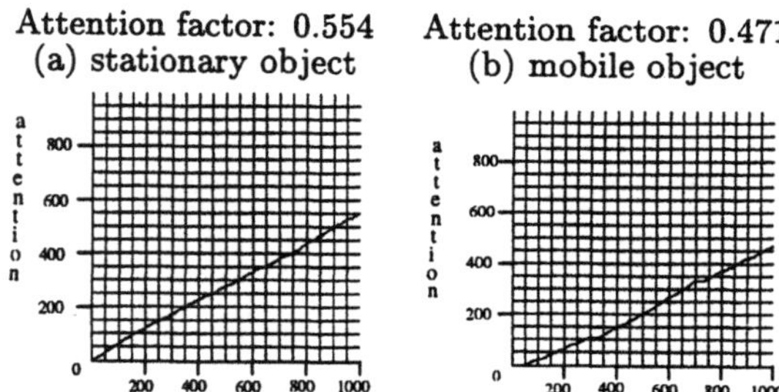

Figure 3: Maintaining interest in a single object.

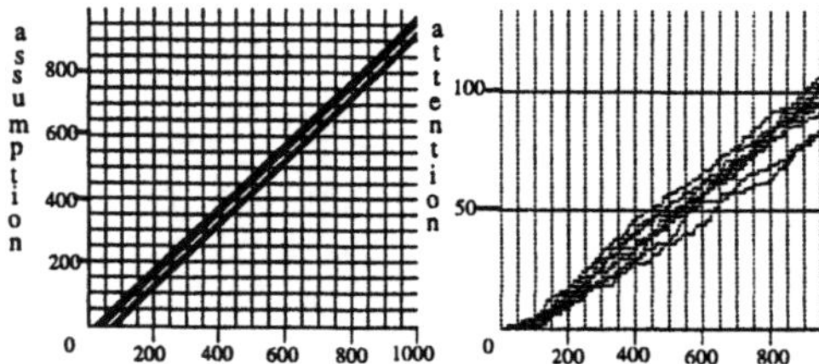

Figure 4: Locating new objects in the scene.

of elements in F were initialised to small values, the initial starting point for the simulated camera, and the location of any objects were set to random grid points on the 10 × 10 grid. The program was tested for three properties: (i) maintaining interest in visual objects already located, (ii) locating new objects in a scene and (iii) forming a compromise between maintaining interest in located objects while monitoring the scene for the arrival of new objects.

4.1 Maintaining Interest in Located Objects

Consider first the problem of maintaining interest in a single stationary object. Maintaining interest in located objects is achieved using the fuzzy membership function f_d. Using this as the input that alters the matrix F provides the model with the ability to maintain interest as demonstrated in the following diagrams. Figure 2 shows a visual representation of the saccade points. Where shown, the numbers in the grid refer to the intensity of saccades to that point in the scene. The attention factor is the number of saccades that terminated upon an object normalised by the number of saccades taken over a run. Attention can be seen as a measure of useful saccades. Figure 3 shows the associated attention graph. Each vertical delineation denotes 50 T and the horizontal delineations mark 50 visual hits upon the object. Figure 2 is a map of the environment showing the intensity with which specific parts of the 10 × 10 scene were monitored through the saccades. The numbers within the box give the frequency that the particular element of the scene is examined.

Suppose that the object is mobile and can move to any of the surrounding grid points every $100T$. As can be seen from Figure 3b the system maintained interest in both stationary and slowly moving objects, with little loss in the attention factor.

4.2 Locating New Objects in a Scene

The second important task was to locate new objects in the scene and this was achieved using the fuzzy membership f_a. Figure 4 shows the success of this mode of the simulation.

A random number of stationary objects were placed at random within the scene and the graphs illustrate the speed with which they were located. The graph labelled *assumption* indicates the system's knowledge of the scene of interest. After each location of an object the program now stores the value at which the object was last seen and assumes that at any time this is where that object is located. This is compared to the true value of the object's location to assess the validity of the assumption.

Table 1: Knowledge and Attention factors for 10 randomly placed objects.

Knowledge factors		Attention factors	
0.914	0.915	0.106	0.101
0.978	0.960	0.094	0.087
0.926	0.913	0.101	0.113
0.951	0.976	0.088	0.107
0.965	0.926	0.112	0.101

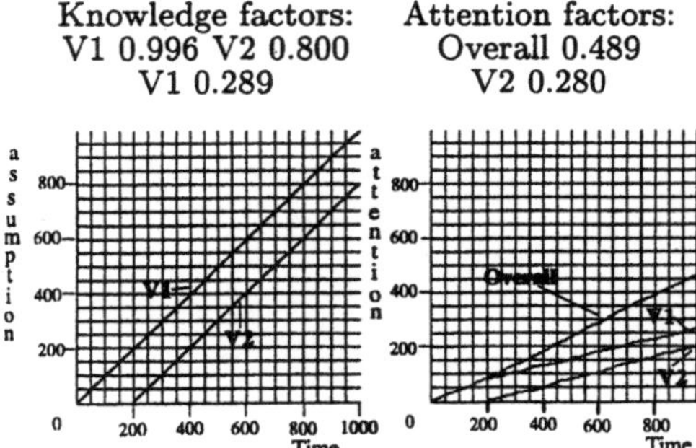

Figure 5: The compromise problem for two stationary objects V1 and V2.

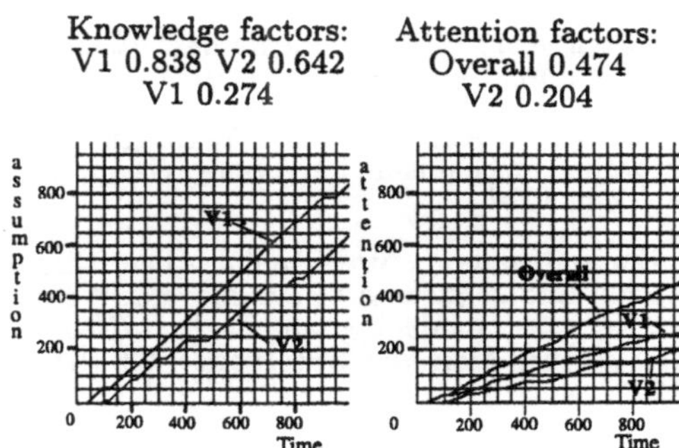

Figure 6: The compromise problem for two mobile objects V1 and V2.

The *knowledge factor* indicates on a unit scale the time for which the objects were correctly located, where 0 would indicate that the object was never found and 1 would indicate the object was located at time $T = 0$. The model was not made aware of how many objects were to be introduced. The *attention graph* shows that each object was allocated a similar proportion of the system's attention.

4.3 Maintaining Interest in Located Objects While Monitoring for New Arrivals

By making use of the previous work and results for maintaining interest and seeking new objects in the field of view and combining the fuzzy memberships f_d and f_a it was possible to produce an initial model that formed the compromise between maintaining interest in known objects while monitoring the scene of interest for new objects. This success is indicated in Figures 5 and 6 for stationary and mobile objects respectively.

In both cases two vehicles were placed at random within the scene and the task was to monitor all objects in the scene while ensuring all objects in the scene were located. In the first case the vehicles were stationary, while in the second the vehicles moved to a random surrounding grid point every $100T$. In both cases the results were successful. The overall attention factors, which measure the efficiency of the saccade trajectories, are comparable to the appropriate results for a single vehicle. The knowledge factor indicates that while occasionally the system lost the true object locations, they were quickly recovered.

5 Conclusion and Further Work

This work indicated the need for an adaptive vision system that would allow an AGV to monitor its environment for important tasks such as path planning and trajectory control and for applications such as security or domestic use. A fundamental task for such a system would be the ability to maintain interests in located objects while monitoring the scene for previously unlocated objects. The model developed in this work and the results subsequently produced indicate that a good compromise between these two tasks has been achieved.

With this work completed the next objective is to use an embedded fuzzy microcontroller system and host computer to implement this model and produce a 'smart' camera system that would be able to carry out independently detection and tracking tasks.

Further work is necessary on developing the model. In particular the camera has no occlusion, i.e. it is assumed that the camera can visually access all areas of interest within the environment. In a real environment this would not be the case. A solution to this would be to use multiple 'smart' cameras that can interact through a host computer and, through an appropriate strategy, would be able to exchange, request or transmit information and work in co-ordination.

References

[1] A. Abu-Alola, N. Gough, Q. Mehdi, and P. Musgrove. Application of a genetic algorithm to an actuation system for robotic vision. In *Proc. IEE, Int. Conf. on Control*, 1994.

[2] R. Carpenter. Eye-motion machinery. *Physics World*, 2, 1989.

[3] R. Ditchburn. *Eye movements and Visual Perception.* Oxford University Press, 1973.

[4] N. Gough, A. Abu-Alola, and A. Gough. Push-pull actuation mechanisms for robotic vision. In *Proc. Melecon*, 1994.

[5] S. Hacisalizhade, L. Stark, and J. Allen. Visual perception and sequences of eye movement fixations: A stochastic modelling approach. *IEEE Trans. on Systems, Man and Cybernetics*, 22(3):474–481, 1992.

[6] R. Monty and J. Senders. *Eye movements and psychological processes.* Lawrence Earlbaum, pages 93–94, 1974.

[7] D. Norton and L. Stark. Scanpaths in eye movements during pattern perception. *Science*, 171:308–311, 1971.

[8] T. Wang, Q. Mehdi, and N. Gough. A human imitation controller for autonomous guided vehicles. In *Proc. of 12th Int. Conf. on CAD/CAM Robotics and Factories of the Future*, 1996.

[9] T. Wang, Q. Mehdi, and N. Gough. A hybrid intelligent approach to navigation and control of AGVs. In *Proc. of 11th Int. Conf. on Systems Engineering*, 1996.

[10] B. Zuber. *Models of Oculomotor Behavior and Control.* CRC Press, Boca Rotan, FL, 1981.

On VLSI Implementation of Multiple Output Sequential Learning Networks

A. Bermak[1] and H. Poulard[2]
[1] Laboratoire d'Analyse et d'Architecture des Systèmes - CNRS,
7 avenue du Colonel Roche - 31077 Toulouse FRANCE.
[2] ACTIA, 25 Chemin de Pouvourville - 31432 Toulouse FRANCE.
Email: {bermak,poulard@laas.fr}

Abstract

In this paper we propose a hardware implementation of a binary neural network architecture obtained from a new efficient constructive algorithm. This algorithm is particularly interesting because it can treat boolean as well as real valued classification problems with an arbitrary number of outputs. The networks obtained consist of binary neurons organized in two hidden layers. The first layer is implemented on a systolic architecture which represents a good tradeoff between speed and area. Due to the particular computation performed by the second hidden layer, its implementation is straightforward and well-suited to the systolic architecture. A limited number of logical gates is needed for its implementation. The output neurons are also easy to implement but require a small size memory.

1 Introduction

Since the beginning of neural network development, many researchers have attempted to provide efficient algorithms for learning in neural networks. Software simulations have been very useful for investigating the capabilities of neural network models, while hardware implementations are essential to take full advantage of the inherent parallelism of a neural network. During the last decade a few dedicated designs have been reported in the literature. The usual choice between digital and analog techniques is critical in neural network implementation. On the one hand, an analog device requires less hardware than the digital one. Analog techniques allow realization on a single chip whereas an entire board is needed with digital methods. However, analog devices suffer from the disadvantages of a low precision and difficulty in modifying the synaptic weights. On the other hand, a digital implementation offers a high level of precision but is very expensive in terms of silicon area. The design presented in this paper is a digital hardware allowing us to take advantage of this design strategy without detrimentally affecting area resources.

Integrated networks are built with a new constructive algorithm which brings in a number of advantages relative to the other algorithms. First, the networks obtained consist of binary units, well-suited for digital technology. Second, it seems to be the first constructive algorithm capable of learning boolean and real valued training sets without preprocessing, and furnishing multiple output networks with a regular connectivity (two hidden layers). The simplicity of the process performed by the second and the output layer leads to a very simple design. The first layer is more complex. It has been implemented using a systolic architecture — which constitutes a good trade-off between area and speed — while the second layer has been implemented using only a few logical gates, as well as the output neurons.

2 The Constructive Algorithm

Although the backpropagation [7] or its various derivatives are still the most widely applied algorithms for neural networks design, these methods still suffer from two major drawbacks: prohibitive computation time and *a priori* choice of the structure. To overcome these problems, new kinds of methods have been investigated over the last years

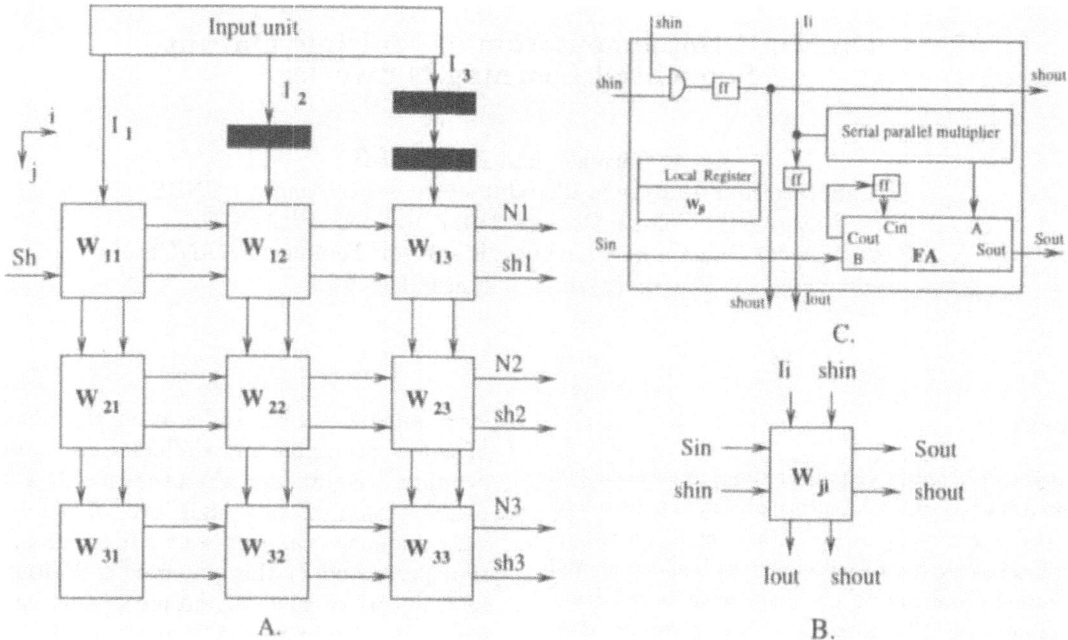

Figure 1: (A) 3×3 systolic array for implementing the fist layer. (B) Input/Output communication for a processing element P_{ij} (C) Schematic rendition of P_{ij}. FA denotes a full adder with carry save and ff denotes a flip-flop for which the clock input has been omitted for simplicity.

in the form of constructive algorithms [4]. Generally, these methods make use of binary units in order to construct the network and this is a significant advantage for VLSI implementation. With these techniques, the network is incrementally constructed by adding units when required according to various strategies. Adaptation of the connections requires training of each unit separately. Many strategies have been proposed but no algorithm matches the three following requirements: convergence on real valued inputs (necessary for a wide range of applications), multiple outputs (for multiple category classification tasks) and simple connectivity (for implementation facilities).

Recently, a multiple output constructive strategy addressing real valued mappings has been reported. It is a natural extension of the sequential learning algorithm [3] to multiple output network meeting the three preceding requirements. A brief overview is given in [5] and a detailed description together with simulations can be found in [4]. In this method, each unit is built up so as to exclude a cluster of patterns belonging to the same class as in the sequential learning. The classes are defined by the different output vectors. Because the number of classes is greater than two, these patterns will always be located in the positive halfspace defined by the hyperplane, *i.e.* the unit's response will be 1 for these patterns and 0 for all of the others. With such a construction process, it can be shown that the training sets for learning the output neurons consist of linearly separable mappings. The output neurons can simply be trained by using the new efficient algorithm proposed in [6], which is the basis of the method employed for learning each unit in the hidden layer. For highly complex mappings, computation of the internal representation becomes highly consuming in terms of time and storage. This problem has been solved by constructing a second hidden layer [4] by direct computation of the weight based on an analysis of hypercube structures. Thus, output neurons are simply logical ORs of output subsets of the second layer. This version offers several advantages for VLSI implementation. Due to the particular function performed (see [4] for details) the output of the second layer neurons can be computed sequentially by a recursive logical equation. Suppose that $k-1$ neurons have been constructed in the first hidden layer. Then the second hidden layer will be made up of k neurons. If we denote the neuron outputs in the first hidden layer by x_i and those in the second by y_i, we get the following logical equations:

$$\begin{aligned}
y_1 &= x_1 \\
y_2 &= \overline{x}_1 \,.\, x_2 \\
y_3 &= \overline{x}_1 \,.\, \overline{x}_2 \,.\, x_3 \\
&\vdots \\
y_{k-1} &= \overline{x}_1 \,.\, \overline{x}_2 \,.\, \overline{x}_3 \,.\, \ldots \,.\, \overline{x}_{k-2} \,.\, x_{k-1} \\
y_k &= \overline{x}_1 \,.\, \overline{x}_2 \,.\, \overline{x}_3 \,.\, \ldots \,.\, \overline{x}_{k-2} \,.\, \overline{x}_{k-1}.
\end{aligned}$$

In order to avoid a particular case in the computation of y_k, a neuron is added in the first hidden layer such that its output is always 1. In this case, one gets

$$y_k = \overline{x}_1 \,.\, \overline{x}_2 \,.\, \overline{x}_3 \,.\, \ldots \,.\, \overline{x}_{k-2} \,.\, \overline{x}_{k-1} \,.\, x_k.$$

Then if one defines z_i such that

$$\begin{aligned}
& & z_1 &= 1 \\
\forall i \in \{2, \ldots, k\} & & z_i &= z_{i-1} \,.\, \overline{x}_{i-1}.
\end{aligned} \qquad (1)$$

computation of y_i is straightforward

$$\forall i \in \{1, \ldots, k\} \qquad y_i = z_i \,.\, x_i. \qquad (2)$$

Each neuron i of the second hidden layer will be activated only for the patterns belonging to the ith excluded cluster. Each neuron is also dedicated to a subset of a given class. Suppose that the classes have been coded on p bits, the number of output neurons. The target of each pattern is also a p-dimensional vector and a class is made up of the patterns which have the same target vector. Let us define the $k \times p$ matrix $T = (t_{ij})$ where the ith row is the target vector of the class corresponding to the ith excluded cluster. Then one can easily show that the output of the neuron j is given by the logical equation

$$o_j = t_{j1}.y_1 + t_{j2}.y_2 + \ldots + t_{jN}.y_N \qquad (3)$$

We will focus now on the hardware implementation for this network.

3 Hardware Implementation

The systolic architecture was adopted to reduce the complexity of the design of a large scale multiprocessor arrays, by using only localized communications which simplify the design of the interconnections between processors [2]. The proposed architecture is a two-dimensional systolic array. This array consists of $N_I \times N_N$ processing elements (PE) where N_I is the number of inputs and N_N the number of neurons. For clarity, the array considered in Figure 1A is a 3 × 3 PEs and it is assumed that we have the same size as the neural network. Note that the systolic architecture proposed can easily be extended to any arbitrary artificial neural network (ANN). In the case where the size of the systolic array is larger than the neural network then the elements in excess are simply set to zero.

Each processor in this architecture needs the storage of synaptic weights in local registers. This local storage is adopted to achieve a high computational speed by overcoming the problem of the delay required for memory access. The serial-parallel multiplier is used in each processor in order to perform the partial products necessary for computing the weighted sum $N_j = w_{ji} I_i$. The weight coefficient w_{ji} is stored in parallel with a fixed number of n bits while the inputs can be of an arbitrary word-length m. This is one of the great advantages of our architecture because it allows for a variable-precision calculation to be obtained. The input vector I_i is serially fed along the upper edge of the array, from the least significant bit to the most significant one. When a processor PE(i,j) receives a bit of the input I_i, it computes the product of the resident w_{ji} by I_i. This product is then added to the partial sum received from the processor located on left. The result is transmitted to PE on the right one clock cycle later. The inputs are transmitted vertically through register D. The process taking place on a row of the array is repeated one cycle later on the next row. The weighted sum N_j is sequentially produced by the row j of the systolic architecture. The least significant bit of N_1 is processed by the first row after N_I clock cycles. All results are collected on the right side of the array just after the activation units.

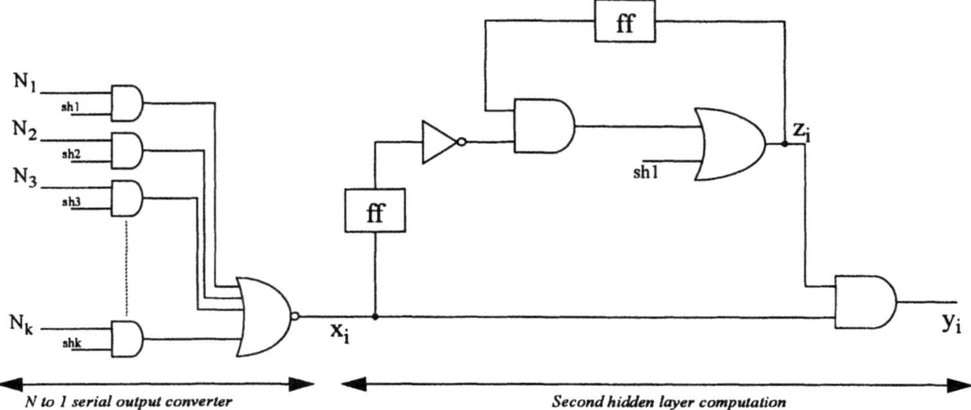

Figure 2: N to 1 serial output converter and Second hidden layer circuit. The signals n_i are the systolic output and sh_i are the control signals.

As shown in Figure 1A, a control signal sh is generated from a control unit [1] to the processor PE(1,1) and is then systolically propagated to all PEs. This signal is on after $(m + n - 1)$ clock cycles and hence it is used for detecting the sign bit of the weighted sum by simply operating an AND gate between the array output and the control signals as shown in Figure 2. The design of the N to 1 serial output converter has been optimized at the logical level in order to reduce the costs. For this the logical function $\overline{(\overline{N_{i-1}sh_{i-1}})(\overline{N_i sh_i})}$ was rewritten as $(N_{i-1}sh_{i-1}) + (N_i sh_i)$ since for all i, $sh_{i-1}sh_i = 0$. This allows us to save $2N_N$ transistors.

The quantity z_i computed through the recursive equation 1 can be easily implemented since the systolic architecture sequentially produces one and only one output x_i at the clock cycle T_i. In this case, the output of the second layer y_i is obtained by simply operating a logical AND between the output x_i available at the clock cycle T_i and z_i. As can be shown in Figure 2, this layer is implemented with two flip-flops and three gates making up 44 transistors only.

Equation 3 yields the computation required for the output neuron. This can be done as shown in Figure 3, by storing all coefficients t_{ji} corresponding to a neuron output j in a same column register of k bits. Note that for this output layer p registers of k bits each are needed. The matrix T shown in Figure 3 stands for these registers. Computations in this layer begin just after the control signal sh_1 has been received from the systolic architecture. The first output y_1 is received from the second hidden layer and multiplied with all the stored coefficient t_{j1} to perform the first partial product in each neuron output. The product is then added to the partial sum stored in the ff_1. At the next clock cycle, all p registers are shifted and output neurons execute the second recursion of Equation 3. Processing is ended as the last output N_k is computed. The signal sh_k is on and all the outputs are stored in the p output register.

4 Concluding Remarks

The complexity of the architecture proposed in this paper is mainly due to the systolic array for the first hidden layer. This systolic array of 4×4 processing elements has been designed using a standard cell $0.7\mu m$ CMOS process and consists of 60832 transistors in an active area of 7.3 mm^2. This design implements 16 neurons but can be easily extended to any arbitrary ANNs by connecting in cascade several chips. Implementation of the second hidden layer and of the output neurons is simple and particularly well-suited to the sequential production of the outputs of the systolic array. When the response of the last neuron has been obtained, the

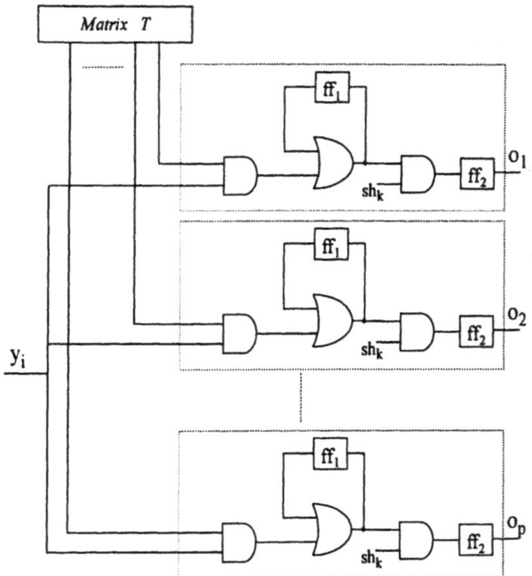

Figure 3: Output layer implementation.

second layer has already effected all computations and all the output neurons simultaneously give their output. Due to the systolic array and the specificity of the networks implemented, this architecture appears as a very good tradeoff between parallelism and silicon area. Although this VLSI implementation is designed for a particular type of neural networks, the generality of the constructive algorithm allows a wide range of applications to be considered.

References

[1] A. Bermak and D. Martinez. A variable-precision systolic architecture for ANN computation. In *Fifth International Conference on Microelectronics for Neural Networks and Fuzzy Systems*. Lausanne, Feb. 1996.

[2] S. Kung. *VLSI array processors*. Prentice Hall, New York, 1988.

[3] M. Marchand, M. Golea, and P. Rújan. A convergence theorem for sequential learning in two-layer perceptron. *Europhysics Lett.*, 11:487–492, 1990.

[4] H. Poulard and N. Hernandez. Two efficient constructive algorithms. Submitted paper, http://www.laas.fr/~poulard/papers/, 1996.

[5] H. Poulard and N. Hernandez. A constructive algorithm for real valued multi-category classification problems. In *Artificial Neural Networks and Genetic Algorithms*, Wien, 1998. Springer-Verlag.

[6] H. Poulard and S. Labrèche. A new algorithm for learning threshold unit. Technical report, Laboratoire d'Analyse et d'Architecture des Systèmes, 1996. Submitted paper, http://www.laas.fr/ poulard/papers/.

[7] D. Rumelhart, G. Hinton, and R. Williams. *Learning internal representations by error propagation*, volume I, chapter 8. MIT Press, Cambridge, MA, 1986.

Automated Parameter Selection for a Computer Simulation of Auditory Nerve Fibre Activity using Genetic Algorithms

C. P. Wong and M. J. Pont
BTSP: Speech and Hearing, Department of Engineering, University of Leicester,
University Road, Leicester, LE1 7RH ENGLAND
Email: {CPW2, MJP9}@le.ac.uk

Abstract

The Meddis computational model of auditory nerve fibre activity [9, 10] is widely used as a research tool in the study of auditory processing. The model is governed by a set of control parameters that allows it to simulate responses derived from physiological observations. In this paper, we describe a novel method for automatically determining the parameters required to simulate a range of rate-intensity responses from auditory nerve fibres using this model. A genetic algorithm [4] is employed to explore possible parameter combinations and to determine a 'best fit' solution. Two sets of experiments used to demonstrate the flexibility of the technique are described. Some possible wider applications of the technique are discussed.

1 Introduction

Recently, computer simulation has emerged as an extremely powerful tool for investigating the neural mechanisms underlying speech perception (e.g. [5, 7, 8, 9, 10]). It is now possible, using results from physiological and anatomical studies, to build detailed software models and explore their potential to explain observed behavioural phenomena (e.g. [12, 13, 14]).

One particularly widely-used computer simulation, and the one upon which we focus in this paper, has been described in detail by Meddis [9, 10]. Briefly, the model is primarily concerned with the production, movement and loss of 'transmitter substance' in the synapse between the inner hair cell and auditory nerve (AN) fibre (Figure 1). Transmitter substance is assumed to be stored in a free pool q near the synaptic junction and to be released across the membrane into the cleft c. The permeability factor k of the membrane is a non-linear function of the instantaneous amplitude of the stimulus signal. A fraction of transmitter substances in the cleft is subject to loss through diffusion, while another fraction is taken back into the cell. The loss of transmitter substance from the system is replenished in the free pool by a 'factory' at a set rate. In summary, the occurrence of an output 'spike' in the AN fibre is a probabilistic function of the residue of transmitter content in the cleft due to the instantaneous stimulus amplitude.

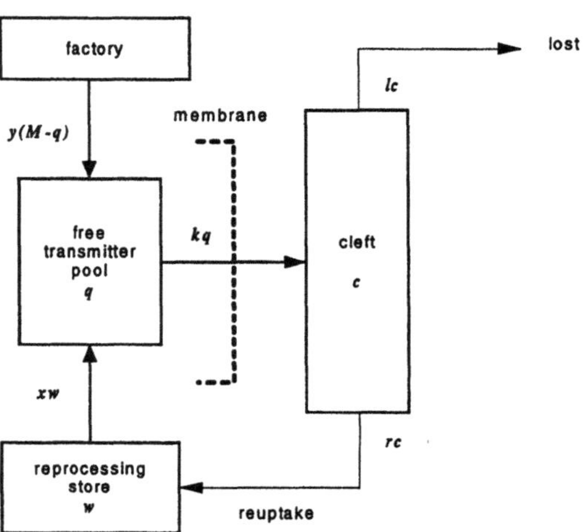

Figure 1: Summary of the operation of the Meddis model.

Conventionally, parameters for the Meddis model are determined using a 'hill climbing' approach: that is, by changing one parameter and observing its effect on the simulated fibre activity. If the new parameter set produces a simulation better able fit the empirical data, then the change is retained and used as the basis for further modifications. If the new parameter produces a worse simulation, it is reversed and a new combination is tried. This process of trial and error can make the selection of parameters required to simulate a particular fibre a time-consuming process.

These problems are, of course, not unique to the Meddis model, but are an important factor in the development of any substantial computer simulation of biological systems. As such simulations are extended in terms of both the areas they cover, and the resolution of the models themselves, it will — we suggest — become increasingly impractical to determine the necessary parameters by trial and error. In this paper, we describe a novel automatic approach which is ultimately intended to allow the selection of parameters for a range of computer simulations of auditory processing. The technique is based on a genetic algorithm [4]. Here, we demonstrate the effectiveness of the approach by employing it to determine automatically the 'best' set of parameters for use with the Meddis model.

We briefly introduce genetic algorithms below, and describe the implementation we have used. We then detail two sets of experiments carried out to demonstrate the effectiveness and flexibility of this technique. We conclude by describing some extensions and other possible applications of this approach.

2 Genetic Algorithms

A genetic algorithm (GA) is a search technique based on the mechanics of adaptive behaviour of natural systems [5, 6]. It is guided towards finding an 'optimal' solution by speculating on new points in the search space based on the performance of previous points. In natural systems, the performance of an organism is measured by its ability to survive against other organisms. In a GA, this performance measure is a fitness function devised according to the task the algorithm was set out to optimise. The existence of an organism is maintained by evolving its genetic structure to produce better offspring. The GA mimics this behaviour by applying a set of genetic operators to explore subsequent points in the search space.

In genetic algorithms, candidate solutions in a search space are called 'individuals' and are grouped together to form a 'population'. The GA evaluates a population and generates a new one, repeatedly, with each successive population known as a 'generation'. In a basic GA, we start off with an initial population, $G(0)$, and for subsequent generation of $G(t)$, the algorithm generates a new one, $G(t+1)$. This process is repeated until a condition is satisfied whereby the GA is deemed to have found an acceptable solution.

In order to use GAs to optimise a set of parameters to fit a model of the real world, such parameters will have to be mapped into a suitable representation that the GA can manipulate. Generally, individuals in a population are represented as a fixed-length string of binary digits. Holland [6] suggested that GAs work best with low-cardinality symbol sets and long string lengths; that is, a long binary string. A real valued parameter is discretised and mapped linearly between a user-defined minimum and maximum. The choice of the upper and lower limits is important in that a narrow range allows values of a higher precision, but risks missing better solutions beyond that range. On the other hand, a wide range will suffer from quantisation errors. In the case of a multi-parameter model, the binary string for each parameter is concatenated to form a single string.

3 Applying Genetic Algorithms to Auditory Problems

In this paper, we describe two experiments based on a simulation of the AN-fibre rate-intensity response characteristics produced with the Meddis model. In each experiment, seven parameters of the Meddis model were encoded into a binary string by linearly mapping each real value to an unsigned 32-bit integer. The binary strings were concatenated to form a 224-bit string to construct a multi-parameter coding. The range and precision of the mapping was controlled by specifying an interval of valid real values to be used in the model. Table 1 summarises the range of intervals for the Meddis model. The parameter ranges were estimated from typical values used in [5, 9, 10].

Table 2 summarises the GA control parameters. A total of 50 generations were 'evolved' with the GA and the best parameter set was chosen from the

Table 1: Parameter ranges used in experiments. Parameters h and M are kept constant as they are scalar factors and play no particular role in dictating the AN-fibre function.

Parameter	Range
A	0 - 100
B	1000 - 1,000,000
g	1 - MaxInt
y	1 - 100
l	1 - MaxInt
x	1 - MaxInt
r	1 - MaxInt
h	50,000
M	1

MaxInt = 2^{16} = 65535

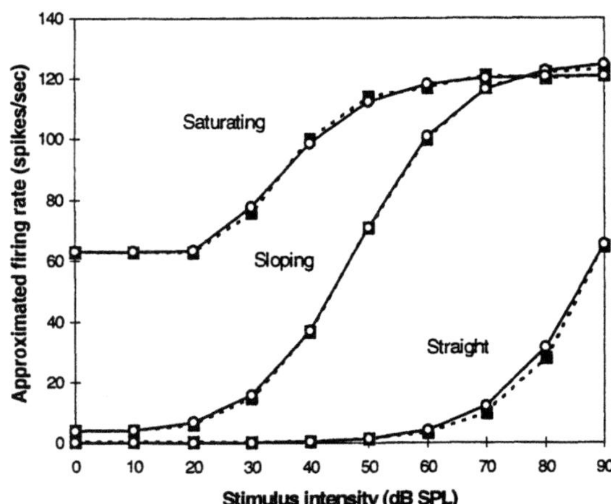

Figure 2: Rate intensity functions of auditory nerve fibres of various types (□ dotted line) matched with the simulated curves (○ solid line) produced using the parameters selected by the genetic algorithm.

individual with the highest overall fitness. The simulations were carried out on an 486-IBM PC compatible computer, running Microsoft Windows 3.1. In each case, the simulations took approximately 12 hours to complete.

Table 2: Genetic algorithm parameters used throughout the experiments.

GA control parameters		
No of individuals	nPop	200
No. of Crossover sites	nCross	2
Prob. of Crossover	pCross	0.8
Mutation rates	pMutate	0.005

Experiment 1. Fibres in the mammalian auditory nerve can be divided into three broad categories on the basis of their spontaneous discharge rates [16]. These categories can be described as saturating, sloping and straight. Fibres of saturating type generally have the lowest thresholds and the highest spontaneous discharge rates of the three groups. Fibres of straight type generally have the highest thresholds and the lowest spontaneous rates. Fibres of sloping type have intermediate threshold and spontaneous rate characteristics.

Our first experiment was conducted in order to demonstrate that it was possible for the GA-based technique to determine a set of parameters for a broad range of different AN fibres. Specifically, we used the method described above to search for parameters to simulate a set of three rate-intensity functions of saturating, sloping and AN fibres, as recorded by Winter and Palmer [16]. The results are shown in Figure 2.

In all three cases, the match is very good. Note, however, that in an earlier study [17], using a 16-bit string representation, it was necessary to use two different parameter ranges to obtain these results. The original range (used for the HSR and MSR curves) did not provide a good match for the LSR curve. Here, two ranges are required because — when a single (large) range is used in all cases, the resolution of the binary string is insufficient to provide a good match in all three cases. In this study, however, using a longer bit-string representation, we are able to work with just a single set of parameter ranges.

Experiment 2. In Experiment 1, described above, we demonstrated that the use of genetic algorithms provides a flexible automatic technique for selecting the parameters required to reproduce a given rate-intensity response in a computational model of auditory nerve fibre activity. This is a useful re-

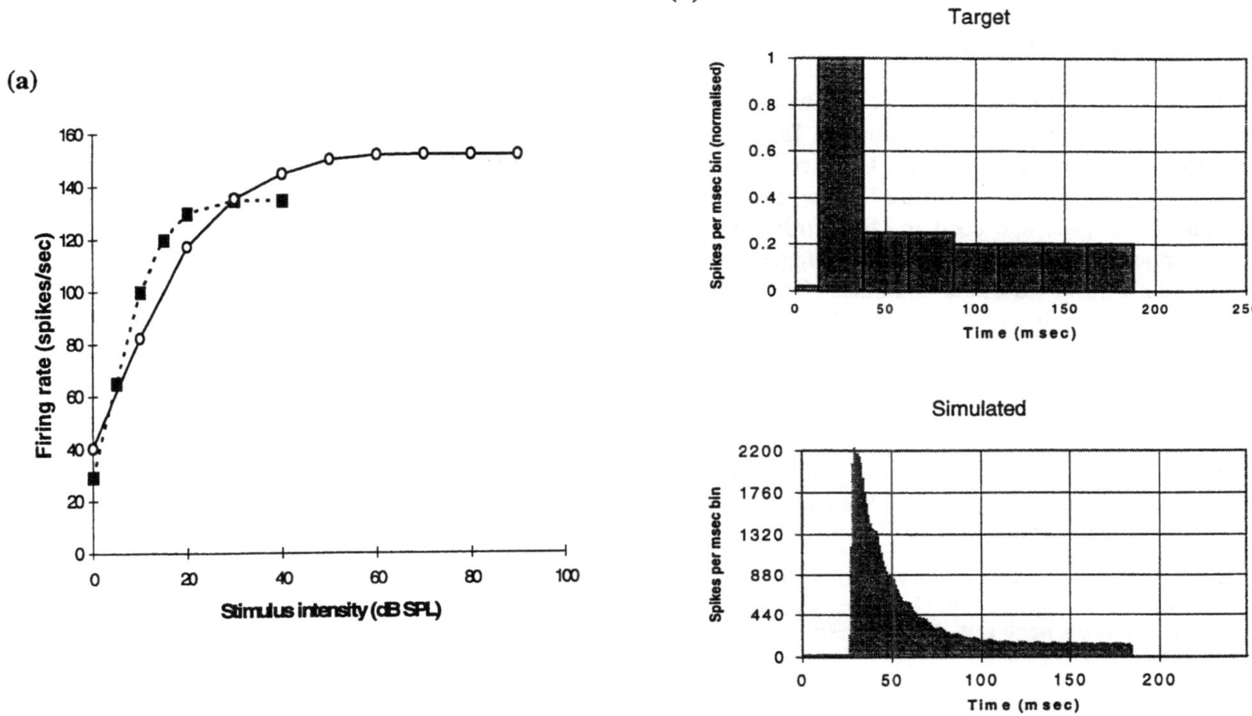

Figure 3: Parameters matching both a rate of intensity curve from an arbitrary fibre (a) and the corresponding PSTH response (b).

sult; however, in addition to the rate-intensity response, AN fibres also generate other characteristic behaviours. For example, many AN fibres phase-lock in the presence of an applied tonal stimulus (e.g. [2]). Also, in the presence of a sustained tone, the rate of firing is not constant but adapts (e.g. [15]). Clearly, if the selection of parameters is to be useful, the resulting simulation must be able to satisfy a set of different objective functions.

In Experiment 2, described here, we considered the selection of parameters required to match both a specific rate-intensity response, and a corresponding peri-stimulus time histogram (PSTH) response from the same fibre. In this experiment, the parameters and methodology were exactly as described for Experiment 1. The results are shown in Figure 3. While the match in simulated and recorded rate-intensity responses is less than perfect, the PSTH responses are seen to have the usual form.

4 Discussion

This paper has demonstrated a technique using genetic algorithms to automatically select parameter sets to a model of AN fibre activity to simulate the required rate-intensity functions recorded from physiological studies. The Meddis model of extracellular AN fibre activity was used in the experiments, simply due to its computational efficiency and availability. The results of the experiments described here illustrate that the technique works effectively. We see no reason why the approach could not be adapted for use with other auditory (and non-auditory) models.

5 Acknowledgements

This work is supported in part by an award from the Lord Dowding Fund for Humane Research. We thank Ray Meddis and Lowel O'Mard (Loughborough University) for providing the computer model of the inner hair cell used in this study, and for a great many helpful comments and suggestions. We also thank Colin Paterson (Leicester University) for providing the original GA code adapted for use in this study.

References

[1] K. A. De Jong. *An analysis of the Behaviour of a class of Genetic Adaptive Systems.* PhD thesis, University of Michigan, 1975.

[2] E.F. Evans. Cochlear nerve and cochlear nucleus. In W.D. Keidel and W.D. Neff, editors, *Handbook of Sensory Physiology.* Springer-Verlag, New York, 1987.

[3] J. M. Fitzpatrick, J. J. Grefenstette, and D. Van Gucht. Image restriction by genetic search. In *Proc. of IEEE Southest Conf.*, pages 460–464, 1984.

[4] D. E. Goldberg. *Genetic Algorithms in Search, Optimisation and Machine Learning.* Addison-Wesley, Reading, MA, 1989.

[5] M.J. Hewitt, R. Meddis, and T. Shackleton. A computer model of a cochlear-nucleus stellate cell: Responses to amplitude-modulated and pure-tone stimuli. *Acoust. Soc. Am.*, 91:2096–2109, 1992.

[6] J.H. Holland. *Adaption in Natural and Artificial System.* University Press, Michigan, 1975.

[7] D.O. Kim, S. Goshal, S.L. Khant, and K. Parham. A computational model with ionic conductances for the fusiform cell of the dorsal cochlear nucleus. *Acoust. Soc. Am.*, 96:1501–1514, 1994.

[8] P. K. Kuhl and J. D. Miller. Speech perception by the chinchilla: Identification functions for synthetic VOT stimuli. *J. Acoust. Soc. Am*, 63:905–917, 1978.

[9] R. Meddis. Simulation of mechanical to neural transduction in the auditory receptor. *Acoust. Soc. Am.*, 79:702–711, 1986.

[10] R. Meddis. Simulation of auditory-neural transduction: Further studies. *Acoust. Soc. Am.*, 83:1056–1063, 1988.

[11] J. K. Moore. The human auditory brain stem: A comparative view. *Hear Res.*, 29:1–32, 1987.

[12] M. J. Pont and R. I. Damper. A computational model of afferent neural processing from the cochlea to dorsal acoustic stria. *J. Acoust. Soc. Am.*, 89:1213–1228, 1991.

[13] M. J. Pont and R. I. Damper. *Exploring the role of the Dorsal Cochlear Nucleus in the Perception of Voice-Onset Time*, volume 2. JAI Press, London, 1992.

[14] M.J. Pont. *The Role of the Dorsal Cochlear Nucleus in the Perception of Voicing Contrasts in Initial English stop consonants: A Computational Modelling Study.* PhD thesis, Department of Electronics and Computer Science, 1990.

[15] L.A. Westerman and R.L. Smith. Rapid and short-termadaptation in auditory responses. *Hear. Res.*, 15:249–260, 1984.

[16] I.A. Winter and A.R. Palmer. Intensity coding in low-frequency auditory-nerve fibres of the guinea pig. *Acoust. Soc. Am.*, 90:1958–1967, 1991.

[17] C.P. Wong and M.J. Pont. Automatic selection of parameters for a computer simulation of extracellular auditory nerve fibre activity. In *Proceedings of the ESCA Workshop on the Auditory Basis of Speech Perception*, pages 61–64, Keele, UK, 1996.

Automatic Extraction of Phase and Frequency Information from Raw Voice Data

S. McGlinchey and C. Fyfe
Department of Computing and Information Systems, University of Paisley, UK.
Email: mcgl0ci@student.paisley.ac.uk, fyfe0ci@paisley.ac.uk

Abstract

We use a simple network which uses negative feedback of activation and simple Hebbian learning to self-organise in such a way as to produce a feature map which has the property of identifying the relative proportions of the components of the input data. Thus it evaluates the angular properties of the input data space and ignores the magnitude of the input data. When used on unprocessed voice data, the network is shown to extract both the phase and the frequency information from the raw data. We show how to use this network for classification of vowels.

1 Introduction

We have recently [2] introduced an artificial neural network which self organises to find a mapping of the input data which preserves neighbourhood relations. The difference between this mapping and the well-known Kohonen mapping is that the mapping is scale invariant. An example of both mappings on a two dimensional uniform input distribution on the square $\{(x,y) : -1 \leq x \leq 1, -1 \leq y \leq 1\}$ is shown in Figure 1. If we crudely define a topology preserving transformation as one in which similar inputs are mapped to similar output neurons while simultaneously ensuring that only similar inputs are mapped to each output neuron we may state that both networks create a topology-preserving mapping between inputs and outputs. However the Kohonen network creates an approximation to a Voronoi tessellation of the input data space while the scale invariant feature map ignores information contained in the magnitude of the inputs and extracts information about the relative magnitude of the components of the input vector i.e. each neuron captures a pie-slice of the input data.

In this report, we use the scale invariant feature map to extract information from voice data: our conjecture was that we could use the mapping to find a neural network which would reliably identify vowels regardless of the amplitude of the sound waves (volume with which the vowel was spoken). We show that the network self-organises to extract both frequency and phase information from voice data which is presented to the network solely as raw unprocessed data in the time domain and use this information in a network which will identify vowel sounds.

2 The Network

Consider a network with N dimensional input data and having M output neurons. Then the activation of the ith output neuron is given by

$$act_i = \sum_{j=1}^{N} w_{ij} x_j \tag{1}$$

Now we invoke a competition between the output neurons by selecting the neuron whose weight vector is closest to the input vector as the winner.

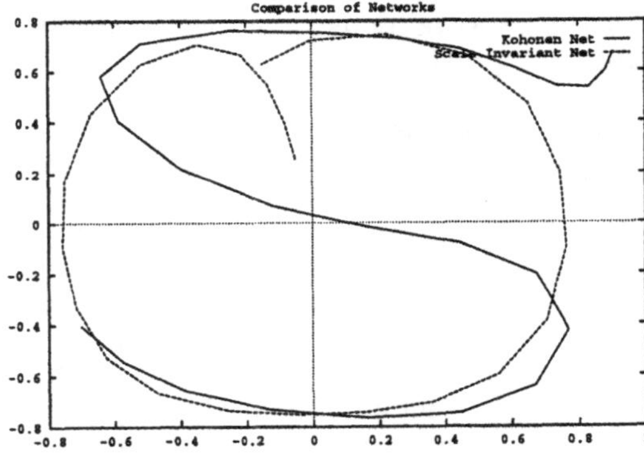

Figure 1: Convergence of the Kohonen network and the scale invariant network.

After a winner, e.g. the pth neuron, is selected it is deemed to be maximally firing (=1) and all other output neurons are suppressed(=0). Its firing is then fed back through the same weights to the input neurons as inhibition.

$$x_j(t+1) \leftarrow x_j(t) - w_{pj}.1 \text{ for all } j \quad (2)$$

where p is the winning neuron. Now the winning neuron excites those neurons close to it, i.e. we have a neighbourhood function $\Lambda(p,j)$ which satisfies $\Lambda(p,j) \leq \Lambda(p,k)$ for all $j,k : \| p-j \| \geq \| p-k \|$ where $\| \, . \, \|$ is the Euclidean norm. In the simulations described in this paper, we use a Gaussian whose radius is decreased during the course of the simulation. Then simple Hebbian learning gives

$$\begin{aligned} \Delta w_{ij} &= \eta_t \Lambda(p,i).x_j(t+1) \\ &= \eta_t \Lambda(p,i).(x_j(t) - w_{pj}) \quad (3) \end{aligned}$$

where we have used $x_j(t)$ as the activation of the jth input neuron at time t and w_{ij} is the weight between this and the ith output neuron. For the pth winning neuron, the network is performing simple competitive learning but note the direct effect the pth output neuron's weight has on the learning of other neurons. This algorithm introduces competition into the same network used in [1] to perform a principal component analysis and in [3] to perform an exploratory projection pursuit.

3 Self Organisation on Voice Data

Typically such networks are trained on data which has been preprocessed by, for example, Fourier transforming the data to the frequency domain and often using the more computationally expensive cepstral coefficients (see e.g. [4]). We, however, wish to test the network by using as crude voice data as possible as inputs.

The input data to the network then is raw voice data sampled at 8KHz and subjected to no preprocessing. The network has been tested using different numbers of inputs from 10 to 160 where, for example, 64 inputs represents 64 consecutive inputs from the data stream (equal to 8 ms of data). It was found that at least 70 inputs (8.75 ms) were necessary to include a complete waveform of all of the selected phonemes. Each presentation of the data consists of a randomly chosen starting point and the following consecutive inputs. The results have been qualitatively similar for each size of network. We use 30 output neurons with a one dimensional neighbourhood function.

The neighbourhood function used was a simple exponential exp $\frac{-x^2}{r}$ where x is the 'distance' between neurons and r determines the width of the function. An initial radius of 30 was selected and this was reduced during training at a rate of 1 every thousand training cycles to a minimum value of 4. The learning rate was given an initial value η_0 of 0.1 and was decayed during the first 10000 iterations by

$$\Delta\eta = \frac{-1}{10000} * \eta_0.$$

Consider first one network which is trained on 8 speakers saying the word 'far'. The trained weights are shown in Figure 2. The top diagram shows the weights into the first three neurons: we can see

- the neurons are extracting the frequency information from the input data in the weight vector, showing a waveform pattern.
- the weights into the first neuron differ very little from those into the second and this in turn differs very slightly from those into the third neuron — this is a prerequisite for any network which is claiming to retain neighbourhood relations. The variation between neurons shows a shift in the phase of the wave pattern.

The second half of the diagram shows the weights into three neurons which are not neighbours. Clearly each neuron is extracting the same frequency information from the raw data but has learned to respond to different phases of this data. A diagram with all 30 output neurons would show a complete coverage of all phases at this frequency.

4 Vowel Classification

We now propose a network which will identify any vowel phoneme from its raw data samples. We use a network such as shown in Figure 3. During the learning phase, each set of output neurons (Net A, Net B, ...) is trained on a different set of vowel data. Each learns the frequency information associated with that vowel and individual neurons within that set will respond maximally at any particular time depending on the correspondence between the phase of the signal and the weights into the neuron. However during the vowel identification phase the network is fully connected - each input is connected to all output neurons in all networks, Net A, Net B, ... When any vowel is presented to the network,

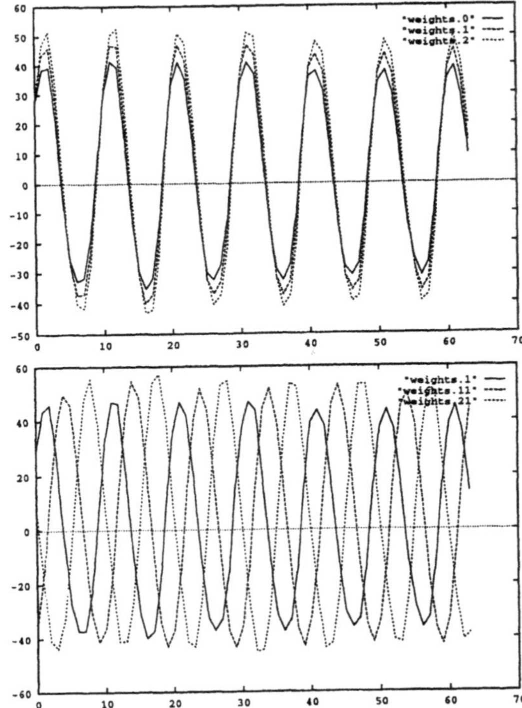

Figure 2: Top diagram: the weights of the first three neurons trained on the vowel sound of 'far'. Bottom diagram: the weights of the first, eleventh and twenty-first neurons on the same data.

the vowel can be identified by noting which of the output networks responds optimally since there is a ripple of activation across that network as particular neurons respond maximally to the particular phase of the inputs. Such a coherent ripple cannot be seen in the other output nets.

During vowel classification, sounds were fed forward through all eight networks, producing eight winning neurons. The network whose winner had the highest activation was chosen as an overall winner. This procedure was repeated a number of times, taking the input vector from a different section in the raw sound data. The network that wins most competitions wins overall. Random selection of input vectors produced less reliable classification, and better results were obtained by selecting input vectors sequentially from the sound data, ensuring than the sound waves are fed forward in all possible phases.

E.g. if a vowel sound is composed of n samples, and we wish to use x components in the input vector, and a phase length of p, then we must select a random starting sample, s, in the range:

$$0 \rightarrow n - x - p.$$

The ranges of samples shown below must then be fed forward:

1 :	range s	$\rightarrow s + x - 1$
2 :	range $s + 1$	$\rightarrow s + x$
3 :	range $s + 2$	$\rightarrow s + x + 1$
4 :	range $s + 3$	$\rightarrow s + x + 2$
..	....	....
p :	range $s + (p - 1)$	$\rightarrow s + x + (p - 2)$

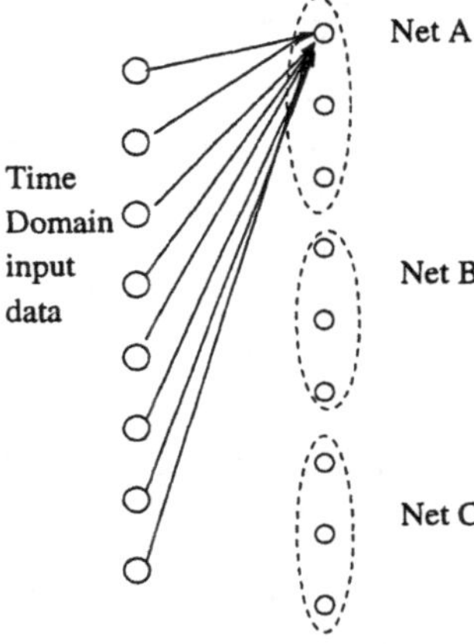

Figure 3: Each group of output neurons comprises a single network trained on examples of a single vowel. Each output neuron responds maximally to a single type of input data.

For this experiment, a phase length of 100 was selected, ensuring that at least a complete wavelength of all sounds was fed forward. Each sequence of 100 tests was repeated ten times, each time from a random starting position. Therefore each vowel sound was presented a total of 1000 times.

Eight words were recorded and the vowel sounds were taken from them. All of the training and test data was spoken by the same person. The voice of the speaker was such that some of the words had common vowel phonemes e.g. 'bat' and 'far' and also 'boot' and 'put' and these sounds are assumed to be equivalent. Table 1 shows the results of tests that were run on the training data and Table 2

Table 1: Number of successes when the network is tested on the training data.

	Inputs							
	bat	bed	beep	boot	but	far	pit	put
bat	10	0	0	0	1	10	0	0
bed	0	10	0	0	0	0	0	0
beep	0	0	10	0	0	0	2	0
boot	0	0	0	8	0	0	0	0
but	0	0	0	0	9	0	3	0
far	0	0	0	0	0	0	0	0
pit	0	0	0	0	0	0	5	0
put	0	0	0	2	0	0	0	10

Table 2: Number of successes when the network is tested on data not seen during training.

	Inputs							
	bat	bed	beep	boot	but	far	pit	put
bat	2	0	0	0	0	10	0	0
bed	0	9	0	0	0	0	6	0
beep	0	0	10	0	0	0	0	0
boot	0	0	0	8	0	0	0	0
but	0	1	0	0	10	0	2	0
far	8	0	0	0	0	0	0	0
pit	0	0	0	0	0	0	2	0
put	0	0	0	2	0	0	0	10

shows the same tests performed on new data. It can be clearly seen that both tables show comparable results, indicating that the networks have been trained to recognise features of the sound without modelling the noise.

All of the selected phonemes could be identified by the network, however, the only phoneme that was not reliably classified was from the word ‘pit’. All other phonemes showed nine or ten out of ten correct classifications. Typically, these classifications were won with around 80 to 90% confidence i.e. 80 to 90% of phase positions gave a correct classification.

This system is speaker dependant and future work will focus on training the networks with a variety of voices to achieve more general, speaker independent recognition. This will require the addition of more neurons since the model will have to learn varying waveforms for each phoneme. We also hope to extend the model to cater for fricative and plosive phonemes.

References

[1] C. Fyfe. PCA properties of interneurons. In *From Neurobiology to Real World Computing, ICANN 93*, pages 183–188, 1993.

[2] C. Fyfe. Radial feature mapping. In *International Conference on Artificial Neural Networks, ICANN95*, Oct. 1995.

[3] C. Fyfe and R. Baddeley. Non-linear data structure extraction using simple Hebbian networks. *Biological Cybernetics*, 72(6):533–541, 1995.

[4] T. Kohonen. *Self-Organising Maps.* Springer, 1995.

A Speech Recognition System using an Auditory Model and TOM Neural Network

E. Hartwich and F. Alexandre
CRIN-CNRS / INRIA Lorraine
BP 239, F-54506 Vandœuvre-lès-Nancy
Email: {hartwich, falkex}@loria.fr

Abstract

This paper is devoted to a neurobiologically plausible approach for the design of speech processing systems. The temporal organization map (TOM) neural net model is a connectionist model for time representation. The definition of a generic neural unit, inspired by the neurobiological model of the cortical column, allows the model to be used for problems including the temporal dimension. In the framework of automatic speech recognition, TOM has been previously tested with conventional techniques of signal processing. An auditory model as front-end processor is now used with TOM, in order to test the efficiency and the accuracy of a physiologically based speech recognition system. Preliminary results are presented for speaker-dependent and speaker-independent speech recognition experiments. The interest of auditory model is the possibility to develop more valuable processing and communication strategies between TOM and the front-end processor, including afferent and efferent information flow.

1 Introduction

Among the various applications of artificial neural networks (ANNs), speech processing is certainly one of the most active domains. This implies that the classical functioning of ANNs had to be extended to temporal aspects, like in recurrent networks [8] or in time-delay neural networks (TDNNs) [15]. The neuronal model TOM [6] is a neurobiologically plausible temporal neural network, whose basic unit refers to neural assemblies and involves different kinds of specialised links. TOM is designed for every problem involving temporal dimension, and initially tested on speech processing [7]. Its modular architecture allows to connect several maps in order to encode different functionalities or modalities. For the auditory modality, our purpose is to make use of processing schemes inspired with the mammalian auditory pathway. We have firstly focused on the afferent auditory pathway. A preliminary architecture, neurobiologically plausible, composed of an auditory model and TOM neural network, is tested on isolated word recognition experiments.

2 TOM, the Temporal Neural Net Model

A TOM involves a set of *super-units*. These super-units are closer to the cortical column model [2, 4] than to the MacCulloch and Pitts formal neuron. Super-units correspond to neural assemblies, and each *unit* in the assembly has a particular temporal functionality. A super-unit involves different kinds of links (see Figure 1):

- Each super-unit is associated to one stimulus of the input space, and so reacts to a given feature. The corresponding spatial mask is carried out by feedforward links.
- The neighbourhood links define a spatial topology preserving map. Two close super-units represent two close stimuli. These links and feedforward links are learned by competitive learning algorithms, like Kohonen SOM [11], neural-gas [13] and growing neural-gas [9] algorithms.
- Units in the same super-unit share the feature encoded by their super-unit, but each of them represents a particular temporal context for this feature. Intra-map links bind units in different super-units one to another, and so allow to modelling of temporal sequences. The intra-map links set of a given unit is called temporal receptive field.

A unit can be activated only if the feature encoded by its super-unit is present, and if one unit of the temporal receptive field was active at the previous

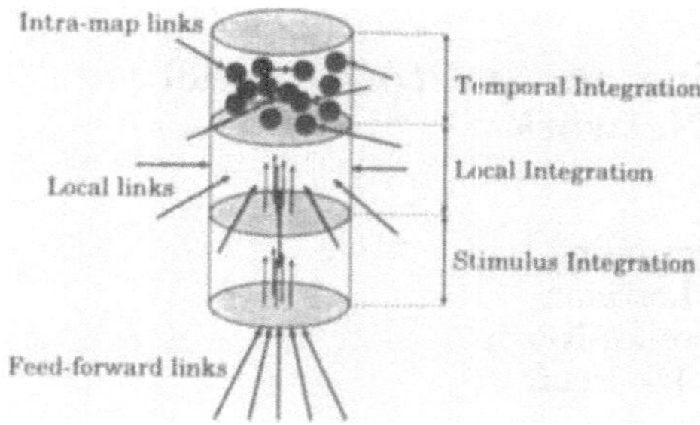

Figure 1: A super-unit.

time step. So the learning procedure of the temporal receptive field consists of creating new units and binding these units according to presented sequences.

TOM algorithms, presented in detail in [7], are dependent on the preprocessing scheme. In the case of speech recognition, the time step is a function of the sampling frequency and then defined *a fortiori*. It does not necessarily reflect the temporal structure and characteristics of the studied phenomenon. For example, with the TOM model, the time is uniformly processed for plosive sounds and for stationary sounds like vowels. Nevertheless, it's quite clear that the few milliseconds around a plosive burst are more important for the classification processing than the same time window in the steady-state part of a vowel sound. To take into account such effects, temporal distortion parameters have been introduced in TOM. Self-organized learning of these parameters may emphasize relevant events in the input signal and gives to the model temporal distortion robustness.

These developments have been achieved in the framework of speech recognition, and particularly to improve classification and recognition of spoken digits. In order to provide generality to the model, most specific processings have to be treated by the preprocessing stage. We present in the next section the auditory model used as a front-end processor.

3 Architecture of the Auditory Model

The auditory model we used consists of two stages. The periphery model represents the processing steps from outer ear to auditory nerve. This stage provides firing probabilities as a function of time for a set of channels. In order to reduce the data rate before entering the TOM model, a spike generation stage is added and a 'pseudo mean-rate' is finally computed. The global architecture is depicted in Figure 2.

3.1 Periphery Model

The periphery model is largely inspired by the models of Seneff [14] and Gao *et al.* [10]. Middle ear and basilar membrane filtering characteristics are approximated by a set of 39 independent channels. These channels cover the frequency range from 150 to 7000Hz. Each channel consists of a filter section, followed by a non linear stage intended to reflect the transformations that occur at the Inner Hair Cells (IHC)/auditory nerve fibres junctions.

Each filter of the filter bank is realized by a bandpass filter, followed by a lowpass filter to sharpen the frequency response of the bandpass filter on the high-frequency side of the filter. Moreover, signal is pre-emphasized by the high-frequency promotion in the gain of bandpass filter.

The transduction section models the transformation from basilar membrane vibrations to auditory nerve fibres responses and predicts the firing probability of 39 tonotopically-organized auditory neurons. This section incorporates effects such as dynamic range compression, short-term adaptation (STA) and automatic gain control (AGC). These nonlinearities are processed by a four-stage transduction mechanism, and implemented as a cascade of independent modules in each channel (see Figure 2).

Firstly, the half-wave rectification (HWR) expresses the fact that spike sequences in auditory nerve fibres are only initiated on one-half cycle of the basilar membrane motions. The model for the instantaneous half-wave rectification is defined as:

$$y(t) = \begin{cases} 1 + \alpha \arctan(\beta x(t)) & x(t) > 0 \\ 0 & x(t) \leq 0 \end{cases}$$

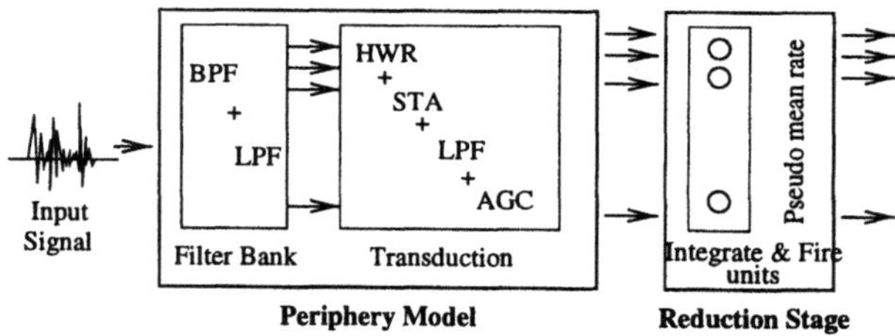

Figure 2: Schematic diagram of the auditory model (See the text for the description of the labels).

The STA stage, which can be considered to represent the release of neurotransmitter into the presynaptic cleft, consists of two separate mechanisms expressed by:

$$dC(t)/dt = \begin{cases} \mu_a [S(t) - C(t)] - \mu_b C(t) & \text{if } C(t) < S(t) \\ -\mu_b C(t) & \text{if } C(t) \geq S(t) \end{cases}$$

where $C(t)$ (resp. $S(t)$) is the concentration of neurotransmitter within the region (resp. in the source region). The quantity $\mu_a [S(t) - C(t)]$ controls the probability of firing of the auditory fibre. These outputs of STA stage are then lowpass filtered (LPF), to simulate synchrony reduction with increasing frequency.

The last component is the AGC and takes into consideration the refractory period of auditory fibre [3]. The probability of firing of the ith auditory fibre (or channel) at time t is :

$$P_i(t) = \lambda_i(t) \left(1 - \sum_{j=1}^{ref} P_i(t-j)\right)$$

where, ref is the refractory period exprimed in time step (relatively to the sampling frequency) and $\lambda_i(t)$ the ith channel output after lowpass filtering.

3.2 Reduction Stage

The output of the periphery model is the firing probabilities of each channel for each time step. The time step depends on the sampling frequency, so the data rate needs to be reduced before processing by the classification and recognition system. This stage can be considered as the first processing step of the central nervous system, although our model is very simple.

It consists of tonotopically organized array of 39 integrate-and-fire units. One unit receives input either from a single channel, or from a set of adjacent channels. In our experiments, each unit is connected to a single channel. In the same way, output of units could be fed back to adjacent units, but such recurrent connections have not been used here. The activity $A(u_i)$ of the unit u_i, connected to the peripheral model i^{th} channel, is initially 0 and evolves according to :

$$\frac{dA(u_i)}{dt} = O_i(t) - \lambda A(u_i),$$

where $O_i(t)$ is the ith channel output at time t, and λ the dissipation describing the leakiness of the integration. When $A(u_i)$ reaches a threshold, the unit emits a spike, $A(u_i)$ is reset, and u_i comes into refractory period, during which it is insensitive to input. We compute then a pseudo mean-rate by summing the spikes in each channel over a 20 ms time window length. This reduction stage is somewhat simple, insofar as we lose the fine temporal structure of spikes emission.

4 Application to Speech Recognition

Experiments have been performed with two different front-end processors. In the model I, input speech is sampled at 16 kHz and hamming windowed. A time window of 32 ms is used to compute an MFCC vector [5] with 12 coefficients. One vector is given every 4 ms. In order to reduce the number of vectors (and in this way to avoid variability), an average of 3 successive vectors is provided as input to the TOM map at each time step. Model II consists of the auditory model previously described and of the same TOM architecture

For both models, the classifier is a single TOM map connected to a sequence detecting map (Figure 3). Preprocessing scheme apart, the receptive field size of the first map is the only difference between the two models, and there are respectively 12 and 39 coefficients computed at each time step for model I and model II. 400 super-units are used, and their feedforward link vectors are learned with a two dimensional Kohonen connectivity. Intra-map learning process starts with the initial parameters for each unit of *decay* = 0.95 and *thres_act* = 0.5. Two experiments have been performed with these architectures. They used data speech from the TI Digit database with a corpus of male voices. Experiment 1 is made with 11 speakers, and the training set includes 100 digits (10 occurences per

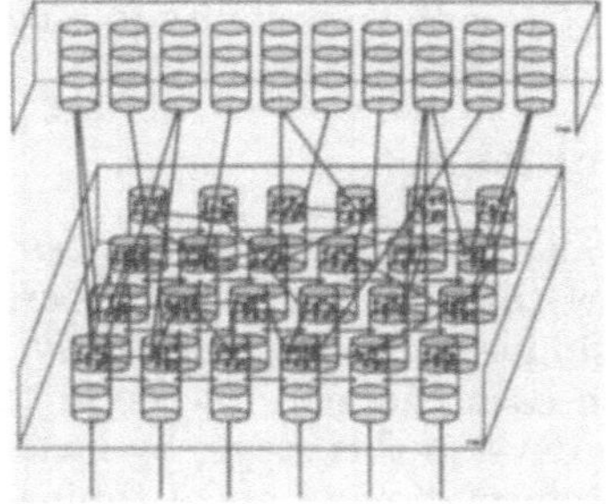

Figure 3: TOM architecture for spoken digit recognition.

Table 1: Recognition rate on TI Digit database.

	Model I	Model II	HMM
Exp 1	99.6%	96.6%	99.8%
Exp 2	98.5%	86.4%	98.6%

speaker), while the testing data contains 100 other digits from the same speakers. Experiment 2 uses a training set of 55 speakers (two occurrences of each digit per speaker), and the testing data involves 340 digits from 17 other speakers. Results are collected in Table 1. We give comparative recognition rates reached by a first-order hidden Markov model (HMM), remembering this model also used MFCC-like preprocessing scheme. Model I performances draw near to the HMM results [7], while the results obtained with the auditory model are the less accurate. The stronger degradation with Experiment 2 mainly comes from the preprocessing step. While the MFCC tends to smooth inter-speaker variability, the auditory model preserves supra-segmental information (like energy and pitch) in an explicit fashion, since we obtain a detailed time-of-firing versus tonotopic representation. But this representation contains most of the speech cues and could thus be used for other speech features processing. For example, auditory models showed great capabilities on speech enhancement and are particularly robust on noisy data [10]. Furthermore, we did not make use for the moment of temporal processing of spatio-temporal firing patterns in the auditory model, whereas it is known that temporal analysis is performed throughout the mammalian auditory structures [12]. The temporal structure of the auditory nerve fibres responses takes part in the coding of features such fundamental frequency, amplitude modulation or formant resonances. The advantage of the auditory model over classical techniques like the MFCC is the possibility to incorporate such temporal processing.

5 Conclusions

Using an auditory model led to loss of accuracy, but we saw that the reduction processing scheme was very simple, since we used solely a pseudo firing rate on each channel. In the future, one challenge is to develop this stage, by including the fine temporal information of spike timing. Models of cochlear nucleus cells seem to be candidates for this purpose. Functionally different cells process the auditory nerve ascending information by at least two different processing schemes. The first processing way offers a goodspectral representation, while the second is devoted to temporal encoding (see [1] for a functional model). Although each one could be used independently for specialized processing (e.g. pitch extraction or source separation), the difficulty is to efficiently combine these information streams in the more general goal of speech recognition. Another improvement of our neurobiologically plausible approach would be to take into account efferent processing schemes, and therefore enable co-operation between bottom-up and top-down process in a fully integrated architecture. A more practical view of our work will lead to us adding a speaker adaptation and to test our model on noisy speech.

References

[1] W. A. Ainsworth. Auditory mechanisms for speech perception. In *Proc. of Eurospeech'95*, pages 171–178, Madrid, Spain, 1995.

[2] F. Alexandre, F. Guyot, J. P. Haton, and Y. Burnod. The cortical column: a new processing unit for multilayered networks. *Neural networks*, 4:15–25, 1991.

[3] F. Berthommier. *Intégration neuronale dans le système auditif. Modélisation de réseaux neuronaux temporo-dépendants*. PhD thesis, Université Joseph Fourier - Grenoble I, 1992.

[4] Y. Burnod. *An adaptive neural network: The cerebral cortex*. Masson Paris, 1988.

[5] S. B. Davis and P. Mermelstein. Comparison of parametric representation for monosyllabic word recognition in continuously spoken sentences. *IEEE Transactions on acoustics, speech, and signal processing*, ASSP-28(4):357–366, 1980.

[6] S. Durand and F. Alexandre. Spatio-temporal mask learning : application to speech recognition. In D. W. Pearson, N. C. Steele, R. F. Albrecht (editors), *Artificial Neural Nets and Genetic Algorithms*, pages 132–135, Springer-Verlag, Wien, April 1995.

[7] S. Durand and F. Alexandre. Tom, a new temporal neural net architecture for speech signal processing. In *IEEE International Conference on Acoustic Speech and Signal Processing*, Atlanta, USA, 1996.

[8] J L. Elman. Finding structure in time. *Cognitive Science*, 14:179–211, 1990.

[9] B. Fritzke. A growing neural gas network learns topologies. In G. Tesauro, D.S. Touretzky, and T.K. Leen, editors, *Advances in Neural Information Processing Systems 7*. MIT Press, Cambridge MA, 1995.

[10] Y. Gao, T. Huang, S. Chen, and J. P. Haton. Auditory model based speech processing. In *Proc. of ICSLP*, pages 73–76, Alberta,Canada, 1992.

[11] T. Kohonen. *Self-Organization and Associative Memory*. Springer Series in Information Sciences. Springer-Verlag, third edition, 1989.

[12] G. Langner. Periodicity coding in the auditory system. *Hearing Research*, 60:115–142, 1992.

[13] T. M. Martinetz and K. J. Schulten. A "neural-gas" network learns topologies. In T.Kohonen, K. Mäkisara, O. Simula, and J. Kangas, editors, *Artificial Neural Network*, pages 397–402. North-Holland, Amsterdam, 1991.

[14] S. Seneff. A joint synchrony/mean-rate model of auditory speech processing. *Journal of Phonetics*, 16:55–76, 1988.

[15] A. Waibel, T. Hanazawa, G. Hinton, K. Shikano, and K J. Lang. Phoneme recognition using time-delay neural networks. *IEEE Transaction on Acoustics, Speech and Signal Processing*, 37(3):328–339, 1989.

Fahlman-Type Activation Functions Applied to Nonlinear PCA Networks Provide a Generalised Independent Component Analysis

M. Girolami and C. Fyfe
Department of Computing and Information Systems, University of Paisley,
High Street, Paisley, Scotland, PA1 2BE
Email: {giro0ci, fyfe0ci}@paisley.ac.uk

Abstract

It has been shown experimentally that Oja's nonlinear principal component analysis (PCA) algorithm is capable of performing an independent component analysis (ICA) on a specific data set [7]. However, the dynamic stability requirements of the nonlinear PCA algorithm restrict its use to data which has sub-gaussian probability densities [6]. The restriction is particularly severe as this precludes the application of the algorithm from performing ICA on naturally occurring data such as speech, music and certain visual images. We have shown that the nonlinear PCA algorithm can be considered as minimising an information theoretic contrast function and develop a more direct link between ICA and the algorithm function [6]. To remove the sub-gaussian restriction and enable a generalised ICA which will span the full range of possible data kurtosis, we propose the use of Fahlman type activation functions [2] in the nonlinear PCA algorithm. We show that variants of these functions satisfy all the dynamic and asymptotic stability requirements of the algorithm and successfully remove the sub-gaussian restriction. We also report on simulations which demonstrate the blind separating ability of the nonlinear PCA algorithm with the Fahlman type functions on mixtures of super-Gaussian data (natural speech).

1 Introduction

Karhunen and Joutsensalo [8] show that Oja's nonlinear PCA algorithm (1) can be derived from a cost function (2) which minimises the mean square error incurred in approximating an L dimensional vector x in terms of nonlinear expansion coefficients, which are functions of the inner product of the original vector x and the corresponding basis vector $w(i)$, where the M basis vectors are $w(1), \ldots, w(M)$: $M \leq L$. The cost function can be written in matrix format as a function of W where W is the matrix whose columns are the individual weight vectors $W = [w(1), w(2), \ldots, w(M)]$

$$W_{k+1} = W_k + \mu \left[x_k - W_k f(W_k^T x_k\right] f(x_k^T W_k) \quad (1)$$

$$J(W) = E\left\{\| x - Wf\left(W^T x\right) \|^2\right\} \quad (2)$$

$E\{\}$ denotes expectation and $\|\|$ denotes the L_2 or Euclidean norm giving a measure of the error vector length. Karhunen *et al.* first realised that the nonlinear PCA algorithm had signal separating properties and applied this to sinusoidal estimation [8]. Fyfe *et al.* [3] detailed the benefits of pre-whitening data, that is covariance diagonalisation, for neural exploratory projection pursuit. It was found that data pre-whitening was the key to successful blind separation of source signals using the nonlinear PCA algorithm [7]. The independent component analysis problem can be considered as a model where a source vector s has independently identically distributed components and a mixture is formed by linear multiplication with the matrix A such that $x = As$. With the data vector x and no *a priori* knowledge of A or s, save that the components of s are independent, ICA attempts to find the basis vectors A. Blind separation of sources can be considered as another facet of ICA where the identification of the original source data s is the goal. We shall refer to ICA and blind separation in the same manner in this paper, noting that Karhunen *et al.* have proposed a fully neural implementation of ICA by using a second layer of neurons which inverts the separating matrix W [7].

The derivation of (1) from (2) shows that the nonlinear PCA algorithm is an approximative stochastic gradient descent algorithm for the minimisation of the mean square representation error. For nonlinear PCA where the expansion coefficients are nonlinear versions of the vectors projected on to the learned subspace, it is clear as to the function of the algorithm. However, a clear intuitive link to ICA for the algorithm is not possible from the cost function of (1). Recently, we have shown [6] that the cost function $J(W)$ can also be considered as an approximation to the contrast function for ICA developed by Comon [1]. This then gives a far clearer indication as to the blind separating properties of the nonlinear PCA algorithm as it can then be considered as maximising (or minimising, depending on the normalised fourth order marginal cumulant value of the source data) the sum of squares of fourth order marginal cumulants. The contrast function is given below, and is a measure of the mutual information at the network output

$$J(W) = (L + \text{sign}(f()) * 6) - 2 * \text{sign}(f()) \sum_{i=1}^{M} K_i^{(4)} \tag{3}$$

where sign() gives the sign of the nonlinearity used at the output neurons and $f(u) = \pm u^3$.

2 Stable Activation Functions for the Nonlinear PCA Learning Algorithm

There are a number of constraints placed upon the form of the nonlinearity $f()$, based on the asymptotic and dynamic stability of the nonlinear PCA algorithm, these are detailed in Oja [9]. We will merely state these here as we wish to ensure that the activation functions proposed satisfy these criteria for the algorithm to demonstrate separating properties and stable convergence. However, detailed derivation and discussion of these criteria can be found in [9].

1. The function $f()$ is odd so that $f(u) = -f(-u) : \forall u$ and is at least twice continuously differentiable $\forall u$.

2. $f()$ should be a non-saturating function which grows less than linearly $f(u) \leq u : \forall u$.

3. The function must satisfy $E\{u^2\}E\{f'(\alpha u)\} - E\{f^2(\alpha u)\} < 0$ for the weight matrix W to be an asymptotically stable stationary point of (1) and a separating matrix which rotates the input x into the factorable output $u = Wx$.

4. For dynamic stability the function must be increasing for all u, this can be seen by viewing (1) if $f()$ is not increasing then the anti-Hebbian decay term will be positive and so the stabilising effect will be removed and instability will follow. This can be written as $E\{uf(\alpha u)\} = \alpha E\{f^2(\alpha u)\}$ where α is a scalar value greater than zero.

Oja [9] shows that the hyperbolic tangent function satisfies the asymptotic stability requirements for separation of sub-gaussian (negatively kurtotic) data vectors. Although the function saturates, for data scaled to operate within the active region of the function, this provides stable operation of (1). The simple polynomial negative cubic does not satisfy criterium 1 as it is a decreasing function and the positive cubic grows faster than linearly and so introduces instability into the learning dynamics. The form $f(u) = u^3$ has also been found to be unstable experimentally [8]. The truncated Taylor series expansion of the *tanh*() function is $tanh(u) \cong u - \frac{u^3}{3} + 2\frac{u^5}{15}$. We neglect the fifth and higher order terms. As the data is zero mean then the expectation will remove the first term in the expansion leaving only $E\{tanh(u)\} \cong E\{\frac{-u^3}{3}\}$. Using *tanh*() in (1) it is clear that this will satisfy the contrast (3), and J will be minimised by maximisation of the rightmost term. This function is then suitable for performing ICA on input data whose original source is negatively kurtotic which satisfies the range of kurtosis $[-2, \ldots, 0[$.

The major problem arises with data which is positively kurtotic and possesses kurtosis in the range $]0, \ldots, \infty[$. No suitable nonlinearity has been reported in the literature and indeed Karhunen has restricted the use of (1) to sub-gaussian data only [5]. Wang *et al.* develop the bigradient algorithm for robust PCA and MCA, it also overcomes this apparent shortcoming of the nonlinear PCA algorithm [5]. We now show that (1) can be employed in performing ICA across the complete range of kurtosis, however, we first of all need to consider the quickprop algorithm [2].

3 Fahlman Activation Functions

Fahlman develops the quickprop algorithm based on Newtons method, it is a heuristic scheme which considers the network weights as almost independent. By approximating the error surface as a quadratic polynomial function of each of the weights, the polynomial coefficients can be evaluated, and so the weight parameters can be moved to the minimum of the parabola describing the error surface. Fahlman notes that when a sigmoid is close to its maximum or minimum limits the slope of the function is near zero and so the corresponding local error is almost zero, thus the corresponding weight change is insignificant. To alleviate this problem, Fahlman uses in his quickprop algorithm activation functions of the form,

$$\Phi(u_i) = \beta u_i + (1 + e^{-\gamma u_i})^{-1} \quad (4)$$

$$\Phi(u_i) = \beta u_i + tanh(\gamma u_i) \quad (5)$$

These functions then are no longer saturating and so the problem of insignificant weight changes at the limits of the activation function is alleviated somewhat. We however note with interest the form of (5) and find a number of desirable properties for use in the nonlinear PCA algorithm to perform ICA on positively or negatively kurtotic original data, by changing the form of (5) slightly to

$$\Phi(u_i) = \beta u_i - sign(K_s^{(4)}) tanh(\gamma u_i) \quad (6)$$

where $\text{sign}(K_s^{(4)})$ indicates the sign of the fourth order normalised marginal cumulant of the original vector components s_i. The four stability criteria listed in the previous section all require to be met, inspection of (6) shows that criteria 1, 2, and 4 are satisfied when $\text{sign}(K_s^{(4)}) > 0$. We now require to consider condition criterium 3 which will indicate the separating performance of the nonlinearity (6) and the shape of the source data probability density function (PDF) which will yield this asymptotic stability. By applying criterium 3 to sources which are super-gaussian we can identify a match with the specific form of data PDF. For $f(u) = u - tanh(u)$ if the PDF of u is positive then $E\{u^2\}E\{f'(\alpha u)\} - E\{f^2(\alpha u)\} < 0$. From this we can see that a separating matrix or a permutation thereof will be an asymptotic stationary point of the learning if using (6) the input data marginal distributions are all super gaussian, that is has positive kurtosis in the range $]0, \infty[$. If we consider the form of the Fahlman type activation function for super-gaussian zero mean data then

$$\begin{aligned} E\{f(u)\} &= E\{u - tanh(u)\} = E\{u\} - E\{tanh(u)\} \\ &= E\{u\} - E\{u - u^3/3 + 2u^5/15\} \cong E\{u^3/3\} \end{aligned}$$

So we have a positively cubic term dominating in the nonlinearity and the nonlinear PCA algorithm will stochastically minimise the contrast function of (3), which is equivalent to maximisation of the sum of squares of fourth order marginal cumulants [1]. The use of this form of function as has been shown will provide dynamic stability in learning and converge to an asymptotically stable point which is a separating matrix for the input mixture x. We now report on simulations carried out on super-gaussian data and show that the sub-gaussian restriction on the nonlinear PCA algorithm has been removed.

4 Simulations

As has been reported in [4, 5, 7, 8, 9] the nonlinear PCA algorithm has signal separating utility for sub-gaussian data. We shall consider now the separation of mixtures of super-gaussian data with the nonlinear PCA algorithm which has not been considered possible until now [5]. We use data generated by natural speech as this has a high positive value of kurtosis and the effects of separation can be checked not only numerically but also visually and audibly. We sampled five seconds of speech from five male and female speakers, sampling was carried out at 8 Khz. The original value of kurtosis for each speech sample was computed and the minimal value of the contrast (3) was calculated. The mixing was carried out by a well conditioned matrix. From Figure 1 we can see that the individual kurtosis has been decreased due to central limit effects brought about by the mixing of the densities; in the limit, the kurtosis will tend to zero. The absolute value of the contrast has dropped significantly indicating the rise in mutual information between the mixed components. We now use a 5×5 nonlinear PCA network and employ the standard nonlinear learning algorithm, however we shall now use the Fahlman type function (6) as the network nonlinearity.

The contrast develops rapidly with seven passes through the data set required to reach the asymptotic limits of convergence, yielding almost perfect separation (98% of original contrast). Another interesting point should be noted regarding the output of the network. Due to the indeterminacy of the ICA problem, the matrix product WA will equal PD where P is a permutation and D is a scaling matrix. This is seen at the network output and analysis of the converged separated value at each of the output neurons shows a uniform distribution over the five original values.

It is anticipated that the implementation of the ICA learning with a hierarchical form of network may constrain formation of the outputs to a speci-

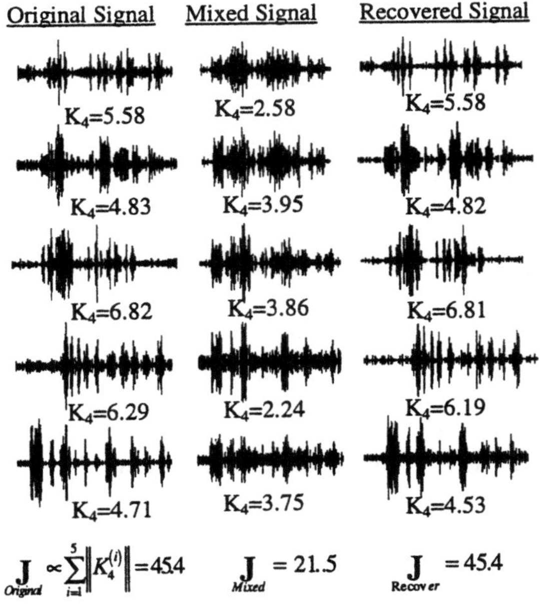

$$\mathbf{W} = \begin{bmatrix} -0.24 & 1.11 & 0.63 & 0.29 & 0.15 \\ 0.74 & 0.21 & -0.09 & 0.48 & -0.53 \\ 0.63 & -0.54 & 0.17 & -0.06 & 1.11 \\ 0.29 & 0.86 & 0.81 & -0.68 & 0.07 \\ 0.15 & 0.41 & -1.08 & -0.55 & 0.24 \end{bmatrix} \qquad \mathbf{WA} = \begin{bmatrix} 0.06 & 0.26 & \boxed{12.7} & 0.46 & 0.07 \\ 0.44 & -0.13 & 0.08 & \boxed{17.2} & -0.03 \\ \boxed{14.8} & -0.18 & 0.04 & -1.60 & 0.44 \\ -1.13 & 0.19 & -0.06 & 0.45 & \boxed{10.7} \\ -0.13 & \boxed{-5.4} & -0.39 & 0.39 & 0.48 \end{bmatrix}$$

Figure 1: Original, Mixed, Unmixed Voice Traces and Converged Weight Matrices.

fied form. Figure (1) also shows the final values of the matrix product WA, and noting that there is only one dominant term in each column it is clear that this is a permutation and scaling matrix.

5 Conclusions

We have shown that, contrary to current belief regarding the applicability of the nonlinear PCA algorithm to solely sub-gaussian densities [5], the use of Fahlman type activation functions allows the algorithm to perform ICA on both sub and now super gaussian data. We have shown that these functions satisfy the requirements for asymptotic and dynamic stability as laid out in [9] and have confirmed this with simulations on positively kurtotic data. The nonlinear PCA algorithm and a network with the Fahlman type nonlinearities will be capable of performing a stable ICA on data which covers the full range of possible kurtosis. Utilising *a priori* knowledge of the nature of the original data an informed choice of the sign in the nonlinearity can be made thus making the network respond optimally to the particular higher order statistics of the source data. We are currently working on the development of a method by which the requirement for *a priori* knowledge of the type of kurtosis the source data has can be dispensed with, this will be reported in a future publication.

References

[1] P. Comon. Independent component analysis, a new concept ? *Signal Processing*, 36:287–314, 1994.

[2] S.E. Fahlman. Faster-learning variations on back-propagation: an empirical study. In *Proc. Connectionist Models Summer School*, pages 38–51. Morgan-Kaufmann, 1988.

[3] C. Fyfe and R. Baddeley. Non-linear data structure extraction using simple Hebbian networks. *Biological Cybernetics*, 72(6):533–541, 1995.

[4] M. Girolami and C. Fyfe. Higher order cumulant maximisation using nonlinear Hebbian and anti-Hebbian learning for adaptive blind separation of source signals. In *Proc. IWSIP-96, IEEE/IEE International Workshop on Signal and Image Processing*, pages 141–144. Elsevier, 1996.

[5] M. Girolami and C. Fyfe. Kurtosis extrema and identification of independent components: A network approach. In *Proceedings ICASSP-97*, 1997. To appear.

[6] M. Girolami and C. Fyfe. Stochastic ICA contrast maximisation using Oja's nonlinear PCA algorithm. *International Journal of Neural Systems*, 1997. In press.

[7] J. Karhunen. Neural approaches to independent component analysis and source separation. In *Proc. ESANN96*, 1996.

[8] J. Karhunen and J. Joutensalo. Representation and separation of signals using nonlinear PCA type learning. *Neural Networks*, 7(1):113–127, 1994.

[9] E. Oja. The nonlinear PCA learning rule and signal separation-mathematical analysis. Technical Report A26, Helsinki University of Technology, 1995.

Blind Source Separation via Unsupervised Learning

B. Freisleben[1], C. Hagen[2] and M. Borschbach[1]
[1]Department of Electrical Engineering and Computer Science (FB12), University of Siegen
Hölderlinstr. 3, D-57068 Siegen, Germany,
[2]Department of Computer Science (FB20), University of Darmstadt
Julius-Reiber-Str. 17, D-64293 Darmstadt, Germany
Email: freisleb@informatik.uni-siegen.de, hagen@iti.informatik.th-darmstadt.de

Abstract

In this paper, a two-layer neural network is presented that organizes itself to perform *blind source separation*, i.e. it extracts the unknown independent source signals out of their linear mixtures. The convergence behaviour of the network is analyzed, and experimental results of separating historical speeches of four different speakers are presented.

1 Introduction

Blind source separation (BSS) [2] is aimed at extracting the individual, unobservable, statistically independent source signals out of given linear noisy mixtures of them. BSS is performed as part of *independent component analysis* (ICA) [1], a novel signal processing technique useful in a variety of applications areas.

From the mathematical point of view, BSS is a method to find a separation matrix which transforms the mixture signals into statistically independent signals being good estimations of the original source signals. There are some numerical algorithms [1, 3] using higher order statistics for estimating the separation matrix. Recently, neural networks have been proposed (see [4] for a survey) to perform BSS. The unsupervised neural approaches are based on the observation that introducing a nonlinear function in well known neural networks for *principal component analysis* (PCA) [5] allows us to perform BSS if the input vector and the distribution of the source signals meet some conditions. This is done in order to deal with the higher order moments required.

In this paper, an unsupervised BSS learning rule for a two-layer neural network is presented. The learning rule is hierarchical in the sense that output unit i receives contributions from all units j with $j < i$. The convergence behaviour of the neural network is mathematically analyzed and demonstrated in practice. It will be shown that the network is capable of performing BSS for a set of acoustic source signals, consisting of (excerpts of) historical speeches by Churchill, Kennedy, Luther-King and Armstrong.

2 Blind Source Separation

Figure 1 illustrates the steps of an ICA algorithm.

The first box represents the unobservable (linear) transformation of the unknown source signal vector $\vec{s}$ into the observed signal vector $\vec{u} = \vec{M} \cdot \vec{s}$. The $m \times n$-mixing matrix $\vec{M}$ is assumed to have full column rank, i.e. $\vec{u}$ has the dimension m, with $m \geq n$.

The second box represents a *prewhitening* step which transforms $\vec{u}$ into $\vec{x} = \vec{D}^{-1} \cdot \vec{F}^T \cdot \vec{u}$, such that $\vec{x}$ has as its covariance matrix $\vec{C}_{xx} = E(\vec{x}\vec{x}^T)$ the identity matrix $\vec{I}_n$. Prewhitening simplifies the BSS task and can be achieved by any non-neural or neural PCA algorithm [5].

The third (dashed) box represents the BSS part: finding an (orthogonal) separation matrix $\vec{B}$ to transform $\vec{x}$ into an output vector $\vec{y}$ whose components have a maximal degree of statistical independence (measured by so called *contrast functions* [1]). Each output y_i, $i = 1, \cdots, n$ can be regarded as an approximation of one of the source signals $\pm s_j$, $j = 1, \cdots, n$.

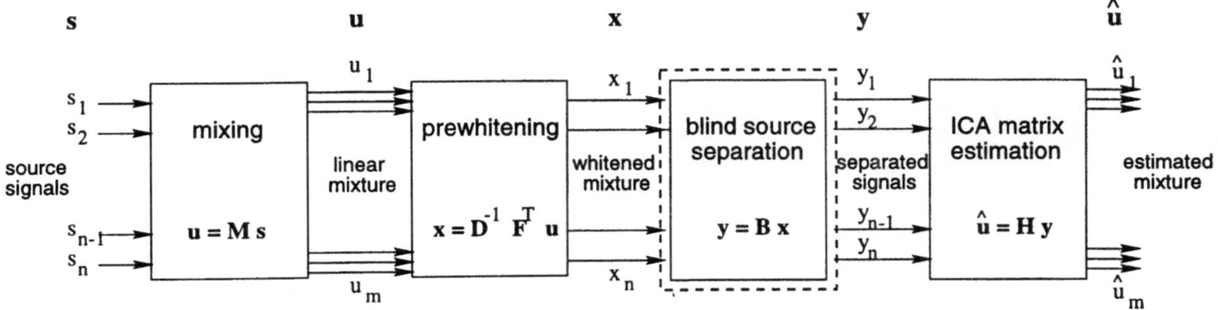

Figure 1: Illustration of the ICA Algorithm.

The fourth box represents the last step missing to complete ICA, namely the estimation of the mixing matrix, the so called ICA matrix $\vec{H}$, whose columns form the ICA basis. The output of the last step is the estimated observed mixture vector $\vec{u} = \vec{H} \cdot \vec{y}$.

All BSS algorithms only work under the assumption that the distributions of the source signals are *non-Gaussians*, except at most one, because a linear mixture of Gaussians is again a Gaussian and cannot be separated. In many practical situations, the source signals are either *sub-Gaussians* or *super-Gaussians*, i.e. distributions with densities flatter or sharper than that of Gaussians, respectively. Therefore, the existing approaches proposed for performing BSS can only be applied to one of these distributions.

3 A Neural Network for BSS

The neural network presented in this section assumes that the input vectors $\vec{x}$ are already prewhitened and that the source signals have *super-Gaussian* distributions. It consists of an input and an output layer, each consisting of n units. Both layers are fully connected to each other.

The connection weights are the n-dimensional vectors $\vec{w}_i, i = 1, \cdots, n$ which form the columns of the $n \times n$ matrix $\vec{W}$. The output of the i-th output unit $y_i, i = 1, \ldots, n$ is

$$y_i = \vec{w}_i^T \vec{x} \quad \text{or} \quad \vec{y} = \vec{W}^T \vec{x} \tag{1}$$

where $\vec{x} = (x_1, \cdots, x_n)^T$ is the input vector and $\vec{y} = (y_1, \cdots, y_n)^T$ is the output vector.

The weight vectors $\vec{w}_i, i = 1, 2, \cdots, n$ are updated according to the learning rule

$$\begin{aligned} \tilde{\vec{w}}_i(t+1) &= \vec{w}_i(t) + \\ &\alpha(t)\, (\vec{w}_i(t)^T \vec{x}(t))^3 \left(\vec{I} - \textstyle\sum_{j=1}^{i-1} \vec{w}_j(t) \vec{w}_j(t)^T\right) \vec{x}(t) \\ \vec{w}_i(t+1) &= \tilde{\vec{w}}_i(t+1) / |\tilde{\vec{w}}_i(t+1)| \end{aligned} \tag{2}$$

In the following, we will show that the weight matrix $\vec{W}^T$, adapted using the learning rule (2) converges to the separation matrix $\vec{B}$. The simplest way to prove the convergence of a stochastic algorithm is to find it being a gradient of a cost function. The cost function used in our proof is the contrast function Ψ_{super} [3] which is defined on the unit sphere $\mathcal{S}$ of n-dimensional unit norm vectors:

$$\Psi_{super}(\vec{w}_i) = \frac{E\left((\vec{w}_i^T \vec{x})^4\right)}{4} \tag{3}$$

Assuming that all source signals $s_i(t), i = 1, \ldots, n$ are super-Gaussians, it will be shown that a gradient ascent of Ψ_{super} has its local maxima when the weight vectors $\vec{w}_i$ are the rows $\pm\vec{b}_k$ of the (orthogonal) separating matrix. The gradient ascent of Ψ_{super} yields:

$$\begin{aligned} \vec{w}_1(t+1) &= \vec{w}_1(t) + \alpha(t) \cdot \frac{\partial}{\partial \vec{w}_1} \Psi_{super} \\ &= \vec{w}_1(t) + \alpha(t) \cdot y_1^3(t) \vec{x}(t) \end{aligned} \tag{4}$$

This is exactly the learning rule for the first weight vector $\vec{w}_1$. In addition, the vector has to be normalized after each step in order to keep it on the

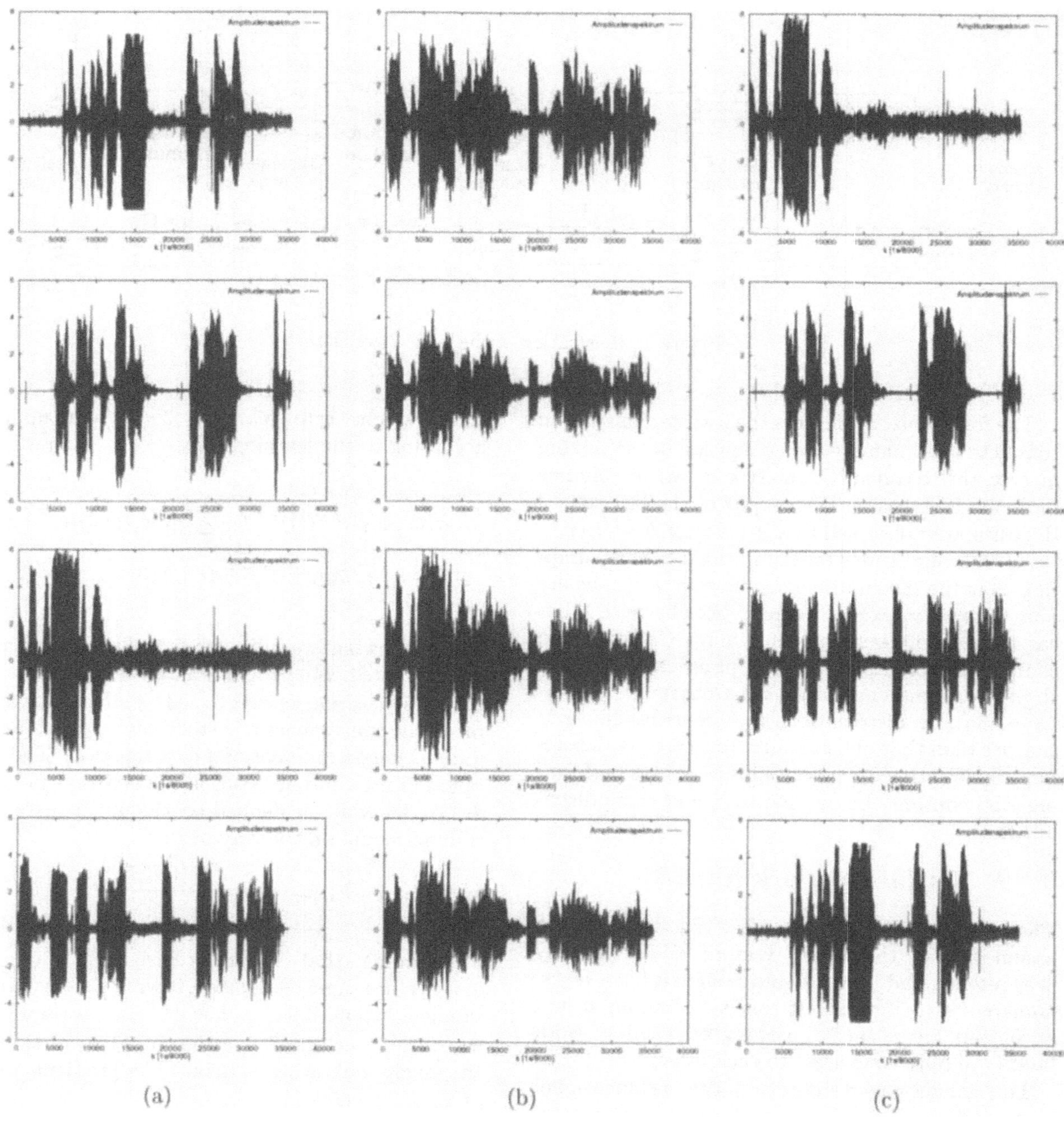

Figure 2: Source signals (a), linear mixtures (b), and separated signals (c).

unit sphere $\mathcal{S}$. Therefore, we can conclude that the first weight vector $\vec{w}_1$ converges to one of the vectors $\pm\vec{b}_{k_0}$ for some $k_0 \in \{1, \cdots, n\}$.

To find the other local maxima $\vec{b}_j, j \neq k_0$, we have to consider that the separation matrix $\vec{B}$ is orthogonal. This condition, combined with the maximization of Ψ_{super}, leads to the Lagrange function

$$L(\vec{w}_i, \lambda_{i1}, \ldots, \lambda_{i,i-1}) = \\ 1/4 \cdot E\left((\vec{w}_i^T \vec{x})^4\right) + \sum_{j=1}^{i-1} \vec{w}_i^T \vec{w}_j \, \lambda_{ij} \tag{5}$$

with λ_{ij} being the Lagrange multipliers.

The optimal values λ_{ij}^* are found by setting the gradient of (5) to zero

$$\lambda_{ij}^* = -E\left((\vec{w}_i^T \vec{x})^3 \cdot \vec{w}_j^T \vec{x}\right) = -E\left(y_i^3 \cdot y_j\right) \tag{6}$$

and the gradient ascent yields:

$$\tilde{\vec{w}}_i(t+1) = \vec{w}_i(t) + \\ \alpha(t)\,(\vec{w}_i(t)^T \vec{x}(t))^3 \left(\vec{I} - \textstyle\sum_{j=1}^{i-1} \vec{w}_j(t) \vec{w}_j(t)^T\right) \vec{x}(t)$$

which together with the required normalization is equivalent to the proposed learning rule (2).

4 Experimental Results

The network has been implemented in *C* on a DEC Alpha workstation under Digital UNIX.

In the BSS example presented in the following, 4 independent sound signals, sampled at a rate of 8 kHz were used (see Figure 2, column (a)). The four signals represent excerpts of four historical speeches given by the following speakers: (1) W. Churchill, (2) J. F. Kennedy, (3) M. Luther-King, and (4) N. Armstrong. They are stationary and have super-Gaussian distributions.

The 4-dimensional input vectors for the neural network, which represent 35 392 samples of the linear mixtures, were constructed by the orthogonal mixing matrix $\vec{M}$ shown in Figure 3. The mixtures are shown in column (b) of Figure 2.

In the training mode, the set of 35392 input vectors was presented 3 times to the network. The weights of the connections were initially set to random values taken from the interval [-1, 1]. An initial learning rate $\alpha(0) = 0.95$ (see Equation (2)) has been used and successively decreased after each simulation step according to $\alpha(1000) = 0.004 \cdot \alpha(0)$.

$$\begin{pmatrix} -0.246559 & -0.311630 & -0.436252 & 0.898941 \\ -0.199433 & -0.377810 & -0.385189 & -0.416525 \\ 0.426000 & -0.377810 & 0.813212 & 0.412838 \\ -0.228142 & -0.309001 & 0.463215 & 0.354782 \end{pmatrix}$$

Figure 3: Mixing matrix $\vec{M}$.

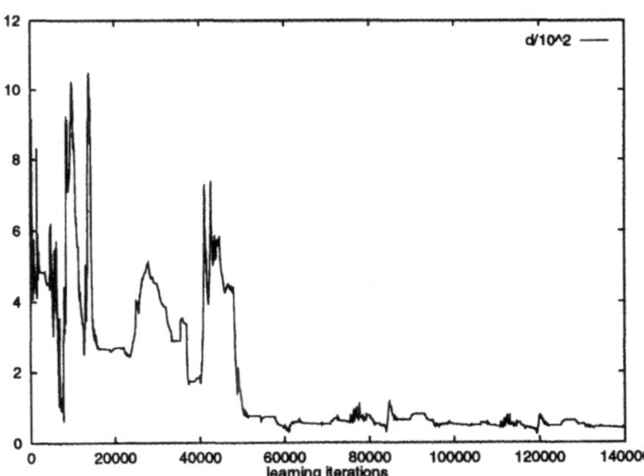

Figure 4: Convergence behaviour.

The convergence behaviour is illustrated in Figure 4, where the x-axis indicates the number of simulation steps (a complete weight update for a single input vector), and the y-axis indicates the error d resulting after each simulation step. The error d represents the distance of the weight matrix to an idealized separation matrix with no difference between the separated and the original signals; it has been introduced in [1].

After the weight vectors have converged to the columns of the separating matrix, the network is capable of separating the original source signals from the mixtures used as the input, as shown in column (c) of Figure 2. The acoustic quality of the separated signals is quite good. It is impossible for humans to distinguish them from the original signals.

5 Conclusions

In this paper we have presented a self-organizing neural network which performs blind acoustic signal separation. Assuming that the input vectors are prewhitened and the source signals are all ***super-Gaussians***, it was shown that the weight vectors of the network converged to the rows of the separating matrix.

There are several areas for future research, such as investigating the suitability of using other network architectures or nonlinear unsupervised learning rules for performing BSS, and finding methods for separating signals with both sub-Gaussian and super-Gaussian distributions.

References

[1] P. Comon. Independent Component Analysis – A New Concept. *Signal Processing*, 36:287-314, 1994.

[2] P. Comon, C. Jutten, J.Herault. Blind Separation of Sources, Part II: Problem Statements. *Signal Processing*, 24:11-20, 1991.

[3] N. Delfosse, P. Loubaton. Adaptive Blind Separation of Independent Sources: A Deflation Approach. *Signal Processing*, 45:59-83, 1995.

[4] J. Karhunen. Neural Approaches to Independent Component Analysis and Source Separation. *Proc. of the 4th European Symp. on Artificial Neural Networks*, Bruges, Belgium, 1996.

[5] J. Karhunen, E. Oja, L. Wang, R. Vigario, J. Joutsensalo. A Class of Neural Networks for Independent Component Analysis. Report A 28, Helsinki University, 1995.

Neural Networks for Higher-Order Spectral Estimation

F.-L. Luo and R. Unbehauen
Lehrstuhl für Allgemeine und Theoretische Elektrotechnik
Universität Erlangen-Nürnberg
Cauerstraße 7, 91058 Erlangen, Germany

Abstract

This paper deals with neural network approaches for higher order spectral estimation. The emphasis is put on how to use analog neural networks to perform in real-time major computations required in the ARMA model based bispectral estimation and the fourth order cumulant based Pisarenko's harmonic method. The proposed approaches are useful for the real-time signal processing with higher order spectral estimation.

1 Introduction

The use of higher order spectra is becoming increasingly widespread. Higher order spectra are defined in terms of higher order cumulants. Particular cases of higher order spectra are the third order spectrum (also called bispectrum) which is the Fourier transform of the third order cumulants, and the trispectrum (the fourth order spectrum) which is the Fourier transform of the fourth order cumulants.

A large number of theories and applications of higher order spectral estimation have been published [5]. Similar to power spectrum estimation techniques, higher order spectral estimation techniques have split into two camps: nonparametric and parametric methods. Because the nonparametric higher order spectral methods are subject to the same problems that plague nonparametric power spectrum methods, that is, high variances and low resolution, the current emphasis is on the parametric higher order spectral methods. Parametric higher order spectral methods first estimate the parameters of an underlying data generating model from the sampled data and then compute the higher order spectrum by use of the model and the estimated parameters.

There are two principal factors for the parametric higher order spectral estimation methods, one is the estimation performance (estimation variances and resolution), the other is the computational complexity. Based on the available power spectral estimation techniques, a number of high resolution higher order spectral estimation techniques have been reported. The ARMA model based method and Pisarenko's harmonic method are two examples. However, these methods are in general computationally intensive, as a result, it is very difficult to implement these algorithms in practical applications, especially, in real-time applications. To attack this problem, this paper proposes neural network approaches for higher order spectral estimation. We will concentrate on the ARMA model based bispectral estimation method and Pisarenko's harmonic method. With the asynchronous parallel and distributed processing, continuous time dynamics, global interconnection of elements, and high speed computational capability, the proposed neural networks can perform in real-time the major computations required in the ARMA model based bispectral estimation method and Pisarenko's harmonic method. The solution provided by these neural networks can approximate the exact solution with arbitrarily small error by appropriately selecting parameters of the networks.

2 ARMA Model Based Bispectral Estimation Method

Consider a zero-mean real stationary signal $x(n)$, its second, third and fourth order cumulants are given as follows [5].

(1) The second order cumulants are

$$C_2(m) = E[x(n+m)x(n)] \tag{1}$$

(2) The third order cumulants are

$$C_3(m,k) = E[x(n+m)x(n+k)x(n)] \tag{2}$$

(3) The fourth order cumulants are

$$\begin{aligned} C_4(m,k,l) &= E[x(n+m)x(n+k)x(n+l)x(n)] \\ &\quad -C_2(m)C_2(k-l) - C_2(k)C_2(l-m) \\ &\quad -C_2(l)C_2(m-k) \end{aligned} \tag{3}$$

The bispectrum $s_{2,x}(f_1, f_2)$ and trispectrum $s_{3,x}(f_1, f_2, f_3)$ of $x(n)$ are defined as the Fourier transform of the third order cumulants and fourth order cumulants, respectively, that is,

$$s_{2,x}(f_1, f_2) = \sum_{m=-\infty}^{+\infty} \sum_{k=-\infty}^{+\infty} C_3(m,k) e^{-j2\pi m f_1 - j2\pi k f_2} \quad (4)$$

$$s_{3,x}(f_1, f_2, f_3) = \sum_{m=-\infty}^{+\infty} \sum_{k=-\infty}^{+\infty} \sum_{l=-\infty}^{+\infty} C_4(m,k,l) e^{-j2\pi m f_1 - j2\pi k f_2 - j2\pi l f_3} \quad (5)$$

Based on the ARMA model [1, 2], $x(n)$ can be written as

$$x(n) = -\sum_{i=1}^{p} a_i x(n-i) + \sum_{i=0}^{q} d_i \epsilon(n-i) \quad (6)$$

where $\epsilon(n)$ is a non-Gaussian, independent identically distributed (i.i.d.) sequence with zero-mean, $E[\epsilon(n)\epsilon(n+k)] = \sigma^2\delta(k)$ and $E[\epsilon(n)\epsilon(n+k)\epsilon(n+m)] = \gamma\delta(m,k)$. In addition, we assume that the above ARMA model is causal, nonminimum phase and free of pole-zero cancellations. Using this model, the bispectrum can be written in the form

$$s_{2,x}(f_1, f_2) = \gamma \frac{\sum_{m=-q}^{q} \sum_{k=-q}^{q} \beta(m,k) e^{-j2\pi m f_1 - j2\pi k f_2}}{\sum_{m=-p}^{p} \sum_{k=-p}^{p} \alpha(m,k) e^{-j2\pi m f_1 - j2\pi k f_2}} \quad (7)$$

where $\beta(m,k)$ and $\alpha(m,k)$ are related to the ARMA parameters via

$$\beta(m,k) = \sum_{i=0}^{q} d_i d_{m+i} d_{k+i} \quad (8)$$

$$\alpha(m,k) = \sum_{i=0}^{p} a_i a_{m+i} a_{k+i} \quad (9)$$

According to [1], we have,

$$\gamma\beta(m,k) = \sum_{i=-p}^{p} \sum_{l=-p}^{p} \alpha(i,l) C_3(m-i, k-l) \quad (10)$$

$$(-q \le m, k \le q)$$

From (8)-(10), we know that the bispectrum can be obtained by determining the coefficients a_i (for $i = 1, 2, \ldots, p$). It has been shown [6, 8] that the coefficients a_i satisfy the linear equations

$$Ca = C_1 \quad (11)$$

where C is a matrix consisting of $C_3(m,k)$ and

$$a = \begin{pmatrix} a_p \\ a_{p-1} \\ \vdots \\ a_1 \end{pmatrix}, \quad C_1 = \begin{pmatrix} C_3(q+1, q-p) \\ C_3(q+1, q-p+1) \\ \vdots \\ C_3(q+1, q) \\ C_3(q+1, q-1) \\ \vdots \\ C_3(q+p, q-p) \\ C_3(q+p, q-p+1) \\ \vdots \\ C_3(q+p, q) \end{pmatrix} \quad (12)$$

However, it is very difficult to solve (11) in real-time because it involves a matrix inversion and a matrix multiplication. In the next section, we will give a neural network approach for solving (11) in real-time.

3 A Neural Network Approach

A neural network for solving (11) is presented in Figure 1 which is derived from the linear programming neural network model [7].

The input-output relationships of the neurons in the left-hand and right-hand parts of this network are denoted by $g(u)$ and $f(u)$, respectively. The left-hand part has N neurons and the right-hand part has M neurons. R_i and C_i (for $i = 1, 2, \ldots, N$) are the input resistance and capacitance of the i'th neuron in the left-hand part, respectively (for mathematical convenience, here we let $R_i = R$ and $C_i = C$). $T = \{T_{ji}\}$ (for $j = 1, 2, \ldots, M; i = 1, 2, \ldots, N$) is the connection strength matrix. $B = [b_1, b_2, \ldots, b_M]^T$ is the bias current vector of the right-hand part, $v_i(t)$ and $q_j(t)$ (for $i = 1, 2, \ldots, N; j = 1, 2, \ldots, M$) are the neuron outputs of the left-hand and right-hand parts of the network, respectively. $u_i(t)$ (for $i = 1, 2, \ldots, N$) is the neuron input of the left-hand part. In terms of

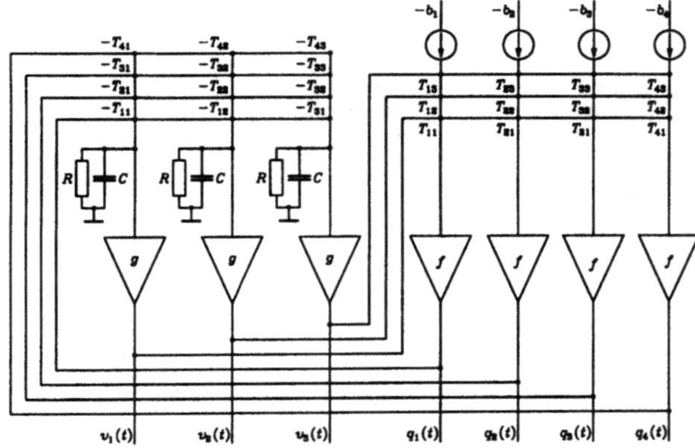

Figure 1: A neural network for solving (11).

Kirchhoff's laws, we have the following relationship

$$C\frac{du_i(t)}{dt} = -\sum_{j=1}^{M} T_{ji} f(\sum_{k=1}^{N} T_{jk} v_k(t) - b_j) - \frac{u_i(t)}{R} \tag{13}$$

For this neural network, we define an energy function

$$\begin{aligned} E(t) = & \sum_{j=1}^{M} F(\sum_{i=1}^{N} T_{ji} v_i(t) - b_j) \\ & + \sum_{i=1}^{N} \frac{1}{R} \int_0^{v_i(t)} g^{-1}(v) dv \end{aligned} \tag{14}$$

where $F(\cdot)$ is an indefinite integral of the function $f(u)$. The following theorem describes the dynamics and the stability of this neural network.

Theorem 1. *If $E(t)$ is bounded from below and $g(u)$ is a monotonically increasing function, this neural network is asymptotically stable.*

The proof of this theorem can be referred to [4].

In order to use the proposed neural network to solve (11), we select

(1) T=C and B=C_1, that is, the available matrix C and vector C_1 are taken directly as the connection strength matrix and bias current vector of the network, respectively.

(2) $f(u) = K_1 u$ (u has the unit of current, K_1 has the unit of resistance) and $g(u) = K_2 u$ (u has the unit of voltage, K_2 is the voltage gain).

Under these conditions, (14) becomes

$$E(t) = \frac{1}{2} K_1 \parallel C_1 - C^T V(t) \parallel^2 + \frac{1}{2RK_2} \parallel V(t) \parallel^2 \tag{15}$$

where $V(t) = [v_1(t), v_2(t), \dots, v_N(t)]^T$ is the output vector of the left-hand part of the network.

Equation (15) shows that $E(t)$ is non-negative (bounded from below). Together with the fact that $g(u)$ is a linear function with positive gain, we know that such a constructed neural network satisfies the sufficient conditions of the above theorem. Hence, this neural network is stable and will provide the solution

$$a_n = (\frac{1}{RK_1K_2} I + C^T C)^{-1} C^T C_1 \tag{16}$$

which is the best approximation to the exact solution of (11) in the sense of the Euclidean norm (that is, under the LS criterion). The time required by this neural network to provide the solution is within a few characteristic time constants of the network. The error between a_n and the exact coefficient vector a determined by (11) can be made arbitrarily small by appropriately selecting other parameters such as R, K_1 and K_2 of the network.

In summary, the above bispectral estimation approach based on the ARMA model and neural networks includes mainly the following three steps:

(1) Estimation of the third order cumulants from the available samples of $x(n)$. In terms of the definition (2), $C_3(m,k)$ is usually estimated by

$$C_3(m,k) = \frac{1}{L-I} \sum_{n=1}^{L-I} x(n)x(n+k)x(n+m) \tag{17}$$

where $I = \max\{\mid m \mid, \mid k \mid\}$ and L is the number of samples.

(2) Using the proposed neural network to solve (11) so as to obtain the coefficients a_i (for $i = 1, 2, \dots, p$).

(3) Computation of the bispectrum from the available coefficients a_i.

The above shows that in this approach the ARMA model and the neural networks are, respectively, used to provide a high resolution and real-time performance.

4 Fourth Order Cumulant Based Harmonic Retrieval and its Neural Network Approach

For the harmonic retrieval problem dealt with in [3, 4], the methods based on higher order cumulants are much more effective than that based on the second order cumulants in the additive colored noise case [5]. In this case, the received signal can be represented as

$$x(n) = \sum_{i=1}^{P} \alpha_i cos(\omega_i n + \theta_i) + \epsilon(n) \tag{18}$$

$$(n = 1, 2, \ldots, L)$$

where $x(n)$ and $\epsilon(n)$ are the measured samples and noise samples, respectively. α_i, ω_i and θ_i denote the amplitude, frequency (normalized) and initial phase (uniformly distributed over $[0, 2\pi]$) of the i'th sinusoid. $\epsilon(n)$ is zero mean and has variance σ^2. $\epsilon(n)$ is assumed to be colored Gaussian noise.

With this in mind, we can prove that all the third order cumulants are identically zero and the fourth order cumulant is

$$C_4(m,k,l) = -\frac{1}{8}\sum_{i=1}^{P} \alpha_i^4 (cos(\omega_i(m-k-l)) + cos(\omega_i(k-m-l)) + cos(\omega_i(l-k-m)))$$

If we let $m = k = l$, then we have

$$C_4(k) = -\frac{3}{8}\sum_{i=1}^{P} \alpha_i^4 cos(\omega_i k) \tag{19}$$

Comparing (19) with (2) of [3], we know that Pisarenko's method can immediately be applied just by replacing the autocorrelation function $R(k)$ with the fourth order cumulants $C_4(k)$. Consequently, the neural network approaches proposed in [3] can be used to provide the eigenvector corresponding to the smallest eigenvalue of the fourth order cumulant matrix C_4,

$$C_4 = \begin{pmatrix} C_4(0) & C_4(1) & \cdots & C_4(N-1) \\ C_4(-1) & C_4(0) & \cdots & C_4(N-2) \\ \vdots & \vdots & \ddots & \vdots \\ C_4(-N+1) & C_4(-N+2) & \cdots & C_4(0) \end{pmatrix} \tag{20}$$

Note that the fourth order cumulant matrix C_4 is positive semidefinite (for $N \geq 2P + 1$), that is, the smallest eigenvalue is zero. As a result, the connection strength matrix of the neural network proposed in [3] should be $C_4 + KI$ instead of C_4, where K is a positive constant.

We can summarize the above approach as follows.

1. Estimate the fourth order cumulant matrix C_4 (equation (20)) of the size $N \times N$ from the measured samples.

2. Use the neural network proposed in [3] to compute the eigenvector corresponding to the smallest eigenvalue of the estimated matrix C_4. All the parameters of the network are selected as proposed in [3] except that the connection matrix is taken as $C_4 + KI$.

3. Compute the roots of the polynomial formed by the elements of the above eigenvector. This polynomial will have $2P$ roots located at $exp(\pm j\omega_i)$ (for $i = 1, 2, \ldots, P$).

Finally, it should be noted that neural networks, which can be used to provide in real-time not only the desired eigenvector but also the roots of the related polynomial, are being developed.

5 Conclusions

We have proposed neural network approaches for the ARMA model based bispectral estimation and fourth order cumulant based harmonic retrieval. It has been shown that the proposed neural networks are stable and can perform the major computations required in these higher order spectral estimation methods in real-time. The error of the solution provided by the proposed neural networks from the exact solution can be made arbitrarily small by appropriately selecting parameters of networks. All of these demonstrate that the proposed neural networks are very suitable for real-time applications of high order spectral estimation. With real-time radar clutter classification as an application example, we have made extensive simulations and these simulation results have demonstrated the accuracy of the above analyses and effectiveness of the proposed neural network methods. Concerning this application example and related simulation analyses, we will report in another paper.

6 Acknowledgements

This paper is supported by the German Research Society (DFG).

References

[1] G. B. Giannakis. On the identifiability of non-Gaussian ARMA models using cumulants. *IEEE Trans. on Automatic Control*, 35(1):18–26, 1990.

[2] G. B. Giannakis and A. Swami. On estimating non-causal nonminimum phase ARMA models of non-Gaussian processes. *IEEE Trans. on Acoustic Speech and Signal Processing*, 38(3):478–495, 1990.

[3] F. L. Luo and R. Unbehauen. Neural network approach to Pisarenko's frequency estimation. In *Proceedings of ICANNGA'95*, pages 261–264, 1995.

[4] F. L. Luo and R. Unbehauen. *Applied Neural Networks for Signal Processing.* Cambridge University Press, New York, 1997.

[5] C. L. Nikias and J. M. Mendel. Signal processing with higher-order spectral. *IEEE Signal Processing Magazine*, 10:10–37, July 1993.

[6] A. Swami and J. M. Mendel. ARMA parameter estimation using only output cumulants. *IEEE Trans. on Acoustic Speech and Signal Processing*, 38(7):1257–1265, 1990.

[7] D. W. Tank and J. J. Hopfield. Simple 'neural' optimization networks: A/D converter, signal decision circuit, and linear programming circuit. *IEEE Trans. on Circuits and Systems*, 33:533–541, 1986.

[8] X. D. Zhang and Y. L. Zhou. A novel recursive approach to estimation MA parameters of casual ARMA models from cumulants. *IEEE Trans. on Signal Processing*, 40(1):2870–2873, 1992.

Estimation of Fractal Signals by Wavelets and GAs

H. Cai and Y. Li
Dept. of Automation, Tsinghua University,
Beijing 100084, P.R.C

Abstract

The $1/f$ family of fractal signals constitutes an important class in signal processing. In many applications a precise estimation of the fractal parameter is needed, which is not easy in many environments. This paper is a study on estimating the fractal parameter, i.e. Hurst index, under the wavelet domain and using genetic algorithms. The results show that this method is robust and converges quickly. Due to the intrinsically parallel nature of genetic algorithms, this method is especially useful in real time applications, e.g. estimation of parameter in self-similar traffic in ATM networks.

1 Introduction

The $1/f$ family of stochastic processes constitutes an important class of fractal signal models for a variety of data (for which they are inherently well suited [3]). These intrinsically scale-invariant processes have a number of interesting characteristics, among which is a much more persistent long term correlation structure that is usually used in ARMA processes. Naturally, $1/f$ processes are well used for modelling a large number of phenomena, such as geophysical and economic time series, biological signals, noise in electronic devices, frequency variation in music, and incoming packets in local area traffic (LAN) [4]. Following Mandelbrot [6], a considerable body of literature is devoted to understanding both the physical origins and the ubiquity of behavior in real data.

There are many ways of estimating the fractal parameter of $1/f$ processes, such as the R/S method, the variance-time method, etc. Our method is based on Wornell's work [8] on orthonormal wavelet basis expansion for the $1/f$ process in terms of a collection of uncorrelated variables. In Wornell's pioneering work it had been shown that estimation of fractal parameters in wavelet domain has the good quality of robustness and precision. But the maximum likelihood (ML) method used there spent quite a lot of time to converge to the real parameter, and this limits its use in real time applications. In this paper we use genetic algorithms to estimate the fractal parameters, which can converge to the real parameters in a few iterations, much less than that used in the ML method.

2 $1/f$ Signal

The $1/f$ processes are generally defined [3] as processes whose power spectra are of the form

$$S(\omega) \propto \frac{\sigma_x^2}{|\omega|^{2H+1}} \tag{1}$$

where H is the so called Hurst index; an important parameter in fractal processes. It is generally convenient to extend the notion of $1/f$ processes to include nearly $1/f$ processes that are defined [9] as having power spectra bounded according to

$$\frac{k_1}{|\omega|^{2H+1}} \leq S(\omega) \leq \frac{k_2}{|\omega|^{2H+1}} \tag{2}$$

where k_1 and k_2 satisfy $0 < k_1 \leq k_2 < \infty$ but are otherwise arbitrary. These processes exhibit only a constant percentage deviation from standard $1/f$ processes and retain the characteristics of long-term dependence associated with $1/f$ processes.

3 Wavelet Transform of $1/f$ Signal

In this section, we review the orthonormal wavelet theory needed in this paper. For more detailed discussion of orthogonal wavelet theory, refer to [1, 5].

3.1 Basic Theory of the Wavelet Transform

An orthonormal wavelet transform of a signal $x(t)$.

$$x(t) \leftrightarrow x_n^m \tag{3}$$

is defined through the synthesis/analysis equations

$$x(t) = \sum\sum x_n^m \Psi_n^m(t) \tag{4}$$

$$x_n^m = \int x(t)\Psi_n^m(t)dt \qquad (5)$$

and has the property that all basis functions are dilations and translations of a single function [1]:

$$\Psi_n^m(t) = 2^{m/2}\Psi(2^m t - n) \qquad (6)$$

where m and n are the dilation and translation indices respectively. For more detail refer to Mallat's algorithm [5].

3.2 Wavelet Representation of $1/f$ Processes

The work [9] constructs a class of nearly $1/f$ processes using wavelet expansions in terms of uncorrelated transform coefficients having the variance progression

$$\text{var } x_n^m = \sigma_n^2 2^{-\gamma n} \qquad (7)$$

where γ is the exponent of the nearly $1/f$ spectrum. This unique property of $1/f$ processes in wavelet transform domain is what we use in the following section.

3.3 The Parameter Estimation Problem

Let us suppose that we have observations of a zero-mean with zero-mean additive Gaussian noise $w(t)$ that is statistically independent of $x(t)$, so,

$$r(t) = x(t) + w(t), \quad -\infty < t < \infty \qquad (8)$$

Suppose we have got the wavelet coefficients of $r(t)$ by DWT, we have

$$r_n^m = \int_{-\infty}^{\infty} \Psi_n^m(t)r(t)dt. \qquad (9)$$

Exploiting the wavelet decomposition's role as a whitening filter for the $1/f$ process, and using the fact that the w_n^m are independent of the $x_n^m$and are decorrelated for any wavelet basis, the resulting observation coefficients

$$r_n^m = x_n^m + w_n^m \qquad (10)$$

can be modelled as mutually independent, zero-mean Gaussian random variables with variance

$$\text{var } r_n^m = \sigma_m^2\beta^{-m} + \sigma_w^2, \qquad (11)$$

where we have defined

$$\beta = 2^\gamma. \qquad (12)$$

It is the parameter set

$$\Theta = (\beta, \sigma^2, \sigma_w^2) \qquad (13)$$

we want to estimate! Wornell *et al.*[9] used an estimate-maximize (EM) algorithm to estimated these parameters, we plot a trajectory of the evolution of the parameters in Figure 1. From this we can see that the method is slow to converge. This paper will present a genetic algorithm based parameter method, which is both robust and efficient. Next we review the basic structure of genetic algorithms.

4 Genetic Algorithms

A genetic algorithm (GA) is a parallel global search technique that emulates natural genetic operators. Because it simultaneously evaluates many possible (high fitness) points in parameter space, it is likely to reach the global minimum (or maximum). Genetic algorithms (GAs) have been applied to a large number of applications, like function optimization, search of parameters, machine learning, etc. [2] . In GAs, the parameters that must be optimized are often coded as a binary string (a string of length L composed of '0' or '1'). Information is exchanged between individuals through 'schemata' in the hope that more 'fit' offsprings are produced. Next is the general structure of GAs:

```
Initialize population
Calculate the fitness function of each individual
Selection
REPEAT
 Crossover
 Mutation
 Calculate the fitness function of each individual
 Selection
UNTIL (termination condition satisfied)
```

The *selection* operator is based on the principle of 'survival of the fittest'. A specific individual having a fitness value above the average level will have more chance of being selected than those individuals having fitness value below the average level.

The *crossover* operator is used to produce new offspring from their parents, at the mean time by exchanging the information between them. Selection and crossover direct the search toward the better areas of the parameter space according to current gained knowledge.

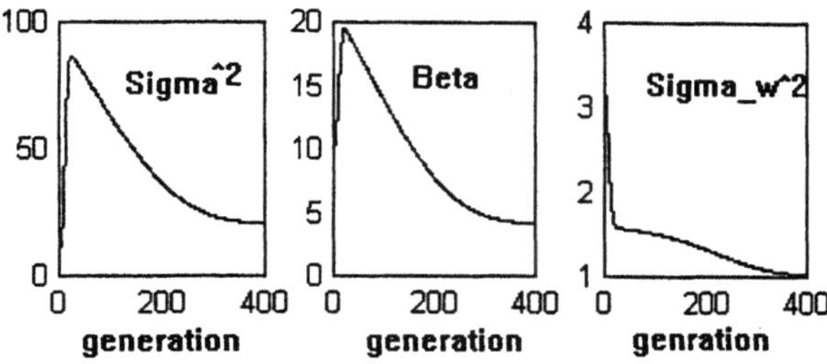

Figure 1: Iteration of method in [9].

The *mutation* operator is used to bring about new information at the bit level, so that the GA can search new areas otherwise not accessible when searched using only selection and crossover. The three basic operators are implemented iteratively until the algorithm converges to a global extrema.

5 Estimation of the Hurst Index by Genetic Algorithms

Our problem of estimating fractal signals can be formalized as follows:

Given a series of sample data $V_m = \text{var}\, r_n^m$, try to minimize $\sum_{m=1}^{M}(V_m - \widehat{\sigma}_m^2)$, where $\widehat{\sigma}_m^2 = \widehat{\sigma}^2\widehat{\beta}^{-m} + \widehat{\sigma}_w^2$. The parameter set $\widehat{\Theta} = (\widehat{\beta}, \widehat{\sigma}^2, \widehat{\sigma}_w^2)$ is the estimated value of $\Theta = (\beta, \sigma^2, \sigma_w^2)$. We encode the three parameters $\widehat{\Theta} = (\widehat{\beta}, \widehat{\sigma}^2, \widehat{\sigma}_w^2)$we want to estimate with 15 bits each, so each string is 45 bits long. We use modified elitist strategy here; that is the best individual in every population has the probability '1' to survive to the next population. [7] has proved that GAs that always maintain the best solution in the population converge to the global optimum. The parameters of the GA are: $P_c = 0.99$, $P_m = 0.005$; we plot the process of parameter estimation using GA in Figure 2.

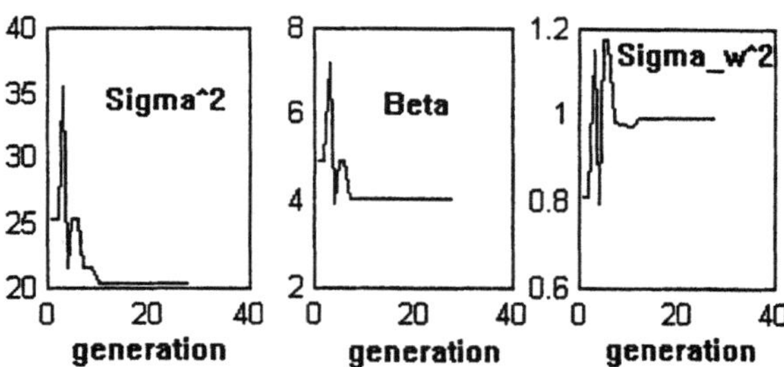

Figure 2: Iteration of our new method.

6 Conclusion

In this paper, we proposed a new method of estimating fractal signals, using the variance coefficients in wavelet domains and genetic algorithms. This is a function optimization problem. We show that since the function is multimodal, GAs are well suited to this problem. Because of the implicit parallel nature of GAs and the possibility of implementing GAs in parallel with VLSI, this method is more applicable to real time applications than the original EM method.

However, we think that there is more work to be done in this area. For example, the proper selection of GAs will have considerable effect on the convergence of parameter estimation processes, including our problem. The proper choice of GA parameters will depend both on empirical study through more experiments and on the deep understanding of GA mechanisms.

References

[1] I. Daubechies. Orthonormal bases of compactly supported wavelets. *Commn. Pure Appl. Math.*, 41:909–996, November 1988.

[2] D. E. Goldberg. *Genetic Algorithms in Search, Optimization and Machine Learning.* Addision-Wesley, Reading, MA, 1980.

[3] M. S. Kesher. 1/f noise. *Proc. IEEE*, 70:212–218, March 1982.

[4] W. E. Leland, M. S. Taqqu, W. Willinger, and D. V. Wilson. On the self-similar nature of ethernet traffic. *IEEE/ACM T. Networking*, 2(1):1–15, 1994.

[5] S. G. Mallat. A theory for multiresolution signal decomposition: The wavelet representation. *IEEE T. PAMI*, 11:674–693, July 1989.

[6] B. Mandelbrot. Some noises with 1/f spectrum, a bridge between direct current and white noise. *IEEE T. IT*, 13:289–298, April 1967.

[7] G. Rudolph. Convergence analysis of canonical genetic algorithms. *IEEE T. NN*, 5(1), January 1994.

[8] G. W. Wornell. A karhunen-loeve-like expansion for 1/f processes via wavelets. *IEEE T. IT*, 36:859–861, July 1989.

[9] G. W. Wornell and A. V. Oppenheim. Estimation of fractal signals from noisy measurements using wavelets. *IEEE T. SP*, 40(3), March 1992.

Classification of 3-D Dendritic Spines using Self-Organizing Maps

G. Sommerkorn[1], U. Seiffert[1], D. Surmeli[1], A. Herzog[1],
B. Michaelis[1] and K. Braun[2]

[1] Institute for Measurement Technology and Electronics, Otto-von-Guericke University of Magdeburg,
PO Box 4120, D-39016 Magdeburg, Germany

[2] Federal Institute for Neurobiology Magdeburg, Brenneckestrasse 6, D-39118 Magdeburg, Germany

Email: andreas.herzog@e-technik.uni-magdeberg.de

Abstract

This work in progress shows a method for classifying dendritic spines by their shape. Focal points are the extraction of features from three-dimensional spine data and the following classification of the spines. Hence there will be only little reflection of biological aspects of this problem. Feature extraction based on moments and spherical coordinates will be discussed. Furthermore, this paper shows and describes a modified kind of self organizing maps (SOM)), which is used for the classification of the dendritic spines.

1 Introduction

Biological examination often depends strongly on image processing and image analysing methods. The same is true for research concerning dendritic spines in the brain. Different forms of learning processes, such as filial imprinting, are accompanied by changes in the density of spines. Domestic chicks that have been imprinted to acoustic stimuli show reduced densities of spine synapses on a distinct neuron type [8]. Preliminary qualitative examination revealed that spine synapses show not only changes in number but also in shape and size. Theoretical considerations implicate that such changes result in changes of efficacy of synaptic transmission [5].

Thus, the quantitative and objective measurement of these parameters, as well as the classification of spine types gives information about changes of network properties in relation to learning and memory formation.

With the aid of a confocal laser scanning microscope operating at the resolution limit (mean spine-size: $0.5-2.0\mu$m long and $0.1-0.5\mu$m diameter) the three-dimensional images are recorded slice by slice [2]. These data contain both: spines and dendrites (Figure 1). A pre-processing phase repairs the noisy and degraded images and separates the spines from the dendrites [4].

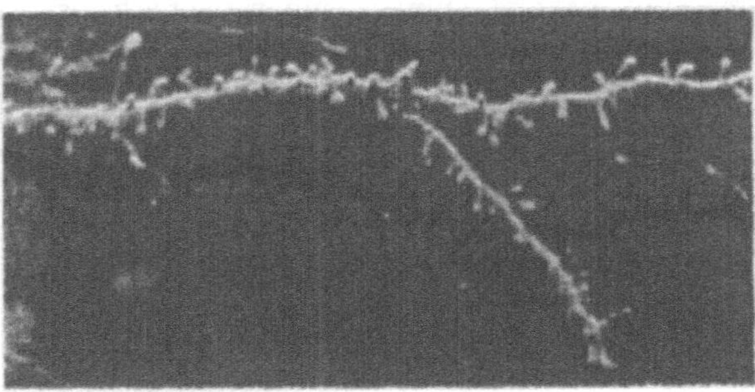

Figure 1: Dendrite with spines after image reconstruction.

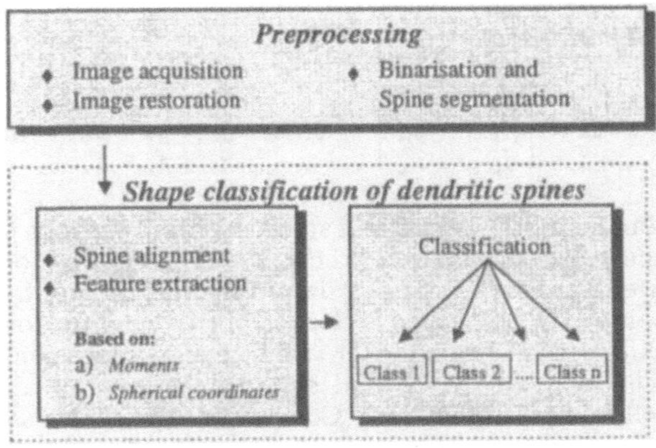

Figure 2: Basic idea.

After this pre-processing the binarised spines are available as a three-dimensional point cloud (sample voxel). Some of the voxels are indicated to be connected to the dendrite (spine foot). This is the starting point for the present paper. The information about the spines has to be sufficient for the desired shape-classification. In Figure 2 the basic idea for the realised spine classification is shown. In this paper suitable image processing classification methods for the description of morphological changes of spines are analysed. Forthcoming papers will consider the concrete biologically relevant results.

2 Feature Extraction

For an accurate classification we have to find good shape-descriptors as extracted features. The first step is to make all spines comparable with respect to their position. All spines are brought into the same position, it follows that the calculated features become independent from translation and rotation of the raw data.
The principal component analysis (PCA) ascertains the directions $\vec{e}$ (eigenvectors) of the most significant variances (eigenvalues λ):

$$\mathbf{C}\vec{e} = \lambda\vec{e} \tag{1}$$

where

$$\mathbf{C} = \mathbf{D}\mathbf{D}^{\mathrm{T}} \tag{2}$$

is the covariance-matrix, with the three-dimensional centred spine-data (n voxel)

$$\mathbf{D} = \begin{bmatrix} \tilde{x}_1 & \cdots & \tilde{x}_n \\ \tilde{y}_1 & \cdots & \tilde{y}_n \\ \tilde{z}_1 & \cdots & \tilde{z}_n \end{bmatrix} \quad \mathit{with} \quad \begin{array}{l} \tilde{x}_i = x_i - \overline{x} \\ \tilde{y}_i = y_i - \overline{y} \\ \tilde{z}_i = z_i - \overline{z} \end{array} \tag{3}$$

Furthermore, we define a spine axis from the foot to the centre of gravity of the spine. The eigenvectors are re-ordered corresponding to the direction of the spine axis (spine axis is close to positive z-direction, x-dimension has a higher variance as y-dimension). The next step is to find suitable features for the following classification. Conventionally, one would start out thinking about obvious features such as length, different radii etc. However, it has been found that limiting the description to those kinds of features, imposes undesirable restrictions on the variability of describable shapes of spines. Due to this and the sensitivity of the voxel data to noise corruption, generic integrating shape descriptors are needed, like the two variants, which are considered in the following.

2.1 Moment Based Feature Extraction

In several papers (e.g. [1, 7]) geometric moments were used for pattern recognition in a plane. Analogous features were derived for spatial objects. Depending on their order, moments are assumed to be suitable shape descriptors, because their computation includes all pixels (voxels) for each moment. The following equation shows the general computation of three-dimensional moments:

$$m_{pqr} = \int_{-\infty}^{\infty}\int_{-\infty}^{\infty}\int_{-\infty}^{\infty} x^p y^q z^r f(x,y,z)\,dx\,dy\,dz \tag{4}$$

with $p, q, r = 0, 1, 2, \ldots$.
The order of a moment results from the sum of p, q and r. $f(x,y,z)$ describes the density function. The computation of the moments from the spines can be simplified immensely, because all spines consist of discrete binarised voxels, which means that spines have a constant density function. The so calculated moments have an enormous dynamic range. Therefore it is necessary to normalise them. This can be done easily by adapting the spines to a unit cube with the scale factor c resulting from the greatest length (other normalisation methods are possible and have been investigated) among the three dimensions.
Equation 4 can now be rewritten as:

$$M_{pqr} = \sum_{i=1}^{n} \left(\frac{x_i - \overline{x}}{c}\right)^p \left(\frac{y_i - \overline{y}}{c}\right)^q \left(\frac{z_i - \overline{z}}{c}\right)^r \tag{5}$$

with $p, q, r = 0, 1, 2, \ldots$.
These are finally the features for the moment based classification. The interpretation of all moments (especially of higher order) is very difficult. Moments of lower order describe important features as variance, skewness parameters and various symmetries. Therefore, only 29 low order moments are used for classification.

2.2 Spherical Coordinates Based Feature Extraction

Another variant consists of scanning the hull of each aligned spine in a specific sequence. Hence the raw data are used. First it is necessary to convert the cartesian coordinates into their spherical counterparts. After this the spine can be scanned by a tracing pointer along a fixed path, i.e. for each discrete angle in space the length of the furthest voxel

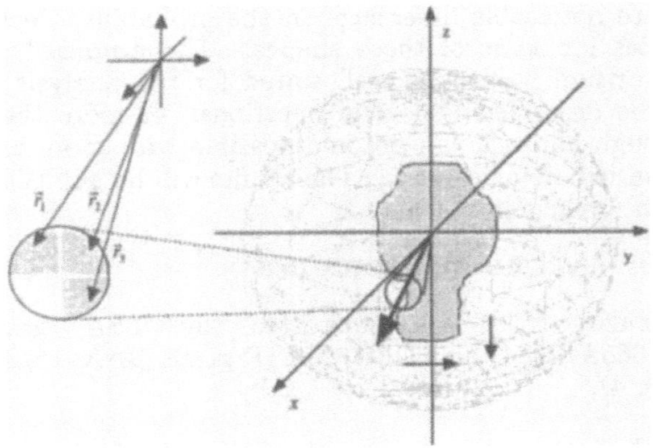

Figure 3: Scanning of the spine ($|\vec{r}_2| < |\vec{r}_1| < |\vec{r}_3| \Rightarrow |\vec{r}_3|$).

in a certain overlapping neighbourhood, here forming a pyramid, is recorded as a feature (see also Figure 3).

3 Implementation of the Classification System

Considering the task at hand, to find among the dendritic spines clusters of characteristic shape and track, prompts a concentration on unsupervised classifiers. This is due to the fact that neither the number nor the properties of appropriate classes are known in advance. As the result is to be implemented online, as few iterations as possible through the data should be required. There are, of course, examples of manual classification of spines by shape, which could be used as reference for the performance of a classifier, yet the classification should not be limited by previous human experience.

For greater flexibility in the approach to clustering a *SOM* has been investigated.

Self-Organizing Maps [6] provide unsupervised partitioning of the input data into several classes according to their distribution, and thus seem to be a predestined artificial neural network to meet the above requirements. In the original *SOM*-algorithm the size (number of neurons) of the Kohonen layer has to be defined in advance. This may be very difficult, especially if there is no further information about the current data set. The number of classes within a considered part of a dendrite is not known a-priori. If the size of the map is too small, the resolution of the classes is insufficient. On the other hand, if the size is huge, the net itself and the resulting number of classes are unnecessarily large. Considering biological plausibility and the noted inadequacy of spine data, parameters have been chosen such that reasonably few classes are formed.

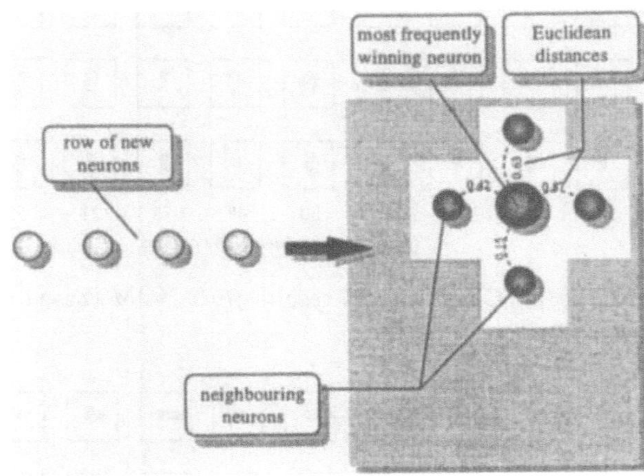

Figure 4: Example for the insertion of a row of neurons between the most frequently winning neuron and the neighbour having the closest weight vector to it in *SOM*.

A size-adaptive Kohonen layer would be a suitable solution to this problem. Based on an approach of Fritzke [3] a modified *SOM* has been implemented. The net grows from an initially small size of 2x2 neurons until a special stopping criterion depending on the mean Euclidean distance over all classes is met.

For *SOMs*, the results depend strongly on parameters such as the number of available neurons and a training regime for learning rate and neighbourhood.

4 Results

Both feature sets were used for the modified *SOM*. A first visual inspection supports the suitability of the feature sets for a description of the spine shapes, when compared to manual classification. The best representatives for all classes for all classifications are depicted in Figure 5 and Figure 6. In each column, a spine is shown as projection on the $x-y$, $x-z$ and $y-z$ planes, respectively. For both feature sets, the *SOM* established 9 classes after an initiali-

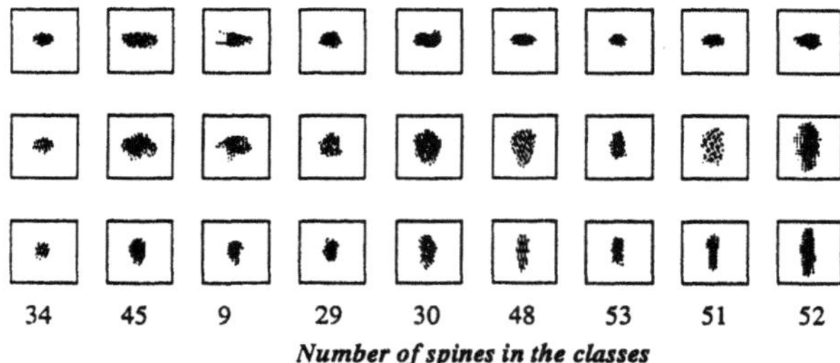

Figure 5: Classification results with SOM (based on moments).

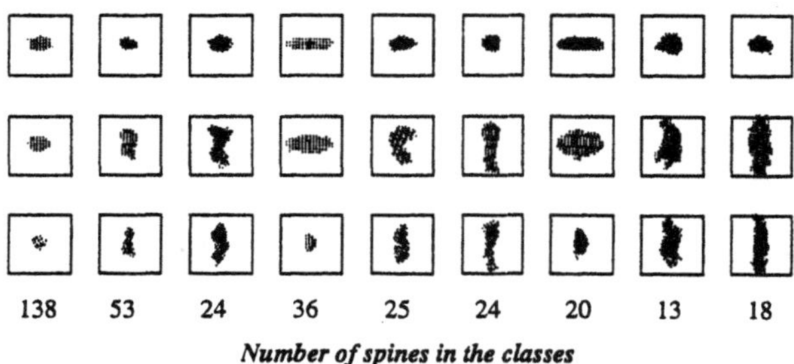

Figure 6: Classification results with SOM (based on spherical coordinates).

sation with 4 classes (neurons). The moment-based feature set comprised 29 geometric moments of up to ninth $(p+q+r)$ order, while there were up to 648 features (10° scanning) from spherical scanning. Scanning rates in the range between 10° and 45° have shown similar results.

The continuous transitions between the shapes of spines that could be expected from a biological subject has been confirmed. For the same feature set both classifications build classes around similar prototypes of shapes, that seem to occur more frequently than others.

5 Conclusion

Due to the problems interpreting the individual geometric moments, investigations are carried out to identify the most significant among the computed, which will both accelerate and sharpen the classification. Alternatively or in addition, the importance of individual features for the classification will be examined and their influence adjusted accordingly. A suitable algorithm for the subdivision of spines by their shape has been developed. The investigation indicates a similar range of spine shapes for socialised and isolated chickens. Nevertheless there are noticeable differences in the probability densities for some of those shapes. The proposed algorithm was found well suited for the analysis of the described biological questions. A more thorough analysis will become possible with more and varied biological data. The results will be published in forthcoming papers.

6 Acknowledgements

This work was supported by LSA grant (665A/2384) and DFG/BMBF grant (INK 15/TP A 4).

References

[1] R. B. Bailey and M. Srinath. Orthogonal moment features for use with parametric and non-parametric classifiers. *IEEE Trans. Pattern Analysis and Machine Intelligence*, 18(4):389–399, April 1996.

[2] K. Braun, W. Zuschratter, J. Wang, R. Watzel, A. Hess, and H. Scheich. Changes in synaptic size and shape during auditory imprinting in domestics chicks and footshock avoidance conditioning in gerbils. In *Proc. Workshop on Neural Networks, 16-18 October 1995, Würzburg, Germany*, 1995.

[3] B. Fritzke. A self-organizing network with constant neighbourhood range and adaption strength. *Neural Processing Letters*, 1(5):1–5, 1995.

[4] A. Herzog, G. Krell, B. Michaelis, K. Braun, J. Wang, and W. Zuschratter. Restoration of three-dimensional quasi-binary images from confocal microscopy and its application to dendritic trees. In *Proc. Intl. Biomedical Optics Symposium (BiOS'97), San Jose, CA*, 1997. (to appear).

[5] C. Koch and A. Zador. The function of dendritic spines: Devices subserving biochemical rather than electrical compartmentalisation. *J. Neurosci*, 13(2):413–422.

[6] T. Kohonen. *Self-Organization and Associative Memory (2nd ed.).* Springer-Verlag, New York, 1988.

[7] A. P. Reeves, R. J. Prokop, S. E. Andrews, and F.P. Kuhl. Three-dimensional shape analysis using moments and fourier descriptors. *IEEE Trans. Pattern Analysis and Machine Intelligence*, 10(6):937–943, November 1988.

[8] E. Wallhäuser and H. Scheich. Auditory imprinting leads to differential 2-deoxyglucose uptake and dendritic spine loss in the chick rostral forebrain. *Dev. Brain. Res.*, 31:29–44.

Neural Network Analysis of Hue Spectra from Natural Images

C. Robertson and G. M. Megson
Department of Computing Science, University of Reading,
Whiteknights, Reading, Berkshire, RG6 6AY, UK.

Abstract

This paper describes a straightforward method of analysing the colour spectra of natural images using a hue histogramming technique. Examples of post-processed first moment data from these histograms are then analysed using simple feedforward neural networks. These networks are shown to provide a good level of generalisation and can therefore be used for classification.

1 Introduction

As Gibson [2] has argued, colour is the most important aspect of how we humans interpret the world around us. Merleau-Ponty [4] even quotes Paul Cézanne (an expert colour theorist in his own right) as saying "colour is the place where our brain and the universe meet". Why then do the vast majority of scene-analysis techniques concentrate on monochrome images ?

Images analysed in traditional scene analysis have been in shades of grey, often taken from so-called 'mini-worlds' of geometric blocks in controlled lighting. This has tended to mean that scene analysis has involved a great deal of line finding and geometric model matching but little colour model fitting. This has been due in some part to the expensive nature of digital hardware and analogue to digital signal translation. Recent advances in hardware, however, have meant that cheap, reliable, high quality digital cameras are widely available and can be used to create databases of images at low cost. This means that the concept of colour scene analysis can now be properly addressed.

The colours in general images tend to be widely spread throughout the colour space with the result that they are not necessarily useful for classification. This can be alleviated (as discussed in Ohta [5]) by reducing the number of parameters in that space by normalisation, in our case by intensity. This creates a 2-D histogram of the colour space spanned by the image that may then be analysed. In the example images used for this paper, this histogram proved to be very revealing and has enabled a simple decision surface to be drawn between the contents of three different types of natural image.

This paper first describes the acquisition of digital images for the natural scene database and their processing to form the hue spectrum histograms. Details of the histogram post-processing are then given as well as a full explanation of the neural network learning algorithms used. Results are then presented which show that elements in natural scenes can be differentiated with a high degree of accuracy.

2 Image Collection

2.1 Hardware

The commercially available digital camera used was constructed around a 0.5cm CCD with $250,000$ pixels arranged in a 320×240 matrix with 3 samples per pixel. The recording system was 24-bit and included hardware JPEG encoding. Exposure was based on a TTL centre-point photographic element for light metering giving an exposure range of +5EV to +18EV with a fuzzy logic element which could adjust by −2EV and +2EV. Manual exposure was also possible but was not used in this series of tests. Shutter speed was equivalent to 1/4000 sec. Images were uploaded to the computing platform via a 19200 baud serial link.

2.2 The Images

The selection of natural images chosen was made such that there would be reasonable distances between their possible hue ranges and so that they would represent the majority of possibilities in any natural scene.

Clouds Many different types of cloud were photographed under a wide range of lighting and weather conditions. Some images include small areas of sky and some 'bleed-through' of sky due to their composition. By the very nature of clouds it was not possible to judge exact distances so a wide range was used.

Foliage In this instance foliage is a general term for both grass and many different kinds of trees. These were photographed under a wide range of lighting conditions and in varying stages of growth and sample distance. This was done in order that fixation did not take place and skew the data sample towards one kind of foliage spectrum. Some overlap with both sky and cloud also took place.

Sky Clear sky was sampled at many points from horizon to zenith and at many different times of the day. Each image is cloudless to the best possible extent although some overlap with the cloud images was inevitable.

Examples Typical examples of hue spectra appear in the Appendix, as discussed in the next section.

3 Hue Spectra

3.1 Hue Spectra Results

Images direct from the camera come in a compressed form as vectors in (R,G,B) space. While this is useful for faithful reproduction of the sampled scene it is less useful for simply identifying the colours present in an image since intensity (overall brightness of the colours) is bound up in the measurement. In order to normalise these vectors they were first shifted into a spherical coordinate space using the standard transformation:

$$\rho = (r^2 + g^2 + b^2)^{\frac{1}{2}} \tag{1}$$

$$\theta = \tan^{-1}\left(\frac{g}{r}\right) \tag{2}$$

$$\phi = \cos^{-1}\left(\frac{b}{\rho}\right) \tag{3}$$

The benchmark used was a standard colour image [1] which contained a full range of colours and would give a useful limits for subsequent plots. It can be noted that there is some degree of similarity with the standard C.I.E chromaticity diagram [1], which was to be expected.

Images once sampled were colour-reduced using the median cut algorithm where the (R,G,B) colour cube is iteratively bisected, as described in [7].

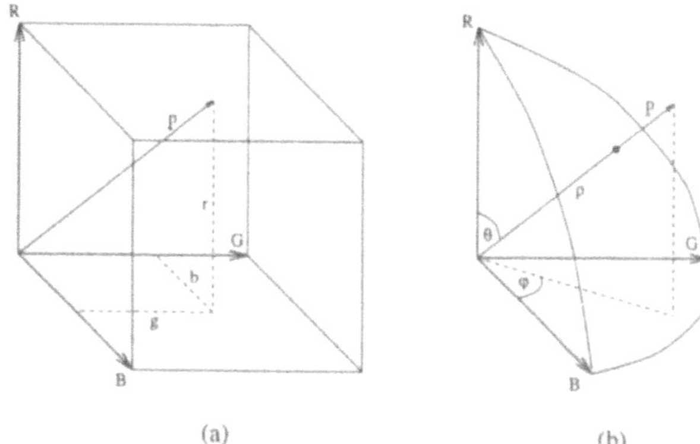

Figure 1: (a) (R,G,B) space, (b) (ρ, θ, ϕ) space.

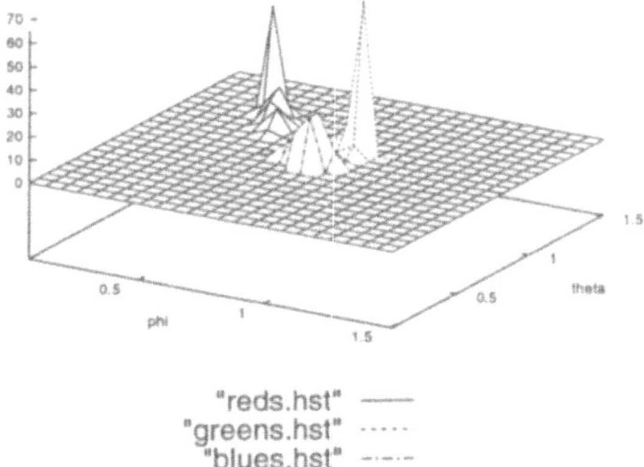

Figure 2: Benchmark hue spectrum plot.

3.2 First Moment Results

The first moment in each dimension of the spectrum-space was calculated for all of the example images. The first two dimensions, ϕ and θ are shown plotted in Figure 3, which shows the intersect of the major axis with the zero-plane. They were calculated as follows:

$$\bar{\phi} = \frac{1}{N} \sum_{j=1}^{j=N} \phi_j.\rho_j \tag{4}$$

$$\bar{\theta} = \frac{1}{N} \sum_{j=1}^{j=N} \theta_j.\rho_j \tag{5}$$

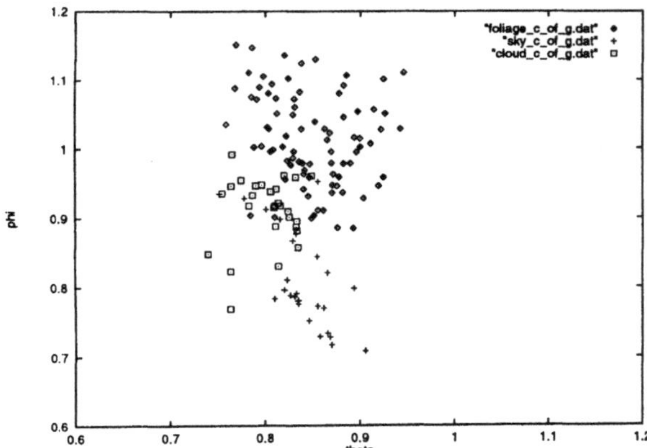

Figure 3: First moment in two dimensions.

It can be seen that the results for foliage are spread over a fairly wide range and partly mix with both sky and cloud. This is to be expected since the images of foliage occasionally include both of these. The group describing sky is perhaps the best defined of the three and intersects only with cloud, as discussed in Section 2.2.

4 Neural Network Analysis

Neural networks were used on this dataset because:

- The decision boundary for the data is not simple. Neural algorithms bring with them a model-wide non-linearity that is particularly useful for modelling complex boundaries.
- The kind of neural networks used provide input to output mapping inherently in their design. This is particularly useful in this context.
- The outputs give *confidence* output rather than a set classification. This is particularly useful if the network is to be embedded in a real-time system.
- A hardware implementation of this system would be inherently fault-tolerant. Extensive damage has to take place before the distributed learning is affected.
- VLSI implementability. Real-time systems can be built which exactly mimic the parallel nature of the networks.
- The network can be grafted directly into a recognition system as described in [6].

4.1 Moments Data

Moments data was used in 2-D format, as shown in Figure 3. The output data for training was the usual orthogonal bases: Foliage 100; Sky 010; Cloud 001.

4.2 Training Parameters and Algorithms

The following algorithms were used, with several parameter settings and at several different levels of iteration and several topologies : quickpropagation; backpropagation with momentum. These algorithms are widely used and can be found in [3, 8] with further theory on optimisation.

Each network was allowed to learn for 5000 iterations.

Quick-propagation Results

By far the best results came from the quickprop algorithm, which is an optimised version of standard backpropagation.

The learning rule takes the standard backpropagation form:

$$\Delta w_{i,j}(t+1) = \eta \delta_j o_i + \mu \Delta w_{i,j}(t) - \nu w_{i,j}(t) \quad (6)$$

as discussed in [3]. In this test the parameters were used as follows:

$$\eta = 0.5 \;\; \mu = 1.75 \;\; \nu = 0.0001,$$

where ν, the weight decay term, is obviously kept small.

Quick-propagation also computes the gradient of the weight space in the direction of each weight, afterwards a direct step to the error minimum is attempted by:

$$\Delta(t+1)w_{i,j} = \frac{S(t+1)}{S(t) - S(t+1)} \Delta(t) w_{i,j} \quad (7)$$

where

$w_{i.j}$ is the weight between units i and j,

$\Delta(t+1)$ is the actual weight change,

$S(t+1)$ is the partial derivative of the error function by $w_{i,j}$

$S(t)$ is the partial derivative.

Using a $2-4-3$ topology, the trained network had a predictive capability of 100% on the training data of 72 images and 94.4% on the remaining testing data. Precisely the same results were found with a $2-5-3$ topology.

Backpropagation with Momentum Results

In this test, the standard momentum learning rule was used:

$$\Delta w_{i,j}(t+1) = \eta \delta_j o_i + \mu \Delta w_{i,j}(t) \tag{8}$$

with parameters $\eta = 0.6$ and $\mu = 0.3$.
When using backpropagation with momentum and a $2-4-3$ topology, 100% was achieved on the training set with 91.5% on the testing set. With a $2-5-3$ topology 100% was again achieved on the training set with 91.5% on the testing set.

Discussion

It is inherent in neural network learning that over-specific learning can destroy the generalisation capabilities of the network so examples of this sort have been avoided and the best results only given. Similarly, overly specified networks with many degrees of freedom have been avoided since they are prone to very complex and contorted decision boundaries which, although are fine for training set tests fair very badly with unseen (test) data. On close examination, the misclassified examples proved to be those which are just inside the decision boundaries of groups.

5 Summary and Further Work

This paper has shown that under certain circumstances, elements in natural images can be classified using the first moment of their hue spectra in conjunction with an unsophisticated neural network. As well as the first moment data descriptor there are many different ways of classifying the spectrum shape with statistical parameters. Two other shape classifiers were used in this paper for reinforcement in the neural network, the second and third moment (skew and kurtosis). This proved to have little effect in classification however. Neural algorithms were also applied directly to hue spectra histograms but the classifications resulting from this were due mainly to overtraining (or under-generalising) due to the large number of degrees of freedom, therefore their predictive capabilities were not as strong as those of more simple networks. The main problem with the first moment dataset is the degree of overlap. Although this is predicable in the case of the sky cloud boundary it can also be noticed that there is overlap in the cloud foliage and sky foliage interface. This may be due in part to the fact that none of the images were taken to be 'canonical' and so contain certain elements from the others.

6 Appendix

This section contains some representative example hue spectrum histograms. The histograms have been sampled at a 25×25 resolution which is representative of their 256 colour (8-bit) depth. The phi-theta axes are from 0 to $\frac{\pi}{2}$ and the height axis is in % of image pixels.

- Cloud

  The majority of points are in the monochrome (central) region, as would be expected since clouds are mostly white to dark grey. The colour variation across the entire cloud image data-base was quite large due to weather and light conditions.

- Foliage

  The foliage spectrum extends over the green part of the spectrum and has a broad base meaning that a range of greens are present in the image, extending into both brown and yellow. Colour variation in the entire foliage image database was surprisingly low. This is probably due to the light-energy collection efficiency of foliage pigmentation being relatively stable at set wavelengths.

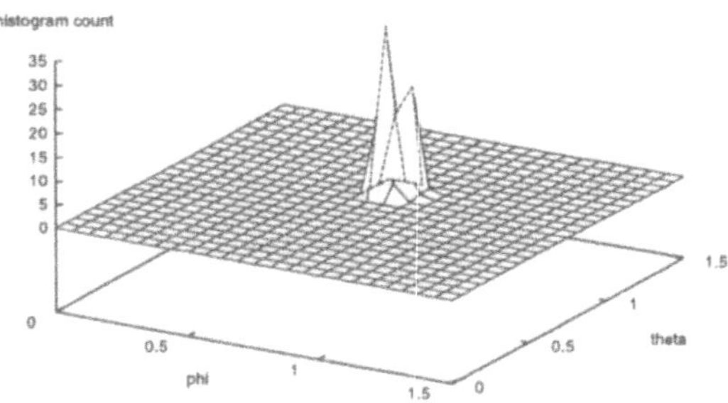

Figure 4: Example cloud spectrum.

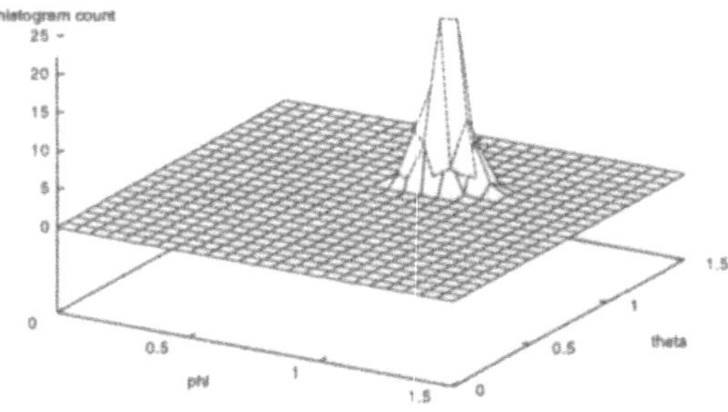

Figure 5: Example foliage spectrum.

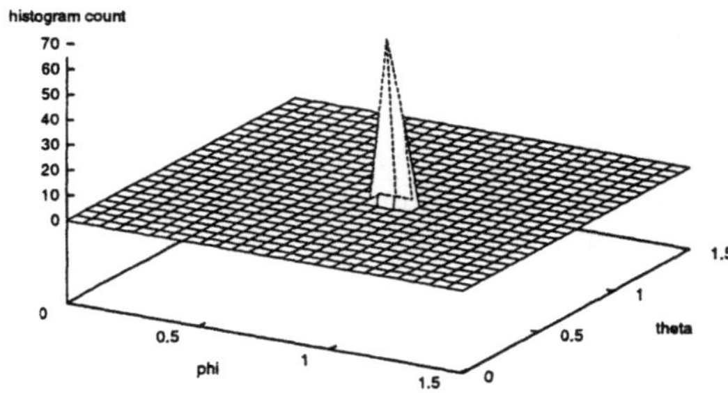

Figure 6: Example sky spectrum.

- Sky

  It seems obvious from the spectrum that the majority of pixels in the image are in the blue region. However, the question arises as to the proximity of the central peak to the centre of the spectrum. This is due to the low saturation (14) level in this particular image. Light from the sky tends to vary widely in its white content with times in the day and position.

References

[1] D. H. Ballard and C. H. Brown. *Computer Vision.* Prentice Hall Inc., New Jersey, 1982.

[2] J. J. Gibson. *The Senses Considered as Perceptual Systems.* Houghton-Mifflin, 1966.

[3] S. Haykin. *Neural Networks - A Comprehensive Foundation.* MacMillan College Publishing Company, New York, 1994.

[4] M. Merleau-Ponty. *Eye and Mind.* NorthWestern University Press, Evanston, 1964.

[5] Y. Ohta. *Knowledge Based Interpretation of Outdoor Natural Colour Scenes.* Pitman, London, 1985.

[6] C. Robertson and G. M. Megson. Parallel segmentation network with topological post-processor. Technical Report RUCS/96/TR/003/A, University of Reading, 1996.

[7] A. Watt. *Fundamentals of Computer Graphics.* Addison-Wesley, Wokingham, UK, 1989.

[8] A. Zell et al. Snns manual version 4.1. Technical Report 6/95, Institute of Parallel and Distributed High Performance Systems (IPVR), 1996. available by ftp from University of Stuttgart.

Detecting Small Features in SAR Images by an ANN

I. Finch, D. F. Yates and L. M. Delves
Department of Computer Science, University of Liverpool, Liverpool, L69 3BX
Email: ian@csc.liv.ac.uk

Abstract

Synthetic aperture radar (SAR) images are intrinsically noisy, and processing them attracts a high computational overhead. This paper relates to developments, involving the use of an ANN to reduce the overhead, in respect of earlier work by the authors on the identification of small objects. It describes how the ANN is utilised, and how it was trained using an artificially created training set.

1 Introduction

Synthetic aperture radar (SAR) is a high resolution imaging technique, which utilises one or more microwave transmitters and receivers. It is usually operated from an airborne or spaceborne platform [1], but can be used at ground level [3]. One significant advantage of the technology is that SAR images can be captured when optical images cannot, at night for example, or when there is cloud cover.

A characteristic feature of SAR technology is that, due to the method of formation, images have a signal to noise ratio of 1:1. As a result, the techniques needed for the successful analysis of a SAR image differ from those normally used for image processing; they are inevitably much more computationally intensive [4]. This is certainly so in the case of the type of analysis addressed here, namely, the identification of regions in a SAR image that correspond to groups of small, related objects. Typical of such objects are airfield runway lights and electricity pylons for which corresponding regions in a SAR image are each only a few pixels (3–12 approximately) in size. Hereafter, the term *target* will be used to refer to such a region.

The high level of noise of which SAR technology is redolent and the small size of targets renders impossible their identification by means of image enhancement; an alternative technique is required. In earlier work by the authors [2] an approach based upon the use of three types of contextual knowledge was proposed. The types: IS (individual specific) knowledge, GS (group specific) knowledge, and MG (multiple group) knowledge, correspond to the three general patterns of occurrence of a target/targets in a SAR image, that is, as an individual, as a single group, or as several groups. The approach was implemented as a three-stage process; each stage making use of one kind of contextual knowledge. Of the stages, the first two, those relating to IS and GS knowledge respectively, were the most computationally expensive, and although effective in identifying the required targets, the approach fell short of providing a real-time facility. This paper describes how an artificial neural net was employed to learn the requisite IS knowledge in respect of an electicity pylon, and in so doing has contributed to achieving the aim of the real-time identification of 'lines (runs) of pylons' by reducing the time overheads associated with both of the first two stages of the approach.

2 The Three-Stage Approach

In order to place the role played by the ANN into context, and to provide a basis for assessing the improvements gained by its use, it is appropriate to consider briefly the three-stage approach as constituted prior to the introduction of the ANN.

Each stage in the approach made use of one type of contextual knowledge. The first stage used *IS* knowledge to identify potential targets, and the second attempted to form these into appropriate groups using GS knowledge which relates to factors such as the regular spacing of targets in a group. Failure to associate a target with a group resulted in its being discarded. The third stage then used MG knowledge to conjoin, where appropriate, those groups of targets identified at the second stage.

The specific IS knowledge employed in the first stage relates to the size, intensity, and shape of a target; a region corresponding to a pylon is small, bright and approximately circular. Some examples of pylons are given in Figure 1. Its brightness

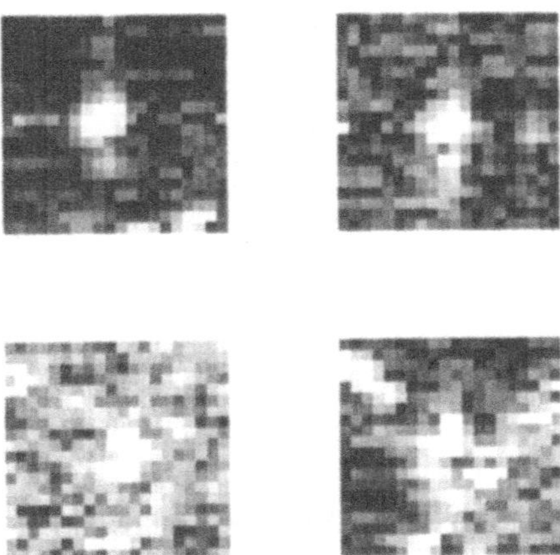

Figure 1: Examples of pylons in a SAR image.

is a consequence of the size of the pylon in relation to other surrounding physical features and the large number of microwave scatterers (sharp edges) it possesses. Its circular shape arises because although a pylon is essentially square in section, its appearance is 'smeared' as a result of its brightness. In order to utilize this knowledge to identify a potential target, the following algorithmic steps were employed:

1. Discard all pixels of sufficiently low intensity;
2. From those that remain form 'regions' by conjoining contiguous pixels;
3. Discard each region that is too big, too small, or is not approximately circular;
4. Replace each region that remains by a single point at its centre of mass to give a potential target.

3 Utilising an ANN

When contemplating improvements to the processing speed of the above approach, it must be noted that the second stage must necessarily involve an exhaustive search strategy. This is so because there may be more than one line of pylons registered in an image, or there may be no pylons at all, and it is clearly necessary to distinguish these situations. Given this and the fact that the third stage is relatively time-inexpensive, it is appropriate to seek improvements to the first stage. Speeding up the first stage would, per se, reduce the time overhead

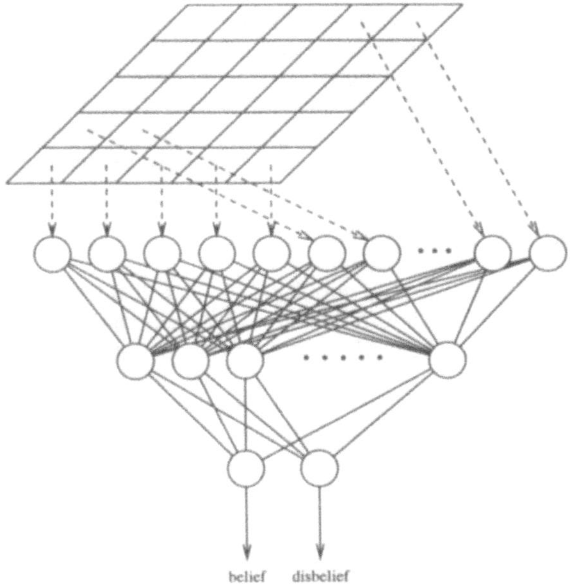

Figure 2: The ANN used for pylon detection.

of the entire approach. Moreover, further overall reductions would ensue, if fewer 'false alarms' (erroneously identified targets) were to be generated by the first stage, since the processing performed at stage two would thereby be lessened. It was, therefore, decided to supplant stage one with an ANN.

Given the 'scale' of a SAR image, that is, the area represented by a pixel, together with knowledge of the physical size of pylons, it is possible to deduce the value of p such that a target is contained in a $p \times p$ matrix of pixels. Correspondingly, an ANN was designed such that it would accept the intensity of each pixel in such a matrix as input, and as output, produce an indication of whether or not that matrix contains a target. The processing of an entire SAR image was to be achieved by abstracting all possible $p \times p$ matrices of pixels and supplying each, in turn, to the ANN. For the images discussed in this paper, $p = 5$, and thus the ANN used to process them possessed 25 input nodes. Every one of these was connected to each of 20 nodes which constituted the single hidden layer, and in turn, each of these was connected to the two output nodes. The architecture of the net is shown in Figure 2. Both output nodes were required to deliver a value in the range [0, 1]; one value representing the extent of the belief that the supplied region contains a target, the other the extent of the disbelief. The difference between the two gives a 'certainty factor' ranging from -1 (absolute disbelief) to $+1$ (absolute belief). If the certainty factor that is produced for a given region exceeds a user-supplied 'certainty threshold', a potential target is registered at the centre of the matrix.

4 Training the ANN

When the data to which an ANN will ultimately be applied is noisy, it is often the case that the ANN can be trained, at least in part, using clean data. However, this is not so for SAR imagery; all real data is noisy. The approach taken, therefore, was that of utilising a training set consisting entirely of artificial data.

As a starting point, the earlier characterisation of a pylon as a small, bright, approximately circular region can be refined somewhat. The pylon can be regarded as a 'star-shaped' region, with the centre pixel having the highest intensity, and the intensity of points along the 'arms' decreasing with distance from the centre. This, of course, is a generalisation, and may not be true in any single case because of the high level of noise, as an examination of Figure 1 will show.

The characteristics of the artificial targets that were generated for purposes of training were based upon these observations. Each target was constituted from 9 pixels in the approximate shape of a cross; each pixel assuming one of three levels of intensity. In each 'cross', only the central pixel was assigned the highest level, whilst pixels on each of the four arms were associated with the two lower levels; those furthest from the centre possesing the lowest intensity. Unfortunately, the relative brightness of pixels lying in different regions of a real target vary too greatly to be encoded in the artificial targets. To overcome this difficulty, a variety of values were used, in the range (0, 1]. This gave the ANN an understanding of 'pylon-ness', without its having to focus too tightly on any particular set of values.

In order to produce training counter-examples, that is, artificial regions which do not contain pylons, it was first noted that the majority of pixels in the corresponding regions of real images, have approximately the same intensity. Consequently, to simulate this, pixel-matrices of constant intensity were used. Figure 3 depicts part of the training set derived; artificial targets occurring on the left-hand side and counter-examples on the right.

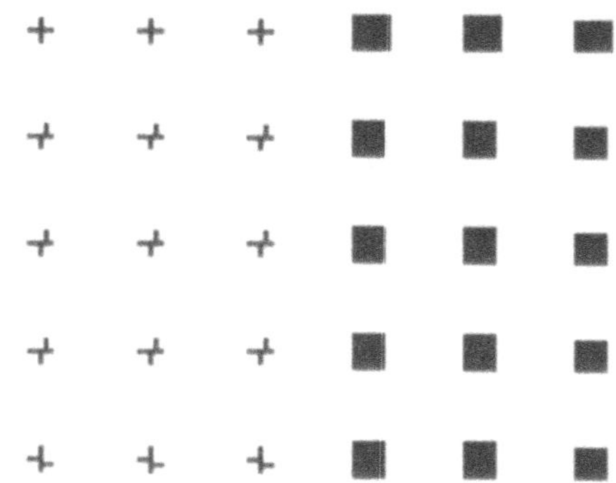

Figure 3: Part of the artificial training set.

The ANN was initialised by assigning to each of its links, a weight selected at random from a uniform distribution on $[-0.5, 0.5]$. The training set contained 50 artificial targets and 50 counter-examples, which were presented alternately to the ANN. After each pattern was presented, the weights in the ANN were adjusted using backpropagation. The network converged reasonably quickly, being able to correctly classify every pattern in the training set as a target or a counter-example, after the entire set had been presented to it 100 times.

5 Results

In order to provide an indication of the level of performance that can be achieved by the ANN descibed above, it is appropriate to present details of the speed and precision of the original three-stage approach. For such purposes, the analysis of a SAR image of a small part of Bedfordshire, U.K. that contains ten targets (see Ordnance Survey map No. 1002), is considered. On the image, which consists of 922×696 pixels and is shown in Figure 4, the following details of the performance of the original three-stage approach were recorded.

In the first stage of the approach, application of an intensity threshold resulted in 93.5% of the pixels being dicarded. Aggregation of the remainder led to the definition of 4200 distinct regions, and after application of constraints on size and shape, 252 potential targets were ultimately identified, see Figure 5. The second stage discarded 241 of these, thereby identifying eleven real targets, all of which it deemed to lie on a single run. There was no need to invoke stage three in this case. An elaboration of these results can be found in [2].

Figure 4: The SAR image.

Figure 5: Pre-processing with the original algorithm.

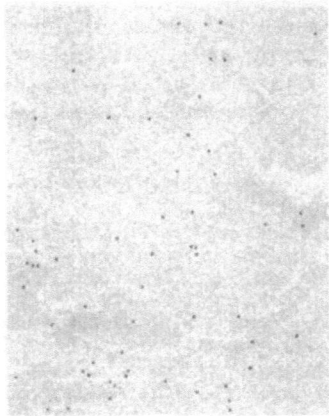

Figure 6: Pre-processing using the ANN.

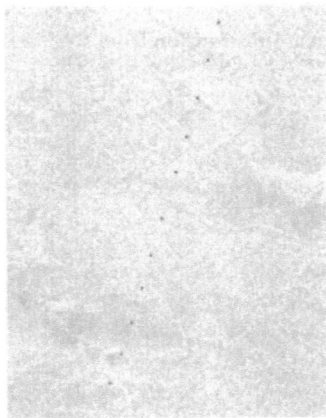

Figure 7: The line of pylons.

When the ANN is used in place of the first step in the approach, as discussed in this paper, only 61 potential targets are identified, see Figure 6. However, these do include all of the real targets (a fact which can be verified by checking the OS map). There are therefore only one quarter of the false alarms that the original approach recorded. When this set of potential targets is supplied to stage two of the approach, the correct single run of ten pylons, that depicted in Figure 7, results.

Furthermore, there is a significant increase in speed. Using the ANN in place of the first step, reduced the time taken for this stage to about two thirds of the time for the original approach. The much lower number of false alarms then resulted in a great speed increase for the second step; about one sixth of the original time. The total time taken was thus reduced to approximately one third of the time needed for the original approach (because the second stage takes twice as long as the first).

6 Conclusions

This paper has shown that artificial data can be used to produce a training set for an ANN when no suitable real data can be found. Furthermore, it has been demonstrated that the artificial data can be generated from only an approximate rather than an exact analysis of the type of data to be supplied to the ANN. As well as providing an example of the use of these techniques, the resulting system for identifying pylons in SAR images is extremely useful, and outperforms the system on which it is based, in terms of both speed and quality of result.

References

[1] J. C. Curlander and R. N. McDonough. *Synthetic Aperture Radar: Systems and Signal Processing.* John Wiley and Sons, New York, 1991.

[2] I. Finch, D. Yates, and M. Delves. Detecting lines of pylons in sar images. In *Proceedings Third International Conference on Satellite Remote Sensing*, volume 2958, pages 152–163, Washington, 1996. SPIE.

[3] K. Morrison and J. C. Bennett. Development of a ground-based synthetic aperture radar remote sensing facility. In *Polarimetric SAR Workshop.* SCEOS, Sheffield, 1994.

[4] C. J. Oliver. Information from SAR images. *J. Phys. D: Appl. Phys.*, 24:1493–1514, 1991.

Optimising Handwritten-Character Recognition with Logic Neural Networks

G. Tambouratzis
Institute for Language and Speech Processing,
22 Margari Str., Athens, 115 25, Greece.
Email: giorg_t@ilsp.gr, tambour@dsclab.ece.ntua.gr

Abstract

This article studies the implementation of a hand-written character recognition task using neural networks. Two logic neural network models are employed to classify the Essex dataset, which comprises real-world hand-written characters. To reduce the underlying dataset variation, several pre-processing approaches are investigated. This allows the comparison of the network models on the basis of their classification accuracy for datasets with different characteristics.

1 Neural Network Description

The neural networks used in this article belong to the family of logic neural networks. These networks operate on the input pattern by partitioning it into tuples which consist of n pixels. The classification of the pattern is achieved by (i) decomposing it into its constituent tuples, (ii) performing a classification operation on each tuple and (iii) combining the results obtained for all tuples in order to calculate the result for the entire pattern. Logic neural networks possess an easily implementable and readily scaleable structure which is closely related to that of Random-Access Memory (RAM) circuits [1]. At the same time, they combine a very high functionality with exceptionally short training and classification times and are thus well-suited to real-world recognition tasks.

The WISARD has been established as one of the first neural network models to be implemented in hardware and become commercially available [2]. It is formed by a set of discriminator nodes [1]. Each discriminator consists of functions which sample tuples of n binary pixels from the input retina and comprise 2^n memory locations, one for each combination of pixel values. Every location consists of a single bit, where a binary number is stored. This number reflects whether the corresponding pixel combination has occurred within the training set.

1. Initialisation Phase:
   (a) set all tuple values to 0.
   (b) insert an equal number of units to each function using randomly-selected addresses.
2. Training Phase: For each pattern j and each function k
   (a) locate address a_j designated by the n-tuple and address a'_j with a non-zero content and maximum Hamming distance from a_j,
   (b) increase the content of a_j by 1 and decrease the content of a'_j by 1.

Figure 1: The SOLNN learning rule.

The SOLNN self-organising logic neural network [6] also consists of discriminator nodes. However, in contrast to the WISARD, in the SOLNN several bits are provided for each memory location. This allows the storage, for each tuple combination, of a frequency-of-occurrence entry, which is considerably more precise than the binary entry stored in the WISARD. In order to update the information in the SOLNN nodes, an adaptation rule is used [7]. This rule is described in Figure 1.

According to that rule, no new units are introduced in any function during learning. Before the start of the training phase, a given number of information units are inserted in randomly-selected addresses. During training, these units are moved between addresses within each function [6], to record the occurrence of patterns. The distribution constraint is employed in the SOLNN learning phase to ensure the separation of distinct pattern classes, by requiring a minimum degree of similarity between the patterns assigned to each discriminator.

The SOLNN has been shown to successfully perform self-organisation tasks, an initially untrained network evolving during learning so that it separates the training set into classes which are in accordance to the clusters existing in the input space [7].

2 The Character Recognition Task

The Essex dataset of hand-written characters is employed in the classification/recognition experiment. This dataset comprises classes corresponding to the 26 letters of the English alphabet and the 10 digits. In this experiment, the digit classes are used, to allow for an extensive study of the networks' behaviour. The first 20 samples from each class, which are depicted in Figure 2, indicate the high variation within each class. This dataset has been used to develop a toolkit for the automated recognition of hand-written forms [3], and its size (over 3900 training and 1800 testing digit images) is sufficient to draw sound conclusions regarding the networks' behaviour. Initial results on this task have been presented in [5].

Each character is represented as a 42×50 array of binary pixels. The characters occupy varying portions of the retina, consist of lines of varying widths and contain noise. Hence, each class comprises patterns with a high degree of variation. To reduce that variation and thus improve the network performance, a number of pre-processing sequences are applied. These sequences are based on window-type convolution operators and consist of two main phases, the first intended to reduce the existing noise and the second to normalise the digit images. The most effective pre-processing methods used are summarised as follows:

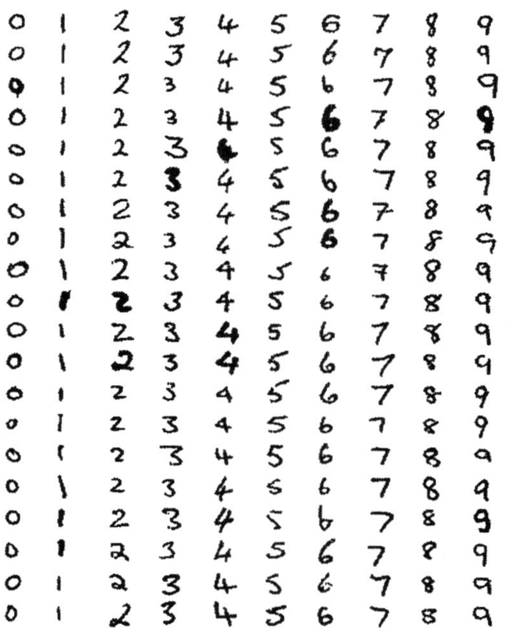

Figure 2: Samples of training patterns.

1. Method 1:
   (a) median filtering using a 3×3 square window for a single iteration,
   (b) centering of the digit on the retina,
   (c) uniform scaling along the x and y axes so that the digit fully occupies one of the retina dimensions.
2. Method 2:
   (a) median filtering using a 3×3 square window for a single iteration,
   (b) centering of the digit on the retina,
   (c) different scaling factors along the x and y axes so that the digit fully occupies both retina dimensions.
3. Method 3 differs from method 2 solely in applying the median filtering operator for several iterations, to reduce the existing variation.
4. Method 4:
   (a) median filtering using a 3×3 square window,
   (b) skeletonisation of the character area,
   (c) centering of the digit on the retina,
   (d) uniform scaling along the x and y axes so that the digit fully occupies one of the retina dimensions
   (e) opening and/or closing of the character area with a 3×3 square window, so that the area of the character becomes equal to a predetermined fraction (set to 30%) of the retina area.

This last method attempts to normalise both (1) the total retina area occupied by the character and (2) the character's outer dimensions, and thus may be expected to be more accurate than the other ones.

3 Experimental Results

The neural networks employed to perform the classification task are (i) the WISARD, (ii) the self-organising SOLNN (denoted as SOLNN-u) and (iii) a supervised version of the SOLNN (denoted as SOLNN-s). The WISARD and the supervised-SOLNN networks consist of 10 discriminator nodes, one node being provided for each digit class. To allow the unsupervised SOLNN to successfully cluster the dataset without any external guidance, a degree of redundancy is required [6], which is achieved by using a 40-node network. The networks are presented with all training patterns in the learning phase. Due to the WISARD training algorithm, each pattern needs to be presented to this network only once. When training the SOLNN, the dataset is presented for 60 iterations, to allow the extraction of the frequency-of-occurrence information.

The classification accuracy of the two supervised networks (expressed as the percentage of correctly classified characters) is depicted in Table 1 for the original Essex dataset (denoted as method 0) and the pre-processed datasets (denoted by the corresponding method label). For each set of results, the standard deviation of the recognition rate throughout the digit classes is noted in italics. The deviation indicates whether a network recognises all classes with a similar degree of accuracy (the variation being close to zero), or whether some classes are recognised less successfully (the variation being relatively high). As expected, both network models perform significantly better using pre-processed datasets rather than the original Essex dataset, while pre-processing method 1 is found to generate the best results. The recognition rate of the WISARD is slightly higher than that of the supervised SOLNN (86.4% rather than 85.3%).

Table 1: Classification results of the supervised networks (the recognition rate is expressed as a percentage, the standard deviation is denoted in italics and method 0 indicates use of the original Essex dataset).

	Training Set	Testing Set
WISARD / method 0	86.7% (9.4)	58.3% (17.0)
WISARD / method 1	99.96% (0.1)	86.4% (6.5)
WISARD / method 2	100.0% (0.0)	84.5% (7.5)
WISARD / method 4	100.0% (0.1)	68.9% (11.1)
SOLNN-s / method 0	90.6% (6.0)	65.8% (11.8)
SOLNN-s / method 1	99.5% (0.4)	85.3% (8.4)
SOLNN-s / method 2	94.75% (3.4)	78.33% (12.7)

However, the decision margin (expressed as the difference between the two highest node responses divided by the winning node response), which reflects the confidence with which a pattern is classified, is much higher for the SOLNN, typically being 40% to 45%, compared to less than 20% for the WISARD. Using the original Essex dataset, which has a higher intra-class variation, the supervised SOLNN is clearly superior to the WISARD for both the training and testing sets, as reflected by the higher recognition rate and the smaller standard deviation. This is due to the SOLNN adaptation rule which allows (i) the storage of the frequency-of- occurrence of each pattern feature as well as (ii) the removal of data referring to non-characteristic samples of each class. In contrast, the learning process of the WISARD prevents the removal of knowledge that is stored during training.

For the datasets derived with methods 1 and 2, the two networks exhibit a similar performance. Pre-processing methods 3 and 4 prove less effective and the recognition rates suffer accordingly. In particular, the 4th pre-processing method is markedly inferior, due probably to the skeletonisation operator, which tends to accentuate the discontinuities in the characters.

The use of supervised networks ensures that each digit class is fully separated from all other classes, allowing the successful extraction its characteristics. However, the SOLNN model is capable of self-organising on a given dataset. As shown in Table 2, when the SOLNN is allowed to self-organise on pre-processed Essex datasets in the absence of external guidance, it achieves a meaningful clustering result. The unsupervised SOLNN results have been obtained by presenting all training patterns to the network for 60 iterations. Following the completion of this learning phase, each training pattern is presented to the network and assigned to the highest-responding discriminator node. Each node is considered to represent the digit class with the highest fraction of training patterns assigned to it, this labelling being used to generate the final classification result.

Table 2: Classification results of the unsupervised SOLNN, for pre-processed training and testing sets.

	Training Set	Testing Set
SOLNN-u / method 1	74.9 %	70.6 %
SOLNN-u / method 2	78.6 %	64.9 %

Table 3: Classification results of the WISARD network, using random and directional sampling for the n-tuples.

	random sampling	horizontal sampling	vertical sampling
method 1 (train)	100.0 %	98.5 %	92.7 %
method 1 (test)	86.4 %	60.5 %	51.4 %
method 2 (train)	100.0 %	99.2 %	92.1 %
method 2 (test)	84.5 %	64.6 %	43.5 %

In unsupervised learning, the clusters generated depend on the similarity between patterns, rather than externally-provided labels. Therefore, the performance of the unsupervised SOLNN, quoted in Table 2, is justifiably lower than that of the supervised networks. If the classification was randomly performed, a 10% recognition rate would be expected. Yet, for method 1, the unsupervised SOLNN classifies correctly over 70% of the previously unseen test patterns. The effect of the mapping of the input retina onto the n-tuples has also been examined. In Table 3, the random mapping of the retina pixels, which has been used to generate the results summarised in Tables 1 and 2, is compared to 'directional' sampling (in which neighbouring retina pixels are used in each n-tuple). Two directional sampling schemes are used, the n-tuple consisting of pixels along a horizontal or vertical line. It can be seen that random sampling leads to much better results than directional sampling, probably due to the fact that in characters, the values of neighbouring pixels are not independent, and thus localised sampling leads to loss of information in the n-tuples.

It is worth noting that for both supervised and unsupervised networks, the dataset generated using method 1 gives the best recognition results. This shows that method 1 performs the most efficient pre-processing. The results obtained using method 1 are very similar to these quoted by Amiri *et al.* [3] for the same dataset using n-tuple-type networks and processing the dataset with window-based operators. More recently, Lucas *et al.* [4] have presented improved results, with recognition rates ranging from 90.6% to 91.4 %, for a pre-processing version of the Essex dataset. This pre-processing involves chain-coding each character and generates much better results than window-type operators similar to those used in this article. It is thus probable that by using similar chain-coded data as inputs to the unsupervised SOLNN, the network may be able to generate improved results.

4 Conclusions

It has been found that, when applied to a hand-written character recognition task, the supervised SOLNN and the WISARD possess a similar performance provided that the dataset is pre-processed to reduce the underlying variation. When the pre-processing step is omitted, the supervised SOLNN has a classification performance clearly superior to that of the WISARD, due to its learning algorithm. The SOLNN is also able to self-organise when presented with the dataset, achieving a meaningful clustering result, even when the pre-processing operation used is not particularly successful. It has been shown that the choice of pre-processing operation can affect considerably the performance of both supervised and unsupervised networks. It also seems that window-based operators are less successful than other pre-processing techniques in reducing the dataset variation.

References

[1] I. Aleksander and H. Morton. *An Introduction to Neural Computing.* Chapman and Hall, 1990.

[2] I. Aleksander, W. V. Thomas, and P. A. Bowden. Wisard: A radical step forward in image recognition. *Sensor Review*, pages 120–124, July 1984.

[3] A. Amiri, A. C. Downton, S. J. Hanlon, C. G. Leedham, S. M. Lucas, and D. Monger. Oscar: A visual programming toolkit for off-line hand-written form recognition. In *Proceedings of the 4th International Workshop on Frontiers in Handwriting Recognition*, pages 441–448. Taipei, Taiwan, December 1994.

[4] S. Lucas and A. Amiri. Statistical syntactic methods for high-performance ocr. *IEE Proceedings on Vision, Image and Signal Processing*, 143(1):23–30, 1996.

[5] G. Tambouratzis. Applying logic neural networks to hand-written character recognition tasks. In *Proceedings of the ICTAI'96 Conference*, pages 268–271. Toulouse, France, IEEE Press, 16-19 November 1996.

[6] G. Tambouratzis and T. J. Stonham. Evaluating the topology-preservation capabilities of a self-organising logical neural network. *Pattern Recognition Letters*, 14(11):927–934, 1993.

[7] G. Tambouratzis and D. Tambouratzis. Self-Organisation in Complex Pattern Spaces Using a Logic Neural Network, *Network: Computation in Neural Systems*, volume 5, pages 599–617. 1994.

Combined Neural Network Models for Epidemiological Data: Modelling Heterogeneity and Reduction of Input Correlations

M. H. Lamers[1,2], J. N. Kok[1] and E. Lebret[2]
[1]Leiden University, Computer Science Department, P.O. Box 9512, 2300 RA Leiden, The Netherlands
[2]National Institute for Public Health and The Environment (RIVM).
Email: lamers@wi.leidenuniv.nl

Abstract

We consider an epidemiological dataset, concerned with predicting pulmonary response to air pollution. To gain more knowledge of nonlinear effects and interactions in the data, nonlinear neural network techniques were applied to model the data. Initially, we modelled the data with standard feedforward network models. Based on the epidemiologic effect of heterogeneity in response, we propose a novel combined neural network modelling strategy to improve prediction quality. Also, we propose the use of a neural network strategy for reducing correlation between covariants to improve modelling quality. The results presented are promising when compared to standard feedforward network modelling.

1 Introduction

One way of learning to understand the cause of diseases is to search for associations between exposures to various conditions and incidence of the studied disease or effect in a population. Such associations may provide testable hypotheses of causal links between exposure and its effects, even in situations that would be impossible to simulate in a laboratory. When exploring relations for which not much biological understanding is yet available, such analyses may be of great value to initiate further research. Considering this, it seems that neural network models can be useful tools for epidemiological studies, in which associations between exposure and disease are of interest.

We have analyzed a dataset concerned with air-pollution epidemiology using neural network methods. Our findings may be illustrative for analyses of epidemiological datasets using similar techniques. The distinguishing features of the epidemiological problem we studied are weak, possibly nonlinear, associations between exposure levels and measured effect, interactions between covariants and high levels of noise in the data. These characteristics are common of epidemiological data, but constitute challenging problems for neural network models, particularly because of the weakness of the featured associations and relatively high levels of noise.

Often in epidemiological studies, parametric models are fitted to the data at hand. Common parametric models are linear regression models, which assume a linear relationship between the covariates in the model and the outcome variable, and parametric Poisson regression, which usually assumes that the logarithm of the outcome variable is linearly related to the covariates included in the model. Although the latter model assumes a nonlinear relation between covariates and outcome, it does also place constraints on the form of this relation. Nonparametric models, however, do not make such assumptions concerning the underlying form of the relation. Any form of relation between the covariates and the outcome variable can be represented by the model. Some advocate that modelling complex relations is essentially impossible without introducing carefully designed biases, in the form of a priori assumptions about the processes studied, into the model [2, 6, 7, 8]. Nonetheless, nonparametric methods, or *black box* methods, remain useful tools in epidemiological research. They enable the exploration of exposure-response relations without a priori assumptions such as linearity.

The interactions between covariates are an important issue in epidemiological studies. Multiple covariates often affect the response variable. Changes in the value of one covariant may even modify the effect another covariate has on the outcome, which is known as *effect modification*. Modelling such interactions requires flexibility and often nonlinearity of the model.

For example, we are interested in relations between ambient levels of air pollution and changes in daily peak expiratory flow (PEF) measurements for children. By modelling the acquired data with a nonparametric feedforward network [3], changes in modelled PEF as a result of changing the values of particular covariates, or combinations of covariates, can be visualised, even if complex interactions are present in these relations. Exposure-response plots relating several pollutants and confounding covariates, such as temperature, to childrens PEF in a nonlinear feedforward network are shown in this paper.

An effect encountered also in epidemiological datasets is that of *heterogeneity*. Similar, or even equal values of the covariates may yield different outcome values in multiple observations. One reason for this effect in epidemiological studies may, for example, be differences between studied individuals. Reactions to similar changes in exposure may differ among individuals, without the presence of confounding issues such as age, sex or race, but for the fact that they are simply different individuals. We propose a method that combines learning vector quantization (LVQ) networks [3, 4] and feedforward networks to deal with issues of heterogeneity, which is discussed in Section 3.

Experiments aimed at reducing correlations between input variables are also discussed in Section 3. Through the application of non-linear autoencoding networks, correlation between time lagged measurements of the same covariant is reduced. Results of experiments employing the less correlated variables in an estimation task are presented.

Section 2 presents examples of neural network modelling of epidemiological data performed by others. Work we have done on modelling pulmonary response to air pollution, and methods we devised, are discussed in Section 3. Conclusions and final thoughts are summarized in Section 4.

2 Other NN Applications

This section describes two examples of previous applications within epidemiological studies. Both experiments involve the application of feedforward network models to fairly large datasets.

A study by Moseholm *et al.* [5] applied nonlinear feedforward network models to estimate pulmonary response of subjects with asthma to gaseous air-pollution, weather, and medicine intake. Two datasets ($n = 918$, and $n = 955$) gathered from different cities were used and an independent test set ($n = 422$) was reserved for testing purposes. Networks with two hidden layers were trained and interactions between the effects of pollutants NO_2 and SO_2 were found in multi-variate exposure-response plots where all other covariates were held at their average value. Nonlinear relations between weather and pulmonary response were found by analyzing such plots. Correlations between observed and estimated PEF values in the test set of approximately 0.52 were reported (ideally, this value is 1).

Another approach was taken by Waschulzik *et al.* [9]. They applied simple perceptron network models (no hidden units) with semi-linear activation functions to the task of predicting the survival or death of patients who suffered acute myocardial infarction, given their age, sex and medically diagnosed conditions. 2856 training patterns were available, and 1429 patterns were set aside for testing. Use of network models without hidden layers enabled the analysis of connection strengths in the network as related to different covariants. These analyses found strong reliance of the network models on clinical conditions for the prediction of survival, and much less on the non-clinical variables.

3 Air Pollution Effect Models

In co-operation with the National Institute for Public Health and The Environment, we applied artificial neural network models to data concerned with air pollution effects on children's pulmonary function. We are interested in learning how the evening pulmonary function changes with ambient levels of air pollution, when correcting for the confounding effects of weather conditions and time lag in pulmonary response.

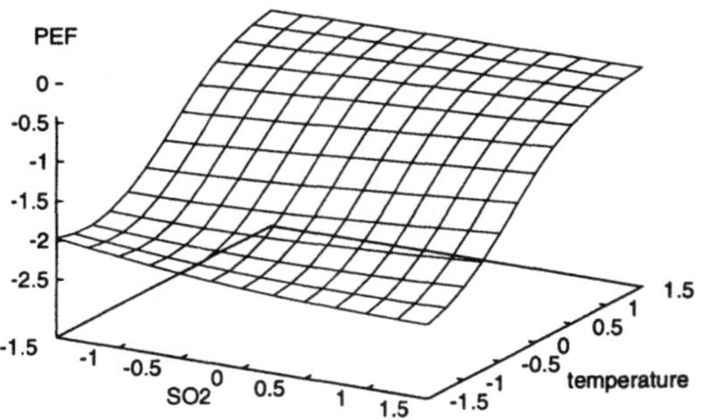

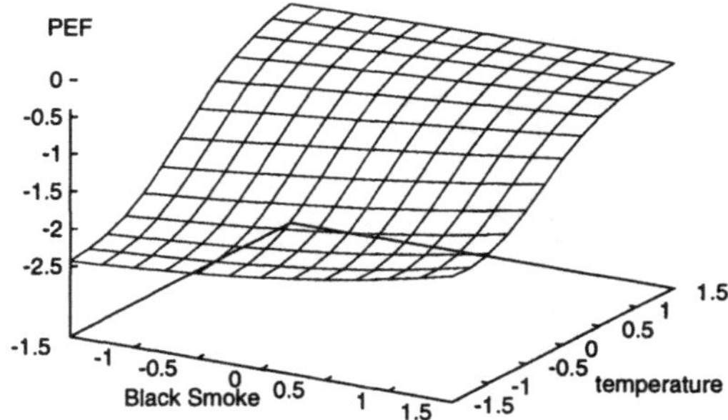

Figure 1: Exposure-response plots from one of the feedforward network models. All other input variables are set to their average value, day-of-the-week is set to Sunday. Both plots show nonlinear response. PEF decreases with increasing SO_2 levels and also with falling temperature (left). Interaction between BS and temperature seems to reverse the response to BS from positive at low temperatures to slightly negative at high temperatures (right). Correlation between observed and estimated PEF values in the test set was 0.627 averaged over multiple fits of the model, with standard deviation 0.057.

The data was collected in the winter of 1991-1992 at measurement sites of the Dutch National Air Quality Monitoring Network and from schoolchildren aged 7 through 12 in the city of Utrecht. Peak expiratory flow (PEF) measurements were performed three times daily by the children in their homes as a measure of pulmonary response. Daily measurements of SO_2, NO_2 and Black Smoke (BS, a particulate pollutant), their previous day values and five-day average values were included as variables in the model. To correct for learning effects in making PEF measurements, a variable was included representing the measurement day number. Daily minimum temperature and day-of-the-week indicator variables were included to correct for weather conditions and weekday effects such as weekend lifestyle patterns. One and two day lagged PEF measurements are included to correct for time lag in pulmonary response, and bring the total number of input variables to 26. Pollution levels and temperature are transformed to mean 0 and unit standard deviation. PEF measurement are scaled to a resolution of 1, and translated such that a childs median PEF value is closest to 0. A total of 620 patterns are available: 496 for fitting the models, 124 are kept aside for testing. Quality of model fit is expressed as correlation coefficients between observed and estimated PEF values, to make results comparable to other epidemiological findings.

Feedforward network models with a single hidden layer of 5 units were trained to estimate the evening PEF given the input variables. Hidden units have sigmoidal transfer functions, the output unit is linear. Training was conducted using standard backpropagation learning on randomly selected 75% of the training patterns, and ceased when performance deteriorated on the remaining 25%. Results are summarized in Figure 1.

Analyses of the estimation quality for separate patterns showed large differences, even for observations with similar input conditions. This could be the effect of heterogeneity in response to air pollution among different children. Reasons for such heterogeneity may be the presence of for instance short illness or even asthma among a subgroup of the studied children.

Attempting to identify possible heterogeneity among the patterns, we search for subgroups with differing response to exposure conditions. These groups are subsequently modelled with individual feedforward network models. We divide the data into two groups based on quality of estimation

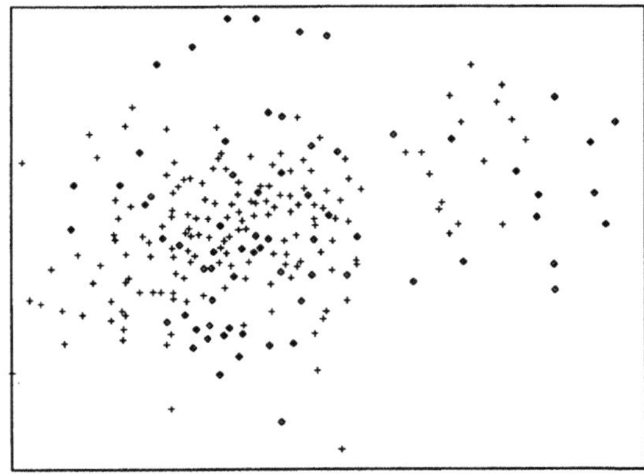

Figure 2: A 2-dimensional Sammon mapping of the 250 prototype vectors of a pattern-based LVQ model shows little clustering of the subgroups. This may indicate the absence of pattern-based heterogeneity, although it is not excluded by this visualization.

for each pattern by a feedforward network model. Training of this network is performed similarly to earlier experiments.

All patterns with squared estimation error below a threshold value form one subgroup, all others form the other subgroup. The threshold value is determined from inspecting the error distibution over all patterns. After the subgroups are formed, an LVQ model with 250 prototype vectors is trained to model the classification of the patterns to either of both groups. Using the LVQ model all patterns are re-classified into one of both groups, and separate feedforward networks are trained on the patterns in each group to estimate children's PEF. To process an input pattern, it is first classified by the LVQ classifier into one of the groups, and subsequently processed by the corresponding feedforward network, to yield an estimate of the PEF value.

Since heterogeneity may be the result of differing responses of children, caused by the presence of asthma for example, we also investigated the result of assigning children to subgroups, instead of individual patterns. All patterns of a child are assigned to either one of the subgroups based on the mean squared estimation error over all patterns of that child.

Correlation coefficients between observed and estimated PEF values in the test set (n=124), and their standard deviation over 10 fits of the models are for the linear model 0.623 (0.145), feedforward network 0.627 (0.057), pattern-based split 0.637 (0.064) and child-based split 0.631 (0.059). Figure 2 illustrates a distribution of LVQ prototype vectors. Both combined models yield higher mean correlation coefficients, although not significantly. Results indicate no child-based heterogeneity.

As noted earlier, correlations between input variables of our models are large. Most prominent are correlations between time lagged measurements of the same covariant, between a covariant and its averaged value over a larger time frame and between covariants that are inherently correlated through time. This last type of mutual association is evident between measurements of ambient SO_2, NO_2, Black Smoke and temperature levels.

Such strong correlations between input variables result in large sections of input-space (the space spanned by the input variables) to be practically void of observations. Input patterns in those sections correspond to unfeasible observations, because of restrictions on input value combinations imposed by the correlations. If the underlying correlations were known, combinations of input variables could be expressed in fewer synthesized variables, thus decreasing dimensionality of input space, and excluding improbable portions thereof. Improved covering of the input space with observations would most likely increase modelling quality. Also, although exposure levels are expressed in new variables, exposure-response plots will be less affected by the problem of input patterns being located in improbable portions of input space. Finally, new synthesized variables express previously complex profiles of exposure in fewer variables, and form more concise descriptions of exposure conditions, thus facilitating interpretation of the models.

Using DeMers' *non-linear dimensionality reduction* (NLDR) neural network [1], we described the eight time-lagged PEF input variables in three new synthisized variables. The method eliminates units in a designated hidden layer of an nonlinear autoencoding feedforward network by penalizing variance of their activation. This way, it forces autoencoding

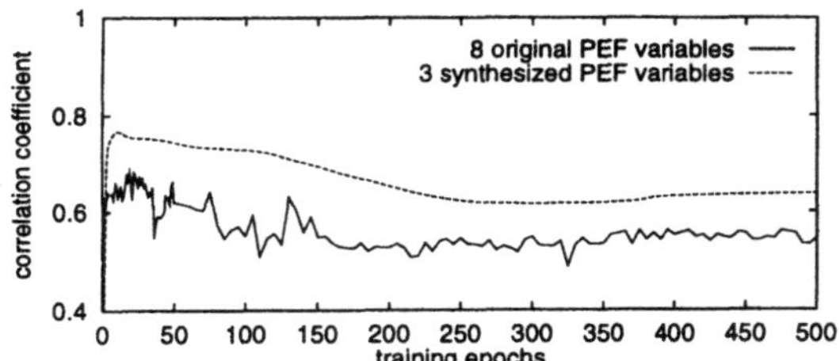

Figure 3: Learning curves showing the correlation coefficient between observed and estimated PEF values on the test data (n=124) as training of a feedforward network proceeds. One 'epoch' signifies one cycle through all training patterns (n=496). Curves are shown for both the 26 original input variables, including 8 PEF variables, and for input patterns with 3 synthesized PEF variables (21 input variables).

of the input patterns through a 'tightening' bottleneck layer, resulting in a low-dimensional activation patterns at that hidden layer for each input pattern. Tightening of the bottleneck hidden layer is ceased when restoration error at the output layer exceeds some user-defined tolerance level.

We applied the NLDR network to the 8-dimensional profiles of time-lagged PEF variables that were part of our previous models. The network reduced the bottleneck hidden layer to three units, before the restoration error became unacceptable. Replacing the eight PEF variables with the corresponding three activation values of the bottleneck layer in each pattern, we again trained standard feedforward network models to the same task as before: estimating the evening PEF. Results were compared to those for the original input patterns, and summarized in Figure 3. Replacing the original eight PEF variables with three synthesized ones increases correlation between observed and estimated PEF values of the test data. Further analyses are necessary to gain understanding of the new synthesized variables.

4 Conclusions

Neural network methods can be useful tools in analyzing epidemiological data. Exposure-response plots of feedforward network models fitted to the data give insight into nonlinear relations and interactions between covariants. Our results indicate nonlinearity and possible heterogeneity in the studied dataset. Our work demonstrated the development of new neural network methods suited for problems that arise within other areas of research — epidemiology in our case. Methods for exploring heterogeneity in response using a combination of LVQ and feedforward networks is fruitful. A model using nonlinear dimensionality reduction networks to reduce correlation between covariants, and hence improve modelling quality, also yields promising results.

References

[1] D. DeMers and G.W. Cottrell. Nonlinear dimensionality reduction. *NIPS 5*, pp.580-587, 1993.

[2] S. Geman, E. Bienenstock, and R. Doursat. Neural networks and the bias/variance dilemma. *Neural Computation*, 4:1-58, 1992.

[3] J. Hertz, A. Krogh, and R.G. Palmer. *Introduction to the Theory of Neural Computation*, volume 1 of *Santa Fe Institute Studies in the Sciences of Complexity*. Addison Wesley, 1991.

[4] T. Kohonen. The self-organizing map. *Proceedings of the IEEE*, 78(9):1464-1480, 1990.

[5] L. Moseholm, E. Taudorf, and A. Frøsig. Pulmonary function changes in asthmatics associated with low-level SO_2 and NO_2 air pollution, weather, and medicine intake. *Allergy*, 48:334-344, 1993.

[6] B.D. Ripley. Statistical aspects of neural networks. In *Séminaire Européen de Statistique*, 1992.

[7] D.A. Savitz. In defense of black box epidemiology. *Epidemiology*, 5:550-552, 1994.

[8] P. Skrabanek. The emptiness of the black box. *Epidemiology*, 5:553-555, 1994.

[9] T. Waschulzik *et al.* Evaluation of an epidemiological data set as an example of the application of neural networks to the analysis of large medical data sets. *4th Conf AI in Medicine Europe*, pp.466-476, 1993.

A Hybrid Expert System Architecture for Medical Diagnosis

L.M. Brasil[1], F.M. de Azevedo[1], J.M. Barreto[2]
[1] Biomedical Engineering Research Group (GPEB), Dept. of Electrical Engineering
[2] Dept. of Informatics and Statistics, Federal University of Santa Catarina, Brazil
Email: {lourdes,azevedo}@gpeb.ufsc.br, barreto@inf.ufsc.br

Abstract

This paper deals with a new methodology for the development of an expert system (ES) using a hybrid architecture. This architecture simplifies the knowledge acquisition phase, by providing some sort of learning corresponding to the training phase of the neural network. So, it is possible to start with the nucleus of a knowledge base and the system will improve during the learning phase using examples.

1 Introduction

Hybrid architectures for intelligent systems are a new field of artificial intelligence (AI) research concerned with the development of the next generation of intelligent systems. Current research interests in this field focuses on integrating the computational paradigms of symbolic manipulation and artificial neural networks (ANN), both having crisp and fuzzy values, and exploring the similarities of the underlying structures of these two methods of knowledge manipulation. Moreover, it focus on several applications where intelligent hybrid systems may and can play an important role.

Expert systems (ESs) appeared about twenty years ago as an application of the symbolic manipulation paradigm of AI. Traditionally, ESs have used symbolic artificial intelligence techniques and knowledge bases to simulate to action of human experts, and their main goal has been to solve specific problems in a given domain. These ESs are called rule-based expert system (RBES).

ANN are composed of several similar units. These processing units can be considered as a model of biological neurons. Therefore, every unit presents several inputs, some excitatory and some inhibitory, that are combined to generate an output (note that if the neuron is dynamical its output depends also on the preceding activation state of the neuron). So, a network is characterized by units (the neurons) and how they are connected (the topology). Moreover, algorithms are used to change the connection weights (the learning rules). Thus, these three aspects constitute the connectionist paradigm of AI [1]. ESs implemented in this a way are called neural network based expert systems (NNES). These systems are, generally, developed using a feedforward network trained by a backpropagation-like learning algorithm [10].

Building a traditional ES leads to several problems. The first of them is the process of knowledge acquisition (KA). In fact, knowledge extraction of a domain expert, which is one of the stages of KA, is a hard task. It is time consuming because normally human beings, even knowing how to solve a problem, have difficulty in explaining how they reached the solution of a specific problem. Another difficulty of the process of KA is the choice of the model for the knowledge representation to code human reasoning [2]. Production rules are one of most common knowledge representation schemes used. It is simple and direct, but it relies on a rich knowledge base. Moreover, it necessitates many updates. In the connecionist paradigm, the problems have been basically the choice of the input data to the ANN, the number of neurons in the hidden layer and the topology used. Finally, when the connectionist approach is used, it is not generaly very easy to obtain the explanation of how the network arrived at a conclusion. The efforts in this direction assume localized representation. So, a hybrid expert system (HES) is proposed to deal with the problems mentioned above [3, 4], and is presented in the next section.

2 Methodology

In this work, a methodology is proposed for building NNES to deal with the problems mentioned in Section 1. The main idea is that, in general, the domain expert has difficulty in specifying all rules mainly when imprecision is pervasive to the problem and fuzzy techniques are to be used. In this case, it is often difficult to chose the membership function. Nevertheless, he is able to supply examples of real cases. So, the knowledge engineer use the rules that were supplied by the domain expert to implement a basic structure of a NNES. Subsequently, the NNES is refined through a training algorithm that uses the set of available examples.

2.1 Knowledge Acquisition Stage

The KA task represents the first stage in developing the proposed methodology, i.e., the basic rules and the set of examples. The main goal of this stage is to extract the knowledge of the expert by rules in a short time such that it can also supply a series of examples of real cases. Moreover, the rules could be improved to add a way to capture the uncertainties associated with human cognitive processes. The model proposed uses fuzzy logic. The theory of fuzzy logic provides a great mathematical framework to represent this kind of knowledge [3, 4, 6, 10].

2.2 Assembly Stage of the Basic NNES

The implementation of the NNES uses neurons representing concepts. The rules relating these concepts are used to establish the topology of the ANN and employing a graphicl tool known as AND/OR graphs helps in the development of the basic structure of this system (Figure 1). In other words, AND/OR graphs, which represent concepts and connections, indicate the number of neurons in the input and output layers. They also show the existence of intermediate concepts and their connections which are translated in the intermediate layer of the neural network [4]. Moreover, the NNES also foresees the possibility of different kinds of variables in its input, where they represent different types of concepts, as quantitative, linguistic, or boolean valued or a combination of these [6, 8, 10]. In this way, the basic NNES is obtained.

Neural Mathematics Model

The mathematical model of the neuron is given by: $X(t)$ = n-dimensional input vector,

$$X(t) = [x_1(t), x_2(t), \cdots, x_i(t), \cdots, x_n(t)]^T \in \Re^n \quad (1)$$

$y(t)$ = scalar output of each neuron and $y(t) \in \Re^1$ N : nonlinear mapping function, $X \to Y, x(t) \mapsto y(t)$, where:

$$X : Z^+ \in \Re^n \quad (2)$$

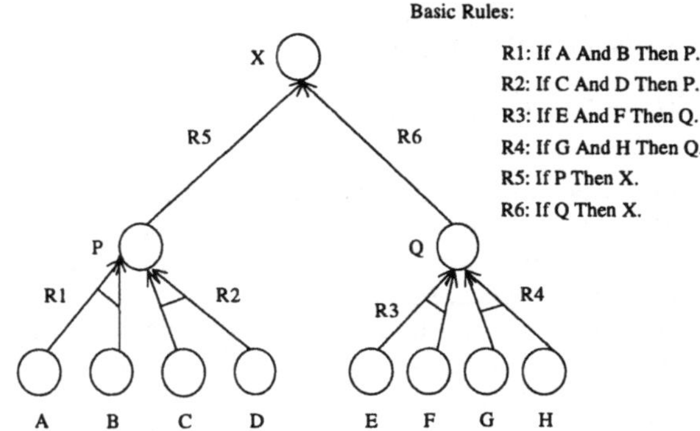

Figure 1: Organization of the rules base as a network.

$$Y : Z^+ \in \Re^1 \quad (3)$$

This mapping can be noted as N and so:

$$y(t) = N[X(t) \in \Re^n] \in \Re^1 \quad (4)$$

Mathematically, the neural nonlinear mapping function N can be divided into two parts: a function called confluence [6] and a nonlinear activation operation. The confluence function is the name given to a general function having as arguments the synaptic weights and inputs. A particular case widely used is the inner product. This mapping yields a scalar output $u(t)$ which is a measure of the similarity between the neural input vector $X(t)$ and the knowledge stored in the synaptic weight vector $W(t)$. So, $u(t) \in \Re^1$ and $W(t)$ is given by,

$$W(t) = [w_0(t), \cdots, w_i(t), \cdots, w_n(t)]^T \in \Re^{n+1} \quad (5)$$

Redefining $X(t)$ to include the bias $x_0(t)$ we have:

$$u(t) = X(t) \odot W(t) \quad (6)$$

The nonlinear activation function Ψ maps the confluence value $u(t) \in [-\infty, \infty]$ to a bounded neural output. Then, the nonlinear activation operator transforms the signal $u(t)$ into a bounded neural output $y(t)$, that is,

$$y(t) = \Psi[u(t)] \quad (7)$$

$$y(t) = \Psi[X(t) \odot W(t)] \quad (8)$$

Applying the Equations (1), (5), (6), and (8) to a multilayer ANN (e.g.,three layers), we have:

$$Y(t) = \Psi_3[W_3(t) \odot \Psi_2[W_2(t) \odot \Psi_1[W_1(t) \odot X(t)]]] \quad (9)$$

where i is non-linear activation operator, $\odot$ is the confluence operator, and $W_1(t)$, $W_2(t)$, and $W_3(t)$ are the synaptic weight vectors for the input, hidden and output layers, respectively.

If we express the neural input signals in terms of membership functions over the interval [0,1], rather than in absolute amplitudes, then we can perform mathematical operations on these signals using logical operations such as AND/OR, according to [6].

Let us express the inputs x_1 and x_2 over [0,1]. Then we define the generalized AND (T-norm) as a T mapping function and generalized OR (T-conorm) as a S mapping function [11]:

$$T : [0,1] \times [0,1] \rightarrow [0,1] \quad (10)$$

$$S : [0,1] \times [0,1] \rightarrow [0,1] \quad (11)$$

$$y_1 = [x_1 \, AND \, x_2] \equiv [x_1 \, T \, x_2] = T[x_1, x_2] \quad (12)$$

$$y_2 = [x_1 \, OR \, x_2] \equiv [x_1 \, S \, x_2] = S[x_1, x_2] \quad (13)$$

Then for OR neurons, in (6), by replacing the $\odot$-operation by the T-operation, and the $\sum$-operation by the S-operation, we get

$$u(t) = S_{i=1}^{n}[w_i(t) \, T \, x_i(t)] \, \epsilon \, [0,1] \quad (14)$$

$$y(t) = \Psi[u(t)] \, \epsilon \, [0,1] \quad (15)$$

Then for AND neurons, in (5), by replacing the $\odot$-operation by the operation of the algebraic product, and the $\sum$-operation by the T-operation, we get

$$u(t) = T_{i=1}^{n}[w_i(t) * x_i(t)] \, \epsilon \, [0,1] \quad (16)$$

$$y(t) = \Psi[u(t)] \, \epsilon \, [0,1] \quad (17)$$

2.3 NNES Refinement Stage

The NNES refinement stage is made through a learning algorithm using the examples of real cases as training set. This learning algorithm allows structural changes of the network through inclusion and/or exclusion of neurons and/or connections. This approach leads to a localized knowledge representation where neurons represent concepts and connections represent relations among concepts [2, 4, 6].

3 Learning Algorithm

The learning algorithm proposed provides modifications not only in the connection weights, but also in the network structure. It generates and/or eliminates connections that had not been in the fuzzy basic rules that were given by the expert. Moreover, it can also occasionally generate more concepts that were not in the fuzzy basic rules. So, the system translates as rules the new fuzzy basic rules that the expert was not able to provide during the development process of the basic ANN, in shape of new concepts and/or connections. Eventually, the number of neurons in the hidden layer must be modified to control of the generation and/or elimination of hidden intermediate concepts. In this case, genetic algorithms (GA) can be used to optimize the hidden layer [3, 4, 5, 8, 9].

GA are based on the work of Holland [7] who took inspiration from the evolution of a population subject to reproduction, mutation and crossover in a selective environment. The implementation of a GA generally involves the following cycle: Evaluate the fitness of all of the individuals in the population. Create a new population by performing operations such as crossover, fitness proportionate reproduction and mutation on the individuals whose fitness has just been measured. Discard the old population and iterate using the new population. One iteration of this loop is called a generation.

So, following this idea, we choose a GA to optimize the size of the hidden layer and determining weights to be set to zero [4, 5]. This can be justified by the following main facts: GAs can avoid local optima and provide near-global optimization solutions and are easy to implement. Nevertheless, when a GA is applied with this goal, in hidden layer of a ANN, we must take care of respecting a maximum and a minimum number of neurons of this layer. In fact, too many neurons generally have as effect a decrease of the generalization capabilities of the network and implies a long learning phase. On the other hand, too few neurons can be unable to learn, with the desired precision, the task. So, there is an intermediate number of neurons that must be put in the hidden layer, to avoid the problems mentioned above. Then, the network must be sufficiently rich to solve a problem and as it must also be adequately simple to solve a problem well and not consume long training [4].

4 Results and Discussions

The primary goal of this project was to develop a methodology for an ES using a hybrid architecture. The intrinsic capacities of a NNES, as well as the learning by examples paradigm were explored. So, the elicited knowledge of a domain expert during the KA phase was useful to define the NNES basic structure. Moreover, it has also aided in training of the NNES through a set of examples. This methodology has shown, in the preliminary studies performed, very promising by leading to an easier KA phase than expected if it was performed using symbolic techniques alone.

A supervised feedforward topology was used to implement the NNES because the input-output data used for the NNES was already well-known. The training of the NNES was performed using the classic backpropagation algorithm, with some alterations. One of them occurred in incorporating logic operators AND/OR in place of the weighted sum, e.g., it was replaced by Max/Min and algebraic product aggregation operators of fuzzy logic. Nevertheless, it was observed that in the backward pass the error propagation between the output and the hidden layer has reached expected values. However, the aggregation operators failed when applied to the connection between the hidden and the input layer, e.g., the network saturated. A solution is to use the formalism of weight adjustment of the generalized delta learning rule.

Another achieved goal was with regard to optimization of the topology to be adopted for the NNES. The optimization of the hidden layer was supported by GA. Nevertheless, when it is applied with this goal, we must take care of respecting a maximum and a minimum number of neurons of this layer. Then, in the maximization case the adopted solution considered the following points: the values for crossover and mutation rates was chosen empirically, e.g., the value of 0.6 for the crossover operator and 0.1 for the mutation operator. Using these values, the system reached a good performance. However, during the mutation process a population of chromosomes codifying a too big hidden layer to be an acceptable solution was created. To eliminate this effect, we considered the peculiarities of the example treated. The minimization case of the hidden layer has not been implemented. In this case, the physician can assert with a certain degree of trust that for a determined disease a certain symptom is predominant with regard to others. Then, the minimun value of chromosomes generated by the selection process and genetic operators should be decreased to maximum equal to number of important symptoms related by the expert. Doing so, we hope that the knowledge will not be altered with regard to a determined subject.

In future, we intend to use in the validation process of this approach a concrete example applied to a medical classification task, which is extremely difficult to solve. So, it will be used as Decision Support Systems to classify epileptic crises. This system will have about 80 to 100 rules combined with their membership degrees. These data will be elicited by physician experts, mainly at the University Hospital of the Federal University of Santa Catarina. Moreover, we intend also to develop a RBES in which it can explain how the NNES reached a given conclusion.

5 Acknowledgements

The first author acknowledges the CAPES (Coordination Foundation of High Level Personnel Improving) for the material support in the development of this work.

References

[1] F.M. de Azevedo. *Contribution to the Study of Neural Networks in Dynamical Expert Systems.* PhD thesis, Institut d'Informatique, FUNDP, Namur, Belgium, 1993.

[2] L.M. Brasil. Aquisição de conhecimento aplicada ao diagnóstico de epilepsia. Master's thesis, UFSC, Engenharia Biomédica, Florianópolis, Brasil, 1994.

[3] L.M. Brasil, F.M. de Azevedo, R. Garcia Ojeda, and J.M. Barreto. Cooperation of symbolic and connectionist expert system techniques to overcome difficulties. In *Proc. of II Congresso Brasileiro de Redes Neurais*, pp. 177–182, Curitiba, Brazil, 1995.

[4] L.M. Brasil, F.M. de Azevedo, R. Garcia Ojeda, and J.M. Barreto. A methodology for implementing hybrid expert systems. In *Proc. of The IEEE Mediteranean Electrotechnical Conference, MELECON'96*, pp.661–664, Bari, Italy, 1996.

[5] R. Garcia, F.M. Azevedo, and J.M. Barreto. Genetic algorithms in the optimal choice of neural networks for signal processing. In *Proc. of The 38th. Midwest Symposium on Circuits and Systems*, v. 2, pp.1361–1364, Rio de Janeiro, Brazil, 1995.

[6] M.M. Gupta and D.H. Rao. On the principles of fuzzy neural networks. *Fuzzy Set and Systems*, 61(1):1–18, 1994.

[7] J.H. Holland. *Adaption in Natural and Artificial Systems.* MIT Press, Reading, MA, 1975.

[8] H. Ishibuchi, R. Fujioka, and H. Tanaka. Neural networks that learn from fuzzy if-then rules. *IEEE Trans. on Fuzzy Systems*, 1(2):85–97, 1993.

[9] D.J. Janson and J.F. Frenzel. Training product unit neural networks with genetic algorithms. *IEEE Expert*, 1(1):26–33, 1993.

[10] S. Mitra and S.K. Pal. Logical operation based fuzzy MLP for classification and rule generation. *Neural Networks*, 7(2):353–373, 1994.

[11] H.J. Zimmermann. *Fuzzy Set Theory - and Its Applications.* Kluwer Academic Publishers, Reading, MA, 1991.

Enhancing Connectionist Expert Systems by IAC Models through Real Cases

N.A. Sigaki[1] F.M. de Azevedo[1] J.M. Barreto[2]
[1] Biomedical Engineering Research Group (GPEB), Dept. of Electrical Engineering
[2] Dept. of Informatics and Statistics, Federal University of Santa Catarina, Brazil
Email: {nancy, azevedo}@gpeb.ufsc.br, barreto@inf.ufsc.br

Abstract

This work presents a study of learning (case-based) in an interactive activation and competition (IAC) connectionist model. In this type of neural network, the basic learning mode may be classified as *rote learning*, and no iterative algorithm is used. The knowledge elicitation corresponds directly to the connection weights and its values are obtained by a type of engineering called *connection engineering*. In a way it is similar to the knowledge engineering in that it obtains functioning rules for an expert system. In this sense, an example of differential diagnosis in rheumatology is used to study the learning performance of a neural network with the introduction of real clinical cases, presented by a expert doctor. These clinical cases are used as a source of additional knowledge that represent relations between diseases and symptoms.

1 Introduction

The first implementations and the very notion of an expert system utilized the symbolic paradigm of artificial intelligence. Many systems were developed using this methodology, some of the most famous being Mycin, Dendral, Xcon, etc. More recently, with the renewed interest in neural networks, they were employed to implement expert systems, the work of Gallant [5] being one of the pioneering works. He utilized a feedforward neural network and its learning capacity. However, despite being expert in very special cases due to the distributed knowledge representation that occurs in these networks, to extract an explanation of the reasoning is a difficult task.

More recently de Azevedo [3] studied some variations of the IAC model. The IAC neural network approach was proposed by McClelland and Rumelhart [7]. Models of this nature have been studied by Grossberg [6]. De Azevedo has also proposed some modifications [1, 2, 3, 4] to the original Rumelhart's model. The original architecture of this model is composed of the processing units arranged in pools that interact among them. This interaction occurs within pools in which units are organized. There are excitatory connections among units of different pools and inhibitory connections among units of the same pool. These excitatory connections, acting between different processing units are generally bi-directional, leading to an interactive processing. The method is interactive in the sense that the processing in each pool influences and is influenced by the processing in the others. The inhibitory connections are generally transferred from each unit in the pool to every other unit in the pool. In this way, it causes competitive processing.

The network presents two kinds of competitive units: visible units (those that receive inputs from the exterior) and hidden units (those that cannot receive exterior inputs).

In IAC models, the time is discrete because the processing is divided into a sequence of steps, or cycles. Each cycle starts with the units having activation values as determined in the previous cycle. The new activation values are considered only in a new cycle. The updating procedure of each unit is synchronous.

There are some differences between the Rumelhart's model and those proposed by Grossberg and by de Azevedo. The main difference of Grossberg's models is in the formulation of the equations. In relation to de Azevedo model, the main difference occurs in the connections. The de Azevedo's model allows connections among all units, i.e., allows the existence of direct relations among all pools and, eventually, it is not necessary to have the hidden pool. Moreover, the synaptic weights belong to the range $[-1, 1]$. They represent a fuzzy value or the strength between concepts represented by units.

These network models have been proposed by

Barreto and de Azevedo [1, 2, 4] as useful in the implementation of expert systems, specially to solve problems in the medical area. They have shown good results. The idea is the exploration of the interactive activation and competition mechanisms, that characterize those models imitating the expert reasoning. These are a new connectionist expert system that traditionally have been implemented using a feedforward static network trained with a set of examples.

2 Case Based Reasoning

In IAC models, the basic learning mode may be classified as *rote learning*. Therefore, these are a *connection engineering* process where the connection weights are set *a priori* and stored on the neural network structure. Thus, there is not a real learning phase.

It is evident that a learning algorithm could be implemented for these kind of networks. Such an algorithm could be based on Hebb's law. However, the development of this algorithm is not the main goal of this work.

In any case, either learning rule or connection engineering, it seems natural to use real cases as knowledge sources. The knowledge about symptoms, the diseases associated to and laboratory tests, may be improved with the addition of data related to real clinical cases.

The objective of this work is the study of the performance of the system as a function of the real cases, previously diagnosed by an expert. These cases are present on the network structure. It is conjectured that, in the presence of such real clinical cases, the network dynamics become richer. An equilibrium point can be obtained more easily by an analogy to the closest real case.

This is the basic idea of the system called case based reasoning (CBR). In this type of system, the methodology is based on the evidence that humans use past cases stored in memory to solve certain problems. That is, they adapt known solutions to the current situation.

3 Methodology

The methodology used on the study of the network performance starts by the creation of the synaptic matrix. This solution takes into consideration the symptoms and disease relations (eventually also the relations with laboratory tests). The network response is observed for a test case set to be diagnosed differently from those used in the creation of the system. Clinical cases, one for each disease, are introduced into the synaptic matrix and again, the system is evaluated for a test set. This procedure is repeated, always followed by an evaluation of the network performance.

4 Case Study

An example of differential diagnosis in rheumatology is used in this work to exemplify and study the approach. It considers 4 diseases, 17 symptoms, and 24 known cases.

The diseases are: Rheumatoid Arthritis adult form (RA), Systemic Lupus Erythematosus (SLE), Psoriatic Arthritis (PA) and Arthritis of Gout (Gout or AG, for short).

- RA — is a chronic inflammatory disease of unknown etiology, affecting primarily the joint synovial membranes, though other tissues and organs may also be involved. There is evidence that a disordered immunological system plays a part in pathogenesis.

- SLE — is a chronic generalized inflammatory disorder of unknown cause affecting skin, joints, kidneys, nervous system and often other organs of the body. It is usually classified with the collagen or connective tissue disorders, and autoimmunity seems to be involved in the pathogenesis.

- PA — is an inflammatory, seronegative arthritis occurring in patients with psoriasis. There are no accurate figures for the overall incidence of Psoriatic Arthritis in the whole population but approximately 7% of patients with psoriasis develop it. It presents itself usually in the form of a chronic Poliarthritis very similar to the RA.

- AG — The Gout represents a group of genetic diseases and acquired metabolic disorders ordinarily identifiable by hyperuricemia. When clinically manifest, Gout presents itself as an acute inflammatory arthritis, with accumulation of sodium urate deposits as tophi, uric acid nephrolithiasis, or renal failure. These manifestations can occur in any combination.

In the initial phase, these four diseases have similar symptomatology. Any symptoms exclusive of one disease are often absent (although some of these specific symptoms appear generally later). This makes a correct diagnosis difficult during this phase if based exclusively in clinical data, and complementary tests are required. The following symptoms were considered: fever, arthralgia, arthritis, morning stiffness, myalgia, subcutaneos nodules, butterfly rash, raynaud's phenomenon, photosensitivity, alopecia, renal manifestations, central nervous system manifestations, pulmonary manifestations, rheumatoid hand, psoriatic lesion, tophi and podagra.

5 System Implementation

To implement the system, a program was written in Visual Basic in a windows environment. Having initialized the program, we see the presentation screen and several options of the paradigms of the neural network. We are interested only in the IAC network. After several choices, one has the implementation of this paradigm taking into consideration the Rumelhart, Grossberg and de Azevedo versions. The program allows the creation of a new network. We can choose the number of pools and the number of neurons for each pool. For the Rumelhart and Grossberg version one can create a hidden layer. There is, also, the possibility to change the parameters, such as: alpha, gamma, min, max, rest, decay and estr, where:

- Alpha — this parameter scales the strength of the excitatory input to units from other units in the network;
- Gamma — this parameter scales the strength of the inhibitory inputs to units from other units from the network;
- Min — the minimum activation parameter;
- Max — the maximum activation parameter;
- Rest — the resting activation level, which tends to settle in the absence of external input;
- Decay — the decay rate parameter, which determines the strength of the tendency to return to resting level;
- Estr — this parameter stands for the strength of external input (i.e., input to units from outside the network). It scales the influence of external signals relative to internally generated inputs to units.

The system is not limited in number of pools and neurons as the original software parallel distributed processing (PDP) (proposed by Rumelhart and Grossberg [7]), but by the quantity of available memory. Figure 1 shows a window of the program in execution. In this case, each pool of neurons is implemented in one window, that can be brought to the foreground when required by using the mouse. Therefore, we navigate through the representation of pools. The system allows us to plot the activation levels of a pre-defined number of neurons. In this way, if the user follows the transitions of the neural network during the consultation phase, it is supposed that we can understand the possible relations that can exist between the different diseases and symptoms and, eventually, laboratory tests. This understanding can become the explanation of the reasoning. This point is important given that connectionist systems are criticized by their inability to explain how they arrived at a conclusion. Note that, Gallant has shown, using a simple example, how this can be done [5].

6 Results

Simulation has been performed taking into consideration de Azevedo's model. The results, up until now, show the viability of this approach. One of the principal advantages of this approach is that the expert systems implemented in this way are able to indicate some supplementary data to arrive at a conclusion. Examples of this are the three steps of the normal process of diagnosis: anamnesis, clinical examination and (if considered in the system) laboratory tests. Given the network performance

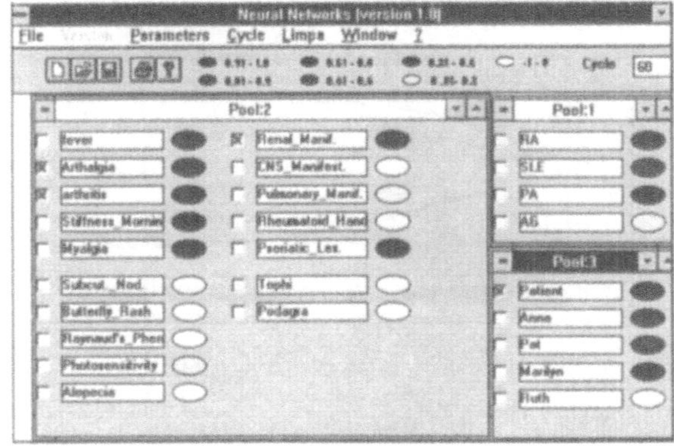

Figure 1: The computer system.

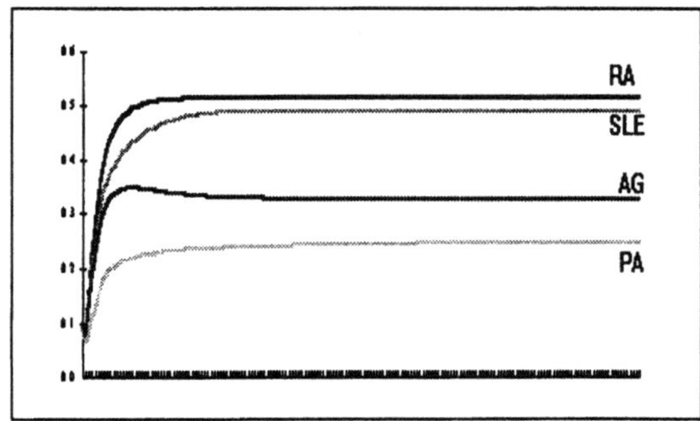

Figure 2: Transients of the activation levels of the four diseases: network considering only the symptoms and diseases relations.

with the introduction of more real clinical cases, it is observed that it can reach a good result compared to the performance of the network formed only by the symptoms and disease relations. The example shows the case of a patient with three symptoms: Arthralgia, arthritis, and renal manifestations. Figure 2 shows the transient of the values of excitation of the units representing the four diseases. The four diseases mentioned above differ only in the membership degree value stored in neural network structure. However, in this case the patient disease is undefined.

After the introduction of real clinical cases, a new state is shown in Figure 3. In this case, the renal manifestation symptom is revealed in all but the AG diseases.

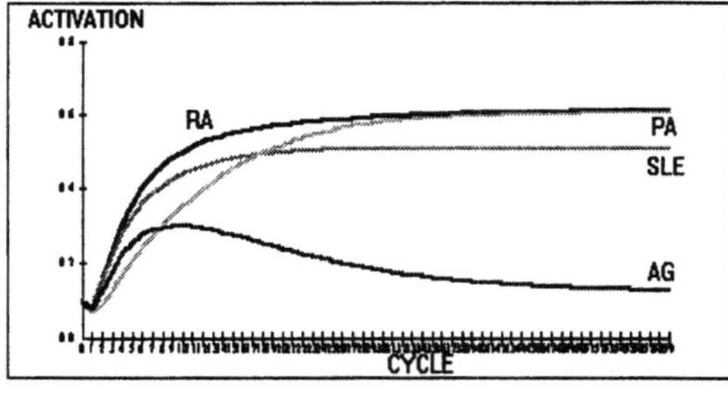

Figure 3: Transients of the activation levels of the four diseases: a network considering also the real clinical cases.

7 Conclusions

It was not the main goal of this work to determine a diagnosis, but to show that the network improve its performance with the introduction of real clinical cases. Figure 2 corresponds to the network transient when only relations between symptoms/diseases are included. It can be remarked that stationary points present a low separation of suggested diagnostics. Figure 3 corresponds to the same network with the real cases added. In the simulation performed it is observed that a different behaviour corresponds to a better functioning.

8 Acknowledgements

The first author acknowledges the CNPq for the material support in the development of this work.

References

[1] F.M. de Azevedo, J.M. Barreto, L.R. Epprecht, W.C. de Lima, and C.I. Zanchin. A neural network approach for medical diagnosis. In *Proceedings of the ISMM International Conference, on Mini and Microcomputers in Medicine and Healthcare*, Long Beach, Ca, U.S.A, 1991.

[2] F.M. de Azevedo, J.M. Barreto, E.K. Epprecht, L.R. Epprecht, and C.I. Zanchin. Two approaches in case-based connectionist expert systems. In *Proceedings of the IASTED International Conference, on Artificial Intelligence Applications and Neural Networks*, Zurich, Switzeland, 1991.

[3] F.M. de Azevedo. *Contribution to the Study of Neural Networks in Dynamical Expert Systems.* PhD thesis, Institut d'Informatique, FUNDP, Namur, Belgium, 1993.

[4] J.M. Barreto and F.M. de Azevedo. Connectionist expert systems as medical decision aid. *Artificial Intelligence in Medicine*, 5(1):515–523, 1993.

[5] S.I. Gallant. Connectionist expert systems. *Communications of the ACM*, 31(1):152 – 169, 1988.

[6] S. Grossberg. A theory of visual coding, memory, and development. In *Formal Theories of Visual Perception.* Wiley, New York, USA, 1978.

[7] J.L.McClelland D.E. Rumelhart and PDP Group. *Parallel Distributed Processing Vols 1 and 2.* MIT Press, Cambridge, Massachusetts, 1986.

A Schema Theorem-Type Result for Multidimensional Crossover

M.-E. Balázs[1,2]
[1] Department of Mathematics and Computer Science, 'Babeş-Bolyai' University, Cluj, Romania
[2] Worcester Polytechnic Institute, Department of Computer Science, 100 Institute Rd., Worcester, MA 01609, USA

Abstract

Most of the genetic algorithms (GAs) used in practice work on linear chromosomes (e.g. binary strings or sequences of some other types of symbols). However some results have been published revealing that for certain problems multidimensional encoding and crossover may give better results than the one dimensional (linear) ones [1, 2, 3]. While some theoretical results have been obtained, no clear criteria are known for deciding the suitable dimensionality of the encoding to be used for a given problem.

In this paper we consider a class of problems for which we define a multidimensional encoding and a corresponding genetic operator. We show that for a genetic algorithm (GA) using this encoding and operator we can obtain theoretical results similar to (under certain conditions even better than) those known for linear encoding. We demonstrate these theoretical results using a set of test examples.

1 Multidimensional Encoding and Crossover

1.1 Multidimensional Problems

We define multidimensional search problems as follows:

Definition 1. *Given the (finite) sets $A_1, A_2, ..., A_n$ by an n-dimensional (search) problem attached to a predicate P defined over $R = R(A_1, A_2, ..., A_n)$ we mean the problem of finding an n-ary relation $\rho^* \epsilon R(A_1, A_2, ..., A_n)$ such that $P(\rho^*)$ is true.*

Our goal in the following is to study some aspects of using genetic algorithms for solving n-dimensional problems. To be able to do this we assume that for every n-dimensional search problem considered there is a *fitness function* $f : R \rightarrow \Re^+$, which measures the 'quality' of any given relation. For our purposes the nature of this function is irrelevant.

1.2 The Common Approach

In using GAs to solve problems of the above mentioned type the current practice is to choose a (binary) encoding for each of the dimensions of a relation and concatenate them. One-point crossover of two relations then means choosing a crossover point, cutting both of the codes in that point and exchanging their 'tails', producing two offsprings. Other, multi-point, crossover operations, have also been defined [4, 6], however none of them are able to provide the type of improvement we are looking for.

Intuitively the drawback of these approaches is that they don't take into consideration the dimensionality of the problem, i.e. they don't try to preserve sub-relations of same arity of the original relations [1, 3].

1.3 n-dimensional Encoding and Crossover

In this section we propose another approach to using GAs for solving multidimensional search problems. Similar approaches have been proposed in [1, 2, 3] and [8]. Our approach is based on the use of multidimensional matrices for encoding and a crossover operator specific to this encoding. This was first defined in [1] and is similar to the encoding used by algorithm Z1 in [3].

The basis for all the above mentioned approaches is that an n-ary relation ρ over $A_1, A_2, ..., A_n$ can be represented by an n-dimensional matrix $M(\rho) = (m_{i_1,i_2,...,i_n})$ defined over $\{0, 1\}$ as follows:

Let $A_i = \{a_i^1, a_i^2, ..., a_i^{k_i}\}$ for $i = \overline{1, n}$. Then

$$m_{i_1,i_2,...,i_n} = \begin{cases} 1 & \text{if } (a_1^{i_1}, a_2^{i_2}, ..., a_n^{i_n}) \epsilon \rho \\ 0 & \text{otherwise} \end{cases}$$

For the two-dimensional case this corresponds to the adjacency matrix associated with the graph corresponding to the relation.

To make our discussion clear we first present our results for two-dimensional problems, the generalization to n-dimensional ones being discussed in a separate subsection.

Two-dimensional Encoding and Two-dimensional One-Point Crossover

Let $A = \{a_1, a_2, ..., a_m\}$ and $B = \{b_1, b_2, ..., b_n\}$ be finite sets, and let us consider the two dimensional search problem attached to a given predicate P over $R(A, B)$. To solve this problem using a genetic algorithm we have to choose a representation for the relations in P. We propose that the representation of a relation $\rho \epsilon R(A, B)$ be the corresponding $M(\rho)$ matrix as defined above. In this case a two-dimensional matrix representing a relation will be a chromosome processed by the genetic algorithm. Let us now define a crossover operator for a GA using the above considered encoding.

Definition 2. *Let $\rho_1, \rho_2 \epsilon R(A, B)$ be two relations and $M(\rho_1)$ and $M(\rho_2)$ the corresponding chromosomes. Let us further consider two positive integers k_1 and k_2 such that $1 \leq k_1 \leq m$ and $1 \leq k_2 \leq n$. The two-dimensional crossover of the two chromosomes using crossover point (k_1, k_2) produces two new chromosomes $M_1 = (m^1_{i,j})$ and $M_2 = (m^2_{i,j})$ defined by*

$$m^1_{i,j} = \begin{cases} r_{i,j} \text{ for } i < k_1 \wedge j < k_2 \text{ or } i \geq k_1 \wedge j \geq k_2 \\ q_{i,j} \text{ for } i < k_1 \wedge j \geq k_2 \text{ or } i \geq k_1 \wedge j < k_2 \end{cases}$$

$$m^2_{i,j} = \begin{cases} q_{i,j} \text{ for } i < k_1 \wedge j < k_2 \text{ or } i \geq k_1 \wedge j \geq k_2 \\ r_{i,j} \text{ for } i < k_1 \wedge j \geq k_2 \text{ or } i \geq k_1 \wedge j < k_2 \end{cases}$$

respectively.

The Two-Dimensional Schema Theorem

In order to give our first result we need some more definitions similar to those given for the classical Schema Theorem [4, 6].

Definition 3. *A two-dimensional schema of size $m \times n$ is a matrix of dimension $m \times n$ over $\{0, 1, *\}$.*

As in the Schema Theorem '' means 'either 0 or 1'. We call the positions having values 0 or 1 specific positions.*

A two-dimensional schema represents a set of matrices. More precisely: a two-dimensional schema with specific positions $S(\sigma) = \{(i_1, j_1), (i_2, j_2), ..., (i_k, j_k)\}$ defines the following matrix set:

$$\sigma = \{(a_{i,j})_{i=\overline{1,m}, j=\overline{1,n}} \mid a_{i,j} = \sigma_{i,j}, \forall (i,j) \epsilon S(\sigma)\}$$

By analogy with the one-dimensional Schema Theorem we also need the following definitions:

Definition 4. *The defining matrix of a schema σ is the smallest sub-matrix of σ that contains all its specific positions ($\Delta(\sigma)$).*

Definition 5. *If the dimensions of $\Delta(\sigma)$ are d_1 and d_2 respectively, then (δ_i, δ_j) is the defining size of σ, where $\delta_i = d_1 - 1$ and $\delta_j = d_2 - 1$.*

Similar to the Schema Theorem we have the following result:

Theorem 1. *(Two-dimensional Schema-Theorem) If σ is a two-dimensional schema of size $n_1 \times n_2$, then*

$$n(\sigma, t+1) \geq n(\sigma, t) \cdot \frac{f(\sigma, t)}{\overline{f(t)}} \cdot [1 - p_c \cdot p_d - \omega(\sigma) p_m],$$

where $p_d = \frac{\delta_i \cdot n_2 + \delta_j \cdot n_1 - \delta_i \cdot \delta_j}{(n_1 \cdot n_2 - 1)}$ is the disruption probability of a schema σ.

Proof. In most of its parts the proof of this theorem is identical to that of the one-dimensional Schema Theorem. Thus we will only insist on the single significant difference which is the estimation of the survival probability of a schema from one generation to the next one.

It is easy to observe that there are $\delta_i \cdot n_2 + \delta_j \cdot n_1 - \delta_i \cdot \delta_j$ crossover points in an $n \times m$ chromosome that disrupt a schema of dimension $\delta(\sigma) = (\delta_i, \delta_j)$.

This allows us to state that the survival probability of a schema σ of defining size $\delta(\sigma) = (\delta_i, \delta_j)$ is given by

$$p_s(\sigma) = 1 - \frac{\delta_i \cdot n_2 + \delta_j \cdot n_1 - \delta_i \cdot \delta_j}{n_1 \cdot n_2 - 1}.$$

The remaining part of the proof is identical to the proof of the one-dimensional Schema Theorem. □

This theorem only shows that a GA using the chosen coding and the above defined crossover operator behaves in a similar way to the one using linear coding and crossover.

Crossover: Two-dimensional versus One-dimensional

In the previous subsection we showed that a genetic algorithm using the two-dimensional encoding and crossover operator defined earlier behaves the same way as one using a linear encoding and single-point crossover.

In the following we shall study in what situations the use of this encoding and crossover can be advantageous. To do this let us consider a schema defined by two specific positions: (i_1, j_1) and (i_2, j_2), where $i_1 < i_2$ and $j_1 < j_2$. We shall compare the probability with which this schema will be disrupted on one hand if the schema is rearranged by rows and a one-point linear crossover is applied, on the other hand if the two-dimensional crossover is used (a similar analysis can be done if we rearrange the schema by columns).

We illustrate our reasoning on the following example. Let us consider the schema:

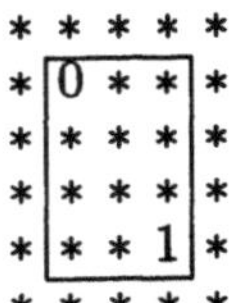

If we rearrange this schema by rows we obtain the following linear schema:

***** | * 0** * | * *** * | * *** * | * **1 * | *****

By simple counting it is easy to show that the specific positions in this schema will be separated by two-dimensional crossover for 15 crossover points and by one-dimensional crossover for 17 crossover over points.

The above example indicates that in some situations the disruption probability of two-dimensional crossover is less than that of the one-dimensional one. Let us find the conditions under which this is true.

The number of crossover points for which two-dimensional crossover disrupts a two-dimensional schema defined by the specific positions (i_1, j_1) and (i_2, j_2) is $a = n_2 \cdot \delta_1 + n_1 \cdot \delta_2 - 2\delta_1 \cdot \delta_2$, where (δ_1, δ_2) is the defining size of the two-dimensional schema. On the other hand, the number of crossover points for which one-dimensional crossover disrupts the corresponding one-dimensional schema (obtained by rearranging the two-dimensional schema by rows) is $l = n_2\delta_1 - \delta_2$.

In order to find under what conditions will the two-dimensional encoding and crossover be advantageous we have to determine when will the inequality $a < l$ hold. This inequality means

$$n_2 \cdot \delta_1 + n_1 \cdot \delta_2 - 2\delta_1 \cdot \delta_2 < n_2\delta_1 - \delta_2,$$

Since $\delta_2 \geq 0$, this inequality can only hold if $\delta_2 > 0$ and $n_1 - 2\delta_1 - 1 < 0$, which is equivalent to

$$\delta_2(n_1 - 2\delta_1 - 1) < 0.$$

that is

$$\delta_1 > \frac{n_1 - 1}{2}.$$

We can interpret this condition by saying that two-dimensional encoding and crossover performs better on 'column-schemata' than the one-dimensional one. Alternatively, we can deduce that by rearranging a two-dimensional schema by rows and applying one-dimensional crossover the GA will be biased towards 'line-schemata'. We may arrive at a similar conclusion if we rearrange the two-dimensional schemata by column.

1.4 Multidimensional Encoding and Crossover

In this section we only present the results obtained for the n-dimensional case. The reasoning behind these results is similar to the two-dimensional case, the difference consisting only in the complexity of the computations needed.

To give an intuitive idea of how crossover is defined in the n-dimensional case we illustrate it graphically in Figure 1. for the three-dimensional case.

It is easy to show that the disruption probability for the n-dimensional case is given by

$$p_d = \frac{\sum_{i=1}^{n} \delta_1 ... \delta_{i-1} d_i \delta_{i+1} ... \delta_n - (n-1) \prod_{i=1}^{n} \delta_i}{\prod_{i=1}^{n} d_i - 1}$$

where d_i is the i-th dimension of the chromosome.

Based on this the n-dimensional Schema Theorem (using similar definitions and notations as above) can be formulated as follows:

Theorem 2. *If σ is an n-dimensional schema of size $d_1 \times d_2 ... \times d_n$, then*

$$n(\sigma, t+1) \geq n(\sigma, t) \cdot \frac{f(\sigma, t)}{\overline{f(t)}} [1 - p_c \cdot p_d - \omega(\sigma) p_m],$$

The comparison to one-dimensional encoding and crossover presented in the previous section can be easily extended to the n-dimensional case.

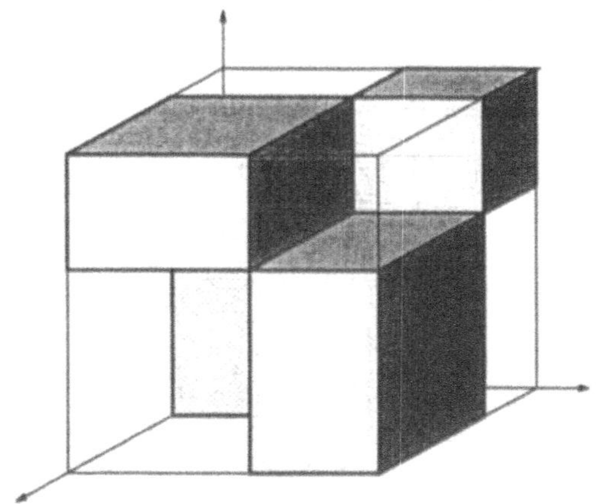

Figure 1: An intuitive idea of how crossover is defined in the n-dimensional case.

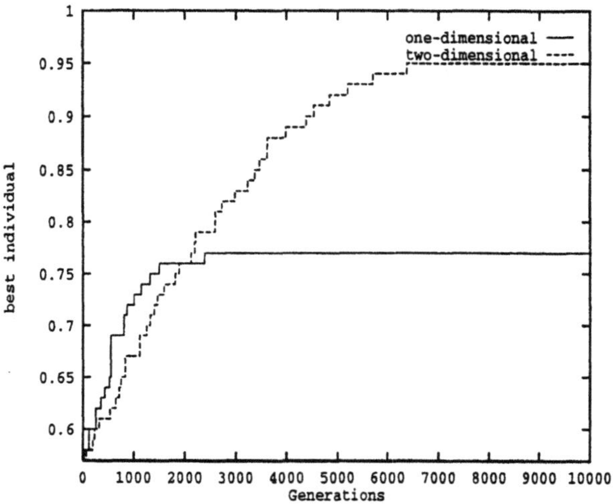

Figure 2: 'X' Relation.

2 Experimental Results

To demonstrate our theoretical results we performed a set of experiments on selected two-dimensional problems. All of these problems required to find homogeneous binary relations, that is relations $\rho \subseteq A \times A$, where A is a finite set $(A = \{a_1, a_2, ..., a_n\})$. The relations to be found were chosen such that reordering by rows or columns of the adjacency matrix doesn't provide an advantage to linear crossover.

For the experiments we used a steady-state GA with a population size of 100, crossover rate of 0.6 and mutation rate of 0.001/footnoteWe chose these values for the parameters of the GA to compare our results with the ones in [3]. The fitness function was simply calculated by counting the positions in a chromosome that matched the positions in the target chromosome.

In all the problems we considered the GA using two-dimensional encoding and crossover gave better results than the GA using one-dimensional encoding and crossover, both in the convergence rate of the fitness of best individual and in the quality of the solution.

For lack of space in Figure 2. we present the graph comparing the performances of the two-dimensional case to the one-dimensional case for one of the problems considered. The problem is to search for the relation (we consider the set A ordered by the indices given above):

'x' Relation: $\rho_{1st} = \{(a_i, a_j) \mid i = j \epsilon A\}$

3 Conclusions and Future Work

In this paper we introduced a multi-dimensional crossover operator and proved a Schema-Theorem type result for the genetic algorithm using it. We also showed that, in certain situations, using this operator is more advantageous than using the classical, one-dimensional one. Finally we demonstrated these theoretical results on a set of examples.

In the future we shall concentrate on trying to apply the operator defined here in real examples and compare the results obtained with results produced using other types of encoding and crossover operators.

Other experiments we performed indicated that two-dimensional one-point crossover may perform better than two-dimensional multi-point crossover (as opposed to the results presented in [3]). This encourages us in trying to develop an analysis of multidimensional crossover similar to the one given in [5].

References

[1] C. A. Anderson, K. F. Jones, and J. Ryan. A two-dimensional genetic algorithm for the ising problem. *Complex Systems*, 1991.

[2] M. E. Balázs. *Genetic Algorithms Theory and Applications.* PhD thesis, Babeş-Bolyai University, Romania, 1994.

[3] T. N. Bui and B. R. Moon. On multi-dimensional encoding/crossover. In *Proceedings of the Sixth International Conference on Genetic Algorithms*, 1995.

[4] L. Davis, editor. *Handbook of Genetic Algorithms*. Van Nostrand Reinhold, 1991.

[5] K. A. De Jong and W. M. Spears. A formal analysis of the role of multi-point crossover in genetic algorithms. *Annals of Mathematics and Artificial Intelligence*, 5(1):1–26, 1992.

[6] D. A. Goldberg. *Genetic Algorithms in Search, Optimization and Machine Learning*. Addison-Wesley, 1989.

[7] J. H. Holland. *Adaptation in the Natural and Artificial Systems*. Ann Arbor: University of Michigan Press, 1975.

[8] A. B. Khang and B. R. Moon. Toward more powerful recombinations. In *Proceedings of the Sixth International Conference on Genetic Algorithms*, 1995.

Möbius Crossover and Excursion Set Mediated Genetic Algorithms

S. Baskaran[1,2] and D. Noever[2]
[1] Institut fuer Theoretische Chemie, Waehringerstr. 17, 1090-Wien, Austria
[2] Biophysics Branch ES76, National Aeronautics and Space Administration
George C. Marshall Space Flight Center, Huntsville, AL-35812 USA
Email: subbiah@darwin.msfc.nasa.gov

Abstract

A non traditional highly disruptive crossover operator, called Möbius crossover, is introduced in the context of Excursion Set Mediated Genetic Algorithm (ESMGA). The new operator and the algorithm are applied to two GA deceptive problems and the results are reported.

1 Introduction

Within the dynamical requirements of a genetic algorithm(GA), crossover is a reproduction technique used to generate new population members. To this effect, many crossover operators have been designed for binary as well as real coded GAs, such as traditional n-point, uniform, discrete crossovers, and their properties studied. In addition, non traditional highly disruptive recombination operators have also been introduced and studied in GA optimization [3]. In the same spirit, in this paper, we introduce a novel crossover operator called Möbius crossover, and study its properties. Early results are presented in its application to GA deceptive problems. The results are compared with those obtained by a traditional simple GA [5].

2 Excursion Set Mediated Genetic Algorithm

Excursion Set Mediated Genetic Algorithm (ESMGA) [2, 8] was developed by combining the power of the geometrical properties of excursion sets in random fields [1] with the conventional sampling potential provided by the schema theorem [6]. Genetic optimization in ESMGA proceeds by inducing an hierarchy of excursion sets in the fitness landscape under optimization. Using these excursion sets, ESMGA in addition to allocating trials to fit individuals as in a simple GA, further allocates and maintains these trials to members which are propagating towards better fitness excursion subsets. Since excursion subsets or coordinate intervals at very high fitness levels (assuming maximization) confine the global optimum (excursion theorem [2]), this double sorted allocation certainly improves GA adaptation and exploration and in principle ends the search in excursion subsets containing the global optimum. Because of this, successful ESMGA dynamics centers around the assumption that during search all the excursion subsets are completely explored. Since mutation operator's exploratory ability is limited to a local vicinity of an excursion subset, we need robust and highly disruptive crossover operator to achieve global exploration of unconnected excursion subsets for a given fitness level. Thus designing efficient crossover operators with broader exploratory power is central to the successful GA dynamics embedded in ESMGA.

3 Möbius Crossover

Figure 1 shows the operational schematics of the Möbius crossover, and a conventional crossover in their one point version.

In this operator, crossover is done between population strings that are assumed to be distributed on a Möbius strip [9]. Since a Möbius strip does not posses orientation, one of the strings loses its original decoding direction. This orientation reversal happens with a certain probability based on how far it is separated from its mating partner. The separation metric can be anything but in the present study a random number between 0.0 and 1.0 decides the separation. A more meaningful metric will be their fitness values. Strings with similar fitness values will be closer and those with different fitness values will be considered further away on the Möbius strip. This introduces a form of crowding.

Möbius crossover in total has three components: an orientation reversal, a relative genome sliding, and a final segment exchange between the partners.

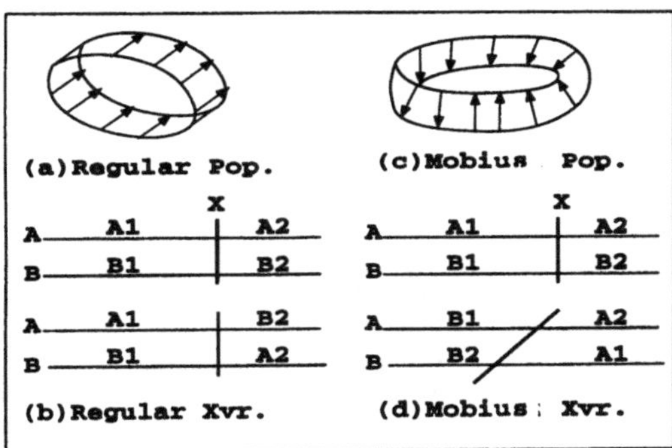

Figure 1: Genetic population on a regular(a) and a Möbius strip(c). Below are shown the corresponding regular(b) and Möbius swap(d) component of crossovers. Note that the polarity has reversed for some genomes on the Möbius strip. The sliding component is not shown.

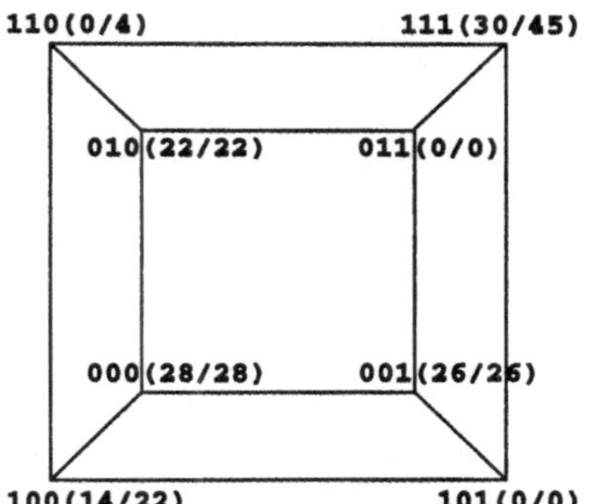

Figure 2: The Hamming cube schematics of a fully deceptive 3-bit function. The optimum is isolated and all local paths lead to the local optimum (000). The number within simple brackets are the fitness values: one above the slash is for DCF-1, and the one below is DCF-2 fitness.

The amount of relative sliding is determined by the sliding factor. The segment exchange part as shown in Figure 1(d) is similar to 1 point crossover except the exchange will be reflective of a Möbius strip.

The sliding component in essence provides a platform to experiment with the redistribution of the original epistasis built among the variables because of the initial fixed-focus encoding. The sliding factor takes values from zero to the maximum genome string length. This is a source of non-linearity in the genetic information exchanged during crossover. In the limit of no orientation reversal, no sliding, and no swapping, the crossover reduces to its 1 point analog.

Möbius crossover has shown robust characteristics when used with ESMGA on a range of test function landscapes.

4 Test Deceptive Functions

Most GA test functions do not show GA deceptiveness. However application of GA to these functions have met with varying degree of success [3, 5, 10]. In a fitness function, GA deception is present, whenever low-order schema fitness averages favor a particular local optimum, while the global optimum is located at its compliment.

Figure 2 shows the Hamming cube of order 3 deceptive basis function used in the construction of the two 30 bit long deceptive functions for our study. The two functions will be called DCF-1 and DCF-2.

If we assume the point 111 is the global optimum in the basis landscape, then for full deception, we require that all order-one and order-two schemata lead away from that point, 111. This translates into the following relations:

for one-bit schemata:

$$f(**0) > f(**1), f(*1*), f(**1).$$

and for the two-bit schemata:

$$f(*00) > f(*01), f(*10), f(*11).$$
$$f(00*) > f(01*), f(10*), f(11*).$$
$$f(0*0) > f(0*1), f(1*0), f(1*1).$$

Both functions, DCF-1 and DCF-2 are constructed from the 3-bit monomers by distributing the bits of these monomers over a 30 bit long string. In both cases, schema averages should drift the population towards deceptive optimum.

The function DCF-1 was introduced by Goldberg in his studies on the messy GA [5] on deceptive functions. DCF-1 is deceptive and has very low fitness difference between global and sub global optima ($f_{111} - f_{000} = 2$). This offered extreme difficulties for the low population limit (= 300) used in our study in contrast to the high population size (= 2000) used by Goldberg [5]. Because of this fact we introduced a reward factor equal to the average landscape fitness (= 15) for every correctly solved subproblem. This modification to the fitness function, although it did not help the simple GA to solve the problem, certainly enabled ESMGA to obtain the optimum. This is done as an experiment to understand how a deceptive function can be made GA easy. However the method provided here can be used only for Partial String Partial Evaluation (PSPE) type fitness functions [5].

The function DCF-2 is introduced in the study for the following reason: It is fully deceptive satisfying all the above schemata requirements, but in addition possesses a large fitness difference ($f_{111} - f_{000} = 15$) between the global and the sub global optimums, very similar to the order 4 deceptive function designed in [10]. For the low population limit, this should make the optimization approachable for a simple GA.

Because of schema theorem, a simple GA should face optimization difficulties depending on the order in which bits belonging to the sub-functions are distributed on the longer string in these functions. We consider three bit ordering schemes, tight, loose and random as used in [5].

The tight ordering refers to keeping bits belonging to a sub-function adjacent as in

$$1\ 2\ 3 \quad 4\ 5\ 6\ \ldots 25\ 26\ 27\ 28\ 29\ 30.$$

The loose ordering refers to keeping bits poorly arranged as in

$$1\ 4\ 7\ 2\ 5\ 8\ 3\ 6\ 9\ \ldots 24\ 27\ 30.$$

Whereas the random ordering refers to arranging the bits according to a preset table. We used the same random ordering table as in [5].

In the tight ordering, the building block length is small (= 2). If GA uses large population size, the traditional GA's positional bias will favor short schematas. But in moderate population sizes (e.g., 300 as in present study) this is highly unlikely and we obtained different results.

For the loose ordering, because schema length is longer, simple GA should experience deception and drift away from 111 settling in 000. For the random ordering, since the schema defining length is intermediate, it should solve a proportionate number of sub-functions.

All our experiments were conducted as described in [5]. We used Goldberg's simple GA [4]. Tournament selection was used for both simple and ESMGA. Mutation was completely disabled. The crossover probability was set as $p_c = 1.0$. For ESMGA Möbius crossover was used with the reversal probability set at .0001, and the slide factor set at 15. This reversal probability value corresponds to a dilute Möbius which is almost comparable to the non reversal in normal GA (1 point traditional crossover). However, in our experiments, the population size was set at 300 unlike Goldberg's setting of 2000.

5 Results and Discussion

From the experiments, we computed the maximum number of sub-functions solved and the average number of best building block, 111, in the population as a function of generation. In all the figures the simple GA has taken a steep course while ESMGA has shown its characteristic cascaded dynamics. The averages incrementally build up in ESMGA each time exploring steadily the available excursion subsets. This innovative exploratory phenomenon is totally absent in the simple GA.

ESMGA solved DCF-2 for all the three orderings. Figure 3 shows the maximum sub-functions solved for the loose ordering. The corresponding average evolution shown in Figure 4. Figures corresponding to tight ordering are not shown. The simple GA, although it solved as much as nine sub-functions, never could fully converge to the optimum. The reason is very obvious. Simple GA followed the schema averages fostered by the schema theorem and explored the landscape within that context and obtained convergence within smaller number of generations. On the other hand, ESMGA respected schema averages but also worked in conjunction with the excursion theorem. Although it converged at the end of each excursion cycle, its exploration continued to higher and higher fitness levels until no further improvement is possible. This cascaded dynamics in ESMGA, where each cascade embeds a miniature GA, is shown clearly in all the figures.

Figure 5 shows maximum sub-functions solved for the random ordering. The cascaded dynamics showed slow growth in the beginning and steady growth afterwards, in contrast to loose ordering where it is steady in the beginning and slow afterwards.

Experiments with DCF-1 did not offer promising results for loose and random ordering for the current parameter setting when the reward is set at 0. Although Möbius offers a wide choice of reversal probability and slide factor combinations, we did not try all of them. An extensive study should reveal combinations that solve the problem.

Here a word about Goldberg's messy GA [5] in solving all the three orders is worth mentioning:

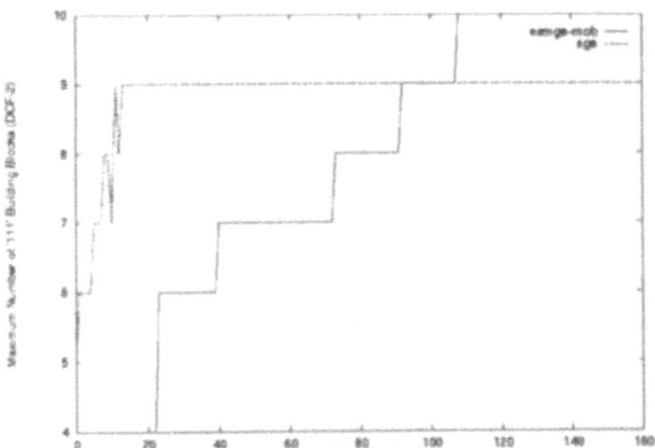

Figure 3: Evolution of maximum sub-functions solved for loose ordering DCF-2.

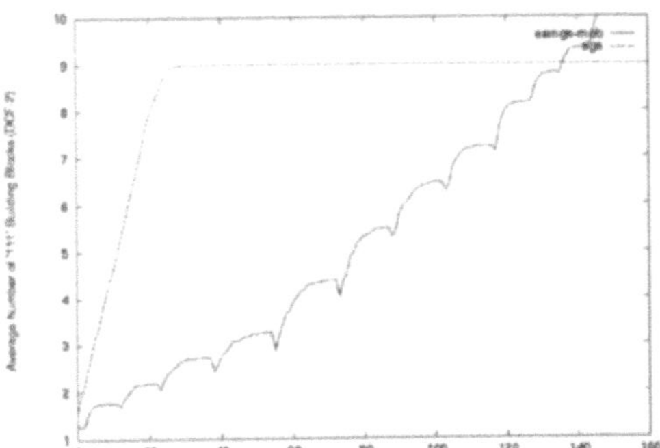

Figure 4: Evolution of average number of 111 building block for loose ordering DCF-2.

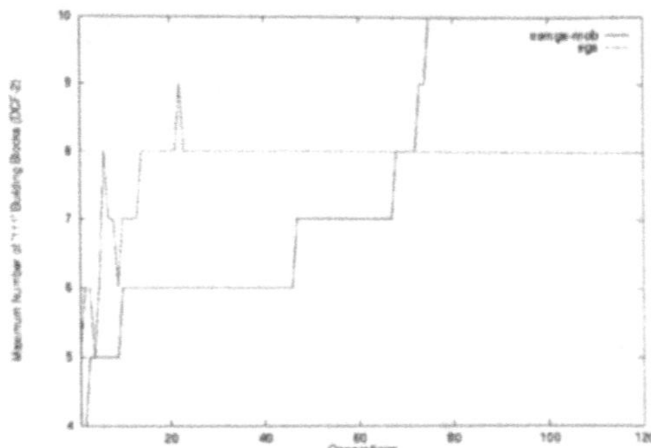

Figure 5: Evolution of maximum number of sub-functions solved for random ordering DCF-2.

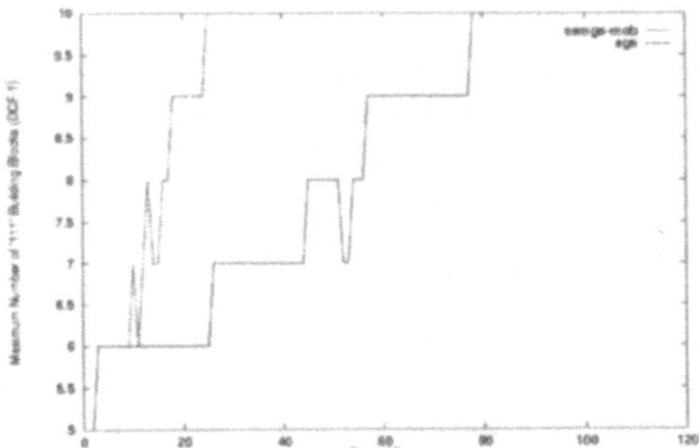

Figure 6: Evolution of best building block 111 for tight ordering DCF-1 with reward factor = 0.0.

messy GA maintains a large population (= 3000) of small primordial strings. The probability of 111 in these strings is very high. The evaluation scheme (PSPE) evaluates only fully contained sub-functions, so poor schematas are eliminated. In the juxtapositional stage, short strings are combined and evaluated and at this stage there will be maximum number of 111 blocks. So the probabilities multiply in promoting the 111 block to the final string which is evaluated for the problem. This makes messy GA very suitable for solving PSPE type fitness functions.

Simple GA solved only the tight ordering case even with reward equal to the average landscape fitness. At this level of reward, however, ESMGA solved all the three ordering cases. In order to show the difference between ESMGA and GA dynamics, we show in Figure 6 the maximum sub-functions solved for the tight ordering with zero reward. ESMGA progressed in a stepwise fashion exploring the landscape and solved the problem. It has taken more generations to avoid premature convergence. Whereas the simple GA followed very quickly the schemata averages and solved the problem. In this case, schemata averages helped simple GA because of the short schema defining length, but in other cases the same misled its flow ending in deceptive attractors [10]. This resulted in poor exploration of the landscape.

For rewards lower than the landscape average fitness, ESMGA consistently solved more sub-functions than the simple GA. Figure 7 shows evolution of the maximum sub-functions solved for loose ordering. The delayed construction of building blocks is one of the characteristics of ESMGA optimization. ESMGA explores the landscape in a stepwise manner delaying the takeover by any furious building block structures constructed by the schemata alone. Because of the second level exploration within excursion sets of a given level towards

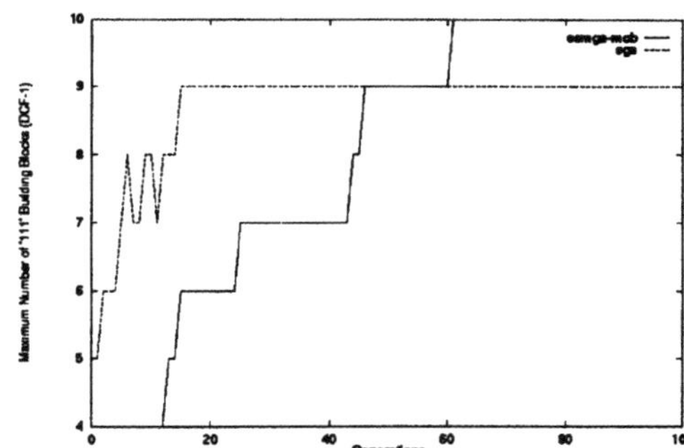

Figure 7: Evolution of maximum number of sub-functions solved for loose ordering DCF-1 with reward factor = 15.0.

optimal subsets (schemata within schemata), this false takeover is delayed as much as possible. A mathematical analysis of the dynamics can be done using Hidden Markov Field Models.

6 Conclusions

In this paper, we introduced a new type of crossover called Möbius crossover, and applied that to solving GA deceptive functions. A sub-function reward was also considered successfully to alleviate GA deception but only in the case of PSPE type fitness functions. Future versions of ESMGA will incorporate genetic control parameters in the evolving genomes, and make them evolve along with the problem. This should in principle enable it to explore all the excursion subsets and hence trap the global optimum in the given domain.

References

[1] A. D. Adler. *The Geometry of Random Fields.* John Wiley & Sons, New York, 1981.

[2] S. Baskaran and D. Noever. *Excursion Sets and a Modified Genetic Algorithm*, pages 55–567. Addison-Wesley, Reading, MA, 1993.

[3] L. J. Eshelman. *The CHC Adaptive Search Algorithm*, pages 265–283. Morgan Kaufmann, 1991. G.Rawlins, ed.

[4] D. E. Goldberg. *Genetic Algorithms in Search, Optimization, and Machine Learning.* Addison-Wesley, Reading, MA, 1989.

[5] D. E. Goldberg. Messy genetic algorithms: Motivation, analysis, and first results. *Complex Systems*, 3:493–530, 1989.

[6] J. H. Holland. *Adaptation in Natural and Artificial Systems.* University of Michigan Press, Ann Arbor MI, 1975.

[7] G. E. Liepins and M. D. Vose. *Deceptiveness and Genetic Algorithm Dynamics*, pages 36–50. Morgan Kaufmann, 1991. G.Rawlins, ed.

[8] D. Noever, S. Baskaran, and P. Schuster. Understanding genetic algorithm dynamics using harvesting strategies. *Physica*, D79:132–145, 1994.

[9] J. R. Weeks. *The Shape of Space.* Marcel Dekker, Inc., New York, 1985.

[10] L. D. Whitley. *Fundamental Principles of Deception in Genetic Search*, pages 221–241. Morgan Kaufmann, 1991. G.Rawlins, ed.

The Single Chromosome's Guide to Dating

M. Ratford, A. Tuson and H. Thompson
Department of Artificial Intelligence, University of Edinburgh,
80 South Bridge, Edinburgh, U.K.
{miker,andrewt,ht}@dai.ed.ac.uk

Abstract

In nature, sexually reproducing organisms do not mate indiscriminately — the choice of mate has an impact upon their offspring's fitness. The investigation described here shows that, for a wide range of problems in the literature, using sexual selection proved to be a robust method for enhancing genetic algorithm performance. In addition, this investigation provides evidence for which parameters are important for a successful implementation.

1 Introduction

In genetic algorithms (GAs) [6], the crossover operator plays a central role. Informally, the rationale behind this is that the recombinative process combines portions of the two parent solutions, in the hope of combining sections of each solution associated with high fitness together, to produce solutions of higher fitness. Standard implementations of the GA select the second parent either at random, or on the basis of fitness. However, using different criteria to select the second parent may make crossover more effective by encouraging recombination between strings that have useful information to exchange, thus making GAs more useful optimisers.

In nature, sexually reproducing organisms do not mate indiscriminately. The reason for this is simply that half of the offspring's genes will come from the other parent, and the choice of mate will therefore have a large impact upon their children's fitness, and hence the survival of the parent's own genes. Therefore organisms effect some form of sexual selection such as choosing a mate to control the balance between out-breeding and inbreeding. Rephrasing this in GA terms, crossover between solutions that are too similar should be discouraged as no useful search takes place; but there should be some similarity as information exchange is more likely to be meaningful.

Work has been performed on using mate choice in GAs: [10] described how the use of a 'seduction function' based upon an aesthetic measure to select the second parent can improve GA performance; [7] discussed the various ways that sexual selection can be used in evolutionary computation; and [4] showed that 'incest prevention', the prohibition of crossover between identical or very similar strings, can prevent premature convergence and give improved results.

An investigation of the potential of using sexual selection, and how it should be implemented in GAs is described in this paper.

2 Implementation

The publicly available GA implementation PGA was used as a testbed for this investigation. Due to lack of space, full details of the implementation are given in [9]. The first parent was selected in the usual fashion; with modifications being made in the selection of the second parent. A seduction function [10], based upon the similarity of each of the prospective second parents to the first parent, was then used to select the second parent. The process is summarised by the pseudo-code below (adapted from [10]):

```
until (termination condition met) do {
   parent1=select(population);
   parent2=seduce(parent1,population);
   operate(parent1,parent2,child1,child2);
   merge(child1,child2,population); }
```

2.1 Similarity Metrics

Three types of a similarity (or distance) metric were investigated: *Hamming distance* which measures the similarity in genotype space; *Euclidean distance* which measures the similarity between solutions in phenotype space; and finally the number of building blocks common to both chromosomes. In each case the similarity metric is normalised to a value between 0 and 1 by dividing the raw value by the maximum possible.

Seduction Functions

The similarity measure, *distance*, needs to be somehow processed to provide a measure of seduction, *sed*, such that chromosomes with high seduction values make better 'suitors'.

Two forms of the seduction function were investigated, with the parameters: *centre*, *width* and *depth*. A bell curve was defined as:

$$sed = \left(\frac{width}{width + (distance - centre)^4}\right)^{depth}$$

and a simpler, and less computationally expensive function was defined as:

$$sed = \begin{cases} 1 - depth + \frac{distance \times depth}{centre} & distance \leq centre \\ 1 - depth \times \frac{distance - centre}{1 - centre} & distance > centre \end{cases}$$

Incest prevention [4] was also implemented: a *threshold* was provided, so that the final seduction measure was defined as:

$$\begin{matrix} thresholded \\ seduction \end{matrix} = \begin{cases} 0 & seduction \leq threshold \\ seduction & seduction > threshold \end{cases}$$

It is possible that using distance alone to select a second parent could be inefficient, allowing unfit parents to contaminate the reproductive process. Thus three options for a final *attraction* measure were used: seduction only, the mean of seduction and normalised fitness, and seduction multiplied by *normalised fitness* defined as:

$$normalised\ fitness = \frac{\#\ solutions - fitness\ rank + 1}{\#\ solutions}$$

where the *fitness rank* is 1 for the fittest chromosome, 2 for the second fittest, and so on.

Selecting The Mate

As calculating the 'overture' from every other chromosome in the population each time an individual was selected for breeding would clearly be unmanageable, tournament [1], marriage [11], and courtship selection [9] were used.

3 The Test Problems

This investigation tested the performance of sexual selection over a wide range of problems in the literature, to provide a strong test of this technique.

- De Jong's test functions (DJ1, 2, 3, and 5) [2].
- Modified binary F6 (BF6) [12].
- Deb's test functions (DEB1, 2, 3, and 4) [3].
- Himmelblau's function (HIMM) [5].
- Maximising 1s (MAX).
- Maximum contiguous block (MCBn).
- The Royal Road (RR) [8].

4 An Initial Study

An initial attempt was made to find trends for each parameter value to provide a general feel for what was and was not important. The test functions used were DJ3, DJ5 and BF6. Due to space constraints only a summary of results is given in this paper. Detailed results are available in [9].

The GA was initially run with and without mate selection, covering as wide a range of the parameter space as practicable. Each run had a population of 50 and a genotype size of 32. Each set of parameters was run 20 times. Each run was terminated when the maximum possible fitness was reached, the population had converged, or 100,000 function evaluations had been made. These results were then examined to detect trends for each parameter choice. A *t-test* was used to determine significance.

4.1 Does Sexual Selection Work?

The use of seduction only, rather than a combination of seduction and fitness, produced significantly better quality solutions for DJ3 and BF6.

4.2 The Seduction Function

There was no difference between either of the two functions (bell or simple) used.

4.3 The Similarity Metric

The results of using phenotypic or genotypic similarity measures varied. The phenotype was slightly better for DJ3, no different for DJ5, and the genotype measure did better for BF6; the choice is problem specific. The building block measure did extremely badly.

4.4 The Selection Method

Courtship selection outperformed marriage selection for DJ3 and BF6 with no difference for DJ5.

4.5 Incest Prevention

Using sexual selection in conjunction with incest prevention using a seduction threshold gave better results each time.

5 Expanding the Study

The study was then expanded to include all of the test problems, to give a convincing demonstration of sexual selection's abilities, and to check the earlier conclusions.

To see how good the current model was, each function was run 50 times with an educated guess at a good, standard genetic algorithm: a steady-state GA, with mutation rate 0.001, string length 30, two-point crossover, and rank-based selection. The functions were then rerun with the same algorithm together with an educated guess at a good sexual selection technique: a simple curve, seduction as an attraction measure, Hamming distance, incest prevention, and marriage selection. The results obtained (Tables 1 and 2) were heartening, with 11 out of the 13 functions having a significant (in **bold**) improvement in solution quality.

The fact that this technique was not tuned indicates that it is fairly robust to its parameter settings. Also, it was interesting to note that, unlike in Section 4.4, marriage selection proved to be the better choice. This demonstrates the dangers of making general statements from only a few problems.

5.1 Incest Prevention Revisited

Is it just incest prevention that is causing the improvement? When the same tests were performed with a seduction threshold of 0.1 and selection based on ranked fitness only, the results were actually worse than for those tests using a standard GA *without* sexual selection. Thus either the initial conclusions (in Section 4.5) about its worth were wrong or incest prevention, as used here, is doing its work as an intrinsic part of sexual selection — follow-up results using sexual selection without a seduction threshold were inconclusive, providing better results than those from using sexual selection with incest prevention for seven test functions and worse results for five. The small size of the differences suggest that the shape of the seduction curve renders explicit incest prevention unnecessary.

5.2 Varying the Aesthetic

The runs were repeated using a phenotype measure of similarity, where the algorithm did generally slightly worse than when using a genotype measure.

This may, at first, appear slightly surprising, as in genotype space the fitness landscape will be more disjointed and ragged. However crossover will often have more chance of success between chromosomes that are genotypically similar, as phenotypically similar chromosomes may be either side of a *Hamming cliff*, where a small difference in natural numbers is equivalent to a large difference in the binary representation. In these cases crossover will be a poor tool for local search.

The results when the building block aesthetic

Table 1: Results for a standard GA.

Test	Opt	Fitness Mean	Fitness σ	Evals Mean	Evals σ
MAX	×25	29.34	0.7982	714.00	110.21
DJ1	×0	99.9637	0.0685	938.00	305.49
DJ2	×1	999.948	0.0626	846.00	244.08
DJ3	×20	54.12	0.9398	15790.00	28218.95
DJ5	×1	496.528	3.0968	982.00	263.76
BF6	×2	0.761324	0.2664	850.00	245.78
MCB5	×0	19.54	3.8184	33786.00	16758.13
RR	×11	0.9804	0.8182	79118.00	40414.82
DEB1	×15	0.999925	0.0003	1002.00	298.46
DEB2	×8	0.986123	0.0445	1022.00	291.40
DEB3	×7	0.999649	0.0006	998.00	315.09
DEB4	×8	0.993056	0.0157	974.00	267.69
HIMM	×1	199.386	0.9354	926.00	285.40

Table 2: Results for a GA with sexual selection.

Test	Opt	Fitness Mean	Fitness σ	Evals Mean	Evals σ
MAX	×49	**29.98**	0.1414	718.00	95.70
DJ1	×1	**99.9926**	0.0178	1406.00	1217.75
DJ2	×1	**999.951**	0.0746	1266.00	290.92
DJ3	×46	**54.92**	0.2740	4074.00	11308.56
DJ5	×8	498.08	1.7311	1202.00	281.57
BF6	×8	**0.928968**	0.1749	3258.00	13974.38
MCB5	×2	**26.36**	2.8978	59586.00	20139.21
RR	×15	1.1332	0.8639	73506.00	43758.20
DEB1	×34	**0.999979**	9.3e-5	1230.00	256.35
DEB2	×11	**0.991279**	0.0417	1270.00	373.07
DEB3	×15	**0.999957**	0.0002	1182.00	230.74
DEB4	×6	**0.994211**	0.0151	1186.00	394.74
HIMM	×1	**199.924**	0.2239	1278.00	361.43

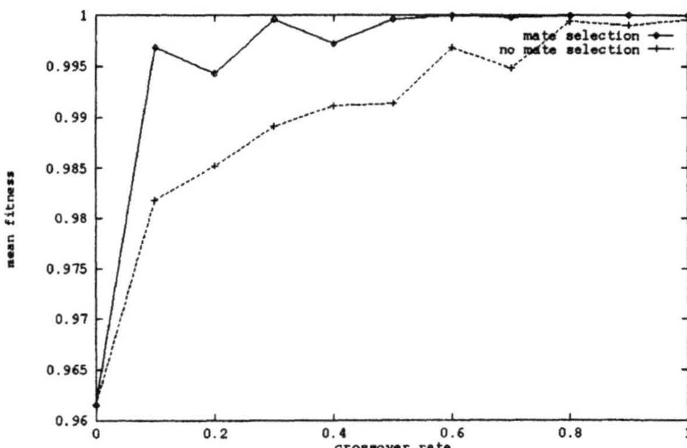

Figure 1: Mean fitness against $p(Xover)$, DEB4.

were used were poor, worse than not using mate selection in most cases.

5.3 The Effect of Crossover Rate

Varying the crossover rate showed that the higher the crossover rate, the better the results achieved, although in more evaluations. This result applied when using a standard GA, but the rate of increase was much higher with sexual selection (Figure 1); i.e. sexual selection makes crossover more effective.

6 Conclusion

On the basis of the results obtained here, sexual selection appears to have made crossover a more effective search operator. Significant increases in solution quality were obtained for most of the problems. Although of course some care has to be taken on how it is implemented, it appears that the approach is fairly robust to its settings in many cases, and some useful guidelines to the application of this technique have arisen from this investigation. This bodes well for its application to problems in the real-world.

7 Acknowledgements

Thanks to the EPSRC for their support of Michael Ratford and Andrew Tuson via studentships with references 95411068 and 95306458.

References

[1] A. Brindle. *Genetic Algorithms for Function Optimization.* PhD thesis, University of Alberta, 1981.

[2] K. A. De Jong. *An Analysis of the Behavior of a Class of Genetic Adaptive Systems.* PhD thesis, University of Michigan, 1975.

[3] K. Deb. Genetic algorithms in multimodal function optimization. Master's thesis, University of Alabama, Department of Engineering Mechanics, 1989. TCGA Report No. 89002.

[4] L. J. Eshelman and J. D. Schaffer. Preventing premature convergence in genetic algorithms by preventing incest. In R. K. Belew and L. B. Booker, editors, *Proceedings of the Fourth International Conference on Genetic Algorithms.* Morgan Kaufmann, 1991.

[5] D. M. Himmelblau. *Applied Nonlinear Programming.* McGraw-Hill, 1972.

[6] J. H. Holland. *Adaptation in Natural and Artificial Systems.* Ann Arbor: The University of Michigan Press, 1975.

[7] G. F. Miller. Exploiting mate choice in evolutionary computation: Sexual selection as a process of search, optimization, and diversification. In T. C. Fogarty, editor, *Proceedings of the 1994 AISB Workshop on Evolutionary Computing.* Springer-Verlag, 1994.

[8] M. Mitchell, S. Forrest, and J. H. Holland. The Royal Road for Genetic Algorithms: Fitness Landscapes and GA Performance. In *Toward a Practice of Autonomous Systems: Proceedings of the First European Conference on Artificial Life*, Cambridge, MA, 1991. MIT Press.

[9] M. Ratford. The Single Chromosome's Guide To Dating. Master's thesis, Department of Artificial Intelligence, University of Edinburgh, 1996.

[10] E. Ronald. When Selection Meets Seduction. In Larry J. Eshelman, editor, *Proceedings of the Sixth International Conference on Genetic Algorithms*, San Francisco, CA, 1995. Morgan Kaufmann.

[11] P. Ross. Personal communication, 1996.

[12] J. D. Schaffer, R. A. Caruna, L. J. Eshelman, and R. Das. A study of control parameters affecting online performance of genetic algorithms for function optimization. In J. D. Schaffer, editor, *Proceedings of the Third International Conference on Genetic Algorithms.* Morgan Kaufmann, 1989.

A Fuzzy Taguchi Controller to Improve Genetic Algorithm Parameter Selection

C.-F. Tsai[1,2], C. G. D. Bowerman[1], J. I. Tait[1] and C. Bradford[1]
[1] School of Computing and Information Systems, University of Sunderland, U.K.
[2] Department of Industrial Engineering Management, Tamsui Oxford University College, R. O. C.
Email: {cs0cts, cs0cbo, cs0jta}@cis.sunderland.ac.uk

Abstract

The selection of operators and parameters for genetic algorithms (GA) depends upon the situation, and the choice is usually left to the users. Identifying the optimum selection is very time consuming and, therefore, it is important to develop a system which can assist the users in their selections. In our fuzzy Taguchi controller, we present a hybrid system, which combines the Taguchi method with fuzzy logic, to select near optimum settings for the design parameters. The Taguchi method selects an optimal orthogonal array from experimental design theory, to reduce the number of experiments required to study the parameter space. Our controller uses this array to determine the selection for fuzzy membership in the dynamic selection process. It then applies fuzzy logic to evaluate the beneficial genes which affect the GA performance. We use the hybrid procedure to produce evidence from simulations and this information is then used to refine the GA behaviour. The system utilises a fuzzy matrix to rearrange the sequence of gene groups within the chromosome and applies a fuzzy knowledge base to tune the GA parameter selection. This provides a simple and easy method to assist users to direct their search and optimisation in an efficient way.

1 Introduction

In the past, many researchers have attempted to utilise a meta-GA or neural network to find an optimal selection for the GA parameters. But if the system applies a meta-GA to solve the problem, the problem becomes recursive (how to optimise the meta-GA's parameters?) and also time consuming [5]. If the researcher uses a neural network to find a single situation, it still needs a lot of data to train the system and it is also very difficult to determine the size of the hidden layer in an efficient way. In most situations, users do not have a lot of data, or enough lead-time to react to the situation so, in practice, neural networks usually cannot be used to produce an optimum solution. Our system attempts to reduce the search space in a logical and efficient way. Hence, we apply the Taguchi method to determine an optimal array to meet the situation; and also to select the fuzzy membership. This can reduce the search time significantly. The system also applies the evolution evidence, which are the (average fitness)/(best fitness) or (worst fitness)/(average fitness) etc. to tune the GA parameter by fuzzy reasoning [2]. As the fuzzy reasoning can map one situation onto two (or more) memberships, it can effectively reduce the required search space by half (or more). In this controller, there are two functions to control the GA performance. Firstly, the system uses a fuzzy matrix to refine the chromosome structure. Secondly, the system can tune the selection of GA parameters over evolutions. Both functions use evolution evidence to direct the fuzzy reasoning.

2 System Architecture

The experimental tasks for the optimisation of the economic design of multiple control charts (EDMCC) consist of four main components: a control chart selector which identifies the process situation and selects the appropriate control charts for the problem; a GA manager which selects the optimal operator and parameter settings for the GA evolutions; a process parameter controller which refines the chromosome structure and assigns suitable weights for each individual objective function; and a genetic algorithm which is a simple implementation of the GA (see Figure 1) [3, 4].

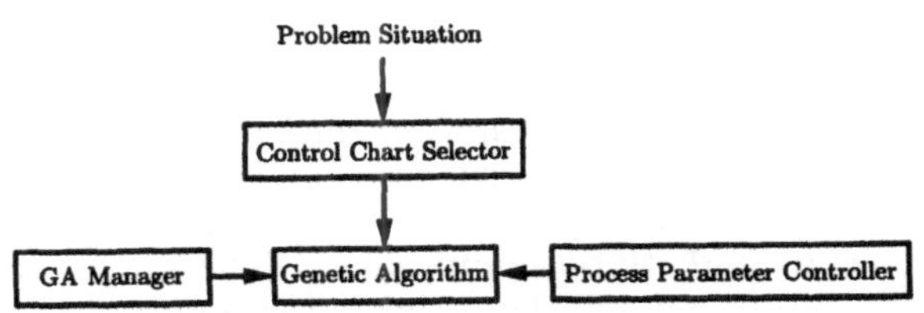

Figure 1: The system architecture.

Table 1: Three parameters with two level orthogonal array.

Exp No.	Parameter 1	Parameter 2	Parameter 3
1	Level 1 (15)	Level 1 (0.85)	Level 1 (0.05)
2	Level 1 (15)	Level 2 (0.95)	Level 2 (0.001)
3	Level 2 (20)	Level 1 (0.85)	Level 2 (0.001)
4	Level 2 (20)	Level 2 (0.95)	Level 1 (0.05)

This experimental framework has two main objectives: the first is to apply the genetic algorithm to find the better solutions for EDMCC. The second objective is to utilise three proposed approaches to improve the behaviour of genetic algorithms. The operations of this experimental framework are divided into three stages. In the initial stage, the control chart selector identifies the product type in the production process and the failure model of the machine operation and then selects the production and maintenance policies accordingly. Once the situation has been defined, the control charts are then chosen for the EDMCC.

3 The Taguchi Method

The Taguchi method is a useful approach to parameter optimisation which has been employed successfully in quality control since its introduction in 1985. It provides the designer with a systematic and efficient method for conducting experiments to determine a near optimum setting for the design parameters. The Taguchi method designs several orthogonal arrays (OA), using experimental design theory, to enable the parameter space to be studied with the minimum number of experiments. The designer evaluates the situation and then chooses an experimental design which can find an optimal solution for the experiment. Taguchi designed a group of orthogonal arrays which can match with a variety of problem demands. Table 1 shows a two level orthogonal array with three Parameters (e.g. Population Size : 15–20, Crossover Rate : 0.85–0.95; Mutation Rate : 0.05–0.001).

Each row in the array represents an experiment and the columns contain the values of the parameters. Once all the experiments have been finished, a final calculation for each parameter level will be done, to find its optimum value. The main procedures to applying the orthogonal array to design the best combinations for the selection of the GAs operators are as follows:

1. Select the GA's operators and parameters to be considered.
2. Define the number of levels for each GAs operators and parameters.
3. Choose an appropriate orthogonal array.
4. Assign the GAs operators and parameters for row and level values for columns.
5. Perform the experiments by the combination of each row.
6. Analyse result and choose several alternatives for fuzzy reasoning tuning.

4 Chromosome Refinement

The GAs chromosome consists of groups of variables, which are represented by groups of genes. The gene structure significantly affects the GAs performance, and an improper choice for the chromosome structure will often result in a poor GA performance. It is therefore very important to design a good structure for the chromosome. Hence, we employ a fuzzy Taguchi controller to adjust the order of the genes within the chromosome. In our research, the controller applies gene identification, grouping and ordering.

The gene identification evaluates the relative effect of each gene on the fitness function. Previous research usually did this by applying traditional approaches, such as: expert systems using domain knowledge; machine learning by simulations; or statistical analysis. In our Gene Identification, we attempt to identify the beneficial genes within the chromosome structure in three stages. First, we calculate the importance ratio of the genes to the total fitness. Then, we select control variables from the fitness value, using a fuzzy matrix. Finally, we analyse the relationships between the importance ratio and the control variables by the level of process policy.

In gene grouping, the system employs cluster analysis to define the grouping relationships between genes. Initially, the control parameter will be grouped in a way to reduce the search space. Then, the system attempts to identify subgroup relationships within an individual group.

After gene identification and grouping, we apply gene ordering to optimise the gene sequence of the chromosome. The system then employs the importance ratio and relationship value, which is extracted from the results of cluster analysis, to arrange the gene sequence within each gene group. Over evolutions, the system attempts to recognise the beneficial genes and fix values for these genes as well as their positions, reducing the effective search space.

5 Fuzzy Taguchi Controller

The fuzzy Taguchi controller is a hybrid methodology which combines the Taguchi method with fuzzy reasoning techniques. In our research, this controller selects an orthogonal array based on the external situation and conducts a small number of experiments. Next, fuzzy mapping modules transform evolution evidence into items within a fuzzy knowledge-based system. Then the fuzzy membership module applies the Taguchi method to select several alternative memberships for the GA's simulations. Finally the system performs fuzzy reasoning to adjust the input parameter values (i.e. the chromosome structure and the GA's operator parameters) using defuzzification [6].

The fuzzy membership module utilises triangular membership functions, the minimum intersection operator and the correlation-product inference procedure. Defuzzification of the output was conducted by applying the fuzzy centroid method [1]. The transformation functions use evolution evidence derived from simulations, mapping the crisp values to fuzzy values and then converting these values back to crisp values again, using a reasoning process. These functions are described as follows (see E1–E5):

E1: Slope of Membership = Membership Value / Variable Value

E2: Left Part Membership (LPM) = (Input Value – Minimum Value) * Slope

E3: Input Value of LPM = (Membership Value / Slope) + Minimum Value

E4: Right Part Membership (RPM) = 1 – ((Input Value – Minimum Value) * Slope)

E5: Input Value of RPM = (Membership Value / Slope) + Minimum Value

Table 2: The comparison of the static and fuzzy refinement on DeJongs function

DeJong	F_1	F_2	F_3
Static	0.003	0.0016	-3.13
Fuzzy	0.002	0.0012	-3.6
DeJong	F_4	F_5	
Static	-2.05	0.0020023	
Fuzzy	-3.16	0.0020012	

Table 3: The comparison of the static and fuzzy refinement on the EDMCC.

EDMCC	E_1	E_2	E_3	E_4
Static	770	730	759	727
Fuzzy	738	727	735	725
EDMCC	E_5	E_6	E_7	E_8
Static	726	725	723	723
Fuzzy	724	723	723	723

6 Results

This section described the experimental results of fuzzy refinement. This approach can cover the weakness of the orthogonal array method which has difficulty considering the relationships between the GAs operators and parameters. Table 2 shows the average performance of fuzzy refinement is better than the orthogonal array setting.

The experimental results for the EDMCC functions (Table 3) also show that the fuzzy refinement approach is better.

When we verified the approach of the fuzzy refinement on DeJongs test-beds and the empirical test-beds of the EDMCC, the overall results showed that using the fuzzy refinement method can select optimal GA parameters and hence improve the GA's searching behaviour.

7 Conclusion

GA users often spend a lot of time selecting the parameters. Therefore, the aim of this research is to design a system which can improve the selection of operators and parameters. In our research, we developed a fuzzy Taguchi controller, which identifies the external environment and directs the chromosome controller and parameters controller to improve the GA's performance. This is an attempt to integrate the selections of operators/parameters and process parameters for the process control application. We designed several combinations of the operators and parameters and process parameters using the fuzzy Taguchi method and the results showed that the searching behaviour of the GA improved significantly. Hence, this approach can act as a useful guideline for users working on the optimisation of process control. Through these simulations, we have attained several combinations of the operators parameters and process parameters (chromosome structure) for different situations. We have also attempted to classify and refine these combinations so that they can become generic procedures that can assist future guidance. We also used a heuristic approach to deduce alternatives to these combinations. This effort also considers the structure to take place within chromosome. In the future, we are going to develop several methodologies which can measure the effect of tuning the fuzzy membership selections on the GA's performance.

8 Acknowledgements

This work was conducted using the SUGAL genetic algorithm, written by Dr. Andrew Hunter at the University of Sunderland, England.

References

[1] M.A. Lee. Dynamic control of genetic algorithms using fuzzy logic techniques. In *Proceedings of the Fifth International Conference on Genetic Algorithms*, pages 76–82. Morgan Kauffman, 1993.

[2] G. Taguchi and S. Konishi. *Orthogonal Arrays and Linear Graphs*. American Supplier Institute, Dearborn, MI, 1987.

[3] C.F. Tsai, C.G. Bowerman, and J.I. Tait. Fuzzy refinement in genetic algorthims for the economic design of control chart. In *Conference on Agile and Intelligent Manufacture Systems*, October 1996.

[4] C.F. Tsai, C.G. Bowerman, and J.I. Tait. A intelligent adaptive system for improving the behaviour of simple genetic algorithms. In *EXPERSYS-96*, October 1996.

[5] B.C. Turton. Optimization of genetic algorithms using the taguchi method. *Journal of System Engineering*, pages 121–130, 1994.

[6] L.A. Zaddeh. QSA/FL-quantative systems based on fuzzy logic. In *Stanford AAAI Symposium on Limited Rationality*, pages 111–114, 1989.

Walsh Functions and Predicting Problem Complexity

R. B. Heckendorn
Department of Computer Science, Colorado State University
Fort Collins, Colorado 80523 USA
Email: heckendo@cs.colostate.edu

Abstract

Theorems are given establishing the epistatic bounds for problems that can be stated as mathematical expressions. Examples of the application of the theorems and techniques for controlling epistasis are presented.

1 Introduction

Many problems can be cast as the optimization of a mathematical model. In the preparation of the model for optimization by a genetic algorithm, the parameters are encoded into a bit string for processing by the algorithm resulting in a function whose domain is in bitspace and whose range is in the reals. More formally:

$$f_{ga} = f_{model}(f_{decode}) : \mathcal{B}^L \to \mathcal{R}$$

where $\mathcal{B}^L$ is an L dimensional bitspace.

Many features of f_{ga} tend to make it more difficult for a genetic algorithm to solve. Epistasis, linkage, hyperplane organization (deception) and interpartition conflict are some of these features. This paper presents several new measures of epistasis. Theorems are given establishing the bounds of epistasis for problems that can be stated as mathematical expressions (proofs available from the author). How this insight might be used to reduce problem epistasis is discussed and empirical evidence to demonstrate the application of the theorems is presented.

2 Walsh Sums and Function Order

In order to measure the degree of interaction between bits in a function, $f : \mathcal{B}^L \to \mathcal{R}$ (where $\mathcal{B}^L$ is the L bit binary space), it is helpful to break the function down into a linear combination of the interactions. For an L bit function there are 2^L possible interactions. A Walsh polynomial [1, 3] is a classic and useful linear decomposition. Any function $f : \mathcal{B}^L \to \mathcal{R}$ can be broken down into the Walsh polynomial as follows:

$$f(x) = \sum_{i=0}^{2^L-1} w_i \psi_i(x)$$

where $\psi_i(x) : \mathcal{B}^L \times \mathcal{B}^L \mapsto \{1, -1\}$ is the ith Walsh function of x and $w_i : \mathcal{B}^L \mapsto \mathcal{R}$ is the ith Walsh coefficient. A Walsh function does a bit by bit parity check between two bit strings x and i. If they share an even number of 1 bits in the same position the function returns a 1 otherwise it returns a -1.

In order to apply Walsh functions to the problem of epistasis we need to group the Walsh coefficients into a more manageable measure. Let a Walsh sum be denoted by W_b where:

$$W_b = \sum_{i:bc(i)=b} |w_i|$$

and $bc(i)$ is the bit count of i. That is W_b is the sum of the *absolute value* of all of the Walsh coefficients for bit patterns i with exactly b bits set to 1. For example:

$$\begin{aligned}
W_0 &= |w_0| \\
W_1 &= |w_1| + |w_2| + |w_4| + \ldots + |w_{2^{(L-1)}}| \\
W_2 &= |w_3| + |w_5| + |w_6| + \ldots + |w_{3 \cdot 2^{(L-2)}}| \\
&\vdots \\
W_L &= |w_{2^L-1}|
\end{aligned}$$

Notice that for an L bit function there are $L+1$ Walsh sums and that W_k is the sum of $\binom{L}{k}$ Walsh coefficients. Since the absolute value of the Walsh coefficients is used, any nonzero Walsh coefficient will force the corresponding Walsh sum to be nonzero. Therefore the nth Walsh sum could be an effective measure of the magnitude of n-bit interactions.

Let the *order* of a function, denoted $\Omega(f)$, be defined as the largest i such that $W_i \neq 0$. In the special case where all W_i are 0, that is $f(x) = 0$, $\Omega(f) = 0$. Intuitively, the order of a function is the size of the largest set of interdependent bits. So $\Omega(f)$ is a measure of the maximum level of epistasis of f and W_i measures the magnitude of the i-bit interdependence.

Another useful linear decomposition of f is by spectral functions. f can be decomposed into the sum of at most $L+1$ spectral functions S_i as

$$f(x) = \sum_{j=0}^{L} S_i(x)$$

where

$$S_b(x) = \sum_{i:bc(i)=b} w_i \psi_i(x)$$

Notice that the only possible nonzero Walsh sum for S_i is W_i and that $S_0 = w_0$. Spectral functions will enable us to more easily express ideas about functions with limited degrees of interaction.

3 Function Order and Models

If we know the mathematical expression for function f do we know anything about $\Omega(f)$? The answer is yes as seen in the next set of theorems:

Theorem 1 (Polynomial Complexity Theorem). *Let P_n be a polynomial with degree $n : n \geq 0$ such that $P_n : \mathcal{B}^L \rightarrow \mathcal{R}$ then*

$$\Omega(P_n) \leq n$$

Extraction is often the basis of composing parameters for functions with domain $\mathcal{R}^n$ from strings in $\mathcal{B}^L$ that are processed by genetic algorithms. Let $x[n_1, n_2] : \mathcal{B}^L \times Integer \times Integer \rightarrow \mathcal{B}^L$ be the extraction operator which extracts bits in positions n_1 through n_2 ($n_2 \geq n_1$) from string x and placing them in the least significant bit portion of the string with zero fill. This operation can be performed by masking and shifting. The extraction theorem shows that Ω for a function is limited by the number of bits extracted for its parameter.

Theorem 2 (Extraction Theorem).

$$\Omega(f(x[n_1, n_2])) \leq \min(\Omega(f(x)), n_2 - n_1 + 1)$$

The next theorem will allow us to unite the previous two theorems and apply them to f_{ga}.

Theorem 3 (Polynomial Composition Theorem). *Let $P_n(x_0, x_1, \ldots x_{n-1})$ be a polynomial in $x_0, x_1, \ldots x_{n-1}$ that takes $\mathcal{R}^n \rightarrow \mathcal{R}$ such that*

$$P_n(x_0, x_1, \ldots, x_{n-1}) = \sum_{all\ terms} a_{k_0 k_1 \ldots k_{n-1}} x_0^{k_0} x_1^{k_1} \ldots x_{n-1}^{k_{n-1}}$$

with $k_i \geq 0$ and let $f_0, f_1, \ldots f_{n-1}$ be functions such that $f_i : \mathcal{B}^L \rightarrow \mathcal{R}$ then

$$\Omega(P_n(f_0(x), f_1(x), \ldots f_{n-1}(x))) \leq \max_{all\ terms} k_0 \Omega(f_0(x)) + \ldots + k_{n-1} \Omega(f_{n-1}(x))$$

4 Predicting Complexity

By complexity I mean the epistatic component of problem difficulty. Although there are certainly easy problems with high epistasis there is a general correlation between epistasis and difficulty of solving a problem with a simple genetic algorithm. Empirical studies [2] show that functions with higher Ω tend to have higher levels of deception and, under a simple genetic algorithm, are less frequently solved to optimality and converge to less optimal answers. So even though high epistasis is not the complete measure of problem difficulty, it provides a mechanism for the creation of more difficult problems.

Barring recombination operators that understand about the structure of the decoding function, the genetic algorithm only sees f_{ga}. Therefore, the function that is really being solved is f_{ga} and it is the difficulty of this *composite* function that estimates the difficulty of solving the problem with a genetic algorithm.

The decoding function can be represented as a vector of functions

$$\vec{f} = (f_0, f_1, \ldots f_{n-1})$$

where $f_i : \mathcal{B}^L \rightarrow \mathcal{R}$ and maps the bitstring to a real argument for the model function. Therefore, for a chromosome x in $\mathcal{B}^L$:

$$f_{model}(f_{decode}(x)) = P_n(f_0(x), f_1(x), \ldots f_{n-1}(x))$$

for model functions that are polynomials of n variables. In these cases the Polynomial Composition Theorem applies. This means given just the mathematical models and extraction functions we can derive an upper bound for the degree of bit interaction in f_{ga}. We may even be able to apply this understanding to control the level of complexity and thereby improve performance.

Table 1: Walsh sums for $f(x) = x$ and x^5.

Order	x	x^5
0	127.5	1.811e+11
1	127.5	4.299e+11
2	0	3.457e+11
3	0	1.088e+11
4	0	1.229e+10
5	0	3.731e+08
6	0	0
7	0	0
8	0	0

In the next section we will give some examples of the predictive power of the theorems.

5 Test Driving the Theorems

In the following examples we present tables of Walsh sums for W_0 through W_8. The first column is the order of the Walsh sum. The remaining columns are the Walsh sums for that order for various functions. All functions presented in this section are evaluated using a string length of 8.

Table 1 shows that the nth power of a linear function has an Ω of n. The second column is the Walsh sum for $f(x) = x$. The third column is $f(x) = x^5$. For both functions $x \in [0, 2^L - 1]$. Notice how $\Omega(f)$ is limited to the maximum power of the polynomial by the Polynomial Complexity Theorem.

A very useful invariant of functions is found in odd and even orderness. The product of any two spectral functions S_i, S_j results in a function that is the sum of a series of all odd ordered spectral functions or all even ordered spectral functions up to and including order $i + j$. Such functions are called odd ordered or even ordered functions. The product of an odd ordered function with an even ordered function is an odd ordered function. The product of two odd or two even ordered functions is an even ordered function.

Since a linear function has an Ω of 1, it must be the sum of the two spectral functions S_0 and S_1. But S_0 is just the average of all of the function values. By adjusting the average of the linear function to zero we can get $S_0 = 0$ leaving the function equal to some S_1'.

In Table 2 we have mapped the range $[0, 2^L - 1]$ to the range $[-(2^{(L-1)} - .5), (2^{(L-1)} - .5)]$. This technique is called argument centering. For the linear function x over this argument range, the average of x is now 0. Therefore, the linear function is now equal to a single nonzero spectral function S_1 and is an odd ordered function. Since odd powers of S_1 have only odd ordered Walsh sums, x^5 becomes the odd ordered function we see in the second column of Table 2.

Table 2: Walsh sums for x^5.

Order	x^5 Without Centering	x^5 With Centering
0	1.811e+11	0
1	4.299e+11	1.923e+10
2	3.457e+11	0
3	1.088e+11	1.409e+10
4	1.229e+10	0
5	3.731e+08	3.731e+08
6	0	0
7	0	0
8	0	0

Table 3: Walsh sums for $\cos(x)$.

Order	Without Centering	With Centering
0	0.6386	0.6366
1	0.512	0
2	0.1363	0.6459
3	0.01511	0
4	0.0007757	0.01548
5	1.837e-05	0
6	2.019e-07	1.634e-05
7	9.314e-10	0
8	1.444e-12	4.692e-10

Keeping a f_{model} restricted to a polynomial may seem limiting but actually it is quite powerful since all continuously differentiable functions can be expressed as a Taylor series expansion. For example, the Taylor series expansion about 0 for cos is:

$$\cos(x) = 1 - x^2/2! + x^4/4! - x^6/6! + ...$$

The Walsh sums for $\cos(x)$ for $x \in [0, \pi/2]$ are presented in Table 3. By centering the argument x so that $x \in [-\pi/2, \pi/2]$ the powers of x in cosine, which are all even, force the function to become even ordered in Table 3. One may not always be in position to change the range of arguments to a function but there is often more flexibility than may seem at first.

Let's exercise the theorems on a more complex

model. Suppose:

$$f_{model}(x, y, z) = x^2 + \cos(y)\sin(z)$$

and that the decoding functions for the chromosome string s are:

$$x = s[1,2], \quad y = s[3,5], \quad z = s[6,7]$$

where the extraction function centers the argument and scales the value to the range $[-\pi, \pi]$. Our analysis of $\Omega(f_{model})$ can proceed as follows:

1. f_{model} has 2 terms so Ω is the max of the Ω's of the individual terms.

2. $\Omega(x^2) = 2$

3. $\Omega(\cos(y))$ is limited by y having a width of 3 bits. Since y is centered and cos is an even ordered function under centered arguments only W_0 and W_2 will be nonzero. Therefore $\Omega(\cos(y)) = 2$. Note that if y included just one more bit $\Omega(\cos(y)) = 4$.

4. $\Omega(\sin(z))$ is limited by z having a width of 2 bits. A Similar argument to above shows only, W_1 will be nonzero. Therefore $\Omega(\sin(z)) = 1$. Again if z included just one more bit $\Omega(\sin(z)) = 3$.

5. So the computation of $\Omega(f_{ga})$ is as follows:

$$\begin{aligned}\Omega(f_{ga}) &\leq \max(\Omega(x^2), \Omega(cos(y)) + \Omega(sin(z)) \\ &= \max(2, 2+1) \\ &= 3\end{aligned}$$

If the arguments to cosine and sine each had just one more bit then $\Omega(f_{ga}) = 7$. This demonstrates that the choice of encoding length can be important and could be engineered to reduce interactions. Table 4 shows that our estimate is correct.

Table 4: $x[1,2]^2 + \cos(x[3,5])\sin(x[6,7])$.

Order	Walsh Sum
0	5.483
1	0.1083
2	4.386
3	0.8885
4	3.331e-16
5	2.22e-16
6	1.11e-16
7	0
8	0

6 Conclusions

In this paper I have shown that in the case where a model function can be expressed as a polynomial and each parameter encoding can be expressed as a polynomial applied to an extraction of bits from a bit string, the degree and magnitude of bit interactions can be predicted. I have shown that this information could be used to design model and encoding functions with lower epistasis. I have shown that by use of Taylor series the meaning of polynomial can be expanded to some common infinitely differentiable functions. Finally, I developed some techniques such as argument centering and odd/even series truncation that could be used to reduce bit interactions.

References

[1] D. E. Goldberg. Genetic algorithms and walsh functions: Part I, a gentle introduction. *Complex Systems*, (3):129–152, 1989.

[2] R. B. Heckendorn, D. Whitley, and S. Rana. *Nonlinearity, Hyperplane Ranking and the Simple Genetic Algorithms*, Foundations of Genetic Algorithms, 4, Morgan Kaufmann Publishers, 1997.

[3] C. Reeves and C. Wright. *An Experimental Design Perspective on Genetic Algorithms*, Foundations of Genetic Algorithms, 3, Morgan Kaufmann Publishers, 1995.

Migration through Mutation Space: A Means of Accelerating Convergence in Evolutionary Algorithms

H. Copland and T. Hendtlass
Centre for Intelligent Systems, Swinburne University of Technology,
P.O. Box 218 Hawthorn 3122. Australia.

Abstract

In this paper a multiple subpopulations technique for evolutionary algorithms is proposed. Each subpopulation is distinguished by the mutation radius and mutation probability assigned to it, with mutation radius being a function of mutation probability. Mutation probabilities across the subpopulations range from 0.005 to 0.75. There is no crossover between subpopulations in the normal course of breeding, and the mechanisms of elite migration from a higher mutation subpopulation to the adjacent subpopulation of lower mutation is used to introduce new genetic material. The evolution of artificial neural networks for solving a variety of problems is demonstrated, with convergence times typically half as long as a standard evolutionary algorithm.

1 Introduction

In biological evolution, viable subpopulations of a species that become isolated from the main breeding pool of the species frequently give rise to an entirely new species incapable of breeding with their 'unevolved' ancestors. At times, the descendents of the isolated subpopulations eventually migrate or otherwise return to the ancestral territory and outcompete their ancestors. For example, the black rat outcompeting the brown rat where both species exist, though their ancestor is common.

This process can be accelerated – in terms of genetic drift away from the ancestral gene pool – in a high (but non-lethal) mutation environment, as is presently the case with the voles inhabiting the Chernobyl region [2]. Should members of the subpopulation be in contact with the ancestral population during an intermediate evolutionary stage where the subpopulations can still interbreed, their mutated characteristics can rapidly be assimilated by the ancestral population since most mutations are dominant in expression.

In evolutionary algorithms, a high mutation rate is often associated with poor convergence to an acceptable error threshold. This is a result of the damage which the high mutation environment causes to the 'genome' of the fitter solutions. However, high mutation is an excellent way to achieve the rapid exploration of a solution space.

A means of achieving the required stability of the evolving gene pool while benefiting from the exploration benefits of high mutation is to divide the total population into subpopulations that each possess a differing mutation probability and mutation radius. The ability of the subpopulations to interbreed is then constrained by a fitness-based mechanism of 'elite migration'. This is achieved by copying the most fit individuals from the higher mutation environment to the adjacent lower mutation environment.

This differs considerably from the approach to coarse-grained parallel genetic algorithms (also known as *island models*) as done by Tanese [4], who established processor-based subpopulations within a parallel genetic algorithm, and allowed intermittent interbreeding of the fitter solutions between all adjacent subpopulations. Subpopulations have only minor changes in parameters. Major changes in mutation or crossover rates showed no net benefit. There was no bias established to move individuals to a lower mutation subpopulation, such as exists in the work being reported here, and no overall underlying structure to the relationship between the adjacent subpopulations.

2 Mutation Environments and Migration

The highest mutation probability implemented is 0.75, which is 0.10 below the highest mutation probability normally found to contribute positively to a solution. Populations which have mutation probabilities exceeding 0.85 are essentially conducting a random search; there generally appears to be no meaningful legacy of fitness preserved for the next generation.

This approach to mutation is considerably different to the mutation technique implemented by Davis [1] which involved varying creep mutation to effect mutation radius by fixed amounts, while mutation probability typically remained either 0.10 for both the basic creep and small creep mutations, and 0.20 for the large creep mutation. This has a superficial similarity in that the creep mutation radius is smallest with the lowest mutation probability and largest with the highest mutation probability. However, the mutation radii are fixed increments or decrements, and, in comparison, the mutation probability is low. Moreover, there is no migratory behaviour across the mutation space as the individuals subjected to the different mutation operators are not isolated.

The mutation space in this migratory algorithm can be viewed as a 'mutation field'. Where, for example, three subpopulations of low, medium and high mutation are used, this results in a mutation field as illustrated in Figure 1.

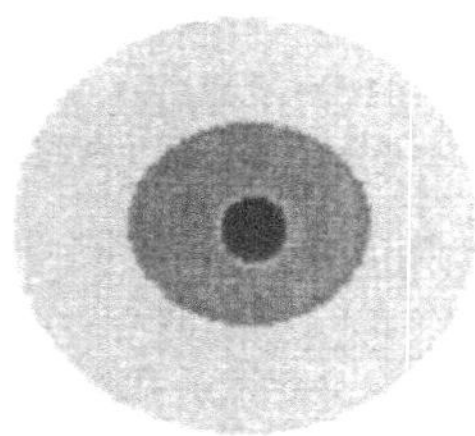

Figure 1: This diagram represents the 'mutation field' of the total population. The lighter the colouring, the higher the probability of mutation occuring. The differences in mutation radius are evident. As a consequence, the search of the solution space is simultaneously broad and narrow.

For the purposes of this research, three subpopulations were established. These were deemed low, medium and high mutation environments, with respective mutation probabilities of 0.005, 0.25 and 0.75 for the weight values. Architectural mutation is considerably different and discussed in the next section. The function for relating mutation probability to mutation radius was:

$R = \pm(B + n^k \times n)$ where
R is the
mutation radius.
B is the base mutation radius (typically 1%).
n is a constant.
k allows exponential increase of R.
n is the mutation probability.

Consequently, the higher the mutation probability, the larger the maximum mutation radius. Values for n used for three different evolutionary trials were 3, 3, and 2, with corresponding values for k of 1, 2 and 2. No significant difference was found between them, and the results presented here are for the n, k combination of 3, 1.

Elite preservation was used, with the most fit 10% of each successive generation being preserved between generations. It was this preserved elite that was 'migrated' (copied) to non-preserved (i.e. non-elite) slots of the adjacent subpopulation of lower mutation. Migration, then, was down the mutation gradient with continual reintroduction of fitter individuals. This continual reintroduction of the fitter individuals prevents loss of advantageous traits through unfortunate crossover.

3 Breeding of Artificial Neural Networks

The ANNs were four layer nonlinear feedforward networks with the number of nodes in the two hidden layers restricted to a maximum of six. No local heuristic such as backpropagation was implemented. The number of actual nodes in the hidden layers varied between individual networks. Input and output nodes were problem dependent in number. The networks were not maximally connected, though there was no constraint on the evolution of maximally connected networks. A tanh output transform was used for the hidden and output nodes. The input was scaled to between −0.8 and +0.8.

During breeding, using uniform crossover, each node of the 'child' network was drawn from either 'parent' at a probability of 0.5. The choice of parents was based on performance, with those having least error (being fitter) having correspondingly higher breeding probabilities. A node was crossed with its connections to other nodes in the previous layer and weight values intact.

Mutation of the architecture was based on the presence or absence of nodes, and the presence or absence of connections between nodes. This is a binary decision, and so has a mutation probability but no radius in the sense that changes to a weight value, for example, have a radius. The mutation probability of a node shifting its state is one third the mutation probability of the subpopulation in question. A highly unstable architecture impacts greatly on convergence, effectively increasing convergence times by an order of magnitude. A highly unstable architecture results where more than half the individuals in the subpopulation have more than half of their nodes change state — off to on or on to off — in each generation. As no local heuristic is in operation, when a node is switched on its weights are assigned random values which exist within a specified range.

The mutation probability for a connection existing with respect to a given node ranged from 0.70 to 0.85. Generally, the fewer the number of inputs, the higher the probability of connections existing should be to ensure faster convergence.

4 Feedback Mechanisms

For the best search of a solution space, a broad search about the best found (i.e. fittest) point is usually optimum. Recombination through crossover and mutation achieve this. However, even with elite migration, the subpopulations drift apart during the course of evolution in terms of the region of solution space they are exploring. In general, the low mutation group becomes fittest once convergence has begun, and the fitness of the best point decreases as the mutation gradient is ascended.

For the evolution of a network to solve XOR, typical figures for error in the best individuals in each subpopulation after convergence are 2.5% (low mutation), 7.5% (medium mutation) and 12.5% (high mutation). While obtaining convergence naturally takes longer for more complex problems, and the convergence errors differ, the performance ratio between the subpopulations appears to remain much the same.

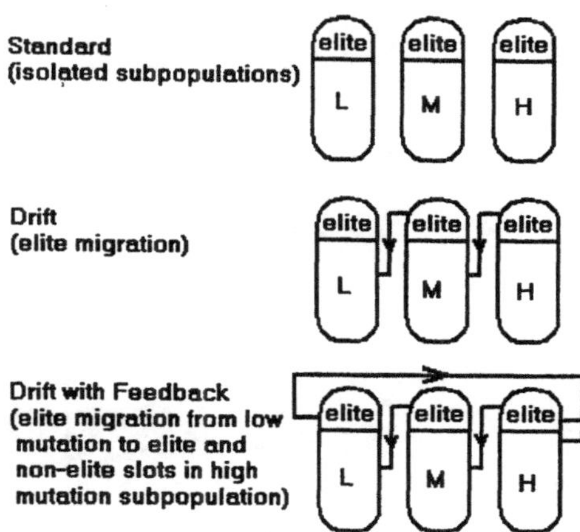

Figure 2: Standard structure of subpopulations, structure with drift, and drift with feedback.

The higher the mutation probability of the subpopulation, the more likely it is be 'trapped in the past' in terms of fitness. The probability of a mutation being beneficial is low, and where mutation is high the population tends toward much poorer fitness than a lower mutation environment. Consequently, hundreds of generations can pass in a high mutation subpopulation without improvement in the elite, whereas in a lower mutation subpopulation the improvement is more gradual, certainly slower, but the fitness can better that of the higher mutation subpopulation over hundreds of generations. When this happens, the higher mutation subpopulation's contribution to accelerating the convergence via enhancing the breeding pool of the adjacent subpopulation rapidly diminishes. To counter this, a feedback mechanism was introduced. The mechanism implemented is to copy the elite of the lowest mutation subpopulation to one (or more) of the preserved elite slots of the highest mutation subpop-

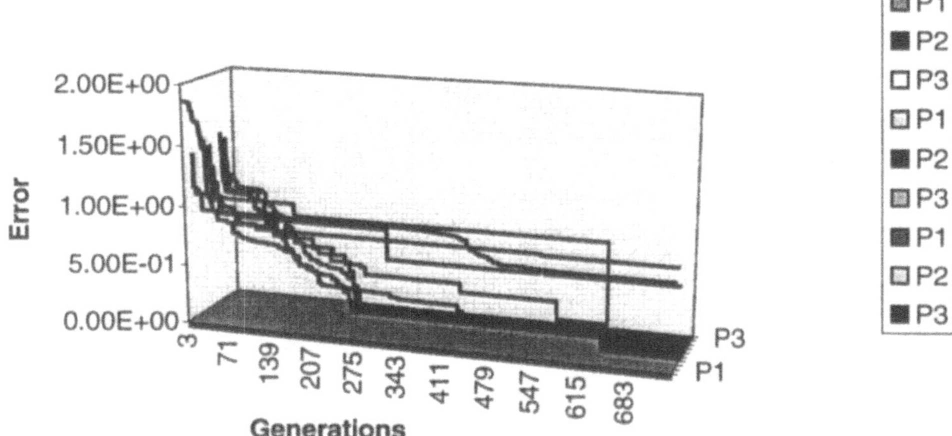

Figure 3: 10 trail average performance for elite members of subpopulations (XOR).

ulation. This allows rapid exploration around the overall fittest point, with elite migration as before.

5 Results

ANNs were evolved to solve a variety of problems: XOR, 4 bit parity, iris categorization and the classification of the letter 'F' from the Frey and Slate [3] character recognization problem.

Since three subpopulations of 40 were used, comparisons are made to a standard evolutionary algorithm population of 120 unless otherwise indicated (as in the graph of generational performance).

In Table 1, the results for six typical evolutionary runs for the XOR problem are given. In only one instance did the standard evolutionary algorithm evolve a network to solve the problem to within an error less than 2.5%. However, in all instances with drift, a solution was arrived at. In one instance with both drift and feedback the evolutionary algorithm failed to converge on the solution. The convergence times with feedback are highly variable: a result of the instability introduced by the feedback mechanism.

Performance at solving the 4 bit parity and iris problems was also, on average, superior to the standard technique. In Table 2, a single exception exists with the 4 bit parity problem in that the standard evolutionary algorithm did slightly better overall than when drift alone was introduced. Results are mean average for 10 trials on each problem.

Table 1: Typical error and generations for XOR (Random seeds were 200, 400, 600, 800, 1000 and 2019.).

Standard	Standard	Drift	Drift	Drift & Feedback	Drift & Feedback
Error	Gens	Error	Gens	Error	
15.52%	750*	2.40%	638	1.91%	226
13.85%	750*	2.45%	553	2.49%	462
17.00%	750*	2.12%	421	2.37%	679
18.38%	750*	2.34%	592	2.50%	750*
2.29%	721	2.48%	527	1.85%	317
16.63%	750*	2.18%	476	2.00%	590

* Failed to converge in 750 generations.

Preliminary results for the Frey and Slate problem indicate correct classification in excess of 80%, for a network evolved using drift and feedback.

Figure 3 shows the average evolutionary 'path' followed by the standard, drift, and drift plus feedback evolutionary algorithms. In this diagram, the forward three lines (which do not reach the 'zero' error represented by the floor of the graph) represent the performance of the differing subpopulations (1 through 3) where no drift or feedback was introduced. The second lot of three lines, which do reach the floor of the graph, correspond to the case where drift was permitted between subpopulations. The final three lines, which reach the floor of the

Table 2: Mean average error at 750 generations (or convergence).

	XOR	4 Bit Parity	Irises
Standard	13.90%	23.2%	4.12%
Drift	2.33%	24.3%	2.96%
Drift&Feedback	2.20%	15.4%	2.78%

graph first, represent the performance where both drift and feedback were introduced.

6 Conclusions

The technique has already demonstrated a good advantage over conventional techniques in terms of time to convergence for a limited range of problems. Consequently, it has application in restricted population scenarios and preliminary indications are that still better results are achieved for a wide range of problems with more subpopulations to 'refine' characteristics acquired in the higher mutation subpopulations.

References

[1] L. Davis. Adapting operator probabilities in genetic algorithms. In *Proceedings of the Third International Conference on Genetic Algorithms*, pages 61–69, George Mason University, June 4th-7th 1989.

[2] S. Dickman. Chernobyls voles: A spring of genetic surprise. *New Scientist*, (1990):14, 12th August 1992.

[3] R. Frey and D. Slate. Letter recognition using Holland-style adaptive classisfiers. *Machine Learning*, 6:161–180, 1991.

[4] R. Tanese. Distributed genetic algorithms. In *Proceedings of the Third International Conference on Genetic Algorithms*, pages 434–439, George Mason University, June 4th-7th 1989.

Dual Genetic Algorithms and Pareto Optimization

M. Clergue and P. Collard
University of Nice Sophia Antipolis, I3S Laboratory,
bat 4,250 av. A.Einstein, 06560 Valbonne, France
Email: clerguem@alto.unice.fr , pc@alto.unice.fr

Abstract

This paper deals with an important class of optimization problems, the multiobjective problems. A new genetic algorithm, called the dual genetic algorithm, is presented. Through two theoretical problems, we show that this approach appears to be efficient for multiobjective optimization.

1 Introduction

In general, real world problems are expressed in terms of many criteria, or objectives, often competitive, which should be satisfied simultaneously. Multiobjective optimization (MOO) involves some different reflections with regard to classical optimization techniques, including genetic algorithms (GA), which often take care of a single objective. In the multiobjective problem (MOP) context, there is no trivial way to say that a solution is better than another and should be preferred. Pareto optimality is a way to do this. A Pareto optimal solution for a multiobjective problem is a solution whose objective vector components could not be improved simultaneously. More formally, consider, without loss of generality, the minimization of the n components of a vector function $\mathbf{f}$. A decision vector x_u is said to be Pareto optimal if and only if there is no x_v for which $v = \mathbf{f}(x_v) = (f_1(x_v), ..., f_n(x_v))$ *dominates* $u = \mathbf{f}(x_u) = (f_1(x_u), ..., f_n(x_u))$, that is, there is no x_v such that: $\forall i, v_i \leq u_i \quad \wedge \quad \exists i \mid v_i < u_i$. The vector u is said to be *non-dominated.*

1.1 Genetic Algorithms and MOPs

As we will see, scalar optimization methods, like hill climbing, simulated annealing or classical GAs, aren't well suited for MOO. The power of GAs is that they are not uniquely a scalar optimization method. They can be easily adapted by incorporating the notion of Pareto optimality.

Non Pareto Approaches

When powerful and well studied scalar methods are provided, the temptation is great to reduce anything to a scalar problem and then to apply these methods. This is done naturally in the MOO context by combining all the objectives in a single scalar value, by means of a linear combination for example. By this way the MOP is reduced to a single objective problem and classical GAs can be applied.

Two remarks have to be made about this method. Firstly, the combination gives more or less importance to each objective. This involves some ***a priori*** knowledges of the problem. Secondly, points in concave regions of a trade-off surface cannot be found by optimizing a linear combination of the objectives, for any set of weights, as pointed out by Fonseca and Fleming [2].

To avoid the design of the combination, one could optimize each objective separately. With this point of view, Schaffer [5] proposed a GA extension called Vector Evaluated GA (VEGA). In VEGA, the next generation is selected by n fractions, each one according to each of the objectives. Although VEGA and other similar population based methods aim to obtain non-dominated solutions, they don't make a direct use of the Pareto optimality.

Pareto Approaches

Goldberg [3], first proposed a direct use of the Pareto optimality. His method is a ranking method based on a Pareto partial order. The non-dominated individuals are assigned rank 1, and then removed; the new non-dominated individuals are assigned rank 2, and so forth. Fonseca and Fleming [2] proposed a similar method based on another partial order defined by the number of dominating individuals.

Horn and Nafpliotis [4] gave another method, called niched Pareto GA (NPGA), not based on a population ranking, but on a tournament selection technique: the two candidates are compared to a random comparison subset of the population; if a candidate is dominated by the comparison set and the other not, the latter wins; if neither or both are dominated the tie is broken by a niching technique.

The difficulty for the GA is now to converge toward a set of equivalent solutions, the Pareto optimal ones, and to stabilize it in the population. The solutions may be so different, formally may belong to incompatible *schemata*, that these tasks are impossible. This is why we introduce dual genetic algorithms. Theoretically, this extension of GAs allows a wider class of solutions to be present in the same stable population.

2 Dual Genetic Algorithms

A schema is a way to identify a subset of individuals sharing some characteristics. Schemata link individuals having the same allele at given loci.

Unfortunately, schemata are unable to represent some sets of individuals. For instance the set $\{00, 11\}$ doesn't correspond to a schema. This is why we introduce the notion of relational schemata (to avoid confusion we call classical schemata positional schemata or P-schemata, as they link individuals having given values at given positions) and their implicit implementation, the dual genetic algorithms.

2.1 Relational Schemata

Relational schemata, or R-schemata, as defined by Collard and Escazut [1], are another type of equivalence class that can be defined over the search space. Instead of having equivalence relations based on the value at given positions, R-schemata are equivalent classes based on relations between loci. They are defined over the ternary alphabet $\{\mathtt{X}, \overline{\mathtt{X}}, *\}$, where the two symbols $\mathtt{X}$ and $\overline{\mathtt{X}}$ represent two complementary variables: if $\mathtt{X}$ is bound to 0 then $\overline{\mathtt{X}}$ is bound to 1 and vice versa. For instance, the set $\{00, 11\}$ corresponds to the R-schema $\mathtt{XX}$ and the set $\{01, 10\}$ corresponds to the R-schema $\mathtt{X}\overline{\mathtt{X}}$.

2.2 Dual Genetic Algorithms

Dual genetic algorithms (DGAs) are a quite simple implementation of the notion of R-schemata through a new encoding of the binary string developed by Collard and Escazut [1]. A ***head-bit*** is added to the string and manages the interpretation of the rest of the string. When this bit is set to 0, the rest of the string remains unchanged. If it is set to 1, the rest of the string is interpreted as its binary complement.

In a more formal way, let us consider λ-bit strings. The search space is then $\Omega = \{0,1\}^\lambda$. The product space $\langle\Omega\rangle = \{0,1\} \times \Omega$ is defined as the ***dual space***. We defined a mapping T, called ***transliteration***, from the dual space to the basic search space by:

$$\forall\omega \in \Omega, T(0\omega) = \omega \text{ and } T(1\omega) = \overline{\omega},$$

where $\overline{\omega}$ is the binary complement of ω.

We apply genetic operators on the dual space. Strings from the dual space are mapped by ***transliteration*** to strings from the basic space and then the fitness function is applied.

A central property is that P-schemata over the dual space may represent P-schemata as well as R-schemata over the basic space. For instance, if $\lambda = 3$, the P-schemata $\underline{0}1 * 1$ and $\underline{1}0 * 0$ of $\langle\Omega\rangle$ are associated with the P-schema $1 * 1$ of Ω. On the other hand, the P-schema $\underline{*}1 * 0$ becomes the R-schema $\mathtt{X} * \overline{\mathtt{X}}$.

As the DGA is a GA that operates on the dual space, any works concerning classical GA apply there, especially methods for MOO.

3 Minimal MOPs

Minimal multiobjective problems (MMOPs) are a class of problems with two objective functions and solutions coded with binary chromosomes of length 2. We constructed two problems from this class. The first one ranks solutions, from the highest to the lowest: $[\{00, 10\}, \{01, 11\}]$. So, the best trade-offs are $\{00, 10\}$. This set forms a stable predicate or equivalently, is strictly represented

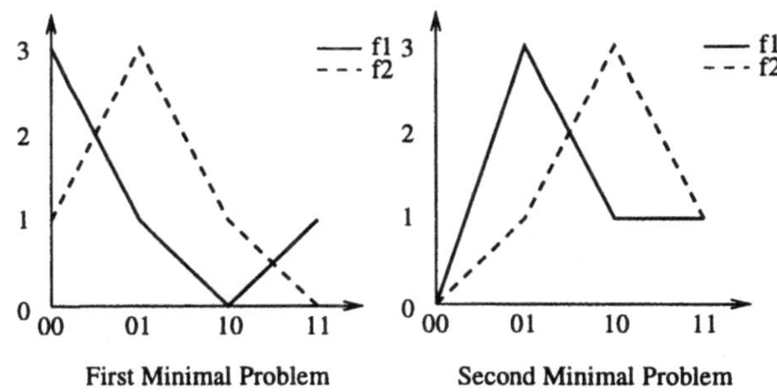

Figure 1: Instances of the first and the second minimal problem.

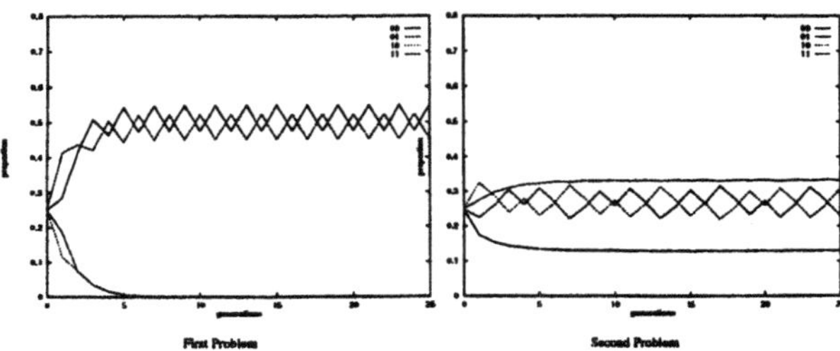

Figure 2: Pareto ranking SGA.

by the P-schema $*0$. The second problem ranks the solutions, still from the highest to the lowest: $[\{01,10\},\{11\},\{00\}]$. The set of the best trade-offs is now $\{01,10\}$. It can't be strictly represented by a P-schema, but it can by the R-schema $\mathrm{x}\overline{\mathrm{x}}$. Instances of these problems are shown in Figure 1.

We realized a simulation of the dynamic of a SGA, with equations found in [6], that we adapted for a niched Pareto ranking method, for the two problems (Figure 2). For the first problem, the SGA converges rapidly toward a population composed only of the two best trade-offs. The population is stable, the oscillations are due to the sharing method. On the second problem, the SGA fails to find the best trade-offs. The most represented individual is not one of the two best trade-offs.

A DGA associated with the same niched Pareto ranking method, for both problems, converges toward the sets of best trade-offs (Figure 3). In the first problem, the final population contains the individual set $\{000,001\}$, transliterated to $\{00,01\}$. In the second problem, the final population contains the stable individual set $\{001,101\}$, transliterated to $\{01,10\}$.

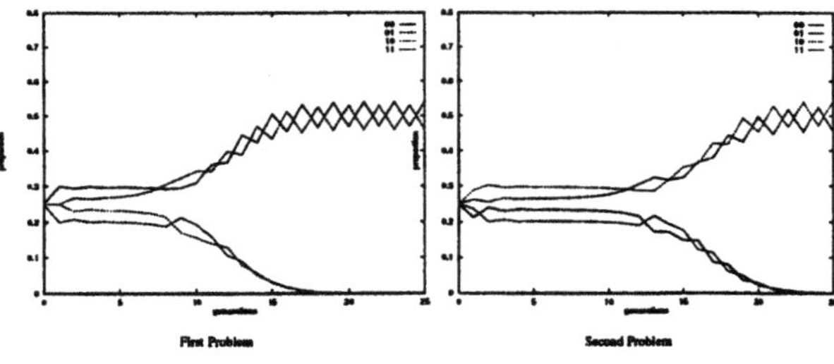

Figure 3: Pareto ranking DGA.

4 Multiobjective Trap Problems

We construct multiobjective trap problems (MOTPs) in order to obtain a kind of Pareto deception. As their single objective homologous, MOTPs are problems with two optima, a local one and a global one. The difference is, in the case of multiobjective problems, the optima is in fact a set of points. Figure 4 shows the two functions involved. They are defined on the normalized unitation of the chromosomes, i.e. the number of bits set to 1 divided by the total number of bits, and they return values between 0 and 1. These two functions have to be minimized. There are two Pareto sets. The right one, which lies between 0.9 and 1, is the local one: all of the points from it are dominated by those of the left one, which lies between 0 and 0.1. The problem is parameterized by two values. The first one, r, measures the importance of the global Pareto set with regard to the other. The second one, x_b, measure the relative importance of the basin of attraction of the two Pareto sets.

Firstly, we tested the two approaches, the SGA one and the DGA one, with several values for r and x_b in order to find their domain of efficiency on the plane $(r,x_b), r \in [0,1], x_b \in [0.1,0.9]$. There is a particularity for $r=0$. Indeed, for this value the two Pareto sets have to be considered equivalent. The DGA is able to find the two sets at the same time, as we will show it.

Domains of Efficiency

Figures 5 and 6 are the results of the experimentations. For each couple $(r,x_b), r \in [0,1], x_b \in [0,1]$, the value drawn is the average of the proportion of the population in the global Pareto set after 200 generations for 100 successive runs. On Figure 5,

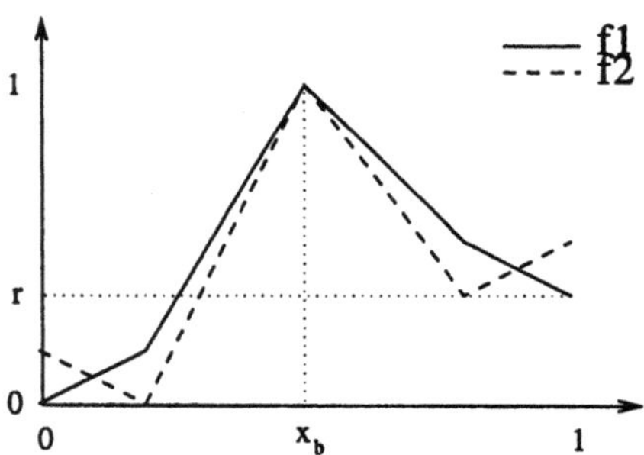

Figure 4: Multiobjective trap problems.

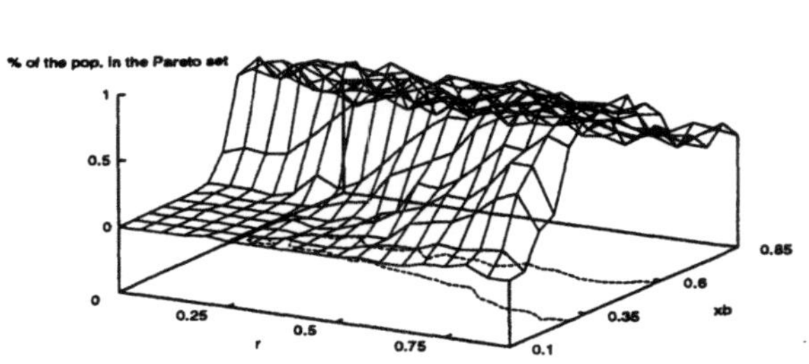

Figure 5: Proportion of the population in the global Pareto set with SGA.

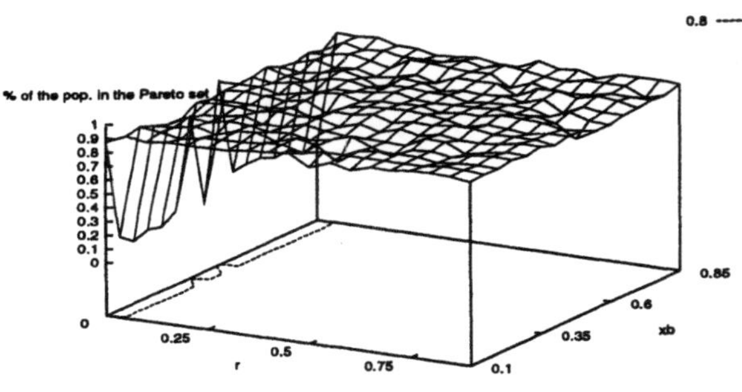

Figure 6: Proportion of the population in the global Pareto set with DGA.

one can see that SGA is unable to find the good Pareto set for low values of r and x_b. This is due to the importance of the basin of attraction of the bad set. Similarly, for x_b near 0.9, the proportion of the population in the Pareto set falls under 0.8. For this region, the basin of attraction of the global Pareto set is too important, and many individuals get 'lost' in it.

For the DGA, as we can see on Figure 6, these remarks don't apply. On the totality of the plane $(r, x_b), r \in [0,1], x_b \in [0.1, 0.9]$, the good set is found and a good proportion of the population is in it.

Disjoint Pareto sets

The case $r = 0$ is interesting as the two Pareto sets are then equivalent. In fact, there is a unique Pareto set, composed of two disjoint subsets. These two subsets are disjoint in the sense that each element of a subset is at least at a Hamming distance superior to 1 away from each element of the other subset.

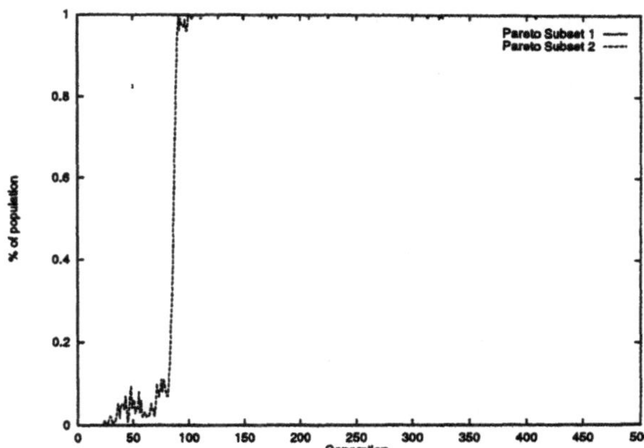

Figure 7: Proportion of the population in each of the two Pareto subsets with SGA.

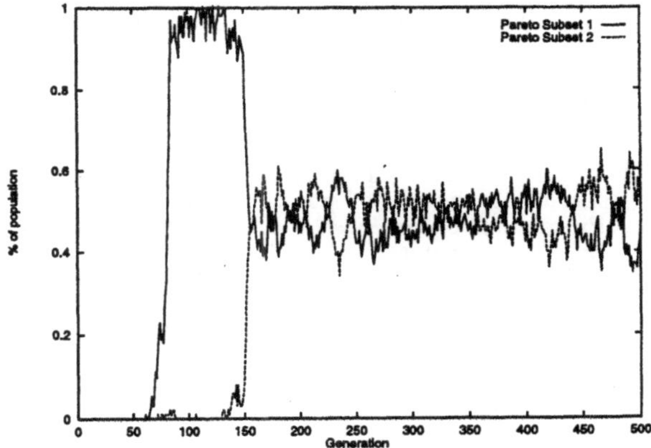

Figure 8: Proportion of the population in each of the two Pareto subsets with DGA.

A SGA, as shown in Figure 7, is able to converge in a single subset. As in MMOP, it is unable to make chromosomes from incompatible schemata coexisting in the same population. So, after a while, the entire population goes to one of the subsets. A DGA, thanks to R-schemata, is able to do it. As we can observe in Figure 8, the population firstly converge to the first subset. After a while, a part of the population goes in the second subset. Then, the two subsets are equally represented.

5 Further Work

The experimentations proposed here are obviously *ad hoc* problems, constructed to demonstrate interesting characteristics of DGAs. Our future work will be to use these successful associations on more complex problems and on real engineering problems.

6 Acknowledgments

This work was partly supported by a Brite-Euram project (EC contract BRPR-CT96-0282).

References

[1] P. Collard and C. Escazut. Relational schemata: A way to improve the expressiveness of classifiers. In *Proceedings of the Sixth International Conference on Genetics Algorithms*, 1995.

[2] C. M. Fonseca and P. J. Fleming. An overview of evolutionary algorithms in multiobjective optimization. *Evolutionary Computation*, 3(1):1–16, 1995.

[3] D.E. Goldberg. *Genetic Algorithms in Search, Optimisation and Machine Learning.* Addison Wesley, Massachussets, 1989.

[4] J. Horn and N. Nafpliotis. Multiobjective optimization using the niched pareto genetic algorithm. In *Proceedings of the First IEEE Conference on Evolutionnary Computation*, 1994.

[5] J. D. Schaffer. Multiple objective optimization with vector evaluated genetic algorithms. In *Genetic Algorithms and Their Application: Proceedings of the First International Conference on Genetic algorithms*, 1985.

[6] M. D. Vose. Modeling simple genetic algorithms. In *Foundations of Genetic Algorithms 2*, 92.

Multi-layered Niche Formation

C. Fyfe
Department of Computing and Information Systems,
The University of Paisley,
Paisley, UK.

Abstract

Recently an abstraction of genetic algorithms has been developed in which a population of GAs in any epoch is represented by a single vector whose elements are the the probabilities of the corresponding bit positions being equivalent to 1. The process of evolution is represented by learning the elements of the probability vector. We have previously extended this to model homeotic genes which are environmentally driven and turn other genes on and off. In this paper we incrementally develop the algorithm on a set of standard problems used to compare methods for the simultaneous optimisation of conflicting criteria within a single population.

1 Introduction

In this paper we consider the problem of simultaneous multiobjective optimisation involving conflicting criteria. This type of problem is of growing importance in e.g. classifier systems which require a set of rules which must be collectively optimised to perform a task. It is well known that, with a finite population, the simple GA settles on a single optimum. Strategies such as fitness sharing, mating restrictions and progressively setting preferences have been shown to be effective but each requires some form of global information outwith the confines of the simple GA.

2 PBIL Algorithm

A recent innovation in genetically motivated optimisation algorithms is the population-based incremental learning (PBIL) [1] algorithm which abstracts out the GA operations of crossover and mutation and yet retains the stochastic search elements of the GA. The algorithm is

- Create a vector whose length is the same as the required chromosome and whose elements are the probabilities of a 1 in the corresponding bit postion of the chromosome. When a binary alphabet is used for the chromosomes, the vector is initialised to 0.5 in every bit.
- Generate a number of samples from the vector where the probability of a 1 in each bit position of each vector is determined by the current probability vector.
- Find the fittest chromosome(s) from this population.
- Amend the probability vector's elements so that the probability of a 1 is increased in positions in which the fittest chromosomes have a 1.

The process is initialised with a probability vector each of whose elements is 0.5 and terminated when each element of the vector approaches 1 or 0. The update of the probability vector's elements is done using a supervised learning method

$$\Delta p_i = \eta(E_{best}(chromosome_i) - p_i) \qquad (1)$$

where $E_{best}(chromosome_i)$ is the mean value of the ith chromosome bit taken over the fittest chromosome(s) and p_i is the probability of a 1 in position i in the current generation. η is a learning rate.

Therefore to transmit information from generation to generation we have a single vector — the probability vector — from which we can generate instances (chromosomes) from the appropriate distribution. Notice that

$$\Delta p_i \to 0 \Longleftrightarrow p_i \to E_{best}(chromosome_i) \qquad (2)$$

Therefore if all the fittest chromosomes in each generation have a 1(0) in the ith position, the probability vector will also tend to have a 1(0) in that position.

3 The Structured GA

A gene which switches other genes off and on is known as a homeotic gene. For example there are environmentally-driven homeotic genes which are

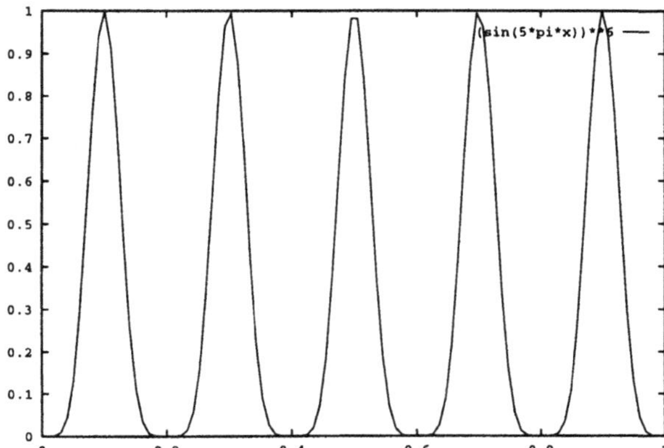

Figure 1: The graph of $F_1(x) = sin^6(5\pi x), 0 \leq x < 1$.

known to occur in nature and determine such major phenotypical properties as the organism's sex or number of wings [2].

The structured genetic algorithm [3] described this structure as a two-layered genetic algorithm. Genes at either level can be either active or passive; high level genes either activate or deactivate the sets of lower level genes i.e. only if gene a_1 is active will genes (a_{11} a_{12} a_{13}) determine a phenotype. Therefore a single change in a gene at a high level represents multiple changes at the second level in terms of genes which are active. Genes which are not active (passive genes) do not, however, disappear since they remain in the chromosome structure and are carried invisibly to subsequent generations with the individual's string of genes. Therefore a single change at the top level of the network can create a huge difference in the expressed phenotype whereas with the simple GA we would require a number of changes in the chromosome to create the same effect. As the number of changes required increases, the probability of such a set of changes becomes vanishingly small.

4 Structured PBIL

The PBIL algorithm provides us with a method for adjusting the elements of the probability vector corresponding to a simple GA. We have augmented [4] the method to mirror the SGA structure.

Let us consider a vector in which element p_0 corresponds to the homeotic gene; we consider it to be a simple switch but with a stochastic element in that the value of p_0 is the probability that the phenotype corresponding to gene 1 (in bit positions p_1 to p_m) being expressed and $1 - p_0$ is the probability that that corresponding to gene 2 in bit positions p_{m+1} to p_{2m} is expressed. Then we generate a random number from the uniform distribution in [0,1] and if it exceeds p_0 evaluate the fitness of the gene corresponding to positions p_1 to p_m; otherwise we evaluate the fitness of the gene corresponding to positions p_{m+1} to p_{2m}. The learning rule is

$$\begin{aligned}\Delta p_0 &= \alpha_0(E_{best}(chromosome_0) - p_0)\\ \Delta p_i &= \alpha(E_{best}(chromosome_i) - p_i) * \\ &\quad E_{best}(chromosome_0),\ 1 < i < m\\ \Delta p_i &= \alpha(E_{best}(chromosome_i) - p_i) * \\ &\quad (1 - E_{best}(chromosome_0)), m + 1 < i < 2m\end{aligned}$$

where α and α_0 correspond to small learning rates which can be adjusted during the course of the simulation. The last two equations are similar to the equation for the PBIL. The sole difference is that each change in the probabilities is gated by the expected value that the appropriate gene's phenotype was expressed during that generation.

The first equation causes the homeotic gene's probabilities to be changed. Notice that unless this gene's probabilities reach 0 or 1, either of the two 'lower-level' genes has a non-zero probability of being expressed. It is simple to ensure that this is always the case by putting bounds on the maximum and minimum values of p_0, however it has been found that a better solution is to have a lower learning rate α_0 for this gene which allows it to track the environment but at a slower rate.

5 Multiple Competing Criteria

We now extend the use of the SPBIL on a representative of the set of problems of multiple competing criteria. We use as an example of such problems the first of Deb's test problems quoted in [7]. Consider the function

$$F_1(x) = sin^6(5\pi x)$$

This function has 5 peaks when defined for $0 \leq x \leq 1$ which occur at 0.1, 0.3, 0.5, 0.7 and 0.9 as shown in Figure 1. We follow Ryan's coding: a 30-bit chromosome is used to represent x. Ryan follows the usual practise of initialising the chromosomes equally within the space of possible values, however such a method is not guaranteed to work since, in real problems, several solutions can be arbitrarily close to one another. Also, as the dimensionality of the data increases, the method's expense increases exponentially. We simply wish to begin with the probability vector of all the realisable genes (the lowest level genes) initialised with a common probability of 0.5. Further we do not wish to build in pre-knowledge of the solution in that the number of realisable genes are equal to the number of optimal solutions (5 in this case). One possible extension to the SPBIL is shown in Figure 2. We now consider a multi-layered sequence of homeotic genes. The top level represents bit 0 which determines whether bit 1 or bit 2 is turned on. Both bit 1 and bit 2 are further homeotic genes so that bit 1 determines whether bits 3 or 4 are turned on and bit 2 determines whether bits 5 or 6 are turned on. Similarly bits 3 - 6 are also homeotic genes and determine which of the realisable genes (each of 30 bits) are activated. There are therefore 8 sets of realisable genes each of which may determine the fitness of the chromosome. The learning rules are as above:

$$\begin{aligned}
\Delta p_0 &= \alpha_0(E(ch_0) - p_0) \\
\Delta p_1 &= \alpha_1(E(ch_1) - p_1)E(ch_0) \\
\Delta p_2 &= \alpha_1(E(ch_2) - p_2)(1 - E(ch_0)) \\
\Delta p_3 &= \alpha_2(E(ch_3) - p_3)E(ch_1).E(ch_0) \\
\Delta p_4 &= \alpha_2(E(ch_4) - p_4)(1 - E(ch_1)).E(ch_0) \\
\Delta p_i &= \alpha_3(E(ch_i) - p_i)E(ch_3).E(ch_1).E(ch_0), \\
&\quad 8 \le i < 38 \\
\Delta p_i &= \alpha_3(E(ch_i) - p_i)(1 - E(ch_3)).E(ch_1).E(ch_0), \\
&\quad 38 \le i < 68
\end{aligned}$$

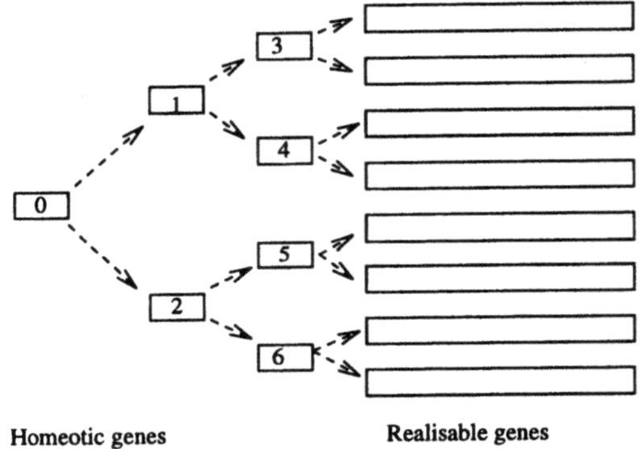

Figure 2: Homeotic gene 0 determines whether homeotic genes 1 or 2 will be realised. They in turn determine which of genes 3-6 will be realised. Genes 3-6 determine which of the lowest level genes will be realised.

where the gating expectations are taken over the fittest chromosomes; the expectations in lines after the third can be thought of as conditional expectations. The obvious extension is made to the other genes. We use lower learning rates for the higher level homeotic genes. With this set of learning rules, we found that convergence to the optimal solution was possible but that it required a very low learning rate i.e. there is a very slow convergence to the optima. If the learning rate increases a subset of the probability vectors converge to the optimal values but since that subset will typically have very accurate realisations on each generation, the whole population will become dominated by the niche solution which is most accurate. This fairly quickly feeds back to the homeotic genes and predisposes them to always select the most fit sub-population.

In order to maintain diversity in the population, we introduce an environmental feedback mechanism: when an instance of the realised genes is one of the fittest chromosomes i.e. one used in upgrading the probability vector, we added a small bias which is used to handicap the gene when its fitness is next calculated. This models the fact that competition for resources in a population is greatest between similar individuals. So in each generation we calculate

$$\Delta bias_i = -0.016, \tag{3}$$

for the best genes

$$\begin{aligned}
\Delta bias_i &= +0.002, \forall i \qquad (4) \\
bias_i &= bias_i + \Delta bias_i \\
fitness_i &= F_1(x) + bias_i
\end{aligned}$$

Using this method we can have an extremely high learning rate and accurate convergence to all optima. At the optima, probability elements representing the homeotic genes will be approximately 0.5 while all probability elements representing realisable genes will either be tending to 0 or 1. Using a training run of only 200 iterations and a learning rate α_3 of 0.2 we got the results in Table 1. The similarity between the use of a bias and that method known as a 'conscience mechanism' in artificial neural networks is obvious.

Now we wish to extend the above method to cope with Deb's other functions. Consider function F_2 defined by

$$F_2(x) = \exp(-2\log(2) * (\frac{x - 0.1}{0.8})^2) * \sin^6(5\pi x) \tag{5}$$

Table 1: Converged values represented by the probability vectors after only 100 generations when the learning rate, α, was 0.2.

Gene	1	2	3	4
Value	0.100	0.500	0.300	0.700
Gene	5	6	7	8
Value	0.300	0.902	0.900	0.900

Experimental findings have shown that with the system defined above the fitness of the highest peak is predominant: all realised genes converge to that peak leaving the other peaks unexplored. What we wish to do is to allow only one niche population to occupy the highest peak and force the other populations to occupy the lower peaks. To do so we use a system very similar to Kohonen's learning vector quantization (LVQ) [5]. At each update of the probability vector, we increase the probability vector relative to the best vector as before but also include a means of moving all (including the winning one) away from the best:

$$\Delta p_i = -\alpha_0(E_{best}(chromosome_i) - p_i) * \ldots$$
$$\Delta p_i = \alpha(E_{best}(chromosome_i) - p_i) * \ldots$$

Clearly we require $\alpha_0 < \alpha$ or we would get no convergence to any peak. A ratio of 0.1 was found sufficient in the experiments described herein. With this simple addition, all peaks were consistently found.

Functions F_3 and F_4 did provide more of a challenge to the algorithm in that we often found more than one population occupying the broader based peaks. They are defined by

$$F_3(x) = \sin^6(5\pi(x^{\frac{3}{4}} - 0.05))$$
$$F_4(x) = \exp\left(-2\log(2) * (\frac{x - 0.08}{0.854})^2\right) * \sin^6(5\pi x)$$

It was found that the simple expedient of causing the learing rate to decrease to 0 during the course of the experiment was sufficient to cause any required degree of accuracy. This method is obviously suggested by the needs of many stochastic approximation algorithms such as the Robbins-Monroe algorithm [6] which requires the learning rate to satisfy

$$\alpha(k) \geq 0, \sum_k \alpha(k) = \infty, \sum_k \alpha^2(k) < \infty \qquad (6)$$

We find ourselves in some difficulty in reporting our results in that

- some authors have been interested in how long their niche populations stayed on the peaks. Ours appear to stay indefinitely in all the simulations described above.
- other authors (e.g. [7]) have been interested in the degree of accuracy of their solutions. We appear to be able to achieve any required degree of accuracy using the learning rate decay described in this section.

We will have to content ourselves with stating that using the methods described in this paper we can cause convergence of the realised genes to the correct optimal values to any degree of accuracy required.

6 Conclusion

The PBIL algorithm has an advantage over the GA in that it converges more quickly to optimal solutions. However it also has the additional property that it enables a novel approach to describing the convergence properties of GAs.

We have seen that the SPBIL algorithm provides a new way of thinking about the convergence of the SGA and have given an experiment in which a single population using a homeotic 'switching' gene has caused divergence into two distinct populations. The SPBIL algorithm can be modified to ensure that genetic diversity always exists within the population by ensuring that no probability exceeds 0.9 or is less that 0.1; by ensuring that the homeotic gene does not reach the extreme values, we can ensure that there exist organisms in any environment which will not be overspecialised to the environment.

The SPBIL algorithm also allowed simultaneous optimisation of conflicting criteria. A simple extension of the SPBIL was found to be necessary when using a multi-layered gene in order that the niche populations equally represent all optima.

Finally when we use a population of more than 1 to update the probability vector, it is possible to use a weighted expected value in the learning equations in which the weights are proportional to the fitness of the appropriate phenotypes.

References

[1] S. Baluja and R. Caruana. Removing the genetics from the standard genetic algorithm. In *Proceedings of the Twelfth International Conference on Machine Learning*, 1995.

[2] J. Cohen and I. Stewart. *The Collapse of Chaos.* Penguin, 1995.

[3] D. Dasgupta and D. R. McGregor. A more motivated genetic algorithm: The model and some results. *Cybernetics and Systems*, 25(3):447–469, 1994.

[4] C. Fyfe. Developing and understanding niche formation in dynamically-changing environments using structured evolutionary algorithms. In *International Symposia on Soft Computing and Intelligent Industrial Automation*, 1996.

[5] T. Kohonen. *Self-Organising Maps.* Springer, 1995.

[6] J. M. Mendel. *A Prelude to Neural Networks: Adaptive and Learning Systems.* Prentice Hall, 1994. (Ed).

[7] C. Ryan. Racial harmony and function optimization in genetic algorithms - the races genetic algorithm. In *Proceedings of EP95.* MIT Press, 1995.

Using Hierarchical Genetic Populations to Improve Solution Quality

J. R. Podlena and T. Hendtlass
Centre for Intelligent Systems, Swinburne University of Technology
P.O.Box 218, Hawthorn, Australia 3122
Email: {jrp,tim}@brain.physics.swin.oz.au

Abstract

A multi-population genetic algorithm is proposed which is hierarchical in nature. This allows the algorithm to solve problems which consist of smaller tasks contributing to the solution of an overall problem. The algorithm feeds the entire pool of individuals between local populations (solving the smaller problems) and a global population (solving the overall task). Results on categorisation tasks are presented.

1 Introduction

Many real world problems can be broken down into a set of smaller problems. The 'divide and conquer' philosophy is often useful in these situations, and can be implemented in a wide variety of ways. This paper discusses the use of hierarchical genetic algorithms (HGA) which follow this philosophy. Problems in which the cost or fitness function can be broken down into different components are solved by using a separate genetic population optimising each of these components. An overall genetic population drawn from these lower level populations then calculates and optimises the combined performance of the individuals. Lower level optimising populations are transferred to and from the overall population until a suitable solution is found. Methods for, and the consequences of, these population interchanges are discussed.

This algorithm is applied to a character recognition problem. The recognition of single individual characters are optimised in the lower level populations, and the full recognition of the alphabet is optimised in the overall population. Results are compared with the standard genetic algorithm in terms of computational expense versus solution quality.

2 The Hierarchical Genetic Algorithm

In solving problems for which the solution must contain a number of lower level solutions which must work well both individually and as a group, it is common sense to optimise these lower level solutions both on an individual and a group basis. In using genetic populations to solve such problems, it also makes sense to separate local and global optimisation into different populations. This is a departure from the standard parallel GA where multiple populations are influenced by the same cost function [3]. In the HGA the lower level populations must optimise local cost functions while the overall cost function must also be optimised within the global population. Two methods for implementation were proposed and investigated, and these are described below.

The first implementation method involves using a standard GA for the local iterations which uses a fitness in breeding which is based on both the local fitness of the individual and also on the value returned from a 'combinational' GA. This combinational GA has an initial population of individuals extracted from the local populations and randomly combined with each other. (The combinational GA can then proceeded to breed these using the crossover of components only (no mutation) to find a team member's fitness value by finding the optimal team.) One problem with this, however, is that each individual component must be evaluated, yet since this is survival of the fittest, many of the original components are lost. This means that the individual components' team fitness must be stored from the first iteration. This could cause a great deal of noise if, for example, a combination

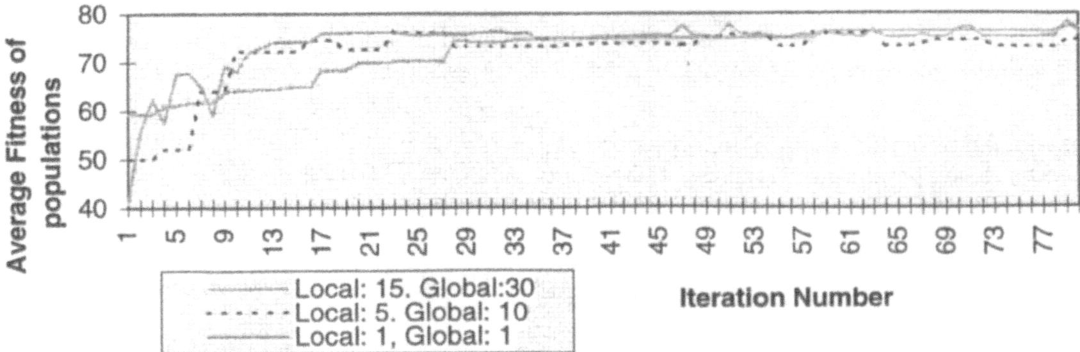

Figure 1: Fitness versus iterations for three runs of the algorithm with differing population transfer intervals.

of a good component with other bad components occurred early on in the run. The second problem related to this is with the actual combination of fitness.

Consider three populations A, B and C, where $A_i \in$ a set of all the possible values in set a, $i = (1, \ldots, \text{populationsize})$, etc.

Let $l(a)$ be a measure of how the individual fitness $(f_i(A_i)) \in a$ is related to its team member fitness $(f_t(A_i))$. $l(a)$ is therefore a measure of how close the following condition is met:

$$\forall f_i(A_j), f_i(A_k) : f_i(A_j) > f_i(A_k) \text{ and }$$
$$f_t(A_j) > f_t(A_k)$$

where j and $k = (1, \ldots, | a |)$. If $l(A_i)$ is high then we can say that the correlation between an individual component's fitness and that of its team member fitness is high. In the case where this is too high, there is little point in calculating the team fitness of the individual because this will obviously be shown in the local fitness. If $l(a)$ is too low, then the two fitness values will be in opposition to one another, and force the algorithm into endless oscillation. The problem is further complicated if, for example, $l(a)$ is much greater than either $l(b)$ or $l(c)$. Since this condition can not be known at run time, the choosing of relative weightings in calculating the overall fitness of an individual becomes non-trivial. If the weightings are fixed and uniform, the effect of noisy data from the combinational GA's calculation of individual team fitnesses could again cause destructive oscillation.

The second method of implementation of the hierarchical genetic algorithm uses two separate GA algorithms, one for individual populations and one for the overall population. This removes the problems of global fitness calculation by giving both local and global GAs their own separate fitness evaluation. It also reduced the chance of oscillation by restricting the transfer of fitness to simply the seeding of populations. It was this method which was used for the results discussed in this paper.

3 Testing Problems

Two categorisation problems were selected for testing the HGA, the iris data classification problem [1] and Frey and Slates' letter recognition problem [2]. In the iris data problem, four parameters are used to distinguish between three classes of iris flowers. In the Frey and Slate problem, each member of the training and testing set contains sixteen numerical attributes which are used to determine which of the letters of the alphabet it characterises. In solving both of these problems, each categorising chromosome is represented as a polynomial separating the division between, for example, the 'A' character and the 'B' to 'Z' characters. This separation is described in the following example:

For the category A recogniser applied to any example:

if A polynomial <0 **then**

categorise this example as category A

else

categorise this example as **not** category A

end if

The polynomial is encoded into the chromosome of the lower level population handling this categorisation task. After each iteration, each lower population is sorted in terms of local fitness. The global population is formed by combining similarly ranked polynomials from each lower level categorisation population. For example, lower level populations with individuals $A_1 - A_n$, $B_1 - B_n$, and $C_1 - C_n$ would combine to form a global population $A_1B_1C_1 - A_nB_nC_n$. This combination of three polynomials is then used to categorise input examples in the following manner:

if polynomial A < polynomial B **and** polynomial A < polynomial C **then** category A
if polynomial B < polynomial A **and** polynomial B < polynomial C **then** category B
if polynomial C < polynomial A **and** polynomial C < polynomial B **then** category C

4 Results on the Iris Categorisation Problem

Table 1 displays the results of trials performed on the iris classification data. All trials ran for a constant amount of computer time, and results were averaged over the three different methods to give the percentages displayed. It can be seen that the use of hierarchical genetic populations is superior to both the use of individual populations to find a set of classifiers (local GA) or single populations to find a single classifier (global GA). The HGA used 15 local population iterations, then formed the global population which ran for 30 iterations, which subsequently seeded the local populations and continued the process. An investigation as to the effects of different iteration lengths between local and global population transfers is outlined in Figure 1.

5 Optimal Values For Local and Global Iterations

Empirical evidence found on using a subset of the Frey and Slate data (see Figure 1) shows that as the frequency of population transfer increased, the oscillation during the run increased. Transferring the populations every local and global iteration gives a fitness profile that oscillates far more widely, but finds the near optimal solution faster (analogous to an under-damped system).

6 Results on Frey and Slate Data Set

The HGA was tested on the Frey and Slate data set using one generation for each transfer of populations between the local and global populations. Tests were made for both equal numbers of iterations (150) and for equal CPU time between the HGA and standard GAs on a local and global task level. The results of the comparison runs for 150 iterations can be seen in Figure 2.

The HGA out-performed the conventional GA using local or global populations only, with 70.9% correct classifications after 150 iterations compared to 12% for the standard local GA and 48.6% for the standard global GA. For the same amount of CPU cycles, the HGA finds 66.9% classifications compared to 11% for the local GA and 48.6% again for the global GA.

7 Conclusion

In using 'divide and conquer' philosophy, it seems that it is wise not to lose sight of the bigger task. In using a HGA with a single iteration between population transfers between local and global populations, both local and global methods can be utilised without losing information which can aide the search for the problem solution. The results as presented show

Table 1: The resulting average correct categorisations over a fixed equal time period for i) a standard local GA using single classifiers for each class, ii) a standard global GA using one classifier for the whole set of classes, and iii) the hierarchical GA.

method:	Categories correct on test data
Local GA	89.3
Global GA	64.0
Hierarchical GA	94.7

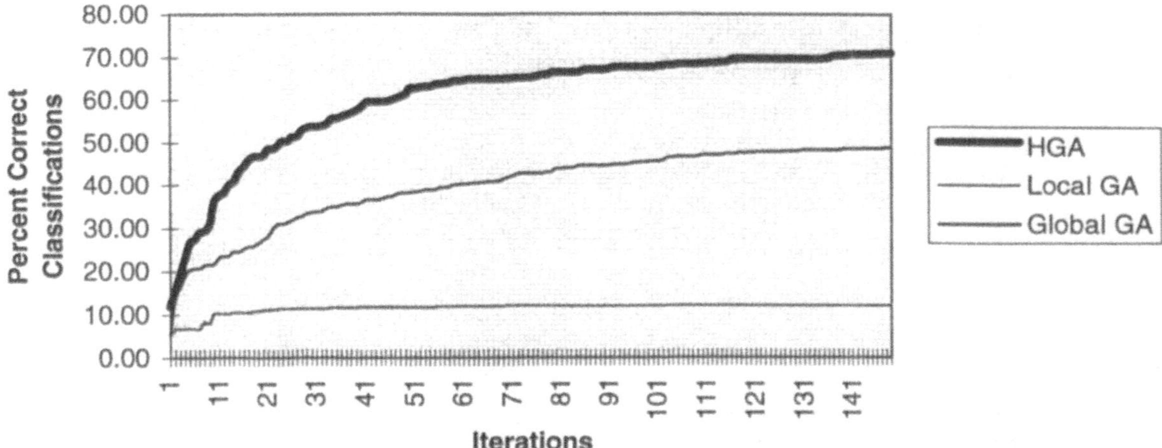

Figure 2: Hierarchical GA fitness versus Local and Global GAs.

that for the test problems using the HGA accelerates the search and therefore increases the quality of the final solution in comparison to the use of the standard GA.

References

[1] B.G. Batchelor. *A Practical Approach to Pattern Design Recognition.* Plenum Press, New York, 1974.

[2] P.W. Frey and D.J. Slate. Letter recognition using Holland-style adaptive classifiers. *Machine Learning*, 6:161–182.

[3] D. Levine. A parallel genetic algorithm for the set partitioning problem. Technical Report ANL-94/23, Mathematics and Computer Science Department, Argonne National Library, May 1994.

A Redundant Representation for use by Genetic Algorithms on Parameter Optimisation Problems

A. J. Soper and P. F. Robbins
School of Computing and Mathematical Sciences, University of Greenwich,
Wellington Street, Woolwich, London SE18 6PF, UK.
Email: A.J.Soper@greenwich.ac.uk

Abstract

In this paper we describe a redundant representation for use by genetic algorithms on continuous, parameter, optimisation problems and subject it to crossover. We examine the schemata induced by the representation and test its performance against a gray coded, binary representation and a real-coded representation acted on by the recombination operator BLX-0.5.

1 Introduction

Representations with redundancy have been little considered to date in genetic algorithm (GA) research for parameter optimisation problems; binary representations supporting the inversion operator provide one example which met with some success [2]. They have arisen more frequently in optimisation problems as for example in the evolution of the weights for multilayer perceptrons [12], where neural networks equivalent up to a permutation of their hidden nodes are normally represented by different genotypes, and representations for sequencing problems [10]. Radcliffe [8] has given minimal redundancy as a design principle and further advocates that where redundancy exists the GA should be able to 'fold (it) out' so that redundant solutions are not treated as unrelated. Redundancy has been regarded as detrimental because it enlarges the search space and produces additional noise when equivalent representations are not recognised during mating - the offspring may bear little resemblance to either parent. However, randomness in the form of higher mutation rates has been shown to be advantageous in studies using small population sizes exploring the effects of varying a GA's parameters [11] and especially so when higher cardinality alphabets [9] or real codings [13] are used. Fogarty [7] has shown that an initial high mutation rate which decreases over time can improve a GA's performance.

We present a highly redundant representation for parameter encoding acted on by crossover which is shown to perform well against a gray coded representation, which has no redundancy, and a real-coded representation subject to the BLX-0.5 operator of Eshelman and Schaffer [5], which processes interval based schemata. We follow this last reference and consider a parameter's values as a range of integer values, using the same number of points as the binary code for the purpose of comparison.

2 The Representation

The representation proposed is intended for parameter optimisation problems where the fitness function is a continuous and gradually varying function of the parameters. There has been much debate on which representations work best for this type of problem; gray coding normally performs better [3] than the standard binary coding having the property that adjacent values are only separated by unit Hamming distance. This provides for a more effective local search of the parameter space as single mutations can provide transitions to adjacent points. Real-coded GAs and evolution strategies [1] respect locality by using mutations that produce offspring near their parents with higher probability.

A redundant representation for integer coding is the unitation representation. This consists of a sequence of 0s and 1s, with integer value equal to the sum of the binary digits; it is clearly possible to reach adjacent values by changing one bit. The representation has been used in theoretical investigations [4], its practical usefulness being limited by the variation of the redundancy over the possible range 0 to l, where l is the length of the bit sequence. There are ${}^{l}C_{p}$ possible representations for a phenotype with value p, the ratio of the number at the centre to that at the ends of the range rising rapidly with l and producing a strong bias towards its centre.

The representation considered here consists of a sequence of integers

$$(a_1, a_2, ..., a_N)$$

each taking values in a set of increasing cardinality. If the cardinality at position r is $card_r$, then the possible values (alleles) are the members of the set

$$0, 1, ..., card_r - 1.$$

The parameter value (phenotype) is the sum of the values at each position. The possible phenotypes lie in the range 0 to the sum of the cardinalities minus N. Clearly the phenotypes are again not represented fairly. To remove this inequality define (a bias)

$$B = \sum_{r=1}^{N-1} (card_r - 1),$$

and choose

$$card_N = B + ParameterRange + 1,$$

where the parameter to be encoded takes integer values from 0 to *ParameterRange* (the maximum value of the binary equivalent). The phenotype now lies in the range 0 to $ParameterRange + 2B$. The frequency of genotypes will be uniform in the range B to $ParameterRange + B$. We therefore adjust our decoding to the phenotype by subtracting B from the sum of the allele values and constrain the phenotypes to lie within the parameter range as described below. The mutiparameter case is handled by juxtaposing the different parameter representations to form a single chromosome.

Considerations of fairness determine $card_N$ in terms of the first N-1 cardinalities which, including their number, have still to be chosen. In the simulation results presented here, the cardinalities found to give the best results after considerable experimentation have been presented. Factors guiding our choices were (i) the necessity of including loci with small cardinalities for effective local search, and (ii) sufficient redundancy that offspring with any parameter value should be possible in the first generation, from random mating of the randomly generated, initial generation. All the parameters, which all had the same range for each test function, were assigned the same cardinalities.

Our representation can be considered as based on a real-coded representation where numerical parameters are directly juxtaposed on a chromosome and subject to crossover. High mutation rates are normally applied to the real-coded parameters. In our extension, the redundancy produces noise under crossover which decays over time as the population converges.

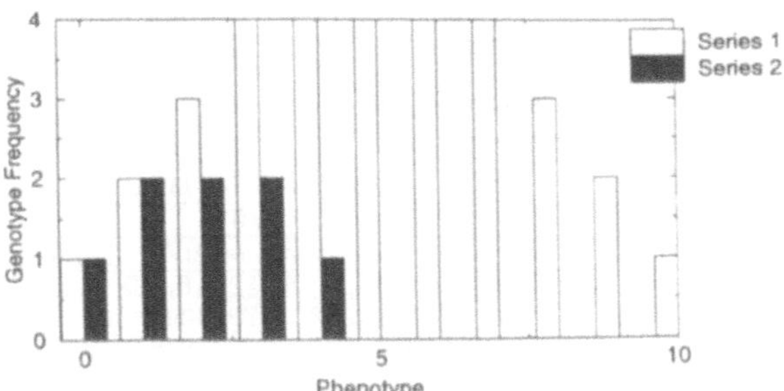

Figure 1: Schemata genotype frequencies.

2.1 Schemata Induced by the Representation

For a single parameter, we examine the possible phenotypes for particular schemata. Consider first schemata of order 1 and the number of genotypes for each possible parameter value, assuming a population containing each genotype once. Such schemata with the first (smallest cardinality) gene specified will give a wide and unbroken range of parameter values represented, while if the highest gene is specified, a narrower range of parameter values is present. As more genes are specified the ranges become increasingly narrow, but are always unbroken intervals for a flat population.

A simple example (where we ignore any bias for simplicity) of a chromosome containing three genes with cardinalities (2, 4, 8), where the possible allele values at each gene are $\{0,1\}$, $\{0,1,2,3\}$ and $\{0,1,...,7\}$ respectively is used for illustration. The frequency distributions of genotypes versus phenotypes, for the schemata (0, *, *) and (*, *, 0) are given by series 1 and 2 in Figure 1.

Each schema, for a flat population, defines a frequency distribution of genotypes over the parameter range, non-zero over a subinterval and peaked at its centre about which it is symmetric. During simulations it was observed that the alleles at loci with higher cardinality, specifying the narrowest distributions converged first.

Table 1: Parameters used with the new representation.

Function	Cardinalities per parameter $card_r$. Numbers in brackets denote a repeat count	bpp	fadx
f1	2(3), 3, 4, 5, 6, 7, 8, 12, 16, 32, 64, 1174	10	0.2
f2	2(5), 3, 4, 6, 8, 16, 32, 64, 128, 192, 256, 4800	12	0.2
f3	2, 4, 8, 16, 32, 64, 1144	10	0.5
f4	2, 4, 8, 16, 32, 313	8	0.5
f5	2, 4, 8, 16, 32, 64, 128, 256, 512, 1024, 2048, 4096, 8192, 147441	17	0.5
f6 and f6 scaled	2(7), 3, 6, 8, 12, 16, 24, 32, 48, 64, 96, 128, 192, 256, 384, 512, 768, 1024, 1536, 2048, 3072, 4096, 6144, 8192, 12288, 16384, 24576, 32768, 49152, 65536, 98304, 131072, 196608, 262144, 5111771	22	0.2

bpp bits per parameter
fadx fraction of differing alleles exchanged

Table 2: Performance tests.

Mean number of trials to find the optimum						
Function	New Represn.	sem	Gray code/ HUX	sem	BLX-0.5	sem
f1	1054	27	1089	25	874	20
f2	7013	589	9065	591	4893	357
f3	1496	117	1169	27	2005	119
f4	952	14	1948	97	933	24
f5	3750	313	1396	38	5561	588
f6	27973	2825	6496	725	14736	1998
f6 scaled	26307	2756	159325	21162	21635	1160

3 Empirical Comparisons

3.1 Test Harness

We used CHC [6] as a test bed, which has a conservative selection mechanism and is suitable for highly disruptive recombination operators, for two reasons:

1. Crossover applied to our representation will be disruptive in the sense that the offspring may be very different from both parents because of the redundancy of the representation, even if the parents are similar,
2. Results are available for CHC on a set of both continuous and non-continuous functions for systems processing both interval and binary schemata.

Eshelman and Schaffer [5] compared the performance of BLX-0.5 (blend crossover), a recombination operator that exploits the parameter intervals determined by the parents against a gray coded binary representation subject to HUX on DeJong's test suite supplemented by more difficult test cases.

We adapted CHC as follows: (i) HUX was applied to our chromosomes directly - however the fraction of differing allele values exchanged between the parents was reduced for tests on some of the functions, (ii) incest prevention was determined by the Hamming distance of gray-coded representations of the parameter values, (iii) new values of parameters for restarts were generated by flipping a fraction 0.35 of the bits of the gray-coded equivalent of the best individual, and the new chromosomes by setting half the first $N-1$ loci for each parameter to the same values as the best and randomly for the remainder, choosing the Nth allele's value to give the new parameter value. The above use of the gray coded representation follows that used for embedding BLX-0.5 in CHC (David Schaffer, personal communication). A population size of 50 was used throughout.

Out of range offspring can be produced by mating. If this occurred, the same parents were remated with the fraction of alleles exchanged during crossover decreased by a factor of 0.9 until legal offspring resulted.

We used the DeJong's functions $f1 - f5$ plus $f6$ [5], a sine envelope sine wave. It is known that f6 has a periodicity that makes it advantageous to use binary coding [5]; a scaled version was therefore also used in which the axes were stretched by a factor of 3 over the same number of sampled points. 50 runs were performed for each function and the average number of evaluations required to find the optimum was recorded in each case.

3.2 Performance Results

The cardinalities and fraction of different alleles exchanged under crossover is shown in Table 1 with the number of bits per parameter used in the gray code. Table 2 shows the mean number of trials needed to find the optimum. The results for HUX/gray code and BLX-0.5 are quoted from reference [5] for functions f1-f6, those for the scaled version of f6 are the authors' own.

4 Discussion of Results

The performance of our representation lay between that of gray/HUX and BLX-0.5 for each function. BLX-0.5 performed better than gray/HUX on the continuous functions $f1$, $f2$, $f4$ and the scaled version of $f6$, all of which have gradual variations suitable for an operator processing interval schemata. Because of the nature of the schemata of our representation, it is not surprising that its relative performance against gray/HUX tends to follow that of BLX-0.5.

For the gradually varying functions, the new representation tended to do better with more genes per parameter, when it has greater redundancy, and with fewer differing alleles exchanged during crossover. Performance was fairly sensitive to the set of cardinalities chosen.

The results show that a highly redundant representation for integers, acted on by crossover, can provide for efficient search on parameter optimisation problems.

References

[1] T. Back, F. Hoffmeister, and H. Schwefel. A survey of evolution strategies. In R. K. Belew and L. B. Booker, editors, *Proc. Fourth Intl Conference on Genetic Algorithms*, pages 2–9, San Mateo, CA, 1991. Morgan Kaufmann.

[2] J. D. Bagley. The behaviour of adaptive systems which employ genetic and correlation algorithms. *Dissertation Abstracts International*, 28(12), 1967.

[3] R. A. Caruana and J. D. Schaffer. Representation and hidden bias: Gray vs. binary coding for genetic algorithms. In *Proceedings of the 5th International Conference on Machine Learning*, pages 153–161, San Mateo, CA, 1988. Morgan Kaufmann.

[4] K. Deb and D. E. Goldberg. Analyzing deception in trap functions. In L. D. Whitley, editor, *Foundations of Genetic Algorithms 2*, pages 187–202. Morgan Kaufmann, San Mateo, CA, 1993.

[5] L. Eshelman and J. D. Schaffer. Real-coded genetic algorithms and interval-schemata. In L. D. Whitley, editor, *Foundations of Genetic Algorithms 2*, pages 187–202. Morgan Kaufmann, San Mateo, CA, 1993.

[6] L. J. Eshelman. The CHC adaptive search algorithm: How to have safe search when engaging in non-traditional genetic recombination. In G. J. E. Rawlins, editor, *Foundations of Genetic Algorithms*, pages 205–218. Morgan Kaufmann, San Mateo, CA, 1991.

[7] T. C. Fogarty. Varying the probability of mutation in the genetic algorithm. In J. Schaffer, editor, *Proc. of the Third Intl. Conference on Genetic Algorithms*, pages 51–60, San Mateo, CA, 1989. Morgan Kaufmann.

[8] N. J. Radcliffe. Forma analysis and random respectful recombination. In R. K. Belew and L. B. Booker, editors, *Proc. Fourth Intl Conference on Genetic Algorithms*, pages 222–229, San Mateo, CA, 1991. Morgan Kaufmann.

[9] C. Reeves. Using genetic algorithms with small populations. In S. Forrest, editor, *Proc. of the Fifth Intl. Conf. on Genetic Algorithms*, pages 92–99, San Mateo, CA, 1993. Morgan Kaufmann.

[10] P. F. Robbins. The use of a variable length chromosome for permutation manipulation. In D. W. Pearson, N. C. Steele, and R. E. Albrecht, editors, *Artificial Neural Nets and Genetic Algorithms*, pages 144–147, Wien, 1995. Springer-Verlag.

[11] J. D. Schaffer, R. A. Caruana, L. J. Eshelman, and R. Das. Study of control parameters affecting online performance of genetic algorithms for function optimization. In J. Schaffer, editor, *Proc. of the Third Intl. Conference on Genetic Algorithms*, pages 51–60, San Mateo, CA, 1989. Morgan Kaufmann.

[12] D. Whitley, T. Starkweather, and C. Bogart. Genetic algorithms and neural networks: Optimizing connections and connectivity. *Parallel Computing*, 14:347–361, 1990.

[13] A. H. Wright. Genetic algorithms for real parameter optimization. In G. J. E. Rawlins, editor, *Foundations of Genetic Algorithms*, pages 205–218. Morgan Kaufmann, San Mateo, CA, 1991.

A Genetic Algorithm for Learning Weights in a Similarity Function

Y. Wang and N. Ishii
Department of Intelligence and Computer Science, Nagoya Institute of Technology,
Gokiso-cho, syowa-ku, Nagoya, 466, Japan
Email: [wang,ishii]@egg.ics.nitech.ac.jp

Abstract

One large problem when employing a similarity function to measure the similarities between new and prior cases is to determine the weights of the features. This paper proposes a new method of learning weights using a genetic algorithm based on the similarity information of given examples. This method is suitable for both linear and nonlinear similarity functions. Our experimental results show the computational efficiency of the proposed approach.

1 Introduction

Similarity measure plays a central role in case-based reasoning and analogical reasoning. Most case-based and analogical reasoning systems represent cases by features and employ a similarity function to measure the similarities between new and prior cases. One large problem here is to determine the weights of the features. Because it is hard to give suitable weights to the features, learning weights is preferable. Recently, many learning methods [1, 4, 5, 8] have been proposed for this problem under various conditions. In this paper, we propose a different approach. We assume the similarity information is given by examples and propose a new method of learning weights using a genetic algorithm based on the similarity information. Our method is suitable for both linear and nonlinear similarity functions. Experimentation is performed and the results show the computational efficiency of the proposed approach.

2 Problem Description

The similarity information of given examples is divided into two kinds: One is called *qualitative similarity information (QSI)* which indicates whether case A is similar to case B (denoted by $A \sim B$) or not. The other is called *relative similarity information (RSI)* which represents, when $A \sim B$ and $A \sim C$, if the similarity between A and B is similar to that between A and C (denoted by $(A \sim B) \approx (A \sim C)$), or if A is more similar to B than to C (denoted by $(A \sim B) > (A \sim C)$), or not. Let $w = (w_1, w_2, \ldots, w_n)$ denote a weight vector where $w_i (i = 1, 2, \ldots, n)$ is the weight for attribute a_i. Let $s_w(x, y)$ be a similarity function which calculates the degree of similarity between case x and y using the weight vector w. Our problem is, given the similarity information of examples, a similarity function $s_w(x,y)$, real number α and ε, to find a weight vector w which satisfies:

Qualitative Similarity Condition(QSC): $\forall A, B$, if $A \sim B$ then $s_w(A,B) \geq \alpha$ else $s_w(A,B) < \alpha$.

Relative Similarity Condition(RSC): $\forall A, B, C$ where $A \sim B$ and $A \sim C$,

if $(A \sim B) \approx (A \sim C)$ then $|s_w(A,B) - s_w(A,C)| < \varepsilon$.
if $(A \sim B) > (A \sim C)$ then $s_w(A,B) - s_w(A,C) \geq \varepsilon$.
if $(A \sim B) < (A \sim C)$ then $s_w(A,C) - s_w(A,B) \geq \varepsilon$.

For any similarity function, it is clear that there may not be only one w which satisfies QSC and RSC. Because most similarity functions have the property: $\forall x, y$ and any w, $\exists w' = (w'_1, w'_2, \ldots, w'_n)$, $s_w(x,y) = s_{w'}(x,y)$ where $w' = w / \sum_{i=1}^{n} w_i$, we add the condition $\sum_{i=1}^{n} w_i = 1$ to w to reduce the search space.

If the similarity function is linear then linear programming is one method of solving this problem. But if the similarity function is not linear then other methods should be studied. Up to now, no other method has been proposed for this case. This paper presents a method of learning weights using a genetic algorithm (GA), which is suitable for both linear and nonlinear similarity functions.

3 The Genetic Algorithm

In this section, we describe the GA proposed for the learning of weights based on the similarity information of given examples.

3.1 Coding

A weight vector is encoded by a string of real numbers where the ith real number is the weight assigned to the ith attribute. An example of an individual is depicted below:

attribute	a_1	a_2	a_3	a_4
weight	0.25	0.30	0.10	0.35

⇑
an individual

3.2 Evaluation

The fitness value of an individual is the satisfaction degree of similarity information, i.e. $(N_{QSC} + N_{RSC})/(M_{QSC} + M_{RSC})$, where N_{QSC} is the number of case pairs which satisfy QSC using the weight vector w represented by the individual, N_{RSC} is the number of sets of three cases which satisfy RSC using w, M_{QSC} is the maximum number of case pairs included in QSI and M_{RSC} is the maximum number of sets of three cases included in RSI. The fitness value of a population is the maximum value of fitness of individuals in the population.

3.3 Selection

All individuals in the population are sorted by their fitness. Let p be the population size. In our method, all individuals are selected for mating. We perform crossover to the ith individual and the $(i+1)$th individual to generate $[(p+1)/2]$ individuals of the next population, where $i = 1, 2, \ldots, [(p+1)/2]$. If the fitness of the offspring is less than that of one of its parents, then add the parent whose fitness is higher to the next population instead of the offspring. This method helps to maintain the quality of the population. Also, we perform crossover to the jth individual and the $(p+1-j)$th individual to generate $[p/2]$ individuals of the next population, where $j = 1, 2, \ldots, [p/2]$, then the average fitness of individuals in the next population could be increased because the fitness of one parent is higher and the fitness of another is lower.

3.4 Crossover

Two selected individuals are called parent-A and parent-B, and we assume the fitness of parent-A is higher or equal to that of parent-B. Every crossover results in one offspring by the following steps:

1. For a gene w_a in parent-A and the gene w_b at the same position as w_a in parent-B, if $w_a = w_b$ then set w_a to the same position in the offspring.

2. Let the number of unset genes in the offspring be m.

   If $m < 4$ then for every unset position in the offspring, set the average value of genes at the same position in parents to the offspring.

   If $m \geq 4$ then: select $[m+1]/2$ genes from parent-A randomly and set them in the offspring. Let the sum of these genes and the genes set by step 1 be $sum1$ and then, let the number and sum of genes in parent-B at the unset positions be n' and $sum2$. For any unset position in the offspring, if $sum2 = 0$ then set $(1 - sum1)/n'$ to the position else set $g(1 - sum1)/sum2$ where g is the gene in parent-B at the unset position, so that the sum of all genes in the offspring is 1.

3.5 Mutation

In our method, mutation means the exchange of one gene with another, or replace g_1 and g_2 by g_1' and g_2' where $g_1 \neq g_1'$ and $g_1 + g_2 = g_1' + g_2'$.

3.6 Differentiation of the Same Individuals

After crossover and mutation, for variety of individuals in the population, mutation will operate on the individual which is the same as another until it is different from all other individuals.

4 Experimentation

In this section, we discuss the settings used in our experiment, the experimental method and results.

4.1 Similarity Function

Two similarity functions are used in our experiments. One is linear and the other is nonlinear.

Linear Similarity Function

Definition 1. *The degree of similarity between case x and y using the weight vector w is denoted by $s_w(x, y)$ and is defined as follows:*

$$s_w(x,y) = \sum_{i=1}^{n} s_i w_i \Big/ \sum_{i=1}^{n} w_i$$

where s_i is the degree of similarity between v_i^x and v_i^y which are values of attribute a_i in x and y respectively, and w_i is the weight assigned to a_i.

This is a well-known linear similarity function.

Nonlinear Similarity Function

We have proposed a nonlinear similarity function [10, 11] based on fuzzy integral.

Definition 2. *The weight of a set f is defined as $w_f = \sum_{v \in f} w_v$, where w_v is the weight assigned to v.*

Definition 3. *The fuzzy measure g_w on a set f based on a weight vector w is as follows:*

1. $g_w(\phi) = 0$;
2. $g_w(f) = 1$;
3. $g_w(f') = w'_f / w_f$, *where* $f' \subset f$.

Definition 4. *The fuzzy integral of a set f on fuzzy measure g_w is denoted by $\int f \circ g_w$ and defined as follows:*

$$\int f \circ g_w = \bigvee_{f' \subset f} \left\{ \left[\bigwedge_{v \in f'} v \right] \bigwedge g_w(f') \right\}$$

where $\bigvee$ and $\bigwedge$ express the upper limit and the lower limit.

Detailed discussion on fuzzy measures and fuzzy integrals can be found in [9].

Definition 5. $s_w(x, y) = \int f(x, y) \circ g_w$

where $f(x, y) = \{s_1, s_2, \ldots, s_n\}$.

We use these two similarity functions in our experimentation where values of any attribute are in $[0, 1]$ and s_i is calculated by $|v_i^x - v_i^y|$.

4.2 GA Parameters

The population size is $2N_w$ where N_w is the number of attributes. The start population is initialized randomly. The probability of mutation is 10%. The GA be terminated after $10kN_w$ individual (i.e. weight vector) evaluations, where k varies on the set $\{1, 10, 25, 50, 100\}$.

4.3 QSC and RSC Parameters

$\alpha = 0.5$ and $\varepsilon = 0.01$.

4.4 Experimental Method

1. Select a similarity function randomly.
2. Decide a weight number N_w(i.e. attribute number) randomly, where $4 \leq N_w \leq 20$. Randomly generate a weight vector w.
3. Generate 100 cases randomly.
4. For each N_c which varies on the set {25, 50, 75, 100}, use w and $1 \sim N_c$th cases to generate similarity information(QSI and RSI). Then use the GA described above to learn weights based on the similarity information. Let w_{GA} denote the learning result (the best weight vector) and f_{best} denote the best fitness value (maximum satisfaction degree of similarity information). We calculate sd, the standard deviation from $w - w_{GA}$.

For 100 runs, we calculate average, minimum and maximum values of f_{best} and sd for each k(see GA parameters) and N_c.

4.5 Results and Discussion

Table 1 and 2 summarize the experimental results. Table 1 shows f_{best} and Table 2 shows sd. In each table, there are three values for each k and N_c: The first is the average value, the second is the minimum value and the third is the maximum value. For example, for $k = 25$ and $N_c = 50$, the average, minimum and maximum value of f_{best} in Table 1 are 0.9928, 0.9616 and 1 respectively. The average, minimum and maximum value of sd in Table 2 are 0.0008, 0 and 0.0058 respectively.

The optimal value of f_{best} is 1. From Table 1, we can see that, for $k \geq 10$, the average value of f_{best} is greater than 0.97, and for $k \geq 25$, it is greater than 0.99 and the maximum value of f_{best}

Table 1: Satisfaction degrees.

k	N_c 25	50	75	100
1	0.8315	0.8263	0.8237	0.8240
	0.7318	0.7075	0.7057	0.6918
	0.9886	0.9496	0.9733	0.9479
10	0.9758	0.9771	0.9768	0.9774
	0.8982	0.8529	0.8684	0.9353
	0.9992	0.9998	0.9997	0.9995
25	0.9924	0.9928	0.9929	0.9928
	0.9421	0.9616	0.8917	0.9419
	1.0000	1.0000	1.0000	0.9999
50	0.9963	0.9975	0.9976	0.9968
	0.9751	0.9775	0.9643	0.9705
	1.0000	1.0000	1.0000	1.0000
100	0.9985	0.9988	0.9994	0.9988
	0.9826	0.9780	0.9914	0.9820
	1.0000	1.0000	1.0000	1.0000

Table 2: Standard deviations.

k	N_c 25	50	75	100
1	0.0192	0.0184	0.0175	0.0176
	0.0033	0.0046	0.0038	0.0064
	0.0685	0.0708	0.0445	0.0694
10	0.0025	0.0022	0.0022	0.0021
	0.0001	0.0001	0.0000	0.0000
	0.0212	0.0243	0.0235	0.0135
25	0.0009	0.0008	0.0007	0.0008
	0.0000	0.0000	0.0000	0.0000
	0.0066	0.0058	0.0149	0.0091
50	0.0005	0.0003	0.0002	0.0004
	0.0000	0.0000	0.0000	0.0000
	0.0037	0.0050	0.0054	0.0027
100	0.0003	0.0002	0.0001	0.0001
	0.0000	0.0000	0.0000	0.0000
	0.0031	0.0049	0.0009	0.0019

is 1, that means f_{best} reached the optimum value in some runs.

The optimal value of sd is 0. From Table 2, we can see that, for $k \geq 10$, the average value of sd is less than 0.003, and for $k \geq 25$, it is less than or equal to 0.0009 and the minimum value of sd is 0, that means sd reached the optimum value in some runs.

From these results, we can say that our GA can find optimal or very close to optimal weight vectors so that it can be of real practical use.

5 Conclusions

We have described a genetic algorithm that learns weights in a similarity function based on the similarity information of given examples. The computational efficiency of the proposed method is confirmed by the experimentation. Our further works will include comparing our GA with other GAs, testing the computational efficiency when the given similarity information is not complete or includes errors, other techniques for learning weights, etc.

References

[1] D. W. Aha, 'Incremental Instance-based Learning of Independent and Graded Concept descriptions', *Proc. of the 6th Intl. Workshop on Machine Learning*, 387–391, 1989.

[2] J. E. Baker, 'Adaptive Selection Methods for Genetic Algorithms', *Proc. of an Intl. Conf. on GAs and Their Applications*, 101–111, 1985.

[3] T. Bäck, 'Selective Pressure in Evolutionary Algorithms: A Characterization of Selection Mechanisms', *Proc. of the 1st IEEE conf. on evolutionary computation*, 57–62, 1994.

[4] J. P. Callan, T. E. Fawcett and E. L. Rissland, 'CABOT: An Adaptive Approach to Case-Based Search', *Proc. of IJCAI'91*, 803–808, 1991.

[5] C. Cardie, 'Using Decision Trees to Improve Case-Based Learning', *Proc. of the 10th Intl. Workshop on Machine Learning*, 25–32, 1993.

[6] J. D. Kelly and L. Davis, 'A Hybrid Genetic Algorithm for Classification', *Proc. of IJCAI'91*, 635–650, 1991.

[7] K. Satoh and S. Okamoto, 'Learning Weights in A Similarity Function from Distance Information'(in Japanese), *Journal of Japanese Society for AI*, 11, 3, 238–245, 1996.

[8] C. Stanfill and D. Waltz, 'Toward Memory-Based Reasoning', *Communications of the ACM*, 29, 1213–1228, 1986.

[9] M. Sugeno, 'Fuzzy Measure and Fuzzy Integral'(in Japanese), *Trans. of the Society of Instrument and Control Engineers*, 8, 2, 218–226, 1972.

[10] Y. Wang, N. Inuzuka and N. Ishii, 'A Method of Similarity Metrics Using Fuzzy Integration', *Proc. of the 3rd Pacific Rim Intl. Conf. on AI*, 2, 1028–1034, 1994.

[11] Y. Wang, N. Inuzuka and N. Ishii, 'Similarity Metrics on Frame Knowledge Expressions'(in Japanese), *Journal of Japanese Society for AI*, 10, 5, 778–785, 1995.

Learning SCFGs from Corpora by a Genetic Algorithm

B. Keller and R. Lutz
School of Cognitive and Computing Sciences, The University of Sussex,
Brighton, UK.
Email: {billk,rudil}@cogs.susx.ac.uk

Abstract

A genetic algorithm for inferring stochastic context-free grammars from finite language samples is described. Solutions to the inference problem are found by optimizing the parameters of a covering grammar for a given language sample. We describe a number of experiments in learning grammars for a range of formal languages. The results of these experiments are encouraging and compare very favourably with other approaches to stochastic grammatical inference.

1 Introduction

Grammatical inference [5] is a fundamental problem in many areas of artificial intelligence and computer science, including speech and language processing, syntactic pattern recognition and automated programming. Although a wide variety of techniques for automated grammatical inference have been devised (for surveys see [1, 4]) most are subject to limitations which severely restrict their range of application. Inference may be limited to regular grammars or require information about negative (ungrammatical) as well as positive (grammatical) strings.

A goal of the present work is to provide a general method for inferring a wide class of grammars on the basis of just positive information about the target language.

Genetic algorithms [6] are a family of robust, probabilistic optimization techniques that offer advantages over specialised procedures for automated grammatical inference. A number of researchers have already described applications of genetic algorithms to language identification problems with some success [3, 7, 8, 10, 12, 13, 15, 16]. However, with the exception of recent work reported by Schwem and Ost [12], the problem of inferring *stochastic* grammars has not been addressed. This is surprising in view of the many practical applications to tasks including speech recognition, part-of-speech tagging, optical character recognition and robust parsing. While Schwem and Ost recognize the importance of the stochastic inference problem, their approach is restricted to the inference of stochastic regular grammars. The present work tackles the more general problem of inferring stochastic grammars for the class of context-free languages.

2 Stochastic CFGs

A *stochastic context-free grammar* (SCFG) is a variant of ordinary context-free grammar where each production has an associated probability. The probabilities associated with all productions expanding the same non-terminal symbol must sum to one. The set of production probabilities will be referred to as the *parameters* of the SCFG. The language $L(G)$ generated by a SCFG G is the set of all strings of terminal symbols derivable from the start symbol of G (typically, S). The parameters define a probability distribution over strings in $L(G)$. For a string $\alpha \in L(G)$, the probability of a parse tree for α is the product of the probabilities of all productions used in its construction. The probability $P_G(\alpha)$ of the string α is the sum of the probabilities of all of its parses. The SCFG shown in Figure 1 generates the language $\{a^n b^n | n \geq 1\}$, where $P_G(ab) = 0.6$, $P_G(aabb) = 0.24$, and so on.

3 Corpus-Based Inference

A *corpus* C for a language L is a finite set of strings drawn from L, where each string $\alpha \in C$ is associated

$$
\begin{array}{llll}
\mathrm{S} \rightarrow \mathrm{A\,B} & (1.0) & \mathrm{B} \rightarrow b & (1.0) \\
\mathrm{A} \rightarrow a & (0.6) & \mathrm{C} \rightarrow a & (1.0) \\
\mathrm{A} \rightarrow \mathrm{C\,S} & (0.4) & &
\end{array}
$$

Figure 1: SCFG for the language $a^n b^n$ $(n \geq 1)$.

ab	595
aabb	238
aaabbb	97
aaaabbbb	49
aaaaabbbbb	14
aaaaaabbbbbb	5

Figure 2: A Corpus for the Language $a^n b^n$.

with an integer f_α representing its *frequency of occurrence.* An example of a corpus for the language $\{a^n b^n | n \geq 1\}$ is shown in Figure 2. The frequency of the string *ab* is 595, the frequency of *aabb* is 238, and so on. Given a corpus C as training data, the inference problem is to identify a SCFG that models the corpus as accurately as possible while generalizing appropriately to the wider language from which the corpus was drawn. For a stochastic grammar, a natural measure of accuracy is the likelihood of the corpus data given the grammar. The best grammar under this measure is that SCFG $\hat{G} = argmax_G P(C|G)$ where

$$P(C|G) = \prod_{\alpha \in C} P_G(\alpha)^{f_\alpha}$$

Unfortunately, the most accurate model in this sense will *over-fit* the training data. What is actually required is the grammar $\hat{G} = argmax_G P(G|C)$ that is most likely given the training data. It is not clear how to calculate $P(G|C)$ directly, but from Bayes rule we see that maximising $P(G|C)$ just corresponds to maximising $P(C|G) \times P(G)$. This poses the problem of fixing an appropriate prior probability distribution over grammars. In principle there are many different priors that could be chosen, but it seems reasonable to assume that we should prefer smaller or simpler grammars to larger, more complex ones. Our choice of prior is related to the *minimum description length* principle of Risannen [11] as well as earlier work on inductive inference due to Solomonoff [14]. We first fix a probability distribution over (parameterized) productions, such that shorter rules are more probable than longer rules. The prior probability of a grammar is taken to be the product of the prior probabilities of all of its rules.

In practice, it is not convenient to compute $P(G|C)$ directly. The genetic algorithm uses an objective function F given by

$$F(G) = \frac{-K_C}{\log P(C|G) + \log P(G)} \tag{1}$$

Minimizing the denominator in Equation (1)(ignoring sign) just amounts to maximising $P(G|C)$. The numerator $-K_C$ is a problem (corpus) dependent normalization factor that yields fitness values in the range [0;1].

4 The Genetic Algorithm

For a given a corpus, grammatical inference is performed in the following steps:

1. construct a covering grammar that generates the corpus as a (proper) subset;
2. set up a population of individuals encoding parameter settings for the rules of the covering grammar;
3. repeatedly apply genetic operations (crossover, mutation) to selected individuals in the population until an optimal set of parameters is found.

The covering grammar is a large context-free grammar in Chomsky Normal Form (CNF) over a fixed set of non-terminals. The grammar contains every rule of the form $A \rightarrow BC$ (A, B and C non-terminals) and every rule of the form $A \rightarrow a$ (A a non-terminal, a a terminal symbol that appears in the corpus). Restricting attention to CNF grammars over finite sets of terminal and non-terminal symbols guarantees a finite number of productions. A CNF grammar with n nonterminals and m terminals has $n^3 + nm$ productions. Note that there is no loss of generality: for any SCFG (not generating the empty string) there is a weakly equivalent CNF SCFG that assigns the same probability distribution.

The population is organized as a two dimensional grid, with opposing sides of the grid identified (i.e. individuals inhabit the surface of a torus). Each individual has exactly eight neighbours and encodes a complete set of parameters for the covering grammar. Within a genome, parameters are encoded as n-bit blocks. Because the parameter values are not independent of one another, we do not encode them directly. Instead, each n-bit block is treated as encoding a numerical *weight* which is then normalized. The probability p_j of production r_j is given by w_j/W, where W is the sum of all weights associated with rules expanding the same non-terminal as r_j.

In general, the covering grammar has many more rules than the target SCFG. The representation is therefore biased in favour of rules having zero weight by reserving a small number of initial bits in each block. If each initial bit is set to 1, then the remaining bits are decoded to obtain the rule weight. Otherwise the rule weight is taken to be zero. The number of reserved bits controls the amount of bias in favour of zero weight, while the number of remaining bits controls the size of the weights and thus the precision of the rule probabilities.

For the experiments described in the following section we have found that between one and three initial bits and 7 'weight' bits is sufficient. The actual number of initial bits is determined automatically in proportion to the size of the covering grammar. The larger the grammar, the more bits are used. The genetic algorithm repeatedly executes the following steps:

> Select a random member of the population for breeding, and choose the fittest of its eight neighbours as the second parent.
>
> Breed by applying crossover and mutation to produce two children.
>
> Replace the weakest parent by the fittest child.

A characteristic of our representation is that the probability of any given rule does not depend solely on local properties of the genome (i.e. the state of the relevant n-bit block). In general, it will also depend on the weight encodings associated with rules expanding the same non-terminal symbol. Further, the fitness of a given individual may be crucially dependent on the state of weight encodings that are widely separated within the genome. In short, our representation is one which exhibits high epistasis and where global properties of the genome are in many ways more important than local properties. This presents a problem, because such global characteristics unlikely to be preserved under the classical, one-point crossover operator.

We have experimented with a number of alternatives to the classical crossover operator. Good performance has been achieved using a novel genetic operator which we refer to as *and-or crossover*. This works by inspecting corresponding positions in the parent's genomes and then performing the logical operations of *and* and *or*. Two children are built up bit-by-bit, with one child selected (with some crossover probability) to receive the value returned by the *and* operation, and the other the value of the *or*. Best results are obtained when the crossover probability is itself randomly generated at the start of each breeding phase. The genetic algorithm employs a standard, point-wise mutation operator, which is performed with low probability. The mutation rate is set inversely proportional to the length of the genome.

5 Experimental Results

A number of experiments were conducted in learning grammars for the following formal languages: the language of all strings of equal numbers of *a*s and *b*s (EQ); the language a^nb^n ($n \geq 1$); the language of balanced brackets (BRA1); balanced brackets with two sorts of bracketing symbols (BRA2); and two and three symbol palindromes (PAL1 and PAL2). For each experiment, a training corpus was produced automatically using a hand-crafted SCFG for the target language. On the order of 16,000 strings were randomly generated up to a pre-specified maximum length. The number of non-terminals in the covering grammar was fixed as the number used in the hand-crafted SCFG. For each problem, the population size was set to twice the number of parameters.

With the exception of PAL2 (three symbol palindromes), ten runs each of the genetic algorithm were performed on each task. For PAL2, the current implementation requires considerable processor time and for this reason only three runs were executed. A run of the genetic algorithm was terminated as 'successful' if a SCFG was found with fitness above a threshold value of 0.93. Experience has shown that grammars attaining this fitness are almost invariably correct in the sense that they generate the target language exactly, and assign appropriate probabilities to the strings. Runs of the genetic algorithm that failed to attain the threshold value were terminated after a maximum number of select-breed-replace cycles.

For each learning task, Table 1 gives the number of non-terminals used in the covering grammar, the number of parameters (rules), the success rate (number of runs that attained the threshold fitness value) as well as the maximum fitness value found on the best and worst runs. The first four tasks (EQ, a^nb^n, BRA1 and BRA2) presented little difficulty. For EQ and BRA2, one run in each case failed to produce an adequate grammar. The

Table 1: Results on language learning tasks.

L	NTs	Params	Success	Best	Worst
EQ	3	33	9/10	0.971	0.679
$a^n b^n$	4	72	10/10	0.979	0.941
BRA1	3	33	10/10	0.956	0.951
BRA2	5	145	9/10	0.957	0.622
PAL1	5	135	2/10	0.950	0.871
PAL2	7	364	1/3	0.937	0.892

S → C B	(0.51875)	C → b	(1.0)
S → D A	(0.48125)	D → A S	(0.252066)
A → a	(1.0)	D → A C	(0.066116)
B → S C	(0.24778)	D → a	(0.681818)
B → b	(0.752212)		

Figure 3: Near-miss grammar for PAL1.

relatively poor fitness values attained on these runs (0.679 and 0.622 respectively) suggest the presence of local maxima around which the population has converged.

The results for the two palindrome languages initially appear less encouraging. For PAL1, only two runs attained the threshold fitness value, while for PAL2 only one of the three runs was terminated successfully. However, even on the worst runs in each case the algorithm found grammars with quite high fitness. Furthermore, it should be noted that the threshold fitness value represents a somewhat arbitrary measure of success. In particular, failure to attain this threshold does not imply that the algorithm has failed to find a grammar with a correct (or nearly correct) set of rules. For example, it is possible that the grammar generates the target language exactly, but with a non-optimal probability distribution.

Inspection of the grammars produced for all runs of the PAL1 learning task showed that the algorithm had performed rather better than suggested. Figure 3 shows the grammar ranked fifth best (with a fitness of 0.897355) out of all those produced by the algorithm on this task. Aside from the presence of one spurious production D → A C, which has a low associated probability (0.066116), the grammar is otherwise correct. Similar comments apply in the case of PAL2. In this case the second-ranked grammar actually achieved a fitness of 0.925312, narrowly missing the threshold value.

6 Conclusion

The approach to grammatical inference described in this paper differs from previous work using genetic algorithms in addressing the problem of corpus-based inference of stochastic context-free grammar. This difference makes direct comparison of our results with those of other researchers difficult. However, the experiments that we have conducted are typical of those in other studies and the results reported in this paper appear promising. The approach also compares well with other (non-genetic) techniques for stochastic grammatical inference, for example the work reported by Lari and Young [9] using the Inside-Outside algorithm [2].

The main limitation of our approach is the cost involved in evaluating the fitness of each candidate solution, which requires parsing every string in the corpus in all possible ways. We are currently investigating ways of overcoming this problem, including the possibility of a massively parallel implementation of our algorithm.

References

[1] D. Angluin and C. Smith. Inductive Inference: Theory and Methods. *Computing Surveys*, 15(3):237–269, 1983.

[2] J. K. Baker. Trainable grammars for speech recognition. In *Proceedings of the Spring Conference of the Acoustical Society of America*, pages 547–550, 1979.

[3] D. Dunay, F. Petry, and B. Buckles. Regular language induction with genetic programming. In *Proceedings of the First International Conference on Evolutionary Computing*, pages 396–400, 1994.

[4] K. S. Fu and T.L. Booth. Grammatical inference: introduction and survey. *IEEE Transactions on Pattern Analysis and Machine Intelligence*, 8:343–375, 1986.

[5] E. M. Gold. Language identification in the limit. *Information and Control*, 10:447–474, 1978.

[6] J. H. Holland. *Adaptation in Natural and Artificial Systems.* University of Michigan Press, Ann Arbor, 1975.

[7] W.-O. Huijsen. Genetic grammatical inference: Induction of pushdown automata and context-free grammars from examples using genetic algorithms. Master's thesis, Dept. of Computer Science, University of Twente, Enschede, The Netherlands, 1993.

[8] M. M. Lankhorst. Grammatical inference with a genetic algorithm. In *Proceedings of the 1994 EUROSIM Conference on Massively Parallel Processing Applications and Development*, pages 423–430, 1994.

[9] K. Lari and S.J. Young. The estimation of stochastic context-free grammars using the inside-outside algorithm. *Computer Speech and Language*, 5:237–257, 1990.

[10] S. Lucas. Biased chromoso[illegible] for grammatical inference. In *Proceedings of Natural Algorithms in Signal Processing*, 1993.

[11] J. Risannen. Modelling by shortest data description. *Automatica*, 14:465–471, 1978.

[12] M. Schwem and A. Ost. Inference of stochastic regular grammars by massively parallel genetic algorithms. In *Proceedings of the Sixth International Conference on Genetic Algorithms*, pages 520–527, CA, 1995. Morgan-Kaufmann.

[13] S. Sen and J. Janakiraman. *Learning to construct pushdown automata for accepting deterministic context-free languages*, pages 207–213. 1992.

[14] R. J. Solomonoff. A formal theory of inductive inference. *Information and Control*, 7:1–22; 224–254, 1964.

[15] P. Wyard. Context-free grammar induction using genetic algorithms. In *Proceedings of the Fourth International Conference on Genetic Algorithms, ICGA'92*, pages 514–518, CA, 1991. Morgan Kaufmann.

[16] H. Zhou and J.J. Grefenstette. Induction of finite automata by genetic algorithms. In *Proceedings of the 1986 IEEE International Conference on Systems, Man and Cybernetics*, pages 170–174, 1986.

Adaptive Product Optimization and Simultaneous Customer Segmentation: A Hospitality Product Design Study with Genetic Algorithms

E. Schifferl
Institute for Tourism and Leisure Studies,
Vienna University of Economics and Business Administration,
Augasse 2-6, 1090 Vienna, Austria
Email: elisabeth.schifferl@wu-wien.ac.at

Abstract

Successful product development depends on many factors. Among the most important factors are identification and satisfaction of customers' perceived needs, the accessibility, size and growth rate of the target market, and, of course, production costs. Since the early 1970s marketing researchers achieved remarkable results in developing methods to measure consumer preferences of multiattributed products. Additionally, market segmentation methods have been an important issue in strategic marketing research. This study, however, concentrates on a new method of product design optimization. It is shown how genetic algorithms are used to simultaneously discover optimal multi-attributed products for different customer preferences. For that purpose we chose an interactive version of the genetic algorithm where genetic operators like selection, mutation and crossover are applied as usual. The use of the interactive genetic algorithm is most suitable, where measures of utility are difficult or impossible to specify mathematically. Imprecise optimization in terms of a priori unknown individual consumer decision rules and preferences is an important issue for marketing researchers. The interactive genetic algorithm tries to solve design problems in that the consumer plays the role of the objective function during data collection.

1 Introduction

In recent marketing research literature individual customer preferences in product design studies are often determined by use of conjoint analysis [2]. The most important application of this technique for the hospitality industry is considered to be the design of a new hotel concept Courtyard by Marriott demonstrated by Marriott Corporation [8]. Genetic algorithms (GA) for product design studies were applied by [1]. They found optimal designs based on utility part-worths of a preceding conjoint analysis. A segmentation was not intended. The problem of discovering individual decision rules with a GA was treated by [5]. The interesting results are again based on previously conducted traditional market research studies. The present paper addresses the problem of optimal design of holiday homes in former log cabins in alpine areas. The main issue of this design study is the consideration of different customer preferences for a simultaneous market segmentation which is not based on any preceding surveys. In the course of a preliminary questionnaire survey, number and types (in terms of capacity, location, equipment, price, etc.) of possible holiday homes were investigated. Similar to the study of Courtyard by Marriott, it turned out that a large number of attributes and attribute levels were to be involved in the product optimization process to describe all possible types of holiday homes. Unlike traditional optimization procedures, the interactive genetic approach does not underlie any mathematical restrictions, like e.g. linearity or distribution constraints. It is based on one premise, which says that only those products will be successful in the market that best fit the customers' perceived needs. The idea of segmentation comes from nature, where the same classes of animals live in different ecosystems and different species naturally evolve. In this optimization problem, one product faces different consumer preferences. Thus, the optimal combination of product attributes has to be found for each customer segment.

2 Objectives

The application of GAs for product optimization tasks seems to be interesting since it can perfectly be used as an adaptive data collection tool [7]. This approach is not based on preliminary surveys but the optimization and segmentation process takes place right at the time of data collection. Figure 1 shows the screen shot of the data collection program. The respondent has to evaluate a certain number of products which are described in the two

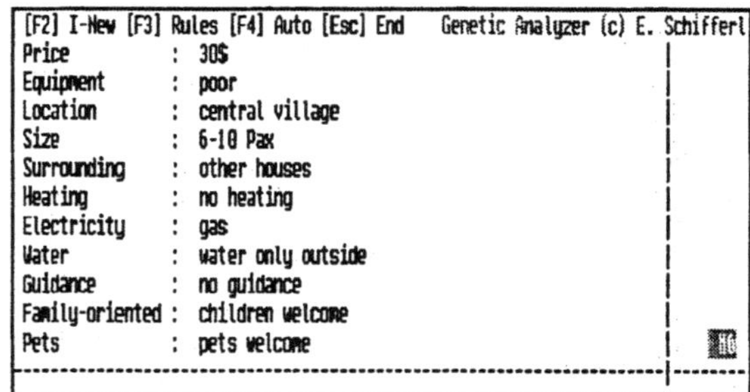

[F2] I-New [F3] Rules [F4] Auto [Esc] End Genetic Analyzer (c) E. Schifferl
Price : 30$
Equipment : poor
Location : central village
Size : 6-10 Pax
Surrounding : other houses
Heating : no heating
Electricity : gas
Water : water only outside
Guidance : no guidance
Family-oriented : children welcome
Pets : pets welcome

Figure 1: Product description and evaluation entry.

```
RD RD RD RD RD RD RD RD RD RD RD
[1]
13121312412 =  63 ( 1)
25326212312 =  45 ( 1)
14337231221 =  46 ( 1)
45337112412 =  40 ( 1)
25111323412 =  54 ( 1)
12133233221 =  53 ( 1)
32231322221 =  51 ( 1)
25126323221 =  45 ( 1)
23136221411 =  54 ( 1)
12434132212 =  49 ( 1)
45132312321 =  46 ( 1)
44227321312 =  40 ( 1)
24132231122 =  54 ( 1)
14233122411 =  54 ( 1)
43214131211 =  50 ( 1)
```

Figure 2: Representation of possible solutions.

text columns.

As especially in the hospitality industry, the rather complex products show a large number of attributes (left column in Figure 1) and attribute levels (right column) that are to be measured, one runs the risk of extremely high requirements on the evaluation methodology and on the respondent's evaluation task respectively. The present study extends the procedure of an individualized conjoint approach in that it presents a product optimization model without preselection or reduction of attributes. Unlike function optimization tasks, the search space is limited and relatively small - depending on the number of attributes and attribute levels. Thus the computing time is reasonable which is very important in the case of interactive computer interviewing. However, the results presented in this paper arose from a computer simulation run. For that purpose the underlying product decision rule is assumed to be non compensatory and is embedded into an additive model. The objective of this study is as follows:

1. It is shown, how GA can be used as a data collection tool.
2. It is shown, how this methodology can be used for simultaneous product optimization and customer segmentation without prespecification of the number of segments.

3 Methods

This method is based on a simple genetic algorithm [3] and starts by randomly generating an initial population of N individuals, each representing an individual solution to the search problem. Each individual in the population is assigned a fitness value (F). According to the fitness value the individuals are selected for breeding. The parent selection procedure a method based on ranking with replacement used. Genetic operators of crossover, mutation and reproduction are applied probabilistically. The procedure continues until a predetermined termination condition is satisfied, in this case until a certain number of generations. There are many kinds of GA that differ from this basic model in some detail or another.

3.1 GA as a Data Collection Tool

The population of individuals ($N = 20$) corresponds to a set of products. The individuals are represented as fixed length character strings (they act like chromosomes in a natural evolutionary process). For specifying the representation scheme the string length was denoted by the number of attributes ($L = 11$) that characterize log cabins. For the encoding of the attribute levels a metric alphabet was chosen. The size of the alphabet (K) depends on the number of attribute levels (see Figure 2, where one string corresponds to a product description as shown in Figure 1).

It is guaranteed that each string is within the search space of the optimization problem. In the case of interactive computer interviewing the objective function is replaced by a human operator who evaluates each product in a population according to its idiosyncratic probability to buy ($F_{max} = 100\%$). Figure 2 shows the assigned fitness values after the 'equals' sign. (For more details on GA as interviewing tool see [6].)

3.2 Model Development Objective Function

In the simulation run, ideal customer preferences are predetermined and the fitness value is calculated as a function of these preferences. The functional form assumes that each attribute has an independent utility value that is weighted and summed up to get the overall fitness value of the product. The weighting is as follows: For each attribute in a population that matches the ideal attribute K a value $100/L$ is calculated, each attribute that misses K by ± 1 is assigned a value of $50/L$, a difference of $K \pm 2$ accounts for $20/L$, all other attributes are assigned a value of $0/L$. The fitness value is the sum of all attribute values. If all attributes meet the ideal preferences the fitness value is 100, which corresponds to the probability to buy of 100%. This way, the fictitious customer is prevented from trading off one attribute against another.

Genetic Operators

After these preparatory steps the following procedures are performed iteratively. A sampling mechanism probabilistically selects individuals for genetic operations and is thus determining the population diversity and the selective pressure. The crossover probability (p_c) was set at 0.6, mutation probability (p_m) was set at 0.1 and reproduction probability (p_r) was set at 0.3. The underlying sampling method implicates a relatively high selective pressure but for the sake of the respondent's time and evaluation ability a fast convergence was intended. Experimental runs showed that due to the selective pressure a smaller mutation rate leads to longer periods stuck in a local optimum. A higher crossover rate on the other hand had hardly effect on the convergence speed. For an overview of sampling mechanisms see [4].

Selection Procedure

The selection procedure in this experiment is determined as follows: A user-defined value $\mu = 1/3 * N$ determines the size of the subpopulation. The products in the population are ranked according to their fitness value so that μ represents the best individuals in the population. A random value r between 1 and μ is defined that determines individual r to be selected for genetic operations.

3.3 GA for Customer Segmentation

The more consistent the customers behave the quicker a convergence will be achieved. In simulation runs with only one customer the algorithm did not converge before 70 generations on average. As it can be assumed, customers have different preferences and a global optimum representing the one and only optimal product can never be reached. For this purpose, the genetic algorithm was modified and extended by a segmentation tool, which is shown in Figure 3.

1. The segmentation starts after a certain number of generations. It turned out that segments converge quicker when the system is already trained, though unstructured.

2. Before every new training an intermediate step determines which segment(s) should be trained. The customer is presented a subpopulation with the best N/S products out of each segment, with S being the number of segments. After the evaluation task, the system checks the summed fitness value (group fitness value) of the different subpopulations and compares it to the equivalent fitness values of the original subpopulations.

3. A segment is trained, when the new group fitness value is equivalent to or exceeds the old group fitness value.

4. If none of the new group fitness values is equivalent to or exceeds any of the old group fitness values, a new segment is initialized with randomly generated N individuals.

5. If any of the new group fitness values is equivalent to or exceeds more than one of the old group fitness values, then these segments are joined.

Step 2 to 5 are repeated until the number of segments remains stable and each segment presents almost optimal products.

4 Findings

It was assumed that the segments are equally strong, thus each fictitious customer had an equal chance to be chosen and to train the model. In a simulation run with three different kinds of customers that were selected for the evaluation task randomly, it turned out that three segments could be found soon after the start of the segmentation procedure (see Figure 4). After only few wrong assignments, the system remained stable until the termination criterion of 50 generations (the first 10 generations ran without segmentation) was reached.

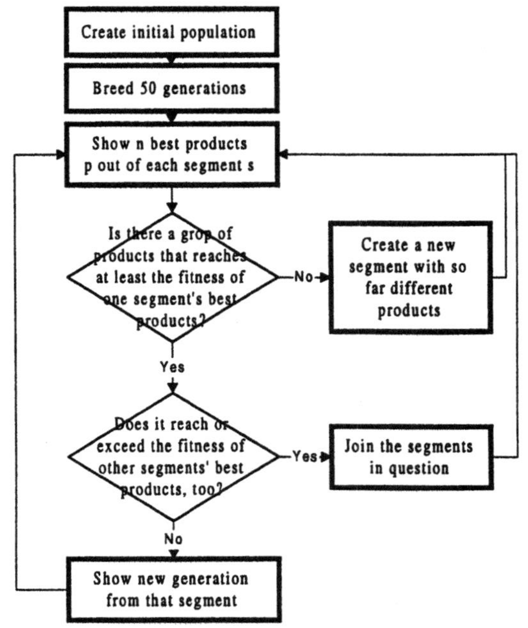

Figure 3: Flow chart of segmentation procedure.

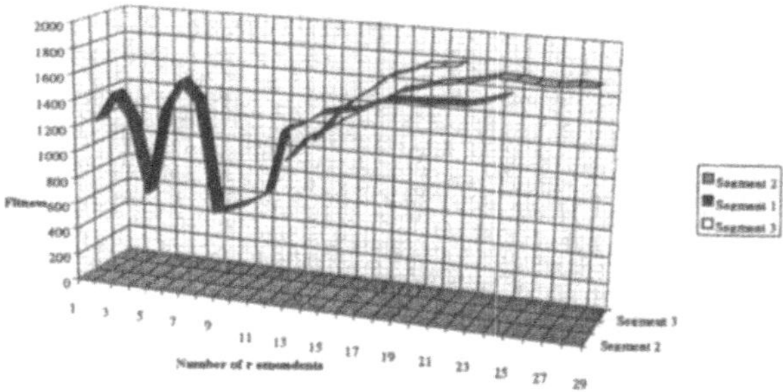

Figure 4: Segmentation procedure within 50 generations.

All segments could reach an average overall fitness value of more than 1,600. None of the segments reached the overall optimum of 2,000. Since optimization runs with only one customer converged not before average 70 generations, the results of the segmentation runs were very promising.

It seems that the application of GA in marketing research is helpful in many ways. Unlike traditional product optimization models there is no methodological limit of the number of attributes and attribute levels. The only limit is the respondent himself when the evaluation task becomes too complex. However, the time spent for evolving a solution, which, of course, is the researcher's time, is relatively short. Another very important advantage of this marketing research model is that once data are gained, they can be used as a dynamic instrument to discover changing customer needs and preferences occurring over time.

References

[1] P. Balakrishnan and V. Jacob. Triangulation in decision support systems: Algorithms for product design. *Decision Support Systems*, 14:313–327, 1995.

[2] P. Green and A. Krieger. Conjoint analysis with product-poisoning applications. In J. Eliashberg and G. L. Lilien, editors, *Handbooks in Operations Research and Marketing Science*, pages 467–515. Elsevier Science Publishers, 1993.

[3] J.H. Holland. *Adaptation in the Natural and Artificial Systems.* Ann Arbor: University of Michigan Press, 1975.

[4] Z. Michalewicz. *Genetic Algorithms+Data Structures=Evolution Programs (2nd ed.).* Springer-Verlag, Berlin, 1992.

[5] J. Oliver. Discovering individual decision rules: An application of genetic algorithms. In S. Forrest, editor, *Proceedings of the Fifth International Conference on Genetic Algorithms*, pages 216–222, San Mateo, CA, 1992. Morgan Kaufmann.

[6] E. Schifferl. Optimizing product development procedures with genetic algorithms: A hospitality case study. In J. Beracs, A. Bauer, and J. Simon, editors, *Proceedings of 25th EMAC Conference*, 1996.

[7] J. Smith. Designing biomorphs with an interactive genetic algorithm. In *Proceedings of the Fourth International Conference on Genetic Algorithms*, pages 535–538, San Mateo, CA, 1991. Morgan Kaufmann.

[8] J. Wind, P. Green, D. Schifflet, and M. Scarbrough. Courtyard by marriott: Designing a hotel facility with consumer based marketing models. *Interfaces*, 19 (January-February):25–47, 1989.

Genetic Algorithm Utilising Neural Network Fitness Evaluation for Musical Composition

A. R. Burton and T. Vladimirova
Department of Electronic and Electrical Engineering, University of Surrey, Guildford, GU2 5XH, England.
Email: A. Burton@ee.surrey.ac.uk

Abstract

The aim of the paper is to propose a means by which neural network fitness evaluation can be applied to a genetic algorithm (GA), and an application of this system to musical rhythm composition. An adaptive resonance theory (ART) neural network is trained using binary information representing classification patterns. By comparing new genetically derived individuals to clustered data, a measure of fitness of the new patterns is determined; the patterns of higher fitness values then being used in successive generations to further improve the overall population fitness. A proposed application for this system is described — a genetic composer that utilises clustered representations of rhythm styles to interactively generate rhythm patterns to the user's general stylistic requirements.

1 Introduction

Genetic algorithms (GAs) are computational methods which use procedures based upon the laws of natural selection to increase the value of a population as a whole by combining those individuals with high value [4]. GAs have been seen to perform better than other optimisation processes such as random walks or enumerative calculus-based methods [4]. GAs combine the probabilistic nature of random search procedures with the guided nature of deterministic search methods to direct the search towards areas of the search space of higher value. A GA utilises encodings of the data through which it is searching, rather than the data itself. This implies that GAs are blind to the application [4], and are thus applicable to a great number of subject areas. One particular subject area is that of musical composition and synthesis, for example [1, 5, 6]. Given an encoded representation of musical phrase or sound, a GA can generate further examples based upon aspects of the original fragment and the means by which fitness is evaluated.

The fitness functions used by GAs in musical applications vary according to the desired results, for example adherence to formal rules [6], or user input to guide population evolution [1]. Both of these methods have produced successful results, but at certain costs. Rule-based fitness regimes tend to limit the scope of compositions, often breaking the rules of composition results in more musically interesting pieces than adhering them. Algorithms that depend on the user to assign fitness to every result are time consuming and sequentially context dependent.

2 Neural Networks and Genetic Algorithm Fitness Evaluation

Neural networks have been applied as fitness evaluators in genetic and evolutive applications, to a varying degree of success [2, 3, 8]. Neural fitness evaluators are used in applications where the knowledge of how fitness should be assigned to any individual is heuristic, uncertain, or in applications where there are many degrees of freedom in the information encoded in each individual. Neural network fitness evaluators require time consuming training, which assumes that knowledge of how the input set will map to a certain output is known in advance. Also, once a training set is learnt, there is no scope for new data to change the way the system evaluates fitness. This may prevent some new genetically produced individuals from being rejected as low fitness, when in actual fact they may be of high fitness, but due to the training method employed, cannot be identified as such. These individuals of potential high fitness may be of use in classifying other similar individuals — of a type that may not have been included in the training set.

An ART [7] network can be used as a fitness evaluator to solve these problems. ART neural networks are self-organising networks, utilising unsupervised learning and clustering algorithms to be able to recognise patterns. One advantage of unsupervised learning is that no *a priori* knowledge is assumed regarding the classification of any pattern — the network determines how a specific input maps to an output. There is also no limit to the number of patterns that can be classified — if a pattern is detected that cannot be clustered with any other patterns, a new cluster is created.

ART systems control the similarity of patterns and the classification clustering using vigilance. This is a threshold which determines whether or not a certain pattern is similar to already classified patterns so that a new cluster can be created if required. One of the most important features of adaptive resonance networks is their ability to resolve the stability-plasticity dilemma. This allows the network to record new information as it arrives, yet remain stable with regard to information previously processed — thus data not in the initial training set does not cause loss or corruption of either the new or old training data.

Three attributes of ART networks — unsupervised learning, unlimited pattern classification groupings, and adaptability of vigilance — make this network architecture suitable to GA fitness evaluation.

3 The ART Neural Network as a Fitness Evaluator

Prior to running the GA with the ART fitness evaluator, the neural network is trained using example training data. The training data used for the experiments to verify the ART fitness evaluator were taken from binary array representations of letters of the alphabet. For example the letter 'T' was represented as shown in Figure 1.

Similar representations were used for four more letters P, M, E and Y. Each training pattern was transformed from the array representation as above

111110
001000
001000
001000
001000
000000

Figure 1: Example of a binary test array.

to a binary string by joining the beginning of one row to the end of the previous row. Hence, the above array becomes:

111110001000001000001000001000000000.

This is an example of the individuals used to create a cluster that represents one classification. Other examples of training data for a classification differed by one or two bits. Experiments employing a different bit length used a different resolution to represent the same symbols.

The ART fitness evaluator was implemented using a GA based upon Goldberg's simple genetic algorithm (SGA) [4]. Single point crossover and single bit mutation with a user-specified probability was employed. The ART network fitness evaluator operates as follows:

1. Each individual is presented as an input to the ART network.
2. The network determines the 'winning' cluster — the maximum value of the matrix product of the input vector (the chromosome) and the bottom-up connection weights of the ART network.
3. The vigilance test is carried out to determine the degree of match between the individual and the cluster.
4. If the vigilance test passes, the individual is added to the cluster. If the vigilance test fails, then the remaining clusters are tested in a similar way.
5. If no existing cluster represents the individual closely enough, a new cluster is added.
6. Fitness is assigned as the degree of similarity to individuals already represented by the cluster — a high fitness is assigned to individuals which show a closer match. Individuals which form a new cluster are assigned maximum fitness.

The use of an ART neural network as a GA fitness evaluator has been validated by the experimental work carried out so far. The effects of cluster formation has been investigated dependent on network vigilance, and the GA parameters population size, mutation probability, and chromosome length. The number of input units used for the ART network is equal to the chromosome length of any individual.

For each test condition, the GA was run several times under the same conditions. This is due to the fact that as the genetic process contains a certain degree of probabilistic operations, then one single result will not be representative of the behaviour of our system. However, if the same conditions are used over a number of runs, then it is possible to average the results to determine the trends in the recorded data.

Cluster creation over the first ten iterations of a GA as a function of population size, mutation probability and vigilance parameter was investigated. Figure 2 shows one set of results of these experiments — the cluster creation behaviour for a population size of 16 individuals.

It can be seen from these results that for higher values of vigilance, the number of clusters that are created by the fitness evaluator is very large compared to the fact that high vigilance parameters require that a very close, or exact, match be found between individuals being evaluated for fitness and those individuals represented by the cluster. Such a trend was observed for all values of population size, although smaller populations would display a slower rate of cluster creation than higher population sizes. Additionally, the rate and amount of clusters created is independent of mutation probability.

The average and maximum fitness values displayed by a number of individuals in a population of constant size was also investigated. The average fitness is the mean fitness of the population at any given iteration of the algorithm, and the maximum fitness is the greatest fitness observed of any individual in any iteration. These values were observed over the first ten iterations of the GA, as a function of chromosome length, mutation probability and vigilance parameter. The average and maximum fitness values for a population of sixteen-bit individuals, with a mutation probability of 5%, are shown in Figures 3 and 4.

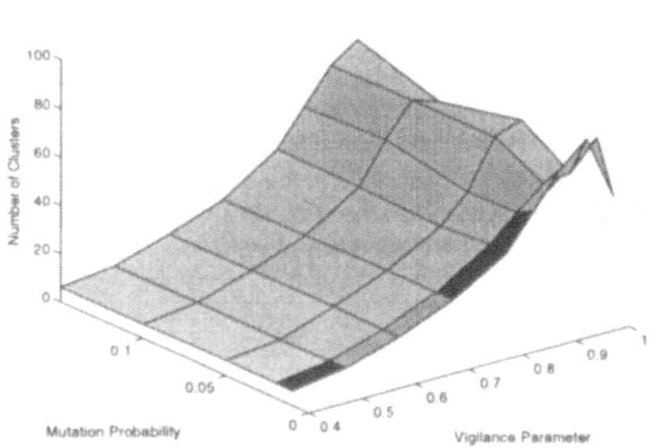

Figure 2: Example of cluster creation results.

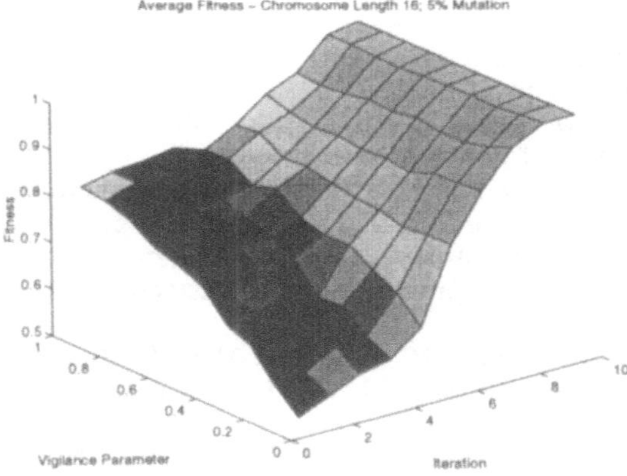

Figure 3: Example of average fitness results.

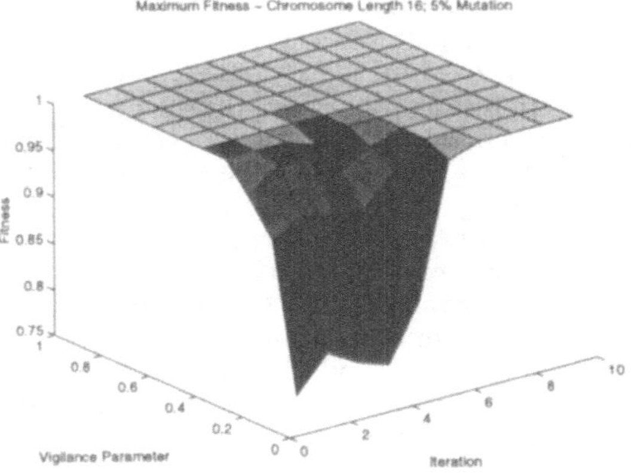

Figure 4: Example of maximum fitness results.

Use of a high vigilance parameter results in a higher average fitness across the population at the beginning of the ten iterations. However, the fact that a high vigilance parameter also results in a very large number of clusters implies that a compromise is required between having a high fitness population and a controlled cluster addition rate. Maximum fitness values are reached faster given a high vigilance parameter — this is also due to the high cluster creation rate. A solution to these cluster creation problems lies in better control of how clusters are added to the fitness evaluator, and is the subject of current work.

Also observed from the results obtained from testing this system is that the average and maximum fitness surfaces obtained were independent of the length of chromosome used in the representation.

4 A Genetic Algorithm Based Interactive Drum Machine

The specific purpose of the proposed system is to act as an interactive drum machine for a musician or composer. Given initial parameters such as a rhythm type, tempo, time signature, the GA constructs a population of individuals, where each represents a rhythm pattern. The representation used for these input patterns will consist of a binary array of data, one dimension representing beats, or discrete time steps, the other dimension representing individual voices. For example the rhythm pattern shown in Figure 5 would be represented by the three row array shown in Figure 6.

The fitness evaluation network uses information obtained from existing MIDI (Musical Instrument Digital Interface) files. This process creates clusters of stylistically similar rhythm patterns.

Depending on the similarity measure of the genetically produced rhythm patterns to the appropriate rhythm pattern cluster, a certain fitness value can be assigned to that pattern. If the rhythm pattern fails the vigilance test, then a new cluster can be created to represent this pattern.

Patterns with higher fitness values can be used as the starting point for the next generation of individuals, after subjecting the population to typical GA breeding processes. Further iteration of the algorithm will result in populations containing individuals that will approach the requirements of the user. By experimentation, the optimum selection/reproduction and genetic operation schemes will be determined, such that the algorithm can reach the desireed result efficiently. More complex crossover operators will be implemented, to allow a more diverse population to be created for subsequent generations. Methods including two point crossover, multiple point crossover and crossover masking [4] will be investigated. Mutation will be, as before, single bit, random, bit-inversion mutation. Additionally, the different means of selection and cluster addition will be implemented depending on the results of tests carried out at the time.

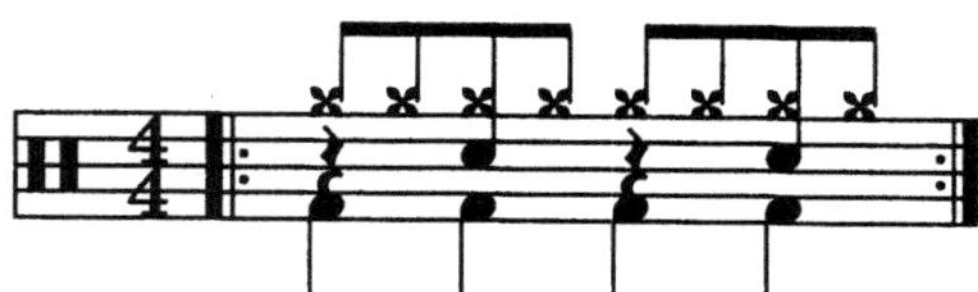

Figure 5: Example of rhythm pattern.

11111111
00100010
10101010

Figure 6: Binary rhythm representation array.

5 Concluding Remarks

Results so far suggest that the ART network clusters binary patterns of genetically produced patterns, and can assign fitness to these patterns as a function of the degree of similarity between an individual pattern and the cluster to which it is closest.

An ART network, when used as a GA fitness evaluator, possesses distinct advantages over the other neural methods of fitness evaluation — the ability to add new classifications when existing clusters do not sufficiently represent an individual. This feature will be applied to an evolutive musical composition system. However, high vigilance parameters have been seen to produce large numbers of clusters. Work in progrss will introduce a system to control cluster creation.

The main areas of focus in developing the rhythm composition system using ART fitness evaluation are the size and type of input representation used, control of cluster creation, using weightings of certain bits of input patterns to signify beat strength, and variation of the vigilance parameter to determine the degree o similarity required between cluster patterns and input patterns.

References

[1] J.A. Biles. Genjam: A genetic algorithm for generating jazz solos. In *Proceedings of the International Computer Music Conference*, 1994.

[2] J.A. Biles, P.G. Anderson, and L.W. Loggi. Neural network fitness functions for a musical iga. Techni-

cal report, Rochester Institute of Technology, 1996. http://www.it.rit.edu/jab/SOCO96/SOCO.html.

[3] C.H. Dagli and S. Sittisathanchai. Genetic neuroscheduler for job shop scheduling. *Computers and Industrial Engineering*, 25(1–4):267–270, 1993.

[4] D.E. Goldberg. *Genetic Algorithms in Search, Optimisation and Machine Learning.* Addison-Wesley, 1989.

[5] D. Horowitz. Generating rhythms with genetic algorithms. In *Proceedings of the 1994 International Computer Music Conference*, pages 142–143, 1994.

[6] R.A. McIntyre. Bach in a box: The evolution of four-part baroque harmony using the genetic algorithm. In *Proceedings of the IEEE Conference on Evolutionary Computation*, pages 852–857, 1994.

[7] A. Nigrin. *Neural Networks for Pattern Recognition.* MIT Press, 1993.

[8] G. Schneider, J. Schuchhardt, and P. Wrede. Amino acid sequence analysis and design by artificial neural network and simulated molecular evolution –an evaluation. *Endocytobiosis and Cell Research*, 11(1):1–18, 1995.

Analyses of Simple Genetic Algorithms and Island Model Parallel Genetic Algorithms

T. Niwa and M. Tanaka
Distributed Systems Section,
Mathematical Informatics Section, Electrotechnical Laboratory,
1–1–4 Umezono Tsukuba-shi Ibaraki-ken, 305 JAPAN
Email: {niwa,mtanaka}@etl.go.jp

Abstract

H. Asoh and H. Mühlenbein investigated empirically the relation among the mean convergence time, the population size, and the chromosome length of genetic algorithms (GAs). In this paper, from the mathematical point of view, the relation they revealed is convincing. Our analyses of GAs make use of the Markov chain formalism based on the Wright-Fisher model, which is a typical and well-known model in population genetics. We also give the mean convergence time under genetic drift. Genetic drift can be described by the Wright-Fisher model. We determine the stationary states of the corresponding Markov chain model and the mean convergence time to reach one of these stationary states. Furthermore, we derive the most effective mutation rate for the standard GAs and also the most effective migration rate for the island model parallel GAs with some restrictions. These rates are coincide with known empirical results.

1 Introduction

Genetic algorithms (GAs) are adaptive methods based on the genetic processes of biological organisms which were introduced by Holland [5]. They succeeded to solve many kinds of search, optimization, and machine learning problems [3].

It is natural to study behaviors of GAs theoretically, because we want to know performances of GAs compared with other methods and how GAs converge to good solutions; it is expected that situations are different from random search methods. Many researchers have studied GAs theoretically [2, 7, 11, 12].

In population genetics and GAs, genetic drift is well known phenomenon. In [1, 4, 6], genetic drift has been studied with computer simulations. Kimura gave the mathematical analysis for the population genetics through diffusion models [8].

In this paper, we derive the most efficient mutation rate for standard GAs and the most efficient migration rate for island model parallel GAs. Section 2 is devoted to analyze the mean convergence time on Wright-Fisher model (W-F model) which is a model of simple GAs, and we obtain the Asoh's proposition. In Section 3 we describe the density function through mathematical analysis of stationary states. In Section 4 we consider the island model parallel GAs, which is often used in parallel GAs. For the details of our analyses on the Markov chain model, see [13].

2 Genetic Drift and Mean Convergence Time for W-F Model

Genetic drift is the random fluctuation of gene frequencies subjected by probabilistic transition from generation to generation in finite population size. It tends to localize genes to particular genes (convergent states of Markov chain). This tendency is against mutation, which makes genes disperse to various genes. In this section, we consider the mean convergence time of W-F model without mutation using standard Markov chain analysis.

In this case, the convergence of W-F model is driven by genetic drift. There are only two alleles, i.e. 0 and 1, in the W-F model. In the general case, i.e. more than two alleles, we pick up a particular allele to use W-F model. If a total population size is n then the state of the population is uniquely specified by the number of 0s, so we define the state as the number of 0s.

On the other hand, the mean convergence time of simple GAs is proportional to the population size, and to the logarithm of the length of chromosomes [1]. We also inspect these propositions mathematically.

Table 1: Mean convergence time at large population.

initial state	mathematical analysis by Equation (1)	numerical analysis in [1]
$p=1/2$	$1.386n$	$1.4n$
$p=1/4$	$1.125n$	$1.0n$
$p=1/8$	$0.754n$	$0.7n$

2.1 Large Population Limit

In [1, 4, 6], genetic drift has been studied with numerical experiments. They showed that the mean convergence time is proportional to the population size of a model. On the other hand, we can show, in the continuous limit such as the large population limit, the mean convergence time is proportional to the size of population. According to Kimura [8], we have the following relation on the mean convergence time τ;

$$\tau \simeq \sum_{j=0}^{\infty} \frac{P_{2j}(1-2p) - P_{2j+2}(1-2p)}{(j+1)(2j+1)} n \tag{1}$$

where $P_*(\cdot)$ are the Legendre polynomials, and p is the gene frequency at the initial state. This shows that the mean convergence time is proportional to the population size. As shown in Table 1, the values calculated by above equation agree with those of numerical experiments [1]. And, we have shown that the right hand side of the Equation (1) is equal to $-2\{p\log p + (1-p)\log(1-p)\}n$.

2.2 Long Chromosomes Case

Generally, chromosomes consist of many loci. Asoh and Mühlenbein showed that the mean convergence time with uniform crossover is proportional to the logarithm of the length of chromosomes [1]. However, this was resulted in numerical computation. Therefore we want to know rigorous results depending on the length of chromosomes. In order to understand the behavior of GAs, at first, we investigate the chromosomes with ℓ loci. Now, we show that the analytic results in the case where the population size is two.

The probability that a certain locus converges before time step t is $1-2^{-t}$. If each chromosomes have ℓ loci, the probability that all of loci converge before time step t is $(1-2^{-t})^\ell$, and the probability that all of loci converge at just time step t is

Table 2: Mean convergence time at long chromosomes with uniform crossover.

length of chromosomes	correct value	numerical analysis [1]
1	2.0000	2.00
2	2.6667	2.67
4	3.5048	3.50
8	4.4211	4.42
16	5.3774	5.37
32	6.3552	6.36
64	7.3440	7.34
128	8.3384	8.34
256	9.3356	9.34
512	10.3341	10.33
1024	11.3335	11.33

$(1-2^{-t})^\ell - (1-2^{1-t})^\ell$. Note that time step t is discrete. Thus the mean convergence time E is given by

$$\begin{aligned} E &= \sum_{t=1}^{\infty} t\left\{\left(1-\frac{1}{2^t}\right)^\ell - \left(1-\frac{1}{2^{t-1}}\right)^\ell\right\} \\ &= \sum_{t=0}^{\infty}\left\{1-\left(1-\frac{1}{2^t}\right)^\ell\right\} . \end{aligned} \tag{2}$$

On the other hand, let the continual approximation of E be $\tilde{E}$,

$$\begin{aligned} \tilde{E} &= \int_0^{\infty}\left\{1-\left(1-\frac{1}{2^t}\right)^\ell\right\} dt + \frac{1}{2} \\ &= \frac{1}{\log 2}\left(\log\ell + \gamma + \frac{1}{2\ell} - \sum_{i=1}^{\infty}\frac{B_{2i}}{2i\ell^{2i}}\right) + \frac{1}{2} \end{aligned} \tag{3}$$

where γ is Euler's constant, ψ is the digamma function, and B_n are Bernoulli numbers. The details of the above calculation are described in a previous paper [10]. The continual approximation of E is proportional to $\log\ell$.

When the population size is more than two, the convergence time E becomes too complex to analyze. For the reference, Table 2 shows the comparison between correct value and numerical experiments [1].

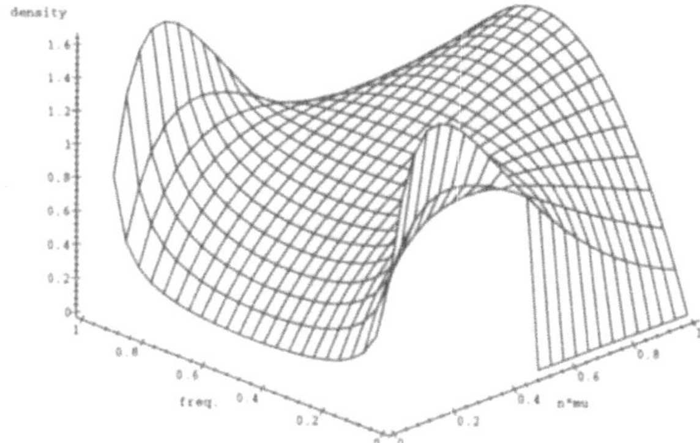

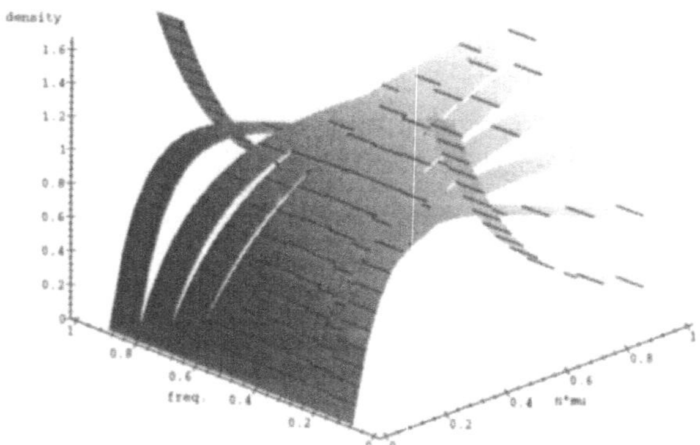

Figure 1: The shape of the density function of the stationary states: the large population limit such that $\frac{\Gamma(4n\mu)}{\{\Gamma(2n\mu)\}^2}\left\{\frac{i}{n}\left(1-\frac{i}{n}\right)\right\}^{2n\mu-1}$.

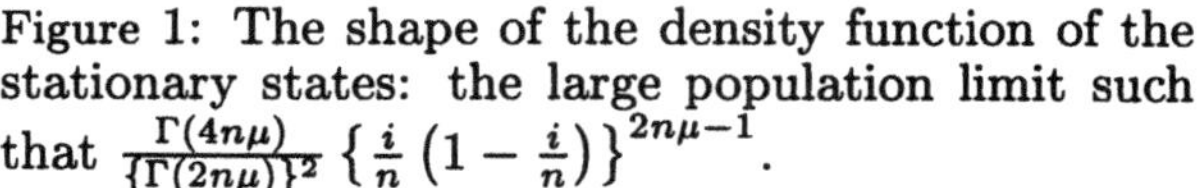

Figure 2: The shape of the density function of the stationary states: Markov chain model. The population size is $n = 10$. The height stands for the density of the stationary states, the width is of the gene frequency and the depth means the product of mutation rate and the number of population.

3 Stationary State

Consider the W-F model with mutation. The transition matrix Q is given as

$$Q_{ij}=\binom{n}{j}\left(\mu+\frac{1-2\mu}{n}i\right)^{j}\left(1-\mu-\frac{1-2\mu}{n}i\right)^{n-j}. \tag{4}$$

Since the largest eigenvalue of Q is 1, we get the density function of the stationary state by normalizing the eigenvector for the eigenvalue 1. By extremity approximation with population size n to infinity, the density function of stationary state is given as [8];

$$\frac{\Gamma(4n\mu)}{\{\Gamma(2n\mu)\}^2}\left\{\frac{i}{n}\left(1-\frac{i}{n}\right)\right\}^{2n\mu-1}. \tag{5}$$

Figure 1 shows the shape of the density function of stationary state in the large population limit, and Figure 2 shows the same function with the population size $n = 10$. Because Figure 2 is quite similar to Figure 1, this extremity approximation is good in the case where population size n is small, such as $n = 10$.

From Equation (5), we see that the mutation rate $\mu = 1/2n$ makes the density function flat. This shows that this mutation rate makes the GAs work well. Because, if the mutation rate is large, GAs are hard to get stationary results, and if the mutation rate is small GAs become easy to converge to a certain value which might not be an optimal result. When the mutation rate μ is $1/2n$ all of the states have the same probability. This means GAs represent the difference between states according to their fitness.

In this consideration, we didn't take the influence of the chromosome length into account. This is because, the genes of each locus behave as described above. Furthermore, since the reciprocal effect spanning plural loci, i.e. epistasis, depend on the problem to be solved, we cannot describe it without knowing the fitness function. $\mu = 1/2n$ is the standard value for mutation rate to decide GA-parameters.

4 Island Model Parallel GAs

Parallel GAs have been investigated since GAs were introduced, and the island model parallel GAs is the typical model of parallel GAs [14]. In this model, the global population is divided into several subpopulations, and one processor is allocated to each subpopulation. Each processor runs the simple GA independently. Inter-processor communication occurs during the migration phase at regular intervals (i.e. migration interval). During migration, a fixed proportion of each subpopulation is selected and sent to another subpopulation. In return, the

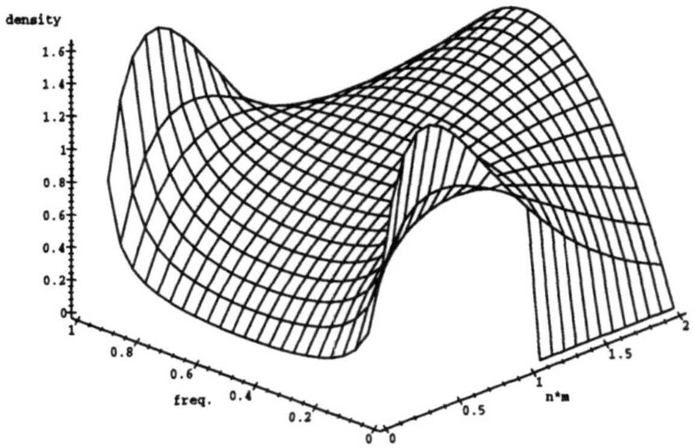

Figure 3: The shape of the density function of the stationary states: the large population limit of island model parallel GAs, where subpopulation size is n, and mean frequency of whole population is 1/2; $\frac{\Gamma(2nm)}{\{\Gamma(nm)\}^2}\left\{\frac{i}{n}\left(1-\frac{i}{n}\right)\right\}^{nm-1}$.

same number of migrants are received and replace individuals selected according to some criteria.

By extremity approximation with subpopulation size n to infinity, the density function of stationary state is given as [8]:

$$\frac{\Gamma(2nm)\left(\frac{i}{n}\right)^{2nm\overline{x}-1}\left(1-\frac{i}{n}\right)^{2nm(1-\overline{x})-1}}{\Gamma(2nm\overline{x})\Gamma(2nm(1-\overline{x}))}, \quad (6)$$

where n is the number of individuals in one subpopulation, m is migration rate or the ratio of migrants for each generation in subpopulation, and $\overline{x}$ is the mean value of i/n of whole populations. Figure 3 shows the shape of this function.

From Equation (6), the migration rate m is double of mutation rate μ of the Equation (5) when $\overline{x}$ is 1/2. That means the migration rate $m = 1/n$ makes the density function uniform, as same as the case of $\mu = 1/2n$. This shows that the migration rate $m = 1/n$, which means one migrant per generation, makes parallel GAs work well. This situation is similar to the case of the mutation rate of standard GAs. However, there are some differences between them; First, mutation is ignored in the Equation (6). Second, we assume $\overline{x}$ (the mean value of i/n of whole populations) is 1/2. Because of these differences, our expectation would have a little error. In fact, the migration rates which are used in the several researches on parallel GAs are larger than $1/n$.

Although Manderick *et al.* [9] tried to define the most effective migration rate, they could not define them. This means that the little difference of migration rate is not so effective. And, since the smaller migration rate needs less communication overhead cost, we expect the migration rate $1/n$ is the same or more effective. We are going to check this by simulations.

5 Conclusions

We considered the mean convergence time subjected to genetic drift and gave reference values of mutation and migration.

We want to know the performance of GAs compared with other methods. Therefore theoretical and experimental studies of GAs must be performed. The roles of mutation, crossover, and selection must be made clear and controllable. Mutation and crossover are effective to tend to increase the diversity in population. Convergence by reproduction affects the tendency of decreasing the diversity. So we want to know the critical point that is to balance between increasing and decreasing the diversity, to make the searching process of GAs effective. Even though the W-F model is a very simple model, it has some remarkable features. From Figure 1, when the mutation rate has the value $\mu = 1/2n$, the density of the stationary states tends to be uniform. Furthermore, in the case of island model parallel GAs, the migration rate $m = 1/n$ makes the density uniform. These values might be a key point to determine the mutation rates and the migration rates.

6 Acknowledgements

The authors would like to give thanks to Dr. N. Otsu, Director of the Machine Understanding Division and Chief of Real World Computing (RWC) Project Team, Dr. K. Ohta, Director of the Computer Science Division, Dr. M. Suwa, Director of the Information Science Division, Dr. B. Manderick, and Dr. H. Asoh for their continual encouragement and valuable discussions and useful comments.

References

[1] H. Asoh and H. Mühlenbein. On the mean convergence time of evolutionary algorithms without selection. In *Parallel Problem Solving from Nature 3*, 1994. to appear.

[2] A. E. Eiben, E. H. L. Aarts, and K. M. Van Hee. Global convergence of genetic algorithms: a markov chain analysis. In *Parallel Problem Solving from Nature*, pages 4–12, 1990.

[3] D. E. Goldberg. *Genetic Algorithms in Search, Optimization & Machine Learning.* Addison-Wesley, Reading, Mass., 1989.

[4] D. E. Goldberg and P. Segrest. Finite markov chain analysis of genetic algorithms. In *Proceedings of the 2nd International Conference on Genetic Algorithms*, pages 1–8, 1987.

[5] J. H. Holland. *Adaptation in Natural and Artificial Systems.* Univ. of Michigan Press, Ann Arbor, Mich., 1975.

[6] J. Horn. Finite markov chain analysis of genetic algorithms with niching. In *Proceedings of the 5th International Conference on Genetic Algorithms*, pages 110–117, 1993.

[7] K. A. De Jong and W. M. Spears. A formal analysis of the role of multi-point crossover in genetic algorithms. *Annals of Mathematics and Artificial Intelligence*, 5:1–26, 1992.

[8] M. Kimura. Diffusion models in population genetics. *J. Appl. Prob.*, 1:177–232, 1964.

[9] B. Manderick and P. Spiessens. Fine-grained parallel genetic algorithms. In *Proceedings of the 3rd International Conference on Genetic Algorithms*, pages 428–433, 1989.

[10] T. Niwa and M. Tanaka. On the mean convergence time for simple genetic algorithms. In *Proceedings of the International Conference on Evolutionary Computing '95*, 1995.

[11] G. Rudolph. Convergence analysis of canonical genetic algorithms. *IEEE Transactions on Neural Networks*, 5(1):96–101, 1994.

[12] J. Suzuki. A markov chain analysis on a genetic algorithm. In *Proceedings of the 5th International Conference on Genetic Algorithms*, pages 146–153, 1993.

[13] M. Tanaka and T. Niwa. Markov chain analysis on simple genetic algorithm. Technical Report ETL-TR-94-13, Electrotechnical Laboratory, 1994.

[14] R. Tanese. Distributed genetic algorithms. In *Proceedings of the 3rd International Conference on Genetic Algorithms*, pages 434–439, 1989.

Supervised Parallel Genetic Algorithms in Aerodynamic Optimisation

D. J. Doorly and J. Peiró
Department of Aeronautics, Imperial College,
Prince Consort Road, London SW7 2BY, U.K.
Email: {d.doorly, j.peiro}@ic.ac.uk

Abstract

This paper describes the application of parallel genetic algorithms (coupled with CFD analysis) to problems of optimal aerodynamic or aerodynamic-structural design of wings and airfoils. The method has been implemented on a variety of parallel architectures, and results to illustrate its application are presented. A common problem with genetic algorithms (GAs) is how to maintain diversity of the gene pool and avoid premature convergence of the population. Subdivision of the population into semi-isolated subpopulations (commonly referred to as 'demes') not only helps significantly in this regard, but is ideally suited to implementation on parallel environments. Considerable further advantages may be obtained when some form of automated supervision is added to direct the operation of the parallel GA. A supervision strategy and its parallel implementation are also considered.

1 A Genetic Algorithm for Airfoil Design

The basic ideas behind the GA have been described in many excellent texts such as [7], and only the parameters specific to this work are described here. Starting with the problem of optimising an airfoil shape, each member of the population of initial trial solutions is encoded as a string (or chromosome), which specifies the particular values of the design variables which define the shape of the individual. A real number encoding is used throughout this work; the encoding specifies the shape (for aerodynamic analysis) and structural properties (for combined aerodynamic and structural optimisation). Airfoil geometries are represented using B-splines; a spline is defined by the vertices of its control polygon, and taking the x-position coordinates of the vertices as fixed, an array which stores the y-ordinates can encodes the airfoil geometry, Figure 1. Alternatives to B-splines include direct surface encoding, and shape modification functions as outlined in [2].

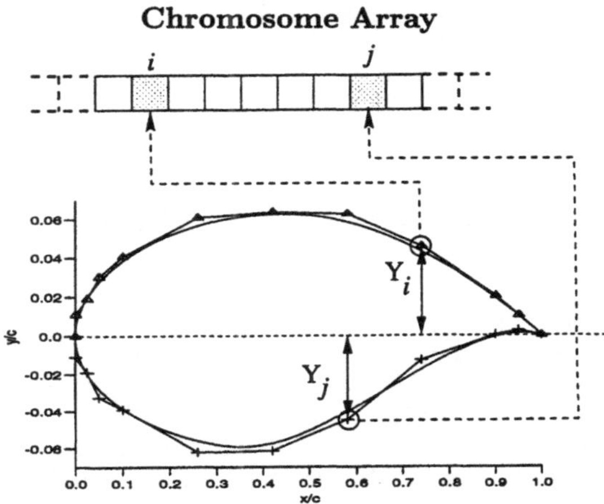

Figure 1: Representation of airfoil geometry.

Using a GA for optimisation, the population of strings or chromosomes is allowed to evolve, by preferentially selecting the fitter individuals for reproduction. In this application area, fitness for reproduction is evaluated by the flow (CFD) and structural solver, where the computational expense of the CFD analysis in particular is usually far greater than that of the other GA operations. The selection schemes used comprise binary tournament and roulette wheel selection [7]. Selection plays a central role in controlling the GA; if over-extreme, it can produce too rapid convergence. Fitness scaling is implemented, and the selection is also treated here as 'elitist'. A uniformly slightly mutated copy of the best of each generation is also carried through to the next generation, and mutation amplitude is varied (effectiveness of mutation 'creep' for this application is described in [2]).

To deal with the large computational effort required for aerodynamic optimisation (which becomes extreme when the more exact Euler or Navier-Stokes equations are solved), parallel com-

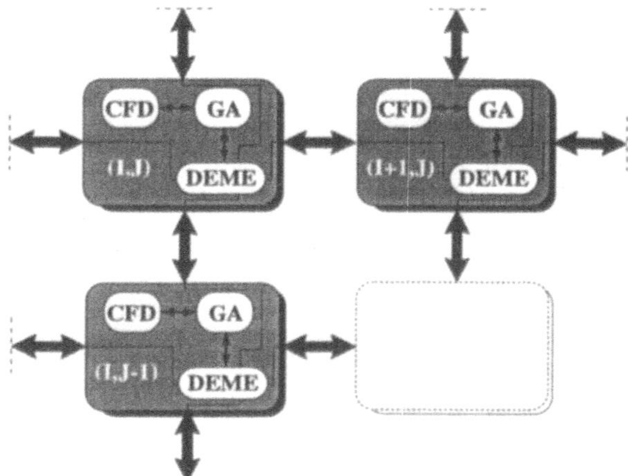

Figure 2: Inter-deme exchange.

puting is often employed to allow reasonable total computing times. Most applications, certainly in aerospace [11, 12, 14], have however employed sequential rather than parallel GAs. When GAs have been used in CFD, parallel computing has generally been used only for the fitness evaluation, i.e. flow solution.

2 The Distributed Genetic Algorithm

Studies (e.g. [13]) have shown that a distributed genetic algorithm (DGA) outperforms the conventional GA, in a variety of applications; work to date suggests this also holds for CFD optimisation [2]. In the distributed GA, the single well mixed (so called 'panmictic') population is replaced by a distributed set of subpopulations or demes. Recombination and genetic exchange between subpopulations is restricted; and the parameters of the basic GA (operator probabilities, selection mode, etc.) are thus augmented by the exchange (migration) strategy between demes and the sizing of the demes. The exchange strategy encompasses variables such as the number of exchanges and their frequency, geographical exchange radius, and the deme topology. The limitation of exchange between demes renders the distributed GA ideally suited for coarse grain parallelisation (Figure 2), on a parallel supercomputer or network of workstations, as reported in [4].

From this work, a comparison of the convergence behaviour obtained using a parallel and a conventional GA, with the same total population and number of evaluations, for the inverse design of a NACA airfoil (with direct surface encoding and a simple panel solution) is reproduced in Figure 3. The sequential GA used a single population with 200 members, whereas for the distributed GA the population was divided into 20 islands, each with 10 individuals on a 5×4 array. The migration between islands ocurred in a 'stepping stone' fashion, i.e. migration occurred only between immediate neighbours [9]. The evolution of the fitness, defined as

$$\left\{ \int [C_{p_1} - C_{p_2}]^2 \, ds \right\}^{-\frac{1}{2}},$$

where C_{p_1} and C_{p_2} denote the computed and target distributions of C_p on the airfoil surface, shows that the use of a distributed GA results in a large improvement in performance above that of the conventional GA. In fact the conventional GA may fail to converge to an acceptable fitness (corresponding to a value of the order of 15) in reasonable time. When multiplied by the speed up factor (for a case where each deme is mapped to a separate processor), the computational saving would thus be further magnified.

3 Implementation as a Parallel Genetic Algorithm

A parallel version of the DGA has been implemented using the MPI standard on a network of workstations and a multiprocessor machine. Instead

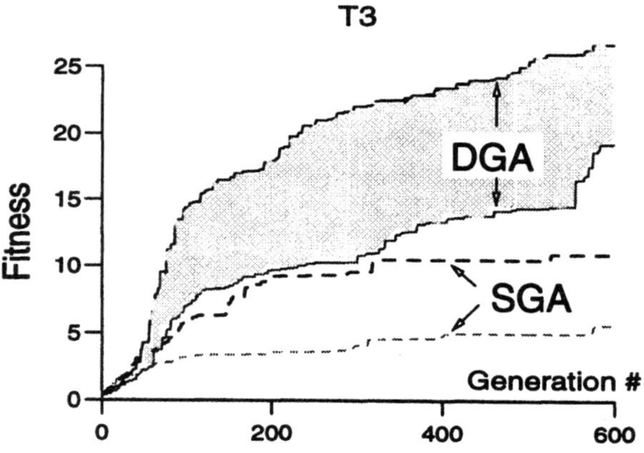

Figure 3: Distributed versus sequential GA.

of allocating a single deme to a processor, it is advantageous to allow several demes to be placed on a processor. This ensures that the number of demes is not restricted to the number of available workstations, and in a heterogeneous network allows faster processors to handle more demes. The parallel genetic algorithm has been coupled to an unstructured mesh based Euler solver [10], to a viscous-inviscid panel solver [6], and to a combined aerodynamic-structural solver [5].

The parallel GA may be made completely independent of the flow solver, interfacing solely via the translation of the chromosome into the surface geometry. This does not imply that the GA should do so; the addition of supervision facilitates allowing the GA to interact with parameters which govern the acuracy/time required by the solver in guiding the fitness evaluations, as well as to introduce suitable heuristics. An example would be pre-evaluation geometry checks to remove very unlikely geometries, so called screening for 'lethal genes'.

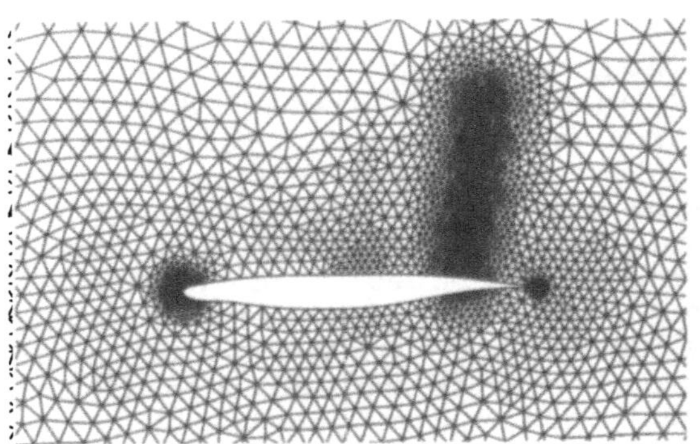

Figure 4(a): L/D airfoil optimisation: sample airfoil mesh.

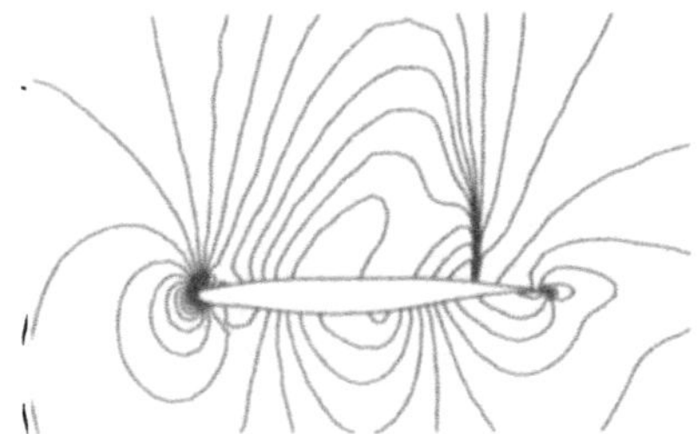

Figure 4(b): L/D airfoil optimisation: flow solution.

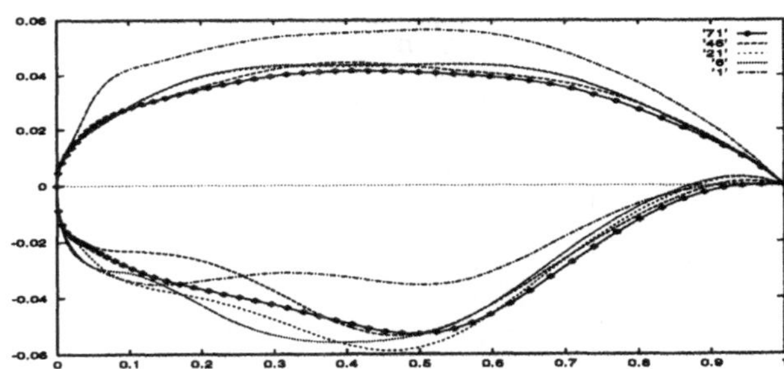

Figure 4(c): L/D airfoil optimisation: history of airfoil evolution.

4 Direct Airfoil Optimisation Results

The algorithm for the optimisation of airfoil geometries in transonic flow was first implemented to utilise up to 16 demes mapped onto a workstation cluster connected via an ethernet network. The coupling of the GA to an unstructured mesh-based CFD solver and a mesh adaptation scheme allows interaction with the flow solution to guide the refinement of the computational mesh. An example of the (adapted) computational mesh and flow solution for the best individual after a few generations are shown in Figures 4(a) and 4(b).

The objective of the design is to optimize the lift over drag ratio at a free-stream mach number of 0.8 and at an incidence of two degress. At this early stage the best individual possesses some good features, but still shows a strong shock on the upper surface, which is also near the peak mach number there and produces a considerable amount of wave induced drag. Figure 4(c) shows the evolution of the 'fittest' airfoil shape during the GA optimization process.

5 Agent Supervision of Parallel GAs

Thus far the improvement in GA performance may be attributed to the improved maintenance of diversity for the parallel distributed population compared to the single population GA. One can observe however that after a certain time, the population on each island converges. Adding an agent to supervise the operation of the parallel GA provides a capability whereby the DGA can adapt more generally than is possible with a sequential GA. Although adaptation can be built into a GA, the use of agents provides a more general framework by decoupling the tasks of higher level supervision from

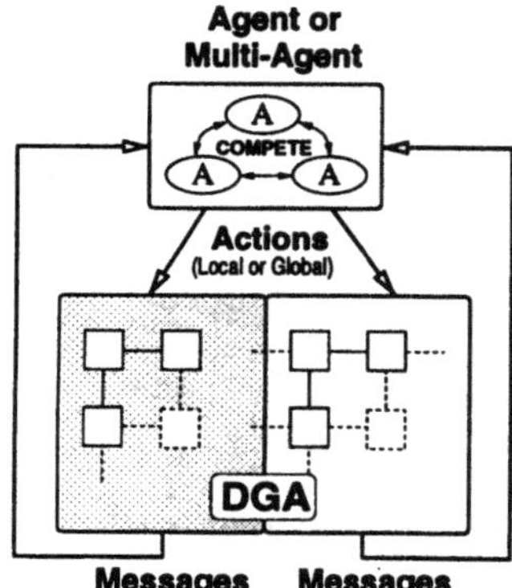

Figure 5: Addition of agent supervision to DGA.

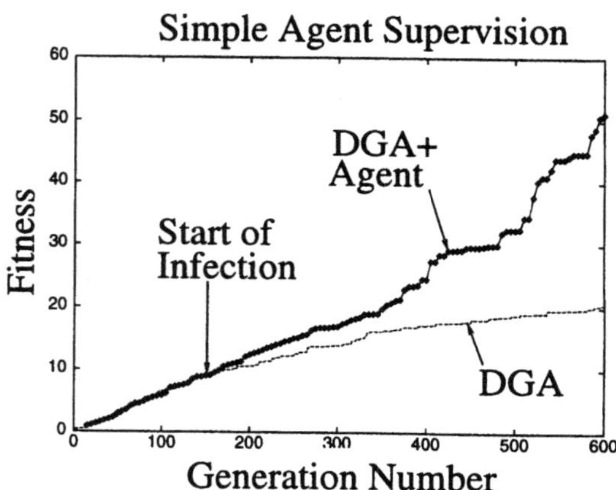

Figure 6: Use of an agent to improve population fitness.

the lower level optimisation. This is particularly so in a distributed computing environment, where agents can direct the operation of the GA on local or global populations, and can additionally direct processing resources. At one extreme, the agent layer may be combined with the DGA software to execute as a single (albeit distributed) entity, or at another, it may run as an entirely separate distributed program, communicating with the DGA by reading external output messages and writing to action inputs.

To illustrate, Figure 5 shows an agent 'layer' directing the operation of a parallel DGA, connected in a stepping stone topology. The agent receives messages from the DGA about its current status, (i.e. generation number, measures of island population convergence, fitness changes etc.), and condition (local mutation and crossover rates/types, reproduction mode etc.). The agent layer can then instruct the DGA to take either global or local action. Examples of such actions could be the introduction of a mechanism to improve diversity between island populations, actions to favour specific local niches within islands, or actions to improve the parallel load balancing by adjusting the deme placing or sizing.

As a simple example, an agent can be used as a vector for infections, with low population diversity encouraging epidemics. When implemented for the problem of inverse design optimisation, greatly improved results are found, beyond the point when the population on all the islands initially converges towards a global 'champion'. This is depicted in Figure 6, which compares the evolution of population fitness using the DGA, and using a modified DGA with the addition of an agent to direct infection. On infected islands, individuals close to the global best have greatly reduced fitness, and undergo increased mutation; the pattern of infection is found to change dynamically.

A learning classifier scheme [1, 8] can be used to train the agent supervisor. Messages from the DGA generate conditions; rules applied to the conditions generate actions. The agent starts with a population of rules, and assigns fitness values to them depending on their average effectiveness for several runs; new rules can be generated by mutating condition or action parts, or combining rules. In this model, agent learning is accomplished by repetitively solving a given type of optimisation problem; the trained agent is then applied to a further series of problems. Current work is aimed at establishing whether an agent supervised DGA can be trained effectively and without excessive computational cost for application to a range of airfoil optimisation problems.

In the preceding, adaptation occurs through repetitively solving a problem or class of problems. A dynamically adaptive agent can be be implemented with a DGA however, by placing several islands in a group under the control of one agent, and other groups of islands under the control of another or others, possibly after initial convergence. If migration between groups is eliminated, the relative improvement over a number of generations may be compared, so that the worse performing agents receive the populations of the best, and try different rule strategies.

6 Conclusion

Parallel genetic algorithms have been applied to airfoil optimisation using different numerical simulation strategies. The approach is very modular and the implementation of different flow and structural solvers is straightforward. The method can handle complicated optimization problems with relative ease but is open to improvement. Encouraging preliminary results, showing improved convergence rates for the DGA, have been obtained through the use of an agent supervision strategy. Further work will concern enhanced parallel GA techniques, learning agents and hybridisation with conventional optimisers.

References

[1] P. Devine, G. Kendal, and R. Paton. *When 'Herby' met ElViS - Experiments with Genetic Based Learning Systems*, chapter 16. John Wiley & Sons, Chichester, 1996.

[2] D. J. Doorly. *Parallel genetic algorithms for optimization in CFD*, chapter 13. John Wiley & Sons, 1995.

[3] D. J. Doorly and J. Peiró. Aerodynamic optimisation using supervised parallel genetic algorithms. In *Proc. 13th AIAA CFD Conference*. Snowmass Co., 1997. to appear.

[4] D. J. Doorly, J. Peiró, T. Kuan, and J-P. Oesterle. Optimisation of airfoils using parallel genetic algorithms. In *Proc. 15th Int. Conf. Num. Meth. Fluid Dyn.* Monterey, 1996.

[5] D. J. Doorly, J. Peiró, and J-P. Oesterle. Optimisation of aerodynamic and coupled aerodynamic-structural design using parallel genetic algorithms. In *Proc. Sixth AIAA/NASA/ISSMO Symposium on Multidisciplinary Analysis and Optimization*, pages 401–409. AIAA, 1996.

[6] M. Drela. *XFOIL, An Analysis and Design System for Low Reynolds Number Aerodynamics.* Number 54 in Lecture Notes in Engineering. Springer Verlag, Berlin, 1989.

[7] D. E. Goldberg. *Genetic Algorithms in Search, Optimisation and Machine Learning.* Addison-Wesley, 1988.

[8] J. H. Holland. *Adaptation in Natural and Artficial Systems.* MIT Press, 1992.

[9] J. Nang and K. Matsuo. A survey of parallel genetic algorithms. *J. SICE*, 33(6):500–509, 1994.

[10] J. Peraire, K. Morgan, and J. Peiró. *Unstructured Grid Methods for Advection Dominated Flows*, volume 787, pages 5.1–5.39. 1992.

[11] C. Poloni. chapter 20. John Wiley & Sons, 1995.

[12] D. Quagliarella and A. DellaCioppa. Genetic algorithms applied to the aerodynamic design of transonic airfoils. *J. Aircraft*, 32:889–891, 1995.

[13] R. Tanese. *Distributed Genetic Algorithms.* PhD thesis, U. Michigan, 1989.

[14] K. Yamamoto and O. Inoue. Applications of genetic algorithms to aerodynamic shape optimisation. In *Proc. 12th AIAA Computational Fluid Dynamics Conference.* San Diego, CA, 1995. AIAA-95-1650-CP.

A Genetic Clustering Method for the Multi-Depot Vehicle Routing Problem

S. Salhi[1], S. R. Thangiah[2] and F. Rahman[2]
[1] Management Mathematics Group, School of Mathematics and Statistics, University of Birmingham, UK.
[2] Computer Science Department, Slippery Rock University, USA.

Abstract

A clustering method based on a genetic algorithm for solving the multi-depot routing problem is proposed. An efficient post optimiser enhanced by reduction tests is embedded into the search to further improve the solutions. Preliminary results, based on a set of problems given in the literature, are encouraging.

1 Introduction

In this paper we introduce an adaptive clustering method based on geometric shapes using genetic algorithms for solving vehicle routing problems with multi-depots. The classical vehicle routing problem (VRP) consists of a set of customers, with known location and demand, and a set of vehicles, with a limited capacity, that are to service the customers from a central location refered to as a depot. The routing problem is to service all the customers without overloading the trucks while minimizing the total distance travelled and using the minimum number of trucks. Heuristic approaches as well as exact methods exist for the VRP [4, 6]. The multi-depot vehicle routing problem (MDVRP) is an extension of the classical vehicle routing problem with vehicles starting from different depots. The constraints of the problem, are similar to the ones of the VRP besides that each vehicle starts and finishes the delivery from the same depot. The primary objective of the problem is to minimize the total number of vehicles used, in addition to minimizing the distance travelled by the vehicles. This problem seems to suffer from a shortage of published work although, in practice, it is unlikely that a distribution system operates from one single depot only.

The MDVRP problem can be formulated as a mixed integer linear program [8]. It can be shown that exact methods are suitable for problems of limited size only. Heuristics seem to offer the best way to find good solutions to this large NP–hard problem. One commonly used technique for solving the MDVRP is a two phase approach; the customers are first allocated to their nearest depots and then for each depot the VRP is solved. Refinements are usually added to improve the obtained solutions. Past work on MDVRP is summarised in [11]. Chao, Golden and Wasil [1] developed a composite heuristic where a slight deterioration in the objective function is allowed in their one point move procedure. Refinements are also added. Renaud, Laporte and Boctor [9] use diversification and intensification in the implementation of their tabu search based method. Salhi and Sari [11] developed a multi-level composite heuristic which is enhanced by suitable reduction tests. The idea is to avoid local optimality by using a different improvement procedure whenever a local optima is encountered. This approach has produced competitive results while requiring only a fraction, say 20%, of the cpu times of the two previous heuristics. New best results are reported in the study of Renaud *et al.*[9] and in Salhi and Sari [11].

In this paper we introduce an adaptive clustering method based upon genetic algorithms to cluster the customers. The clusters of customers are found using a genetic algorithm and then in each cluster a travelling salesman problem (TSP) solution is obtained using a simple insertion heuristic. The routes are improved via a post-optimization method. The adaptive clustering method combined with the post-optimization obtains solutions to a set of problems from the literature that are more promising than those found by alternate heuristic methods.

2 Genetic Algorithms

The genetic algorithm (GA) is an adaptive heuristic search method based on population genetics. The basic concepts of a GA were primarily developed by Holland [5] and described later by Goldberg [3]. The

GA consists of a population of chromosomes that evolve over a number of generations and are subject to genetic operators at each generation. Each chromosome has a fitness value associated with it and a set of best fit chromosomes from each generation survive into the next generation. The genetic operations that the chromosomes are subjected to are crossover and mutation.

For a chromosome that is represented by a binary string, the crossover operation takes two strings, randomly cuts at two points of the strings and exchanges the bit string subset. The mutation operator randomly picks a bit in the chromosome and changes it to its complementary value. The GA is used to adaptively search for a set of attributes for a geometric shape used in routing vehicles. The next section explains the concept of using geometric shapes to adaptively cluster customers using a GA.

3 Adaptive Clustering using a GA

The adaptive clustering method is based upon using a route primitive, similar to the route primitives of Cullen [2], to cluster customers. In the work of Cullen, the attributes of the route primitives were obtained by working interactively with the primitives displayed on the screen. In the adaptive clustering method, the attributes of the route primitives are obtained using the genetic algorithm. For problems like the VRP with multiple depots, that have multiple-objectives and multiple constraints, it is not possible to interactively design routes. Geometric shapes can be used effectively to route vehicles if the shapes have the capability to adapt to the route shapes that result in the minimization of the routing cost.

In this research we map a geometric shape, namely the circle, to the chromosome. Each chromosome is mapped to the attributes of a set of circles used for clustering customers. The GA is used to adaptively search for the attributes of a set of circles that clusters customers using the routing cost as the fitness value for the individual chromosomes. Such an approach, when applied to the vehicle routing with time windows, has shown to be promising [12].

There are three conditions to be considered when assigning customers to be clustered within a circle. Figure 1 depicts an example containing two circles, A and B, and three customers, p_1, p_2 and p_3. A customer location can be inside a circle, on the circumference of a circle or outside a circle. If a customer location is inside a circle, as p_1 is in circle A, or on

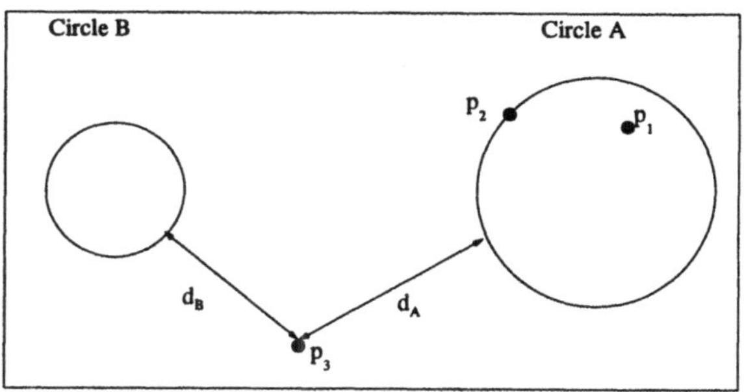

Figure 1: The association of customers p_1, p_2 and p_3 with respect to circles A and B.

the circumference of a circle, as p_2 is on circle A, it is assigned to that respective circle. If a customer is outside of all the circles, then the distance of that customer location from each of the circumference of the circles is calculated and the customer is assigned to the circle whose circumference is closest to it. In Figure 1 the customer location p_3 is outside of the circles A and B. If the distance between p_3 and the circumference of circle A is d_A and between p_3 and circle B is d_B, then p_3 will be assigned to circle A if $d_A \leq d_B$ or to circle B if $d_B < d_A$.

The efficiency of the clustering method is dependent on the placement of the circles such that it minimizes the total distance travelled by the vehicles. The location of a circle is based upon two attributes, namely the origin and its radius. The adaptive clustering method uses the genetic algorithm to search for attributes of a set of circles that minimizes the cost of routing a set of vehicles.

Let us assume N to be the total number of customers and K to be the initial number of vehicles required to service the customers. The clustering method represents the origin of the circle, (x_k, y_k), with a bit string of length B and the radius of the circle, r_k, with length L. The number of bits required to represent the attributes for one circle is $B + L$ bits. The length of the chromosome is the number of vehicles, K, times the length required to represent the attributes of a circle, namely $B + L$. The bit string size for (x_k, y_k) and r_k were derived empirically and set at 8, 8 and 6 bits respectively. This allows (x_k, y_k) to range between (0,0) and (M_x, M_y) where M_x, M_y are the maximum grid coordinate values of the MDVRP. The radius r_k of the circle is allowed to range between 0 and 16. During the clustering process, not all of the customers fall within the circles generated by the chromosome.

The formula for determining the circle, say k, an unassociated customer, say j, should belong to is as follows:

$$C_{kj} = \begin{cases} w_1 \tilde{d}_{kj} \Delta_{kj} + w_2 \tilde{d}_{kj} & \text{if } \Delta_{kj} > 0 \\ \tilde{d}_{kj} & \text{otherwise} \end{cases}$$

$$\Delta_{kj} = \sum_{s \in \text{Circle k}} q_s + q_j - \hat{Q}$$

$$\tilde{d}_{kj} = d_{kj} - r_k$$

where Δ_{kj} denotes the extra vehicle load when assigning customer j to the kth circle, d_{kj} the distance between customer j and the centre of the kth circle, q_j the demand of the jth customer, $\hat{Q}$ the vehicle capacity and (w_1, w_2) are weights $(w_1 = 1 - w_2)$.

The formula is implemented in this manner to place the unassociated customer in a circle that is closest to the customer and that does not lead to an overload with respect to the maximum capacity of the vehicle. The above cost is computed between the unassociated customer and each of the circles and the customer is assigned to the one that has the lowest cost. Once all of the customers are assigned to the circles, the customers within each circle are routed using the cheapest insertion criterion. These routes are then improved using a customer exchange refinement process. The fitness value of the chromosome is the fitness value obtained after the refinement is performed. It is necessary to improve the routes at each of the chromosomes as the solution obtained the chromosome could result in a genetic drift and the search could end up in a region that does not produce good results. The refinement process used to improve the routes consists of moving a customer from one route to another and exchanging customers between routes [12, 13].

Once the genetic search terminates, the best set of clusters obtained during the search is used to route the vehicles and then a post optimiser is introduced to further improve the solution. This post-optimization process will be briefly discussed in the next section.

4 The Post Optimisation Phase

Once the solution is found by the above clustering procedure, we improve the solution using the post-optimiser developed by Salhi and Sari [11]. There are two main points which may be worth describing:

(i) a set of improvement routines which are used in sequence and repeatedly;

(ii) two reduction tests are incorporated to speed up the search without affecting the quality of the solutions significantly.

In (i) the routines used include the swapping of customers between depots and routes, reallocating customers between routes, dropping an entire route, splitting a longer route in smaller ones, combining smaller routes into larger ones, partial combining and splitting, allocating a group of customers which are in chain and involving 3 routes simultaneously. Some of these improvement routines are described in details in Salhi and Rand [10].

In (ii), the aim is to evaluate only those possibilities which are likely to affect the solution when applying the modules described in (i). The first reduction test is built for single depot routing whereas the second is designed for multi-depot routing problems. In the first one, for each customer only those customers situated within its neighbouring sector are considered for evaluation whereas in the second only the borderline customers (those customers which are nearly half way between their nearest depots) with their surrounding customers are considered. For further details, see Salhi and Sari [11].

5 Computational Results and Conclusions

The genetic clustering method was used to solve the first five problems obtained from the literature. The five problems consisted of customers varying in size from 50 to 100 customers with depots ranging from 2 to 5. We compare our results with the ones obtained by the best MDVRP heuristic, namely the heuristic by Renaud *et al.* [9].

The parallel genetic algorithm (PGA), see [7], was used to adaptively search for the set of circles that would minimize the cost of routing vehicles. The parameter values for the number of generations, population size, crossover rate and mutation rate in the PGA were set at 200, 50, 0.6 and 0.001 respectively. The GA search will also terminate if there is a 70% similarity in the chromosomes at any given population.

The initial results obtained by the clustering method indicates that it has the capability to obtain good quality solutions. Although the total distance is larger than the one given by the best heuristic, our method has the advantage of requiring usually

Table 1: Computational results obtained by our genetic clustering and the best MDVRP heuristic.

Problem data				Genetic clustering		Renaud *et al.*	
No	n	m	$\hat{Q}$	cost[NV]	cpu (secs)	cost[NV]	cpu (secs)
1	50	4	80	624.9[10]	61	576.9[11]	66
2	50	4	160	476.7[5]	53	473.5[5]	72
3	75	5	140	884.1[10]	174	641.2[11]	108
4	100	2	100	1062.3[15]	114	1003.9[15]	132
5	100	2	200	768.2[8]	127	750.3[8]	144

one less vehicle. The computation time spent to obtain those solutions is slightly smaller than the ones given by the other authors. The results for these 5 problems are summarised in Table 1 in which n is the number of customers, m the number of depots, and NV the number of vehicles found.

In this study, we have addressed an important distribution problem, the multi depot routing problem. We have designed a clustering GA enhanced by a post optimiser to solve efficiently multi depot routing.

Two reduction tests are embedded into the search to speed up the computation without affecting significantly the quality of the solutions. The proposed approach is tested on benchmark problems varying in size from 50 to 100 customers, and 2 to 5 depots with encouraging preliminary results. Extensive testing using larger problems may be necessary to enhance the usefulness of this approach. The use of a combination of shapes such as circles, ellipses, squares, etc. may be worth considering in future research.

References

[1] I. M. Chao, B. L. Golden, and E. Wasil. A new heuristic for the multi-depot vehicle routing problem that improves upon best-known solutions. *American Journal of Mathematical and Management Sciences*, 13:371–401, 1993.

[2] F. H. Cullen. *Set Partitioning Based Heuristics for Interactive Routing.* PhD thesis, Georgia Institute of Technology, Georgia, 1984.

[3] D.E. Goldberg. *Genetic Algorthims in Search Optimization and Machine Learning.* Addison Wesley, 1989.

[4] B. L. Golden and A. Assad. *Vehicle Routing: Methods and Studies.* North Holland, Amsterdam, 1988.

[5] J. H. Holland. *Adaptation in Natural and Artificial Systems.* University of Michigan Press, Ann Arbor, 1975.

[6] G. Laporte. The vehicle routing problem: An overview of exact and approximate algorithms. *European Journal of Operational Research*, 59:345–358, 1992.

[7] Parallel Genetic Algorithm Package. Argonne national laboratory. USA, 1996.

[8] J. Perl and M. S. Daskin. A warehouse location routing problem. *Transportation Research*, 19B:381–396, 1985.

[9] J. Renaud, G. Laporte, and F. F. Boctor. A tabu search heuristic for the multi-depot vehicle routing problem. Technical Report 94-44, Centre de Recherche sur les Transports, University of Montreal, Canada, 1994. Working Paper.

[10] S. Salhi and G. K. Rand. Incorporating vehicle routing into the vehicle fleet composition problem. *European Journal of Operational Research*, 66:313–330, 1993.

[11] S. Salhi and M. Sari. A multi-level composite heuristic for the multi-depot vehicle fleet mix problem. *European Journal of Operational Research*, 1997. (to appear).

[12] S. R. Thangiah. *Genetic Algorithms for Vehicle Routing Problems with Time Windows.* CRC Press, Florida, 1996.

[13] S. R. Thangiah, I. H. Osman, R. Vinayagamoorthy, and T. Sun. Algorithms for vehicle routing problems with time deadlines. *American Journal of Mathematical and Management Sciences*, 13:322–355, 1993.

A Hybrid Genetic / Branch and Bound Algorithm for Integer Programming

A. P. French, A. C. Robinson and J. M. Wilson
Loughborough University Business School,
Loughborough LE113TU, UK
Email: {a.p.french, a.c.robinson, j.m.wilson@lboro.ac.uk}

Abstract

An approach to combine a genetic algorithm with traditional linear programming based branch and bound for integer programming is described in this paper. Branch and bound provides a systematic search procedure for pure integer programming problems and a genetic approach offers the possibility of rapid movement towards a useful solution. Hence the two approaches look worthy of combination as a way to solve certain {0,1} integer programming problems. The approach has been tested out on satisfiability problems and computational results look promising in certain aspects of speed and solution quality.

1 Introduction

Branch and Bound (B&B) has been the favoured algorithm for integer programming problems for many years since it was first described by Land and Doig [6]. It is the standard tool incorporated in most commercial integer programming optimisation software, e.g. XPRESS-MP [3]. The algorithm takes advantage of the relatively quick time taken to obtain solutions of the linear programming relaxation of the integer programming problem to provide useful information. However, for pure combinatorial problems this approach may not always be best as the relaxation may not be so helpful and long futile searches through the B&B tree may ensue. Heuristic approaches provide a way to move a search on, cutting across a binary tree. However, they lack the search uniformity of an implicit enumeration scheme and consequently may miss solutions. There would seem to be scope for combining the two approaches.

2 The Satisfiability Problem

The satisfiability problem is the problem of assigning truth values (true or false) to a set of logical variables (literals) which occur in a series of logical expressions (clauses) to establish the truth or falsity of the entire collection of logical expressions. The logical expressions may be expressed in a variety of forms, such as conjunctive or disjunctive normal forms for convenience, but we will make use of an integer programming representation as follows.

The satisfiability problem may be expressed as a series of constraints, one per clause, in {0,1} variables of the form:

$$\sum_{j \in J} x_j \leq |J| - 1 \tag{1}$$

where each $x_j (j \in J)$ is a literal (possibly negated) and J is the index set of literals for a particular clause.

As it stands, such a problem has no objective function and all that is required is a feasible solution to the set of constraints. Alternatively, a satisfiability problem containing m clauses and n literals may be formulated as

$$\text{minimise } z = \sum_{i=1}^{m} s_i \tag{2}$$

subject to

$$\sum_{j \in J_i} x'_j - s_i \leq |J_i| - 1 \quad , \quad i \in \{1, 2, ..., m\} \tag{3}$$

$$s_i, x'_j \in \{0, 1\}, \quad i \in \{1, 2, ..., m\}, \quad j \in \{1, 2, ..., n\} \tag{4}$$

and

$$x'_j \text{ is } x_j \text{ or } 1 - x_j$$

If the optimal solution to the problem given by (2) - (4) is such that $z = 0$, then the problem is satisfiable, otherwise it is not.

The satisfiability problem with at least three literals per clause is known to be NP-hard and problems where m/n is approximately 4.3 are known to be hard to solve (Gent and Walsh [5]). A paper by de Jong and Spears [4] describes approaches to solve satisfiability problems using a genetic algorithm.

3 A Hybrid Algorithm

Other authors have attempted to combine B&B with Genetic Algorithm (GA) approaches and work is reported in Cotta *et al.* [2], Nagar *et al.*[7] and Reeves [8].

In our approach we have amalgamated a GA, based on one developed by Beasley and Chu [1] for the set covering problem, with the commercial integer programming software XPRESS-MP [3] which provides the B&B capability. The hybrid algorithm commences in XPRESS-MP and traverses a number of nodes before entering the GA. New potential solutions generated by the GA are added to the B&B tree, and control returned to the B&B engine. Subsequently our algorithm iterates between the two approaches until the solution is obtained or the search abandoned. Thus the hybrid algorithm is using the systematic search procedure of B&B, with the added information from the linear programming relaxation of the problem with a GA approach which attempts to kick start the search into new areas when B&B appears to be getting nowhere.

A number of considerations need to be made for this hybrid approach. These include:

- pool generation,
- pool size,
- fitness measurement,
- switching criteria from B&B to GA,
- pool operations (crossover, mutation rates),
- solution selection from B&B,
- handling of fractional solutions for variables in chromosome structure and GA operations,
- returning solutions from the GA,
- termination criteria.

These considerations will be systematically examined in results presented.

4 Computational Experience

- A series of 10 satisfiability problems was generated with 80 and 100 variables and a number of clauses determined by 4.3 times the number of variables. Each clause contained 3 variables (or complements of variables). Each variable had an equal chance of being selected for a clause and was complemented with probability 0.5. Each problem was satisfiable.

  After considering the various parameters discussed in the previous section, the following settings were found to be beneficial.

- A randomly generated pool was used initially.
- A pool size of 50 was used for the GA.
- Fitness for the GA was measured as the satisfiability or potential satisfiability of each clause aggregated over all clauses and in B&B the objective function used was

$$\text{minimise} \sum C_j x_j \tag{5}$$

  where

$$C_j = C_j^+ - C_j^-$$

  and C_j^+ = number of clauses in which x_j occurs

  and C_j^- = number of clauses in which $\bar{x}_j$ occurs, $j \in \{1, 2, ..., n\}$.

- The leaving criterion from B&B was a composite measure based on the number of variables set by B&B and the current depth of the search.
- Crossover was performed by a fusion operator and prior parent selection by a binary tournament approach. There was limited mutation.
- Hanging nodes in B&B were used to indicate pool solution values, together with randomisation.
- Representation of 1 and 0 in the chromosome structure mapped the binary equivalent, but fractional quantities were starred for further consideration.
- When the GA returned to the B&B phase it brought with it the 5 most fit solutions for consideration by B&B.

Table 1: Results obtained from 30 runs averaged.

Problem	Nodes B&B	Nodes B&B + GA	%
1	1318	848.8	64.4
2	1048	392.0	37.4
3	782	476.2	60.9
4	718	480.3	66.9
5	3728	2076.5	55.7
6	1135	120.3	10.6
7	1224	198.3	16.2
8	2503	290.3	11.6
9	1937	492.0	25.4
10	1709	1022.0	59.8

- As a feasible solution was sought, the search was terminated as soon as one was obtained.

In Table 1, the second column represents the actual number of B&B nodes required by XPRESS-MP. The third column represents the mean number of B&B nodes required by the hybrid algorithm. The fourth column is obtained from the number in the third column expressed as a percentage of the number in the second column.

As can be seen, the hybrid algorithm achieves a good reduction in the number of nodes required to solve the problems.

5 Conclusions

The hybrid algorithm described in this paper is still under development, but early indications provide some hope for solving difficult combinatorial problems, where neither B&B nor a GA approach is the ideal method. The combination of the two techniques seems to add new power.

6 Acknowledgments

The financial support of the Engineering and Physical Sciences Research Council for this research and the software support of Dash Associates, for provision of XPRESS-MP, are gratefully acknowledged.

References

[1] J. E. Beasley and P. C. Chu. A genetic algorithm for the set covering problem. *European Journal of Operational Research*, 94:392–404, 1996.

[2] C. Cotta, J. F. Aldana, A. J. Nebro, and J. M. Troya. Hybridizing genetic algorithms with branch and bound techniques for the resolution of the TSP. In D. W. Pearson, N. C. Steele, R. F. Albrecht (editors), *Artificial Neural Nets and Genetic Algorithms*, Wien, pages 277–280, 1995. Springer-Verlag.

[3] Dash Associates Ltd, Blisworth, Northamptonshire, UK. *XPRESS-MP.*

[4] K. A. de Jong and W. M. Spears. Using genetic algorithms to solve NP-complete problems. In *Proceedings of the 3rd International Genetic Algorithms Conference*, New Jersey, USA, 1989. Lawrence Erlbaum Associates.

[5] I. P. Gent and T. Walsh. Easy problems are sometimes hard. *Artificial Intelligence*, 70:335–345, 1994.

[6] A. H. Land and A. G. Doig. An automatic method for solving discrete programming problems. *Econometrica*, 28, 1960.

[7] A. Nagar, S. S. Heragu, and J. Haddock. A combined branch-and-bound and genetic algorithm based for a flowshop scheduling algorithm. *Annals of Operations Research*, 63:397–414, 1996.

[8] C. Reeves. Hybrid genetic algorithms for bin-packing and related problems. *Annals of Operations Research*, 63:371–396, 1996.

Breeding Perturbed City Coordinates and Fooling Travelling Salesman Heuristic Algorithms

R. Bradwell, L.P. Williams and C. L. Valenzuela
University of Teesside, Middlesbrough, TS1 3BA
Email: christine@tees.ac.uk

Abstract

Standard heuristic algorithms for the geometric travelling salesman problem (GTSP) frequently produce poor solutions in excess of 25% above the true optimum. In this paper we present some preliminary work that demonstrates the potential of genetic algorithms (GAs) to perturb city coordinates in such a way that the heuristic is 'fooled' into producing much better solutions to the GTSP. Initial results for our GA show that by using the nearest neighbour tour construction heuristic on perturbed coordinate sets it is possible to consistently obtain solutions to within a fraction of a percent of the optimum for problems of several hundred cities.

1 Introduction

Weaknesses in many of the standard GTSP heuristic algorithms are easy to spot by examining typical solutions produced by such techniques. Tour construction heuristic algorithms, for example, tend to start off well but are often observed to 'run out of steam' towards the end of the process.

The tour in Figure 1 was produced using a tour construction algorithm called the *nearest neighbour* heuristic algorithm (*NNHA*) [4]. This algorithm proceeds as follows: starting with an arbitrarily chosen starting city, choose for the next city the unvisited city closest to the current one. This new city is then labelled as the current one and the process repeated until all the cities have been chosen. The tour is then closed by returning to the initial city. 'Good parts' and 'bad parts' of the tour are clearly visible in Figure 1, illustrating the reduced effectiveness of the *nearest neighbour* (*NN*) heuristic when only a few cities remain to be incorporated in the tour.

The aim of this paper is to demonstrate the potential of a genetic algorithm (GA) to perturb the city coordinates in such a way that heuristics such as these are 'fooled' into producing better solutions to the GTSP than they otherwise would.

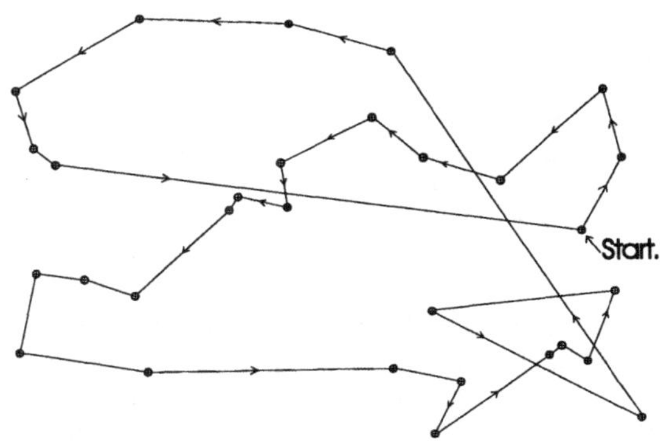

Figure 1: A nearest neighbour solution to a 20 city problem.

The original motivation for this work was to use perturbed city coordinates in conjunction with the cellular dissection algorithms of Richard Karp [8]. The run-time guarantees of Karp's algorithms (they run very fast and scale $O(n \log n)$) make them an attractive proposition for solving very large travelling salesman problems. The success of the *evolutionary divide and conquer* (*EDAC*) algorithm [11, 13, 14] in producing high quality solutions in $O(n \log n)$ time suggests that an approach that endeavours to combine divide-and-conquer with GAs has much potential. The effect we are hoping eventually to achieve with perturbed coordinates is to 'fool' a much simpler and faster version of Karp's algorithm than we used in *EDAC* into producing better solutions.

It would appear, however, that the potential of using a GA to breed perturbed city coordinates is not restricted to the algorithms of Richard Karp. Indeed the general approach could be applied to *any* heuristic algorithm for solving the TSP. The purpose of the present paper is to explore the potential of a GA based on perturbed coordinates on the simplest of heuristic algorithms before we extend the approach to compete with *EDAC*. We have chosen the *NNHA* as our starting point.

The basic idea is to perturb the city coordinates slightly, and use these perturbed coordinates to produce a tour using the chosen heuristic algorithm. The cities in the permutation list resulting from this tour are then moved back to their original position and a 'true tour' is produced.

The use of perturbed coordinate sets for solving the GTSP is not new. Codenotti, Manzini, Margara and Resta [2, 3] incorporated this technique into their version of Iterated Local Search, randomly perturbing the city coordinates of the TSP instance I by small amounts to give a new instance I' every time a locally optimal tour, T, is found on I. T will not normally be locally optimum with respect to I', so local optimization is then performed with respect to I' to give a new tour T'. T' then provides a new starting point for locally optimizing the GTSP with respect to I. In this way the perturbed coordinates provide a simple 'mutation' enabling the local search algorithm to escape local optima. About half of the time is spent applying the heuristic algorithm to the original coordinates and half to the perturbed coordinates. Our approach differs from that of Codenotti *et al* in three very important ways:

- In our study the TSP heuristic is applied *only* to perturbed coordinate sets.
- We use a genetic algorithm to breed perturbed coordinate sets.
- Our approach can be applied to tour construction heuristics as well as to local search heuristics.

The perturbed coordinate sets are the 'chromosomes' in our experiments, and the *NNHA* produces a different tour for each of the virtual coordinate sets in the population at any one time. These tours, which are easily represented by permutation lists, can then be evaluated with respect to the original city coordinates and actual intercity distances. The 'true' tour lengths form the basis for the fitness function of the GA.

All the TSP problems used in our study are uniform random points in a square region of the Euclidean plane.

In order to assess the quality of the solutions obtained by our genetic algorithm, we use a problem specific lower bound known as the Held-Karp lower bound [5, 6, 12]. This technique is known to produce very good estimates of optimal solutions for uniform random points, optimal tour lengths averaging less than 0.8% over the Held-Karp bound [9].

2 The Genetic Algorithm

A simple GA which appeared in [13] is used here. It is derived from the model of [7] and is an example of a 'steady state' GA (based on the classification of [10]). It uses the 'weaker parent replacement strategy' first described by [1]. The GA applies the genetic operators to the perturbed coordinate sets. The fitness values are based on tour lengths, as already explained.

The process for generating an initial population of perturbed coordinates required some thought. The most obvious way to generate a 'random' population at the start of a genetic algorithm is to randomise each x and y coordinate in such a way that all the original cities are effectively free to move anywhere within a region containing the GTSP problem. Intuitively, however, we favoured a scheme which perturbed the city coordinates within some small preset rectangular region surrounding each city, letting a suitable size for this rectangular region be determined by some early experimentation.

Figure 2 shows a snapshot of our GA running on a 20 city problem with the *NNHA*. The rectangular regions surrounding each city delimit the perturbation zones. The tour has been drawn through the cities located at their original positions, but the 'virtual positions' (which are randomly generated within each rectangle) are also visible in the diagram.

3 Perturbation Zone Scaling

How large should the rectangular zones in Figure 2 be? For random uniform points the size of the ideal perturbation zone is likely to be inversely proportional to the density of the cities. More formally if R represents the area of the (square) region, and n the size of the problem then,

$$l = k\sqrt{\frac{R}{n}} \tag{1}$$

relates the required dimension of the side of the perturbation zone, l, to city density through a factor, k, which can be estimated experimentally.

We carried out a set of experiments to establish a suitable value for the parameter k, the perturbation

Table 1: Results of GA and Random Search on 100 City Problem

Heuristic	Mean	Standard Deviation	95% Confidence Intervals
One Point GA	98.21	0.74	98.07, 98.34
Two Point GA	99.08	1.16	98.87, 99.29
Uniform GA	97.77	0.33	97.70, 97.83
Random Search	104.39	1.06	104.01, 104.77
NNHA	119.11	6.91	118.42, 119.80

factor of Equation (1). We ran our GA 30 times for each of a range of values for k on a single 100 city problem (random uniform points), and recorded the mean and standard deviation for each run. For all of our experiments we set the population size to 100, the mutation rate to 0.1% and used one-point crossover (see results section). Each run terminated after 300 generations. $k = 0.6$ gave the best result and was used for the remaining experiments.

4 Results

Uniform crossover was the obvious choice for our main genetic operator [10]. Since the predetermined sequence of the coordinate pairs in the 'chromosomes' is effectively generated at random (i.e. the sequence in which the city coordinates appear on the problem file) the use of either one-point or two-point crossover is unlikely to be any more effective at preserving subtours than uniform crossover. However, if one rearranges the lists of coordinates in an attempt to improve the propagation rate of subtours between parents and their offspring, it is feasible that the more traditional one-point and two-point crossovers will become more effective.

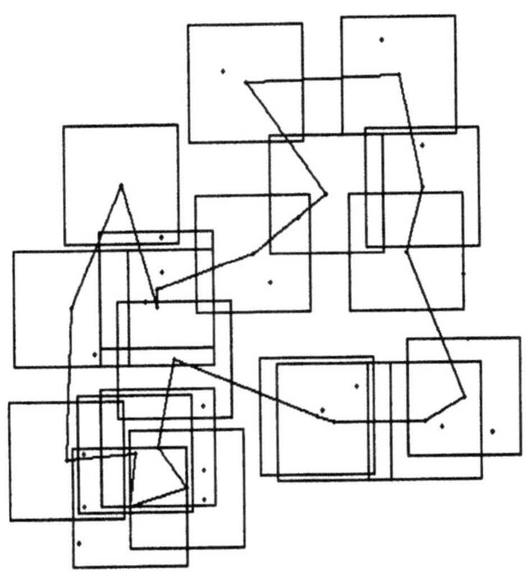

Figure 2: A good tour obtained on a 20 city problem using the GA on perturbed coordinates for the NN heuristic algorithm.

The results of our experiments are presented in Table 1. In order to assess the potential of one-point and two-point crossover, we incorporated a sorting algorithm into our program which, at regular intervals of time, rearranged the virtual coordinate sets in the current population together with the coordinates for the original problem into a sequence matching the current 'best tour so far'. This rearrangement was not expected to fulfil any useful purpose when the GA was using uniform crossover, so the sorting routine was left out in this case.

For each operator we ran 30 trials using the same 100 city problem. Each trial ran for 300 generations on a population of 100 members (a total of 30,000 evaluations plus the initial population). The perturbation parameter used was 0.6, with the mutation rate set to 0.1% for the GA. The random search was based on 30 trials each consisting of 30,100 randomly generated individuals with $k = 0.6$. The *NNHA* was executed 100 times, using each of the 100 cities in turn as the starting city for the algorithm.

Given that the *NNHA* produced a mean of 119.11 on the original coordinate data for the problem, all the other results presented represent a considerable improvement on the basic algorithm. In addition, the GA clearly outperforms the random search. From Table 1 it is possible to see that in our trials the uniform crossover was the most effective, producing on average a result that measures 1.01% above the Held-Karp lower bound of 96.37. The superior performance of uniform crossover seems counter intuitive since one would expect that operators such as two point crossover would preserve subtours more readily.

Initial experiments indicate that significant improvements over *NNHA* are maintained by using our GA for problems up to 500 cities, although run-time overheads for *NNHA* are becoming significant at this level.

Preliminary trials using a version of Karp's algorithm with our GA are encouraging. For a 500 city problem the GA consistently lifts the solution by about 20%.

5 Conclusion

In this paper we have clearly demonstrated that our technique for breeding perturbed coordinates for the TSP heuristics has potential. The method has succeeded in improving the quality of solution produced by the nearest neighbour heuristic algorithm by about 20% on a 100 city problem, to within a fraction of a percent of the optimum. Work is underway to incorporate the technique for breeding perturbed coordinates into Karp's cellular dissection algorithms.

References

[1] D. J. Cavicchio. *Adaptive search using simulated evolution.* Unpublished doctorial dissertation, University of Michigan, Ann Arbor

[2] B. Codenotti L. Margara, G. Manzini and G Resta. *Global strategies for augmenting the efficiency of TSP heuristics*, volume 709. Springer-Verlag, Berlin, 1993.

[3] B. Codenotti L. Margara, G. Manzini and G Resta. Perturbation: an efficient technique for the solution of very large instances of the euclidean TSP. *IFORMS Journal on Computing*, 8(2):125, Spring 1996.

[4] D. S.Johnson. Local optimization and the traveling salesman problem. In *Automata Languages and Programming: 17th International Colloquium Proceedings*, 1990.

[5] M. Held and R. M. Karp. The travelling salesman problem and minimum spanning trees. *Oper. Res*, 18:1138–1162.

[6] M. Held and R. M. Karp. The travelling salesman problem and minimum spanning trees: part ii. *Maths. Programming*, 1:6–25.

[7] J. H Holland. *Adaptation in natural and artificial systems.* The University of Michigan Press, Ann Arbor

[8] R. M. Karp. Probabilistic analysis of partitioning algorithm for the travelling- salesman problem in the plane. *Mathematics of Operations Research*, 2(3):209–224, August 1977.

[9] L. A. McGeoch D. S. Johnson and E. E. Rothberg. Asymptotic experimental analysis for the Held-Karp traveling salesman bound. In *Proceeding 1996 ACM-SIAM symp. on Discrete Algorithms*, 1996.

[10] G. Syswerda. Uniform crossover in genetic algorithms. In *Proceedings of the Third International Conference on Genetic Algorithms*, Hillsdale, NJ, 1996. Lawrence Erlbaum Associates.

[11] C. L. Valenzuela and A. J.Jones. A parallel implementation of evolutionary divide and conquer for the TSP. In *Proceedings of the First IEE/IEEE conference on Genetic ALgorithms in Engineering Systems: Innovations and Applications (GALESIA)*, pages 499–504, Sheffield, U.K., September 1995.

[12] C. L. Valenzuela and A. J. Jones. Estimating the Held-Karp lower bound for the geometric TSP. *European Journal of Operational Research*, to appear.

[13] C. L. Valenzuela and A. J.Jones. Evolutionary divide and conquer (I): a novel genetic approach to the TSP. *Evolutionary Computation*, 1(4):313–333, 1995.

[14] C. L. Valenzuela. *Evolutionary Divide and Conquer: a Novel Genetic approach to the TSP.* PhD thesis, University of London, 1995.

Improvements on the Ant-System: Introducing the $\mathcal{MAX}$–$\mathcal{MIN}$ Ant System

T. Stützle and H. Hoos
TH Darmstadt, FB Informatik, FG Intellektik
Alexanderstr. 10, D-64283 Darmstadt, Germany
{tom,hoos}@informatik.th-darmstadt.de

Abstract

In this paper we present $\mathcal{MAX}$–$\mathcal{MIN}$ Ant System ($\mathcal{MM}$AS) that improves on the Ant system. $\mathcal{MM}$AS is a general purpose heuristic algorithm based on a cooperative search paradigm that is applicable to the solution of combinatorial optimization problems. In the experiments we apply $\mathcal{MM}$AS to symmetric and asymmetric travelling salesman problems. We describe in detail the improvements on Ant system, discuss the addition of local search to $\mathcal{MM}$AS, and report on our computational results, showing that our system also improves over other variations of Ant system.

1 Introduction

Ant system (AS), introduced originally in [1, 2], is a cooperative search algorithm inspired by the behaviour of real ants, that are able to find good solutions to shortest path problems between food sources and their home colony. Ants usually communicate via pheromones, i.e. aromatic substances, that they may lay down in some quantity. An ant's tendency to choose a specific path is positively correlated to the intensity of a found trail. The pheromone trail evaporates over time if no more pheromone is laid down by other ants. If many ants choose a certain path and lay down pheromones, the intensity of the trails increases attracting more ants. The behaviour of ant colonies is imitated to some extent by AS by using simple agents, called *ants*, that may communicate among themselves via a mechanism inspired by the pheromone trails. This allows the application of this search metaphor to the solution of combinatorial optimization problems [3]. Here we present $\mathcal{MM}$AS, an improvement over Ant system at hand of the application to symmetric and asymmetric Traveling Salesman Problems (TSPs) contained in TSPLIB (`ftp.iwr.uni-heidelberg.de/pub/tsplib`).

2 The Ant System for TSPs

A TSP can be represented by a complete weighted directed graph $G = (\mathcal{V}, \mathcal{A}, d)$ where $\mathcal{V} = \{1, 2, \ldots, n\}$ is a set of nodes (cities), $\mathcal{A} = \{(i,j) \| (i,j) \in \mathcal{V} \times \mathcal{V}\}$ a set of arcs, and $d : \mathcal{A} \mapsto \mathbb{N}$ a weight function, associating a positive integer weight d_{ij} with every arc (i,j) which can be interpreted as the distance between i and j. The aim is to find a shortest Hamiltonian Cycle, i.e., a cyclic route of minimal length visiting every node exactly once. For symmetric TSPs, the distances between nodes are all independent of the direction, i.e. for every pair of nodes we have $d_{ij} = d_{ji}$. If this condition does not hold, we have the more general case of an asymmetric TSP (ATSP). TSP is a $\mathcal{NP}$-hard optimization problem which has many applications and is extensively studied in the literature [8]. It also frequently serves to test new algorithmic ideas, one more reason to choose it here as an example application.

To solve TSPs, AS uses pheromone trails τ_{ij} associated with each arc (i,j). Initially, each of the m ants is set on some randomly selected city and starts constructing a tour from there. A tour is successively built by choosing the next node probabilistically according to a probability distribution (induced by normalization) proportional to:

$$p_{ij} \sim \tau_{ij}^{\alpha} \cdot \eta_{ij}^{\beta} \quad \text{if } j \text{ not yet visited, else } 0 \qquad (1)$$

where η_{ij} is a local heuristic function which is defined as $\eta_{ij} = 1/d_{ij}$ in AS. The parameters α and β determine the relative influence of the trail strength and the heuristic information. To keep track of the cities already visited, every ant maintains a tabu list, in which the partial tour is stored. The trails are updated after all ants have constructed a complete tour and have calculated the corresponding tour length L_k. Every ant is allowed to lay down a constant quantity Q of pheromone and the trail intensities are updated according to

$$\tau_{ij}^{new} = \rho \cdot \tau_{ij}^{old} + \sum_{k=1}^{m} \Delta\tau_{ij}^{k} \qquad (2)$$

where ρ is the persistence of the trail, thus $(1-\rho)$ simulates the evaporation. The amount $\Delta\tau_{ij}^{k}$ is equal to Q/L_k if arc (i,j) is used by ant k in its tour, otherwise zero. Frequently used arcs and arcs contained in short tours receive a high amount of pheromone. This can be interpreted as a learning of *good* arcs where arcs leading to short tours receive high amounts of trail and may be selected more often. The two basic steps ***tour construction*** according to (1) and ***trail update*** according to (2) are then simply repeated for a given number of iterations (complete cycles of tour construction and trail updates).

The most important part in AS is the treatment of the trail intensities. If only few trail intensities are very high, the ants are very likely to choose the next arc among those with a very high trail intensity. In practice, the long term effect of the trail intensities is to reduce the size of the effective search space by concentrating the search on a relatively small number of arcs. To characterize the amount of exploration AS still performs, the mean λ-branching factor can be used, for details see [7]. If the mean λ-branching factor is very low, practically no more new tours are built, thus leading to a stagnation of the search.

3 $\mathcal{MAX}$–$\mathcal{MIN}$ Ant System

$\mathcal{MM}$AS was inspired by some of our observations while experimenting with AS. With AS, a modified strategy in which only ants with very good tours are allowed to update the trails, showed promising results. Yet with this variation the major problem was premature stagnation of the search, leading to suboptimal tours. We conjectured that one of the most promising possibilities to improve AS should be to exert more influence directly onto the trail strengths. Thus, we introduced explicit maximum and minimum trail strengths on the arcs, hence the name $\mathcal{MAX}$–$\mathcal{MIN}$ Ant System. The maximum and minimum trail limits are chosen in a problem-dependent way. For the maximum trail limit $\tau_{\max}$ we use $\tau_{\max}(t) = n/L_{\min}$, where $L_{\min}$ is the minimal tour length found during the run of the algorithm. The minimum trail limit $\tau_{\min}$ is fixed to $\texttt{const}/\ \emptyset \cdot n^2$, where $\emptyset$ is the average arc length.

As we use as a lower limit $\tau_{\min}$ the probability that a specific arc is chosen may get very small, but still will never be equal to zero. This leads to a higher degree of exploration. $\mathcal{MM}$AS differs also in one more important issue from AS. We only allow the best ant in each iteration to update the trail intensity according to Equation (2). The trail strengths in $\mathcal{MM}$AS are initialized to $\tau_{\max}$ for all arcs. After each iteration the evaporation will reduce the trail strength by a factor ρ and only the trails on arcs of the best tours are allowed to increase their intensities or maintain them at the upper level. Hence, arcs that do not receive any or very rare reinforcement will continuously lower their trail strength and be selected more rarely by the ants.

3.1 Experiments with $\mathcal{MM}$AS

A very important parameter to influence the performance of $\mathcal{MM}$AS is ρ which determines how fast the trail strength on the arcs may decrease. We present the results obtained with different values for ρ and different number of iterations in some more detail in Table 1. If not indicated otherwise the parameter values are chosen as: $\alpha = \beta = 1$, $Q = 1$, and the number of ants m is chosen equal to the number of cities n. The results are presented for the problem instance `ry48p` as it was small enough to run many experiments. Note, that this instance was solved to optimality only recently in [5], the optimal tour length being 14,422. The values for ρ should be rather high, as usually by increasing ρ better average performance is obtained. The particular parameter values for ρ also depend on the number of iterations allowed, as can be seen by comparing the performance for $\rho = 0.99$ with 1,000 and 2,500 iterations. A careful analysis shows that the average of the branching factor at which the best solution is found in a run is rather low. For the runs with $\rho = 0.99$ and 1,000 iterations the mean branching factor is still too high and thus by allowing more iterations (here 2,500) better results can be obtained. By this example we also can see that the branching factor is an essential descriptive parameter to judge the performance of $\mathcal{MM}$AS and to see whether improvements can still be expected. For higher values for ρ, like $\rho = 0.995$, and more iterations, better average results may be obtained at a rather low standard deviation. Yet, the disadvantage are the considerably higher run times. Note that the average solution quality of $\mathcal{MM}$AS with reasonable parameters for ρ is considerably better than the best results obtainable with AS.

Table 1: $\mathcal{MM}$AS on `ry48p.atsp`, 25 runs. `best` is the shortest tour found in 25 runs, `avg.best` is the average tour length and σ the standard deviation, `iterations` is the number of iterations performed.

ρ	best	avg.best	σ	iterations
0.95	14482	14860.76	230.60	1000
0.98	14422	14693.20	168.92	1000
0.99	14689	14938.68	187.67	1000
0.95	14422	14827.60	235.57	2500
0.98	14422	14666.16	171.03	2500
0.99	14495	14609.48	109.58	2500
0.995	14495	14546.60	22.01	25000

Now we shortly indicate our results corresponding to other parameter choices for $\mathcal{MM}$AS. We observed that after adding the maximum and minimum trail limits, it is significanlty better to allow only the best ant to update the trails. Good parameter values for β and α are $\alpha = 1.0$ and $1 \leq \beta \leq 5$. It is also interesting to note that the best value for β depends on the problem instance, whereas $\alpha = 1$ always seemed to work best. If not indicated otherwise, our results detailed below are obtained with $\alpha = 1.0$ and $\beta = 1.0$. We also investigated the influence of the values for $\tau_{\min}$ and $\tau_{\max}$. It turned out that the particular settings for $\tau_{\min}$ have less influence on the performance of $\mathcal{MM}$AS than the values for $\tau_{\max}$.

3.2 Smoothing of the Trails

As we have seen, the best tours are usually found at rather small values for the mean branching factor. Yet, if the branching factor is too low, only few new tours are built, thus leading to a very small exploration of possibly better tours. This led us to the following idea: If $\mathcal{MM}$AS shows stagnation as indicated by the mean λ-branching factor we have to adjust the trail intensities in such a way that the exploration of new tours is increased again. We adjust the trails according to a ***proportional update:*** The trail intensity on arc (i,j) is increased proportionally to the difference between $\tau_{\max} - \tau_{ij}(t)$.

An advantage of the proportional update is that we do not completely forget the trails learned so far. Its overall effect is that by increasing the trail intensities the probability distribution for the selection of the next node is influenced in such a way that the exploration of new tours is higher. We call this approach ***smoothing*** of the trails as the differences between high and low trail intensities become less pronounced, i.e., smoother. A major advantage of smoothing is that it makes $\mathcal{MM}$AS more robust against premature convergence for a broader range of parameter values (especially ρ). In Table 2 we present the best average tour quality obtained with several extensions of AS, including Ant Colony System (ACS) [6], the best improvement so far over AS. It can be noted that $\mathcal{MM}$AS with additional trail smoothing ($\mathcal{MM}$AS+sm) performs best. For larger problem sizes the differences especially to AS become even larger.

Table 2: Experimental results with 960.000 tour construction on `ry48p.atsp`. `quality` is the percentage deviation from the optimal solution. Ant Colony System 15 runs, others 25 runs.

Version	ρ	avg.best (quality)	σ
AS	0.3	14622.24 (1.38%)	126.03
ACS	–	14565.45 (0.99%)	115.23
$\mathcal{MM}$AS	0.995	14571.68 (1.03%)	90.72
$\mathcal{MM}$AS + sm	0.99	14461.40 (0.27%)	39.27

Table 3: Results on contest problems, Ant Colony System, 15 runs.

Problem	best	avg.best	σ
ry48p.atsp (48)	14422	14565.45	115.23
ft70.atsp (70)	38781	39099.05	170.32
kro124p.atsp (100)	36241	36857.00	521.19
ftv170.atsp (170)	2774	2826.47	33.84
eil51.tsp (51)	426	428.06	2.48
kroA100.tsp (100)	21282	21420	141.72
d198.tsp (198)	15888	16054	71.15

In Table 4 we present the results obtained with $\mathcal{MM}$AS for some problems of the First International Contest on Evolutionary Optimization [4]. We only present results for the maximum number of tour constructions. The parameter settings are $\rho = 0.99$, $\alpha, \beta = 1.0$, and $m = n$ except for `ftv170.atsp`, where we chose $m = n/2$. In Table 3 we reproduced the results obtained with ACS in [6]. With exception of problems `ftv170.atsp` and `d198.tsp`, the average performance of the $\mathcal{MM}$AS achieved over 25 independent runs is better and its standard deviations are generally smaller. The results show the very good performance of the $\mathcal{MM}$AS when compared to the best improvement so far over AS.

Table 4: Results on contest problems. 25 independent runs. 20,000 Iterations for ATSPs, 10,000 for symmetric TSPs.

Problem	best (quality)	avg.best	σ
ry48p	14422 (0.0%)	14461.64	39.27
ft70	38690 (0.04%)	38903.44	149.85
kro124	36416 (0.50%)	36594.36	156.03
ftv170	2826 (1.74%)	2836.40	14.91
eil51	426 (0.0%)	427.2	1.13
kroA100	21282 (0.0%)	21352.05	50.30
d198	15960 (1.14%)	16065.95	73.82

Table 5: Results for symmetric TSPs and ATSP for $\mathcal{MM}$AS + local search. 1000 iterations, $m = 10$.

Problem	best	avg.best	σ
eil51	426 (0.0%)	426.3 (0.07%)	0.48
kroA100	21282 (0.0%)	21285.3 (0.015%)	7.60
d198	15807 (0.17%)	15824.80 (0.28%)	10.96
lin318	42762 (1.74%)	42892.5 (2.05%)	110.7
pcb442	51591 (1.60%)	52114.30 (2.63%)	239.1
p43	5620 (0.0%)	5620.60 (0.011%)	0.516
ry48p	14422 (0.0%)	14464.60 (0.30%)	43.75
ft70	38679 (0.016%)	38713.80 (0.11%)	22.45
kro124p	36235 (0.04%)	36387.40 (0.43%)	104.8

4 A local search $\mathcal{MM}$AS

We now investigate the performance of $\mathcal{MM}$AS when an additional local search is performed. The reason for adding local search algorithms to $\mathcal{MM}$AS is twofold. On the one hand, we want to enhance the performance by adding local search yielding a faster convergence of the algorithm and an earlier detection of high quality solutions. On the other hand, $\mathcal{MM}$AS should be able to construct good initial tours for the following local search phase, guiding the local search procedure towards better solutions. For symmetric TSP we implemented the so called 2-opt heuristic. For ATSPs we used *reduced 3-opt*, which is based on modified 3-opt exchanges that allow reinserting arcs without reversing the direction of a partial tour.

In our experiments we used 10 ants for each problem and, except for problems `lin318` and `pcb442`, all ants performed local search. In these experiments local search was performed in each iteration. In Table 5, results for symmetric TSPs with additional 2-opt and for ATSPs with reduced 3-opt are given. For the larger symmetric TSP `pcb442` (with 442 cities) only the best ant in each iteration is allowed to perform 2-opt and to reinforce the trail.

We also verified that it is better to use local search in addition with $\mathcal{MM}$AS than to use it in addition to AS. We also investigated the performance of greedy random tour construction followed by a local search phase, a simple method for the solution of combinatorial optimization problems. For random tour construction with the $\mathcal{MM}$AS one simply has to $\alpha = 0$. For β the best value seems to be 5. Note that for $\beta \to \infty$ we obtain the nearest neighbour tour construction heuristic. For the problem `ry48p.atsp` we obtained as the best result 14,782 after 1,000,000 applications of the reduced 3-opt.

5 Discussion and Conclusion

In this article we presented $\mathcal{MM}$AS and showed that it performs much better than AS and at least at the same level of performance as ACS. By the addition of minimum and maximum trail limits, $\mathcal{MM}$AS offers a more direct control over the trails strength. Our approach of smoothing the trail led to an effective balance between exploration of new tours and exploitation of the learned trail strength. Additionally, we showed that the performance of $\mathcal{MM}$AS can be increased by adding a local search phase and that $\mathcal{MM}$AS proved to be very valuable in guiding the local search heuristic. In future we want to apply also the more sophisticated Lin-Kernighan heuristic to $\mathcal{MM}$AS to hopefully obtain competitive results with the best algorithms for TSPs. Another issue would be to dynamically adjust β during the run. This seems reasonable as depending on the problem instance, different values for β seem to be best, and higher values for β in the beginning of the algorithm lead to a faster convergence to good tours. To date, $\mathcal{MAX}$–$\mathcal{MIN}$ Ant System seems to be a very promising tool providing an adaptive framework for the solution of combinatorial optimization problems.

6 Acknowledgments

We'd like to thank Marco Dorigo for valuable comments and Christoph Herrmann for careful reading of an earlier draft of this paper.

References

[1] A. Colorni, M. Dorigo, and V. Maniezzo. Distributed Optimization by Ant Colonies. In *Proceedings of ECAL91 - European Conference on Artificial Life*, pages 134–142. Elsevier Publishing, 1991.

[2] M. Dorigo. *Optimization, Learning, and Natural Algorithms*. PhD thesis, Politecnico di Milano, 1992.

[3] M. Dorigo, V. Maniezzo, and A. Colorni. The Ant System: Optimization by a Colony of Cooperating Agents. *IEEE Transactions on Systems, Man, and Cybernetics – Part B*, 26(1):29–41, 1996.

[4] H. Bersini, M. Dorigo, L. Gambardella, S. Langerman and L. Seront. Results of the First International Contest on Evolutionary Optimisation. Technical Report TR/IRIDIA/96-18, IRIDIA, Université Libre de Bruxelles, 1996.

[5] M. Fischetti and P. Toth. An Additive Bounding Procedure for the Asymmetric Travelling Salesman Problem. *Mathematical Programming*, 53:173–197, 1992.

[6] L. Gambardella and M. Dorigo. Solving Symmetric and Asymmetric TSPs by Ant Colonies. In *IEEE Conference on Evolutionary Computation (ICEC'96)*. IEEE Press, 1996.

[7] L. M. Gambardella and M. Dorigo. Ant-Q: A Reinforcement Learning Approach to the Traveling Salesman Problem. In *Proceedings of the Twelfth Iternational Conference on Machine Learning*, pages 252–260. Morgan Kaufmann, 1995.

[8] G. Reinelt. *The Traveling Salesman: Computational Solutions for TSP Applications*, volume 840 of *LNCS*. Springer Verlag, 1994.

[9] T. Stützle and H. Hoos. A detailed report on the *MAX-MIN* Ant System. Technical Report AIDA-96-11, FG Intellektik, TH Darmstadt, August 1996.

A Hybrid Genetic Algorithm for the 0-1 Multiple Knapsack Problem

C. Cotta and J. M. Troya
Departamento de Lenguajes y Ciencias de la Computación
Complejo Politécnico (2.2.A.6), Campus de Teatinos, 29071 - Málaga, SPAIN.
Email: {ccottap, troya}@lcc.uma.es

Abstract

A hybrid genetic algorithm based in local search is described. Local optimisation is not explicitly performed but it is embedded in the exploration of a search metaspace. This algorithm is applied to a NP-hard problem. When it is compared with other GA-based approaches and an exact technique (a branch and bound algorithm), this algorithm exhibits a better overall performance in both cases. Then, a coarse-grain parallel version is tested, yielding notably improved results.

1 Introduction

Genetic algorithms have been traditionally considered as robust techniques, easily applicable to almost any domain. This robustness is not only an advantage, but also a major point of weakness. As shown in recent research, (the so-called No Free Lunch Theorem [13]) no algorithm can be expected to outperform any other one (including random search) when averaged over all possible problems. In other words, an algorithm is as good for a problem (or problem domain) as the problem-dependent knowledge it incorporates.

For that reason, hybrid genetic algorithms (i.e., genetic algorithms using specialised non standard mechanisms) have often been defined to deal with problems that would be very hard for a blind black box search algorithm. These hybrid algorithms have been proved to be very effective tools in many situations (see [3]). Problem-dependent knowledge is incorporated by means of appropriate representations and specialised operators. In that sense, the use of operators performing local search is a widely used technique. These operators frequently play an essential role (e.g., in Mühlenbein's parallel genetic algorithm [7]). In this work, we study a genetic algorithm based in local search. Unlike other hybrid approaches, local optimisation is not achieved via an improvement heuristic (like hill climbing) but by means of a construction heuristic. This implies that the search does not take place in the solution space but in a solution metaspace (the problem space).

2 Problem-Space Search

Problem-space exploration was first suggested by Storer *et al.* [11]. In short, it consists of the intelligent generation of different starting points for a construction heuristic. Since a construction heuristic H is a function that returns a feasible solution for a given problem instance, every new starting point defines a new problem instance for which a solution is generated (see Figure 1). All these solutions the heuristic provides are evaluated with respect to original data. Thus, the alternative problem instances are only used to appropriately modify the behaviour of the heuristic.

New problem instances are generated perturbing original data by means of an appropriate problem-specific procedure. This procedure is parameterized by a perturbation vector D that indicates how to modify each problem datum. Assuming that the heuristic H provides good solutions for the class of problems being solved, it is expected that the magnitude $\| D \|$ of the perturbation vector D needed to find the optimum is not very large.

Notice that, strictly speaking, the search is done in an auxiliary space (the perturbation space) that yields a problem space when composed with original data.

3 The 0-1 Multiple Knapsack Problem

The 0-1 multiple knapsack problem (0-1 MKP) is a well-known member of the NP-hard class [4]. It is also referenced as the 0-1 integer programming problem or the 0-1 linear programming problem.

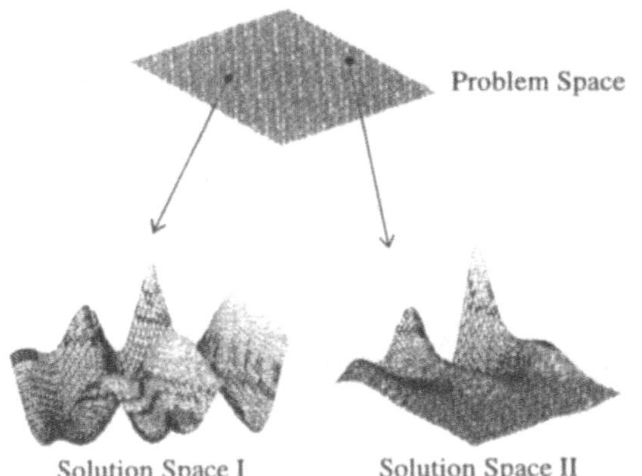

Figure 1: Every point in the problem space defines a different solution space. The goal is to find a fitness landscape that optimally matches the requirements of a given heuristic.

3.1 Definition of the Problem

The 0-1 MKP is a generalisation of the 0-1 simple knapsack problem. In the latter, a set of objects $O = \{o_1, \ldots, o_n\}$ and a knapsack of capacity C are given. Each object o_i has an associated profit p_i and weight w_i. The objective is to find a subset $S \subseteq O$ such that the weight sum over the objects in S does not exceed the knapsack capacity and yields a maximum profit. The 0-1 MKP involves m knapsacks of capacities $c_1, \ldots, c_m$. Every selected object must be placed in all m knapsacks. Moreover, the weight of an object o_i is not fixed, but it has a different value in each knapsack. Again, the goal is to obtain the maximum profit.

3.2 A Construction Heuristic

The greedy approach is a typical construction heuristic for the 0-1 simple Knapsack problem. This algorithm first calculates the profit density $\delta_i = \frac{p_i}{w_i}$ of every object and sorts them by decreasing values of this ratio. Then, objects are successively taken and included in the knapsack if they fit in it. To generalise this heuristic for the 0-1 MKP, the profit density of every object in every knapsack is calculated, and only the lowest value for each object is considered (i.e.,

$$\delta_i = min(\frac{c_j \cdot p_i}{w_{ij}}) = \frac{p_i}{max(w_{ij}/c_j)}, 1 \leq j \leq m).$$

This is better than taking

$$\delta_i = avg(c_j p_i / w_{ij})$$

since the latter could hide extreme values of the weight ratio c_j/w_{ij} for a certain knapsack.

4 The Hybrid Genetic Algorithm

A genetic algorithm for the 0-1 MKP performing problem-space search may be designed using the greedy construction algorithm described above. Each individual in the population represents a perturbation vector defining a new problem instance.

4.1 Perturbing Data

New problem instances may be obtained from a given one modifying the profit p_i of each object. This is an appropriate procedure because all instances generated this way have the same set of feasible solutions, although the evaluation of every solution is different. Thus, each perturbation vector D is a list of n numbers $\{d_1, \ldots d_n\}$ defining a problem instance in which the object o_i has a profit of $p_i + d_i$ and the same original weights. The magnitude of the perturbation is problem- dependent and must be large enough to allow the optimum to be generated. Let x be the amount of profit that has to be added to the object with lowest profit density so it becomes the one with the highest ratio. Taking δ_i from $[-x, +x]$ suffices to ensure that any feasible solution may be generated by the heuristic.

4.2 Experimental Results

Experiments have been carried out with an elitist generational genetic algorithm, using a population size of $\mu = 100$, a crossover rate of $p_c = 0.9$, a mutation rate of $p_m = 0.01$ and proportional selection. The crossover operator is Radcliffe's n-dimensional R^3 operator [9], i.e., it picks at each locus a random value in the range defined by the corresponding parents' alleles. The mutation operator replaces a component of D by a random value.

For comparison purposes with [6] we have chosen a benchmark composed of nine problems taken from the OR-Library by Beasley [1]. Each run of the genetic algorithm involves $2 \cdot 10^4$ evaluations, and a total number of 100 runs are executed for each test problem. Results are shown in Table 1.

These results clearly improve those referenced in [6] (using a binary encoding and a graded penalty

Table 1: Experimental results obtained by the GA.

Problem	n	m	Optimum	Avg.	Opt. found
Knap15	15	10	4015	4015.0	100%
Knap20	20	10	6120	6119.4	94%
Knap28	28	10	12400	12400.0	100%
Knap39	39	5	10618	10609.8	60%
Knap50	50	5	16537	16512.0	46%
Sento1	60	30	7772	7767.9	75%
Sento2	60	30	8722	8716.5	39%
Weing7	105	2	1095445	1095386.0	40%
Weing8	105	2	624319	622048.1	29%

Table 2: B&B results for the generated problems.

$n(=m)$	$p_i \in [0,1]$	$p_i \in [.45,.55]$	$p_i = .5$
	Iterations		
20	67	6925	37373
30	90	37259	295210
40	80	> 250000	-
	Max. queue length		
20	64	2163	7383
30	89	6997	50872
40	77	> 150000	-

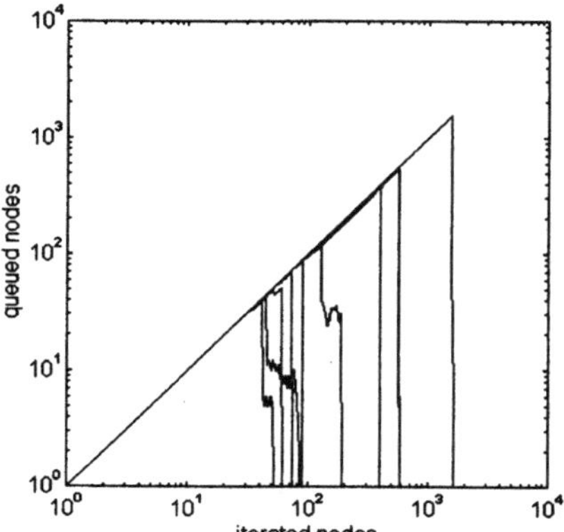

Figure 2: Queued vs. iterated nodes (OR-Library problems).

term in the fitness function). They are much better not only in average but also in the number of times the optimum is found. For example, the larger instances (Sento1, Sento2, Weing7, Weing8) are solved to optimality at least five times more often (with the number of evaluations being a factor of ten lower).

5 The Exact Solution: Branch and Bound

An exact method has been applied to the above test problems to obtain a measure of the effort that it is required to optimally solve them. Since the search space is composed of 2^n points for a problem with n objects (the number of knapsacks does not enlarge it but may make the feasible region be smaller) a simple Branch and Bound (B&B) algorithm (as described in [5]) is unable to solve most of the problems above. Therefore, a more sophisticated version of the algorithm using the linear-programming relaxation [12] has been considered. This B&B algorithm solves the LP-relaxed version of the problem and then branches on the object with highest profit that has a fractional value, generating two subproblems in which that object is forced to be included/excluded respectively.

As shown in Figure 2, about 10^3 subproblems must be solved to obtain the optimal solution for the most difficult problem (Sento2). This is not a very high value, suggesting that the above problems are relatively easy. For that reason, harder problem instances were generated as shown in [5], i.e. randomly picking the weights w_{ij} from the interval $[0, 1]$, setting all knapsack capacities to $\frac{n}{3}$ and using object profits to define three levels of difficulty (from easiest to hardest): $p_i \in [0,1]$, $p_i \in [.45, .55]$, and $p_i = .5$. Results are shown in Table 2.

It can be seen that the B&B algorithm is unpractical for homogeneous ($p_i = .5$) or nearly homogeneous ($p_i \in [.45, .55]$) problems of size around 30-40.

6 A Distributed Version of the Hybrid GA

Finally, a coarse-grain parallel version of the hybrid genetic algorithm following the island model [2] (i.e., k genetic algorithms running in parallel, periodically interchanging some individuals) is tested. This model is used to study the improvement that can be achieved in both the *easy* and *hard* problem instances discussed above.

The experiments carried out so far have been realised with 2 and 4 genetic algorithms. The population size is $\mu = 100/k$ and one individual migrates

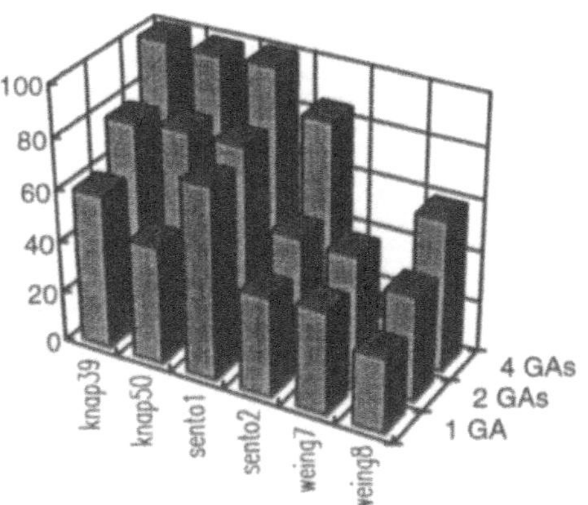

Figure 3: Number of times the optimum was found using a distributed genetic algorithm (results for the OR-Library problems).

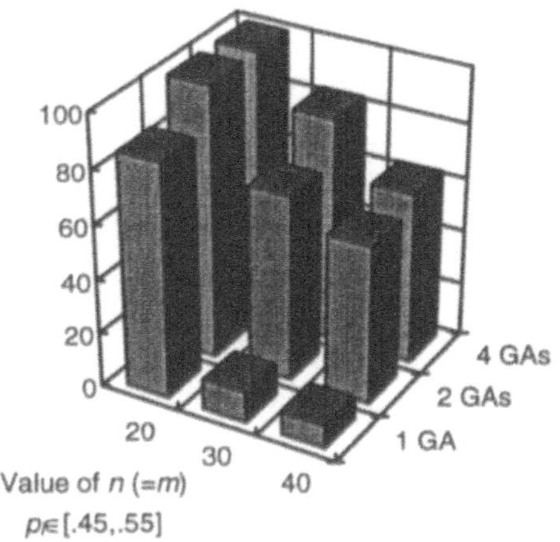

Figure 4: Number of times the optimum was found using a distributed genetic algorithm (results for the randomly generated problem instances).

every 20 generations to a random subpopulation (no fine tuning has been attempted). As shown in Figures 3 and 4, the quality of the results (in terms of how often the optimum is found) is notably improved.

It must be noted that the B&B-*hard* instances were solved to optimality in 100 out of 100 runs both for the easiest ($p_i \in [0,1]$) and the homogeneous case ($p_i = .5$) with a sequential genetic algorithm. Thus, a genetic algorithm seems to be more appropriate than B&B for this kind of problems.

7 Conclusions

A hybrid genetic algorithm using problem-space search has been presented and applied to the 0-1 multiple knapsack problem. This technique is appropriate when a problem-dependent construction heuristic and a perturbing algorithm are available. The resulting hybrid algorithm performs better than other GA-based approaches on problems taken from the literature. Moreover, it consistently solves to optimality randomly-generated problem instances that are very hard for an exact (and specialised) technique. Two important advantages of this method can be stressed: on the one hand, the results are assured to be at least as good as those of the construction heuristic as long as a null perturbation vector is inserted into the initial population. On the other hand, perturbation vectors are usually points in a n-dimensional numerical domain. This allows using techniques oriented to continuous parameter optimisation (e.g., evolution strategies [10]) for the resolution of discrete-domain problems without needing to define artificial encodings. Furthermore, the use of the construction heuristic avoids needing to handle unfeasible solutions.

References

[1] J. E. Beasley. OR-library: Distributing test problems by electronic mail. *Journal of Operational Research Society*, 41(11):1069–1072, 1990.

[2] A. Chipperfield and P. Fleming. *Parallel Genetic Algorithms*, pages 1118–1143. McGraw-Hill Series on Computing Engineering, 1996.

[3] L. Davis, editor. *Handbook of Genetic Algorithms*. Van Nostrand Reinhold Computer Library, NY, 1991.

[4] M. Garey and D. Johnson. *Computers and intractability: A guide to the theory of NP-Completeness*. Freeman and Co., San Francisco, 1979.

[5] E. Horowitz and S. Sahni. *Fundamentals of Computer Algorithms*. Computer Science Press, 1978.

[6] S. Khuri, T. Baeck, and J. Heitkotter. The zero/one multiple knapsack problem and genetic algorithms. In *Proceedings of the ACM Symposium of Applied Computation*. ACM Press, 1993.

[7] H. Muehlenbein. Parallel genetic algorithm, population dynamics and combinatorial optimization. In *Proceedings of the Third International Conference on Genetic Algorithms*, pages 416–421, San Mateo, 1989. Morgan Kaufmann.

[8] C. Peterson and B. Södeberg. Artificial neural networks. In *Modern heuristic techniques for combinatorial problems.* C. R. Reeves editor, Advanced Topics in Computer Science, pages 197–242, 1993, Oxford Scientific Publications.

[9] N. J. Radcliffe. Forma analysis and random respectful recombination. In *Proceedings of the Fourth International Conference on Genetic Algorithms*, pages 222–229, San Mateo CA, 1991. Morgan Kaufmann.

[10] I. Rechember. *Evolutionsstrategie: Optimierung technischer Systeme nach Prinzipien der biologischen Evolution.* Frommann-Holzboog Verlag, Stuttgart, 1973.

[11] R. H. Storer, S. D. Wu, and R. Vaccari. New search spaces for sequencing problems with application to job shop scheduling. *Management Science*, 38:1495–1509, 1992.

[12] W. L. Winston. *Operations Research. Applications and algorithms.* Duxbury Press, Belmont CA, 1993.

[13] D. H. Wolpert and W. G. Macready. No free lunch theorems for search. Technical Report SFI-TR-95-02-010, Sante Fe Institute, 1995.

Genetic Algorithms in the Elevator Allocation Problem

J. T. Alander[1], J. Herajärvi[1], G. Moghadampour[1], T. Tyni[2] and J. Ylinen[2]
[1] Department of Information Technology and Production Economics
University of Vaasa, PO Box 700, FIN-65101 Vaasa, Finland
[2] KONE Elevators Research Center, PO Box 677, FIN-05801, Hyvinkää, Finland
Email: Jarmo.Alander@uwasa.fi, hattty@hatmail.msgw.kone.com

Abstract

The purpose of the work was to test the feasibility of genetic algorithms (GAs) in the landing call allocation problem of an elevator group.

In the first test case the results given by GAs were compared with two current control programs by using an elevator simulator program. In the second test case there were three different buildings which were tested by three different realisations of the same type of passenger traffic flows generated by an elevator simulation program. The results obtained are given as averages of these runs. In the second test case the GA controller was compared with the controller that proved to be better in the first test case.

According to the results, it seems that GAs would be suitable for the elevator allocation problem or to solve similar demanding real-time optimization problems. In the simulation tests performed, the average waiting time decreased achieved by GA-based controller was (evaluated with 99% confidence interval) at most 15-33%, when traffic intensity was 140% and at least 1-13%, when traffic intensity was 40%. Accordingly, when traffic intensity was nominal (100%) average waiting time was decreased by GA-controller at most 15-24% and at least 4-12%.

1 Introduction

The problem with elevator allocation is to determine which elevator serves which landing call in a way that the cost factor indicating the behaviour of elevator group will be minimised. This is a difficult optimisation problem. The cost factor indicating the behaviour of an elevator group and subject of optimisation can be, for example, the average waiting or journey time of the passengers (Figure 1), the number of starts of the elevator, etc. [16]. In this work, the average landing call time was the chosen cost factor.

In traditional control methods, when selecting a suitable elevator for a call, decisions are made case by case with complicated conditional if-then-else statements. This easily results in a situation where

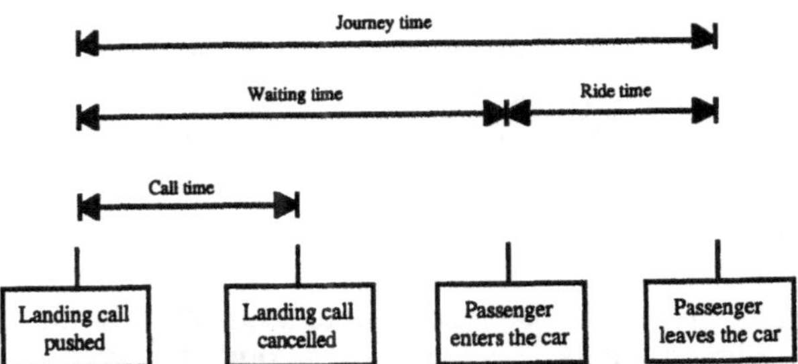

Figure 1: Passenger service definitions related to elevator group.

not all the factors affecting are taken into consideration. As a consequence the control does not operate in the best possible way.

Because the number of alternatives to choose an elevator increases exponentially as the amount of landing calls increases, it is not possible to go systematically through all of the alternatives. This has restricted the use of the traditional enumerative algorithms to small elevator groups without any practical significance. GAs were expected to be appropriate for the allocation problem, because, in general they tend to find good solutions without going through the whole search space. In a real elevator system, the time to obtain the solution to the landing call allocation problem is restricted to one second.

1.1 Related Work

Control has been a popular application area of genetic algorithms with about 500 papers [4], but according to our quite complete GA bibliography [1] there does not seem to be much related work applying GAs to elevator control. The only paper is [10], while [8] applies GA in elevator design. In our previous work we have applied genetic algorithms to optimise elevator control parameters [5, 6]. Applying fuzzy logic and neural networks to the allocation problem has been popular, especially in Japan.

2 Simulation Environment

Because it difficult to gather data of each individual passenger in a real elevator installation, the evaluation of elevator group performance is usually done by a simulator, which simulates the test building itself, elevators, elevator movement and passenger behaviour [12, 16].

The distributed computing environment of the optimization system consisted of several PCs connected by a local area network. One computer acted as a fileserver, one ran the GA-based elevator controller, the third one ran the elevator simulation program.

3 Results

The first test case was a simulated office building designed to handle traffic in a building with 400 persons, for which the calculated average traffic intensity is 61 persons per 5 minutes. This is called nominal '100% intensity'. The size of the elevator group was three cars with a capacity of 8 passengers each. The traffic type was purely outgoing because that traffic type gives most degrees of freedom for an elevator group controller to make decisions. Two reference controllers were used: one for mid rise and one for high rise buildings. Simulation time was one hour for each traffic intensity. This benchmark case was also used in our earlier study [5, 6].

As can be seen in Figure 2, the GA-based controller gave shorter average waiting times than the traditional controller. However, the variance was sometimes, especially at high traffic intensities, slightly higher with the GA control unit (Figure 3). In figures below, dotted lines indicate upper and lower limits and black one the ratio.

The second test case consisted of simulation of three different building types. The first was an office building with 16 floors, three elevators and one entrance floor. This building was called 'typical'. The second was a similar high office building with two elevators. It was called 'duplex'. The third one was a small office building with 10 floors and 3 elevators. The building was called 'small'.

All the buildings were tested by three different traffic realisations of the same type with the better traditional and the GA-based controller. The results are given as averages of the runs (Figures 4-6). The traffic type was outgoing traffic in this case also, and the simulation time one hour for each traffic intensity. In this case, the average waiting time of the GA-based controller was up to 15-33% shorter at very high traffic intensities. There was also less variance with the GA-based controller.

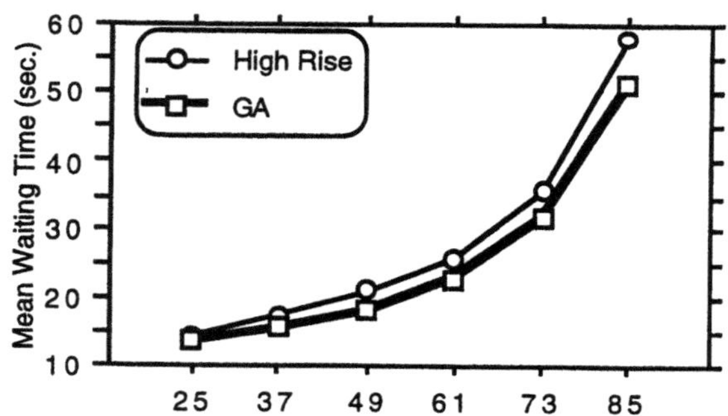

Figure 2: Cell mean of average waiting time at traffic intensities 25-85 passengers/5 min for the GA-based and the traditional (high rise) controllers.

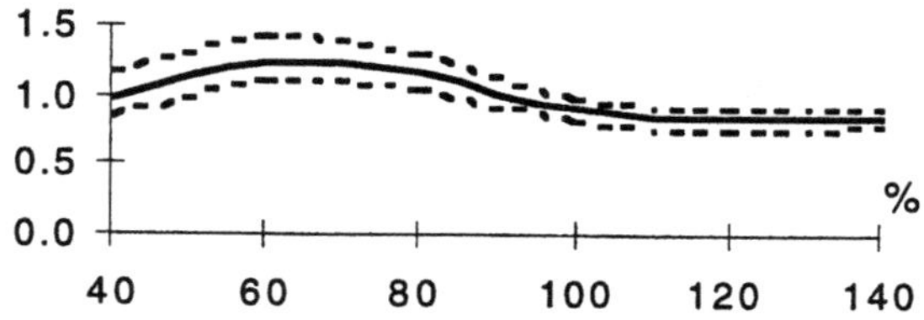

Figure 3: Comparison of waiting time variances evaluated with 99% confidence interval.

4 Conclusion and Future Research

According to the results, it seems that GAs would be suitable for the landing call allocation problem of elevators or to solve similar demanding real-time combinatorial optimisation tasks. In the simulation tests performed the GA-based controller gave better results, measured by almost every indicator applied, than the currently used allocation methods, e.g., the average waiting time obtained by the GA-based controller was at most 15-33% (evaluated with 99% confidence interval) shorter than with the traditional control methods. Because the work is continuing there is room for improvements with the GA-based controller, and hence, we expect results to be even better in the future.

When compared with the more traditional allocation methods, the GA-based approach is more flexi-

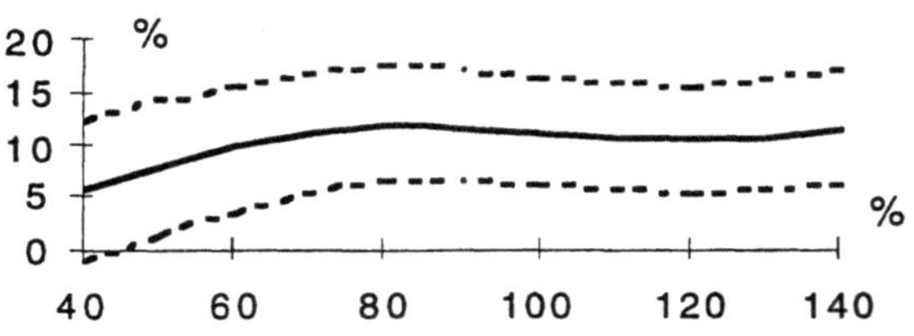

Figure 4: Comparison of average waiting time, evaluated with 99% confidence interval, in 'typical' building (16 floors and 3 elevators).

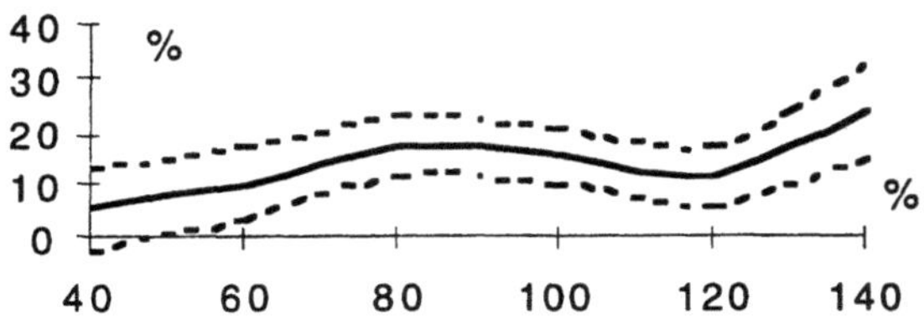

Figure 5: Comparison of average waiting time, evaluated with 99% confidence interval, in 'duplex' building (16 floors and 2 elevators).

ble and its behaviour is easier to tailor by designing a proper fitness function.

In the future, the control programs will be tested using different traffic types and fitness functions. Then, the GA-based controller will be developed by including relevant cost factors in it, for example, the energy consumption, journey time, etc. Moreover, those traffic types and patterns in which the control manages worst will provide a whole new area of research, in which GAs could be used to find program quality problems. More information about the statistical comparisons can be found in [13].

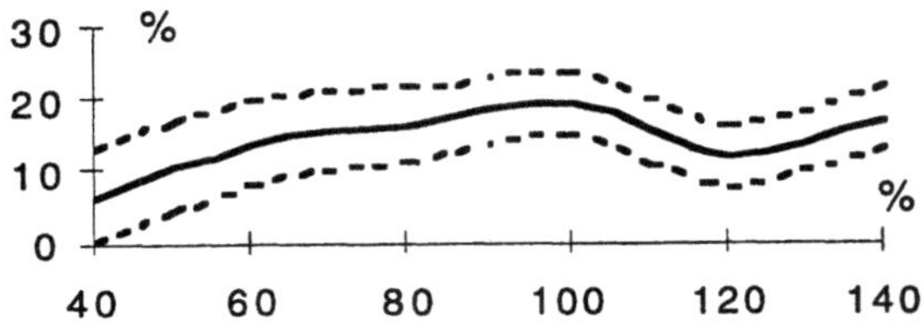

Figure 6: Comparison of average waiting time, evaluated with 99% confidence interval, in 'small' building (10 floors and 3 elevators).

5 Acknowledgements

The work was supported by the Finnish Technology Development Center (TEKES) and KONE Elevators. The authors also want to acknowledge Minna Moisio's and Elizabeth Heap-Talvela's kind help with proof-reading of the manuscript of this paper.

References

[1] J. T. Alander. *An indexed bibliography of genetic algorithms: Years 1957–1993*. Art of CAD Ltd., Vaasa, Finland, 1994.

[2] J. T. Alander, editor. *Proceedings of the Second Finnish Workshop on Genetic Algorithms and their Applications*, University of Vaasa, Vaasa, Finland, 16-18 March 1994.

[3] J. T. Alander. Indexed bibliography of genetic algorithms basics, reviews and tutorials. Technical Report 94-1 BASICS, Department of Information Technology and Production Economics, University of Vaasa, 1995.

[4] J. T. Alander. Indexed bibliography of genetic algorithms in control. Technical Report 94-1 CONTROL, Department of Information Technology and Production Economics, University of Vaasa, 1995.

[5] J. T. Alander, T. Tyni, and J. Ylinen. *Optimizing Elevator Group Control Parameters using Distributed Genetic Algorithms*, pages 105–113. Proceedings of the Second Finnish Workshop on Genetic Algorithms and their Applications. 1994.

[6] J. T. Alander, J. Ylinen, and T. Tyni. Elevator group control using distributed genetic algorithms. In Pearson et al. [14], pages 400–403.

[7] G. C. Barney and S. M. dos Santos. *Elevator Traffic Analysis Design and Control.* Peter Peregrinus Ltd., London, UK, 2nd edition, 1985.

[8] S. E. Carlson, R. Shonkwiler, and M. E. Ingrim. Comparison of three non-derivative optimization methods with a genetic algorithm for component selection. *J. Eng. Des. (UK)*, 5(4):367–378, 1994.

[9] D. E. Goldberg. *Genetic Algorithms in Search, Optimization and Learning.* Addison Wesley, CA, 1989.

[10] R. R. Gudwin and F. A. C. Gomide. Genetic algorithms and discrete event systems: an application. In IEEE [11], pages 742–745.

[11] IEEE, editor. *Proceedings of the First IEEE Conference on Evolutionary Computation*, volume 2, Orlando, Florida, June 1994. IEEE Press.

[12] M. Kaakinen and N-R. Roschier. Integrated elevator planning system. *Elevator World*, pages 73–76, March 1991.

[13] G. Moghadampour. Statistical comparison of elevator controlling systems. Master's thesis, Department of Information Technology and Production Economics, University of Vaasa, Vaasa, Finland, 1996. (in Finnish).

[14] D. W. Pearson, N. C. Steele, and R. F. Albrecht, editors. *Artificial Neural Nets and Genetic Algorithms*, Wien, 1995. Springer-Verlag.

[15] M.-L. Siikonen. *Computer Modelling of elevator traffic and control.* PhD thesis, Helsinki University of Technology, 1989 (in Finnish).

[16] M.-L. Siikonen. Elevator traffic simulation. *Simulation*, pages 257–267, October 1993.

Generational and Steady-State Genetic Algorithms for Generator Maintenance Scheduling Problems

K. P. Dahal and J. R. McDonald
Centre for Electrical Power Engineering,
University of Strathclyde, UK

Abstract

The aim of generator maintenance scheduling (GMS) in an electric power system is to allocate a proper maintenance timetable for generators while maintaining a high system reliability, reducing total production cost, extending generator life time etc. In order to solve this complex problem a genetic algorithm technique is proposed here. The paper discusses the implementation of GAs to GMS problems with two approaches: generational and steady state. The results of applying these GAs to a test GMS problem based on a practical power system scenario are presented and analysed. The effect of different GA parameters is also studied.

1 Problem Description

Generator maintenance scheduling (GMS) is an essential part of the problem of economic operation and control of power systems, and involves finding the optimum timing of the outages of generating units. This is important primarily because other planning activities are directly affected by such decisions. In modern power systems the demand for electricity has increased with related expansions in system size, which has resulted in higher numbers of generators and lower reserve margins making the GMS problem more complicated. A good maintenance schedule increases system operating reliability, reduces generation cost, extends equipment life time and relaxes new installation pressure.

The GMS problem is a complex combinatorial constrained optimisation problem. There are generally two categories of objective functions in GMS based on reliability or economic cost criteria. The problems have the following general constraints to be satisfied.

• Maintenance window — defines the limitation on the earliest and latest times and the duration of maintenance for a unit.
• Sequence constraints — the maintenance of certain units is allowed only after the maintenance of other specified units.
• Non-simultaneous constraints — the simultaneous maintenance of certain units is not allowed.
• Crew and resource constraints — consider the availability of manpower and resources.
• Load constraints — consider the demand and the reliability of power supply.

Mathematically, GMS problems can be formulated as integer programming problems using binary variables associated with answers to 'When does maintenance start?'. The use of these variables instead of the variables associated with answers to 'When does maintenance occur?' reduces the number of variables [1]. The first formulation satisfy the constraints on the periods and duration of maintenance. The answer to the first question automatically provides the answer to the second.

Several deterministic mathematical methods and heuristic techniques are reported in the literature for solving these problems [1, 3, 5]. General solution methods are based on integer programming, branch and bound techniques, dynamic programming, etc. However, such approaches are severely limited by the 'curse of dimensionality' and are poor in handling the non-linear objective and constraint functions that characterise the GMS problem. The heuristic approach uses a trial-and- error method to evaluate the maintenance objective function in the time interval under examination. This requires significant operator input and in some situations it fails to produce even feasible solutions [3].

Genetic algorithms (GAs) provide a new approach to the solution of complex combinatorial optimisation problems. This paper describes the procedure for implementing GAs for solving the GMS problem. Two GA approaches, namely generational and steady state, have been applied to test GMS problems which include features of real systems. The paper discusses the application of GAs to GMS problems using a reliability criteria based on levelling reserve generation [3]. This is achieved by minimising the sum of squares of the reserves over the entire operational planning period.

2 GA Approach

Two basic approaches, known as the generational approach and the steady state approach, may be implemented in the realisation of a genetic algorithm (GA). In each iteration step, called a 'generation', a generational genetic algorithm (GN GA) replaces the population of the previous generation by off-spring which are reproduced by applying genetic manipulation to parents selected from the population of the previous generation according to some selection procedure. The iteration is continued until a termination criteria has been reached. A widely available GA package GENESIS [2] has been used to carry out the numerical tests for GMS problems using this GN GA approach.

A steady state genetic algorithm (SS GA) selects two individuals from the population pool in each iteration step according to some selection procedure. A new off-spring is created by applying genetic manipulation to the selected individuals, and is inserted into the population pool replacing a less fit individual. Hence, the parents and off-spring can co-exist in the population pool for the next iteration step. A SS GA software package GENITOR [4] which uses this steady state structure has been applied to GMS problems.

The ranking selection method, where parents are selected according to their ranked fitness score has been used for both algorithms.

In order to tackle GMS problems using a GA, a candidate solution of a GMS problem is encoded as a one dimensional binary array as follows,

$$\begin{aligned}[X_{1,e_1}, X_{1,(e_1+1)}, \dots, X_{1,(l_1-d_1+1)},\\ X_{2,e_2}, X_{2,(e_2+1)}, \dots, X_{2,(l_2-d_2+1)}, \dots,\\ X_{N,e_N}, X_{N,(e_N+1)}, \dots, X_{N,(l_N-d_N+1)}]\end{aligned}$$

where

$$X_{it} = \begin{cases} 1 & \text{if unit } i \text{ starts maintenance in period } t, \\ 0 & \text{otherwise,} \end{cases}$$

e_i = earliest period for maintenance of unit i to begin,

l_i = latest period for maintenance unit i to end,

d_i = duration of maintenance for unit i,

N = total number of generating units.

This binary string (chromosome) consists of sub-strings which each contain the variables over the whole scheduling period for a particular unit. The size of the GA search space for this type of representation is

$$2^{\sum_{i=1}^{N}(l_i-d_i-e_i+2)}.$$

To take into account the various constraints of GMS problems, we have taken a penalty function approach. The penalty value for each constraint violation increases linearly with the amount by which the constraint is violated. The evaluation function is a weighted sum of penalty values for each constraint violation and the objective function itself, hence

$$\text{evaluation} = \sum_c w_c V_c + w_o F,$$

where w_c and w_o are the weighting coefficients, V_c is the violation of constraint c and F is the objective value. The coefficients are chosen in such a way that the violation of harder constraints gives a greater penalty value than for the soft constraints. In general the penalty value for the constraint violations dominates over the objective function. Feasible solutions with low objective values have high fitness values while unfeasible solutions with high objective values take low fitness measures.

In the test problem described below the crew constraint was assigned a low penalty coefficient. This is because a solution with a high reliability but requiring more manpower may well be accepted for a power utility as the unavailable manpower may be hired.

3 Test Results

A number of small problems have been tested with the proposed GAs with different objectives and constraints. GAs with both generational and steady state approaches yield the optimum solution for small problems when appropriate GA parameters are chosen. Here we present the results of applying a steady state GA to a larger test problem comprising 21 units over a planning period of 52 weeks, which was loosely derived from the example presented in [5] with some simplifications and additional constraints. The data for the test problem is given in Table 1.

Table 1: Data for the test system.

Unit	Capacity (MW)	Allowed period	Outage (weeks)	Manpower required for each week
1	555	1-26	7	10+10+5+5+5+5+3
2	555	27-52	5	10+10+10+5+5
3	180	1-26	2	15+15
4	180	1-26	1	20
5	640	27-52	5	10+10+10+10+10
6	640	1-26	3	15+15+15
7	640	1-26	3	15+15+15
8	555	27-52	6	10+10+10+5+5+5
9	276	1-26	10	3+2+2+2+2+2+2+ 2+2+2+3
10	140	1-26	4	10+10+5+5
11	90	1-26	1	20
12	76	27-52	3	10+15+15
13	76	1-26	2	15+15
14	94	1-26	4	10+10+10+10
15	39	1-26	2	15+15
16	188	1-26	2	15+15
17	58	27-52	1	20
18	48	27-52	2	15+15
19	137	27-52	1	15
20	469	27-52	4	10+10+10+10
21	52	1-26	3	10+10+10

The objective is to schedule the maintenance outages of generators to minimise the sum of the squares of the reserve generation. Each unit must be maintained exactly once and the maintenance for each unit must occupy the required time duration without interruption. The system's peak load is 4739 MW. There are only 20 people available for the maintenance work each week. Due to its complexity the optimum solution for this problem is unknown.

Table 2 presents the results of a number of runs of the GN GA and SS GA taking different values of the GA parameters. The total number of trials for each run was fixed at 100000. The crossover operator used for both GAs is a simple two-point crossover. In GENITOR crossover is applied in each iteration of the SS GA when the exchanged information is unique to each parent. In the GN GA the crossover probability was similarly set to be 1.

The first part of Table 2 demonstrates the test results obtained with varying values of the mutation probability (MP), while taking other GA parameters as constant. The selection bias (SB) and population size (PS) are taken as 2.0 and 50 respectively. Each case presents the outcome of 5 GA runs, using a different random seed.

The results show that the GN GA is more sensitive than the SS GA to variations in MP. For both GAs, lower values of MP are recommended to achieve a better solution.

For each unit $i = 1, 2, \dots, N$, the maintenance window constraint forces exactly one variable in $\{X_{it} : e_i \leq t \leq l_i - d_i + 1\}$ to be one and the rest to be zero. Therefore, a maintenance window feasible genetic structure contains many more '0' bits than '1' bits. For our test problem, only 21 out of 496 bits in the string must be '1' and the rest '0'. Hence the most of the search space represents unfeasible solutions. A high mutation probability increases the chance of changing these '0's into '1's and has the potential to disrupt and degrade the search process. With higher mutation probabilities the GA could not find a maintenance window feasible solution even in 100000 trials. However, with lower mutation probabilities the GA found maintenance feasible solutions.

The selection bias value specifies the amount of preference to be given to the superior individuals in selection of parents. If SB=2, for example, then the selection probability for the best individual is twice that of the mean individual. When SB is close to 1, the distribution of selection probability becomes nearly uniform. In general, if the selection bias is too high, then a superior solution strongly dominates the less fit solutions and this may lead the GA to converge prematurely to a local minimum. Low values of the selection bias cause less preference to be given to the good genetic structures previously found. The second part of Table 2 presents results

Table 2: Effect of GA parameters for 100000 trials.

	Value	GN GA min	GN GA avg	GN GA max	SS GA min	SS GA avg	SS GA max
MP SB=2 PS=50	.001	278	488	885	2987	4441	8051
	.005	342	616	971	679	1495	2105
	.01	7870	1.2e5	3.0e6	1285	1851	2701
	.05	7.1e6	7.6e6	8.1e6	3.5e6	4.6e6	5.2e6
SB PS=50	1.01	2.0e5	4.5e5	7.0e5	818	878	950
	1.25	217	305	378	914	1627	2758
	1.5	227	585	807	703	1056	1925
	2.0	278	488	885	679	1495	2105
PS	25	229	268	339	433	1162	2388
	50	278	488	885	818	878	950
	100	355	642	1132	363	1107	2954
	200	253	349	543	412	1041	2011
	500	6.0e5	6.6e6	7.0e5	166	408	918
With seeding		163	172	184	163	183	203
Best solution		163.62			163.62		
CPU time		1m1.59s			1m53.28s		

found using 4 different bias values for each of the two GAs, with MP chosen to give the best performance from above. Taking selection bias 1.01 for the SS GA and 1.25 for the GN GA gives the best solution. Thus the SS GA gives good results when all individuals in the pool have virtually uniform probability of selection. For the GN GA selection pressure towards fitter individuals leads to better performance.

Table 2 also presents the outcomes of 5 GA runs for different population sizes with other GA parameters chosen to give the best performance from above. For a fixed number of trials, the number of generations in the GN GA decreases as the population size increases. Hence the GN GA performance is poor for the large population size. The SS GA performs better with larger population size.

In order to enhance the performance of the GAs, one of the individuals in the initial population was created meaningfully and the remainder chosen randomly as before. The 'seeded' solution was developed heuristically by ranking the generating units in order of decreasing capacity to level the reserve generation while considering the maintenance window constraints. The results of 5 runs of both GAs are shown in Table 2. The seeding significantly improves the performance of both GAs. During the early iterations, the GAs with random initial population spend most of their time on finding maintenance window feasible solutions. Therefore, the inclusion of a maintenance window feasible solution incorporating domain knowledge in the initial population leads to the improvement of the GA performance. In the real system problem, some previously used solutions may be available for initialisation.

In the GN GA the reproduction of individuals within a generation is independent of the offspring produced in that generation as parents and children do not co-exist in a genetic pool. The influence of new offspring in the reproduction procedure can only occur in the next generation. For each generation a number of individuals equal to the population size are selected for genetic manipulation which helps to preserve the diversity of population in the genetic pool. This reduces the chance of premature convergence.

In the steady state GA, there is no concept of 'generation' and the produced offspring enter to the genetic pool before the next trial. Hence, the influence of the offspring in the reproduction procedure is immediate. As a fitter solution generally replaces a less fit solution in the pool at every trial there is a chance of filling the genetic pool with individuals converging towards the top individual of the pool during the course of GA run. The crossover operator acts to improve the solutions in the initial trials. However, in the later trials the improvements to the offspring are expected to be due to the mutation operator only, as the crossover operator does not change any information between identical parents. Therefore, the improvement of candidate solutions in the pool is faster for initial trials.

The CPU times for both GAs on a DEC Ultrix 5000/260 workstation are shown in Table 2. With GA parameters chosen to give the best performance for each GA, the GN GA takes about a half of the computation time than that of the SS GA to obtain the same result.

4 Conclusions

Two GAs with generational and steady state design were tested for a GMS problem. The effects of varying the mutation probability, selection bias and population size were studied. The test results show that both GAs are sensitive to variation in these parameters and appropriate values must be chosen in order to obtain good solutions. In both cases a low value of mutation probability must be chosen. The GN GA gives better results with a small population. For the SS GA a large population size and virtually uniform selection give the best results. The effect of seeding a heuristically derived individual in the initial pool was investigated for both GAs. Seeding greatly enhances the performance of both GAs in finding better solutions. The obtained results show that the GN GA gives better performance than the SS GA in terms of the speed and the average quality of the solution.

Binary values were used for representing the maintenance start period during the encoding of the GMS problem. However, the problem variables are numeric so representing them directly as a integers rather than bit strings can reduce the size of the GA search space greatly. The use of problem specific knowledge in the formulation of the evaluation function and in the design of the GA operator could reduce the computational time of the GA. Furthermore, domain knowledge may be used to prevent obviously unfit chromosomes, or those which would violate problem constraints from being created within the GA. Further research is in progress to investigate all these and other issues.

References

[1] J. F. Dopazo and H. M. Merrill. Optimal generator maintenance scheduling using integer programming. *IEEE Transactions on Power Apparatus and Systems*, PAS-94(5):1537–1545, September/October 1975.

[2] J. J. Grefenstette. A user's guide to GENESIS version 5.0. available at ftp site: ftp.aic.nrl.navy.mil /pub/galist/src/ga/genesis.tar.z, 1990.

[3] X. Wang and J. R. McDonald. *Modern Power System Planning*. McGraw-Hill, London, 1994.

[4] D. L. Whitley. *GENITOR*. Colorado State University, 1990. available at ftp site: ftp.cs.colostate.edu/pub/GENITOR.tar.

[5] Z. Yamayee and S. Kathleen. A computationally efficient optimal maintenance scheduling method. *IEEE Transactions on Power Apparatus and Systems*, PAS-102(2):330–338, February 1983.

Four Methods for Maintenance Scheduling

E. K. Burke[1], J. A. Clarke[2] and A. J. Smith[1]
[1] University of Nottingham, UK
[2] University of York, UK
Email: jac@minster.york.ac.uk, {ekb,ajs}@cs.nott.ac.uk

Abstract

We had a problem to be solved: the thermal generator maintenance scheduling problem [13]. We wanted to look at stochastic methods and this paper will present three methods and discuss the pros and cons of each. We will also present evidence that strongly suggests that for this problem, tabu search was the most effective and efficient technique.

The problem is concerned with scheduling essential maintenance over a fixed length repeated planning horizon for a number of thermal generator units while minimising the maintenance costs and providing enough capacity to meet the anticipated demand.

Traditional optimisation based techniques such as integer programming [2], dynamic programming [14, 15] and branch and bound [3] have been proposed to solve this problem. For small problems these methods give an exact optimal solution. However, as the size of the problem increases, the size of the solution space increases exponentially and hence also the running time of these algorithms.

To overcome this difficulty, modern techniques such as simulated annealing [6, 12], stochastic evolution [8], genetic algorithms [4] and tabu search [7] have been proposed as an alternative where the problem size precludes traditional techniques.

The method explored in this paper is tabu search and a comparison is made with simulated annealing (the application of simulated annealing to this problem is given in [9]), genetic algorithms and a hybrid algorithm composed with elements of tabu search and simulated annealing.

1 Problem Description

Consider I generating units producing output over a planning horizon of J periods. Each unit $1 \leq i \leq I$ must be maintained for M_i contiguous periods during the horizon. However the starting period denoted by x_i for each unit i is unconstrained even in the case that $x_i = J$ and $M_i > 1$ for some i. Since we are considering a rolling plan, the maintenance period would wrap around to the start of our planning horizon.

The operating capacity of each unit is denoted by C_i. Under no circumstances is it possible for a unit to exceed this limit.

In order to avoid random factors in the problem, such as unit random outages, a reserve capacity variable proportional to the demand is incorporated into the problem description. This problem is classified as a deterministic cost-minimisation problem and can be solved using an optimisation-based technique.

Therefore in period j where $1 \leq j \leq J$, the anticipated demand for the system as a whole will be denoted by D_j and the reserve capacity required by R_j. Fuel costs can also be estimated for each period as a constant, f_j per unit output.

Finally, let p_{ij} represent the generator output of unit-i at period-j, $c_i(j)$ be the maintenance cost of unit-i if committed at period-j and let y_{ij} be a state variable, equal to one if unit-i is being maintained in period-j and otherwise zero.

The objective of the problem is to minimise the sum of the overall fuel cost and the overall cost of maintenance:

$$\text{Minimise} \sum_{i=j}^{J}(f_j \cdot \sum_{i=1}^{I} p_{ij}) + \sum_{i=1}^{I} c_i(x_i)$$

Once the maintenance of unit-i starts, the unit must be in the maintenance state for M_i contiguous periods.

$$y_{ij} = \begin{cases} 0 \text{ if } j = 1, 2, \ldots, x_i - 1 \\ 1 \text{ if } j = x_i, \ldots, x_i + M_i - 1 \\ 0 \text{ if } j = x_i + M_i, \ldots, J \end{cases}$$

If $x_i + M_i > J$ then the maintenance wraps around to the next repetition of the planning horizon. This formulation captures the notion of continual maintenance. It would be possible to arrange matters so that overlap never happens; the difference is minor.

The generator output must not exceed the upper limit; the output of the generator is set to zero during maintenance.

$$0 \leq p_{ij} \leq C_i(1 - y_{ij})$$

The total output must equal the demand in each period,

$$\sum_{i=1}^{I} p_{ij} = D_j \text{ where } j = 1, 2, \ldots, J$$

and the total capacity must not be less than the required reserve.

$$\sum_{i=1}^{I} (1 - y_{ij})C_i \geq (D_j + R_j)$$

Our formulation of the problem is based closely on that of [9].

To simplify the operation of the algorithm, all solutions in the solution space are considered valid. A solution could be infeasible if the demand and reserve constraints cannot be met. In this case the solution is penalised by the addition of a penalty function:

$$\alpha \sum_{j=1}^{J} u_j + \beta \sum_{j=1}^{J} v_j$$

where α and β are tunable parameters and u_j and v_j are derived from the shortfall in output:

$$(\sum_{i=1}^{I} p_{ij}) + u_j = D_j$$

and the shortfall in capacity.

$$(\sum_{i=1}^{I} C_i(1 - y_{ij})) + v_j = D_j + R_j$$

u_j and v_j are not permitted to be negative, thus a feasible solution incurs no penalty function.

Thus any initial solution can be chosen and the optimisation algorithm will be directed towards feasible solutions through the choice of sufficiently high α and β.

2 Problem Solution and Discussion

Simulated annealing [6], genetic algorithms [5], tabu search [7] and a hybrid algorithm composed from elements of simulated annealing and tabu search were implemented to solve this problem.

The problem as described in the previous sections can be reduced to that of finding the optimal values of x_i.

2.1 Implementation of the Methods

Simulated Annealing

Each iteration consists of generating n states in the neighbourhood of the current state. A solution is accepted with probability $\exp^{-\Delta/T_k}$ where T_k is the current "temperature" and if accepted, it replaces the current state from which the new states are derived during the current iteration and so the current solution can change several times during one iteration. T_k is initialised to some value T_0 and varies according to a cooling schedule $T_{k+1} = pT_k$, $0 \leq p \leq 1$.

Once n states have been generated, if the number of solutions which were *accepted* during the last iteration is less than some value then the algorithm terminates.

Genetic Algorithms

The encoding used is a binary representation of each unit's starting period concatenated together. This gives a suitable representation for use with a genetic algorithm.

After generating a starting population consisting of individuals generated at random, for a predetermined number of iterations the genetic algorithm performs roulette-wheel selection to choose a pair of individuals. Single-point crossover is then applied, followed by the application of a mutation operator.

Selection favours the better individuals by ensuring that each individual's probability of being selected is proportional to the fitness of that individual.

The probability of crossover taking place is predetermined and if crossover does occur, single-point crossover is applied from a random position in the individual.

Finally, each bit in a new individual is flipped, again with a predetermined probability.

Tabu Search

During each iteration the solutions in the neighbourhood of the current solution are considered as candidates for the next solution. Two different move functions are considered for use by the algorithm.

The first move function returns solutions where for each generating unit the start period x_i is moved to each possible starting period. In each case, only one start period is moved. $(J-1)*I$ solutions are considered.

The second move function is similar to the first, except for each unit only the periods adjoining the current start period are considered. $2*I$ solutions are considered.

For each function, only one generating unit is selected and its maintenance start period is changed. Thus it is possible to implement the cost function calculation by means of computing the change caused by the move rather than by evaluating the whole cost function.

Note that these move functions are deterministic, in that the neighbourhood is searched in a deterministic way. Non-deterministic (random) move functions were used in an initial implementation of the simulated annealing algorithm but it was found that the use of a deterministic search improved the algorithm's effectiveness.

Two different methods of tabu classification were implemented. The first method compared state of the trial solution to that of the last N solutions, thus the algorithm cannot revisit a previous solution for N iterations. This is referred to as "solution tabu".

The second method compared the move that took the current solution to the trial solution with those made in the last N iterations, thus the algorithm cannot perform a move more frequently than once in N iterations. N is called the tabu list size. This is referred to as "move tabu".

When no improving solution has been found over the last n iterations the algorithm terminates. Typically this value is between 200 and 1000 depending on the desired quality of the solution and the size of the problem since for a larger problem, improving moves are likely to be found with a lower frequency.

An alternative termination criterion was considered: the algorithm stops after computing n iterations. This was found to be more dependant upon the problem size.

Tabu Search / Simulated Annealing Hybrid

The addition of a tabu list to the simulated annealing algorithm ensures that recently visited states are not candidates for selection during each iteration.

The only difference between this algorithm and the simulated annealing algorithm described in Section 2.1 is that once a neighbourhood solution is generated, it only becomes a candidate to be accepted if it does not appear in the list of recently accepted solutions.

2.2 Numerical Examples

To find out the optimal values for the problem parameters, a number of tests were performed. It was found that for this particular problem, best combination of search algorithm and tabu criteria was the first search algorithm and using 'solution tabu' where a solution is tabu if it was visited during the last N iterations.

This is illustrated in Figure 1 where hollow markers indicate that the neighbourhood search function tried each move that could be made by moving the start period to every other possible period for each unit (search algorithm 1). Solid markers indicate that the neighbourhood search function tried to adjust the start period by ± one period for each unit (search algorithm 2). Square markers indicate that the tabu criterion used was solution comparison and round markers indicate that the tabu criterion used was move comparison.

This figure also indicates that the ideal tabu list size is between 30 and 125. Below this range the so-

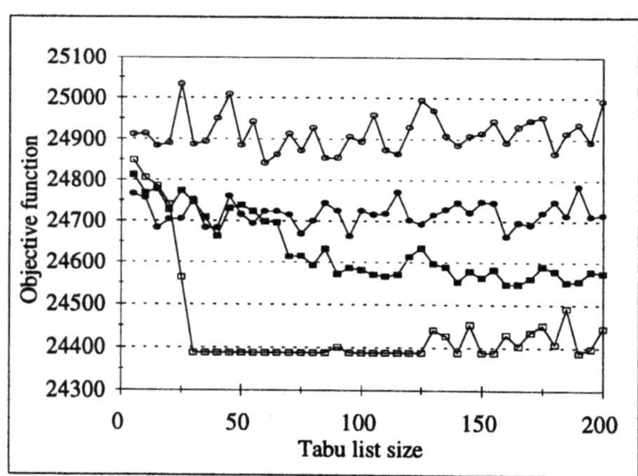

Figure 1: Tabu search.

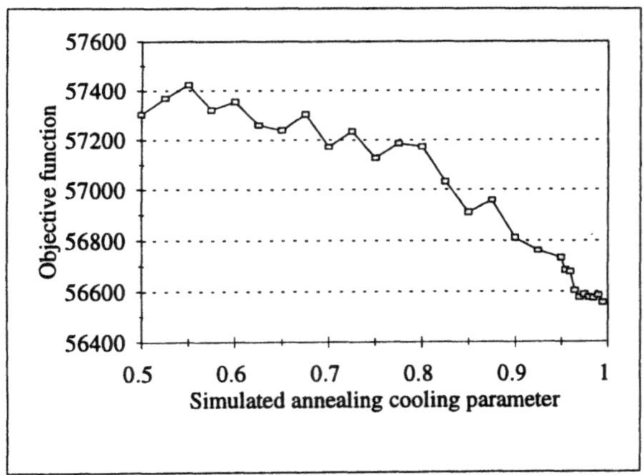

Figure 2: Simulated annealing.

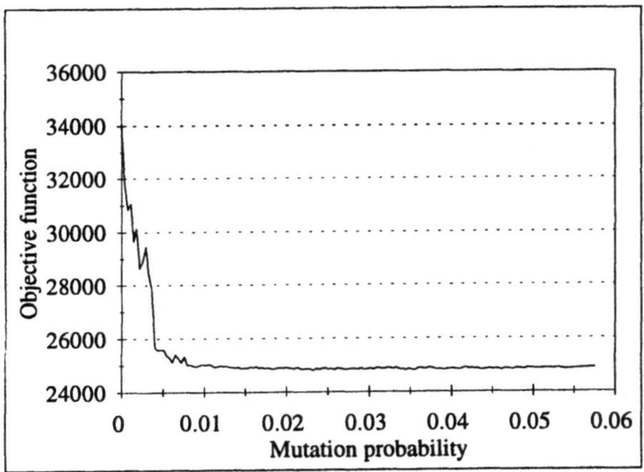

Figure 3: Mutation probability.

lution is stopped from cycling and above this range, ideal solutions are not reached because too many solutions are in tabu and they cannot be reached. Solution aspiration may occur but it will still pick the solution whose cost value is greater than the other tabu solutions.

Simulated annealing has only one tunable parameter, namely the cooling parameter. In Figure 2 it can be seen that the quality of the solution increases as the cooling parameter approaches 1. There is a tradeoff between the quality of the solution and the length of time it takes for the algorithm to terminate since as the cooling parameter increases towards 1 it takes longer for the temperature to drop enough to cause the algorithm to terminate.

To perform tests with the parameter greater than 0.99 would take a very long time, also there are theoretical convergence results for this algorithm. It can be seen that there is little improvement to be gained by setting the parameter higher than 0.97 so for this particular problem the results strongly suggest that this is the best value to use for the cooling parameter.

The genetic algorithm was the worst performing algorithm for solutions with large numbers of feasible solutions but performs slightly better than simulated annealing for problems with a small number of feasible solutions.

Tests were carried out to determine the optimum value for the parameters *GA_crossover_prob* and *GA_mutation_prob* for the hard 5-generator problem. These parameters were changed between ranges 0.0 to 1.0 and 0.0 and 0.05 respectively.

In Figure 3 the mutation results are sorted into ascending order, and plotted against the objective function.

As can be seen, if the mutation parameter is close to zero the quality of the results is severely affected. It is not clear how the solution quality varies as the mutation parameter varies above 0.01. To rectify this, further tests were carried out where the mutation probability was varied in the range 0.01 to 0.2. The results are shown in Figure 4.

Although the variance of the results is quite large, the trend of the results is clearly increasing after the mutation parameter passes 0.05. The optimum mutation probability therefore is in the neighbourhood

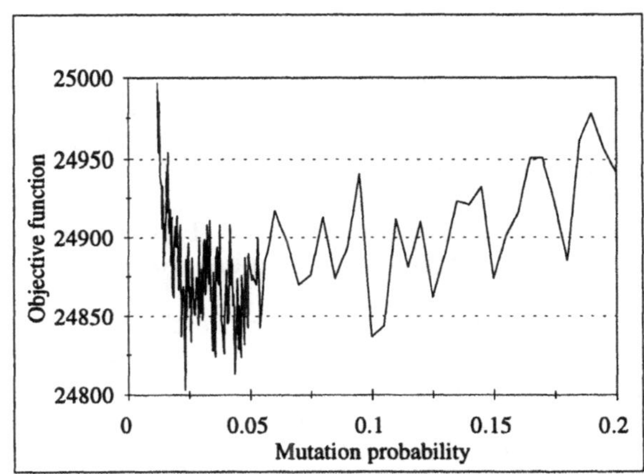

Figure 4: Mutation probability.

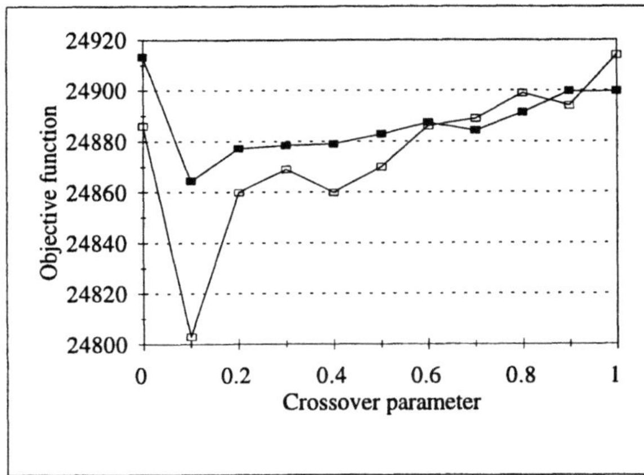

Figure 5: Crossover probability.

of 0.04 which is consistent with DeJong's recommendation [1] that the ideal mutation parameter is 0.0333.

Figure 5 plots the results from varying the *GA_crossover_prob* parameter. Solid markers indicate that the values were obtained by averaging all results regardless of the mutation parameter. Hollow markers indicate the result corresponding to where the mutation parameter was set to 0.04, i.e. the optimum value.

The optimum crossover parameter occurs at 0.1 where both the average and the individual result is better than for the other results. However the variation in the objective function as the crossover parameter is varied is significantly less than that observed as the mutation parameter is varied.

In fact it is suggested [10] that the optimum probability for mutation is much more critical than that for crossover. In [11] this hypothesis was investigated further and it was found that crossover algorithms evolve much faster than mutation alone but mutation generally finds better solutions than algorithms that utilise only crossover. This is certainly true for the maintenance scheduling problem as seen above.

Mutation is important otherwise there is no escape (ever) from certain patterns, i.e. if a bit is 0 in all the population then crossover will not change it.

To conclude, the optimum parameters for the maintenance scheduling problem are 0.1 for the crossover parameter and 0.04 for the mutation parameter, however even the most optimal genetic algorithm performance does not approach the performance of either simulated annealing or tabu search.

The time analysis for genetic algorithms is less complex than that for simulated annealing since the algorithm terminates after a predefined number of iterations according to our implementation, although another stopping criterion could have been chosen. For a population size of n the objective function is evaluated $n+1$ times per iteration. Running genetic algorithms over 1000 iterations with a population size of 15 takes 3.94 seconds.

A number of improvements could be made to this algorithm, such as using an element of local search to improve each individual. Starting with a strong initial population could also prove useful. Finally, using dynamic crossover and mutation probabilities might also yield an improvement. For example, if the best solution has not improved for a number of generations then increasing the probability of mutation would ensure that the diversity of the population is maintained.

In order to compare the performance of the algorithm, the problem was also implemented using simulated annealing and a hybrid simulated annealing and tabu search algorithm where a tabu list is used as an additional move acceptance criteria. These algorithms were implemented in C and run on an IBM RS/6000-530.

For very small problems it is the case that simulated annealing, the hybrid algorithm and tabu search all find the same solution. By enumeration the same solution was found, therefore the solutions obtained by using the three algorithms were optimal.

In the paper by Satoh and Nara [9] three problems of varying complexity are given and a comparison performed between simulated annealing and integer programming. Here a comparison is performed between tabu search, the hybrid simulated annealing/tabu search algorithm and simulated annealing. Table 1 shows the results obtained by using the four methods, where TS 1 uses search algorithm 1 in combination with 'solution tabu' and TS 2 uses search algorithm 2 in combination with 'move tabu'. Times are shown in minutes.

Clearly tabu search generates the best results, at the cost of longer execution time. This can be reduced by using the second search algorithm since less of the neighbourhood is searched on each iteration. There is a tradeoff between two conflicting factors: execution time and solution quality. The combination of the first search algorithm and 'solution tabu' (TS 1) produces better results but at

Table 1: Comparison of the algorithms.

System		TS 1		TS 2	
I	J	Cost	Time	Cost	Time
15	25	199157	58	199846	15
30	40	698020	110	699629	19
60	52	1350328	334	1357315	48

System		SA/TS		SA	
I	J	Cost	Time	Cost	Time
15	25	200255	12	200801	11
30	40	705450	20	706426	16
60	52	1359006	33	1359499	30

System		GA	
I	J	Cost	Time
15	25	201252	24
30	40	711543	31
60	52	1362343	72

the cost of increased computation time wherease the second search algorithm and 'move tabu' (TS 2) yields slightly worse results but in a much faster time. The increase in time overhead involved with using TS 2 instead of one of the simulated annealing algorithms is very small, even for the large problem.

The utility of the tabu concept can also be seen in the simulated annealing algorithms where the addition of a tabu list to the simulated annealing algorithm yields an improvement in the quality of the result at no cost in terms of the execution time. This tabu list is an additional stage during trial solution generation where the trial solution is accepted with probability

$$\exp^{-\Delta/T_k}$$

only if it does not appear in the tabu list (or solution aspiration occurs).

This paper proposes to use tabu search to solve the problem of maintenance scheduling. Although tabu search does not guarantee to find a global optimum, with a suitable choice of search algorithm and tabu criteria, it is expected that the solution found will be sufficiently close to the global optimum. The numerical results show that the proposed method can be applicable to problems in real systems.

References

[1] K. A. De Jong. An analysis of the behaviour of a class of genetic adaptive systems. *Dissertation Abstracts International*, 36(10):5140B, 1975.

[2] J. F. Dopazo and H. M. Merrill. Optimal generator maintenance scheduling using integer programming. *IEEE Transactions on Power Apparatus and Systems*, PAS-94(5):1537–1545, 1975.

[3] G. T. Egan, T. S. Dillon, and X. Morsztyn. An experimental methof of determination of optimal maintenance schedules in power systems using the beanch-and-bound technique. *IEEE Transactions on Systems, Man and Cybernetics*, SMC-6(8):538–547, 1976.

[4] D. E. Goldberg. *Genetic algorithms in search, optimization and machine learning.* Addison-Wesley, 1989.

[5] J. H. Holland. *Adaptation in natural and artificial systems.* University of Michigan, 1975.

[6] S. Kirkpatrick, C. D. Gelatt, Jr., and M. P. Vecchi. Optimisation by simulated annealing. *Science*, 220:671–680, 1983.

[7] C. Reeves. *Modern Heuristic Techniques for Combinatorial Problems.* Blackwell, 1993.

[8] Y. G. Saab and V. B. Rao. Combinatorial optimisation by stochastic evolution. *IEEE Transactions on Computer-Aided Design*, 10:525–535, 1991.

[9] T. Satoh and K. Nara. Maintenance scheduling by using the simulated annealing method. *IEEE Transactions on Power Systems*, 6:850–857, 1991.

[10] J. D. Schaffer, R. A. Caruna, L. J. Eshelman, and R. Das. A study of control parameters affecting online performance of genetic algorithms for function optimization. In *Proceedings of the Third International Conference on Genetic Algorithms*, pages 51–60. Morgan Kaufmann, 1989.

[11] J. D. Schaffer and L. J. Eshelman. On crossover as an evolutionary viable strategy. In *Proceedings of the Fourth International Conference on Genetic Algorithms*, pages 61–68. Morgan Kaufmann, 1991.

[12] V. Černy. Thermodynamic approach to the travelling salesman problem: An efficient simulation algorithm. *Journal of Optimization Theory and Applications*, 45:41–52, 1985.

[13] Z. A. Yamayee. Maintanance scheduling: Description, literature survey and interface with overall operations scheduling. *IEEE Transactions on Power Apparatus and Systems*, PAS-101(8):2770–2779, 1982.

[14] Z. A. Yamayee, K. Sidenblad, and M. Yoshimura. A computationally efficient optimal maintenance scheduling method. *IEEE Transactions on Power Apparatus and Systems*, PAS-102(2):330–338, 1983.

[15] R. H. Zorn and V. H. Quintana. Generator maintenance scheduling via successive approximations dynamic programming. *IEEE Transactions on Power Apparatus and Systems*, PAS-94(2):665–671, 1975.

A Genetic Algorithm for the Generic Crew Scheduling Problem

N. Ono and T. Tsugawa
Department of Information Science and Intelligent Systems
Faculty of Engineering, University of Tokushima
2-1 Minami-Josanjima, Tokushima 770, JAPAN
Email: ono@is.tokushima-u.ac.jp

Abstract

A topic of interest in genetic algorithm (GA) research is the design of GAs for highly constrained optimization problems. We believe that the most general approach which may be taken in dealing with such difficulty is incorporating domain-specific knowledge into GA architecture. To demonstrate this idea, we consider application of GA to a highly constrained combinatorial optimization problem, the generic crew scheduling problem (CSP). We design a steady-state genetic algorithm for solving the CSP. Computational results are given for a number of standard test problems. Experimental results demonstrate the effectiveness of the proposed algorithm and support the applicability of our idea.

1 Introduction

A topic of interest in genetic algorithm (GA [2]) research is the design of GA for highly constrained optimization problems. In order to tackle the difficulty of maintaining feasibility in highly constrained problem, a number of approaches have been proposed by GA researchers. We believe that the most general approach which may be taken in dealing with such difficulty is incorporating domain-specific knowledge into GA architecture [3, 4, 5, 6].

To demonstrate this idea, we consider application of GA to a highly constrained combinatorial optimization problem, the generic crew scheduling problem [1]. CSP is the problem of assigning K crews to tasks with fixed start and finish times such that each crew does not exceed a limit on the total time it can spend working.

We design a steady-state genetic algorithm for solving the CSP. The main characteristics of the GA are: (i) a representation scheme which directly expresses the topological structure of a candidate solution, (ii) a set of domain specific mutation operators, and (iii) an adaptive mutation application scheme. Computational results are given for a number of standard test problems [1]. Experimental results demonstrate the effectiveness of the proposed algorithm and support the applicability of our idea.

2 The Generic Crew Scheduling Problem (CSP)

The generic crew scheduling problem [1] is defined as follows:

- N tasks have to be performed by K ($K \leq N$) crews. Each crew is able to perform every task, but has a limit T on the available working time.
- each task i has the following attributes: (i) the *cost* of performing the task d_i, and (ii) a fixed *start time* s_i as well as (iii) a fixed *finish time* f_i.
- for any two tasks i and j, there is a *transition arc* A_{ij} of cost c_{ij} if it is possible for the same crew to perform the task i and then to perform task j.
- a non-empty sequence of tasks which can be performed by the same crew is called a *crew path*. The objective of the CSP is to find the best collection of K crew paths i.e. those with minimum total cost, such that each task is covered exactly once by a single crew and the total working time needed by each path does not exceed the available working time T.

3 A Genetic Approach

3.1 A Steady-State Model

We designed a steady-state GA for the CSP. Our GA starts with a population of randomly generated initial candidate solutions. Then the GA performs the following steps iteratively until some termination condition is satisfied.

1. Select a pair of parents. Roulette wheel parent selection scheme is used in conjunction with a sigma truncation.

2. From the selected pair of parents, generate a fixed number ($N_o \geq 2$) of their offsprings by applying the available genetic operators. To generate a pair of offsprings, a standard crossover operator and domain specific mutation operators explained below are applied.

3. Among the selected parents and their offsprings, select two new candidate solutions with roulette wheel scheme, and replace the parents by the two solutions eventually selected.

3.2 Representation

We assume that the tasks have been indexed in ascending start time order, and we represent a candidate solution for the CSP by the following vector:

$$X = [a_1, a_2, \ldots, a_N]$$
$$a_i = NIL \text{ or } A_{ik}(i, k = 1, 2, \ldots, N)$$

where A_{ik} is a possible transition arc outgoing from the task i and incoming to k. The vector is interpreted as follows:

- $a_i = NIL$ means that for any crew in charge of task i, the task is the last one to perform.
- $a_i = A_{ik}$ means that any crew in charge of task i must perform task k immediately after task i.
- When no transition arc is in the vector which enters task i, the task i is the first task to perform by a crew in charge of it.

From such a vector, we can extract all of crew paths constituting the corresponding solution. For example, the vector

$$[NIL, A_{23}, A_{35}, A_{46}, NIL, NIL]$$

encodes a solution consisting of the following 3 crew paths.

$$< task_1 >, < task_2, task_3, task_5 >, < task_4, task_6 >$$

The advantages and disadvantages of this representation scheme are as follows.

Even when a solution vector is randomly generated, every task is guaranteed to be performed by some crew, and any two consecutive tasks on a crew path are guaranteed to be executable consecutively. We do not have to consider the correspondence between a crew path and the crew who performs it, and hence we do not have to introduce the crews' indices explicitly. Furthermore, a standard crossover operator is expected to work effectively together with this representation.

The number of crew paths is not always equal to that of available crews. It is possible that the same task is put on multiple crew paths and hence is performed by multiple crews. The working time needed for a crew path may exceed the available working time T.

3.3 Fitness Function

We have to define a fitness function to evaluate each solution and assigns a value which reflects the solution's goodness. To solve the CSP as a maximization problem, we define the fitness function as:

$$f(X) = \frac{C}{c(X) + p(X)}$$

where C is an arbitrary positive constant, $c(X)$ is the cost of candidate solution X, and $p(X)$ is a penalty function, which adjusts the fitness when some constraints are violated by X.

In the CSP, the objective is simply to find a feasible solution with minimum cost. The cost of a solution is the total sum of cost of performed tasks and that of transition arcs involved. Since all of tasks have to be performed, the first term $\sum_{i=1}^{N} d_i$ remains constant for any feasible solution and can be neglected. Thus, the cost associated with the solution X, $c(X)$, is defined as the second term.

Under the above-mentioned representation scheme, the following violations can be made by a candidate solution.

Task Duplication: Each task must be performed exactly once. But the same task can be allocated to multiple crews.

Overtime: All the tasks on a crew path may not be executable within the available time T.

Inequality: The number of crew paths P and that of available crews K can be different.

In order to reflect the degree of violation made by a candidate solution, we use the following penalty function.

$$\begin{aligned} p(X) =& p_{dup}(X) + p_{overtime}(X) + \\ & p_{inequality}(X) \\ p_{dup}(X) =& P \times N_{dup} \\ p_{overtime}(X) =& P \times N_{overtime} \\ p_{inequality}(X) =& P \times |K - P| \end{aligned}$$

where $P = \sum_{i \in Tasks} \max_j c_{ij}$, N_{dup} is the number of tasks performed by multiple crews, and $N_{overtime}$ is the total sum of extra working time for individual crews.

3.4 Genetic Operators

We use 2-point crossover operator at first to generate new candidate solutions. Any candidate solution generated by the crossover operator undergoes a mutation. The CSP is a highly constrained combinatorial optimization problem. If we leave it to a standard random mutation, we can have little hope to find optimal or near-optimal solutions. By incorporating problem specific knowledge into standard mutation, we have devised the following specialized mutation operators:

- $\mathcal{M}_{dup}$ is applied when a task is shared by multiple crew paths. This operator focuses on a randomly selected shared task. By scanning the current solution vector, it collects all arcs entering the shared task, randomly selects one of them, and probabilistically replaces it by an alternative arc or *NIL*.

- $\mathcal{M}_{overtime}$ is needed when a crew path is not executable within the available time T. This operator randomly chooses such a crew path and probabilistically replaces its component arc by an alternative arc or *NIL*.

- $\mathcal{M}_{inequality}$ is invoked when the number of crews K is not equal to that of crew paths P. If K is greater than P, this operator selects a crew path with relatively long working time via roulette wheel selection, and divides it into two paths by replacing a randomly selected component arc by *NIL*. If K is less than P, the operator concatenates two crew paths with relatively short working time, each selected via roulette wheel selection.

- $\mathcal{M}_{swap}$ is applied only when all constraints are satisfied. It tries to improve the solution by swapping two transition arcs constituting the solution.

- $\mathcal{M}_{merge\÷}$ is also applied when the solution is feasible. It tries to improve the solution by merging two crew paths and dividing it into two other paths.

When a solution is selected to undergo a mutation, a single mutation is applied according to the following scheme: we first check for a constraint violation. If a violation is found, then the corresponding mutation is selected adaptively. Mutations $\mathcal{M}_{dup}, \mathcal{M}_{overtime}$, and $\mathcal{M}_{inequality}$ are selected with a probability of $N_{dup}/N_{violations}, N_{overtime}/N_{violations}$, and $N_{inequality}/N_{violations}$, respectively, where $N_{violations}$ is the number of all violations, and $N_{dup}, N_{overtime}$, and $N_{inequality}$ are the number of their corresponding violations. If the selected solution is feasible, then the mutation $\mathcal{M}_{swap}$ is applied to it. The mutation $\mathcal{M}_{merge\÷}$ is applied, when no improvement is made by $\mathcal{M}_{swap}$.

4 Computational Results

To implement these ideas, the genetic algorithm described above was coded in the C programming language on a SPARCstation 20. Population size was set to 50. The number of offsprings, N_o, generated at each GA iteration was set to 20. A crossover rate of 0.6 was employed. Ten independent runs, with random initial seed, were carried out and the best and average solutions found were recorded. Due to the very limited computational resources, a set of 15 benchmark problems was selected from a public library, *OR Library* compiled by J.E.Beasley, and used to investigate the performance of the algorithm. The results obtained are shown in Table 1. In every run, we obtained feasible solutions at its early iterations. The optimal solutions and the best and average ones obtained are shown in terms of their costs, respectively.

5 Yet Another GA Approach

Recently we have applied yet another GA, based on domain-specific operators, to the CSP. We briefly present the GA and preliminary results we have obtained.

Table 1: The crew scheduling problems and computational results obtained.

name of problem	number of crews	optimal solution	best solution	average solution
csp50	31	1872	1872	1872.0
	30	2092	2092	2092.0
	29	2399	2399	2453.3
	28	2706	2706	2769.3
	27	3139	3139	3181.3
csp100	48	3905	3923	3931.4
	47	4107	4108	4129.2
	46	4310	4312	4370.3
	45	4514	4578	4574.4
	44	4812	4821	4854.0
csp200	97	6288	6288	6314.2
	96	6430	6430	6468.1
	95	6583	6611	6642.8
	94	6747	6770	6801.3
	93	6914	6944	6987.6

In this GA, a candidate solution is simply represented by a vector of feasible crew paths:

$$[c_1, c_2, \dots, c_N]$$

where c_i is a feasible crew path, i.e. a non-empty sequence of tasks which can be performed by the same crew. We suppose that any two paths in the vector do not begin with the same task and the crew paths are placed in the vector in ascending order of the indices of their first tasks. Under this representation scheme, only two types of constraint violations can be made: *Task Duplication* mentioned above, and *Task Omission* where a task is performed by no crew.

The same steady-state model and the same form of fitness function $f(X) = C/(c(X)+p(X))$ as those mentioned above are used, but the penalty function $p(X)$ is modified as follows:

$$p(X) = p_{dup}(X) + p_{omitted}(X)$$
$$p_{dup}(X) = P \times N_{dup}$$
$$p_{omitted}(X) = P \times N_{omitted}$$

where $P = \sum_{i \in Tasks} \max_j c_{ij}$, N_{dup} is the number of tasks performed by multiple crews, and $N_{omitted}$ is the number of tasks performed by no crew.

We use a modified two-point crossover. For two parent vectors, $m = [m_1, m_2, \dots, m_N]$ and $n = [n_1, n_2, \dots, n_N]$, this operator generates two offsprings of the same length N. Two crew paths $m_s, m_t (s < t)$ are randomly selected from the first parent, and accordingly two crew paths $n_u, n_v (u < v)$ are selected from the second one such that

$$1st(n_u) \geq 1st(m_s),$$
$$1st(n_k) < 1st(m_s)(k = 1, 2, \dots, u-1),$$
$$1st(m_t) \geq 1st(n_v),$$
$$1st(m_t) < 1st(n_k)(k = v+1, v+2, \dots, N).$$

where $1st(c)$ denotes index of the first task in the crew path c.

Crew paths m_1, m_2, ..., m_{s-1}, m_{t+1}, m_{t+2}, ..., m_n are simply copied into an offspring vector, and $n_1, n_2, \dots, n_{u-1}, n_{v+1}, n_{v+2}, \dots, n_N$ are also copied into the other one. The other crew paths $m_s, m_{s+1}, \dots, m_t, n_u, m_{u+1}, \dots, m_v$ are inserted into one of the two offsprings chosen randomly so that any two paths in the same offspring vector do not begin with the same task.

We also incorporate two types of domain-specific mutations: $\mathcal{M}_d$ applied when a task is covered by multiple crews, and $\mathcal{M}_o$ applied when a task is not covered by any crew.

Preliminary results obtained by this GA are summarized in Table 2.

6 Concluding Remarks

Since most real-world problems are inherently constrained, performing constrained optimization is an important task. Any optimization method, including those based on GAs, must address the central issue of constraint handling. It is important to note that there is no general unique approach for the handling of constraints within the framework of GAs. All proposed approaches have their own merits and demerits.

Table 2: Preliminary results by new GA.

name of problem	number of crews	optimal solution	best solution	average solution
csp50	31	1872	1872	1872.0
	30	2092	2092	2092.0
	29	2399	2399	2405.4
	28	2706	2706	2718.9
	27	3139	3139	3139.0
csp100	48	3905	3905	3957.8
	47	4107	4107	4122.8
	46	4310	4310	4332.4
	45	4514	4514	4578.0
	44	4812	4812	4864.2

Although GAs belong to the class of black-box optimization methods, using domain-specific information should not be ignored. As suggested in this paper, incorporating domain-specific knowledge in genetic algorithms will significantly enhance the performance and hence such knowledge should be use where and whenever available and applicable.

References

[1] J. E. Beasley and B. Cao. An algorithm for the crew scheduling problem. The Management School, Imperial College, 1993.

[2] D. E. Goldberg. Genetic Algorithms in Search, Optimization, and Machine Learning. Addison-Wesley, 1989.

[3] A. T. Rahmani and N. Ono. A Genetic Algorithm for Channel Routing Problem. Proc. 5th International Conference on Genetic Algorithms, pages 494–498. 1993.

[4] A. T. Rahmani and N. Ono. A Genetic Algorithm for the Two-Dimensional General Guillotine Cutting Problem. In *Proc. ICYCS'95 Workshop on Soft Computing*, pages 39–44, Beijing, China, 1995.

[5] A. T. Rahmani and N. Ono. An Evolutionary Approach to Two-Dimensional Guillotine Cutting Problem. In *Proc. 1995 IEEE International Conference on Evolutionary Computing, Vol.1*, pages 148–151, Perth, Western Australia, Australia, 1995.

[6] A. T. Rahmani and N .Ono. Constrained Optimization with Genetic Algorithms: Channel Routing Case, *Journal of Japanese Society for Artificial Intelligence*, 11(3):461–469, 1996.

Genetic Algorithms and the Timetabling Problem

B.C.H Turton
Cardiff School of Engineering
University of Wales,
Cardiff, CF2 1XH
Email: Turton@cf.ac.uk

Abstract

This paper investigates a number of approaches to encoding and crossover to support timetable design using genetic algorithms, thus extending the range of techniques available for solving such problems. Timetabling is used in this paper to refer to organising a weekly lecture timetable, as used in universities. In addition the algorithm is designed to produce a 'good' timetable as defined by a fitness function rather than merely a legal solution. The first approach to encoding timetabling dealt with in this paper uses a 'greedy algorithm' variant and a variety of standard crossover methods. The second encoding method searches a wider space of solutions but requires a new adaptation of existing order and position-based crossover algorithms. Results are compared with a traditional search technique and timetables provided by lecturers. These results demonstrate the effectiveness of genetic algorithms when used to optimise a timetable and introduce a combinatorial crossover operator which can deal with a more general class of problem than the normal order and position based operators. The greedy algorithm version of the genetic algorithm outperformed the other methods, despite the fact it cannot search the whole of the legal solution space.

1 Introduction

Timetabling is a widespread problem and due to the large number of possible combinations, exhaustive searches are prohibitive. The problem is well known as being NP hard. In particular the university lecture timetable has a large number of variables and constraints.

A variety of methods have been tried to either optimise a timetable or assist in optimising timetables. Constraint satisfaction (C-SAT) techniques are used in this paper to form a comparison against which the genetic algorithm methods can be tested [6]. Other techniques that have been tried by other authors include tabu search [7] and simulated annealing [5].

The University of Nottingham keeps a comprehensive list of timetabling references at 'cs.nott.ac.uk/ttp/References'.

Genetic algorithms (GAs) provide a powerful alternative for tackling this type of problem. To date many of the timetabling problems tackled by genetic algorithms have been based on examination timetables [1,2,4]. These problems are characterised by a large number of 'hard' constraints and in some cases extend the time in the examination schedule. A legal solution is the prime objective. School timetabling has been tackled [3] using genetic algorithms, however again the timetable is highly constrained and the emphasis is on finding a feasible solution. Such papers have mainly based the genetic algorithm's search on the number of conflicting lectures or examinations. This paper concentrates on a timetable where 'soft' constraints are the key consideration. In addition to the constraint satisfaction approach a greedy algorithm using an order based GA, with a new fitness function, has been used as the closest reasonable approximation to existing GA techniques for tackling the problem. The key disadvantage to the order-based GA using a greedy algorithm is that it searches a subset of legal space and so is not ideal. Consequently, this paper concentrates on different encoding techniques and operators in order to try and identify a more effective method. Variants of the crossover operators used in scheduling are proposed in order to try and find an improved timetabling algorithm.

2 Timetabling

Four key entities exist in a lecture timetable, lecturers, students, room/time slots and courses. A timetable specifies the location and time of a lecture course along with the lecturer and students attending. Students that take the same set of courses are grouped into 'schemes of study'. Timetabling within the university system is subject to a number of constraints and preferences that are listed below:

- must fit within a weekly plan,
- six time slots are available per day,
- five days are available per week,
- all courses must be allocated,
- each course taught by an appropriate lecturer,
- the lecturer must be available,
- the room must be available,
- the students must be available,
- the room must be large enough for the course,
- multiple 'Schemes of Study' may take a particular course,
- students will have a preferred timetable and must be free to attend laboratories,
- staff prefer lectures at particular times of day,
- ideally the room should not have a significantly greater seating capacity than is required.

Additional factors not considered in this study are the distance between lecture theatres and the number of consecutive lectures.

This problem is expressed by using a formula based on weightings for different time/course slots so that a fitness value can be obtained for a particular timetable. The consequent fitness function is of necessity subjective in nature. The particular problem chosen was one that could reasonably be solved by an administrator so that comparisons could be made with manually produced result. There are 30 time slots available assuming five working days of six time periods. Twenty-nine courses must be timetabled within three rooms. Fifty percent of the time rooms are not available due to external courses. Each course is attended by any or all of three schemes of study. Five lecturers cover the twenty-nine courses. This was based on a simplified version of a first year timetable.

Room	Seats	Scheme	No Students
1	125	1	60
2	70	2	40
3	40	3	20

The algorithm used to determine the figure of merit for the genetic algorithm is as follows:

l–lecturer number $\quad$ t - time/day slot

s–scheme number $\quad$ r - room number

c–course number

T–Timetable array which holds the course number for a particular time/day and room $T_{t,r}\{1,\ldots,29\}$

L–Lecturer day/time weighting array for a particular time slot and lecturer $L_{t,l}\{0,\ldots,5\}$

A–Room availability for a particular time slot and room $A_{t,r}\{0,1\}$

S–Scheme weighting array for a particular scheme and time slot $S_{t,s}\{0,\ldots,5\}$

H–Scheme/course array indicating if a course belongs to a scheme $H_{s,c}\{0,1\}$

C–Course/lecturer array indicating the lecturer than lectures a particular course $C_c\{1,\ldots,5\}$

R–Room weighting array for a room and scheme combination for a course$R_{r,c}\{0,\ldots,3\}$

The following algorithm is used to determine the fitness:

Calculate_Fitness
Fitness ← 0
Select the first course
repeat
  Calculate the value (V_c) for that course (as described in the following equations 1-4)
  Add the value to the fitness total
  select the next course (c←c+1)
until all courses have been evaluated
End Calculate_Fitness

Let the number of schemes participating on the course (N_c) be

$$N_c = \sum_{s=1}^{3} H_{s,c}. \tag{1}$$

The lecturer number for the course can be obtained from the course lecturer/array (C)

$$l = C_c. \tag{2}$$

If a scheme of study attends a course then a value (n) is attached to the course based on the room availability (A), the scheme weighting for a day/time slot (S), the lecturer day/time slot weighting (L) and the weighting array for the room and scheme combination (R).

$$v_{s,c} = \begin{cases} A_{t,r}.S_{t,s}.L_{t,l}.R_{r,c} & \text{if } H_{s,c} \neq 0, \\ 1 & \text{if } H_{s,c} = 0. \end{cases} \tag{3}$$

The individual values for each scheme of study attending a course are then combined to give a fitness value for the course, which is multiplied by the number of schemes of study attending the course.

$$V_c = N_c \sqrt[N_c]{\prod_{s=1}^{3} (v_{s,c})} \tag{4}$$

Finally the fitness of the table is determined by summing over the individual course values

$$\text{Fitness} = \sum_{c=1}^{29} V_c \tag{5}$$

2.1 Encoding the Timetable

This paper develops two approaches based on two different ways of encoding the timetable as a string. The first approach lists all courses in a particular order and uses that order to determine the timetable using a greedy algorithm based on the fitness score, (Figure 1a). The timetabling problem then becomes one of ordering the courses [8]. This approach must have a method of determining the 'best' room/time slot available for a particular course which can be found by exhaustive search. A value for the 'best' course is found in Equation (6).

$$V_c = N_c \sqrt[N_c]{\max \left[\overset{30}{\underset{t=1}{}} \left[\overset{3}{\underset{r=1}{}} \prod_{s=1}^{3} (v_{s,c}) \right] \right]} \tag{6}$$

Calculation of the maximum is in fact efficient as many of the options will evaluate to zero because

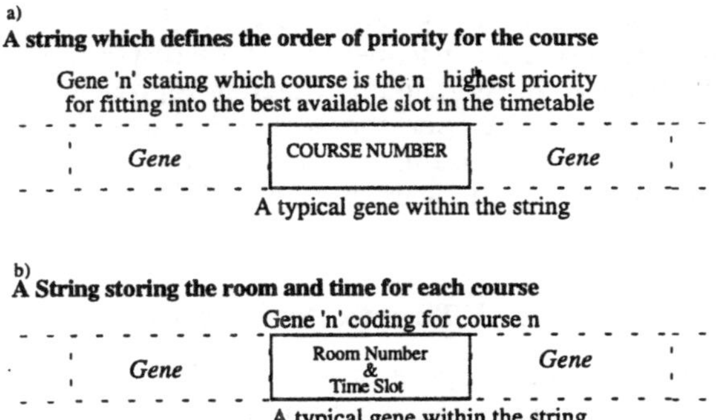

Figure 1: Timetable encoding methods.

they are illegal, in addition the most time consuming operation takes place after one particular time slot has been identified. Consequently the calculation of the course value will not be much longer than a normal (no search) evaluation. Alternatively the string representing the timetable can be encoded by associating a course with the room and time value (Figure 1b). Since all attributes for a course are known for this method no additional search is required and Equation (4) can be used directly. However this approach to encoding the timetable poses a serious problem to standard GA operators. Existing GA operators can produce unreasonable timetables when trying to combine the best features of two timetables. Typically this forces the use of some form of extra filter to remove the unwanted strings [3]. Consequently new genetic algorithm operators are required to complete this task efficiently.

2.2 Genetic Algorithms for Timetabling

The order-based operater, as defined by Syswerda [8], manipulates the ordering of genes and consequently can be used for the first method of encoding suggested by the author. This operator can be described as follows: 'A set of positions is randomly selected, and the order of courses in the selected positions in one parent is imposed on the corresponding courses in the other parent'. Along with the crossover operator, position and order based mutation can be uses. Position-based mutation selects two genes and places the second before the first. Order-based mutation selects two genes and swaps them. This technique cannot cover all possible combinations of room/course/lecture/day/time, as can be shown by a simple counting argument. The second approach which encodes the room and time value for each gene (Figure 1b) allows all possible values and consequently can investigate course timetables not searched by the first approach. However existing GA operators would produce illegal results for this encoding method, so new 'extended' operators, which overcome this difficulty, are required.

2.3 Extended Timetable Operators

The order-based operator was described in the previous section, so the position-based operators will now be described followed by their shortcomings and the proposed new variants.

Position-based crossover as used in this paper is defined by Syswerda [7]. A set of positions is randomly selected and the positions of tasks selected in one parent are imposed on the corresponding tasks in the other parent. Such an operator depends on finding a matching value in parent 2 for every value in parent 1. However for the timetabling problem encoded as shown in Figure 1b, parent 1 may store room/time slots that do not appear in parent 2. Consequently an algorithm must be devised which preserves order or position between two parents where genes match, whilst ignoring non-matching genes. The following variants have been devised.

Algorithm for Variant 1:
Position-Based Crossover Operator
Select positions at random
Child1=create_child(parent2,parent1,positions)
Child2=create_child(parent1,parent2,positions)
end Position-Based operator

create_child(ParentA,ParentB, Positions)
Copy selected Positions in ParentA into the Child
Copy those values in ParentB not already in the Child, into the Child, using the ordering of ParentB, till all empty positions are filled
return Child

Algorithm for Variant 2:
A minor change is made to the create_child function.
create_child(ParentA,ParentB, Positions)
Copy selected Positions in ParentA into the Child
Copy those values in Parent B not at the selected Positions and not already in the Child, into the same positions in the Child. Copy those values in ParentB not already in the Child, into the Child, using the ordering of ParentB, till all empty positions are filled
return Child

The following variant on the order-based operator is also used. A set of positions is randomly selected, and the order of genes at the selected positions in one parent, that are duplicated in the other parent, are imposed on the corresponding genes in that parent.

Algorithm for the Extended Order-Based Operator
create_child(ParentA,ParentB, Positions)
Copy ParentA to Child
Identify values at the selected Positions in ParentB
Ignore those values which do not exist in ParentA
Reorder the selected values in Child to be the same as the order for ParentB
return Child

This description is still not sufficient as the number of selected positions becomes important in this case. Normally each gene has a 50% chance of being selected and used in the crossover. Consequently on average half the genes take part. If some genes are ignored because they do not appear in both strings then the number of selected genes may need to increase in order to compensate for those which are chosen but cannot be used. So to compensate for the failed matches the number of selected genes is increased in proportion to the chances of such a failed match. The validity of this method was checked against an extended version of the travelling salesman problem and proved to be effective. Each of the three extended operators, can be used on the second encoding method.

3 Results

The results of this work are shown in Figure 2. The order-based 'greedy' algorithm proved to be superior to the other algorithms despite the potential for the other operators for searching a larger space. An example is given in Table 1.

Table 1: Timetable from order based greedy algorithm.

	Mon	Tue	Wed	Thur	Fri
1				2:23	
2	3:25	2:17	1:21	1:8,2:28	1:4
3	2:27	1:1	1:5	1:14,3:22	1:19
4	2:7	1:16, 3:12	3:29	1:13	2:26, 3:18
5		2:11	2:6	2:9,3:2	
6	3:15	1:20	1:10	1:24,3:3	

Notation:- Room number: Course number
(Fitness=1866)

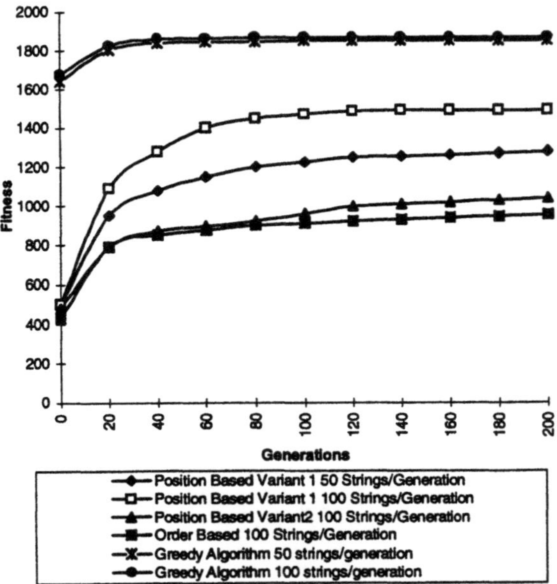

Figure 2: Genetic algorithm results.

Table 2: Results after 10^5 course evaluations.

Method	Fitness 10^5 Evals
Greedy GA 50 strings	1860
Greedy GA 100 strings	1800
CSAT	1570
Lecturer 2 (Manual)	1420
Lecturer 1 (Manual)	1382
V1 Position Based GA 100 strings	1280
Lecturer 3 (Manual)	1209
V1 Position Based GA 50 strings	1200
V2 Position Based GA 100 strings	920
Order Based GA 100 strings	900

GA timetable assumes 29 course evaluations / string.

Results were obtained for the same problem solved by three professionals who normally organise timetables. In addition, constraint satisfaction techniques (CSAT) were used to reduce the size of the search space with a best first search to provide an automated method for comparison. A maximum value for each course is known so the search can be truncated if there is no route for achieving an improved fitness value. It is assumed that a single chromosome requires 29 course evaluations for comparison purposes. The greedy-algorithm based GA required approximately sixty generations, with 50 strings (chromosomes) to reach a result of ~1850.

Table 3: List of courses taken by each lecturer.

Lecturer1	1	2	3	16	17	18
Lecturer2	4	5	6	19	20	21
Lecturer3	7	8	9	22	24	
Lecturer4	10	11	12	25	26	29
Lecturer5	13	14	15	23	27	28

Table 4: List of courses for each scheme of study.

Scheme 1	1 4	5 6	8 9	10 13	14 16	19 20	21 23	24 26
Scheme 2	1 4	7 11	12 13	14 15	17 18	19 20	24 25	27 28
Scheme 3	1 2	3 5	7 8	10 11	13 16	17 20	21 22	27 29

The search based technique was unable to reach this value after 109 course evaluations (Table 2). The problem is defined in Tables 3, 4 and 5.

4 Conclusion

The results obtained in this paper show new variants of existing GA operators that are capable of solving the timetabling problem. Results from the tests run indicate that the genetic algorithm solution is superior, as defined by the numeric criteria, to the timetables devised by the lecturers and to constraint satisfaction search techniques with a similar number of course evaluations. The best solution to the timetabling problem was achieved using an

Table 5: Weighting factors for the timetable.

	Mon	Tues	Wed	Thur	Fri
1	11111 011NYN	11111 111YNN	11111 101NYN	11112 111YYN	11111 110YNY
2	11111 011NYY	31331 333NYN	13133 303YNY	31313 333YYN	22121 220YNY
3	13313 033NYN	51551 555YNN	15155 505YNN	51515 555YYY	33131 330YNN
4	14414 044YYN	51551 555YNY	15155 505NNY	51515 555YNN	22121 220YYY
5	14414 044NNN	41441 444NYN	14144 404NYN	41414 444NYY	11111 110NYN
6	12212 022NNY	21221 222YNN	12122 202YNN	21212 222YNY	11111 110YNN

Weightings- Lecturer 1,2,3,4,5; Scheme of study 1,2,3; Room Availability 1,2,3. Each one character long. Room weighting 0 if illegal otherwise 3 to 1 in rank order

order based operator with the greedy genetic algorithm, not by the extended operators. These results indicate that fairly simple techniques will obtain a solution (1600 for the case shown in this paper) but that producing good results within a reasonable number of evaluations (above 1800 in less than 105 evaluations for this study) requires more carefully designed algorithms. Only the order-based greedy algorithm was able to obtain such a result in less than 105 course evaluations. Further research on new operators for timtabling problems is required in order to identify more effective operators that search a larger space than the greedy genetic algorithm. In addition the type of figure of merit used can be extended, for instance by adding 'distance between lecture theatres' information to discourage moving students around from place to place.

References

[1] E. Burke, D. Elliman, and R. Weare. Specialised recombinative operators for timetabling problems. In *Proc of the AISB*, pages 75–85, Berlin, 1995. Springer-Verlag.

[2] M. W. Carter, G. Laporte, and S. Y. Lee. Examination timetabling: Algorithmic strategies and applications. *J.Opl Res. Soc.*, 47:373–383, 1996.

[3] A. Colorni, M. Dorigo, and V. Maniezzo. Genetic algorithms and highly constrained problems: The time-table case. In *Proc. of the First Workshop on PPSN*, pages 55–59, 1990.

[4] D. Corne and P. Ross. Some combinatorial landscapes on which genetic algorithm outperforms other stochastic iterative methods. In *Proc of the AISB*, Berlin, 1995. Springer-Verlag.

[5] K. Dowsland. A timetabling problem in which clashes are inevitable. *J. Opl Res. Soc.*, 41(10):907–918, 1990.

[6] R. Feldman and M. C. Golumbic. Optimization algorithms for student scheduling via constraint satisfiability. *Comp. J.*, 33(4):356–364, 1990.

[7] A. Hertz. Tabu search for large scale timetabling problems. *European Journal of Operational Research*, 54:39–47, 1991.

[8] G. Syswerda. *Schedule Optimisation Using Genetic Algorithms*, pages 332–349. Van Nostrand Reinhold, 1991.

Evolutionary Approaches to the Partition/Timetabling Problem

D. Corne
Parallel Emergent and Distributed Architectures Laboratory,
Department of Computer Science, University of Reading,
Reading, RG6 6AY, UK. Email: D.W.Corne@reading.ac.uk

Abstract

Recent research has yielded a variety of mainly evolutionary algorithm (EA) based methods for dealing with exam and course timetabling problems continually faced by educational institutions A problem as yet unaddressed in the timetabling research literature, however, is the *partition/timetabling problem* (PTP). This problem most typically arises early during a university term and requires the need to partition groups of students into small tutorial or lab groups, and then timetable these sessions. Even in cases where an institution has effectively 'solved' the main course or exam timetabling problems it faces, the continual need to address combined partition/timetabling problems still costs much staff time and effort. This article describes recent work which addresses the PTP. Three different techniques are tried, and results using these techniques are compared on some real world PTPs. Best results so far are achieved by a two-stage method in which specially developed PTP heuristics are first used to derive a good partition of students into the several tutorial/lab (etc) sessions, and then a relatively standard timetabling EA generates a timetable based on this partition, incorporating operators which may slightly alter the partition.

1 Introduction

Research into timetable optimisation is gathering pace. In particular, recent work has yielded a variety of mainly evolutionary algorithm (EA) based methods for dealing with the exam and course timetabling problems continually faced by educational institutions [1, 3, 4]. An overview of recent techniques can be found in [2]. Most of this research has concentrated on a particular common aspect of the timetabling problem: the assignment of times and/or rooms to a collection of events. Exam and course timetabling involves the need to make such assignments without violating the constraint that a person cannot be in two places at once. More realistically, research in this area considers further objectives such as the need to minimise the number of students who need to take examinations consecutively, for example.

A problem as yet seemingly unaddressed in the timetabling research literature, however, is the *partition/timetabling problem* (PTP). This complex combinatorial problem most typically arises early during a university, college or school term (or semester) and concerns the need to partition several connected groups of students into restricted-capacity tutorial or lab-sessions, *and* then timetable (find times and/or rooms/tutors for) these sessions.

In a typical case, there may be a pool of about 100 students, each with their own individual timetables, and each having about 10 free timeslots. Early in term, lab sessions need to be arranged for one or more of the courses. Each student will then need to attend typically one lab session per week for each of the courses requiring labs. 'Compilers', for example, may have a pool of 40 students, who need to be arranged into 4 tutorial groups. Meanwhile, the 'Software Engineering' course may involve 75 students, who need to be arranged into 10 group-work sessions involving a maximum of 8 students each. Each of these individual lab or tutorial sessions must have an appropriate collection of students assigned to it, and then be timetabled.

Even in cases where an institution has effectively 'solved' the main course or exam timetabling problems it faces (via use of commercial or public domain automated timetabling systems), the continual need to address combined partition/timetabling problems still costs much staff time and effort.

2 The Partition/Timetabling Problem

The partition/timetabling problem (PTP) involves a pool of students S, organised into k (possibly non-disjoint) subsets $S_1, S_2, ...S_k$. Subset S_i will typically represent those students in S who are taking course i.

Owing to capacity constraints on the sizes of laboratory/tutorial sessions, students for each course i will need to be partitioned into $g(i)$ subgroups, $S_i^1, S_i^2, ..., S_i^{g(i)}$, each such subgroup representing an individual lab group. Also, each such subgroup has a collection of allowable rooms (some may require a PC-equipped lab, some may need a multimedia lab, and so on ...), and a collection of allowable timeslots.

Meanwhile, since lab/tutorial allocation is invariably done in the first week or two of term, a variety of background constraints exists concerning the existing timetable: each course i will have assigned times and rooms, while a variety of other courses perhaps not requiring lab sessions, or with already timetabled lab sessions, also have allocated times and rooms which will impinge directly on the timetabling of the labs.

Against this background, the PTP concerns establishing the partition of students $P_i = S_i^1, S_i^2, ..., S_i^{g(i)}$ for each i, *and* assigning suitable times and rooms for the laboratory/tutorial sessions represented by the sets in each partition.

An ill-determined partition can greatly constrain the choice of time and room assignments, and vice versa.

3 Techniques for Addressing the PTP

There are three main styles of technique which seem applicable. A *two-stage* approach would involve first establishing the partitions for each course, and then generate a timetable respecting these partitions. An *all-in-one* method would simultaneously develop the partition and the time/room assignments. Thirdly, a *linked* approach would first develop an initial partition, followed by a timetabling stage within which small adjustments to the partition would be allowed. Further possibilities include *co-evolving* the partition and time assignments; this and other techniques will be explored in later work.

We report on experiments with each of these three techniques on a collection of real-world partition/timetabling problems faced by the Department of Computer Science, University of Reading in 1996 and 1997.

Further details on the methods employed are as follows.

3.1 Two-Stage Approach

The first stage partitioned each student set S_i into the required number of groups, while the second stage solved the resulting well-defined timetabling problem using published EA-based methods [3]. Several potential techniques exist for the first stage, of which three were tried here in different variants of the two-stage approach: 'Random', in which the partitioning was done purely at random (implemented largely as a baseline measure), 'Sequential' (an implementation of the rough, heuristic method commonly used by administrators faced with this task), and 'Evolved', in which an EA was used to evolve a partition with minimal interaction between subsets in the partition.

In this context, the interaction between subsets is the number of timetable-clashes which would result if both were allocated the same time period. The 'Evolved' method therefore used a simple genetic algorithm (steady-state [5], population size 100, binary tournament selection) in which a chromosome was of length A (the total number of student/lab assignments which needed to be made), and each allele expressed the lab assignment for a particular student and a particular course. Standard uniform crossover and genewise mutation were used, and the fitness of a partition was measured in terms of a weighted sum of the subset interaction and the violation of lab capacity constraints.

The 'Sequential' method simply generated a base partition by filling up subsets one by one, trying to choose the next student in a subset to be one maximally similar to those currently in it. The measure of similarity was the number of courses students had in common.

3.2 All-in-One Approach

This involved extending the commonly used direct representation for EA-based timetabling [3] to cater for partition construction. A chromosome had two parts: the standard timetable-representation part, and an extra partition-representation part. The latter involved one chunk of genes for each course requiring partitioning. For example, chunk i contained k genes, one for each student taking course i. A gene's alleles ranged from 1 through $g(i)$, where $g(i)$ was the number of groups into which course i was to be partitioned. Hence, if the 3rd allele of partition chunk 4 was 5, this meant that the third student on the list of students doing course 4 was in tutorial group 5 of that course. At initialisation,

Table 1: Basic details of the problems addressed.

Problem	Students	Lab Pools	Total Labs	Assignments
RUCS96	84	3	17	200
RUCS97a	87	3	17	206
RUCS97b	219	3	24	502

each such chunk was generated randomly, though ensuring that capacity constraints for an individual lab/tutorial groups were respected. Mutation operators on these partition chunks reshuffled the group assignments of a small number (e.g.: 2 or 3) of students

3.3 Linked Approach

Based on the same representation strategy as the all-in-one approach discussed above, this method first installed a pre-generated partition (using one of the methods mentioned in the Two-Stage Approach) in the partition-parts of the chromosomes. Subsequent evolutionary search established a timetable to suit the partition, backed up by a low level of mutation of partition parts.

4 Experiments

4.1 The Problems

In common with many other institutions, the Department of Computer Science at the University of Reading (RUCS) constantly needs to solve hard PTPs in the early weeks of term, following the fixation of the remainder of the course timetable. All of the problems addressed in this paper arose from RUCS, and are referred to in turn as RUCS96, RUCS97a, and RUCS97b. Problems RUCS96 and RUCS97a refer to the timetabling of labs for 2nd year undergraduates. In each case, there was a pool of 90 students, and three courses requiring different numbers of labs. Problem RUCS97b refers to 1st year undergraduates, involving over 200 students and three courses requiring labs. Table 1 summarises the problem details.

For example, problem RUCS96 involved 84 students, 3 courses requiring lab sessions, 17 labs altogether, and a total of 200 laboratory assignments. That is, if x students need to attend labs for one course, y for another, and z for the third course, then $x + y + z = 200$. The number of lab assignments precisely reflects the number of decisions to be made of the form: "which particular lab from course C does student S attend?". The 'Labs' column, of course, gives the number of decisions which need to be made of the form: "what time shall this lab session be set?".

Each of these problems was addressed in the context of the existing relevant course timetable, thus constraining the times that the lab sessions could be assigned to. In addition, sundry further constraints were in force regarding the rooms that labs could occupy, additional constraints on the allowed times, and maximum student capacities for labs of particular courses.

In addition to these, a further six problems were generated by making the originals harder. Hence, the problems P+3 and P+6 respectively refer to removing 3 and removing 6 further timeslots from the allowable times for each lab session.

Table 2: Performance of various techniques on three PTPs.

Method	RUCS96	RUCS97a	RUCS97b
T/R	38.3–0	74–0	34.7–0
T/S	9.7–0	58–0	20–0
T/E	2.8–0	7.1–0	2.7–0
A/R	25–0	79–0	21.8–0
A/S	9.1–0	57–0	15–0
A/E	4.5–0	3–0	4–0
L/R	16.6–2	32–5	22.2–2
L/S	8.8–0	14.9–0	2.5–0
L/E	0.4–0	2.3–2	0–0

4.2 Results

Each of the six techniques was tested on each of these 9 problems. A technique/problem test involved 20 trial runs, for which we record the *capacity conflicts* and *clash conflicts* associated with the best partition/timetable found during a run. Capacity conflicts simply sums the excesses of lab sessions which exceeded their capacity. Clash-conflicts records the total number of cases in which a student is required to do two things at once (either two labs, or a lab and a lecture on the existing timetable).

In Table 2, and the two tables following, the 'Technique' column incorporates the abbreviations 'T' for two-stage, 'A' for all-in-one, and 'L' for linked, along with 'R' for random, 'S' for sequential, and 'E' for evolved. Each entry notes the average number of clash conflicts and the average number of capacity conflicts during the 20 trial runs, separated by '–'. In all cases, the capacity conflicts figure for

Table 3: Performance on three harder PTPs.

Method	RUCS96+3	RUCS97a+3	RUCS97b+3
T/R	58.1–0	89.7–0	63.7–0
T/S	11.7–0	60–0	20.4–0
T/E	2.4–0	8–0	1.8–0
A/R	40–0	85–0	28.7–0
A/S	12.1–0	64–0	15.5–0
A/E	5.5–0	5–0	4–0
L/R	32–4	51.8–6	37.1–0
L/S	11–0	15.1–0	3.1–0
L/E	0.5–0	3.3–2	0–0

Table 4: Performance on three even harder PTPs

Method	RUCS96+6	RUCS97a+6	RUCS97b+6
T/R	62.1–0	96–0	80.5–0
T/S	15.2–0	63.1–0	29.6–0
T/E	3.8–0	11.3–0	3.2–0
A/R	56–0	88–0	39.2–0
A/S	17.8–0	67.8–0	16.4–0
A/E	5.4–3	7.4–0	8–2.5
L/R	38.2–5.4	67–12.5	43.3–7
L/S	15–0	22.1–0	8.1–0
L/E	3.8–0	3.9–3	0.7–0

non-linked and non-evolved methods is always zero, since the initial partitions were constrained to have no capacity constraints, and the technique did not involve any later alteration of the partition.

On the evidence of Table 2, it is fairly clear that methods incorporating random partitioning perform very poorly, while use of heuristic techniques to develop an initial partition are much more successful. In particular, the linked/heuristic technique seems most promising.

Tables 3 and 4 respectively outline results on the '+3' and '+6' versions of both problems.

Table 3 basically shows worse performance all round as we make the problems harder. In particular, the deterioration in performance of methods using 'Random' early partition development is most marked. This seems to indicate the important dependence of the eventual PTP result on the initial base partition.

This effect is magnified again on the '+6' problems in Table 4, where we can most clearly see the benefits of the Linked/Evolved method.

5 Conclusion

This report has introduced the partition/timetabling problem (PTP), a recurring headache of timetable administrators at educational and similar institutions, and considered a number of EA-based approaches to tackling it. Preliminary results are presented for nine separate methods, based on three overall approaches. These indicate that the Linked/Evolved approach seems most promising, whereby partitions are initially developed using a genetic algorithm to evolve the necessary student subgroups with minimal interaction, and to meet capacity constraints. Subsequent EA-based search using the 'direct' timetable representation method [3] then develops the overall timetable, occasionally making alterations to the initially-generated partition.

The test-problem set used here is evidently too small, however, on which to justify general conclusions. Extant partition/timetabling problems are likely to vary very widely in a great many dimensions which may affect the differential performance of algorithms used to address them.

References

[1] E. K. Burke, J. P. Newall, and R. Weare. *A Memetic Algorithm for University Timetabling*, LNCS volume 1153, pages 241–250. Springer-Verlag, Berlin, 1995.

[2] M. W. Carter and G. Laporte. *Recent Developments in Practical Examination Timetabling*, LNCS volume 1153, pages 3–21. Springer-Verlag, Berlin, 1995.

[3] D. Corne, P. Ross, and H-L. Fang. Fast practical evolutionary timetabling. In *Proceedings of the AISB Workshop on Evolutionary Computation*, LNCS volume 865, pages 250–263, Berlin, 1994. Springer-Verlag.

[4] B. Paechter, H. Luchian, A. Cumming, and M. Petruic. Two solutions to the general timetable problem using evolutionary methods. In *Proceedings of the First IEEE Conference on Evolutionary Computation*, pages 300–305, 1994.

[5] D. Whitley. The GENITOR algorithm and selection pressure. In *Proceedings of the Third International Conference on Genetic Algorithms*, San Mateo, 1989. Morgan Kaufmann.

Discovering Simple Fault-Tolerant Routing Rules by Genetic Programming

I.M.A. Kirkwood, S.H. Shami and M.C. Sinclair
Dept. of Electronic Systems Engineering, University of Essex,
Wivenhoe Park, Colchester, Essex CO4 3SQ, UK.
Email: mcs@essex.ac.uk

Abstract

A novel approach to solving network routing and restoration problems using the genetic programming (GP) paradigm is presented, in which a single robust and fault-tolerant program is evolved which determines the near-shortest paths through a network subject to link failures. The approach is then applied to five different test networks. In addition, two multi-population GP techniques are tried and the results compared to simple GP.

1 Introduction

The aims of this paper are to demonstrate the principle of applying genetic programming (GP) [3] to find the shortest or near-shortest path route through simple networks subject to link failures, to assess the approach on a number of test networks, and to explore whether multi-population GP could provide improved results. Traditional centralised methods, such as the simple Dijkstra's Algorithm [1] can be used to identify the shortest path through a static network. Should links fail, Dijkstra's Algorithm could be re-applied to the faulty network and new shortest paths obtained. Alternatively, a distributed restoration algorithm, such as Grover's [2], could be applied. However, the single centralised program described in this paper, found by GP and composed of problem-specific functions, is not in competition with Dijkstra's Algorithm or any other traditional centralised routing algorithms, but rather was 'blindly' evolved to find the near-shortest paths in the given network subject to link failures, and thus is fault-tolerant and robust.

In Figure 1 a notation of the form p(q) is used to identify the links in a network where p is the link number and q is the length of that link. For example in Network 1 of Figure 1, Node 3 is connected to Node 4 via Link 6 whose length is 2 units. Now consider the application of GP to, say, Network 1. Our problem then is to evolve a single program that

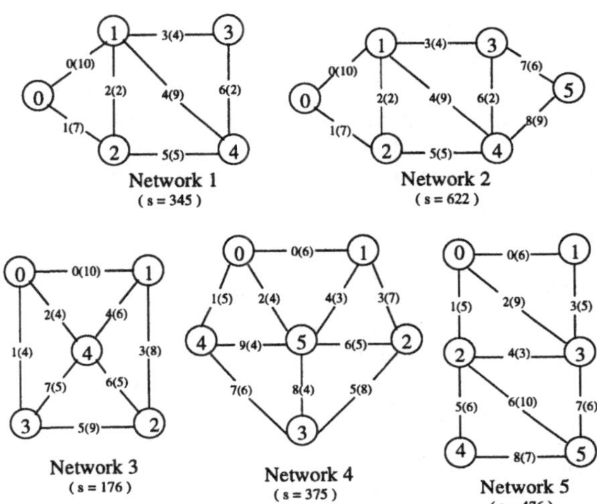

Figure 1: The test networks.

can route information from any node back to Node 0, when there can be up to one link failure. The routes found will be the shortest or near-shortest paths. In other words, a single program is to be able to find simultaneously the shortest or near-shortest path trees through eight networks: the network when all links are fault-free, and seven networks when each of the seven links, in turn, is broken.

This is the first known application of GP to either a telecommunications network routing or restoration problem.

2 Problem Representation

GP usually describes its solution using LISP *s*-expressions which can be redrawn as parse trees containing the problem-specific functions and terminals used. Our novel problem-specific function, used to evolve a solution to the problem, is as follows:

```
(IF-CUR-GO W X Y Z) =
  X, if the Current Node is Node W,
     Node W and Node X are directly
     connected, and Node X has not been
     visited twice before.
  Y, if the Current Node is Node W,
     Node X and Node W are not directly
     connected, or Node X has been
     visited twice before.
  Z, if the Current Node is not Node W.
```

For example, in Network 1, if the information to be routed was at Node 3, the `Current Node`, the function `(IF-CUR-GO 3 1 4 2)` would route the information from Node 3 to Node 1 provided Node 1 has not been visited twice before, else it would be routed to Node 4. This next node then becomes the `Current Node`. Our Function Set [3] is then {`IF-CUR-GO`} and the Terminal Set comprises the nodes {`0,1,2,3,4`}. For our initial experiments on Network 1, a population of 500 programs and Koza's default parameters [3] were used, except that the maximum number of generations was 31. The driving force behind all evolutionary algorithms is the problem-specific fitness function. In this application, the fitness of a potential solution was obtained by calculating the lengths of the paths from Nodes 1, 2, 3 and 4 back to Node 0 in each of the eight possible networks (Network 1, and the seven variants of Network 1 with one link broken); clearly, in this case, the smaller the fitness measure, the better the solution.

3 Initial Results

For our initial experiments on Network 1, different random seeds were used to obtain a different initial population on each occasion. At Generation 30, the fitness of the best individual was 356; this compares with 345 for the optimum solution.

To assess the results, we define a new fitness measure, *Dijkstra fitness* (f_D) which is the difference between the fitness of the solution found by our GP run (f) and the true optimum (found by Dijkstra's Algorithm [1]), which we call s, i.e. $f_D = f - s$. Thus zero Dijkstra fitness corresponds to the true optimum, indicating that the value found by that GP run is equal to s. In order to have a fitness measure which is more representative for comparing several networks, we use another measure which we call P_A (percent above the optimum) and define it as $P_A = (f_D/s) \times 100$. As an example, in our initial experiments on Network 1, the Dijkstra fitness of the best run was 11 and P_A was 3.2.

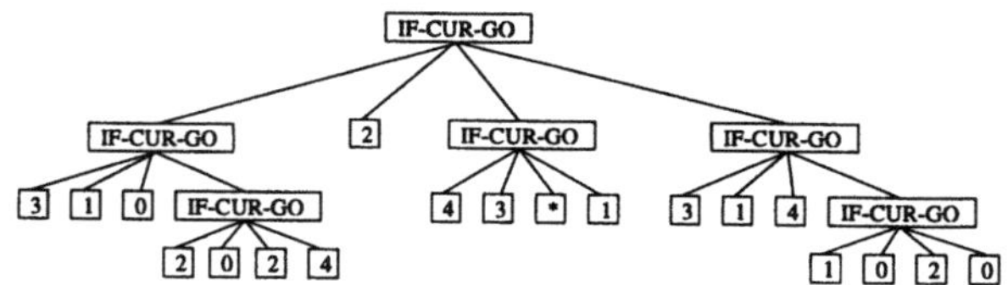

Figure 2: Diagram of fault-tolerant routing parse tree for Network 1.

Applying some judicious manual editing to the LISP code, superfluous genetic material was removed from the parse tree of the best evolved individual from Generation 30. The resulting simplified program is at Figure 2, where the asterisk indicates an element of the parse tree that is never reached and so can equal any value.

4 Robustness of Solution

The robustness of the best evolved solution can be examined by testing its ability to find near-shortest path routings in the same network, but this time with more than one link failure. In a seven-link network there are twenty-one possible networks resulting from breaking two links. Of these, two are discarded as invalid because they isolate at least one node; clearly, no program can solve that network problem. On examination, the best evolved program finds near-shortest paths for seventeen of the remaining nineteen valid networks. Here, the program has a success rate of 89.5% when almost 30% of the links have failed. Considering three link failures, there are twenty-one valid networks out of a possible thirty-five networks. The program finds near-shortest path routes for fourteen networks, a success rate of 66.7% when 43% of the network's links have failed. Clearly for Network 1, there are no valid networks when four or more links have failed.

5 Use of Multiple Populations

Having established proof of principle with our initial experiments, we decided to test our approach on a larger set of networks, as well as explore the use of multi-population GP. Our initial experiments were implemented using `sgpc` (v1.1) [5], a poorly documented and supported GP package. For our subsequent work we moved to the far superior `lilgp` (v1.02) [7]. This package is clearly documented, regularly upgraded, and supports multiple populations with arbitrary exchange topologies.

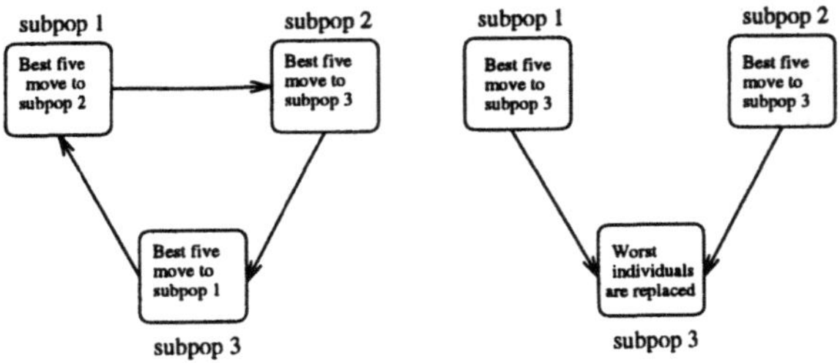

Figure 3: Multi-population (two distinct techniques).

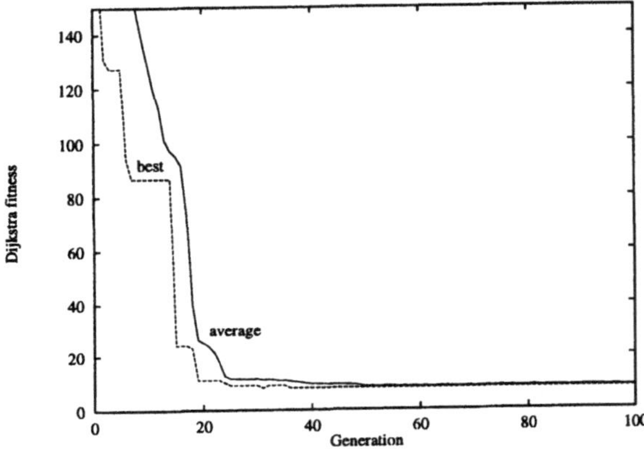

Figure 4: Best and average f_D vs. generations for Network 1.

Premature convergence [4] is a problem that is often faced in the standard uni-population GP (UGP) runs. One effective approach to deal with this problem is to employ multiple populations (MP) within each generation. This relates to a more realistic model of nature than a single large population. Multiple populations of genetic programs have been found to reduce processing time and also explore the search space better, although recent results indicate this is not always the case [4]. The uni-population GP technique is essentially sequential and entails high computational cost for maintaining genetic diversity based on similarity comparison. However, the MP technique maintains separate sub-populations which are allowed to evolve independently. This enables each sub-population to explore different parts of the search space, retain high fitness individuals and control how mixing occurs with other sub-populations. The MP technique can easily be ported to a multi-processor or distributed computing environment.

The two distinct MP techniques used are called MP-Ring Architecture (MP-R) and MP-Injection Architecture (MP-I) are explained with reference to Figure 3 [4]. In the ring architecture each sub-population chooses its five best individuals and sends it to the next sub-population in the ring. Each sub-population takes the individuals sent to it, and uses them to replace its five worst members. This is done after a specified number of generations. On the other hand, the injection architecture is a hierarchical arrangement. In the three sub-population example shown, sub-populations 1 and 2 both send their five best solutions (total ten) to sub-population 3. Sub-pupulation 3 replaces its ten worst individuals with those received from the donors.

6 Additional Test Networks

Four additional networks with varying topologies were arbitrarily constructed in order to further test the ability of GP to discover routing rules for near-shortest paths. The objective is the same, that is to obtain near-shortest paths from every other node to Node 0 in each test network, which can be subject to one link failure. All five networks are shown in Figure 1. Beneath each network s refers to the actual sum of shortest path lengths for that network. The Function Set for all five networks remained {`IF-CUR-GO`} and the Terminal Set comprised the node numbers as before: either {`0,1,2,3,4`} or {`0,1,2,3,4,5`}, as appropriate.

7 Additional Results

For each network fifteen runs were carried out, five for each of the three techniques: namely UGP, MP-R and MP-I. Each run was given 100 generations for uniformity. A population size of 600 was used for all runs. All other control parameters used were Koza's default parameters [3].

Figure 4 shows a plot obtained for Network 1 using UGP and shows the best and average Dijkstra fitness (f_D) for each generation. Note that these additional runs, with a slightly larger population, but many more generations, have improved f_D to 8 and P_A to 2.3. Figure 5 shows the performance of the best runs using the three techniques on Network 2 with the same random seed. The delay in convergence using MP can be clearly seen.

Table 1 shows the performance of GP on the five networks. The paradigm performed extremely well on Network 3 with the GP result being only 0.57%

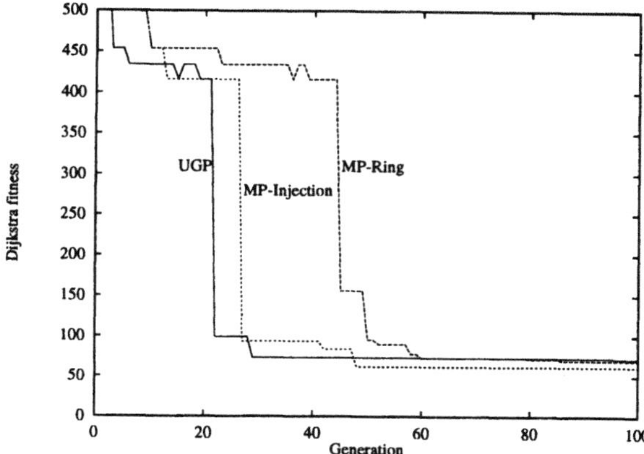

Figure 5: The three techniques on Network 2.

Table 1: GP results for the five networks.

Network Number	Target optimum (s)	Best Fitness obtained by GP	Best % above optimum (P_A)	Technique responsible
1	345	353	2.32	all three
2	622	684	9.97	MP-1
3	176	177	0.57	all three
4	375	603	60.8	all three
5	476	581	22.1	all three

Table 2: Comparison of the three techniques.

Network Number	Uni-population GP			Multipop-Ring			Multipop-Injection		
	Best Dijkstra fitness	Best P_A	Five run average P_A (95% C.I.)	Best Dijkstra fitness	Best P_A	Five run average P_A (95% C.I.)	Best Dijkstra fitness	Best P_A	Five run average P_A (95% C.I.)
1	8	2.32	2.41-3.16	8	2.32	2.31-3.15	8	2.32	2.19-2.91
2	70	11.25	11.19-11.64	70	11.25	11.14-11.56	62	9.97	10.10-11.90
3	1	0.57	0.57	70	0.57	0.57	1	0.57	0.57
4	228	60.8	60.8	228	60.8	60.8	228	60.8	60.8
5	105	22.1	22.0-22.8	105	22.1	22.0-22.8	105	22.1	21.9-22.8

above the optimum. It was under 3% for Network 1, and under 10% for Network 2. For Networks 4 and 5 the performance of GP is certainly poor.

Table 2 lists the comparative performance of the three techniques for each of the five test cases. A 95% confidence interval to cover the five runs within each technique is also computed for better comparison. It can be seen that MP-I has performed better in two networks (1 and 2), but there was little to distinguish the three techniques on the other networks. Overall, the results for Network 3 are nearly ideal, for Networks 1, 2 and 5 are reasonable, but for Network 4, none of the techniques brought the fitness within useful limits.

8 Conclusions and Further Work

The above results demonstrate that genetic programming (GP) has some limited potential, for small networks, to evolve near-optimal fault-tolerant routing rules which are robust enough to be able to solve a high proportion of multiple link failures. Overall, though, this approach lacks adequate performance even for modest-sized networks. In addition, our comparison of uni-population and multi-population GP is arguably inconclusive (*cf.* [4]).

For the future, the use of a context-free grammar [6] to bias the evolution of GP programs for routing holds promise and needs to be explored. Currently, however, the second and third authors are investigating evolving distributed software agents for telecommunications network routing and restoration, rather than centralised routing rules, in an effort to obtain more scalable results.

9 Acknowledgements

The initial experiments on Network 1 were undertaken by the first author as part of an MSc in Telecommunication and Information Systems and subsequently continued by the second author as part of a PhD project, both supervised by the third author. The first author was sponsored by the RAF, and the second by the Government of Pakistan.

References

[1] M. Gondran and M. Minoux. *Graphs and Algorithms.* John Wiley & Sons, Chichester, 1994.

[2] W. D. Grover. The self-healing network: A fast distributed restoration technique for networks using digital cross-connect machines. In *IEEE Global Conference on Communications*, pages 1090–1095, 1987.

[3] J. R. Koza. *Genetic Programming: On the Programming of Computers by Means of Natural Selection.* MIT Press, 1992.

[4] B. Punch, D. Zongker, and E. Goodman. *The Royal Tree Problem, a Benchmark for Single and Multiple Population Genetic Programming*, volume 2. MIT Press, 1996.

[5] W. A. Tackett and A. Carmi. *Simple Genetic Programming in C.* 1993.

[6] P. A. Whigham. Inductive bias and genetic programming. In *Proc. GALESIA'95*, pages 461–466, London, 1995.

[7] D. Zongker and B. Punch. *Lil-GP 1.0 User's Manual.* Michigan State University, 1995.

The Ring-Loading and Ring-Sizing Problem

J. W. Mann[1] and G. D. Smith[2]
[1] Nortel plc, London Road, Harlow, Essex, CM17 9NA, UK.
[2] School of Information Systems, UEA, Norwich, NR4 7TJ, UK.
Email: jasman@nortel.co.uk, gds@sys.uea.ac.uk

Abstract

Ring based structures are often desirable in telecommunication networks since they offer a structure which is inherently fault-tolerant. The simplest such structure consists of a set of nodes connected by links to form a simple cycle. This simple configuration provides high survivability since every pair of nodes is connected by a physically diverse route, and hence no single link failure will disconnect the ring. In the event of failure, all traffic can be diverted as long as the ring has enough capacity. The *ring sizing problem* is then to determine the minimum capacity to handle all traffic while guaranteeing fault protection. In order to solve this problem, the traffic on the ring has to be routed in such a way as to minimise the maximum load on any link, this is termed the *ring loading problem*. In this paper we apply both a genetic algorithm and a simulated annealing algorithm to solve this problem.

1 Introduction

Designing fault-tolerant, or survivable, network configurations is becoming increasingly important since current synchronous optical technology allows telecommunication networks to accommodate extremely high data transmission rates. A failure of one of these high capacity network elements could therefore affect a large amount of traffic. Ring based structures are desirable in telecommunication networks since they offer a structure which is inherently fault-tolerant [10]. A SONET (synchronous optical network — the US standard)/SDH (synchronous digital hierarchy — the European standard) ring is one such structure consisting of a set of nodes connected by high capacity optical links to form a cycle.

There are two protection schemes used within ring configurations; dual protected ring (DPRING) configurations route every traffic request both ways round the ring. In the event of failure, all traffic will still reach its destination. While this scheme guarantees fault protection, it can be wasteful on bandwidth for certain traffic profiles.

Shared protected ring (SPRING) configurations are designed to reduce bandwidth consumption by only sending traffic one way round the ring. In the event of failure, all traffic can be diverted (excluding traffic originating and terminating at a failed node) onto reserved shared protection capacity. To protect all traffic on the ring, it is only necessary to reserve enough capacity to protect the maximum load across a single link. The *ring sizing problem* is then to determine the capacity of the ring including the protection capacity. In order to solve this problem, the traffic on the ring has to be routed in such a way as to minimise the maximum load of traffic routed on any link, this is the *ring loading problem.*

We describe this particular problem for which GA and SA solutions have been developed. Previously the authors have applied GAs and SA to the problem of routing traffic requests through a telecommunications network so as to 'balance' the utilisation distribution [3, 6, 7]. In this paper we are more concerned with routing traffic so as to minimise utilisation. The software developed uses the GAmeter and SAmson toolkits [4, 5] and is described in the following section together with the results of the experiments undertaken.

2 The Ring Loading Problem

The ring loading problem and its application to SONET ring sizing was first presented in [1] which provides a proof that the general ring loading problem is NP-complete.

A telecommunications ring network can be defined by a set of n nodes, N, and m edges (links), E. Associated with each edge $e \in E$ is a cost, $c(e)$. For each distinct $v, w \in N$, we are given an amount of traffic, $t(v, w)$, which must be routed from v to w in the network. This traffic must all be routed on the same path, $P(v, w)$, which is either clockwise or counter-clockwise round the ring. Thus, the ring loading problem can be stated as:

$$\min\{\max_{e\in E}\{\sum_{v,w\in V}\{t(v,w)x_{t(v,w)} + t(v,w)y_{t(v,w)}\}\}\}$$

subject to

$$x_{t(v,w)} + y_{t(v,w)} = 1$$
$$x_{t(v,w)}, y_{t(v,w)} \in \{0,1\} \forall\, t(v,w) \in T$$

where

$$x_{t(v,w)} = \begin{cases} 1 & \text{if traffic is routed clockwise} \\ 0 & \text{otherwise.} \end{cases}$$

$$y_{t(v,w)} = \begin{cases} 1 & \text{if traffic is routed anti-clockwise} \\ 0 & \text{otherwise.} \end{cases}$$

If the traffic $t(v,w)$ can be split so as to send an integral proportion of the traffic clockwise and the rest anti-clockwise, then a polynomial time algorithm can be determined [2]. If the demands can be split arbitrarily, then the ring loading problem is equivalent to solving a linear program and can also be solved in polynomial time [9]. Sending the traffic for each source/destination pair in either one direction or the other makes the ring loading problem computationally very difficult to solve. We apply a GA and SA to this problem by using a representation which maps to the problem naturally.

3 Representation

The software we developed is built on top of the toolkits, GAmeter and SAmson, based on general GA and SA paradigms as described in [8]. These toolkits are built on a Common Toolkit Framework, providing comparable and compatible systems with an open architecture that can be easily transformed and expanded in accordance to the findings of current and future research in the respective fields.

We are aiming to find the path for each pair $(v, w, \in N)$, to carry the traffic between v and w. Thus, the chromosomal representation will have $n \times (n-1)/2$ genes, where each gene should represent the path from v to w. Since the ring loading problem is inherently binary in that the traffic is either sent clockwise or anti-clockwise, we do not need to encode the path taken for each service request, merely the direction taken.

We also include within the evaluation function a term for the length of the paths round the ring. Although the primary aim is to minimise the number

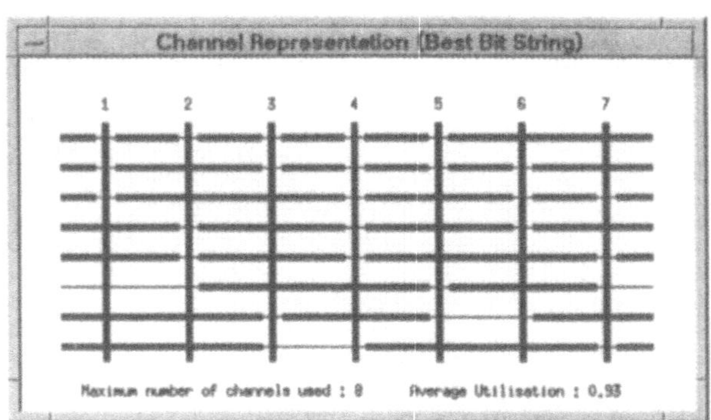

Figure 1: The channel allocation for a ring loading solution for a 7 node ring.

of channels used, this should not be achieved at the sake of sending traffic through more nodes than necessary. This can be achieved by applying suitable weights to both terms in the evaluation function, giving users control over the type of solution obtained. The evaluation is therefore given by

$$Evaluation = (w_1 \times PathCost) + (w_2 \times MaxLoad),$$

where $PathCost$ is the sum of the lengths of all paths. $MaxLoad$ is the maximum load on any link and w_1 and w_2 are the weights which the user can alter to force the application to concentrate on minimising the path costs or the number of channels used.

Both GAmeter and SAmson are fully configurable to provide information about the current progress in a visual or textual manner. We have used this functionality to display solutions to the ring loading problem graphically as shown in Figure 1.

The thick vertical bars are the nodes, in this problem there are seven nodes labelled 1 to 7. Since we are dealing with rings, node 1 is connected to node 2 and node 7. This solution uses 8 channels as depicted by the 8 horizontal lines running through the nodes. The thick horizontal lines show the routes taken by the traffic. For example, the traffic between nodes 1 and 2 takes the shortest route, as shown on the first channel. The traffic between nodes 1 and 3 goes via node 2 as shown on the second channel. The traffic between nodes 1 and 5 also use channel 1, but wrap round the diagram passing through nodes 6 and 7.

Using the toolkits we have experimented with various parameter settings and applied our imple-

Table 1: SA parameter test values.

Initial temperature	10	20	
Cooling Schedule	geometric	arithmetic	imhomogeneous cooling
Temperature duration	constant	adaptive	geometric increases
Neighbourhood operator(s)	flip (F) F, R adaptive	reverse (R) F, R random	move (M) F, R, M adaptive
Accept statistics	all accepted	better accepted	

Table 2: GA parameter test values.

Crossover rate	60	80	
Mutation rate	0.01	0.1	1.0
Pool size	10	25	50
Selection mechanism	roulette	exponential	tournament
Crossover mechanism	single point	multi-point	uniform
Merge mechanism	elitist	new solutions	replace all

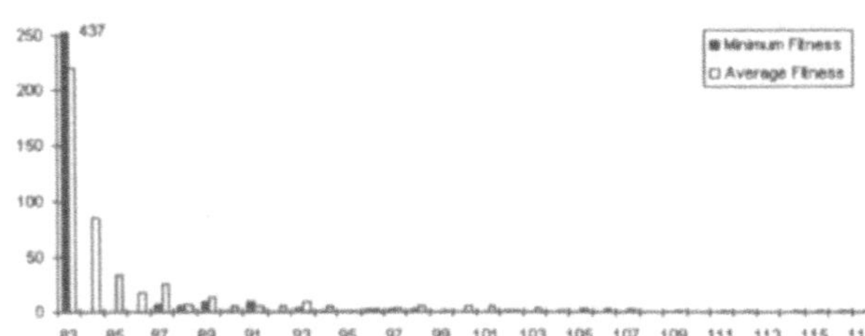

Figure 2: GA results for a 7 node network.

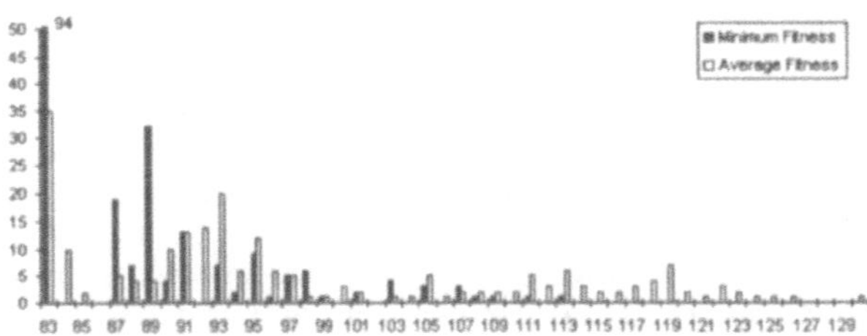

Figure 3: SA results for a 7 node network.

mentations to a set of data files. The next section presents some results achieved.

4 Results

Both the GA and SA were run on four classes of ring networks. The smallest class of network only contained seven nodes and we compared our solutions against that generated manually by an experienced network planner.

For all network classes, both the GA and SA implementations consistently found the best known solutions in many of the experiments. Those solutions that were not the best known solution were still commercially viable.

Since any solutions obtained are generally quasi-optimal solutions, it is more interesting in an industrial setting to test the robustness of such implementations (for example, the traffic distributions may be expected traffic distributions) with respect to the various parameters available within each heuristic.

Both GAs and SA have a number of parameter settings to fine tune and amend. Indeed, in the early years of both algorithms' development, a lot of time was spent determining 'optimal' parameter settings on test problems.

The experiments undertaken in this study attempt to determine how robust these techniques are in terms of parameter settings. With both the GA and SA, common parameter settings were chosen and a series of experiments were made to test various combinations of these parameters. The parameter settings chosen for experimentation in each heuristic are detailed in Tables 1 and 2.

Explanations of the parameters in the tables can be found in [4, 5]. Note that the GA parameter set is larger than that of the SA. The GA contain 486 different parameter combinations while the SA test set consisted of 216 parameter combinations. In all experiments, five runs were carried out on each parameter set and the solution and average solutions from those five parameter sets were recorded. Thus for each *problem*, the GA was ran 2430 times and the SA 1080 times. While these may seem large figures, these are far from exhaustive tests. In the interests of clarity, a summary of the results for two of the network classes tested are described below.

As stated above, the smallest network only consisted of seven nodes and has a known optimal solution. Figures 2 and 3 show the distribution of results for this network from the GA and SA respectively.

Figure 2 shows that GA found the optimal solution at least once out of the five runs a total of 437 times out of 486 experiments and the optimal solution was found in all five runs a total of 221 times. Figure 3 shows that the SA did not fare so well in that the optimal solution was found at least once only 94 times out of 216 experiments and the optimal solution was found in all five runs a total of 34 times.

Concentrating on a larger size network, Figures 4 and 5 show the GA and SA distribution of results respectively. A similar pattern appears with the GA finding the best known solution at least once in the five runs, 320 times out of the 437 experiments and finding this solution in all five runs, 105 times. The SA managed to find this solution 39 times and in 7 experiments reached this solution in all five runs. This pattern of results is repeated for the other network classes.

Note, that the experiments are *not* comparing the quality of solutions obtained by either heuristic, but the robustness with respect to the parameter settings available within either heuristic. We can deduce from these results that the SA is much more sensitive to its parameter settings than the GA. This is partly due to the parameters tested; a crossover rate of 60% does not generally have much difference than that of 80%. However, since SA is a local search-based technique, the neighbourhood function used plays a very important part in the success of any implementation.

With the GA, although many parameter settings worked well, there exists certain combinations of parameters that will generally yield sub-optimal results. Thus, small populations with a very small mutation probability and a convergent merge strategy should generally be avoided. For the SA implementation, the cooling schedule played an important part in the quality of the final solutions. An inhomogeneous schedule generally produced superior results while experiments using an arithmetic-based schedule usually fared worse. The other major factor in SA implementations, the neighbourhood function, showed that a reverse neighbourhood (in this problem) fared worse than the other two when applied singularly.

5 Conclusion

It has been shown that both the GA and SA approach can be used to find solutions of commercial quality for the ring loading problem. Both GAs and SA have a number of strengths; they are relatively easy to understand and implement, they can provide a plurality of solutions with only limited development cost, and perhaps most importantly, the software designed for a particular problem is readily adaptable to small variations of the original specification.

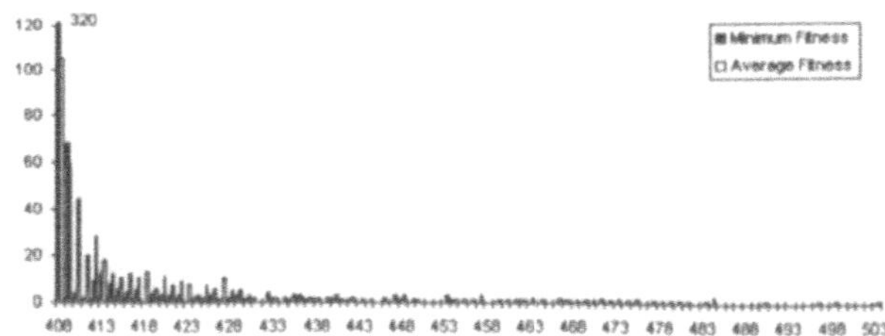

Figure 4: GA results for a 12 node network.

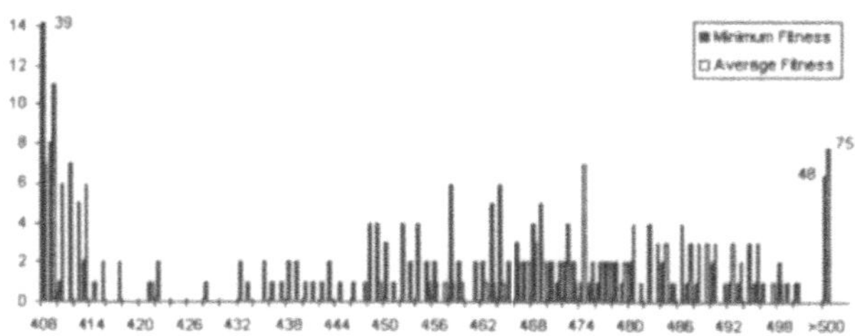

Figure 5: SA results for a 12 node network.

It has also been shown that with a simple representation, the basic GA can perform very well and that 'tweaking' individual parameters is not necessary. The SA, on the other hand, needs care in the choice of parameters that gain the most from the representation. The SA parameters available give the designer more control over the convergent ability of the algorithm than the GA parameters. Whether this is an advantage depends on the objective. If finding the best solution possible is more important, the extra degree of freedom in the SA parameters may help in attaining that objective. On the other hand, if a solution is required with minimal effort, the basic GA with no fine tuning, can be just as effective.

6 Acknowledgements

The research is being undertaken within a Teaching Company Scheme programme, grant reference GR/K40086, between Nortel plc (formerly BNR Europe Ltd) and the University of East Anglia.

References

[1] S. Cosares and I. Saniee. An optimisation problem related to balancing loads on SONET rings. *Telecommunication Systems*, 3(2):165–181, 1994.

[2] A. Frank, T. Nishizeki, N. Saito, H. Suzuki, and E. Tardos. Algorithms for routing round a rectangle. *Discrete Applied Mathematics*, 40:363–378, 1992.

[3] J. W. Mann. Applications of genetic algorithms in telecommunications. Master's thesis, University of East Anglia, 1995.

[4] J. W. Mann. *X-SAmson v1.5 User Manual.* University of East Anglia, 1996.

[5] J. W. Mann, A. Kapsalis, and G. D. Smith. The GAmeter toolkit. In V. J. Rayward-Smith, editor, *Applications of Modern Heuristic Methods*, chapter 12, pages 195–209. Alfred Waller, 1995.

[6] J. W. Mann, V. J. Rayward-Smith, and G. D. Smith. Telecommunications traffic routing: A case study in the use of genetic algorithms. In *Proceedings of the Applied Decision Technologies (ADT'95) - Modern Heuristic Search Methods*, pages 315–325, Uxbridge, 1995. UNICOM Seminars Ltd.

[7] J. W. Mann and G. D. Smith. A comparison of heuristics for telecommunications traffic routing. In V. J. Rayward-Smith, I.H. Osman, C. R. Reeves, and G. D. Smith, editors, *Modern Heuristic Search Methods*, chapter 14, pages 237–256. John Wiley, 1996.

[8] V. J. Rayward-Smith. A unified approach to tabu search, simulated annealing and genetic algorithms. In V. J. Rayward-Smith, editor, *Applications of Modern Heuristic Methods*, chapter 2, pages 17–38. Alfred Waller, 1995.

[9] A. Shulman, R. Vachani, J. Ward, and P. Kubat. Multicommodity flows in ring networks. Technical Report 02254, GTE Laboratories, 1991.

[10] A.R.P. White, J. W. Mann, and G. D. Smith. Genetic algorithms and network ring design. *Annals of Operations Research.* to appear.

Evolutionary Computation Techniques for Telephone Networks Traffic Supervision Based on a Qualitative Stream Propagation Model

I. Servet[1], L. Travé-Massuyès[1] and D. Stern[2]
[1] LAAS/CNRS 7, avenue Colonel Roche 31077 Toulouse Cedex France
[2] CNET 38-40, rue du Général Leclerc 92131 Issy-les-Moulineaux France
Email: servet@laas.fr

Abstract

Evolutionary computation techniques have received a great deal of attention regarding their potential as optimization techniques for complex functions. In this paper, we consider three of them: multiple restart hill-climbing, population-based incremental learning and genetic algorithms. Their binary version and a real-coded variant of each of these techniques are experimented on a real problem: traffic supervision in telephone networks. Indeed, this task need to determine streams responsible for call losses in a network by comparing their traffic values to nominal values. However, stream traffic values are not directly available from the on-line data acquisition system and, hence, have to be computed by inverting a computational model of stream propagation in circuit-switched networks only based on the Erlang's formula plus qualitative knowledge about the network. Then, our stream propagation model inversion has been computed thanks to the previous techniques and using several fitness measures to show how their choice can impact on the final results.

1 Introduction

Evolutionary computation is an umbrella term used to describe computer-based problem solving systems which use models of evolutionary processes as key elements in their design and implementation [6].

Their common base is the simulation of the evolution of *individuals* via processes of *selection, mutation* or *crossover* which depend on the performance, computed by a *fitness measure*, of the individuals.

Hence, before an evolutionary algorithm can be run, a suitable representation for each structure must be determined. It is assumed [3] that a potential solution to a problem may be represented as a set of parameters, known as *genes* that are joined together to form a string of values, often called a *chromosome*.

Besides, these algorithms are often used for function optimization [2] whereas their effectiveness strongly depends on this function. Thus, in systems with epistasis (i.e. high connectivity between genes), such as circuit-switched networks, only empirical studies can show which technique gives better results.

Moreover, whereas binary algorithms are usually considered, a lot of researchers have supported the idea of real genes. Eshelman *et al.* [5] shows that real-coded variants often have an advantage over a binary version (they exploit local continuities in function optimization).

2 Evolutionary Techniques

Three techniques are presented here:

- *multiple restart hill-climbing*: an heuristic and, consequently, a very simple method.
- *population-based incremental learning* that can be seen as an abstraction of genetic algorithms.
- *a genetic algorithm*, a well-known method for solving complex optimization problems.

They have been chosen both for the good results they are able to give and for their increasing complexity.

Then, we have conceived, according to the genetic algorithm real variant principles, a real variant of the two other methods which were initially binary ones.

2.1 Multiple Restart Hill-Climbing (MRHC)

Binary Version

This iterative method, however very simple, sometimes leads to better results than genetic algorithms [1]. The simplest algorithm that described it is:

$V \leftarrow$ randomly generated solution vector
$BEST \leftarrow V$
Loop NB_ITERATIONS
  $MUT\_POSITION \leftarrow$ random bit position
  $N \leftarrow$ mutation of V in $MUT\_POSITION$
  If N is better than $BEST$ **Then** $BEST, V \leftarrow N$

Real-Coded Variant

First, the solution vector V, which is randomly generated, is a real vector. Hence, when optimzing a function, each component of this vector represents the value of the corresponding variable.

Then, $MUT\_POSITION$ corresponds to the number of the variable which has to undergo a mutation. Therefore, a mutation of the ith component $V[i]$ of solution vector V consists, like in genetic algorithms real-coded variants [4], in adding to $V[i]$ a small random real (for instance a real in $[-V[i]/10, V[i]/10]$).

2.2 Population-Based Incremental Learning (PBIL)

Binary Version

Contrary to the previous method, this one does not study a single solution vector but a set of *nb* solution vectors called a *generation*. The object of this method is to create a real-valued ***probability vector*** P which is dedicated to reveal high quality solution vector with high probabilities [3]. Hence, P is updated at each generation and the solution vectors of each generation are created according to its values: each component $V[i]$ of each solution vector V is determined as follows:

Given $random_i$ a random value in [0,1],
If $random_i > P[i]$ then $V[i] \leftarrow 0$ else $V[i] \leftarrow 1$

Therefore, $P[i]$ represents the probability to have '1' at this position in the best solution vector. P evolution, from a generation to an other, is computed thanks to competitive learning mechanisms [1].

Real-Coded Variant

First, we assume that each gene (each variable in function optimization) has a continuous and bounded definition set: each component $V[i]$ of a vector V s assumed to belong to the interval $[Low_i, Up_i]$. Then, each component $P[i]$ of P gives the probability of the ith variable to be greater than $(Low_i + Up_i)/2$.

The probability vector updating is similar to P updating in the binary version but a stage for intervals $[Low_i, Up_i]$ updating is added: when, for a given iteration, the ith component of P is such that:

- **P[i] $\geq$ 0.9**, V[i] is assumed to be greater than $(Low_i + Up_i)/2$. Then, Low_i is updated to $(Low_i + Up_i)/2$ and P[i] is reinitialized at 0.5.
- **P[i] $\leq$ 0.1**, V[i] is assumed to be lower than $(Low_i + Up_i)/2$. Then, Up_i is updated to $(Low_i + Up_i)/2$ and P[i] is reinitialized at 0.5.

P reinitialization is due to the fact that the change of a bound of the interval $[Low_i, Up_i]$ comes to start again the algorithm with a new definition set.

2.3 A Genetic Algorithm (GA)

Binary Version

In this method, an offspring can be the result of gene mutation, as in the two previous techniques, but also the result of two chromosomes undergoing crossover. We have chosen to apply *1-point crossover* [3]. Hence, a genetic algorithm description [1, 3, 6] is:

Generate $POPULATION\_SIZE$ random vectors
Compute fitness of each individual of the population
Loop $GENERATIONS$
  Loop $POPULATION\_SIZE/2$
  • Select probabilistically (according to the fitness measure) 2 vectors ONE and TWO
  • Recombine (thanks to crossover) ONE and TWO to give $CHILD_1$ and $CHILD_2$
  • Perform mutation according to MUT_RATE of $CHILD_1$ and $CHILD_2$
  • Compute fitness of $CHILD_1$ and $CHILD_2$ and insert them in the new generation
  $WORST \leftarrow$ worst vector of the new generation
  $BEST \leftarrow$ best vector of the old generation
  Replace $WORST$ by $BEST$ in the new generation

Real-Coded Variant

As in the real-coded variant of MRHC, the only changes in real-coded GAs concern the way the chromosomes operators are done. The mutation operator has been previously described in Section 2.1.2. Moreover, many real-coded crossover operators can be envisaged. However, to keep the idea of some random choice, we apply a ***blend-crossover*** (BLX-α) [5] which is able to create two offsprings.

Let us consider two parents p_1 and p_2 and their two offsprings c_1 and c_2. Then, for i=1,2, each component $c_i[j]$ of c_i is a random value that belongs, if we assume $p_1[j] < p_2[j]$, to the interval

$$[p_1[j] - \alpha|p_1[j] - p_2[j]|, p_2[j] + \alpha|p_1[j] - p_2[j]|]$$

where α is a user specified GA parameter.

3 Stream Propagation Model

Due to the increasing size of telephone networks and to their high connectivity (several organs are shared by several streams), the task of networks managers becomes more and more complex. Determining which streams are responsible for an overloaded situation is difficult, especially as the disturbances may propagate within the network in a very short time frame.

Real-time network supervision does not require accurate values (approximate values are sufficient enough to incriminate the streams responsible for overloads) but they need to be computed rapidly. Thus, the stream propagation computation cannot be done using a simulation program because of the extremely long computer running time it requires. This is the reason why analytical methods are interesting [7]. The simplest approach for network analysis is the one-moment method where the offered traffic, modelled by a Poisson arrival process is only characterized by its mean. The model presented here uses this method and has been developed under classical assumptions of single moment traffic modeling methods [7, 8]. This model is based on the concept of *blocking* organs and the search of such elements in the network.

Definition 1. *A network element: exchange or circuit group is said to be* blocking *when, according to the network structure, it is likely to experiment traffic loss.*

It can be divided in three main stages [9]:

- *Blocking organ search* which is performed thanks to two qualitative rules: the overdimensioned capacities rule (independent of the offered traffic values) and the loss rate rule which is performed using **Erlang's formula**.

  Definition 2. *Consider an organ whose capacity is N and which routes a traffic stream of intensity A. Then, the probability that the N circuits are busy is:*

  $$E[N, A] = \frac{A^N/N!}{1 + \ldots + A^N/N!}.$$

- *Blocking organs ordering.* This is especially important when overflow from a circuit group to an other is allowed, the offered traffic of the latter depending on the amount of lost traffic on the former.

- *Traffic loss computation* for each blocking organ, which is computed thanks to the *resource availability backpropagation* algorithm [9]

4 Model Inversion

To supervise the French long distance telephone network, we propose to determine streams responsible for call losses by comparing their traffic values to nominal values. However, stream traffic values are not directly available from the on-line data acquisition system and, hence, need to be computed in real time. This is done by inverting the previous model thanks to the techniques presented in Section 2.

It consists of iteratively calling the stream propagation model with stream traffic values updated by one of the previous methods and in using a fitness measure to compute the distance between the observed values and computed values. First, the observed (measured) quantity can be either the number of calls offered to each organ or the stream carried traffic value, which are the only on-line available measures in the French long distance telephone network. Then, the distance between observed and computed values can be:

- Euclidean Distance (ED), defined by

  $$\sqrt{(\sum \text{observed value} - \text{computed value})^2}$$

- χ^2 distance (χ^2), defined by

  $$\sqrt{\sum \frac{(\text{observed value} - \text{computed value})^2}{\text{observed value}}}$$

- Infinite Norm (IN), defined by

  $$max|\text{observed value} - \text{computed value}|$$

These distances have been chosen for their simplicity.

The six methods described in Section 2 have been tried successively with the six previous fitness measures, the three distances being applied to the number of offered calls (C) and the stream carried traffic (T). In a perspective of real-time traffic supervision, each of these methods has been computed 500 iterations only.

4.1 Numerical Results

Let us consider the following network:

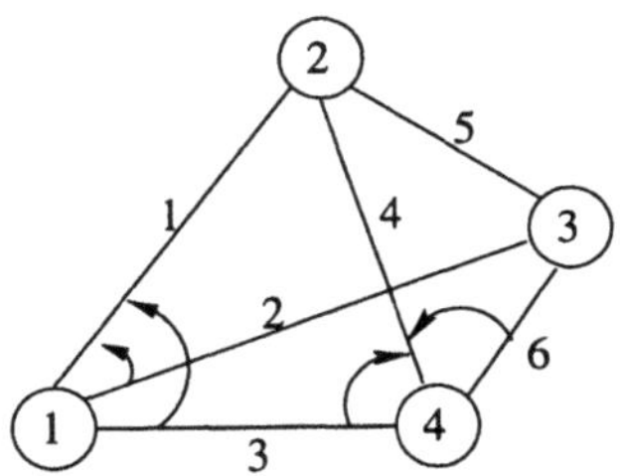

where the switch capacities are infinite and the circuit groups ones are:

Circuit group	1	2	3	4	5	6
Capacity	50	20	30	20	15	50

and the offered traffic of (I, J) streams (going from switch I to switch J)are:

(I,J) stream	(1,2)	(1,3)	(1,4)	(4,2)	(4,3))
Traffic value	15	30	45	10	60

Then, the different algorithms previously presented provide the results in Table 1.

Table 1

		(1,2)	(1,3)	(1,4)	(4,2)	(4,3)
Binary MRHC	IN (T)	13.89	14.94	3.35	3.85	1.80
	IN (C)	3.91	14.94	3.35	3.85	1.80
	ED (T)	7.68	4.28	31.33	1.63	22.33
	ED(C)	27.26	14.94	3.35	3.85	1.80
	χ^2 (T)	5.72	12.53	8.57	2.88	27.56
	χ^2 (C)	27.26	14.94	3.35	3.85	1.80
Real MRHC	IN (T)	13.39	26.49	35.45	8.07	49.93
	IN (C)	0.15	0.65	1.68	3.22	0.98
	ED (T)	14.17	27.84	39.09	9.30	52.20
	ED (C)	14.86	28.73	43.18	9.90	58.71
	χ^2 (T)	14.14	27.84	38.51	9.03	51.00
	χ^2 (C)	14.93	29.87	43.45	9.57	58.93
Binary PBIL	IN (T)	6.14	11.30	24.93	3.59	31.31
	IN (C)	7.63	11.01	27.07	6.42	22.42
	ED (T)	4.75	14.65	32.41	2.89	21.12
	ED (C)	4.78	14.77	32.45	6.91	20.99
	χ^2 (T)	4.78	14.77	32.45	6.91	20.99
	χ^2 (C)	7.62	11.07	26.94	6.40	22.99
Real PBIL	IN (T)	112.9	111.0	218.2	159.7	264.0
	IN (C)	59.25	1.87	15.19	2.25	55.50
	ED (T)	294.9	204.4	457.9	70.50	147.0
	ED (C)	18.37	49.50	51.19	20.50	47.62
	χ^2 (T)	15.00	295.5	40.50	86.50	135.0
	χ^2 (C)	24.00	21.75	36.00	1.75	69.75
Binary GA	IN (T)	4.96	9.99	31.72	6.94	28.81
	IN (C)	4.80	11.95	31.68	6.60	23.84
	ED (T)	4.80	11.95	31.68	6.60	23.84
	ED (C)	4.17	8.54	31.68	6.82	18.83
	χ^2 (T)	4.80	11.95	31.54	8.12	23.84
	χ^2 (C)	4.80	8.54	32.55	5.57	23.84
Real GA	IN (T)	11.27	33.42	30.87	5.83	41.12
	IN (C)	13.55	29.38	38.21	7.58	58.67
	ED (T)	13.29	27.91	40.14	9.37	52.30
	ED (C)	18.34	33.50	36.11	8.43	64.55
	χ^2 (T)	14.33	27.11	37.32	10.90	54.33
	χ^2 (C)	14.69	31.34	40.39	7.98	61.50

5 Conclusion

First of all, although the inverted model gives only approximate values and despite its complexity, non-linearity and non-differentiability, the results computed thanks to evolutionary computation techniques are rather good.

Then, an important conclusion of our study is the influence of distance choice on evolutionary computation techniques performance: in our case, the χ^2 distance with offered calls always gives the best results whereas the results given by the distances applied to the carried traffic are quite bad. Unfortunately, actually, there is no way, except empirical methods of course, to say whether a distance is able to give better results than an other one or not.

Besides, our application clearly show that a real variant, that is closer to human reasoning, can outperform a binary evolutionary technique. Then, the real variant of an heuristic method, MRHC, gives results comparable to the ones given by a standard GA.

6 Acknowledgements

This work has been supported by CNET as part of contract nr. 93 1B 142, project 513.

References

[1] S. Baluja. Population-based incremental learning: a method for integrating genetic search based function optimization and competitive learning. Technical Report CMU-CS-94-163, 1994.

[2] S. Baluja. An empirical comparison of seven iterative and evolutionary function optimization heuristics. Technical Report CMU-CS-95-193, 1995.

[3] D. Beasley, D. R. Bull, and D. R. Martin. An overview of genetic algorithms: Part1, fundamentals. *University computing*, 15(2):58–69, 1993.

[4] D. Beasley, D. R. Bull, and D. R. Martin. An overview of genetic algorithms: Part 2, research topics. *University computing*, 15(4):170–181, 1993.

[5] L. J. Eshelman and J. D. Schaffer. *Real-coded genetic algorithms and interval-schemata*, volume 2. 1992.

[6] J. Heitkokker and D. Beasley. The hitch-hiker's guide to evolutionary computation: A list of frequently asked questions. available by anonymous ftp at rtfm.mit.edu., 1994.

[7] F. Le Gall, J. Bernussou, and J. M. Garcia. *A one-moment model for telephone networks with dependance on link blocking probabilities*, pages 449–458. 1984.

[8] A. Passeron. Notions élémentaires sur le trafic téléphonique. Technical Report DE/ATR/57.84, CNET, Issy-les-Moulineaux, 1984.

[9] I. Servet, L. Travé-Massuyès, and D. Stern. Traffic supervision based on a one-moment model of telephone networks built from qualitative knowledge. In *Proc. IMACS/IEEE CESA'96*. Villeneuve d'Asq (France), 1996.

NOMaD: Applying a Genetic Algorithm/Heuristic Hybrid Approach to Optical Network Topology Design

M. C. Sinclair
Dept. of Electronic Systems Engineering, University of Essex,
Wivenhoe Park, Colchester, Essex CO4 3SQ, UK.
Email: mcs@essex.ac.uk

Abstract

This paper describes the use of a genetic-algorithm (GA)/heuristic hybrid approach in a tool for optical network modelling, optimisation and design (NOMaD) being developed by the author at the University of Essex [11] and, in particular, early results from its application to virtual-topology design.

NOMaD is used as part of the author's own research into the application of GA/heuristic hybrid optimisation techniques to network design, as well as in several research projects, including two Advanced Communications Technologies and Services (ACTS) projects [6]: WOTAN (wavelength-agile optical transport and access network) and OPEN (optical pan-European network).

1 Introduction

The design problems that NOMaD seeks to address will, at least initially, be those presented by multi-wavelength all-optical telecommunications networks, including combined access/core networks at national level and transport networks at the international, as well as consideration of both 'flat' and hierarchical approaches. The two key application areas are:

- Topological design, starting by making the choice of links in both the virtual and physical topologies (the virtual topology ignores the underlying physical reality of which ducts and fibres will actually be used), and then building on that to network extensions (additional node placement), and perhaps eventually node placement for an entire network.
- Routing and wavelength allocation, initially taking account of only the virtual topology, but then recognising the need to include duct and fibre choice.

To begin with, NOMaD will model networks with static traffic demands, but subsequently it is hoped to incorporate dynamic traffic into the network model. Further, although at the start the focus will be on cost optimisation (whilst meeting certain performance and reliability constraints), additional performance metrics will be added later. It is also envisaged that NOMaD will allow the interactive study of network failure scenarios as part of the overall assessment of a design.

2 Overall Architecture

NOMaD is being developed as an overall architecture that will be extended by several people in a variety of directions to serve the needs of their individual projects. Consequently, its users are not just given a single unchangeable piece of software, but rather a framework which they can actively add to for themselves. As a result, NOMaD has to be easy to understand, flexible and extensible. The overall layered architecture of NOMaD is shown in Figure 1.

To aid both understanding and extensibility, the network optimisation, modelling and design toolset is being developed using object technology. The analysis and design is being documented using Booch Notation [1], and the implementation done in the C++ programming language. This process is aided by the use of a CASE tool, Rational Rose/C++ [9], which generates skeletal C++ code directly from the diagrams and specifications created by the author.

To provide flexibility, rather than only providing a C++ interface for users of the NOMaD network toolset, a NOMaD-specific tcl extension is also being developed. Tcl [7] is a scripting language designed to be extended by the incorporation of application-specific commands, coded in C or C++. Consequently, `NOMaDsh` can be used either interactively to access the underlying network toolset — creating networks, modifying them, assessing their

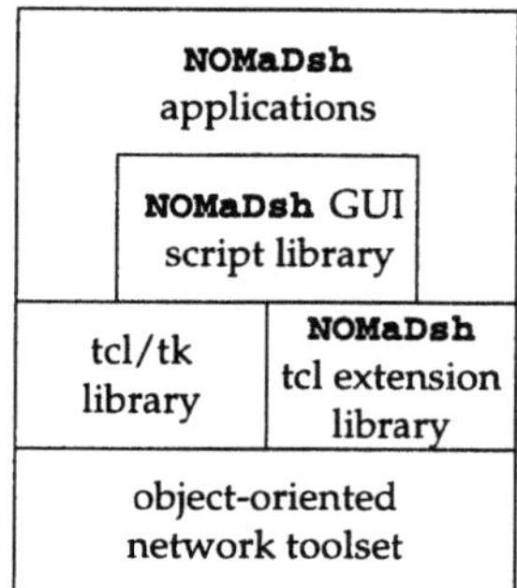

Figure 1: Overall NOMaD architecture.

cost/performance, etc. — or via scripts written using a combination of tcl and the NOMaD-specific commands. In addition, a library of graphical user interface (GUI) scripts is being developed, using the tk toolkit (tcl/tk), that allow `NOMaDsh` scripts to edit and display networks.

Further details on the initial architecture of NOMaD can be found in [11], including the object-oriented network toolset, the tcl scripting layer and the GUI.

3 GA/Heuristic Hybrids

Within the field of GA research, there are two 'schools of thought' with respect to GAs. One approach [3] is to try and develop an algorithm that is robust and general in its application *i.e.* it makes use of no problem-specific information — the focus of the research is on the development of a better GA for all problems. The other [2] is to blend the traditional GA with problem-specific operators, often based on existing heuristic algorithms for a particular problem (or problem domain), such that the combined algorithm performs better than either a 'pure' GA or the heuristic alone. This approach is more pragmatic in nature, with the better solution of particular engineering problems being the primary focus of research.

The GA/heuristic hybrid approach thus adopts a problem-specific encoding, and then develops and employs problem-specific operators that together combine the best existing heuristics for the problem with an overall GA framework. The motivation for this approach is simple, according to Davis: "Although genetic algorithms using binary representation and single-point crossover and binary mutation are robust algorithms, they are almost never the best algorithms to use for any [specific] problem" [2]. It is almost always possible to develop a problem-specific GA/heuristic hybrid that will outperform either the heuristic or a traditional GA alone, and this is one of the approaches that is used in NOMaD.

For a brief survey of work on the application of GAs to network design, see [8], which indicates that even without the incorporation of heuristic-based operators, GAs are often able to match heuristics alone in terms of results, if not in speed. Building on this promise, GA/heuristic hybrids should prove even better for network design.

4 O-O Network Toolset

Clearly, NOMaD's object-oriented network toolset is too large to present in detail here. Consequently, we will focus on those elements that, on the one hand, represent the structures undergoing adaptation, and on the other, implement the overall GA/heuristic hybrid framework itself.

The main `NetworkModel` class diagram is given in Figure 2, which shows the `Network` class, composed not only of `Nodes`, `Links`, and `PathLossSeqs` (ordered sequences of `Paths`), but also several implementation classes used to represent the `Network`'s adjacency matrix, connection matrix (which records `Link` levels in hierarchical networks) and the traffic requirements. To date, the focus has been on the virtual-topology design of both flat and hierarchical networks, but in future, the network model will also incorporate classes to represent a network's physical topology (e.g. `Ducts` and `Fibres`). Within NOMaD, `Network` objects are themselves the structures undergoing adaptation — a highly structured and very problem-specific representation that requires no decoding or interpretation.

Figure 3 is the main `GA` class diagram, illustrating the `GeneticAlgorithm` class, with its `Population` of `Individuals` and its `OperatorPool` of `Operators`. An `Individual` simply consists of a single `Network` object and some additional accounting information. As in Davis [2], the probabilities of individual `Operators` being applied to an `Individual` adapt through the course of a run using a credit-assignment algorithm, rather than being chosen with a fixed probability throughout. Whenever an `Individual` is created whose fitness exceeds that of the best `Individual` so far (`bestIdv`), the improvement in fitness is credited to the `Operator` responsible. In addition, a decreasing fraction of the credit (say, 0.5 and 0.25) is awarded to the parent `Operator` (*i.e.* the `Operator` that created this

Individual's parent, parentOp) and grandparent Operator (*i.e.* grandParOp), to ensure that two- and three-Operator sequences are rewarded appropriately. After a few generations (say, 4), a small proportion (say, 0.15) of the Operator probabilities is reassigned according to the average credit earned per child each Operator has generated. In addition, Operators are given a minimum probability (say, 0.05) to ensure they do not lose the opportunity to gain some credit if they, having decayed to the minimum level, are later found to be useful to further evolve the Population.

A variety of operators have been incorporated within NOMaD (by creating derived classes from either UnaryOp or BinaryOp) including those used in this paper: MutateLink (single-link mutation, ML), SPLinkCO (standard single-point crossover, SPCO), LinkGroupCO (link-group crossover, LGCO), DegreeTwo (degree-two operator, DEG2) and DegreeDec (degree-decrement operator, DEGD). Link-group crossover selects a group of nodes in one network (those whose distance from a randomly-selected node is less than or equal to the distance to a second randomly-selected node) and then exchanges the links used within that group with those used within the same group of nodes in the other network. It was believed that this would lead to better building blocks than a standard single-point crossover with a linear encoding of the network topology, as employed by some earlier authors [5, 10]. The degree-two operator alters a network by adding links to nearest nodes, where necessary, making all nodes at least of degree two (*i.e.* connected to two other nodes), to ensure adequate reliability and avoid a cost penalty. The degree-decrement operator simply deletes a random link from the highest degree node (if there is one of degree greater than two), hopefully thereby reducing the network cost. In general, operators derived from UnaryOp are analogous to mutation in a traditional GA, but can be heuristics that incorporate arbitrary amounts of problem-specific information. Likewise, BinaryOp subclasses are analogous to traditional crossover, although once again, there is considerable potential for problem-specific heuristics.

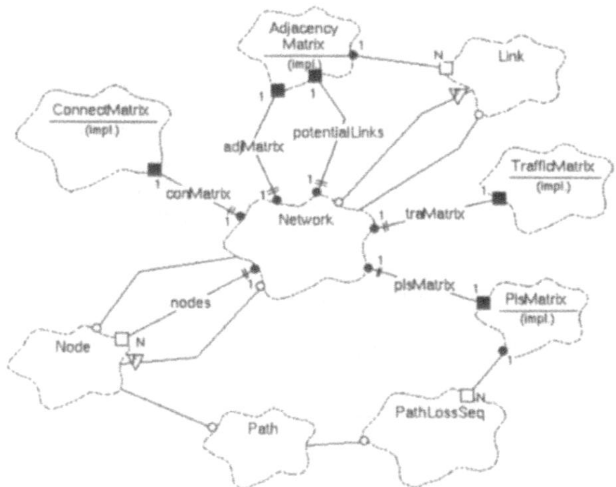

Figure 2: NetworkModel class diagram.

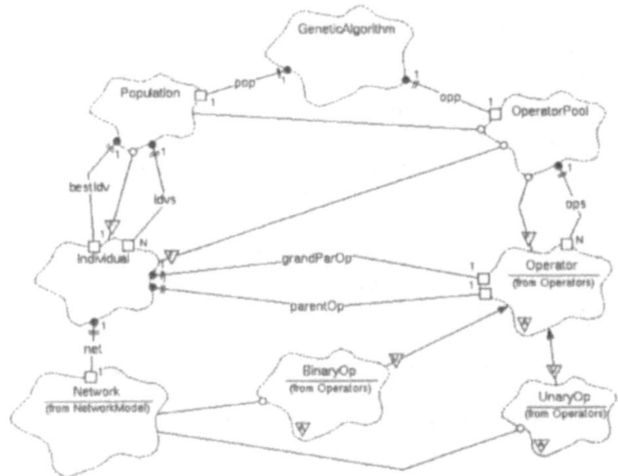

Figure 3: GA class diagram.

5 Experimental Results

To assess the relative performance of the above operators, five test networks were generated using the approach described in [4], although there were further modified to ensure reasonable node separations. Each network had fifteen nodes, used a 1,000km × 1,000km area and carried an overall traffic of 1,500Gbit/s. The network costs were assessed using the model in [10].

Three genetic algorithms were employed. Each had a population size of 200, used tournament selection (of size 4), and was limited to a maximum of 5,000 trials (*i.e.* somewhat more than 25 generations). The operators used, and their initial probabilities (chosen with a few trial runs), are given in Table 2.

Table 1: Results.

Network	GA1		GA2		GA3	
	Cost	Diff.	Cost	Diff.	Cost	Diff.
1	4,974,888	20,906	4,953,982	0	4,965,731	11,749
2	4,785,018	34,002	4,751,016	0	4,756,635	5,619
3	4,449,713	34,311	4,415,402	0	4,426,182	10,780
4	4,619,478	20,579	4,604,944	6,045	4,598,899	0
5	4,461,352	34,021	4,427,331	0	4,430,814	3,483

Table 2: Genetic algorithms.

GA	ML	SPCO	LGCO	DEG2	DEGD
1	0.35	0.45			
2	0.35		0.45		
3	0.25		0.35	0.10	0.10

The three GAs were each applied five times with different random seeds to each of the five networks, and the lowest cost obtained for each group of independent runs. In addition, the lowest overall result for each network from the three different GAs was found, and used to calculate the difference in cost (above the best) of the results found for that network by the other two GAs (see Table 1).

Clearly, for these five test networks and the cost model employed, link-group crossover (GA2) is uniformly better than single-point crossover (GA1). Surprisingly, however, combining the degree-two and degree-decrement operators with link-block crossover only produced an improvement in one case; in all the others, it actually seemed to reduce performance slightly.

6 Conclusions & Further Work

The use of a GA/heuristic hybrid approach in a tool for optical network modelling, optimisation and design (NOMaD) has been described and, in particular, its application to virtual-topology design. The experimental results show that the adoption of a problem-specific crossover can enhance GA performance in this context. However, incorporation of simple heuristic operators has so far provided more mixed results. It is anticipated that, as well as further exploring those operators already developed and the operator-probability adaptation algorithm, future work will investigate other, more powerful heuristic operators to incorporate into NOMaD, as well as addressing wavelength allocation and routing, and physical-topology design. For these latter problems, more difficult than virtual-topology design, the GA/heuristic hybrid approach may better fulfill its promise. Overall, although NOMaD is still being developed, these early results are encouraging, and it is envisaged that NOMaD will continue to play a central role in both network design and GA/heuristic hybrid research at the University of Essex.

7 Acknowledgements

The author is grateful to Katerina Proestaki and Noel Parnis (also of E.S.E., University of Essex) for the many interesting and stimulating discussions during the continuing development of NOMaD. They are both supported by the European Commission under the ACTS programme: Ms. Proestaki on project AC029 (WOTAN) and Mr. Parnis on project AC066 (OPEN).

References

[1] G. Booch. *Object-Oriented Analysis and Design with Applications (2nd Ed.)*. Benjamin-Cummings, 1994.

[2] L. Davis. *Handbook of Genetic Algorithms*. Van Nostrand Reinhold, 1991.

[3] D. E. Goldberg. *Genetic Algorithms in Search, Optimization and Machine Learning*. Addison-Wesley, 1989.

[4] P. S. Griffith, A. Proestaki, and M. C. Sinclair. Heuristic topological design of low-cost optical telecommunication networks. In *Proc. 12th UK Performance Engineering Workshop*, pages 129–140. University of Edinburgh, 1996.

[5] J. Hewitt, A. Soper, and S. McKenzie. CHARLEY: A genetic algorithm for the design of mesh networks. In *Proc. GALESIA'95*, pages 118–122. University of Sheffield, 1995.

[6] A. M. Hill and A. J. N. Houghton. Optical networking. In *European ACTS programme OFC'96*. San Jose, USA, 1996.

[7] J. K. Ousterhout. *Tcl and the Tk Toolkit*. Addison-Wesley, 1994.

[8] A. Proestaki and M. C. Sinclair. Initial survey of heuristics for optical core network design. In *Proc. 13th UK Teletraffic Symposium*, pages 20/1–20/8. University of Strathclyde, Glasgow, 1996.

[9] Rational Software Corporation. *Using Rational Rose/C++ (Revision 2.7)*, 1995.

[10] M. C. Sinclair. Minimum cost topology optimisation of the cost 239 european optical network. In *Artificial Neural Nets and Genetic Algorithms*, D. W. Pearson, N. C. Steele, R. F. Albrecht (editors), Pages 26–29. Wien New York, 1995. Springer-Verlag.

[11] M. C .Sinclair. NOMaD: Initial architecture of an optical network optimisation, modelling and design tool. In *Proc. 12th UK Performance Engineering Workshop*, pages 157–167. University of Edinburgh, 1996.

Application of a Genetic Algorithm to the Availability–Cost Optimization of a Transmission Network Topology

B. Mikac and R. Inkret
University of Zagreb, Faculty of Electrical Engineering and Computing,
Department of Telecommunications, Unska 3, HR-10000 Zagreb, Croatia
E-mail: branko.mikac@fer.hr

Abstract

The paper presents an approach to topology optimization of all-optical telecommunications network, minimizing pair(s) of unavailability — cost values, by means of a genetic algorithm. The solutions satisfy traffic requirements, technological limitations, and the defined routing rules.

1 Introduction

The paper deals with the issues involved in generating an optimum topology of an European core all-optical network — a case study within the framework of the European Commission project COST 239 'Ultra-high capacity optical transmission networks' [3]. The objective of the optimization is the minimization of network unavailability and cost, while satisfying the traffic requirements among the major European cities, meeting current technological limitations in the optical domain, and the defined routing rules. The problem could be defined in another way too: how to minimize the network cost while keeping unavailability within the prescribed requirements, if possible. The goal is not only to have as minimum unavailability as possible, despite the high costs of the network, but to achieve a low cost topology which fulfills the availability requirements, if any. In order to find an optimum topology for n nodes network, one should consider 2^b different solutions (topologies), where $b = n(n-1)/2$. Even for a small number of nodes (in our case study the network comprises 11 nodes and has the set of 3.6 10^{16} different topologies) only a quasi-optimal solution could be obtained, using some heuristic search techniques. This paper presents the application of a genetic algorithm.

2 Assumptions

In order to obtain an acceptable network topology, let us call it a regular topology, different requirements, limitations and routing rules have to be fulfilled. The network topology must fulfill the following requirement: all node-to-node connections should be established through the two shortest, mutually independent paths, primary and spare, the same for both directions of communication, ensuring network survivability in the case of single network element failure, the link or node. A link failure is assumed to be caused by a failure in an optical amplifier, or in the fiber cable, causing an interruption of all services in the cable. The following definition is assumed: a node-to-node connection is available if both directions of the connection are available. The traffic requirements between all pairs of nodes are given. All link capacities are multiples of 2.5 Gbit/s (standard capacity in digital transmission), achieved through a number of wavelengths in one or more different optical fibers on the same optical link. The node pair direct distances are derived from the road distances among major European cities. Because of the accumulated noise and distortions in optical fibers, amplifiers and node elements, the optical path length limitation is fixed at 2000 km. The distances between optical amplifiers are assumed to be 100 km. Component failure and repair rate data for calculating the unavailability of the future all-optical network are taken from the existing data set for mature optical components whereas for new photonic components the calculation is based on estimated data. Steady state unavailability (the asymptotic value of unavailability if time tends to infinity) is considered, assuming constant failure and repair rates. In the total path unavailability calculation, the impact of node unavailabilities is negligible as compared to the unavailabilities of optical links.

3 Network Unavailability Calculation

Network unavailability is defined as the worst case of all node-to-node connection unavailabilities (source-termination unavailability):

$$U = \max_{i,j}\{U_{ij}\},$$

$$U_{ij} = \left(1 - \prod_{k \in pp}(1 - U_k)\right)\left(1 - \prod_{l \in sp}(1 - U_l)\right),$$

where U_k is the unavailability of link from the primary path (*pp*), and U_l is the unavailability of link from the independent spare path (*sp*). In other words, the unavailability model could be described as a serial structure of two parallel ones. Optical link is treated as a non-redundant structure comprising fiber in optical fiber cable and optical amplifiers. For small unavailability values of link elements an approximate formula for the total link unavailability can be used:

$$U_{link} = \lambda_F \, L \, MTTR_F + N_{OA}\lambda_{OA} \, MTTR_{OA}$$

where λ_F is fiber cable failure rate per km, λ_{OA} is failure rate of optical amplifier (OA), N_{OA} is the number of optical amplifiers on the link, L is the link length, $MTTR_F$ and $MTTR_{OA}$ are mean times to repair of fiber (F) and OA respectively ($\lambda_F = 114$ fit/km, $MTTR_F = 21h$, $\lambda_{OA} = 4500$ fit, $MTTR_{OA} = 21h$, fit = number of failures per 10^9 hours).

4 Cost Model

The cost model applied in the network availability optimization was taken from [4]. The total network cost for the set of nodes N is a sum of all link and node costs.

$$C = \sum_{i,j \in N} C_{Lij} + \sum_{i \in N} C_{Ni} = C_L + C_N,$$

where C_{Lij} is the cost of the link between nodes i and j, and C_{Ni} is the cost of the node i. Link cost is a function of link length L_{ij} (km) and link capacity V_{ij} (Gbit/s):

$$C_{Lij} = L_{ij} \, V_{ij}.$$

The link capacity is determined for each link by summing up the contributions from all the primary and spare paths that make use of it. The node cost C_{Ni} is a function of node effective distance N_i (km) (N_i represents the cost of node in equivalent distance terms), and the total capacity of all links incident to the node–V_i (Gbit/s):

$$C_{Ni} = 0.5 N_i \, V_i, \;\; N_i = E + d_i \, F,$$

where d_i is node degree (the number of links incident to node i), E and F constants assumed to be 200 km and 100 km respectively. In order to determine two independent shortest paths for each pair of nodes, weights W_{ij} of links have to be defined, reflecting the influence of node parameters on the path length:

$$W_{ij} = 0.5 N_i + L_{ij} + 0.5 N_j.$$

5 Optimization Procedure using a Genetic Algorithm

Possible solutions are coded as binary strings with $n(n-1)/2$ bits. Each bit in the string corresponds to a link in the fully-meshed topology: 0 stands for the missing link, and 1 for the existing link in the solution. Genetic algorithm parameters and the fitness function should be defined. Two different approaches in defining the selection process were analyzed and tested.

The first approach: In the pre-selection process, all the solutions not satisfying some easy-to-test fundamental requirements are rejected. For example: if a generated graph has a node degree less than 2, or if the number of branches in a topology is less than (graph tree), it can surely be inferred that these solutions cannot satisfy the network requirement–two independent paths between all node pairs. The advantage of this approach lies in reducing the number of topologies to be analyzed in details (cost and unavailability calculation). For 11 nodes, as used in the case study, the number of acceptable topologies is reduced to 10-4 % of all topologies, according to the second preselection rule, as mentioned above. On the other hand, the disadvantages of this approach are: poor diversity of solutions in the population, and the very rough distinction between solutions–a solution is either regular, i.e. acceptable, or irregular, i.e. unacceptable. In the cases where solution limitations are very restrictive, the whole initial population could be rejected, disabling further search. Note that even a bad solution could produce a good offspring.

The second approach: No topology is rejected but penalized, if assumptions or dynamic limitations are not satisfied. The advantage of this approach lies in the great diversity of solutions to be evaluated, increasing the probability of finding different areas of local minima to be tested, in order to select the global one. The disadvantage of the approach lies in an extensive evaluation time. After a number of experiments and analyses performed, using both approaches, the second was accepted and applied in further research. In order to minimize unavailability-cost pairs, two types of optimization alternate. In odd optimization steps, the network cost is minimized. The fitness function in this step is equal to:

$$FF = \frac{1}{(C+PF)\,k\,(1+UP)},$$
$$UP = \begin{cases} \frac{U-U_{lim}}{U_{lim}}, & U > U_{lim}, \\ 0 & U \leq U_{lim}, \end{cases}$$

where k is the penalty slope, U_{lim} is the dynamic unavailability upper bound in an odd step, achieved as minimal in previous steps. PF is the penalty factor defined as follows

$$PF = 2.5\,PathOver\,CapOver,$$

where *PathOver* is the sum of all path length limitation exceedings and distances between the node pairs without primary and/or spare paths. *CapOver* is the sum of capacity demands between the node pairs contributing to the *PathOver*. In even optimization steps, the unavailability is minimized. The penalty is effective for the costs higher than the cost limit C_{lim},-the dynamic cost upper bound reached in previous steps.

$$FF = \frac{1}{U\,k\,(1+CP)},$$
$$CP = \begin{cases} \frac{C+PF-C_{lim}}{C_{lim}} & C+PF > C_{lim}, \\ 0 & C+PF \leq C_{lim}. \end{cases}$$

Note that the genetic content from one step is transferred to the next one. The absolute minimum unavailability, as a reference value, could be determined from the fully meshed network. The optimization target could be to find the same or very close unavailability value with as low cost as possible. The genetic algorithm parameters are chosen as follows: population size = 100, chromosome size = 55, crossover probability = 0.6, mutation probability = 0.05, two point crossover, roulette wheel selection scheme, generation gap = 1, number of generations per step = 200, elitism.

6 Optimization Results

A number of runs were performed. Several quasi-optimal unavailability-cost pair values were obtained. Table 1 shows two selected topologies, the minimum unavailability topology ($MinU$) (Figure 4) and the minimum cost topology ($MinC$) (Figure 3), compared to the reference topology COST 239 (EON) (Figure 1) [5], manually designed grid network (MG) (Figure 2) [2] and fully meshed topology (FM).

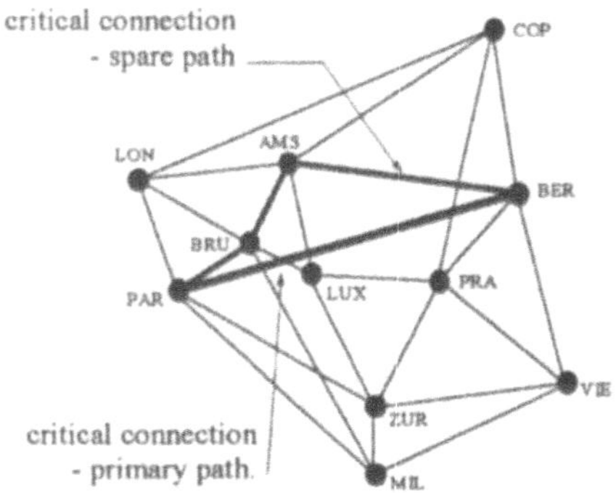

Figure 1: COST 239 case study (EON).

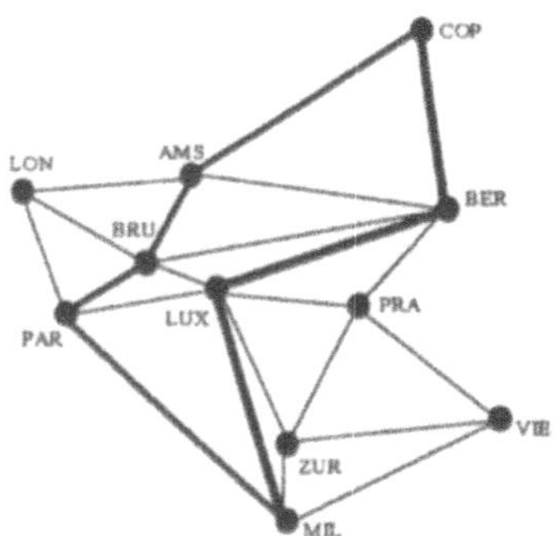

Figure 2: Manual-Grid (MG).

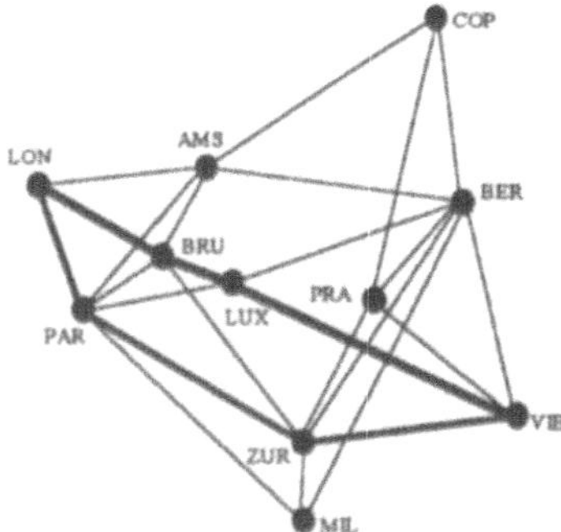

Figure 3: Minimum Cost (MinC).

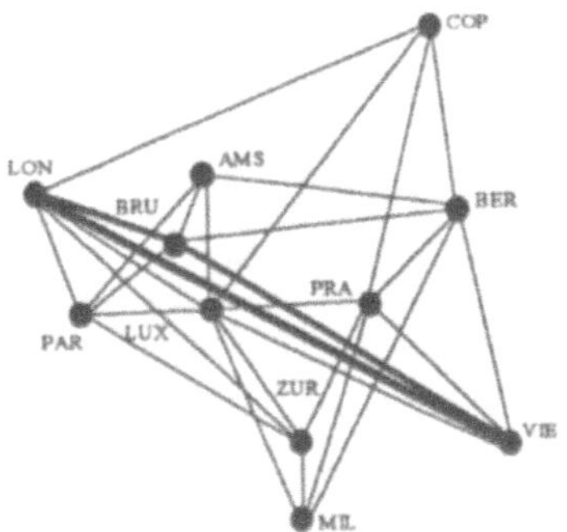

Figure 4: Minimum Unavailability (MinU).

Table 1: The comparison of topology performances.

	FM	EON	MG	MinC	MinU
U x 10^{-5}	2.502	3.789	4.235	3.130	2.502
C x 10^6	4.537	3.765	3.903	3.706	3.793
C_L x 10^6	1.441	1.685	1.711	1.576	1.615
C_N x 10^6	3.096	2.080	2.192	2.130	2.178
TFCL [km]*	44145	14775	11635	14610	19115
No. of links	55	25	22	25	29
dmin	10	4	2	3	4
dmax**	10	5	6	7	8
PathOver [km]	0	50	675	0	0

* *TFCL* - total fiber cable length

** *dmin, dmax* - the minimum and maximum node degree

7 Conclusion

The complex availability/cost optimization problem of an all-optical network was being solved. The number of quasi-optimum topologies were achieved. The minimum unavailability and cost topology performances were presented. The selected topologies, as well as other topologies which have unavailability and cost values very close to those presented, could be a good basis for the optimizations concerning other network performances, minimizing the number of wavelengths, the total fiber (cable) length, etc.

In order to improve the feasibility of all-optical network solutions, other limitations should be introduced, such as maximum node degree, noise accumulation limitations and limitations due to nonlinear effects. In the domain of the genetic algorithm application, running speed could be improved by introducing adaptive searching, homogeneity measurement, and presetting links in learning process.

References

[1] D. E. Goldberg. *Genetic Algorithms in Search, Optimization and Machine Learning.* Addison-Wesley, Reading, MA, 1989.

[2] R. Inkret. All-optical network reliability optimization by means of genetic algorithm (in croatian). Technical report, Dept. of Telecomm., FER, University of Zagreb, 1995.

[3] M. J. O'Mahoney, M. C. Sinclair, and B. Mikac. Ultra-high capacity optical transmission networks: European research project cost 239. *ITA*, 12:33–45, 1993.

[4] M. C. Sinclair. Minimum cost topology optimisation of the cost 239 european optical network. In *Artificial Neural Nets and Genetic Algorithms*, D. W. Pearson, N. C. Steele, R. F. Albrecht (editors), pages 26–29, Wien New York, 1995. Springer-Verlag.

[5] L.G. Tan and M.C. Sinclair. *Wavelength Assignment Between Central Nodes of the COST 239 European Optical Network.* PhD thesis, ESE Dept., University of Essex, 1995.

Breeding Permutations for Minimum Span Frequency Assignment

C. L. Valenzuela[1], A. Jones[2] and S. Hurley[2]
[1] School of Computing and Mathematics, University of Teesside, Borough Road, Middlesbrough TS1 3BA
[2] Department of Computer Science, University of Wales, Cardiff, PO Box 916, Cardiff CF2 3XF,
Email: christine@tees.ac.uk, steve@cs.cf.ac.uk

Abstract

This paper describes a genetic algorithm for solving the minimum span frequency assignment problem (MSFAP). The MSFAP involves assigning frequencies to each transmitter in a region, subject to a number of constraints being satisfied, such that the span, i.e. range of frequencies used, is minimised. The technique involves finding an ordering of the transmitters for use in a sequential (greedy) assignment process. Results are given for several practical problem instances.

1 Introduction

The management of the radio spectrum would be eased if frequencies could always be assigned for a particular purpose in an optimum or near-optimum manner. However, the assignment of radio frequencies for problems of a practical size remains a considerable challenge.

The primary objective in minimum span frequency assignment is to assign radio frequencies to a number of transmitters subject to a number of constraints, such that no interference is suffered, and the range of frequencies used, i.e. the difference between the largest and smallest frequency used, is minimised (this is the *span* of the assignment). The general assignment problem is classified computationally as *NP*-hard. Hence, there is no known algorithm that can generate a guaranteed optimal solution in an execution time that may be expressed as a finite polynomial of the problem dimension.

Computing methods based on exact algorithms and graph theory are successful for small problems but out of the question for large problems. Methods based on the so-called sequential heuristics, which mimic the way the problem might be solved manually, are fast enough for large problems but give results which are well short of the best possible.

The purpose of this paper is to explore the possibility of combining a genetic algorithm with sequential assignment methods. Traditionally, metaheuristics have been applied to an initial solution consisting of an assignment of frequencies to transmitters and then attempting to minimise the number of constraint violations [2, 4, 8]. Here, the iterative transformations are applied to permutations of transmitters. A simple sequential assignment algorithm is then applied to each of these permutations to produce an allocation of frequencies that does not violate any constraints.

1.1 Interference and Constraints

In order to model interference, constraints are imposed on the assignment. Pairs of transmitters can interfere with each other when the assigned frequencies are the same or close together. This can happen when transmitters are at the same location or within a few tens of metres of each other *(co-site interference)*, or when equipment is at a distance of several kilometres or more *(far-site interference)*.

The constraints are due to the following interference mechanisms and can be expressed in terms of equalities or inequalities involving no more than four frequencies.

Co-channel and Harmonic constraints: This is the most important factor in the consideration of far-site interference. A pair of transmitters located at different sites must not be assigned the same frequency, or harmonics of each other, unless they are sufficiently geographically separated. This gives rise to constraints of the form:

$$f_i \neq n f_j \quad \text{for} \quad n = 1, 2, 3, \ldots$$

When $n = 1$ we have a co-channel constraint.

Adjacent channel constraints: When a transmitter and a receiver are tuned to similar frequencies (normally within three channels of each other), there is still the potential for interference. Therefore a number of constraints arise of the following form:

$$|f_i - f_j| > m$$

for some value of m, where m is the number of channels separation.

Co-site frequency separation: Any pair of frequencies at a site must be separated by a certain fixed amount, typically, for a large problem, 250 kHz or 5 channels. If a channel is to be used by a high power transmitter then its frequency separation should be larger, say 500 kHz or 10 channels. The constraint can therefore be of the form:

$$|f_i - f_j| \geq m$$

where m refers to the number of channels separation required between transmitters i and j.

Other types of constraints that can arise include *intermodulation products* and *spurious emissions and responses.* However, in this paper we only consider co-site frequency separation constraints, and co-channel and adjacent channel constraints, as these constraints represent the most important interference problems to avoid. Also, we have assumed that interference can be avoided if there is sufficient channel separation between the frequencies assigned to pairs of transmitters.

1.2 Sequential Assignment Algorithms

Sequential assignment methods mimic the way the problem might be solved manually. They are fast enough for large problems but tend to give results which are well short of the best possible. The transmitters are simply considered one at a time, successively assigning allowable frequencies as we proceed, until either we have assigned all transmitters or run out of frequencies. An important factor affecting the quality of solutions generated by this method is how the next transmitter is chosen. In addition, the initial ordering of the transmitters is important, as is the method by which the next frequency is chosen. We may therefore generate a series of assignment methods based on three components:

- initial ordering,
- choice of next transmitter,
- assignment of frequency.

The simplest way to choose the next transmitter is sequentially, simply picking the next one on the list produced by the initial ordering. A more complicated method, which has proved more effective than sequential selection with the various initial ordering methods, is called ***generalised saturation degree.*** In this method the choice of the next transmitter is influenced by the constraints imposed by all those transmitters that have already been chosen. One could view the more complicated process as a mechanism for correcting those mistakes that have been made by the initial ordering technique.

The simplest assignment technique is to assign the selected transmitter to the smallest acceptable channel i.e. the lowest numbered channel to which it can be assigned without violating any constraints. Variations upon this technique attempt to assign transmitters to channels that are already used in favour of those that are not. A detailed description of sequential assignment methods can be found in [10].

In this paper a genetic algorithm is used to search the state-space of initial orderings. The choice of the next transmitter is made sequentially, and the smallest acceptable channel is assigned to each chosen transmitter.

2 The Genetic Algorithm

A simple genetic algorithm (GA) which appeared in [12] is used for this work. It is derived from the model of [7] and is an example of a 'steady state' GA (based on the classification of [11]). It uses the 'weaker parent replacement strategy' first described by [3]. The GA applies the genetic operators to permutations of transmitters. The fitness values are based on the spans produced when the simple sequential assignment algorithm is applied to each permutation list produced by the GA. The first parent was selected deterministically in sequence, and the second parent was selected in a roulette wheel fashion, the selection probabilities for each genotype being calculated using the following formula:

$$\begin{array}{c}\text{selection}\\\text{probability}\end{array} = \frac{(\text{population size} + 1 - \text{Rank})}{\sum \text{Ranks}}$$

where the genotypes are ranked according to the values of the spans that they have produced, with the best ranked 1, the second best 2 etc.

Mutation: The mutation chosen was to select two transmitters at random from a permutation list, and swap them.

Table 1: Test data characteristics.

	No. transmitters	No. co-site constraints	No. far-site constraints	edge density
test12	12	16	21	0.56
test95	95	90	1124	0.27
test190	190	160	4882	0.28
test410	410	411	22346	0.27
hex481	481	0	97835	0.85

Permutation Crossovers: Permutation crossovers were originally developed primarily for the travelling salesman problem (TSP), where the genotypes consist of lists of cities which are converted to TSP tours. Because TSP tours are circuits, it is irrelevant which city is represented first on the list. The permutation lists represent cycles and an edge in a TSP tour always joins the last city on the list to the first. Thus, for the TSP it is the relative sequence of cities that is important, rather than the absolute sequence. In the frequency assignment problem (FAP), however, the permutation lists making up the genotypes represent lists of transmitters, and intuitively it would seem likely that absolute sequences are important in this case.

The best known permutation operators from an historical standpoint (which are also amongst the simplest to implement) are partially matched crossover (PMX), order crossover (OX) and cycle crossover (CX) [5]. PMX, OX and CX are the chosen crossovers for this study on the FAP. Although many more sophisticated variations of permutation crossover have been developed over the years and these have proven far more successful on the TSP (for example, genetic edge recombination [13] and maximum preservative crossover [6]), unfortunately these variations rely almost entirely on problem-specific heuristics for their improved performance.

Pearson's correlation coefficient was used to assess how effective the various permutation crossovers were at propagating parental qualities to the offspring. Mid-parental values of "span" were correlated with the values for their respective offspring for a 95 transmitter problem. From an initial population of 1000 individuals, 1000 pairs of parents were selected, and from them 1000 offspring were generated using one of PMX, OX or CX and no mutation. The values of the correlation coefficients for all three permutation crossovers proved to be highly significant at the 0.0001% level, with the value for CX the best (0.3937).

3 Results

The test problems used are given in Table 1. The files test*xxx* refer to military frequency assignment problems that arise in irregular networks. The file hex481 refers to a cellular assignment problem which is loosely based on a regular network arising around the Philadelphia area in the USA [1].

Results based on single runs of the GA are presented in Table 3. The GA is compared with the best result obtained from using several state-of-the-art sequential assignment algorithms [10]. The cosite value refers to the value of m used for the cosite frequency separation constraint (see section 1.1).

The 'optimum' column refers to the known optimum span (where available). (The optimum values are obtained either from using exhaustive search techniques where the problem is small enough, or where computed results match theoretical lower bounds [9]). The 'random search' column corresponds to the best span produced by random search through permutation space for each of the problems, where each random search processed exactly the same number of individuals as the corresponding GA.

We can see that the GA, which starts with a population of random permutations of transmitters, produces an improvement over the results from the sequential assignment algorithms, and reaches the

Table 2: Timing results.

Problem	Number of constraints	Time (secs)	Time / # constraints
test12	37	30	0.81
test95	1214	518	0.43
test190	5042	2765	0.55
test410	22757	20019	0.88
hex481	97835	229916	2.35

Table 3: Comparison of results.

Problem	GA			Best Sequential	Random Search	Optimum
	PMX	OX	CX			
test12	22	22	22	24	22	22
test95 (cosite 4)	48	48	48	51	50	48
test95 (cosite 5)	48	49	48	54	52	48
test190	82	84	76	87	84	–
test410	165	165	154	158	170	–
hex481	443	442	426	449	475	426

optimum value in several cases. The run times for the test examples are given in Table 2. In each case the GA was run using a population of 200 for 200 generations.

4 Concluding Remarks

A genetic algorithm has been presented which finds an initial ordering of the transmitters. Each transmitter in this ordered set is then assigned the smallest frequency possible while still satisfying any associated constraints. Of the permutation crossover operations used, cycle crossover gives the best results in all the test cases and gives the optimum solution in four out of six cases. Further work is necessary to test the applicability of the technique to larger problems, and to develop efficient problem specific crossover operators.

References

[1] L. Anderson. A simulation study of some dynamic channel assignment algorithms in a high capacity mobile telecommunications system. *IEEE Transactions on Communications*, 21:1294–1301, 1973.

[2] D. Castelino, S. Hurley, and N.M. Stephens. A tabu search algorithm for frequency assignment. *Annals of Operations Reasearch*, 63:301–319, 1996.

[3] D.J. Cavicchio. Adaptive search using simulated evolution. Unpublished Doctorial Dissertation, 1970.

[4] W. Crompton, S. Hurley, and N.M. Stephens. A parallel genetic algorithm for frequency assignment problems. In *IMACS/ IEEE Internation Symposium on signal Processing, Robotics and Neural Networks*, pages 81–84, France, 1994.

[5] D. Goldberg. *Genetic Algorithms in Search, Optimization and Machine Learning.* Addison Wesley, 1989.

[6] M. Gorges-Schulter. Asparagos: An asynchronos parallel genetic optimization strategy. In *Proceedings of the Third International Conference on Genetic Algorithms*, Hillsdale,NJ, 1989. Lawrence Erlbaum Assocciates.

[7] J.H. Holland. *Adaptation in the Natural and Artificial Systems.* Ann Arbor: University of Michigan Press, 1975.

[8] S. Hurley and D.H. Smith. Fixed spectrum frequency assignment using natural algorithms. In *Proceedings of the First International Conference on Genetic Algorithms in Engineering Systems*, pages 373–378, Sheffield, September 1995.

[9] D.H. Smith and S. Hurley. Bounds for the frequency assignment problem. *Discrete Mathematics.* To Appear.

[10] D.H. Smith and S.U. Thiel. Frequency assignment algorithms. Final report, Radiocommunications Agency Agreement, March 1996.

[11] G. Syswerda. Uniform crossover in genetic algorithms. In *Proceedings of the Third International Conference on Genetic Algoritms*, Hillsdale, NJ, 1989. Lawrence Erlbaum Associates.

[12] C.L. Valenzuela. Evolutionary divide and conquer (i): A novel genetic approach to the TSP. Unpublished Doctoral Thesis, 1995.

[13] D. Whitley. Scheduling problems and travelling salesman: The genetic edge recombination operator. In J.D. Schaffer, editor, *Proceedings of the Third International Conference on Genetic Algorithms*, pages 133–140, San Mateo, 1989. Morgan Kauffman.

A Practical Frequency Planning Technique for Cellular Radio

T. Clark[1] and G. D. Smith[2]
[1] Nortel Technology, London Road, Harlow, Essex, CM17 9NA, UK.
[2] School of Information Systems, UEA, Norwich, NR4 7TJ, UK.
Email: tclar@nortel.co.uk, gds@sys.uea.ac.uk

Abstract

A practical algorithm using a combination of simulated annealing and two problem specific heuristics has been developed for frequency planning in cellular radio networks. It is designed to rapidly generate complete Frequency plans. Experimental work is described that characterises the performance and behaviour of our algorithm.

1 Introduction

The assignment of frequencies to the base stations of a cellular network is a hard combinatorial optimisation problem which is known to be NP-complete [5]. Many different heuristic techniques have been applied to this problem, including; simulated annealing [2, 10], genetic algorithms [6, 8], neural networks [7] and local search [11].

1.1 The Cellular Radio Design Process

The cellular radio design process starts with an estimated traffic demand, a limited portion of the radio spectrum and a geographical region over which the network must operate. Base stations are then located across the region and a propagation model is generated. The propagation model is then used to derive an interference matrix that describes which base stations could potentially interfere with each other. The interference matrix gives an approximation of the real interference situation that can be manipulated by frequency planning algorithms. Once a frequency plan has been produced the amount of interference it produces can be assessed from the interference matrix, or more accurately from the propagation model.

This process is severely constrained. Near optimal locations for base stations can be found, but it may be impossible to actually acquire these sites in practice. Additionally, traffic demand may be too large to produce interference free frequency plans with the bandwidth available.

Traditional frequency planning [9] utilises the principles of hexagonal geometry to rapidly produce frequency plans for cellular networks. However, practical networks rarely possess hexagonal cells so these methods are not applicable to detailed network design.

Graph theory can be used to set bounds upon the frequency planning problem [3]. This enables a description of a planning problem to be evaluated so that a bound can be set upon the number of channels required to generate an interference free asssignment. Iterative algorithms can be designed that use these bounds to produce frequency plans. However, this bound will often be much larger than the number of available channels, a parameter over which the network designer usually has no control.

The frequency planning problem is an instance of the generalised graph colouring problem [5]. Graph colouring techniques can be applied to the frequency planning problem [1] and perform well. These techniques work by ordering the channels to be assigned using a measure of assignment difficulty. A heuristic then assigns frequencies to the channels in the specified order.

Previous work [2, 10] on applying simulated annealing and genetic algorithms [6, 8] to frequency planning has used direct representations of the problem solution with problem specific operators.

The technique described in this paper uses an indirect representation of the problem solution, namely orderings, and heuristics inspired from graph colouring techniques to produce solutions to the frequency planning problem. The orderings are generated by a simulated annealing algorithm.

1.2 Problem Definition

A network is defined as a set of n cells, a demand vector, D and an interference matrix, M. Each cell,

i, has an associated channel demand, $d_i \in D$, where d_i is an integer. A frequency $f_{ik}, 1 \leq i \leq n, 1 \leq k \leq d_i$, must be assigned to the each channel of each cell in such a way that interference is minimised.

Potential interference is described by an interference matrix, M, where m_{ij} indicates the minimum required frequency separation between a channel in cell i and a channel in cell j to prevent interference between the channels. That is $\mid f_{ik} - f_{jl} \mid \geq m_{ij}$ for all i, j, k and l ($k \neq l$ iff $i = j$).

There is a finite set of available frequencies $\mathcal{F}$, and a set of blocked frequencies $\mathcal{B}$. It is usual to define three different values of frequency separation based upon the radio technology being used to implement the network, a co-site separation, a co-channel separation, and an adjacent channel separation.

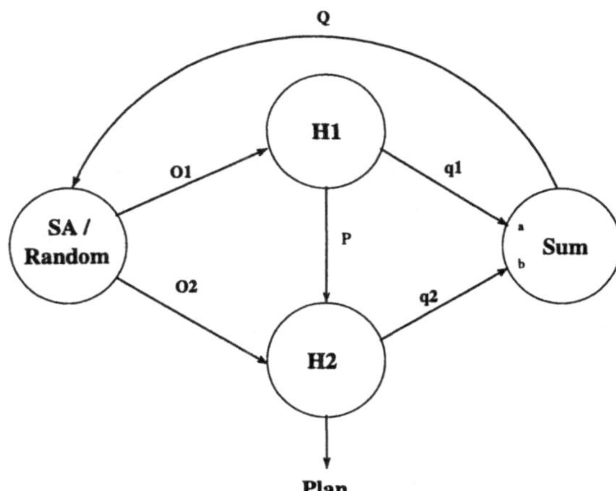

Figure 1: Data flow for frequency planning algorithm. In some experiments O1 ≡ O2 and in some experiments O1 represents multiple orderings.

2 The Frequency Planning Algorithm

Unlike graph colouring approaches [1] and the local search of [11] that utilise orderings of channels to produce frequency plans, the approach described here uses orderings of cells. The cells are repeatedly stepped through in order, so that frequencies are assigned to the network in layers, that is an attempt will be made to assign a frequency to the first channel in each cell before any attempt is made to assign frequencies to additional channels in any cell. This is done as an attempt to spread any interference caused by a frequency assignment across the network so that there is at least one interference free channel available in each cell.

Two assignment heuristics are used, the first heuristic constructs a partial frequency plan that is interference free, the second heuristic then completes the partial frequency plan whilst causing minimal interference.

Figure 1 illustrates the flow of information between the ordering algorithm and the assignment heuristics. An ordering, or multiple orderings, are generated by a simulated annealing algorithm. An ordering, $O1$, is then used by the first heuristic, $H1$, to generate a partial frequency plan, P, and partial quality measure $q1$. The second heuristic, $H2$, takes the partial plan, P, and an ordering, $O2$, to produce a complete frequency plan and a partial quality measure $q2$. An overall quality measure is computed and returned to the simulated annealing algorithm.

2.1 Interference Free Assignment

This algorithm assigns frequencies in layers. An attempt is made to assign a single frequency to every cell in a network before any attempt is made to assign an additional frequency to any single cell. This is done to ensure an even allocation of coverage. A brief description of the core loop of this algorithm is given in Figure 2.

2.2 Assignment Completion

This algorithm assigns frequencies to channels so as to minimise the interference between channels. It works by selecting the frequency used least in

- **for** $i = \mathcal{O}(1)$ **to** $\mathcal{O}(n)$ **do**
  - **if** cell i has an unassigned channel **and** an interference free channel is available **do**
    1. Select lowest available frequency and assign to cell.
    2. Calculate set of blocked frequencies.
    3. Remove blocked frequencies from neighbouring cells.

Figure 2: The core loop of the interference free assignment heuristic. $\mathcal{O}$ represents an ordering, that is a set of n integers.

- **for** $i = \mathcal{O}(1)$ **to** $\mathcal{O}(n)$ **do**
  - **while** cell i has any unassigned channels **do**
    1. Select frequency used least in neighbouring cells and assign to cell.
    2. Record number of constraints violated by assignment.

Figure 3: The core loop of minimal interference assignment heuristic. $\mathcal{O}$ represents an ordering, that is a set of n integers.

neighbouring cells thus minimising the potential interference. A brief description of the algorithm is given in Figure 3.

2.3 Solution Quality

The actual quality of any frequency plan can only be assessed with reference to the propagation model used to produce the interference matrix for the network. It is possible to approximate the amount of interference by counting the number of frequency separation constraints that are violated.

The quality measures produced by the assignment heuristics and used by the ordering algorithm indicate how successfully each heuristic is performing. The measurements are indirectly related to the number of frequency constraint violations. The measure produced by the interference free assignment heuristic, $q1$, is the number of unassigned channels. The measure produced by the minimal interference heuristic is an approximate measure of the frequency constraint violations caused by this heuristic. These two measures are scaled and summed to provide a single quality measure for use by the simulated annealing algorithm.

3 Experiments

Experiments were carried out in order to assess the effect of single and multiple orderings within the two heuristics. This was done by utilising orderings generated by a simulated annealing algorithm and orderings generated randomly.

All experiments reported here consisted of 100 runs of each algorithm variant, where each run was terminated after 5 minutes of computation. The simulated annealing algorithm used a geometric cooling schedule with a cooling rate of 0.90, simple swap and move operations were used as neighbourhood operators.

Experimental results have been presented as distributions of solution qualities. Solution quality is shown on the horizontal axis, whilst the number of trials that achieved this quality is shown as a vertical line.

The algorithms were evaluated on real network design data supplied by Nortel Matra Cellular. The network investigated had eighteen base stations and was a mixture of omnidirectional and tri-sectored antennae giving a total of 48 cells with a total demand for 107 channels. There are 27 frequencies available for use within the network. The co-site separation is two, and the co-channel and adjacent channel separations are one. The solution quality scaling factors, a and b, are set to 1 and 200, respectively.

Seven variants of the algorithm were evaluated:

1. A single random ordering used by both the first and second heuristics.
2. A separate random ordering for each heuristic.
3. A single SA ordering used by both the first and second heuristics.
4. A separate SA ordering used for each heuristic.
5. A single SA ordering for the first heuristic, and a single random ordering for the second heuristic.
6. A single random ordering for the first heuristic, and a single SA ordering for the second heuristic.
7. Multiple SA orderings for the first heuristic, and a single SA ordering for the second heuristic.

The first two experiments evaluated the effect of using a different ordering for each heuristic. Two trials were carried out, a trial with algorithm variant 1, and a trial with algorithm variant 2. Figure 4 indicates that using a different ordering for each heuristic improves performance.

However, the equivalent simulated annealing trials, that is trials with algorithm variants 3 and 4, indicates that simulated annealing can discover a single ordering that performs as well as two different orderings, as shown by the distributions of Figure 5.

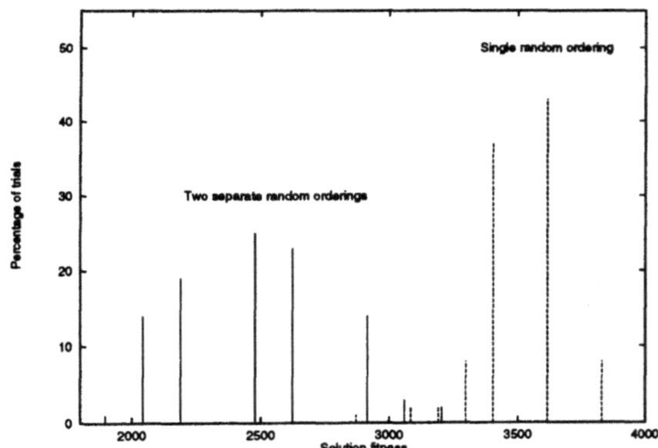

Figure 4: Fitness distributions for random orderings.

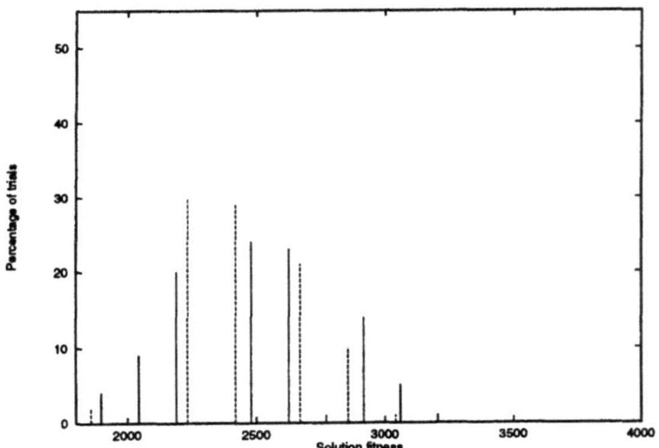

Figure 5: Fitness distributions for simulated annealing orderings.

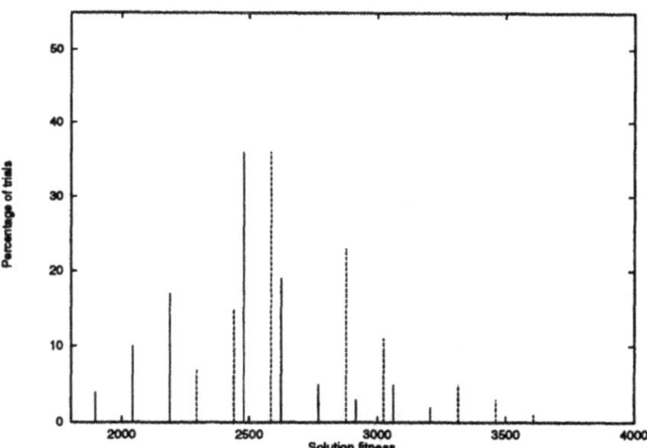

Figure 6: Fitness distributions for combinations of simulated annealing and random orderings.

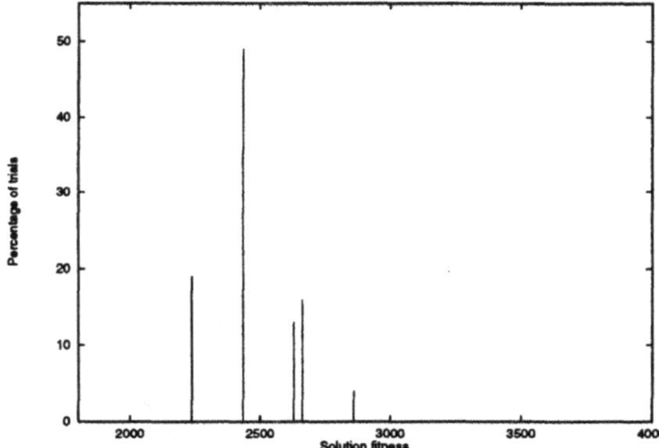

Figure 7: Fitness distributions for multiple intraphase orderings discovered by simulated annealing.

The fitness distribution of variant 2, that is the leftmost distribution in Figure 4 will be used as a baseline distribution against which other results will be compared. Comparing Figure 4 with Figure 5 it is apparent that although simulated annealing can discover a single ordering that performs as well as two different orderings, there is no significant performance difference between using a simulated annealing ordering or two different random orderings.

Figure 6 shows results of experiments with variants 6 and 5. There is still no significant difference between these variants and the baseline distribution.

The layered nature of the assignment heuristic used in the first algorithmic phase is capable of using multiple orderings, that is one ordering for each layer of frequencies assigned by the heuristic. Figure 7 shows results for using multiple orderings, generated by simulated annealing, within the first heuristic, that is algorithm variant 7. Once again there is no significant difference between this distribution and the baseline distribution.

Further work is ongoing to evaluate genetic algorithms in place of simulated annealing. Results will be presented in the completed paper.

3.1 Discussion

The only conclusion that can be drawn from these experiments is that algorithm variant 1 has the worst performance of all the algorithms, and that no significant performance difference can be observed between the remaining variants.

Additionally, simulated annealing provides no performance gain over multiple random orderings. This implies that simulated annealing is unable to conduct a useful search of the problem space which may indicate a degree of non-linearity. Simulated annealing is able to create a single ordering that can perform as well as multiple orderings, but it is unable to exploit the benefits of multiple orderings.

4 Conclusions

Our current approach to the frequency planning problem leads to a large degree of non-linearity. This may be partly due to the non-linear nature of the problem domain, but there is considerable interaction between the two heuristics used to generate the plans which may also cause non-linearity.

However, the frequency plans produced by these heuristics have been assessed using real propagation models and are substantially better, in terms of interference, coverage, and time to produce, than the algorithm described in [4].

5 Acknowledgements

The data used in this study was kindly supplied by Raoul Jacquand of Nortel Matra Cellular.

The research is being undertaken within a Teaching Company Scheme programme, grant reference GR/K40086, between Nortel Technology (formerly BNR Europe Ltd) and the University of East Anglia.

References

[1] A. A. F. Bloemen. Frequency assignment in mobile telecommunication networks. Technical report, Eindhoven University of Technology, September 1992.

[2] M. Duque-Antón, D. Kunz, and B. Rüber. Channel assignment for cellular radio using simulated annealing. *IEEE Transactions on Vehicular Technology*, 42(1):14–21, February 1993.

[3] A. Gamst. Some lower bounds for a class of frequency assignment problems. *IEEE Transactions on Vehicular Technology*, 35(1):8–14, February 1986.

[4] A. Gamst. A resource allocation technique for FDMA systems. *Alta Frequenza*, 57(89–96), 1988.

[5] W. K. Hale. Frequency assignment: Theory and applications. *Proceedings of the IEEE*, 68(12):1497–1514, December 1980.

[6] A. Kapsalis, V. J. Rayward-Smith, and G. D. Smith. Using genetic algorithms to solve the radio link frequency assignment problem. In *Artificial Neural Networks and Genetic Algorithms*, D. W. Pearson, N. C. Steele, R. F. Albrecht (editors), Wien New York, 1995. Springer-Verlag.

[7] D. Kunz. Channel assignment for cellular radio using neural networks. *IEEE Transactions on Vehicular Technology*, 40(1):188–193, February 1991.

[8] W. K. Lai and G. G. Coghill. Channel assignment through evolutionary optimisation. *IEEE Transactions on Vehicular Technology*, 45(1), February 1996.

[9] V. H. MacDonald. Advanced mobile phone service: The cellular concept. *The Bell System Technical Journal*, 58(1), January 1979.

[10] R. Mathar and J. Mattfeldt. Channel assignment in cellular radio networks. *IEEE Transactions on Vehicular Technology*, 42(4):647–656, November 1993.

[11] W. Wang and C. K. Rushforth. An adaptive local-search algorithm for the channel assignment problem (CAP). *IEEE Transactions on Vehicular Technology*, 45(3):459–466, August 1996.

Chaotic Neurodynamics in the Frequency Assignment Problem

K. Dorkofikis and N. M. Stephens
Goldsmiths University, New Cross, London, SE14 6NW, UK
{map01kd,nelson}@gold.ac.uk

Abstract

The frequency assignment problem belongs to the quite difficult to deal with class of NP (nondeterministic polynomial) - hard combinatorial optimization problems [3]. Its computational complexity directs researchers in the field at developing efficient techniques for finding solutions realizing minimum (or maximum) values of an objective function subject to a set of, often conflicting, constraints. To seek an optimal (or near optimal) solution, many methods have been proposed, such as dynamic programming methods, branch and bound methods, etc., and, lately, some heuristic algorithms relating to physical and biological phenomena. They include tabu search, genetic algorithms, simulated annealing and artificial neural networks [4]. We propose a Hopfield neural network model with chaotic neurodynamics to overcome the obstacle of local minima in the energy function and obtain optimal solutions in less iterations than the time-consuming convergent dynamics.

1 The Frequency Assignment Problem

The radio link frequency assignment problem occurs in many civil and military applications. The main objective is to assign radio frequencies to a number of transmitters, subject to a number of constraints, so that no interference occurs. However, it may be impossible to satisfy all the constraints, in which case trying to minimize the number of violated constraints is a more realistic goal.

1.1 Constraint Satisfaction

In the paradigm of the mobile telephony the frequency assignment problem involves again efficiently assigning channels (or frequencies) to all of the links (or nodes) in each radio cell in the cellular radio network, while satisfying the two electromagnetic compatibility constraints:

1. The constraint between two different radio cells (co-channel or adjacent channel constraint); each pair of frequencies selected, one from each cell, must be separated by a precomputed constant which depends on distance and geographic terrain.

2. The constraints between two frequencies in the same cell (co-site constraint); each pair of frequencies must be separated by a constant which is usually larger than the separation required between frequencies for different cells.

1.2 Representation of the Channel Assignment Problem

According to the Gamst and Rave definition [10] the electro-magnetic compatibility constraints in an n-cell network are described by an n x n symmetric matrix called the compatibility matrix (C), and the channel requirements for each cell are described by an n-element vector called the demand vector (D).

Each non diagonal element c_{ij} in C represents the minimum separation distance in the frequency domain between a frequency assigned to cell i and a frequency assigned to cell j. Typically, the cochannel constraint is represented by $c_{ij} = 1$, and the adjacent channel constraint by $c_{ij} = 2$. When $c_{ij} = 0$ cells i and j are allowed to use the same frequency. Each diagonal element c_{ii} represents the minimum separation distance between any two frequencies assigned to cell i, which, for $c_{ii} > 1$, is the cosite constraint.

Each element d_i in D represents the number of frequencies to be assigned to cell i.

Let f_{ik} be the k-th frequency assigned to cell i; then, the electro-magnetic compatibility constraints are represented by

$$|f_{ik} - f_{jl}| > c_{ij}$$

for $i = 1, ..., n$, $j = 1, ..., n$, $k = 1, ..., d_i$ and $l = 1, ..., d_j$ except $i = j$ and $k = 1$.

2 The Hopfield Neural Network Model

In 1982, Hopfield [11] brought together several earlier ideas concerning recurrent networks and present a complete mathematical analysis based on Ising spin models [2].

The Hopfield network consists of a set of n interconnected neurons which update their activation values asynchronously and independently of other neurons; being a recurrent network model, all its neurons are both input and output neurons. The activation values are binary. Originally, Hopfield chose activation values of 1 and 0.

The state of the system is given by the activation values $V = (V_i)$. The net input U_i of a neuron i at cycle $t+1$ is a weighted sum

$$U_i(t+1) = V_j(t).W_{ij} + \mathrm{BIAS}_i, \text{ with } j \neq i$$

where W_{ij} is the weight (or strength) of the connection between neurons i and j, and BIAS_i is the bias of neuron i.

The threshold function is applied to the net input to obtain the new activation values Vi(t+1) at time t+1:

$$V_i(t+1) = \begin{cases} 1 & \text{if } U_i(t+1) > \mathrm{UTP} \\ 0 & \text{if } U_i(t+1) < \mathrm{LTP} \\ V_i(t) & \text{otherwise} \end{cases}$$

where UTP is the Upper Trip Point, and LTP the Lower Trip Point.

When the extra restriction of the symmetric weights is made, the behaviour of the system can be generally described with the Lyapunov energy function given by

$$E = -\frac{1}{2}\sum_{i=1}^{n}\sum_{j=1}^{m} V_i V_j W_{ij} - \sum_{i=1}^{n} V_i \mathrm{BIAS}_i$$

Hopfield and Tank [12] used the Hopfield Network with graded response units in a heuristic solution to the NP-complete travelling salesman problem.

2.1 Representation of the Channel Assignment Problem

The neural network model for the channel assignment problem consists of a n x m matrix, where n is the number of cells and m the number of frequencies. Each neuron in the network can have an output value V_{ij} of either 1 or 0, indicating respectively whether frequency j is assigned to cell i or not.

Adapting the Hopfield neural network to the frequency assignment problem involves introducing specific heuristics so that minimizing the objective energy function guarantees optimum assignment. Takefuji and Lee [14] have pointed out that the sigmoid neuron in Hopfield and Tank model requires much longer computation time on a conventional digital machine, although the two output states of the binary neuron are necessary for solving optimization problems. Moreover, that the motion equation

$$\frac{dU_i}{dt} = -\frac{U_i}{T} - \frac{\partial E(V_1, \ldots, V_n)}{\partial V_i}$$

where T is a time constant, includes the unnecessary decay term $-U_i/T$. They argued the decay term sometimes disturbs network convergence. Funabiki *et al.* proposed the neural network model for the general case [9], based on a $n \times m$ network for the n cells m frequencies problem. The energy function is given by considering the non-decay term Hopfield network motion equation:

$$E = -\frac{A}{2}\sum_{i=1}^{n}\left(\sum_{q=1}^{m} V_{iq} - d_i\right)^2 + B\sum_{i=1}^{n}\sum_{j=1}^{m}\left(\sum_{\substack{q = j-(c_{ii}-1) \\ q \neq j \\ 1 \leq q \leq m}}^{j+(c_{ii}-1)} V_{iq} + \sum_{\substack{p=1 \\ p \neq i \\ c_{ip} > 0}}^{n} \sum_{\substack{q = j-(c_{ip}-1) \\ 1 \leq q \leq m}}^{j+(c_{ip}-1)} V_{pq}\right) V_{ij}$$

together with problem-specific heuristics.

2.2 The Local Minima Obstacle

However, it is argued that the deterministically convergent neurodynamics of the Hopfield model have certain disadvantages, in particular, their inability to overcome local minima in the energy landscape. Our simulations revealed the high dependence of the model on its initial state. To ensure convergence in all problems we had to increase the number of frequencies m, sacrificing near optimal usage of the frequency span.

3 Chaos Neural Networks

On the other hand, contrary to the conventional gradient descent neurodynamics, neural networks with chaotic dynamics (CNN) provide a richer landscape of attractors, despite the simple equations. The fact that it is usually difficult to decide when to terminate the chaotic dynamics is the reason their convergence problems have not yet been satisfactorily solved.

The model of the chaotic neural network, is that the output of neuron i is given by the known sigmoid activation function

$$V_i(t) = \frac{1}{1 + e^{-S_i(t)/\text{STP}}}$$

where STP is a positive steepness parameter, and with $S_i(t)$ given by

$$\begin{aligned} S_i(t) &= kS_i(t-1) \\ &+ a\sum_{\substack{j=1 \\ j \neq i}}^{n}(W_{ij}V_j(t-1) + \text{BIAS}_i) \\ &- Z_iV_i(t-1) \end{aligned}$$

where $i = 1, \ldots, n$, $0 \leq k \leq 1$, $a > 0$ and $Z_i > 0$.

Note in the last equation that the first term corresponds to damping of the neuron membrane, and that the last term represents the chaotic element of the system.

The plethora of methodologies implemented in order to benefit from the chaotic neural networks in combinatorial optimization problems [5, 6, 7, 8, 13, 15] is associated with the decision of when to terminate the chaos neurodynamics without disturbing its behaviour for convergence. Lately, Clen and Aihara [8] have proposed a Transiently Chaotic neural network, by introducing an extra variable into the original input function which corresponds to the temperature in stochastic simulated annealing. This function has the effect of producing successive bifurcations so that the neurodynamics eventually converge from strange attractors to a stable equilibrium point; the behaviour of such a system is interpreted as initially chaotic, that finally almost coincides with the Hopfield network after the aforementioned bifurcation parameter has been decreased enough. Starting from chaotic dynamics utilizes an efficient searching ability for escaping from local minima, and later, when convergent dynamics dominate, the neural network is directed towards a global minimum.

4 Simulation Results and Conclusion

We have implemented a simulated neural network to the frequency assignment problem combining heuristics similar to Funabiki and Takefuji [9] with a refractory strength component similar to Clen and Aihara [8]. Results so far show that the transiently chaotic dynamics exhibit the expected behaviour, utilizing the problem-specific heuristics to converge globally while preserving the frequency span. However, we should stress that, although the proposed approach provides a promising method, further studies, in particular, additional simulations and strict comparison with other methods are needed to evaluate its true ability.

References

[1] D. H. Ackley, G. E. Hinton, and T. J. Sejnowski. A learning algorithm for Boltzmann machines. *Cognitive Science*, 9:147–169, 1985.

[2] D. J. Amit, H. Gutfreund, and H. Sompolinsky. Spin-glass models for neural networks. *Physical Review*, A(32):1007–1018, 1986.

[3] D. J. Castelino, S. Hurley, and N. M. Stephens. A tabu search algorithm for frequency assignment. *Annals of Ops Res*, 63:301–319, 1996.

[4] Euclid cepa 6 project proposal. RTP 6.4 combinatorial algorithms for military applications, project specification. appendix 3, ftp://ftp.win.tue.nl/pub/techreports/CALMA/.

[5] L. Clen. Application of chaotic simulation and self-organizing neural net to power system voltage stability monitoring. In *Second Int. Forums on Applications of Neural Networks to Power Systems*, 7B1. Yokohama,Japan, 1993.

[6] L. Clen and K. Aihara. *Chaotic Simulated Annealing for Combinatorial Optimization*, volume 1, pages 319–322. 1994.

[7] L. Clen and K. Aihara. *Transient Chaotic Neural Networks and Chaotic Simulated Annealing*, pages 347–352. 1994.

[8] L. Clen and K. Aihara. Chaotic simulated annealing by a neural network model with transient chaos. *Neural Networks*, 8(6):915–930, 1995.

[9] N. Funabiki and Y. Takefuji. A neural network parallel algorithm for channel assignment problems in cellular radio networks. *IEEE Trans. Veh. Technol.*, 41(4):430–437, 1992.

[10] A. Gamst and W. Rave. On frequency assignment in mobile automatic telephone systems. In *Proc. GLOBECOM '82 1982*, pages 57–64, May 1978.

[11] J. J. Hopfield. Neural networks and physical systems with emergent collective computational abilities. In *Proceedings of the National Academy of Sciences '79*, pages 2554–2558, 1982.

[12] J. J. Hopfield and D. W. Tank. Neural computation of decisions in optimization problems. *Biological Cybernetics*, 52:141–152, 1985.

[13] T. Kasahara and M. Nakagawa. Parameter-controlled chaos neural networks. *Electronics and Communications in Japan , Part 3*, 78(7), 1995.

[14] Y. Takefuji and K. C. Lee. Artificial neural networks for four-coloring map problems and k-colorability problems. *IEEE Trans. Circuit systems*, 38(3):326–333, March 1991.

[15] J. Tani. Proposal of chaotic steepest descent method for neural networks and analysis of their dynamics. *Trans. Inst. Electron. Inf. Commun.*, J74-A-8:1208–1215, 1991.

A Divide-and-Conquer Technique to Solve the Frequency Assignment Problem

A. T. Potter and N. M. Stephens
Goldsmiths University, New Cross, London, SE14 6NW, UK.
Email: map01ap@gold.ac.uk, nelson@gold.ac.uk

Abstract

The frequency assignment problem is a computationally hard problem with many applications including the mobile telephone industry and tactical communications. The problem may be modelled mathematically as a T-colouring problem for an undirected weighted graph; it is required to assign to each vertex a value from a given set such that for each edge the difference in absolute value between the values at the corresponding vertices is greater than or equal to the weight of the edge. Tabu search, simulated annealing, simulated neural networks and other heuristic algorithms have been applied to this problem. In this paper we describe a divide and conquer technique incorporating heuristics and present results using test data from real problems which show that with respect to both quality of solutions and speed of execution it is superior to simulated annealing and steepest descent algorithms. It is particularly successful in minimising the span and order of the set of frequencies required to solve the problem.

1 Introduction

In recent years the demand on the electromagnetic spectrum due to the wider use of mobile communications has increased dramatically. However the frequency spectrum available is limited and this has led to extensive research into techniques that are able to use the available frequencies in the most economical way. Frequencies must be reused many times and some pairs of allocated frequencies must be separated by a predetermined amount, depending on distance between transmitters and geographical terrain, in order to minimise interference. Ideally we wish to obtain a solution with zero interference using a small set of frequencies of minimal span. The density of the transmitters and restricted frequency set means that minimal interference is a more reasonable aim.

A great deal of work has already been done for the frequency assignment problem which investigates the use of heuristic neighbourhood search and other techniques. Castelino *et al.* [1, 2] investigated the application of genetic algorithms, tabu search and tabu thresholding. The CALMA project [3] resulted in the application and comparison of many heuristic techniques including the aforementioned techniques and neural networks. Each of these techniques has been applied to the *whole* problem throughout computation, although early processing may concentrate on the vertices with large valency. We have investigated a divide and conquer strategy that subdivides the original problem into a number of smaller frequency assignment problems, which are solved using a neighbourhood search algorithm. The sub-problems are significantly smaller and can be solved more effectively. The solutions of the sub-problems are then combined to give a solution of the initial problem.

The algorithm has been implemented using real-life trunc scenario test data supplied by Defence Research Agency Malvern. We compare divide and conquer with standard simulated annealing and with two steepest descent algorithms acting on the same data, investigating speed of execution, interference measures of solutions and span and order of frequencies used in solutions.

We conclude that the divide and conquer strategy is superior. Since it does not exploit any problem-specific knowledge it should prove to be robust enough to deal with a variety of different test data sets.

2 The Frequency Assignment Problem

In order to effect radio communication it is necessary to emit a signal from a transmitter to be picked up by a receiver. Interference occurs when signals combine to produce unwanted frequencies or when a receiver is unable to distinguish between similar frequency signals. For each pair of links a separation value for the frequencies can be computed from terrain data which will ensure no interference.

The aim is to assign frequencies, from the available band-width, to the communication links, subject to the set of constraints, so that minimum interference is suffered. Constraints are of the form

$$|f_i - f_j| >= C_{ij}$$

where f_i and f_j are the frequencies assigned to links i and j respectively and C_{ij} is the separation required. If there is no constraint, C_{ij} is zero. There are two basic types of constraints, co-site and far-site. Co-site interference occurs when transmitters or receivers of two links are closer than two kilometres and lead to larger values of C_{ij} than far-site interference. Co-site constraints are thus harder to satisfy.

3 Mathematical Model of the Assignment Problem

The N links can be considered as the vertices of an undirected weighted graph. The weight of an edge between two vertices i and j is the constraint value C_{ij}. The problem is to assign to each vertex i a frequency f_i from a frequency set F such that, for all i, j, $|f_i - f_j| >= C_{ij}$. In the final analysis, the quality of an assignment is measured by:

1. the number of constraints not satisfied (or the sum of the discrepancies $C_{ij} - |f_i - f_j|$),
2. the difference between the largest and smallest frequencies used by the assignment (span) and
3. the number of frequencies used by the assignment (order).

Ideally we seek a solution that firstly satisfies all the constraints (no interference) and secondly minimises span and thirdly minimises order (releasing frequencies for other applications).

4 Divide and Conquer Strategy

In this section, we describe in more detail the implemented strategy. There are five stages: division of the band of frequencies into sub-bands; allocation of a sub-band to each link; to each link an initial assignment of a frequency within its allocated sub-band; a sequence of local changes to the frequencies; and, finally, a sequence of global changes to frequencies which enhance the overall quality of the solution.

Stage 1: The band of frequencies, F, is partitioned into disjoint, approximately the same order, non-overlapping sub-bands $F_1, F_2.....F_k$, for some k.

Stage 2: Each link is allocated one of these sub-bands. In allocating a sub-band to a link, consideration is given to the satisfaction of the two types of constraints; the *external* constraints between that link and other links which have been allocated *other* sub-bands and *internal* constraints between that link and other links allocated the *same* sub-band. A constraint is *feasible* if it is possible for the links to be assigned frequencies within the appropriate sub-bands which satisfy the constraint. Allocation of the sub-bands uses a greedy algorithm. For any particular link, the sub-band allocated the fewest links is considered first. If all constraints, external and internal are feasible, that sub-band is chosen. Otherwise, the next sub-band is considered. If no sub-band is suitable, the original sub-band is allocated.

Stage 3: There are now k frequency assignment sub-problems - each sub-problem has approximately N/k links to be assigned frequencies from a sub-band of size approximately $|F|/k$. The strategy is to solve each sub-problem in turn repeating the cycle as often as necessary. In solving a sub-problem, links involved with external constraints are assigned first and those only with internal constraints last.

Stage 4: Simulated annealing is used to improve the initial assignment after stage 3. A link and candidate new frequency within the appropriate sub-band are chosen at random. If this candidate improves the quality of the overall assignment it is accepted. If not it is also accepted but with a probability that tends to zero as the number of iterations increases.

Stage 5: Finally, simulated annealing is applied again but this time a candidate frequency is allowed to be any value in the whole band F.

5 Test Data

The test data used in this paper was supplied by D.R.A Malvern. The eight scenarios are constraint based descriptions of real life trunc scenarios used in tactical communications. The tables below gives some indication of their size and complexity. In Table 1, for each scenario (Sc), the values of the number of frequencies ($|F|$), the number of nodes(N) and the number, C, of non-zero constraints is given. Table 2 gives, for each scenario, the percentage of

Table 1

Sc	F	N	C
1	40	158	3099
2	40	60	244
3	80	98	193
4	40	124	215
5	40	158	4058
6	40	12	20
7	80	164	846
8	40	240	779

Table 2

Sc	%c-s	sp(F)	LV	Lc-s	Lf-s	S
1	5.3	169	123	27.1	9.1	253
2	16.4	169	25	27.1	6.1	96
3	35.8	209	18	40.1	8.1	187
4	63.3	399	8	195.9	9.5	199
5	3.3	169	93	27.1	9.1	253
6	0.3	169	6	27.1	0.6	19
7	17.0	209	24	40.1	8.1	312
8	36.5	399	16	195.9	9.5	384

co-site constraints (*%c-s*), the span (*sp(F)*) of the frequency set, the largest valency (*LV*), the largest co-site and far-site values (*Lc-s* and *Fc-s*) of C_{ij} and the value of S where the search space has size 10^S.

6 Results

The eight scenarios were solved using each of the 4 techniques; divide and conquer, basic simulated annealing, steepest descent with one link re-assigned at each iteration and steepest descent with a new random solution generated at each iteration.

The divide and conquer technique out-performed all the other algorithms in all four metrics of solution evaluation. Generally it obtained solutions with less interference using less of the available range and fewer distinct frequencies in less time than the basic simulated annealing and steepest descent techniques. The random technique generally gave solutions that suffered an excessive amount of interference and so this technique has been removed from the comparison figures.

Table 3 gives the test results for the difficult scenario 4 comparing the divide and conquer technique with the simulated annealing algorithm and the steepest descent algorithm. The time limit was 12 seconds and the results are obtained by averaging over 4 test runs with different frequency bands

Table 3

Algorithm	Interference	Order
Divide & Conquer	0.0	31.25
Simulated Annealing	67.65	33.50
Steepest Descent	84.65	33.75

F (with approximately the same span). The table gives the interference measured by the sum of the positive discrepancies and the order of the frequencies. In each case the methods used the full span available.

7 Conclusions

We have described the technique of divide and conquer as applied to the frequency assignment problem and presented the test results. The divide and conquer technique out-performed the simulated annealing and steepest descent techniques on all four metrics of solution evaluation (interference, execution time, span and order). These results were obtained without any fine-tuning of the algorithm parameters which were constant throughout.

The superiority of the divide and conquer method with respect to the span and order has very important practical applications. The frequencies that are not used for the assignment can be freed for use by other users of the electromagnetic spectrum. Since demand on the spectrum is increasing rapidly this advantage will become essential in the future.

8 Further Work

Experiments will be designed and implemented to test the robustness of the divide and conquer technique. We will investigate more sophisticated versions replacing the current simulated annealing algorithms in stages 4 and 5 by tabu search. We intend to use other test data including different scenarios provided by D.R.A. and the 11 CELAR Scenarios. These scenarios were formulated for an international military project seeking to compare heuristic techniques used to solve the frequency assignment problem. We would like to thank Roger Edwards of the Defence Research Agency at Malvern for providing the test scenarios used in this paper.

References

[1] D. J. Castelino, S. Hurley, and N. M. Stephens. A

tabu search algorithm for frequency assignment. *Annals of Ops Res*, 63:301–319, 1996.

[2] D. J. Castelino and N. M. Stephens. Solving frequency assignment problems with tabu thresholding. In *Proceedings of MIC'95*. Berkenbridge, USA, 1995.

[3] The CALMA Project. papers available from ftp://ftp.win.tue.nl/pub/techreports/CALMA/.

Genetic Algorithm Based Software Testing

J. T. Alander, T. Mantere and P. Turunen
Department of Information Technology and Industrial Economics, University of Vaasa
PO Box 700, FIN-65101 Vaasa, Finland
Email: {Jarmo.Alander, Timo Mantere, Pekka.Turunen}@uwasa.fi

Abstract

In this work we are studying possibilities to test software using genetic algorithm search. The idea is to produce test cases in order to find problematic situations like processing time extremes. The proposed test method comes under the heading of automated dynamic stress testing.

1 Introduction

Real-time software is increasingly applied to products in which failure may have severe consequences, thus the requirements for correctness and reliability are getting higher, too. In very reliable sequential programs, the rate of errors should be less than 10 errors/1000 lines of code, to avoid functional failure. Achieving this level is very labourious, because the amount of program testing work grows exponentially with code size.

Testing software manually is slow, expensive and demands inventiveness. Automated testing can reduce both the time and costs needed for performing tests. Exhaustive test data generation is not possible. The most common way of generating test data is random, which is considered weak [11]. For this reason, efforts have been made to optimize test data sets using various methods, for example heuristic methods.

In our study we are trying to identify the situations where the software has the slowest reaction time. The slowest reaction time is identified by having the software tested in difficult input generated by GA.

As the first step, we have tested our approach by a small sequential program consisting of a set of delay loops. In the next step (in a forthcoming paper), we are going to test more complicated real-time software.

1.1 Related Work

The automatic generation of test data using genetic algorithms has been studied by Xanthakis et al. [21], Watkins [20] and Jones et al. [9]. Studies have been mostly based on white-box testing methodology. Here we are using a black-box technique.

Recently there has been a growing interest to use GA based methods to test VLSI circuits [4, 5, 7, 8, 10, 12, 13, 15, 16, 17, 19]. See [2] for further references.

2 Software testing

Dynamic testing techniques execute a program on input data, in contrast to static analysis which uses the program requirements and design documents for visual review. Dynamic testing can be subdivided into two categories of testing techniques, functional and structural. Functional testing, known as the black-box test, aims to test the code by measuring its output or performance without actually viewing the statements of code which are being activated and traversed. In contrast, structural tests are considered white-box or glass-box; the actual code of the program is viewed [18]. The combination of approaches makes testing effective [1, 20].

2.1 The Automation of Testing

Software testing is quite an expensive and time-consuming task. It can be done more efficiently through the automatic generation of test data. Some benefits of automatic testing are [6]:

- testing can be prepared beforehand
- it makes test runs considerably faster
- test runs can be done during night-shift
- the amount of routine work is reduced
- it can be done remotely.

Disadvantages include:

- preparing tests is quite hard
- more knowledge is needed than in non-automatic testing.

Because of these disadvantages the profitability of automatic testing is achieved through repeating the test for the newer version of software or for a different configuration [6]. Normally, the large number of test cases is the problem and this is why automatic testing is done. This problem can't be solved completely by automatic testing and for this reason, we are trying to use genetic algorithms for generating better test data sets.

2.2 Stress Testing

The purpose of stress testing is to identify peak load conditions under which the system fails. The system is subjected to peak loads for key operational parameters: transaction volume, user load, file activity, error rates or their combinations.

3 Testing Environment

In this work we have used ESIM, which is a test automation environment for embedded software development. ESIM uses a workstation for testing embedded software written in the C programming language. Software is compiled with C compiler and linked to the ESIM environment library. The user describes the input and output system (the application-specific hardware). ESIM then simulates the I/O system and the operating system of the application, allowing the user to monitor what is happening in each of them [14].

The GA and the tested program run separately in their own ESIM-tasks, which communicate with each other through simulated ESIM hardware ports.

GA sends inputs to the tested black-box program and measures the response time. Response time is the time it takes for the black-box program to perform the operations caused by the input parameters it received. When the black-box program is completed, it sends a response signal. The response time is the fitness value for the GA.

3.1 A Test Case

Our first test case is a program consisting of 100 randomly generated slowing-down loops. (see `ftp.uwasa.fi cs/report97-1` for the generating program).

The result is a bell shaped response time distribution shown in Figure 2. GA feeds the tested program with a 32 bit string, which is used to select slowing-down loops.

The population size was 40, of which 20 items was always selected to the next generation.

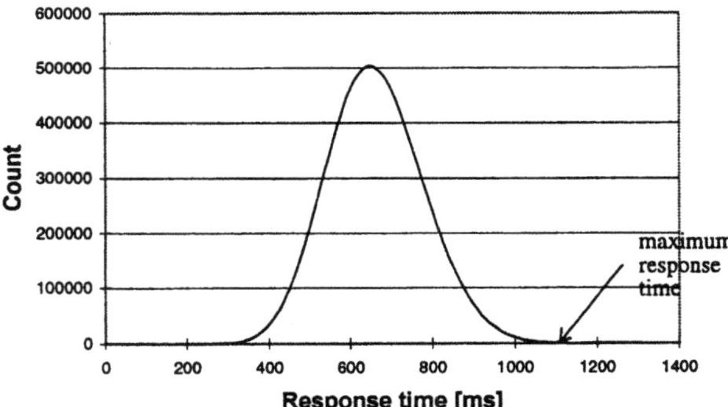

Figure 1: Fitness development via generations.

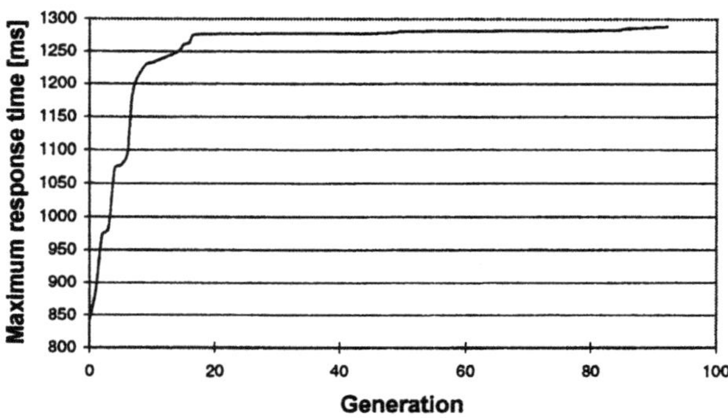

Figure 2: Curve of response times.

4 Results

Figure 1 shows how the worst case develops via generations, when testing our simple program. Figure 2 shows the execution times distribution. From all the 2^{32} possible solutions every 289th was tested in order to obtain the figure. The maximum response time is 1292 ms.

The response time is somewhat non-deterministic, i.e. the same inputs do not always result in the same response time. Figure 3. shows how the response times differ with the same input parameters (= the worst case found). The same inputs were fed in 3000 times, the average deviation was about 5 ms (time between quartiles).

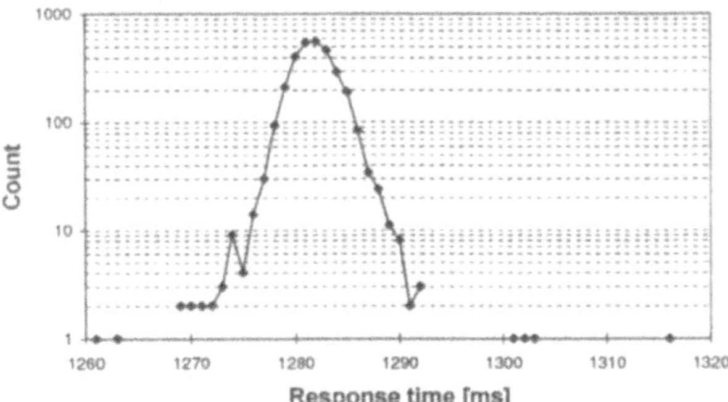

Figure 3: Curve showing the non-determinism of the worst case time response.

5 Conclusion and Future Research

It seems that GA could be suitable for software testing with certain limitations. In white-box testing, problem complexity might cause problems [20]. In black-box testing, the problem might be to find characteristic fitness functions. One possible alternative could be to use the number of warnings from the operating system as a fitness function. If memory-critical software is used, the rate of used memory could be useful a fitness function.

The next step in our project is to evaluate real embedded software, which is a more complex task. Concurrency, continuous operation and the state behaviour of software may introduce non-determinism into the system, which increases the difficulty of software testing [3]. The function of embedded software depends on the scheduling of parallel processes. There are thousands of different scheduling combinations. When the same inputs are given, different response times are obtained. This can interfere with the search made by GA. However, because of the stochastic behaviour of the GA, several test runs are needed in any case, so non-determinism should not be an unsurmountable problem.

In our simple test case there was already some non-determinism, caused by the operating system. This means that the results should be verified in a real environment but it also means that GA might solve the problem in spite of the non-determinism, if it is not too strong.

Because of hardware dependencies, a common solution for validating embedded software is to test it in a simulated or target environment [3]. When software is tested without a target environment, a simulation program is needed. Using simulating software the functioning of a program can be traced. Tracing and repeatability may also make it easier to find errors when the program is black-box tested by a genetic algorithm. If a simulated environment has been used, results based on execution time, are not precise enough. In this case, results must be verified in the real environment. This test could also be called a system test because it tests the whole functionality of the program.

Further information on this work can be found in our anonymous ftp server (`ftp.uwasa.fi`) in directory `cs/report97-1`.

6 Acknowledgements

The work is supported by the Finnish Technology Development Center (TEKES) and ABB Corporate Research. The authors will also gratefully acknowledge the assistance of Mr Jukka Matila from ABB Transmit and Mrs Elizabeth Heap-Talvela and Miss Lilian Grahn for their kind proofreading of the manuscript of this paper.

References

[1] *Proceedings of the 3rd Computer Science Forum*, Baden (Germany), 1996. ABB.

[2] J. T. Alander. Indexed bibliography of genetic algorithms in electronics and VLSI design and testing. Report 94-1-VLSI, University of Vaasa, Department of Information Technology and Production Economics, 1995. (available via anonymous ftp at ftp.uwasa.fi: /cs/report94-1/gaVLSIbib.ps.Z).

[3] A. Auer and J. Korhonen. State testing of embedded software. In *EuroStar -95*, London (UK), 1995.

[4] J. H. Aylor, J. P. Cohoon, E. L. Feldhousen, and B. W. Johnson. Compacting randomly generated test sets. In *Proceedings of the 1990 IEEE International Conference on Computer Design: VLSI in Computers and Processors*, pages 153–156, Cambridge, MA, 17.-19. Sept. 1990. IEEE Computer Society Press, Los Alamitos, CA.

[5] F. Corno, P. Prinetto, M. Rebaudengo, and M. Sonza Reorda. GATTO: a genetic algorithm for automatic test pattern generation for large synchronous sequential circuits. *IEEE Transaction on Computer Aided Design of Integrated Circuits*, 15(8):991–1000, Aug. 1996.

[6] J. Eskelinen. Automatic testing of an embedded software. In J. Jokiniemi and A. Lehtola, editors, *Realtime and Embedded Systems*, Espoo (Finland), 1992. (in Finnish).

[7] T. Hayashi, H. Kita, and K. Hatayama. A genetic approach to test generation for logic circuits. In *Proceedings of the Third Asian Test Symposium*,

pages 101–106, Nara (Japan), 15-17. Nov. 1994. IEEE Computer Society Press, Los Alamitos, CA.

[8] M. S. Hsiao, E. M. Rudnick, and J. H. Patel. Automatic test generation using genetically-engineered distinguishing sequences. In *Proceedings of the 14th IEEE VLSI Test Symposium*, pages 216–223, Princeton, NJ, 28. Apr.- 1. May 1996. IEEE Computer Society Press, Los Alamitos, CA.

[9] M. R. Jones, A. Tezuka, and Y. Yamada. Thermal tomographic methods. *Kikai Gijutsu Kenkyusho Shoho*, 49(1):32–43, Jan. 1995.

[10] T. Lee and I. N. Hajj. Test generation for current testing of bridging faults in CMOS VLSI circuits. In *Proceedings of the IEEE 38th Midwest Symposium on Circuits and Systems*, pages 326–329, Rio de Janeiro (Brazil), 13. -16. Aug. 1996. IEEE, New York.

[11] G. J. Meyers. *The Art of Software Testing.* John Wiley & Sons, New York, 1979.

[12] M. J. O'Dare and T. Arslan. Transitional gate delay detection for combinational circuits using a genetic algorithm. *Electronics Letters*, 32(19):1748–1749, 12. Sept. 1996.

[13] I. Pomeranz and S. M. Reddy. Locstep: A logic simulation based test generation procedure. In *Proceedings of the 25th International Symposium on Fault-Tolerant Computing*, pages 110–118, Pasadena, CA, 27.-30. June 1995. IEEE, Piscataway, NJ.

[14] Prosoft. *ESIM - Testing environment for embedded software, User's Guide Version 2.1 for Windows NT.* Prosoft, Oulu (Finland), 1995.

[15] E. M. Rudnick and J. H. Patel. Combining deterministic and genetic approaches for sequential circuit test generation. In *Proceedings of the 32nd Design Automation Conference*, pages 183–188, San Francisco, CA, 12.-16. June 1995. IEEE, New York.

[16] E. M. Rudnick, J. H. Patel, G. S. Greenstein, and T. M. Niermann. Sequential circuit test generation in a genetic algorithm framework. In *Proceedings of the 31st Design Automation Conference*, pages 698–704, San Diego, CA, 6.-10. June 1994. IEEE, New York.

[17] D. G. Saab, Y. G. Saab, and J. Abraham. CRIS: A test cultivation program for sequential VLSI circuits. In *Proceedings of the International Conference on Computer Aided Design*, pages 216–219, 1992.

[18] I. Sommerville. *Software Engineering.* Addison-Wesley, New York, 1996.

[19] J. Stefanovic and E. Gramatova. RTL level test generation using genetic algorithm and simulated annealing. In *Proceedings of the 2nd Workshop on Hierarchical Test Generation*, Duisburg (Germany), 25-26. Sept. 1995.

[20] A. L. Watkins. The automatic-generation of test data using genetic algorithms. In I. M. Marshall, W. B. Samson, and D. G. Edgar-Nevill, editors, *Proceedings of the 4th Software Quality Conference*, volume 2, pages 300–309, Dundee (UK), 4.-5. July 1995. University of Abertay Dundee, Scotland.

[21] S. Xanthakis, C. Ellis, C. Skourlas, A. L. Gall, S. Katsikas, and K. Karapoulios. Application of genetic algorithms to software testing (application des algorithmes génétiques au test des logiciels). In *Proceedings of the 5th International Conference on Software Engineering*, pages 625–636, Toulouse, France, 7.-11. Dec. 1992.

An Evolutionary/Meta-Heuristic Approach to Emergency Resource Redistribution in the Developing World

A. Tuson, R. Wheeler and P. Ross
OUTLOOK Group, Department of Artificial Intelligence,
University of Edinburgh, 80 South Bridge, Edinburgh EH1 1HN
{andrewt,richardw,peter}@dai.ed.ac.uk

Abstract

The problem of logistics and resource management in disease control projects in the developing world can hardly be understated. One example is the occurance of regional imbalances in supply. A prototype system, based upon evolutionary and 'meta-heuristic' optimisation techniques is described that recommends a plan for the redistribution of available resources to minimise shortages. Evaluation of the system on data from real world situations indicated that the generation of good, feasible redistribution plans is possible even on large datasets. Comparison of the optimisers showed that evolutionary techniques perform poorly on this problem compared to stochastic hill climbing.

1 Introduction

Dealing with aid projects in the developing world can be made difficult by the fact that though an effective policy exists, the actual implementation is often more difficult. For instance, situations may occur where regional imbalances arise: an example would be a treatment site with sufficient diagnostic kits for a disease, but with no drugs to treat the diagnosed cases.

This problem can be described as follows: we have N sites, each of which need to have minimum amounts of M resources in order to operate effectively. Furthermore, each site has different requirements for each resource, and shortages and surpluses of each resource occur at different sites. A resource management system has to give a list of recommendations of the form 'move X amount of resource Z from site A (which has a surplus) to site B (which has a shortage)'. The aim is then to maximise the number of resource targets met (referred to as resource-sites), whilst minimising the number of shipments. This is subject to constraints on feasibility, for example, a site cannot supply more of a resource than it has.

The hardware limitations are also severe — the final system aims to run on a 386 class PC compatible, with a 1 MB memory, and a 20 MB hard drive. The goal is to generate plans for regions with at least 250 relief centres within several hours.

This paper outlines a prototype resource management system to evaluate the feasibility of a system that recommends an *implementable* shipment plan to minimise such shortages in real-world situations.

2 Implementation

The above problem is therefore to find a redistribution plan that minimises shortages and the number (and cost) of shipments required, whilst satisfying the constraints outlined earlier. The solution adopted is now outlined.

2.1 Representation

The representation of the redistribution plan adopted was indirect [1] — the candidate solution encodes not the redistribution plan; but rather instructions to a plan builder. This approach is compact in its memory usage and always generates feasible plans. This 'anytime' characteristic is of particular use as an answer may be required quickly, and a sub-optimal but feasible answer is preferable to no answer at all. The encoding is the order in which sites are considered for supply (i.e. a permutation) by the plan builder. The plan builder is given the order in which the relief sites are to be considered, the current situation as regards resource levels at all the sites, and the desired level of supply, and information on the location of each site.

A brief description of how the plan builder produces a redistribution plan is as follows: the first site of the sequence becomes the current site, then each resource at the current site is considered in turn and if there is a shortage then an attempt to find a supply of that resource is begun. If a site has already been found to supply the current site with another resource, that is tried first (so as to minimise the number of separate shipments made), otherwise the neighbouring sites are considered in order of their distance from the current site, and surplus resources transferred until either the resource target is met, or there are no more sites available. This process is then repeated for the other resources, and then onto the next site in the sequence. It is important to note that if a site has been supplied with a resource, then it cannot supply another site with that resource. This prevents 'daisy-chaining' which can lead to brittle plans, because if one of the shipments in the plan was not to occur, the later shipments are likely to be disrupted.

2.2 Optimisation

It can readily been seen that the plan generated is dependant upon the order that sites are presented to the plan builder. A variety of optimisers were implemented to optimise the initial redistribution plan: steepest-ascent (SAHC), first-ascent (FAHC), and stochastic (SHC) hill climbers with iteration (i.e. random restart if no improvements after a certain number of evaluations); tabu search (TS) [3]; simulated annealing (SA) [5]; threshold accepting (TA) [2]; and an evolutionary algorithm (EA) [4].

The neighbourhood operators used by all of the optimisation methods were permutation-shift and permutation-swap [7] (Figure 1).

For the evolutionary algorithm the crossover operator used was 'modified PMX' [6], which has been found to give good results on some permutation-encoded scheduling problems (Figure 2).

The repair procedure analyses one string for duplicates: when one is found it is replaced by the first duplicate found in the second string.

The evaluation function used in this system was a simple one — a linear combination of the number of targets met (targets_met) and the number of

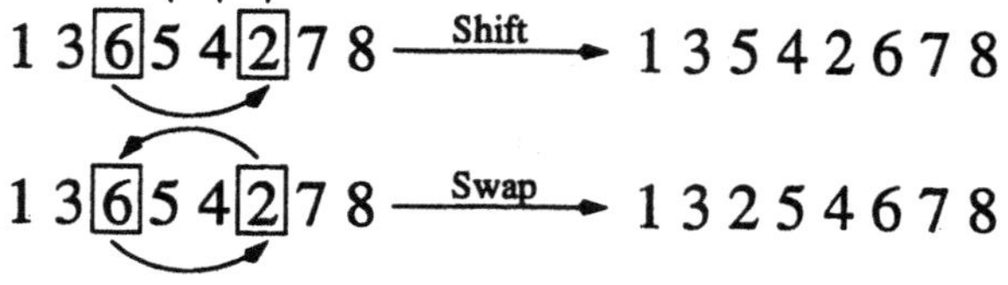

1 3 [6] 5 4 [2] 7 8 —Shift→ 1 3 5 4 2 6 7 8

1 3 [6] 5 4 [2] 7 8 —Swap→ 1 3 2 5 4 6 7 8

Figure 1: The permutation shift and swap operators.

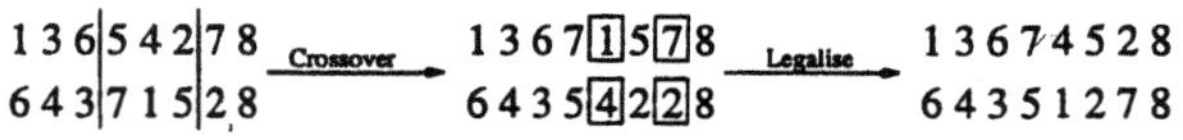

1 3 6 | 5 4 2 | 7 8 —Crossover→ 1 3 6 7 [1] 5 [7] 8 —Legalise→ 1 3 6 7 4 5 2 8

6 4 3 | 7 1 5 | 2 8 —→ 6 4 3 5 [4] 2 [2] 8 —→ 6 4 3 5 1 2 7 8

Figure 2: Modified PMX crossover.

shipments made (shipments); as is shown below:

$$\text{fitness} = 10 \times \text{targets\_met} - \text{shipments} \quad (1)$$

The targets_met was weighted by a factor of 10, so that the system will optimise targets_met before trying to optimise shipments. This was deemed a desirable behaviour by the potential end-users.

3 An Initial Evaluation

The data for the test problems was provided by an expert with experience of the type of situations the system is intended to play a part in, and was based upon actual data.

A small domain consisting of 21 sites and 12 resources was examined first so that a thorough test of the optimisers could be made in a reasonable amount of time. This test dataset had the feature that there were roughly sufficient resources available to supply all of the resource targets.

SHC was used to give a rough idea of the expected performance of the system. The initial solution met about 85-92% of resource targets when processed by the plan builder, compared to 32% before the system is run. Further optimisation increased this figure to 90-92%, and managed to reduce the number of shipments required by 20% — from about 125 to about 105 shipments — in less than 1000 evaluations. This took 7 seconds on a SPARC 5, and 5 minutes on a 386-class PC.

The performance of each of the optimisers, for a range of technique-specific parameter values was then investigated, including the choice of a swap or shift neighbourhood. Fifty runs, each lasting 2000 evaluations were taken in each case (significance was been ascertained using a t-test).

Due to lack of space, not all of the results obtained can be given here (these can be obtained from the authors). Instead the mean and standard deviation (in brackets) of the results obtained, after tuning, in terms of solution quality, and the number of evaluations to solution for each technique are summarised in Table 1.

Some points can be made about the behaviour of the optimisers in general. The type of hill climbing that the optimiser was based upon was found to be

Table 1: A summary of optimiser performance.

Algorithm	Shift Neighbourhood		Swap Neighbourhood	
	Quality	Evals. Reqd.	Quality	Evals. Reqd.
SHC	2321.96 (1.15)	1091.36 (519.64)	2321.68 (1.19)	1042.56 (474.51)
SA	2321.64 (1.53)	1045.92 (527.88)	2321.26 (1.59)	1074.68 (534.29)
TA	2321.82 (1.19)	981.36 (501.17)	2321.58 (1.30)	1053.26 (467.23)
EA	2320.42 (1.17)	1683.02 (247.80)	2319.98 (1.10)	1526.56 (296.02)
FAHC	2319.86 (1.76)	1083.26 (505.30)	2319.98 (1.41)	953.40 (587.77)
SAHC	2317.46 (1.88)	1921.00 (178.89)	2319.52 (1.99)	1404.02 (461.24)
TS(FA)	2319.16 (2.35)	834.26 (427.57)	2319.24 (1.69)	800.56 (583.02)
TS(SA)	2318.28 (1.83)	> 2000 (-)	2319.34 (2.00)	1346.98 (388.00)

important. SHC-based optimisers significantly outperformed FAHC and SAHC based optimisers for both types of neighbourhood; with SHC the method of choice for this problem. Only the EA and SAHC-based optimisers took significantly longer to converge upon a solution than SHC.

The choice of neighbourhood made little difference for SHC and FAHC based optimisers. However, for a SAHC-based optimiser, a swap neighbourhood gave better quality solutions. This was due to swap neighbourhoods being roughly half the size of shift neighbourhoods.

3.1 Hill Climbing

Iterating the hill climber gave a significant improvement for SHC (shift-only), FAHC (both), and SAHC (swap-only).

3.2 Simulated Annealing

Compared against SHC without iteration, solution quality was increased, but not as much as for iterated hill climbing. Low initial temperatures were found to be most effective, which allowed increases in *shipments* whilst preventing decreases in *targets_met*.

3.3 Threshold Accepting

The best results were obtained with a low initial threshold of one. In a similar fashion to SA, it appears that there are gains to be made by allowing a little exploration; but this must be used sparingly.

3.4 Evolutionary Algorithm

The EA performed less well than SHC. The choice of population model was found to be important. A steady state model gave the best performance, followed by a (N+N) ES model. Generational models performed worst of all (worse than SAHC). Also, crossover seemed to have no positive effect, and in some cases a high crossover probability degraded solution quality slightly. Although it is unclear whether this lies with the type of recombination operator used, or with population-based search in general.

3.5 Tabu Search

The level of performance was not significantly affected by the length of the tabu list, remaining comparable to the basic hill climber. The exception was SAHC-based tabu search with a shift-neighbourhood; probably due to the tabu list reducing the size of a large neighbourhood.

4 Additional Constraints

So far, the assumption has been made that the shipment can travel over any distance — this is in fact not true. For the real-world problem this system is aimed at, traversing over the catchment area of more than 3 sites is infeasible.

To see if imposing a constraint on shipment distance would affect the coverage attainable, SHC was used with the constraint that the shipments could not be made further than a maximum value of 3. The coverage attained was around 90%, a slight drop compared with 96% without the constraint.

It would also be desirable for the distance of the shipments to be minimised. The evaluation function was modified to include the total distance covered by the shipments made.

$$\text{fitness} = 10 \times \text{targets\_met} - \text{distance} \qquad (2)$$

The use of the modified evaluation function was found to have no effect upon coverage. A lower distance solution was also attained, though at the expense of using more shipments.

5 Large Dataset Results

The experiments so far have been performed on a relatively small domain. A practical system will have to deal with more than this. So a larger data-set was used, consisting of 283 sites and 12 resources, based on actual data as before with the constraint that a shipment cannot go further than distance 3.

Initially, only 19% of the resource targets were met. The initial solution met about 54% of the targets when processed by the plan builder. The most effective of the optimisers on the small dataset, SHC, was used with the first of the evaluation functions for 2000 evaluations (which took about 10.5 minutes on a SPARC 5 workstation, or 7-8 hours on a 386 PC). This increased the number of resource targets met to 57%. The search was then extended for 20,000 evaluations. This gave a best solution of 58%; though the local optimum was not reached at that stage.

The lower coverage obtained, compared with the small dataset, was found not to be due to any failing of the system itself, but that at least 30% of sites in the data set were inaccessible to any of the sites with surpluses. This was due to a combination of the sites with surpluses being heavily concentrated (spatially) at one end of the dataset, and the maximum shipment distance constraint.

6 Extensions

The problem studied still requires more work to fully reflect the real-world situation; addressing this requires discussion with the end-users to ascertain their requirements. Also, the introduction of directed neighbourhood operators and 'delta-evaluation' [8] could increase the speed of search. Finally, the question of why the EA performed so poorly could be explored. A possible reason may lie in the fact that the crossover operator used manipulates the wrong type of 'building block'; finding out what the appropriate building blocks of this problem are and devising a crossover operator to exploit them may improve EA performance.

7 Conclusion

The system described was able to quickly produce a workable redistribution plan for both small and full-sized versions of the problem which increased dramatically the number of resource targets met. Further optimisation improved upon this further; but this was found to take some time for large problems. Fortunately the 'anytime' characteristic of the system means that this is not a real concern. Experiments with the full-sized dataset indicated that the constraints upon the problem may often prevent all the resource targets being met, even if there are sufficient resources available.

Stochastic hill climbing was found to be the method of choice, with type of neighbourhood operator used being unimportant. In fact SHC-based optimisers were the best performers; though the performance of the Evolutionary Algorithm was disappointing.

8 Acknowledgements

We would like to express our gratitude to the Engineering and Physical Sciences Research Council (EPSRC) for their support of Andrew Tuson via a research studentship (95306458).

References

[1] E. Burke, D. Elliman, and R. Weare. The Automated Timetabling of University Exams using a Hybrid Genetic Algorithm. In *The AISB Workshop on Evolutionary Computing*, 1995.

[2] G. Dueck and T. Scheuer. Threshold Accepting: A General Purpose Optimisation Algorithm Superior to Simulated Annealing. *Journal of Computation Physics*, 90:161–175, 1990.

[3] F. Glover. Tabu Search: A Tutorial. *Interfaces*, 4:445–460, 1990.

[4] J.H. Holland. *Adaptation in Natural and Artificial Systems*. Ann Arbor: The University of Michigan Press, 1975.

[5] S. Kirkpatrick, C.D. Gelatt, Jr., and M.P. Vecchi. Optimization by Simulated Annealing. *Science*, 220:671–680, 1983.

[6] G.F. Mott. Optimising Flowshop Scheduling Through Adaptive Genetic Algorithms. Chemistry Part II Thesis, Oxford University, 1990.

[7] C.R. Reeves. A genetic algorithm for flowshop sequencing. *Computers & Ops. Res.*, 22:5–13, 1995.

[8] P. Ross, D. Corne, and H-L. Fang. Improving Evolutionary Timetabling with Delta Evaluation and Directed Mutation. In Y. Davidor, H-P. Schwefel, and R. Manner, editors, *Parallel Problem-solving from Nature - PPSN III*, LNCS, pages 566–565. Springer-Verlag, 1994.

Automated Design of Combinational Logic Circuits by Genetic Algorithms

C. A. Coello Coello[1], A. D. Christiansen[2] and A. Hernández Aguirre[2]
[1] (coello@depauw.edu) 246 Julian Science Center, Department of Computer Science, DePauw University, Greencastle, IN 46135, USA
[2] ({adc,hernanda}@eecs.tulane.edu), 301 Stanley Thomas Hall, Department of Computer Science, Tulane University, New Orleans, LA 70118, USA

Abstract

We introduce a method, based on a genetic algorithm (GA) approach, to design combinational logic circuits. This problem is quite difficult for a traditional GA, but we have overcome these difficulties and have implemented a computer program that can automatically generate high-quality circuit designs. We describe the important issues to consider when solving this circuit design problem: the importance of the representation scheme, the encoding function, and the definition of the fitness function. We present several circuits derived by our system under various assumed constraints, such as the maximum number of allowable gates and the types of available gates. We compare the solutions produced by our system against those generated by a human designer. We also show that our representation approach, when compared to a standard binary encoding, produces better performance both in terms of quality of solution and in terms of speed of convergence.

1 Introduction

Design is usually considered to be an activity requiring considerable human creativity and knowledge. Even the definition of the term ***design*** itself is quite elusive, since it can be interpreted in several different ways depending on the task to be performed. Although there have been many attempts at developing programs for automated design, such programs are notoriously difficult to build.

In the research reported in this paper, we seek a computer-based tool that can make the design process less tedious for the human designer without sacrificing quality of the design produced. In this paper, we limit our focus to combinational logic circuits, which contain no memory elements. Such circuits contain no feedback paths.

2 Previous Work

A general search technique inspired by natural evolution, called the *genetic algorithm* [4], has been widely used for optimization tasks [3] and is known to be a very powerful tool in certain domains. In our current work we wish to find a way to use the genetic algorithm (GA) as a design tool, with particular emphasis in the design of combinational circuits.

The design process for combinational logic circuits has evolved from its first notions [12] to a standard element of undergraduate computing curricula [11]. Standard graphical design aids such as Karnaugh Maps [5, 13] are widely used and tools suitable for computer implementation have evolved from the Quine-McCluskey Method [9, 10] to freely available tools such as Espresso [1] and MisII [2] and many commercial products.

Louis [7] is one of few sources found in the literature to address the use of GAs for the combinational logic design problem. In his dissertation [8] Louis combines knowledge-based systems with the genetic algorithm, making use of a genetic operator called *masked crossover* that adapts to the encoding, being able to exploit information unused by classical crossover operators. His results, although very encouraging for certain examples, do not seem to have solved the combinational circuit design problem completely. However, his idea of incorporating knowledge about the domain in the genetic operator constitutes a big step toward increasing the power of the GA as a design tool. Unfortunately, the incorporation of knowledge into the GA decreases its usefulness as a *general* search tool. Louis overcomes this problem by defining an operator that he claims to be domain independent, but whose efficiency turns out to depend on the representation used.

Koza [6] has used genetic programming to design combinational circuits. He has designed, for example, a two-bit adder, using a small set of gates (AND, OR, NOT), but his emphasis has been on generating functional circuits rather than on optimizing them. In fact, this is also the case in Louis' research, where the main focus was to provide an easier way to generate functional designs using the GA rather than in optimizing a functional design according to certain metrics. So far, genetic programming has been considered a more powerful tool in such tasks, because the representation it uses is more powerful for structural design in general.

In the work reported here, we are interested not only in producing functional designs, but also in optimizing them according to certain metrics. This is a complicated task for the GA, because we must deal with two difficult problems at the same time.

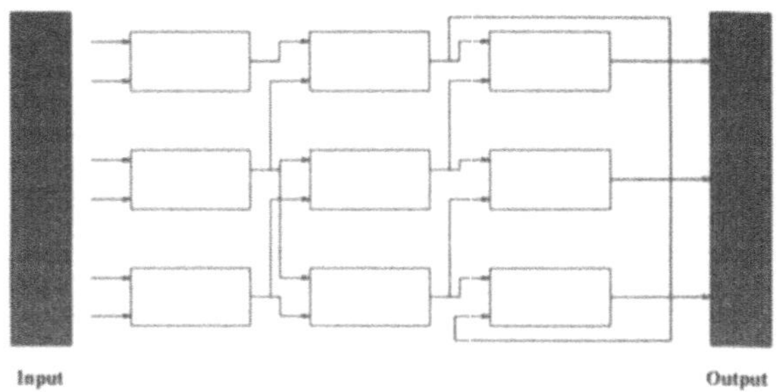

Figure 1: A gate in a two-dimensional template, gets its second input from either one of two gates in the previous column.

3 Statement of the Problem

The problem of interest to us consists of designing a circuit that performs a desired function (specified by a truth table), given a certain specified set of available logic gates. The complexity of a logic circuit is a function of the number of gates in the circuit. The complexity of a gate generally is a function of the number of inputs to it. Because a logic circuit is a realization (implementation) of a Boolean function in hardware, reducing the number of literals in the function should reduce the number of inputs to each gate and the number of gates in the circuit—thus reducing the complexity of the circuit. The algebraic method used to minimize functions is tedious and error prone. Its success depends on our ability to recognize the application of a theorem or a postulate during the minimization process. Such recognition may not be obvious. Furthermore, there is no general set of rules to aid that recognition.

Two popular minimization techniques are the *Karnaugh Map* [5], which is based on a graphical representation of Boolean functions, and the *Quine-McCluskey Procedure* [9, 10], which is a tabular method. Both of these methods are mechanical in nature and their efficiency depends on the designer's abilities.

In this work, we compare the designs produced by a GA with those generated by a human designer using Karnaugh maps. The comparison is in many ways unfair because of differing capabilities of man and machine. For example, a human designer tends to use only the gates NOT, AND, OR and has more difficulties using XOR because the Karnaugh Map does not support the identification of XOR terms as well as it supports 'seeing' simple product terms. The computer, using our GA approach, and not being restricted by human pattern recognition abilities, uses many XOR gates, often disregarding the NOT gate. Our overall measure of circuit optimality is the total number of gates used, regardless of their kind. This is approximately proportional to the total part cost of the circuit. Obviously, we perform this analysis for only fully functional circuits.

3.1 Example 1: A Two-Bit Adder

We want to find the combination of five possible types of gates (AND, NOT, OR, XOR and WIRE) so that our circuit performs the two-bit addition of its inputs, producing a three-bit sum. Our objective is to find a functional design that minimizes the use of gates other than WIRE (essentially a logical no-operation). The circuit can be represented as a two-dimensional array of gates $S_{i,j}$. The index j indicates the *level* of the gate, where gates closest to the inputs have the lowest values of j. (Level numbers increase from left to right in Figure 1.) For fixed j, the index i ranges over gates that are 'next to' each other in the circuit, but such gates are not directly connected to each other. Each gate can get its inputs from any row (any gate in the next smaller level) in the matrix.

3.2 Example 2: A Two-Bit Multiplier

This problem is similar to the previous one, but in this case we want to design a circuit that performs the two-bit multiplication of its inputs, producing a four-bit result. The set of gates available is the same as in the previous example and, again each gate can get its inputs from any gate in the previous level according to our matrix representation.

4 Using the Genetic Algorithm

The first interesting aspect of this problem is the encoding of solutions. Each circuit is encoded in the following way: < *input*1 >< *input*2 >< *gate_type* >, where *input*1 and *input*2 can be any of the gates or inputs at the previous column in the matrix representing the circuit (see Figure 1), and *gate_type* is the gate that we will be using for that particular position of the array. A chromosome is formed with as many triplets of this kind as needed, according to the size of the matrix that the user wants to use to represent the circuit. In our experiments, floating point representation was used, and the order chosen for the gates was column-order, starting from the position $S_{1,1}$ (the top left-most position), followed by the gate below it ($S_{2,1}$) and so on. We also experimented with an alphabet of cardinality n, where n can be defined by the user and will be normally taken as the number of rows allowed in our circuit, according to the matrix encoding adopted in this problem. This representation allows the manipulation of shorter strings, it decreases the complexity of the decoding task, and it provides better solutions.

Another difficulty is the development of a good fitness function. Our approach compared the output produced by the circuit generated by the GA with the desired values according to the truth table, on a bit-per-bit basis. Our fitness function works in two stages. At the beginning of the search, only validity of the circuit outputs is taken into account, but once functional solutions start appearing in the population, we switch to a second fitness function which tries to minimize the number of gates by rewarding those circuits that use more WIREs.

5 Comparison of Results

We compared, for both examples, our results with those produced by an experienced human designer. Populations of 3000 chromosomes were used in all tests, and the GA was run for 200 generations.

5.1 Example 1: A Two-Bit Adder

Figure 2 shows a functional circuit generated by a human designer using Karnaugh maps. Our solution, using a floating point representation that used the integers from 0 to 3, to represent 4 rows of inputs, is shown in Figure 3. This solution is quite efficient, since only 7 gates were required, instead of the 12 gates used by the human designer.

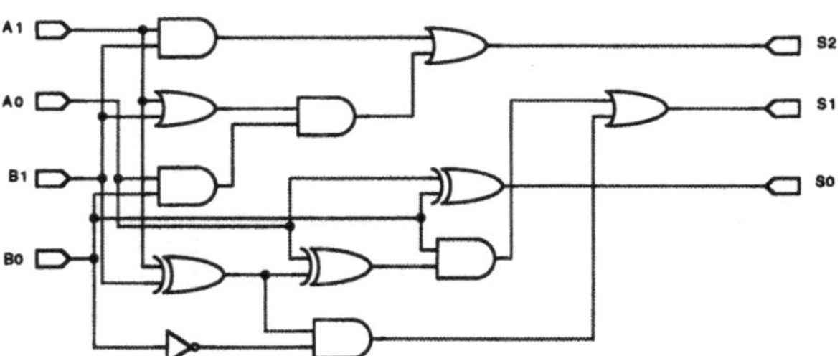

Figure 2: A 2-bit adder generated by a human designer.

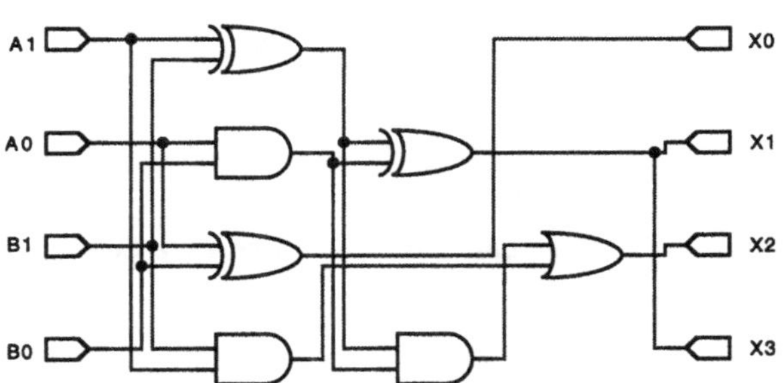

Figure 3: A sample 2-bit adder generated by our GA-based approach.

5.2 Example 2: A Two-Bit Multiplier

This is a more complicated problem in which we compared again our solution with the circuit generated by a human designer (see Figure 4). Once more, the solution generated by the GA using an alphabet of cardinality 8 (using integers from 0 to 7) produced a circuit that uses only 7 gates (see Figure 5) instead of the 16 gates used by the human designer.

6 Conclusions and Future Work

We have demonstrated a fully implemented GA-based approach to the combinational logic circuit design problem, and we have compared the results obtained by our system to circuit designs produced by experienced human designers. The results of our two simple multiple output logic functions are quite encouraging, and it should be mentioned that several other much more complicated examples have also been successfully solved using our approach. We are interested in building a system that can automatically design a wide range of combinational

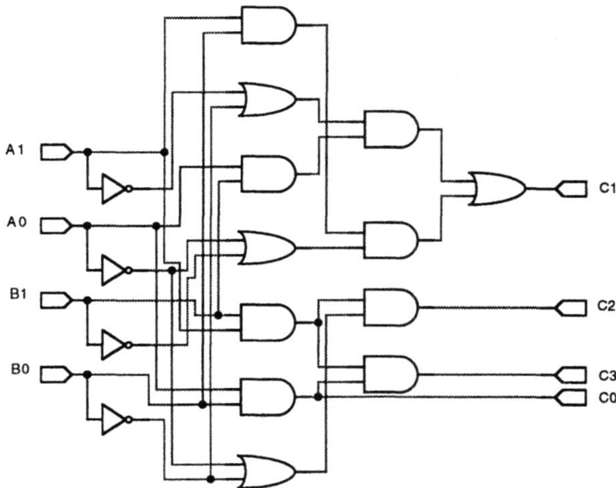

Figure 4: A 2-bit multiplier generated by a human designer.

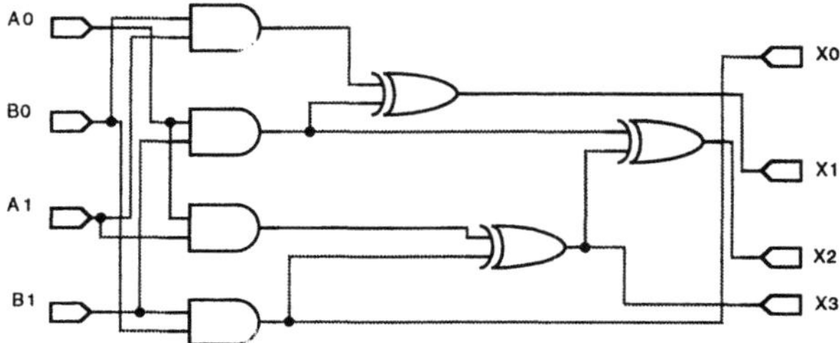

Figure 5: A 2-bit multiplier generated by our GA-based approach.

and sequential logic circuits. We are interested in exploring optimization tradeoffs involving number of levels in the circuit, total number of gates, number of gate inputs, number of integrated circuit packages, part costs, etc. We believe that a GA-based approach has great potential to provide a practical tool for assisting designers of logic circuits.

References

[1] R. K. Brayton, G. D. Hachtel, C. T. McMullen, and A. L. Sangiovanni-Vincentelli. *Logic Minimization Algorithms for VLSI Synthesis.* Kluwer Academic Publishers, 1984.

[2] R. K. Brayton, R. Rudell, A. Sangiovanni-Vincentelli, and A. R. Wang. Mis: A multiple-level logic optimization system. *IEEE Transactions on Computer-Aided Design*, CAD-6 (6):1062–1081, November 1987.

[3] David E. Goldberg. *Genetic Algorithms in Search, Optimization and Machine Learning.* Reading, Mass. : Addison-Wesley Publishing Co., 1989.

[4] J. H. Holland. *Adaptation in Natural and Artificial Systems. An Introductory Analysis with Applications to Biology, Control and Artificial Intelligence.* MIT Press, Cambridge, Massachusetts, 1992.

[5] M. Karnaugh. A map method for synthesis of combinational logic circuits. *Transactions of the AIEE, Communications and Electronics*, 72 (I):593–599, November 1953.

[6] J. R. Koza. *Genetic Programming. On the Programming of Computers by Means of Natural Selection.* The MIT Press, 1992.

[7] S. J. Louis and G. J. Rawlins. Using genetic algorithms to design structures. Technical Report 326, Computer Science Department, Indiana University, Bloomington, Indiana, feb 1991.

[8] S. J. Louis. *Genetic Algorithms as a Computational Tool for Design.* PhD thesis, Department of Computer Science, Indiana University, aug 1993.

[9] E. J. McCluskey. Minimization of boolean functions. *Bell Systems Technical Journal*, 35 (5):1417–1444, November 1956.

[10] W. V. Quine. A way to simplify truth functions. *American Mathematical Monthly*, 62 (9):627–631, 1955.

[11] C. H. Roth, Jr. *Fundamentals of Logic Design (4th Edition).* West Publishing Company, 1992.

[12] C. E. Shannon. A symbolic analysis of relay and switching circuits. *Transactions of the AIEE*, 57:713–723, 1938.

[13] E. W. Veitch. A chart method for simplifying boolean functions. *Proceedings of the ACM*, pages 127–133, May 1952.

Forecasting of the Nile River Inflows by Genetic Algorithms

M. E. El-Telbany[1], A. H. Abdel-Wahab[1], and S. I. Shaheen[2],
[1] Computers and Systems Dept., Electronics Research Institute, Dokki, Giza, Egypt
[2] Computers Engineering Department, Faculty of Engineering, Cairo University.
E-mail: {telbany, ashraf}@eri.sci.eg, sshaheen@frcu.eun.eg

Abstract

The prediction of time series phenomena is a hard and complex task. The selection of a proper statistical model and the setup of its parameters (in terms of the number of coefficients and their values) is also a difficult task and it is usually solved by trial and error. This paper presents a hybrid system that integrates genetic algorithms and traditional statistical models to overcome the model selection and tuning problem. The system is applied to the domain of river Nile inflows forecasting. This domain is characterized by the availability of large amount of data and prediction models. Finally, the results of applying the proposed system are presented and discussed.

1 Introduction

The forecasting of the stochastic systems using the available data is a very complex task. Stochastic mathematical models like autoregressive moving average (ARMA) and its periodic form (PARMA), generally, do not capture all the characteristics of the available data. As a result of the complexity of the statistical techniques, the inaccurate identification of a correct model, and the estimation of parameters, other artificial intelligence techniques have been tried like for example, artificial neural networks which have been applied for *White Nile river* forecasting based on a routing model from the Packwatch (Lake Albert exit) to Wadi Halfa (Lake Nasser entrance) [2]. In this paper, a hybrid system that integrates both statistical models and a genetic based search technique is presented and discussed. Genetic algorithms are used to search for the best prediction model and its parameter settings (*off-line* learning mode) [3]. Also, they are used to adapt the selected prediction model dynamically while observing new data points (on-line learning mode). In the *off-line* mode, the system extracts the appropriate model and its parameter values depending on previous available data. In the *on-line* mode, it adapts the model and its parameter values according to real-time observations.

2 Statistical Models for Forecasting

Statistical models have been used extensively to simulate and predict the behavior of an environment. These models include the regression or '*least squares*' analysis which is a statistical technique for modeling and investigating the relationship between two or more variables. Also, the moving averages approach is helpful in looking for patterns in data when the underlying signal is disturbed by noise. In this sense, moving averages are *linear data smoothers or linear filters* in that they systematically reduce the noise in the observations, thus making the underlying pattern easy to detect. A widely used statistical method is the exponential smoothing. It is mainly used for smoothing discrete time series in order to forecast the immediate future. In this method, the new forecast (for time $t+1$) may be thought of as a weighted average of the old forecast (for time t) and the new observation (at time t), with the weight (called the smoothing constant) assumed to range from 0 to 1, $0 \leq \alpha \leq 1$. Thus

$$\begin{aligned}(\text{new forecast}) = &(1-\alpha) \times (\text{old forecast}) + \\ &(\alpha \times \text{new observation})\end{aligned} \quad (1)$$

The statistic $\tilde{y}_{t+1}$ is represented as a weighted average of all past observations. that is:

$$\begin{aligned}\tilde{y}_{t+1} &= \alpha y_t + (1-\alpha) y_t \quad (2)\\ \tilde{y}_{t+1} &= \alpha y_t + (1-\alpha)\alpha y_{t-1} + (1-\alpha)^2 y_{t-1}\end{aligned}$$

The practical value of α lies between 0.01 and 0.3 [4]. On the other hand, the Box-Jenkins methodology consists of a large class of time series models and a set of procedures for choosing models from this class to fit to data from a given series. The models form a *real-time* forecasting framework. The time series models are called *autoregressive moving average models*, or ARMA models for abbreviation. To encompass the diverse forecasting applications that arise in practice, this class of models has to be, and is, very large. For example, the exponential smoothing, the autoregressive models, and the random-walk models are all special forms of ARMA models. Two important methods are used in ARMA models namely, the *autocorrelation function (acf)* and the *partial autocorrelation function (pacf)* which are used to *identify and estimate* appropriate ARMA model parameters. To achieve greater flexibility in fitting actual time series, it is sometimes advantageous to include both autoregressive and moving average terms in a single model. Different models are identified by the number of autoregressive parameters p, and the number of moving average parameters q. Seasonal ARMA models consist of two parts: terms that incorporate the season-to-season movements, and terms that model the nonseasonal, or within season, movements. Because of this definition, the identification stage requires two separate passes, one to determine the seasonal parameters and another for the nonseasonal parameters. The seasonal part of the model has its own autoregressive and moving average parameters with order P and Q, while the nonseasonal part has orders p and q. The model is then put in the form $ARMA(p,q) * (P,Q)$.

3 Proposed Genetic Based Forecasting System

The implemented forecasting system consists of the following main modules:

- **Data Manipulation and Transformation**: In this module, raw data is processed to eliminate existing noise as much as possible. This processing is described by the following equation:

$$\tilde{y} = y_t - \mu \tag{3}$$

  where μ is the mean value of the samples. Also, missing data points are set to zero value.

- **Genetic Based Search Module**: This module is used for the model selection and parameter tuning operation using genetic algorithms. Each chromosome in the population represents a model and its parameter values (a variable length binary string). Initially, the population is initialized randomly for the parameters numbers p, q and their values. The crossover operator is slightly modified to accommodate for variable length chromosomes. The crossover point is selected to be within the boundaries of the shortest chromosome (single point crossover is adopted). The mutation operator is used in its conventional form i.e. a bit is selected at random (according to the mutation rate) and flipped. The genetic algorithm cycle continues until the prediction error produced by the best chromosome is within a certain error limit. The population size is set to 50, crossover and mutation rates are 0.7 and 0.01 respectively. The generation gap and maximum number of generations are 0.1 and 1000. The initialization of the population is based upon the Box-Jenkins time series ARMA(p,q) models [1]. The identified models are based on the PARMA time series models as shown in the equation below.

$$\begin{aligned} y_{i,j} = & \sum_{k=1}^{p} \phi_k y_{i,j-k} + \sum_{k=1}^{q} \theta_k e_{i,j-k} + \\ & \sum_{L=1}^{P} \Phi_L y_{i-L,j} + \sum_{L=1}^{Q} \Theta_L e_{i-L,j} + e_{i,j} \end{aligned} \tag{4}$$

Equation 4 defines the $PARMA(p,q) * (P,Q)$ mathematical model, where $y_{i,j}$ is the stochastic process under forecasting for step t, and $t = (i-1) * \text{season} + j$. The values of $\phi_k, \theta_k, \Theta_L, \Phi_L$ are the nonseasonal and seasonal autoregressive and moving average coefficients respectively. Assume $e_{i,j}$ to be the white noise with mean zero and variance δ. Also the values of p, q, P, and Q represent the orders of the nonseasonal and seasonal terms.

The fitness function used for ranking the chromosome (i.e. a model) in the population is the *Akaike Information Criteria (AIC)* [5] which is calculated as follows:

$$AIC(p,q) = N \ln(\Omega_e^2) + 2(p+q) \tag{5}$$

where N is the number of trained observations, $(p+q)$ is the number of parameters estimated, and the maximum likelihood is determined according to

$$\Omega_e^2 = \frac{1}{N} \sum_{t=0}^{N} \Omega_e^2(t) \tag{6}$$

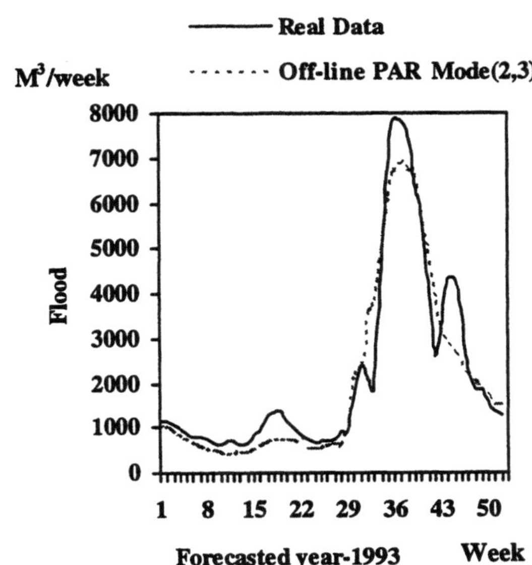

Figure 1: The off-line PAR model forecasts for 52 weeks ahead using genetic algorithms.

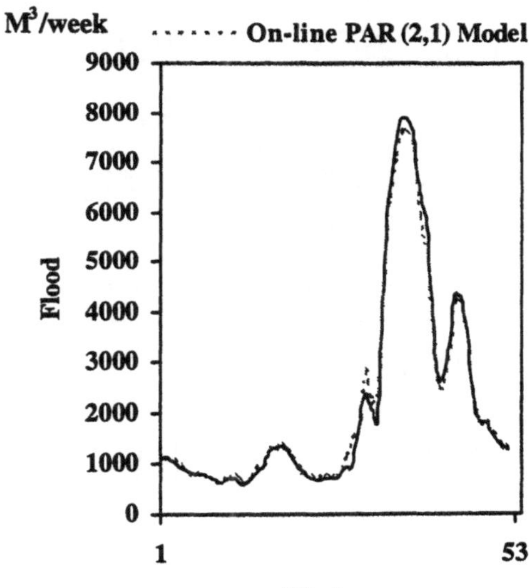

Figure 2: The on-line PAR model forecasts for 1 week ahead using genetic algorithms.

- **Forecasting**: The discovered and identified model that represent the stochastic process is used to predict the future values for a specified horizon. In the off-line learning mode, the forecasting error is based upon a portion of the data that has not been used for training (unseen samples). The forecasted value produced by the system is then compared with the actual value and the forecast error is then calculated. In the on-line learning mode, the current actual reading (data sample) is used as part of the data set to adjust the model parameters in the next time step.

Table 1: The parameters of best models identified.

Model Type	Model Parameters
PAR	$\bar{y} = 0.59\tilde{y}_{t-52} - 0.49\tilde{y}_{t-53} + 0.915\tilde{y}_{t-1} + 0.219\tilde{y}_{t-2} - 0.219\tilde{y}_{t-3}$
PARMA	$\bar{\tilde{y}} = 0.3\tilde{y}_{t-52} + 0.075\tilde{y}_{t-53} + 0.014\tilde{y}_{t-54} + 0.48\tilde{e}_{t-52} + 0.45\tilde{e}_{t-53} + 0.45\tilde{e}_{t-54} + 0.9\tilde{y}_{t-1} - 0.28\tilde{y}_{t-3} + \tilde{e} - 0.32\tilde{e}_{t-1} - 0.16\tilde{e}_{t-2}$

4 Results and Discussion

The Nile river is the main source of water in Egypt for irrigation and daily usage. It is important to predict its inflow to plan the consumption of water in Egypt. The system has been applied to the Nile inflow data at Dongla station (the entrance of Lake Nasser). The measured inflow data from 1985 to 1993 was used. The identified models are shown in Table 1. The results of the forecasting system are shown in Figure 1 and 2 for the off-line mode and the on-line mode respectively. The results of applying the ARMA statistical model —conventionally tuned— are shown in Figure 3. A comparison of the implemented system and the statistical method in terms of the mean absolute forecast percentage error ($MAPFE$) is shown in Table 2. The obtained results show that the proposed system has reduced the MAPFE in all cases as compared to the conventionally selected and tuned model. Also, the MAPFE in the on-line mode has been reduced as

Table 2: Comparison between the implemented system and a statistical model.

Model	MAPFE
Off-line PAR model	22.3%
Off-line PARMA model	24.1%
On-line PAR model	6.5%
On-line PARMA	12.8%
Statistical Method	26.6%

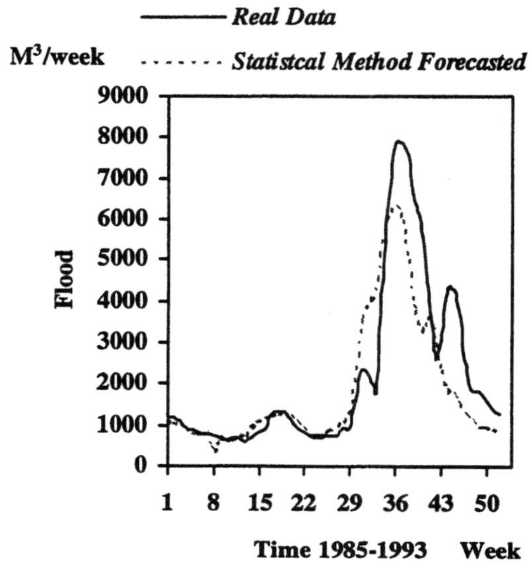

Figure 3: The off-line PARMA model forecasts for 52 weeks ahead using the conventional method.

opposed to the off-line mode since new observations are treated as part of the data set in the next time step, thus avoiding the accumulation of error as the forecasting process proceeds.

5 Conclusion

In this paper, a genetic based system for time series prediction has been presented. The proposed system has been applied to real data to forecast the Nile river water inflows. Genetic algorithms have been used as the search technique for selecting the best model and its parameters. Inspite of its simplicity, it exhibits promising results. Currently, the system is under comparison with other artificial intelligence based techniques like artificial neural networks. Also, a classifier system for the prediction of the river Nile inflow is under development.

References

[1] G. E. P. Box and G. M. Jenkins. *Time Series Analysis.* Holden-Day, Inc., 1976.

[2] A. P. Georgakakos and H. Yao. a routing model for the white nile. Technical report, Georgia Institute of Technology, Atlanta, April 1994.

[3] D. E. Goldberg. *Genetic Algorithms in Search, Optimization and Machine Learning.* Addison-Wesley, 1989.

[4] D. C. Montgomery, L. A. Johnson, and J. S. Gardener. *Forecasting & Time Series Analysis.* John Wiley & Sons, 1992.

[5] H. Tong. *Non-linear Time Series: a Dynamic System Approach.* Oxford University Press, 1995.

A Comparative Study of Neural Network Optimization Techniques

T. Ragg, H. Braun and H. Landsberg
Institute of Logic, Complexity and Deduction Systems
University of Karlsruhe, 76128 Karlsruhe, Germany,
Email: {ragg,braun}@ira.uka.de

Abstract

In the last years we developed ENZO, an evolutionary neural network optimizer which we compare in this study to standard techniques for topology optimization: optimal brain surgeon (OBS), magnitude based pruning (MbP), and unit-OBS, an improved algorithm deduced from OBS. The algorithms are evaluated on several benchmark problems. We conclude that using an evolutionary algorithm as meta-heuristic like ENZO does is currently the best available optimization technique with regard to network size and performance. We show that the time complexity of ENZO is similar to magnitude based pruning and unit-OBS, while achieving significantly smaller topologies. Standard OBS is outperformed in both size reduction and time complexity.

1 Optimization Techniques

Optimizing the topology of neural networks is an important task when one aims to get smaller and faster networks, as well as a better generalization performance. Moreover, automatical optimization avoids the time consuming search for a suitable topology. Techniques for optimizing neural network topologies can be divided into destructive algorithms like pruning or optimal brain surgeon and constructive algorithms, e.g. cascade correlation. Our evolutionary algorithm ENZO [2, 7] is a mixture of both methods, since it allows for reduction as well as for growing structures.

We will show that ENZO surpasses magnitude pruning as well as unit-OBS by evolving topologies with significantly less size while using nearly the same amount of computing time. Moreover, the size of the evolved topologies is even smaller as constructed by the very time consuming incremental optimal brain surgeon algorithm, i.e., we get a better network size reduction using several orders of magnitudes less computing power.

The main criteria for optimizing the network topology is the size of the network, furthermore the time needed for the optimization and the classification error. A problem of pure network reduction algorithms lies in the fact that the smallest network achieving a learning error below a given error limit has not the best generalization performance in general. The tradeoff between network size and generalization capability can be balanced by ENZO using both criteria in the fitness function while MbP and OBS do not consider the generalization and achieve therefore worse generalizing networks. In order to keep the comparison clear, we do not evaluate this essential advantage of ENZO. Therefore, the only criteria is the size of the achieved neural network under the constraint that the learning error (and thereby the classification error) remains under a given error bound. The size of the network is measured by three parameters: number of input units, number of hidden units and number of weights. In typical applications of the multilayer perceptron model there are some redundant or even irrelevant input units which may decrease the generalization capability. Therefore, we are interested in both: reducing the input size and the network size. The first gives hints about the salient parameters in the input representation, the second speeds up the network evaluation and both together improve the generalization capability.

All destructive techniques are used in the same way. The networks are trained until they reach a local minimum, i.e. the mean square error is below a given bound or a maximal number of epochs is exceeded. A candidate c (either a link or a unit) is deleted and then the network gets retrained. If it is not possible to learn the patterns with the given constraints (#epochs mean square error), optimizing stops and the network before the last operation is restored. Different strategies are used to select the candidates:

Pruning: Magnitude based pruning (MbP) searches the weight with the smallest absolute value. This weight is the candidate to delete.

Optimal Brain Surgeon: Optimal brain surgeon (OBS) was introduced by Hassibi and Stork [4]. It uses the Hessian matrix H (second derivate of the error function with respect to the weights) to compute the less important weight, i.e., the one which causes the lowest increases in error (ΔE), if pruned. This is the weight w_q which minimizes $\frac{1}{2}\Delta w H \Delta w$ such that $(\Delta w)_q = -w_q$. A special case of OBS is optimal brain damage [3], which makes the assumption that the Hessian matrix is diagonal. This simplification causes it to delete wrong weights sometimes. Thus it is inferior to OBS and is not further considered here.

Optimal Brain Surgeon with Units: A promising variant to speed up the OBS-algorithm is to prune units or several weights at a time. This variant, unit-OBS, is described in detail in [9]. Deleting several weights at once leads to a generalized equation for $w_{q_1} w_{q_2} \dots w_{q_m}$ minimizing a term similar as for OBS such that $(\Delta w)_{q_i} = -w_{q_i}, i = 1 \dots m$.

2 Evolutionary Algorithm

The basic principles of evolution as a search heuristic may be summarized as follows: the search points or candidate solutions are interpreted as individuals. The optimization criterion has to be one-dimensional and is called the fitness of the individual. Constraints can be embedded in the fitness function as additional penalty terms. New candidate solutions, called offspring, are created by a mutation operator using current members of the population, called parents. Weights (respectively units) can be removed or added depending on their ranking in a list, where the list is sorted with respect to a criterion reflecting the relevance of the weight respectively unit. Several criteria can be chosen, which differ in their accuracy, e.g. size of a weight or information content of an input unit. The selection of the parents is random but biased, preferring the fitter ones, i.e. fitter individuals produce more offspring. Each new inserted offspring replaces another population member with lower fitness in order to maintain a constant population size.

Evolution as a search heuristic offers the possibility of a balance between exploration and exploitation. On the one hand parallelizing the search using a population of search points and by stochastic search steps allows for a explorative search. On the other hand this explorative search is biased towards exploitation by biasing the selection of the parents, preferring the fitter ones.

3 Comparing the Strategies for Minimizing the Topology

Comparing MbP and OBS we may state that both select single connections for elimination. Whereas MbP uses a simple heuristic (select the smallest weight), this selection is done much more carefully by OBS: It approximates the error function by a Taylor approximation of second degree (quadratic polynomial) and computes for this approximation the optimal connection, i.e. the connection with smallest increase of learning error caused through its elimination. As a positive side effect, the according weight matrix is also computed which achieves the minimal increase of learning error on the Taylor approximation which reduces the number of necessary training steps drastically. A problem of this single weight elimination step is its short sightedness: it cannot handle the synergetic effect of eliminating several connections in one step but only the singular effect of eliminating just one weight. This is improved by unit-OBS, where the effect of eliminating whole groups of connections is evaluated.

All three algorithms (MbP, OBS, unit-OBS) are greedy and cannot backtrack from an unfavorable elimination. Since there are many single elimination steps any error resulting from the approximations may cut off favorable areas of the search space and therefore does not necessarily lead to an optimal reduction of the topology. These problems can be handled by ENZO: it does not only eliminate the most promising unit (or connection) but λ promising variants and evaluates their actual increase in error. Moreover, the short sightedness is reduced by following not only one line of development but a whole population of promising lines which are extended by chance according to their fitness.

4 Time Complexity

All four algorithms (MbP, OBS, unit-OBS and ENZO) proceed in the same way: they produce iteratively an offspring from a parent by deleting a unit or some connections and optimize it by gradient descent. The time complexity for producing an offspring can be subdivided into time complexity for the mutation and for the training.

The time complexity for the mutation is determined by the selection operator which specifies the unit or connections for elimination (or addition using ENZO, respectively). For MbP and ENZO, the time complexity of selection is $O(n)$ (with n = number of connections in the topology). Since one learning step has time complexity $O(pn)$ (with p = number of learning patterns), the time complexity for training maximizes the selection and therefore the time complexity for producing one offspring is $O(epn)$ (e = number of learning steps). For OBS and unit-OBS the time complexity of the selection operator is $O(pn^2)$ (cf. [4, 9]). Therefore it maximizes the time complexity of training ($O(epn)$) if $e < n$. This is true since both OBS and unit-OBS compute not only the connection (or unit respectively) with the smallest increase of the learning error but also the weight matrix, which achieves this smallest learning error. Nevertheless some learning steps are necessary, caused by the approximation error of the quadratic polynomial. In our experimental investigations the number of learning steps were typically less than ten. Therefore we may conclude that the time complexity of producing an offspring is $O(pn^2)$ for both OBS and unit-OBS.

Comparing the time complexity for all four algorithms, we have to take into account the number of offspring which are produced. Of course, this depends upon how many units or connections can be eliminated. On a minimal topology none of the algorithms can produce an improved offspring. In typical applications, however, the number of connections can be drastically reduced by less than half. Therefore we may assume, that the number of eliminated connections is $O(n)$. The elimination of a unit can be simulated by the elimination of all its connections. If we consider a weight matrix with $n = mm$ weights, then the elimination of a unit is the elimination of its according row and column, i.e. about $2m$ weights are eliminated. Therefore a unit elimination can be simulated by the elimination of $O(\sqrt{n})$ connections. Consequently we may conclude that we need $O(\sqrt{n})$ eliminations of units. Both ENZO and unit-OBS eliminate first units and then connections. Since the time complexity is proportional to the size of the network, the time consumed for producing the first offsprings is much larger than the time for eliminating single connections when the topology is already nearly minimal (such that no unit can be eliminated).

Therefore, the time complexity is dominated by the elimination of units and the elimination of connections may be neglected for both unit-OBS and ENZO. Summarizing, we get for all four algorithms the time complexities as shown in Table 1.

5 Generalization Behavior

It is well known that the generalization capability may decrease through intensively optimizing on the learning set. This overfitting effect can be handled by limiting the degree of freedom of the network topology [1]. There are two types of limitations: firstly we can reduce the number of free parameters (weights) and secondly we may penalize the size of these parameters by a weight decay factor. Both approaches are important for good generalization behavior. The reduction of the number of weights is automatically obeyed since we reduce the network topology. The penalization of the size (weight decay) has to be optimized beforehand. The weight decay factor has to be chosen so large that the overfitting effect is just suppressed. A larger weight decay factor decreases again the generalization behavior, because in this case the gradient descent priorises the minimization of the size of the weights instead of the learning error. Therefore we optimized in our experimental investigations the weight decay factor beforehand according to the above considerations.

6 Results

For all benchmarks we used fully-connected networks with shortcut connections between all layers, RPROP with weight-decay as learning algorithm [6] using 1.0 as maximal step size. The weight decay is adjusted manually in a series of training runs by increasing its value until overfitting on the validation set does not occur any more. The number of generations, the population size and the number of offspring each generation for ENZO are 30, 20 and 7, respectively.

The purpose of using an artificial benchmark like

Table 1: Time complexity of the four examined optimization algorithms. e is the number of training epochs, p the number of patterns, n the number of weights and λ the number of offspring.

MbP	OBS	unit-OBS	ENZO
$O(epn)O(n)$	$O(pn^2)O(n)$	$O(pn^2)O(\sqrt{n})$	$O(epn)O(\lambda\sqrt{n})$
$= O(epn^2)$	$= O(pn^3)$	$= O(pn^{2.5})$	$= O(\lambda epn^{1.5})$

the TC-Problem, is solely to examine if the optimization algorithm is able to find a minimal topology that still solves the problem with a 100% correctness and the probability to do so, i.e., to determine the dependency from the initialization. The training set consists of all possible T's and C's on a 4×4 pixel matrix, i.e. 17 T's and 23 C's. The topology is 16-16-1 with 288 weights in total. Table 2 shows that only unit-OBS and ENZO were able to determine the minimal topology, where the average results of ENZO are still better.

Breast Cancer is a real world benchmark from the UCI benchmark collection. A tumor is to be classified as benign or malignant based on cell description, e.g. the cell size and shape, gathered by microscopic examination. The network topology is 9-8-4-2 with weights. The data consists of 350 training patterns, 174 validation and 175 test patterns, from which 65.5% are benign. We needed only about 700 training epochs to get satisfactory results and a classification error between 2% and 3%. We used a weight decay of 10^{-2}. Other parameters were set as above.

Thyroid Gland is also a real-world benchmark available at the UCI repository. This task requires a very good classification, because 92% of the patterns belong to one class, thus useful networks should have a classification error of less than 2%. The data consists of 1886 training patterns, 1886 validation and 3428 test patterns. The topology is 21-10-3, with 304 weights. The classification error was for all methods between 1% and 2%, which is in the same range as reported in [8] for an evolved network (284 weights, 1.4% classification error). The smallest network was evolved by ENZO using 15 weights achieving a classification error of 1.0%.

Table 2: Results of the optimization processes for the TC-Problem. The first two columns give the results for the median from 21 runs, i.e., the 11th best run. The other two rows are the values for the best run.

	Pruning	OBS	unit-OBS	ENZO
median topology	14-5-1	13-5-1	11-4-1	9-2-1
median #weights	33	29	23	22
best topology	12-2-1	12-4-1	8-2-1	8-2-1
best #weights	20	22	22	18

Table 3: Results of the optimization processes for breast cancer classification.

	Pruning	OBS	unit-OBS	ENZO
median topology	8-6-2	8-3-2	5-2-2	3-2-2
median #weights	29	29	17	10
best topology	7-6-2	8-3-2	5-1-2	3-1-2
best #weights	26	26	12	8

Table 4: Results of the optimization processes for thyroid gland classification.

	Pruning	OBS	unit-OBS	ENZO
median topology	7-10-3	9-6-3	8-6-3	6-1-3
median # weights	45	42	29	21
best topology	6-10-3	7-4-3	6-3-3	5-1-3
best # weights	26	26	16	15

7 Discussion

Optimization techniques were compared for several benchmarks with regard to their performance in minimizing the network size. Our results show that standard techniques depend heavily on the initialization of the network, as well as on the weight decay term. On the one hand, using weight decay makes it easier to prune minor important weights and thus leads to better results. On the other hand it decreases the degree of freedom for fitting the weights and thereby may increase the minimal network size.

The time consuming optimal brain surgeon algorithm is outperformed by unit-OBS with respect to network size reduction and computing time. Moreover, our results show that ENZO surpasses both magnitude pruning and unit-OBS by evolving topologies with significantly less size while all algorithms have about the same time complexity.

We conclude that using evolutionary algorithms as a meta-heuristic as done in ENZO is concurrently the best available optimization method, with regard to network size and performance. Furthermore, it is scalable in time and simple to parallelize [5]. Extra constraints on network topology or performance are easily integrated into the fitness function. Thus evolution combined with learning appears to be the best framework with which to train neural networks.

References

[1] C. M. Bishop. *Neural Networks for Pattern Recognition.* Oxford Press, 1995.

[2] H. Braun and T. Ragg. ENZO – Evolution of Neural Networks, User Manual and Implementation Guide, http://i11www.ira.uka.de. Technical Report 21/96, Universität Karlsruhe, 1996.

[3] Y. L. Cun, J.S. Denker, and S.A. Solla. Optimal Brain Damage. In *NIPS 2*, 1990.

[4] B. Hassibi and D. G. Stork. Second order derivatives for network pruning: Optimal Brain Surgeon. In *NIPS 4*, 1992.

[5] T. Ragg. Parallelization of an Evolutionary Neural Network Optimizer Based on PVM . In *Parallel Virtual Machine - EuroPVM'96*, Lecture Notes in Computer Science 1156, 1996.

[6] M. Riedmiller and H. Braun. A Direct Adaptive Method for Faster Backpropagation Learning: The RPROP Algorithm. In *Proceedings of the ICNN*, 1993.

[7] J. Schäfer and H. Braun. Optimizing classifiers for handwritten digits by genetic algorithms. In *Artificial Neural Networks and Genetic Algorithms*, D. W. Pearson, N. C. Steele, R. F. Albrecht (editors), pages 10–13, Wien New York, 1995. Springer-Verlag.

[8] W. Schiffmann, M. Joost, and R. Werner. Application of genetic algorithms to the construction of topologies for multilayer perceptrons. In *Artificial Neural Networks and Genetic Algorithms*, . D. W. Pearson, N. C. Steele, R. F. Albrecht (editors), pages 675–682, Wien New York, 1993. Springer-Verlag.

[9] A. Stahlberger and M. Riedmiller. Fast network pruning and feature extraction by removing complete units. In *NIPS 9*. MIT Press, 1997.

GA-RBF: A Self-Optimising RBF Network

B. Burdsall[1] and C. Giraud-Carrier[2]
[1] Mastère 2IA, ENST de Bretagne, BP 832, 29285 Brest Cedex, France
[2] Department of Computer Science, University of Bristol, Bristol, BS8 1UB, England
Email: burdsall@gti.enst-bretagne.fr, cgc@cs.bris.ac.uk

Abstract

The effects of a neural network's topology on its performance are well known, yet the question of finding optimal configurations automatically remains largely open. This paper proposes a solution to this problem for RBF networks. A self-optimising approach, driven by an evolutionary strategy, is taken. The algorithm uses output information and a computationally efficient approximation of RBF networks to optimise the K-means clustering process by co-evolving the two determinant parameters of the network's layout: the number of centroids and the centroids' positions. Empirical results demonstrate promise.

1 Introduction

Radial basis function (RBF) networks (e.g., [8, 9, 13]) are a class of hybrid connectionist models. Whilst they are essentially three-layer feedforward networks, RBF networks differ from classical multilayer perceptrons in three significant ways: there is only one set of trainable weights, from the hidden layer to the output layer; the nodes' activation functions are non-standard (i.e., neither *sign* nor a sigmoid); and learning is effected by both supervised and unsupervised techniques.

In a RBF network, the nodes of the hidden layer encode a set of well positioned centroids, each representing one or part of a class. Generally, the centroids are obtained by K-means clustering (i.e., unsupervised learning), whilst the weights of the output layer are trained by a single-shot process using pseudo-inverse matrices or SVD (i.e., supervised learning). RBF networks have found wide applicability in traditional classification problems as well as in modern fuzzy control systems.

As with other neural network models, experience shows that the performance of RBF networks is greatly affected by their topology, that is, the choice of centroids making up the hidden layer. Too many centroids leads to over-fitting, while too few centroids may prove insufficient to capture intrinsic class divisions adequately. In general, the network's classification accuracy is influenced primarily by the number of centroids used to represent each class and the position of each centroid within its class.

With K-means, the number of centroids per class is set a priori by the user and is typically the same for all classes. Whilst this greatly simplifies the design of RBF networks and is probably adequate for "regular" problem spaces, it is often too restrictive and generally impractical. In complex problem spaces with multiple classes, it is likely that, though many classes consist of only a few compact clusters, other classes may exhibit singularities that require additional centroids. In a hyper-dimensional problem space, such singularities cannot be detected readily by a human user. Another difficulty with K-means is its *random* choice of starting positions for the centroids. Such non-determinism hinders robustness and directly impacts performance. Finally, K-means defines clusters for one class at a time and thus does not take into account potential interactions between classes.

A variety of techniques have been suggested to overcome the limitations of the traditional, K-means-based approach to centroid selection. They range from algebraic solutions inspired by wavelet transforms [10] to symbolic solutions using dynamic regression trees [2] and concept learning algorithms [1] to statistical solutions using means tracking [14], clustering [1] and simulated annealing [6] to genetically-based solutions [12] to more ad hoc techniques such as elimination [4]. The solution proposed here, like [12], is based on the use of an evolutionary strategy for self-optimisation. Unlike [12], where the genetic algorithm (GA) evolves functionally-equivalent canonical parametrisations of RBF networks, the GA here evolves fuzzy centroids that become the basis functions of the network's hidden layer. Both the number of centroids per class and their positions are optimised. Because the GA acts on all of the classes at once and measures fitness as classification accuracy, the method naturally profits from global information about class interactions. Furthermore, robustness is increased by eliminating the non-deterministic choice of starting positions.

2 GA-RBF

To construct an optimal network layout, GA-RBF uses output information and a computationally efficient approximation of RBF networks to optimise K-means clustering by co-evolving the two determinant parameters of the network's topology: the number of centroids and the centroids' positions.

2.1 Encoding

A real-valued, rather than a binary encoded, genetic algorithm (GA) is used, causing the representation to be of a more phenotypical nature. Each chromosome, or individual, encodes one set of centroids' starting positions for each class, as follows.

$$\text{individual} \stackrel{\text{def}}{=} [\{s_1^1, \ldots, s_{k_1}^1\}, \ldots, \{s_1^n, \ldots, s_{k_n}^n\}]$$

where n is the number of classes and k_i is the number of centroids for class i. The k_is vary continually during evolution, thus providing the necessary diversity in the population to optimise both determinant parameters. In the current implementation, a practical upper bound of 7 is placed on the k_is since empirical evidence suggests that most problems require fewer than 7 centroids per class.

2.2 Fitness

GA-RBF's training set is split into a clustering set and an evaluation set. The objective fitness function used for evaluating individuals consists of the application of the K-means algorithm to the clustering set using the starting positions encoded in the individual's genes, followed by a test classification of the evaluation set using the K-means-computed centroids and the nearest-attracting prototype (NAP) classifier [3]. Hence,

$$\text{fitness} \stackrel{\text{def}}{=} \frac{\text{correct\_classifications}}{\text{size\_of\_evaluation\_set}} \times 100$$

The NAP classifier extends the classical nearest-neighbour classifier by using infinite fuzzy support as in RBF networks. Using NAP classifiers rather than RBF networks saves the construction of the networks and the associated computation of the output weights at each generation of the GA. Hence, the NAP classifier is used here as a computationally efficient approximation of a RBF network's performance. Note that although the centroids are labelled during optimisation, they are not labelled once placed in the RBF network upon convergence. The early labelling allows GA-RBF to use output information and NAP to conduct a faster, more informed search for an optimal solution.

2.3 Reproduction

The selection of individuals for reproduction is biased towards fitter individuals, as follows. Let N be the size of the population and $fitness(i)$ denote the fitness of individual i. The probability that i is selected is:

$$p_i = \frac{\text{fitness(i)}}{\sum_{j=1}^{N} \text{fitness(j)}}$$

The p_i's define a probability density function over the population. The parents chosen for reproduction are obtained by sampling the distribution without replacement. Genetic operators are applied to pairs of parents to produce offspring that replace the least fit individuals in the population.

Because a phenotypic representation is used, the traditional genetic operators have to be modified to maintain the integrity of the solution space. The classical crossover operator is replaced by a novel operator, called *gene pooling*, and three different types of mutation are introduced.

Gene-pooling takes two parents and produces a single offspring. Following [5], a population is viewed as a pool of genes rather than a collection of individuals. A gene, here, is a single centroid's starting position. During crossover the genes of two parents are mixed together into a kind of genetic soup. From this mixture, a new individual is formed by random selection of genes. The chromosome size of the offspring is determined by a random number which follows a normal distribution centred on the average parental chromosome size. Non-individuals are eliminated by constraining the mixture of genes to contain at most one instance of each gene. Gene-pooling is a valid form of crossover since, in the adopted representation, the order of genes on a chromosome is irrelevant.

Mutation is applied to offsprings and consists of adding, removing or swapping a centroid in a randomly chosen class in the chromosome. Let K be the maximum number of genes per class (here, $K = 7$). Let i be the randomly selected class and k_i be the number of genes in class i. The probabilities of occurrence of each mutation are given by:

$$\begin{aligned} P_{\text{add}} &\stackrel{\text{def}}{=} \tfrac{k_i}{K} \times \gamma_{\text{add}} \\ P_{\text{remove}} &\stackrel{\text{def}}{=} \tfrac{K-k_i}{K} \times \gamma_{\text{remove}} \\ P_{\text{swap}} &\stackrel{\text{def}}{=} 0.3 \end{aligned}$$

where $\gamma_{\text{add}} = 0.67$ and $\gamma_{\text{remove}} = 0.875$. Mutation by adding and mutation by removing are generally mutually exclusive. Together with gene pooling, they provide a way of optimising the number of centroids per class. The centroids added or swapped during mutation are drawn at random from within the clustering data points for the selected class. They may not already exist in the class.

2.4 Population Control

In addition to the above, a punishment function is used to combat over-fitting and an aging function is used to prevent saturation by super-individuals.

Individuals with more centroids than others generally have higher fitness as they can place their numerous centroids near the training examples. Unfortunately, these individuals also tend to over-specialise on the evaluation set used and to perform poorly on other sets, i.e., over-fitting occurs. To prevent over-fitting, GA-RBF allows individuals to evolve larger number of centroids per class but punishes them for doing so. Here, the punishment introduces a form of natural handicap for highly fit, greedy individuals. It is given by:

$$\text{punishment} \stackrel{\text{def}}{=} \sum_{i=1}^{n} \alpha_i \times \frac{1.38}{n} \times (k_i - 3) \times \text{fitness(i)}$$

where n, k_i and $fitness(i)$ are as before, and α_i is 1 if $k_i > 3$ and 0 otherwise. Only greedy individuals are punished and their punishment increases with both greed and fitness. With the addition of the punishment mechanism, the actual fitness function for GA-RBF becomes:

$$\text{fitness}' \stackrel{\text{def}}{=} \text{fitness} - \text{punishment}$$

Individuals whose fitness far exceed the average fitness of the population are termed super-individuals. Super-individuals reproduce rapidly and come to dominate the population after only a few generations. The propagation of super-individuals leads to rapid exploitation but often results in premature convergence to a local optimum. A reasonable amount of exploration must be maintained. To address this issue, the notion of a life span for individuals has been suggested (see, [7]). GA-RBF follows this idea and defines the life span of individual i, at "birth", to be:

$$\text{Life\_span}(i) \stackrel{\text{def}}{=} \beta \times \frac{\text{fitness(i)}}{\text{Mean}}$$

where $Mean$ is the average fitness of the population and β is some constant. Each time an individual reproduces, its life span is reduced by 1. When the life span reaches 0 the individual dies and is removed from the population. Should a parent die following reproduction, its offspring takes its place rather than that of the weakest individual in the population.

2.5 Network Construction

The GA population is initialised by creating between 30 and 60 random individuals. Starting positions for each class are chosen at random from among that class' data points in the clustering set. The GA population is then evolved until convergence. Upon convergence, the starting positions encoded in the fittest individual are used by K-means to compute the centroids which are then fed into the hidden layer of the RBF network. Finally, the output layer of the network is trained using SVD.

3 Experimental Results

Two datasets from [11] were used to test GA-RBF. The noisy versions contain the original data with substantial noise added. Multiple simulations and cross-validation techniques were used to increase the validity of the results. The results for GA-RBF are compared against those of a NAP classifier with 3 (K-means-generated) centroids per class, user-optimised infinite RBF network with 3 centroids per class, and an optimised multilayer perceptron (MLP). Results are in Table 1 and record predictive accuracy on unseen test data. The fixed number 3 of centroids per class used in the non-evolutionary models results from intensive, user-driven testing. With GA-RBF, the number of centroids per class evolves naturally to settle on a value between 2 and 4. The cost of the evolution process is relatively low as convergence occurs in less than 3 minutes (Sun

Table 1: Experimental Results

Application	GA-RBF	RBF	MLP	NAP
Iris	96.3	96.3	96.0	95.6
Iris (N)	84.4	81.7	83.0	79.9
Diabetes	95.6	95.1	96.0	95.2
Diabetes (N)	89.6	89.0	85.1	86.3
Averages	91.5	90.5	87.5	89.3

Workstation). Moreover, with no a priori knowledge nor manual fine tuning, GA-RBF results in a slight increase of predictive accuracy for the problems considered.

4 Conclusion

This paper presents GA-RBF, a self-optimising RBF network. GA-RBF harnesses the power of genetic algorithms to optimise the k-means clustering process by co-evolving the determinant parameters of the network's topology. GA-RBF uses a real-valued genetic encoding and custom-designed genetic operators. With its fitness based on classification accuracy, GA-RBF also enhances the search with output information. Empirical results show that GA-RBF is a powerful optimiser.

Because it is such a strong optimiser, the GA itself tends to lead to over-fitting. This is particularly problematic with small datasets. In the experiments reported here, the training set was split only once into clustering and evaluation sets. Improved performance could be expected with cross-validation during optimisation. Alternatively, one could consider extending the fitness function with anti-overfitting.

5 Acknowledgements

This work was supported in part by the DES Defence Engineering and Science Group, UK.

References

[1] C. Baroglio, A. Giordiana, M. Kaiser, M. Nuttin, and R. Piola. Learning controllers for industrial robots. *Machine Learning*, 23(2/3):221–249, 1996.

[2] E. Blanzieri, A. Giordiana, and P. Katenkamp. Growing radial basis function networks. In *Proceedings of the Fourth Workshop on Learning Robots*, 1995.

[3] B. Burdsall and C. Giraud-Carrier. Evolving fuzzy prototypes for efficient data clustering. In *Proceedings of ISFL'97*, 1997.

[4] C. Decaestecker. Design of a neural net classifier using prototypes. Technical report, IRIDIA, Université Libre de Bruxelles, 1994.

[5] L. J. Fogel. Evolutionary programming in perspective: The top down view. In J. M. Zurada *et al.*, editors, *Computational Intelligence: Imitating Life*, NY, 1994. IEEE Press, Inc.

[6] B. Lemarié and A-G. Debroise. A dynamical architecture for radial basis function networks. In D. W. Pearson, N. C. Steele, R. F. Albrecht, editors, *Artificial Neural Networks and Genetic Algorithms*, pages 305–308, Wien New York, 1995. Springer-Verlag.

[7] Z. Michalewicz. *Genetic Algorithms + Data Structures = Evolution Programs*. Springer Verlag, 1992.

[8] J. Moody and C. Darken. Learning with localized receptive fields. In *Proceedings of the 1988 Connectionist Models Summer School*, Pittsburgh, PA, 1988.

[9] J. Moody and C. Darken. Fast learning in networks of locally-tuned processing units. *Neural Computation*, 1(2):281–294, 1989.

[10] S. Mukherjee and S. Nayar. Automatic generation of rbf networks. Technical Report CUCS-001-95, Department of Computer Science, Columbia University, 1995.

[11] P. M. Murphy and D. W. Aha. UCI Repository of Machine Learning Databases. University of California, Irvine, Department of Information and Computer Science.

[12] R. Neruda. Functional equivalence and genetic learning of RBF networks. In D. W. Pearson, N. C. Steele, R. F. Albrecht, editors, *Artificial Neural Networks and Genetic Algorithms*, pages 53–56, Wien New York, 1995. Springer-Verlag.

[13] T. Poggio and F. Girosi. Networks for approximation and learning. *Proceedings of IEEE*, 78(9):1481–1497, 1990.

[14] K. Warwick, J. D. Mason, and E. L. Sutano. Center selection for a radial basis function network. In R. F. Albrecht D. W. Pearson, N. C. Steele, editors, *Artificial Neural Networks and Genetic Algorithms*, pages 309–312, Wien New York, 1995. Springer-Verlag.

Canonical Genetic Learning of RBF Networks Is Faster

R. Neruda
Institute of Computer Science, Czech Academy of Sciences
PO Box 5, 18207 Prague 8, Czech Republic
roman@uivt.cas.cz

Abstract

We extend our previous theoretical results concerning functional equivalence of Gaussian RBF networks and test the proposed canonical genetic learning algorithm on two problems. In our experiments, canonical learning achieved the same error threshold about two times faster in comparison to standard GA.

1 Introduction

The problem of universal approximation property has been successfully solved for the main network architectures and assured that MLPs and RBFs are indeed general computational tools. On the other hand, this property failed as a criterion for judging the suitability or superiority of individual network architectures.

This is the reason why other, finer, criteria have to be found. So, it is sensible to study possibilities of reducing the size of weight space by minimizing parameterization redundancy. This can be done by the restriction to minimal subsets of weight space containing at least one representative of each class of *functionally equivalent* weight vectors (i.e., vectors of parameters determining the same input/output function of the network). Hecht-Nielsen [2] pointed out that characterization of functionally equivalent network parameterizations might speed up some learning algorithms. Several authors studied functionally equivalent weight vectors for perceptron-type networks with various activation functions ([1, 4, 7]).

Here, we extend our previous results [5, 6] and study the problem of redundancy of network parameterization for radial-basis-function networks with Gaussian radial function. We propose a *canonical* genetic learning algorithm for this class of networks, test it on two problems and present practical results to approve the expected speedup.

2 Theoretical results

From now on we will consider feedforward networks with one hidden layer containing radial-basis-function (RBF) units with a radial function $\psi : \mathcal{R}_+ \rightarrow \mathcal{R}$ and a metric ρ on $\mathcal{R}^n$ (n is the number of input units) and with a single linear output unit. Such a network computes the function: $f(\mathrm{x}) = \sum_{i=1}^{k} w_i \psi \left(\frac{\rho(\mathrm{x}, \mathrm{c}_i)}{b_i} \right)$.

Definition 1. *A radial-basis-function network parameterization with respect to ψ, n, ρ is a sequence $\mathrm{P} = (w_i, \mathrm{c}_i, b_i;\ i = 1, \dots k)$, where k is the number of hidden units and for the i-th hidden unit the vector $\mathrm{c}_i \in \mathcal{R}^n$ describes the centroid while the real numbers b_i and w_i are widths and output weights, respectively. If additionally, for every $i = 1 \dots, k$ $w_i \neq 0$, and for every $i, j = 1, \dots, k$ such that $i \neq j$ either $\mathrm{c}_i \neq \mathrm{c}_j$ or $b_i \neq b_j$, it is called* a reduced parameterization.

A parameterization P determines a unique I/O function of a ψ, ρ-RBF network with single linear output unit and k hidden units.

Definition 2. *Two RBF network parameterizations* P *and* P′ *are* functionally equivalent *if they determine the same input/output function.*

Definition 3. *Two network parameterizations are called* interchange equivalent, *if $k = k'$ and there exists a permutation π of the set $\{1, \dots, k\}$, such that $w_i = w'_{\pi(i)}$* and *$b_i = b'_{\pi(i)}$* and *$\mathrm{c}_i = \mathrm{c}'_{\pi(i)}$, for each $i \in \{1, \dots, k\}$.*

The standard choice of a radial function is Gaussian and the most popular metrics are those induced by various inner products (such as Euclidean), or the maximum metrics. By $\cdot$ we denote a general inner product on $\mathcal{R}^n$ which induces a norm $\|\cdot\|$ on $\mathcal{R}^n$ by $\|\mathrm{x}\|^2 = \mathrm{x} \cdot \mathrm{x}$; $\mathrm{x} \in \mathcal{R}^n$. ρ then denotes the metrics derived from the norm $\|\cdot\|$ by: $\rho_I(\mathrm{x},\mathrm{y}) = \|\mathrm{x}-\mathrm{y}\|$; for $\mathrm{x},\mathrm{y} \in \mathcal{R}^n$. One of the possible examples is the usual Euclidean metrics $\rho_E(\mathrm{x},\mathrm{y}) = \sqrt{\sum_{i=1}^{n}(x_i-y_i)^2}$, where $\mathrm{x} = (x_1,\dots,x_n)$ and $\mathrm{y} = (y_1,\dots,y_n) \in \mathcal{R}^n$. The symbol ρ_M denotes the maximum metrics: $\rho_M(\mathrm{x},\mathrm{y}) = \max_{i=1\dots n}|x_i - y_i|$.

The following theorem shows that in the cases of ρ_M and ρ_I an input-output function determines the network parameterization uniquely up to a permutation of hidden units.

Theorem 1. *Let n be a positive integer, ρ metrics on $\mathcal{R}^n$ induced by an inner product, or ρ_M, $\gamma(t) = \exp(-t^2)$. Then any two reduced (γ,ρ)-RBF network parameterizations are functionally equivalent if and only if they are interchange equivalent.*

Theorem 1 enables us to describe a canonical representation of a network computing a particular function easily. One of the possible choices is to impose a condition on a parameterization that weight vectors corresponding to hidden units are increasing in a lexicographic ordering on a parameterization. Represent a parameterization $\{w_i, k_i, \mathrm{c}_i; i = 1,\dots,k\}$ as a vector $\mathrm{p} = \{\mathrm{p}_i,\dots \mathrm{p}_k\} \in \mathcal{R}^{k(n+2)}$, where $\mathrm{p}_i = \{w_i, b_i, c_{i1},\dots,c_{in}\} \in \mathcal{R}^{n+2}$ is a weight vector corresponding to the i-th hidden unit. Let $\prec$ denote the lexicographic ordering on $\mathcal{R}^{n+2}$, i.e. for $\mathrm{p},\mathrm{q} \in \mathcal{R}^{d+2}$ $\mathrm{p} \prec \mathrm{q}$ if there exists an index $m \in \{1,\dots,d+2\}$ such that $p_j = q_j$ for $j < m$ and $p_m < q_m$. For a norm $\|\cdot\|$ induced by an inner product, we call a $(\gamma,\|\cdot\|)$ RBF network parameterization p *canonical* if $\mathrm{p}_1 \prec \mathrm{p}_2 \prec \dots \mathrm{p}_k$.

In this terminology, Theorem 1 guarantees that for every (γ,ρ)-RBF parameterization (where ρ is induced by an inner product or ρ_M), a canonical RBF network parameterization determining the same input-output function exists. Thus, the set of canonical representations forms a minimal search set-weight space subset containing exactly one representative of each class of functionally equivalent weight vectors — proposed in [2].

In order to take advantage of the previous result a learning algorithm that can operate only on canonical parameterizations is needed. Unfortunately, neither back propagation nor the more complicated three step learning algorithms of RBF networks (cf. [3]) are suitable, since the analytical solution obtained in each step of the iterative process cannot in principle be limited to a certain weight space subset. This is not so with genetic algorithm whose operations can be changed to preserve the property of being canonical.

The core of canonical GA is the same as usual: In the beginning a population of m canonical parameterizations $\mathcal{P}_0 = \{\mathrm{P}_1,\dots,\mathrm{P}_m\}$ is generated at random. Having population $\mathcal{P}_i$, the successive population $\mathcal{P}_{i+1}$ is generated by means of three basic genetic operations: *reproduction*, *crossover* and *mutation* that are proposed such that they generate only canonical parameterizations.

To generate the *initial population* of canonical parameterizations at random, one has to preserve the property that for each parameterization P holds: $\mathrm{p}_s \prec \mathrm{p}_{s+1}$.

The *reproduction* operator represents a probabilistic selection of parameterization $\mathrm{P}_l \in \mathcal{P}_i$ according to the values of objective function $\mathcal{G}(\mathrm{P}_l)$ which is computed by means of the error function (i.e. the sum of distances between the actual and desired output of the network over all of the patterns from the training set).

Thus, we have $\mathcal{G}(\mathrm{P}) = C - \sum_{j=1}^{z}\|f_{\mathrm{P}}(\mathrm{x}_j) - y_j\|^2$, where y_j is the desired network output, $f(\mathrm{x}_j)$ represents the actual response of the network when the input x_j is presented, and C is the maximal error.

For each individual P_l the value of its objective function $\mathcal{G}(\mathrm{P}_l)$ is computed and then normalized by dividing by the sum of objective function values over all individuals in the population: $p_l = \frac{G(\mathrm{P}_l)}{\sum_{r=1}^{m} G(\mathrm{P}_r)}$. The number p_l then represents the probability with which the parameterization is selected at random. We use the roulette wheel selection together with the elitist mechanism.

The *mutation* operates on two levels—first an element p_s is chosen randomly as a candidate for mutation. Its neighbors p_{s-1} and p_{s+1} then determine the lower and upper border of the range in which the p_s is changed at random.

The *crossover* operator chooses two parameterizations $\mathrm{P} = (\mathrm{p}_1,\dots,\mathrm{p}_k)$ and $\mathrm{Q} = (\mathrm{q}_1,\dots,\mathrm{q}_k)$ in $\mathcal{P}_i$ and generates a new offspring $\mathrm{P}' \in \mathcal{P}_{i+1}$. A position s is found at random such that the parameterization $\mathrm{P}' = (\mathrm{p}_1,\dots,\mathrm{p}_s,\mathrm{q}_{s+1},\dots,\mathrm{q}_k)$ still satisfies the condition: $\mathrm{p}_s \prec \mathrm{q}_{s+1}$.

3 Experiments

In the following we describe our experiments testing the performance of canonical and standard learning algorithms on two problems: the XOR problem and the approximation of $\sin(x)\cdot\sin(y)$ function. The experiments were made on the Pentium 90 PC with 32Mb RAM running Linux. The following figures

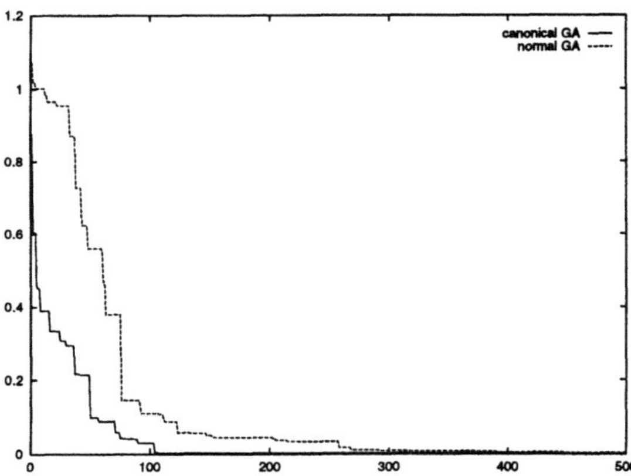

Figure 1: XOR with two RBF units.

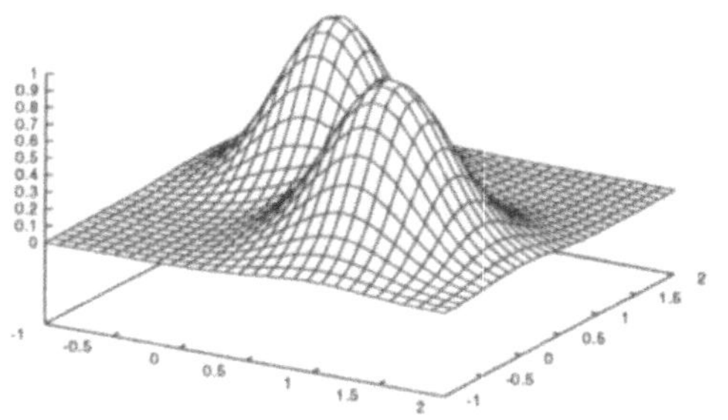

Figure 2: The canonical GA XOR solution.

have been obtained by averaging results of 5–10 runs on every task. The program used is written in C++ by the author and is publicly available.

The first task was a XOR problem defined by four training examples. We used two hidden units in the network, 50 networks in the population and elitist selection for the two best networks. Error values for the first 500 iterations are also plotted on Figure 1.

See also Figure 2 which shows the function realized by the resulting network. Both algorithms were able to successfully learn the given task quite fast. (cf. Table 1); the canonical algorithm was about two times faster.

Running times of both algorithms were roughly identical— about 7 seconds per 1000 iterations.

The second experiment was an approximation

Table 1: XOR: number of iterations necessary to reach a particular error threshold.

$\mathcal{E}$ threshold	canonical GA	standard GA
10^{-1}	50	76
10^{-2}	104	268
10^{-3}	443	705
10^{-4}	934	1553
10^{-5}	1075	1988
10^{-6}	1733	4485
10^{-8}	7623	>10000

Table 2: $\sin(x) \cdot \sin(y)$: number of iterations necessary to reach a particular error threshold.

$\mathcal{E}$ threshold	canonical	standard
10	177	200
5	313	765
3	489	1043
2	1268	—

of the function $f(x,y) = \sin(x) \cdot \sin(y)$ given by a 10×10 mesh of points regularly taken from a $[0; 2\pi] \times [0; 2\pi]$ square. Again, both algorithms with 50 networks in population and 4 RBF networks were used with the same elitist rate as in the previous experiment. The learning speed is shown in Table 2 and Figure 4.

The performance was similar to the previous experiment: the canonical GA was again about twice

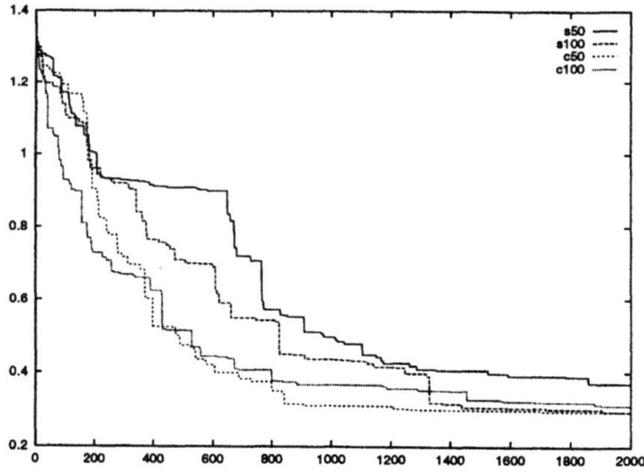

Figure 3: Comparison of error decrease for standard and canonical algorithm with population size of 50 and 100 networks (y scale represents $\log_{10} \mathcal{E}$).

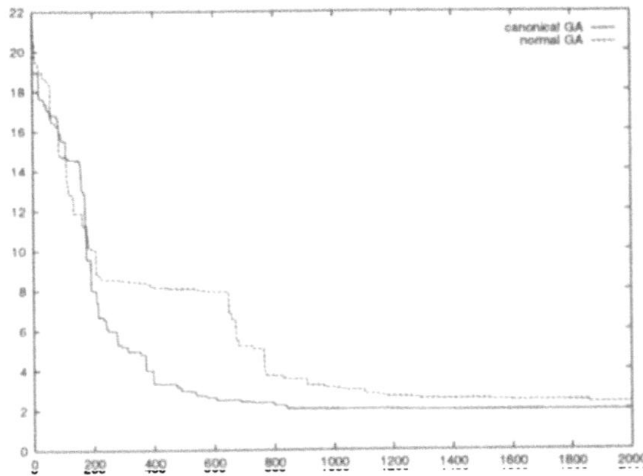

Figure 4: Error decrease for the $\sin(x) \cdot \sin(y)$ problem with four RBF units.

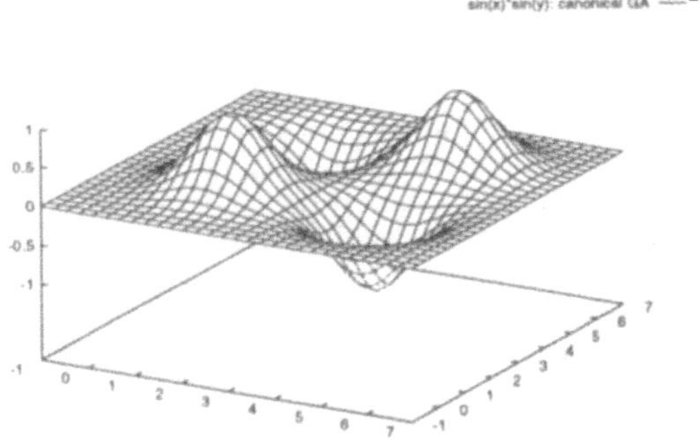

Figure 5: Plot of the function realized by four RBF units trained on the $\sin(x) \cdot \sin(y)$ problem.

faster. The average speed of 1000 iterations was 134 seconds. The resulting network performs the function shown in Figure 5. This function is a good approximation of the original set which can be seen from Figure 6 showing the approximation error.

4 Conclusions

Based on our theoretical results we have proposed and realized the canonical GA for RBF networks. The anticipated speedup was tested and it has been shown that for small/middle tasks the canonical GA is about twice faster. Moreover, the canonical GA does not show any relevant increase in time for one iteration in comparison to standard GA. Thus, the twice better times hold also in real time.

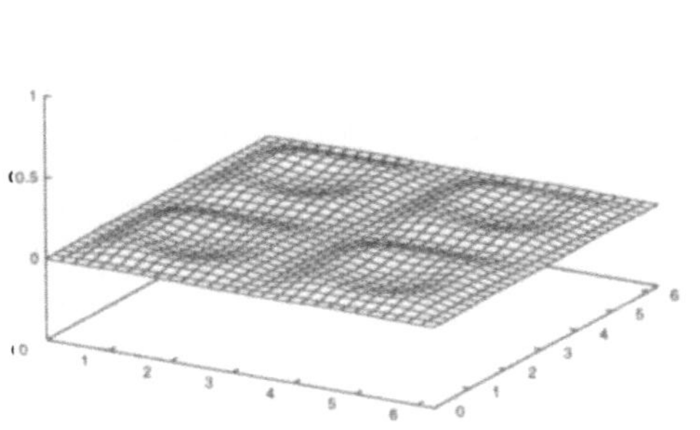

Figure 6: Plot of the square approximation error for the function from previous figure.

5 Acknowledgements

This research was partially supported by GA CR under grants no. 201/96/0917 and 201/94/0729.

References

[1] F. Albertini and E.D. Sontag. For neural networks, function determines form. *Neural Networks*, 6:975–990, 1993.

[2] R. Hecht-Nielsen. On the algebraic structure of feed-forward network weight spaces. In *Advanced Neural Computers*, pages 129–135. Elsevier, 1990.

[3] K. Hlaváčková and R. Neruda. Radial basis function networks. *Neural Network World*, (1):93–101, 1993.

[4] V. Kůrková and P. Kainen. Functionally equivalent feedforward networks. *Neural Computation*, 6:543–558, 1993.

[5] V. Kůrková and R. Neruda. Uniqueness of the functional representations for the gaussian basis functions. In *Proceedings of the ICANN'94*, pages 474–477, London, 1994. Springer.

[6] R. Neruda. Functional equivalence and genetic learning of RBF networks. In D. W. Pearson, N. C. Steele, R. F. Albrecht (editors), *Artificial Neural Networks and Genetic Algorithms*, pages 53–56, Wien New York, 1995. Springer Verlag.

[7] H.J. Sussmann. Uniqueness of the weights for minimal feedforward nets with a given input-output map. *Neural Networks*, 5(4):589–594, 1992.

The Baldwin Effect on the Evolution of Associative Memory

A. Imada[1] and K. Araki[2]
[1] Graduate School of Information Science, Nara Institute of Science and Technology
8916-5 Takayama, Ikoma, Nara, 630-01 Japan
[2] Graduate School of Information Science and Electrical Engineering,
Kyusyu University, 6-1 Kasuga, Fukuoka, 816 Japan
Email: akira-i@is.aist-nara.ac.jp

Abstract

We apply genetic algorithms to the Hopfield model of associative memory. Previously, we reported that a genetic algorithm evolves a network with random synaptic weights to store eventually a set of random patterns. In this paper, we show how the Baldwin effect on the evolution enhances the storage capacity.

1 Introduction

Associative memory is a system in which an incomplete or a noisy input of stored patterns should result in the retrieval in its complete form. The error-correcting capability is due to its distributed storage of the patterns. We are exploring basic behaviors of the associative memory under simple evolutionary processes.

In 1982, Hopfield [6] proposed a neural network model of the associative memory. He used the Hebbian rule [4] to make connection matrix store a set of patterns. He estimated the storage capacity to be at most 15% of the number of neurons with computer simulations. Later the capacity was verified analytically by Amit *et al.* [1] with spin-glass theory. Since then many researchers have been trying to enlarge the storage capacity in various ways. We try it by genetic algorithms (GA).

In an earlier paper, we showed that a network with random connections evolves to store some number of patterns by a GA without any learning rules [8]. We also showed that the GA enlarges the capacity of Hebb rule associative memory by pruning some connections [7]. In these experiments, individuals in the population do not learn the patterns in their lifetime.

The effect of lifetime learning on evolution was first studied as a biological process by Baldwin [2]. This is known as the Baldwin effect. As an analog of this effect, research concerning the combination of GAs and neural networks have addressed the relationship between ***learning on a population level through evolution*** and ***on an individual level during its lifetime.*** Gruau *et al.* show an enhancement of performance of their neural networks by introducing the Baldwin effect on evolution (see also [5, 10] for example).

In this paper, we redesigned our previous GA so that each individual in the population can learn a set of pre-determined random patterns before its fitness evaluation, i.e., the Baldwin effect on the evolution. We found that it enlarges the storage capacity, as well as making the convergence much faster.

The overall goal for this research is to clarify the process of this learning mechanism of the GA.

2 Associative Memory

The Hopfield network which has N neurons can store some number of N-bit bipolar patterns. The pattern is denoted by $\xi^\mu = (\xi_1^\mu, \xi_2^\mu, \cdots, \xi_N^\mu)$, where ξ_i^μ is the *i-th* bit of the *μ-th* pattern which takes the value of either -1 or 1. To store these p patterns, each connection weight has to be appropriately determined. If we use the Hebbian rule, for example, the weight w_{ij} is determined as follows:

$$w_{ij} = \sum_{\mu=1}^{p} \xi_i^\mu \xi_j^\mu.$$

This process of determining w_{ij} is called *learning.*

The retrieval of one of the stored memories is as follows. Each neuron state is asynchronously updated by

$$s_i^\mu(t+1) = sgn\left(\sum_{j\neq i}^{N} w_{ij} s_j^\mu(t)\right),$$

where $s_i^\mu(t)$ is the state of the *i-th* neuron at time t when the *μ-th* pattern is given to the network. For each neuron, the weighted sum of inputs is computed. If this is greater than or equal to zero, the state of the neuron takes the value of 1, and -1 otherwise. In this paper, a neuron is chosen according to a pre-assigned order (once in a cycle) at each step of updating, instead of chosen randomly.

When input is one of the stored patterns which is given small noises, this input pattern relaxes to the stored pattern after several steps of update. In this simulation, however, we do not give any noise to the input. In this case, when one of the stored patterns is given to the network, neuron states should remain unchanged from the start. The patterns are said to be memorized as *fixed points*. We evaluate the ability of the network by the upper bound of the number of patterns possible to store as fixed points.

3 GA Implementation

In this section we briefly describe our GA implementation.

1. At first a weight matrix $W_0(N \times N)$ is produced randomly so that each component of the matrix w_{ij} is chosen from $\{-1, 1\}$. Hence the matrix does not store any patterns at present. This matrix remains unchanged during evolution.

2. A population of chromosomes are produced randomly. The chromosome has a fixed length of $N \times N$ alleles which are chosen from $\{-1, 0, 1\}$, where the probability of choosing either -1 or 0 is set to $\frac{1}{70}$. As we shall see in procedure (3) below, each component of the original weight matrix w_{ij} is multiplied by one of these alleles. Allele 0 implies pruning the connection which is multiplied, while allele -1 implies reversing the excitatory/inhibitory connection.

3. Chromosomes modify the original weight matrix W_0, and produce a population of copies slightly different from W_0. The modification is made as follows:
$$w_{ij}^{(n)} = w_{ij} \cdot c_{ij}^{(n)}$$
where $c_{ij}^{(n)}$ is an allele of the *n-th* chromosome and $w_{ij}^{(n)}$ is the *i-j* component of the *n-th* copy of the original matrix. Our population size is restricted to 256 here because of our computer resources.

4. Each individual phenotype learns a set of predetermined random patterns ξ^μ by the Hebbian learning rule as follows:
$$w_{ij} = w_{ij} + \lambda \sum_{\mu=1}^{p} \xi_i^\mu \xi_{j \neq i}^\mu .$$
We set the value of λ to 0.4. These matrices are used only for fitness evaluation in procedure (5) below, and the learning results do not affect the chromosomes. Hence we can regard this as the Baldwin effect on the evolution.

5. When one of the memorized patterns ξ^ν is given to the network as an initial state, the state of neurons varies from time to time afterwards (unless ξ^ν is a fixed point). In order for the network to function as associative memory, the instantaneous network state must be similar to the input pattern. The similarity is defined by
$$m^\mu(t) = \frac{1}{N} \sum_{i=1}^{N} \xi_i^\mu s_i^\mu(t),$$
where $s_i^\mu(t)$ is the state of neuron i at time t. This $m^\mu(t)$ is referred to as *overlap*. The quality of retrieval of memorized patterns are represented by $\langle m^\mu \rangle$, a temporal average of $m^\mu(t)$ over a certain time interval t_0. We evaluate the fitness value of each network by further averaging these $\langle m^\mu \rangle$'s over all memorized patterns. Namely, the fitness f is
$$f = \frac{1}{t_0 \cdot p} \sum_{t=1}^{t_0} \sum_{\mu=1}^{p} m^\mu(t).$$
In this paper, t_0 is set to $2N$, twice the number of neurons. Note that a fitness of 1 implies that all the p patterns are stored as fixed points, while all other cases have a fitness less than 1.

6. Two parent chromosomes are chosen to be recombined uniformly at random from the upper 40% of the population which is ranked by fitness.

7. Recombinations are made with uniform crossover. We tested several types of crossover including one and two point crossover, and observed that uniform crossover outperformed the others. Furthermore, the

offspring are occasionally mutated, where mutation rotates the value of a randomly chosen allele in a chromosome cyclically, as $(1) \rightarrow (-1), (-1) \rightarrow (0), (0) \rightarrow (1)$. The mutation rate is set to 0.01.

8. Unless the highest fitness value reaches the value of 1 or generation exceeds 12000, the individuals in the best 40% survive to constitute the next generation with their offspring (60%), and the processes from (2) to (8) are repeated.

Both these operations and parameter values were determined mainly on the basis of trial and error.

4 Results and Discussion

To see the Baldwin effect on the evolution of associative memory, we first show two preliminary results. One is evolution from a random matrix without using any learning algorithms, and the other is from the Hebbian matrix. Then we will describe the Baldwin effect on the evolution of random weight matrix using the Hebbian rule as a lifetime learning of each individual. All these simulations are carried out on networks with 49 neurons.

The genetic algorithm *per se* has an ability to make a random weight matrix store some number of patterns. In fact, we succeeded in evolving the random weight matrix to store 8 patterns as fixed points [8]. The GA used in this experiment was similar to the one described above, except for the Hebbian learning process before fitness evaluation. In Figure 1(left), we show the results of the best fitness versus generation for 4, 6, 8, and 9 patterns. These are samples of fastest convergence among several runs with different random number seed, or, when they do not converge, the samples of maximum number of patterns stored as fixed points. For 9 patterns, we did not obtain the perfect solution after repeating 30 runs with different random number seed. Although this capability of storing 8 patterns with 49 neurons is almost the same as in the original Hebb rule associative memory, the structure of the weight matrix is different.

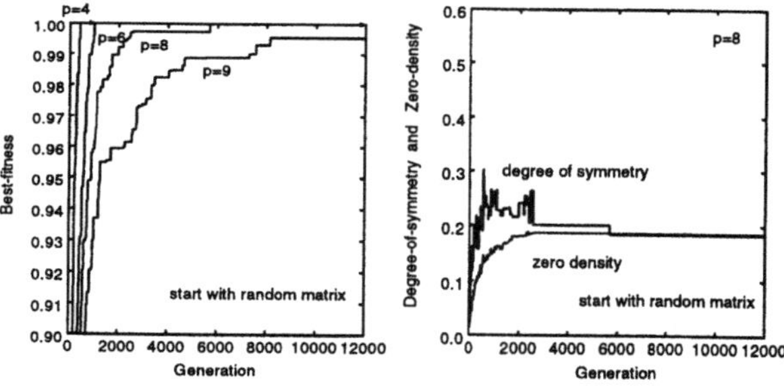

Figure 1: Evolution from random matrix.

The Hebbian matrix is symmetric and all its diagonals are zero. In this experiment, on the other hand, evolutions are started with a fully asymmetric matrix with no zero components. In Figure 1(right), we show the zero-density and degree of symmetry (defined as $\sum_{i=1}^{N}\sum_{j=1}^{N} w_{ij}w_{ji}/\sum_{i=1}^{N}\sum_{j=1}^{N} w_{ij}^2$) of the weight matrix which eventually stores 8 patterns. Both zero-density and degree of symmetry increases from zero and asymptotically approach to the value of approximately 0.2. We conjectured that both the symmetry recovery and pruning connections play an important roll to store the patterns effectively.

Since we use Hebbian learning rule as a lifetime learning in the simulation of the Baldwin effect, it is interesting to see whether the Hebbian matrix itself will be able to increase its capacity through evolution. We applied the GA to the over-loaded Hebbian matrix [7]. As a result, we were able to enlarge the capacity to 16 patterns. In Figure 2(left), we show the evolution for 14, 16 and 17 patterns. We obtain the perfect solution for a maximum of 17 patterns.

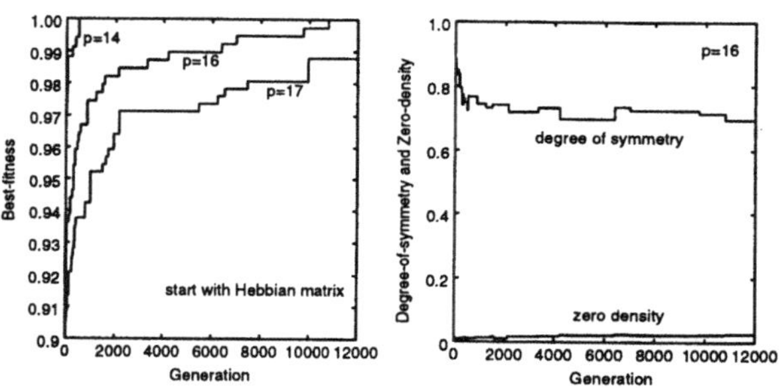

Figure 2: Evolution from Hebbian matrix.

The structure of the obtained matrix is very different from the evolution starting with random matrix. In Figure 2(right), we show the zero-density and degree of symmetry of the weight matrix which eventually stores 16 patterns. Since this matrix learns even number of patterns, some of its com-

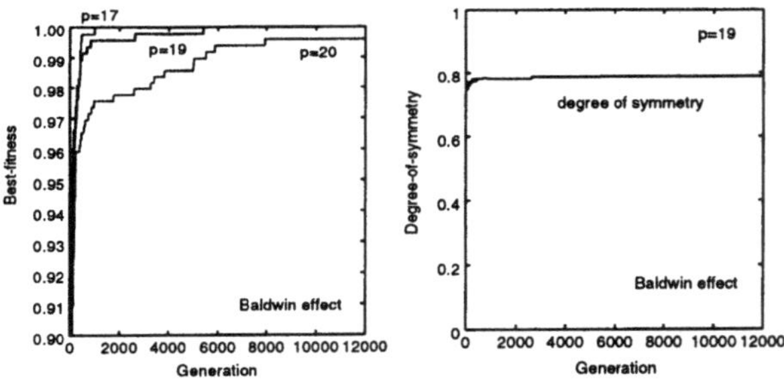

Figure 3: The Baldwin effect.

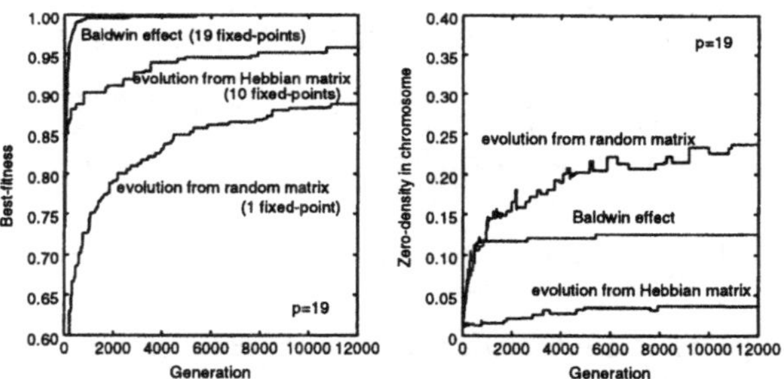

Figure 4: Baldwin versus no-Baldwin.

ponents are zero as a result of Hebbian learning. Here only the zero density introduced by the GA is shown. As shown in the figure, the matrix obtained in this experiment is more symmetric, and the GA prunes much fewer connections than in the above experiment. The degree of symmetry and the zero density approach approximately 0.7 and 0.03, respectively.

In Figure 3(left), we show the evolution from random matrix under the Baldwin effect for 17, 19 and 20 patterns. We can see that the storage capacity is enhanced to 19 patterns. Although the increase in capacity is not so drastic compared to the evolution starting with the Hebbian matrix, the convergence speed is tremendously improved. The perfect solution was obtained at generation 996 for 17 patterns, while the evolution from the Hebbian matrix for 16 patterns needed 10 795 generations to converge. In the Baldwin evolution, as we can see in Figure 3(right), the degree of symmetry remains unchanged from the beginning (around 0.8). This is because learning of the patterns affects individuals only in their lifetime.

To further compare the Baldwin effect with the other two evolutions, we show best-fitness versus generation of the three experiments for the same set of 19 patterns in Figure 4(left). We can see that evolution under the Baldwin effect outperforms the other two. A matrix which stores all these 19 patterns as fixed points emerged at generation 5 408, while the other two evolutions never converged. Temporal expansions of zero density in the chromosome of the best individual are also shown in Figure 4(right). We can see the significant differences among them. We conjecture that the evolution pressures in these GAs are different from each other. In the Baldwin version, the pressure works to enhance learnability rather than to directly increase capacity. That is, individuals which have higher ability to learn tend to survive.

In the experiment of the Baldwin effect, the results of learning of individuals do not change their chromosomes, but only affect the selection after fitness evaluation. However, it is reported that incorporation of learning results into chromosomes also enhances the performance of GAs (see [3, 11] for example). This is known to be the Lamarckian inheritance. We also investigate the effect of *Lamarckian inheritance* on the evolution of associative memory, though a different implementation is required where the chromosomes are made up of components of the weight matrix instead [9]. We obtained similar enhancement of the capacity to the result in this paper.

5 Conclusions

We have described the Baldwin effect on the evolution of Hopfield model of associative memory. The comparison of three evolutions from a random matrix, a Hebbian matrix, and a random matrix with the Hebbian lifetime learning reveals that the learning *on an individual level during its lifetime* enhances the learning ability of the subsequent generations than in the case of only the *learning on a population level.*

However the more precise relationship between the two is still an open question. We leave this to future research.

References

[1] D. J. Amit, H. Gutfreund, and H. Sompolinsky. Statistical mechanics of neural networks near saturation. *Annals of Physics*, 173:30–67, 1987.

[2] B. J. Baldwin. A new factor in evolution. *The American Naturalist*, 30:441–451, 536–553, 1886.

[3] F. Gruau and D. Whitley. Adding learning to the cellular development of neural networks: Evolution and the baldwin effect. *Evolutionary Computation*, 1(3):213–233, 1993.

[4] D. O. Hebb. *The Organization of Behavior*. John Wiley, Chichester, 1949.

[5] G. E. Hinton and S. J. Nowlan. How learning guides evolution. *Complex Systems*, 1:495–502, 1987.

[6] J. J. Hopfield. Neural networks and physical systems with emergent collective computational abilities. *Proceedings of the National Academy of Sciences*, 79:2554–2558, 1982.

[7] A. Imada and K. Araki. Genetic algorithm enlarges the capacity of associative memory. In *Proceedings of 6th International Conference on Genetic Algorithms*, pages 413–420, 1995.

[8] A. Imada and K. Araki. Mutually connected neural network can learn some patterns by means of ga. In *Proceedings of the World Congress on Neural Networks*, volume 1, pages 803–806, 1995.

[9] A. Imada and K. Araki. Lamarckian evolution of associative memory. In *Proceedings of IEEE International Conference on Evolutionary Computation*, pages 676–680, 1996.

[10] D. Parisi and S. Nolfi. *How Learning can Influence Evolution within a Non-Lamarckian Framework.* Santa Fe Institute Series. Addison-Wesley, 1995.

[11] D. Whitley, V. S. Gordon, and K. Mathias. Lamarckian evolution, the baldwin effect and function optimization. In *Proceedings of the 3rd Conference on Parallel Problem Solving from Nature*, pages 6–15, 1994.

Using Embryology as an Alternative to Genetic Algorithms for Designing Artificial Neural Network Topologies

C. MacLeod and G. Maxwell
School of Electronic and Electrical Engineering, Robert Gordon University, Aberdeen, UK.

Abstract

This paper considers the role of the embryological algorithm or embryology in defining artificial neural network architecture. Such an approach is based on the biology and growth of the embryonic nervous system and operates by 'growing' the neural network from a simple to a complex form. The operation of both the embryology and the genetic algorithm are considered, contrasted and compared. A practical algorithm is presented, together with results demonstrating the relevance, application and advantages of the algorithm.

1 Introduction

The problems of artificial neural networks are well documented [5]. In recent times, many researchers have tried to overcome these problems by utilising evolutionary or genetic methods [3]. The main method applied is known as the genetic algorithm. This algorithm is a computer model of the biological process of evolution. The genetic algorithm is generally used in one of two ways:

- to change the weights of a network (that is, as a learning algorithm);
- to choose the topology of the network.

It is the issue of selecting the topology which is discussed here. In general, this problem is further subdivided into changing the network connectivity and choosing the number of neurons in the network.

This paper presents a further classification of how network structure can change and shows how this can be applied to the algorithm known as an embryology. It is argued that the genetic algorithm is not as useful a technique for specifying topology as the embryology.

The embryology proves useful because it grows the network from a simple form to a complex form. This parallels the development of the brain, both in terms of the evolutionary process and the growth of the embryonic nervous system.

The genetic algorithm, on the other hand, generally begins with a large population of complex networks and weeds out those which do not perform well. It lacks the growth aspect which is inherent in the embryology approach.

The following sections provide an introduction to both algorithms, compares the two approaches and presents results from the application of an actual embryology.

2 The Genetic Algorithm

The starting point for the genetic algorithm is a large number of strings which hold information on system parameters. In a neural network, for example, the strings might hold information such as: the number of layers in the network, the number of neurons in each layer and the connectivity between the layers.

The process begins by evaluating the performance of the strings. This is called fitness testing. Each string is given a mark of merit according to its fitness. The population then enters a stage called the breeding cycle. In this stage all the strings are copied; the higher the fitness of the string, the more likely it is to get copied. Due to this process, the population now contains a higher number of fit strings.

The strings are then 'bred' by copying their genetic material at random points. Since there was a higher population of fitter strings, there should be more fit individuals in the new generation of strings (because better strings have a high probability of breeding with other good strings, as there are more of them). In each generation, a number of the weaker strings are eliminated. A random mutation is also introduced to add variation. The whole process is then repeated. As the generations proceed, the population performance tends to get better [2].

3 The Embryology

An embryology is a computer model of the growth of the foetus. The technique starts with a single cell and grows to a complex system. This process of growth is similar to the development of the living embryo. In the natural world, development proceeds through cell division; a single celled egg becomes a multi-celled organism.

When this approach is applied to neural networks, the principles must be adapted since we are dealing with an artificial system. Although several authors have published ideas or schemes for network growth in this way, none have been particularly successful [1,6].

The scheme used here is adapted from work by Dr Richard Dawkins, who used forms which he named 'Biomorphs' in his book 'The Blind Watchmaker'. Dr Dawkins was attempting to illustrate evolution in animal populations using these means. The biomorphs are a basic embryology [4].

The scheme discussed in this paper uses a single string (gene) to hold information. It contains such information as the number of layers in the network and the number of neurons in each layer. The algorithm proceeds by increasing one parameter at a time and then measuring the network fitness. If the fitness increases, then the algorithm keeps the change and proceeds; otherwise, it skips that change and tries the next parameter. In this scheme each string parameter represents a growth strategy. When network fitness does not increase further, then the network stops growing, having found its optimum level.

There are several ways of configuring the algorithm to work; the method above is only one. It is also possible, for example, to select a parameter at random and generate a new network that way. Either way, the network solution 'walks' through the search space from simple to complex forms and can be programmed to stop at the simplest form which fulfils the task. It may be noted that the algorithm has similarities to statistical methods such as the Boltzmann algorithm.

4 Comparison

The basic operation of the genetic algorithm and the embryology differ a great deal. The genetic algorithm is an analogue of the natural situation of having a large breeding population of animals. The embryology, on the other hand, is an analogue of the development of the embryo, or of the evolution of a single fossil line through time.

Since the genetic algorithm represents a snap shot of a population of animals evolving slowly over time and the embryology represents the evolution of just a single species over a large time period, the two can also be considered complementary in some systems and used together. This more complex method will not be discussed further here.

The major advantage of the embryology is that the network grows from a single neuron (or other, simple form) to a complex system. As each stage is reached, the network is tested for fitness and the algorithm terminated when a suitable level is achieved. The method, therefore, settles on the simplest form which can fulfil the task. This is in contrast with the genetic algorithm which always works with large numbers of complex networks. The genetic algorithm may find therefore, the best network, but not always the simplest.

Speed is another advantage of the embryological approach. Since it only uses a single gene to encode the information, it only has to assess one network per cycle, as opposed to many, in the case of the genetic algorithm. It is therefore an efficient search strategy. This efficiency can be further enhanced by structuring the algorithm. This is achieved by grading the effectiveness of the growth strategies and always trying the most effective first. Such techniques further enhance the efficiency of an already structured approach. Again, this is in contrast to the randomness of the genetic algorithm approach.

A further advantage of this approach is that it is well suited for implementation in hardware. Networks may be based on the use of FPGAs; this will allow the connections to be modified as the network evolves. Genetic algorithms are not well suited to this approach.

5 Growth Strategies

The description given in the previous section illustrates the operation of the algorithms which can be used to grow the network. However, to implement these algorithms, the designer must know in which way the network is to grow. From consideration of the possible connection topologies, it is possible to draw up a list of the ways in which a network can evolve. Which of these strategies are used will depend on the problem which the network is trying to solve. Most types of problem only require the use of a few of the strategies. These different growth patterns also have different effects in networks devised for different problems. Careful consideration needs to be employed in selecting strategies, because some can fundamentally change the nature of the network; an example of this is the addition of feedback to a network which changes a pattern classifier into an associative network.

5.1 Change the Number of Layers

The algorithm may add another layer to the network. Care needs to be exercised with this strategy as altering the number of layers in the network changes the operation. It may be shown that three layers is all that is required to produce a generalised network.

5.2 Change the Number of Neurons

The number of neurons in the layer under consideration may be increased or reduced.

5.3 Change the Connectivity

Each neuron may connect to all the other neurons in above layer (a fully interconnected network), or have a reduced number of connections. Removing connections in this way is often referred to as network pruning.

5.4 Asymmetry

It is possible to have one side of the network fully connected and the other side pruned. The success of this approach depends on the mode in which the network is employed. It is useful, however, in studying non-linear systems in input space.

5.5 Sideways Connections

In a synchronous system, neurons may be connected to other in the same layer.

5.6 Skipping Layers

Rather than connecting down to the layer directly below, a connection may skip a layer. Care must be taken when considering the algorithm to implement this, as there are several modes of operation. In certain networks, this form of connection may be shown to be equivalent to a simpler network with more connectivity on the same layer.

5.7 Feedback

Feedback paths may be added to the network. These may skip layers and be symmetrical or asymmetrical. Their introduction changes the nature of the network. Networks which have feedback are known as recurrent and have a simple memory; the current output depends on the previous state of the network. Such networks can perform associative tasks. In adding a feedback path into a network, the possibility of instability is also introduced.

5.8 Bias Units

Bias, if not already used in the network, may be added.

5.9 Sideways Growth

In a three layer network, rather than adding to the top layer, further neurons may be added to the middle layer. This approach is successful because adding a third layer onto two layers changes the structure and operation of the network. The second layer may therefore have to be revisited.

6 Results

In using the embryology to grow a particular type of network, only those growth strategies which are appropriate to use in a particular network are included in the algorithm. In a simple network for pattern recognition, for example, it is not necessary to include strategies which include feedback in the network as a feedforward topology is all that is necessary to perform the function. Pattern recognition is used as an example below.

Choosing the strategies also means deciding on how to apply them. For example, in the case of a recurrent network, it can be shown that there are cases, when the network is symmetrical, of it being unconditionally stable. Therefore the designer might decide to grow the network by adding neurons and feedback paths symmetrically to the network.

Once the appropriate strategies have been chosen, the network may be grown by applying them in a random order and assessing the fitness of the result, or by applying them in a set order. It has been found that application in a fixed order of priority is more efficient than the random approach. The strategy which shows the most effect in reducing the error of the network is applied first. If this fails, then the second best strategy is applied, and so on.

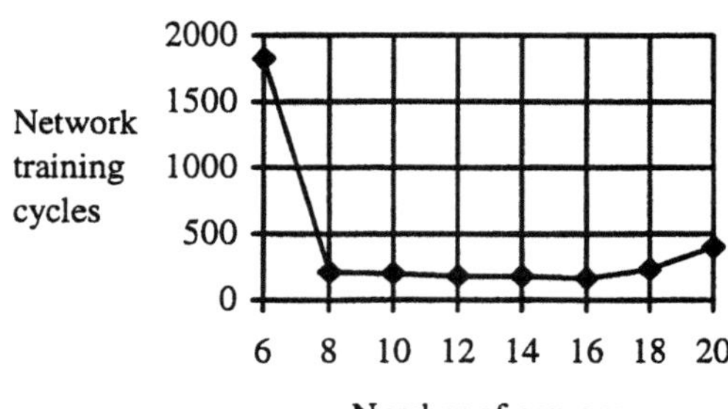

Figure 1: Effect of increasing the number of neurons in a layer.

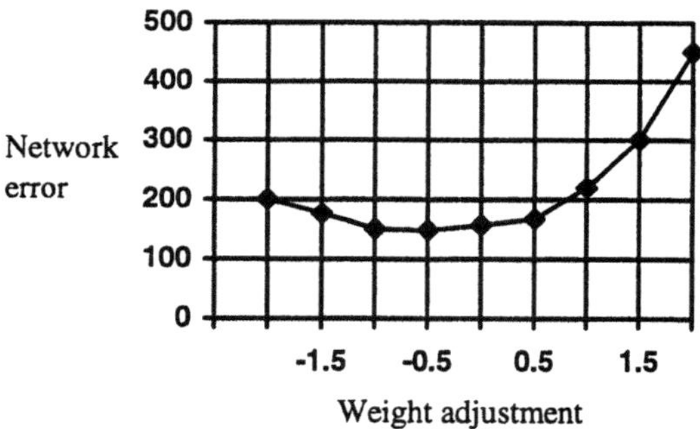

Figure 2: The addition of bias to a network.

Two typical examples of the result of applying one strategy are shown in Figures 1 and 2. They are: adding more neurons in each layer and using different bias units.

These examples are typical of the results which are achieved by the application of a single growth strategy, in that there is a low point on the error curve at a fixed point. In the case of the number of neurons in each layer, the network starts overfitting above 20 neurons and does not give consistent results.

Figure 3 shows the results obtained by applying the algorithm to a feedforward pattern classifier. In this case only four strategies were used. In order of priority, these were:

- change the number of neurons,
- change the connectivity,
- add bias units,
- sideways growth.

The test application was character recognition. The network was trained to recognise all 26 letters of

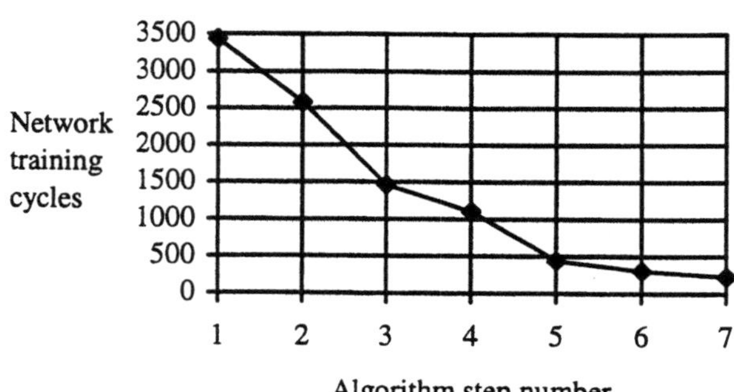

Figure 3: Results from character recognition problem.

the alphabet and had a fixed number of 26 output neurons (one for each letter). Although this is a very simple system, the approach has been shown to work with other problems including recurrent networks.

7 Conclusion

The work outlined above has proved that this approach is a useful optimisation approach when used to design neural networks. It has several advantages over the more traditional genetic algorithm approach. These are:

- it is highly structured,
- it grows from simple to complex,
- it may be optimised to find either the simplest network or the best network,
- it is efficient in terms of time,
- it may be used with FPGAs and other ASICs to implement genetic hardware ANNs.

Much work remains to be done in the field of artificial embryologies and their application to neural networks. However, their advantages mean that it is likely that they will become important techniques in the future.

References

[1] R. Dawkins. *The Blind Watchmaker*. Penguin, 1991.

[2] H. de Garis. Neurite networks: The genetic programming of cellular automata based neural nets which grow. In *Proc. IJCNN93*, pages 2921–2924. IEEE Press, 1993.

[3] D. E. Goldberg. *Genetic Algorithms in Search, Optimization and Machine Learning.* Addison-Wesley, Reading, MA, 1989.

[4] A. N. Kolmogorov. On the representation of continuous functions of many variables by superposition of continuous functions of variable and addition. *American Math. Soc. Trans.*, 28:55–59, 1963.

[5] J. D. Schaffer, D. Whitley, and L.J. Eshelman. Combinations of genetic algorithms and neural networks: A survey of the state of the art. In *Proc. COGANN-92.* IEEE Press, 1992.

[6] F. J. Vico and F. Sandoval. Use of genetic algorithms in neural network definition. In *Proc. IWANN91*, pages 196–203. Springer-Verlag, 1991.

Empirical Study of the Influences of Genetic Parameters in the Training of a Neural Network

P. Gomes[1], F. Pereira[1] and A. Silva[2]
[1] Instituto Superior, de Engenharia de Coimbra
[2]Instituto de Sistemas, e Robótica
{pgomes, xico}@sun.isec.pt, arlindo@isr.uc.pt

Abstract

This paper presents the empirical results achieved in the computer system ROBOTS. This program simulates a virtual world, where agents, called robots, interact with an environment. Each robot is controlled by a neural network. The evolution of the robot behaviour (which is determined by the variation of the weights in the neural network) is done using a genetic algorithm. We describe the conceptual model used in ROBOTS. We also show how the genetic parameters and the environment itself influence the robot's adaptation to the environment.

1 Introduction

In this paper we will study the influence of the parameters of a genetic algorithm on the training of a neural network. In particular, we are interested in the way the behaviour of an artificial agent can evolve using a genetic approach. The agent's behaviour is modelled through the use of a neural network. Development of a neural network is a difficult task. One of the difficulties is training the neural network in order to achieve the desired configuration of weights. This can be done using various algorithms. One of the well known algorithms developed for training neural networks is backpropagation [3]. However this algorithm is dependent on the training examples and is difficult to use in a application where the evaluation function is non-stationary (the fitness landscape of a non-stationary function changes over time) [3]. This problem is due to the local search level of back propagation.

Genetic algorithms are an alternative to conventional algorithms (like backpropagation). As a probabilistic global search procedure [2] they avoid the traps of local minima and are able to converge upon the region where the optimal solution is. Good results in training neural networks with a genetic algorithm have been reported (see e.g. [1,4]).

In order to study the influence of genetic parameters in the training of a neural network, we implemented a computer system called ROBOTS. In the next section we describe this system. Section 3 presents the results obtained in preliminary experiences and shows how the robots behaviour evolved in the virtual world. Finally we present our conclusions and future work in section four.

2 The ROBOTS System

This section presents the conceptual model used in ROBOTS. We start by describing the virtual world used for simulation. Then we present the neural network topology and genetic algorithm.

2.1 Environment

Robots live in a bidimensional virtual environment containing randomly distributed pieces of water and food. Each position in the world may be empty or have one of the following items: water, food, robot, robot and water, or robot and food. Positions with food or water have an associated quantity. Quantities range from 1 to 9.

Robots

The robot behaviour is controlled by a neural network. All robots have the same neural network architecture (only the network weights differ from robot to robot).

A robot has two major properties: energy and mobility. Energy is used by the robot to fight and to survive when it is not moving. Mobility is used for moving around the arena. Energy and mobility are measured in units raging from 0 to 7 for energy, and from 0 to 15 for mobility. Any robot can 'see' the contents of the position where it stands, and three contiguous positions in front of it. Board positions

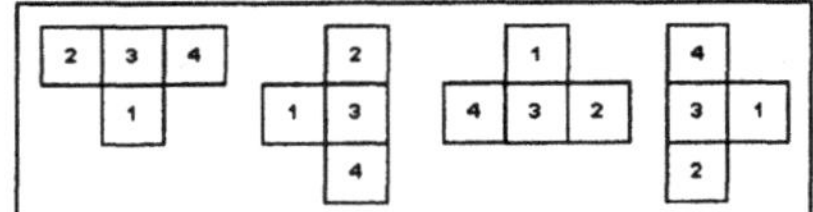

Figure 1: Board positions seen (by the robot) depending on the robot's orientation.

seen by a robot standing in position 1 are presented in Figure 1.

Interactions Between Entities

Relations between entities are ruled in the following way:

- Robot/Food: if a robot is in a position with food, the quantity of food decreases by one and the robot's energy increases by one;
- Robot/Water: if a robot is in a position with water, the quantity of water decreases by one and the robot's mobility increases by one;
- Robot/Robot: if two robots move into the same position, the robot with less energy loses as much energy as half of the opponent's energy. If the energy level drops below zero then the robot dies.

Other Rules

Other rules for life in the arena are:

- each time that a robot moves it loses one unit of mobility;
- if the robot's mobility is zero then it can't move any more;
- each time that a robot doesn't move, it loses one level of energy, except if it is in a position with food or water;
- if the robot level of energy drops below zero then it dies.

2.2 The Neural Network

Robot's behaviour is modelled by a feedforward neural network with two layers. The input layer has 18 neurons, each one corresponding to an input. Inputs correspond to values about the world, which are: perception of the positions in front of the robot (4 positions) and level of energy and mobility that the robot has.

The output layer has nine neurons corresponding to the possible robot moves. The output with the higher value corresponds to the action to be executed.

Since all robots have the same neural network architecture, different behaviours are due to the weights associated with the connections between neurons.

2.3 Genetic Algorithms

The evolution of the robot's behaviour is done using a genetic algorithm. A generation is composed of individuals, where each individual is a robot and is defined by a set of genes. Each gene is a weight of the neural network. Genes are represented by real values. Individuals of the initial population are randomly created (genes are initialised with real values between 0 and 1). The genetic algorithm will apply two genetic operators: selection and mutation. Following [4] we will not apply any kind of crossover operator.

Selection

Selection is performed in a natural way. Individuals of a generation are placed in the artificial world and compete for existing resources, trying to survive. When the number of individuals falls under a given threshold, established by the user, the generation is completed. The surviving robots will be the parents of a new generation.

Mutation

Individuals of the next generation are obtained by making copies of the surviving robots and then mutating some of their genes. The number of individuals in the new generation is defined by the user as the population size. There is always one non-mutated copy of each surviving robot in the new generation. The other copies have mutations in their neural network weights by adding a random value to the weights between -1 and +1. Not every weight is mutated, only a percentage of them. This percentage is defined by the user as the mutation rate. Selection of weights is done by a random process.

3 Experimental Results

In this section we present results achieved with the ROBOTS system. Two types of experiments were done. One is the study of the influence of the population size in the evolution of the robot's behaviour. In the second experiment we will compare the behaviour of individuals from a random population with the behaviour of individuals which have evolved over 1000 generations.

3.1 Influence of the Genetic Parameters

In order to study the influence of the population size in the evolution of the robots, three experiments with different population sizes were performed.

All experiments run for 1000 iterations (each iteration corresponds to a new generation). The arena is a 30x30 grid. Experiment one had the following conditions: population size of 20, a mutation rate of 5%, a percentage of food in the arena of 5%, and a percentage of water of 5%. Experiments two and three had a population size of 30 and 40 individuals respectively. Each generation is completed when the population size reaches 10 live robots.

Data gathered in the experiments comprises: the average life of surviving robots, percentage of used food and water, number of natural deaths, and number of deaths due to robot competition.

In Table 1 we present the statistical values. The results show that the increase in the population size improves the robot's evolution. Robots live more time, and because of that they encounter each other more times increasing the number of deaths caused by other robots. With the increase of the population size it was expected that the usage of water and food would increase too, and that's what happened.

Table 1: Statistical values obtained in the three experiments.

	Exp. 1	Exp. 2	Exp. 3
AMRLD	13.584	13.898	16.201
APFU	5.411	10.922	30.414
APWU	5.122	7.713	27.807
ANND	9.114	18.685	24.95
ANDRC	0.886	1.315	5.05

AMRLD — Average life duration of the surviving robots; APFU — Average percentage of food used; APWU — Average percentage of water used; ANND — Average number of natural deaths; ANDRC — Average number of deaths by robot competition;

Table 2: Statistical values of the population comparison experiments.

Merged Population Number	Random Population	Evolved Population
1	7.5	12.5
2	6	14
3	7.5	12.5
4	8	12
5	6	14
6	3.5	16.5
7	1	19
8	6	14
9	4.5	15.5
10	0.5	19.5

3.2 Evolution of the Robots' Behaviour

We evolved ten different populations over 1000 generations. Individuals of these populations were then compared against individuals of a random population. The process was the following: we created ten new populations by merging the individuals of each evolved population and the individuals of the random one. Evolution of these ten merged populations was then simulated in the virtual world. The simulation runs for 100 iterations. Finally we calculated the average number of surviving robots from each population. The results are shown in Table 2.

The conclusion taken from Table 2 is that populations which evolved over 1000 generation, are better adapted for survival than the random population. The evolution of the neural network performed by the genetic algorithm works, thus improving the robots' survival capacity.

4 Conclusions and Future Work

This work presented a hybrid system using a genetic algorithm to train a neural network. We showed that a genetic algorithm can train a neural network in order to improve its performance, despite the non-stationary evaluation function. The results presented are preliminary, and by the time this paper is written, experiments with the mutation rate and environment parameters will be done.

5 Acknowledgements

This research was partially supported by a MSc. grant from JNICT (PRAXIS XXI BM 6563 95).

References

[1] R. K. Belew, J. McInerney, and N. N. Schraudolph. Evolving networks: Using the genetic algorithm with connectionist learning. In *Proceedings of the Second Artificial Life Conference*, pages 511–547. Addison-Wesley, 1991.

[2] D. E. Goldberg. *Genetic Algorithms in Search, Optimisation and Machine Learning.* Addison-Wesley, Reading, MA, 1989.

[3] J. Hertz, A. Krogh, and R. Palmer. *Introduction to the Theory of Neural Computation.* Addison-Wesley, Reading, MA, 1991.

[4] V. Porto, D. Fogel, and L. Fogel. Alternative neural network training methods. *IEEE Expert*, pages 16–22, June 1995.

Evolutionary Optimization of the Structure of Neural Networks by a Recursive Mapping as Encoding

B. Sendhoff and M. Kreutz
Institut für Neuroinformatik, Ruhr-Universität Bochum,
44780 Bochum, Germany
Email: {bs, kreutz}@neuroinformatik.ruhr-uni-bochum.de

Abstract

The determination of the appropriate structure of artificial neural networks for a specific problem or problem domain remains an open question. One attempt to solve this optimization problem is the application of evolutionary algorithms and the choice of an appropriate coding, the genotype → phenotype mapping. We employ a coding procedure from the class of *recursive coding methods* and apply the optimization process to the problem of prediction and modeling of chaotic time series. The network structure and the inital weight setting are determined by an evolutionary process and the 'fine tuning' of weights is achieved by a standard backpropagation algorithm. We focus on the properties of the coding procedure and the understanding of the network structures in this context.

1 Introduction

Optimization of the structure of neural networks is becoming more and more important both for successful applications of neural networks and for the process of understanding their information processing capacities on a more general level. The application of neural technologies for tasks of increasing complexity requires a more efficient way to determine which kind of network and which structure should be employed in order to increase the overall performance of the network. It would be desirable to be able to make this decision for a whole class of problems, in order to make the process more efficient.

Any understanding of the information processing capabilities of networks, whether they are networks of real neurons or artificial ones, must include the understanding of the organization of the basic elements. We believe that the overall structure of natural neural networks is genetically predetermined via a complicated process of encoding. It is therefore sensible to solve the structure optimization problem along the same line, applying an encoding procedure for the structure of the networks and an evolutionary algorithm to optimize it. After this optimization process a standard learning technique such as backpropagation can be applied to fine tune the weights.

The definition of the structure of neural networks is itself not fixed. A base structure, for example the organization in layers, is often assumed. If we want to avoid any prestructuring, we are confronted with large connection matrices for networks with even a moderate number of neurons. This problem is overcome by encoding the connection information in such a way so that it scales reasonably with the number of neurons, whilst avoiding making any unnecassary assumptions. We will introduce an encoding procedure which is based upon the work of Kitano [4, 5]. This recursive encoding has the advantage over other proposed methods [1, 3] in that it is simple enough to analyse and understand the evolutionary optimization process. In this way it is possible to draw conclusions relating back to the encoding process [8]. The inclusion of some effects observed in biological evolution, such as hidden mutation and redundant encoding, was another reason for applying the recursive encoding method.

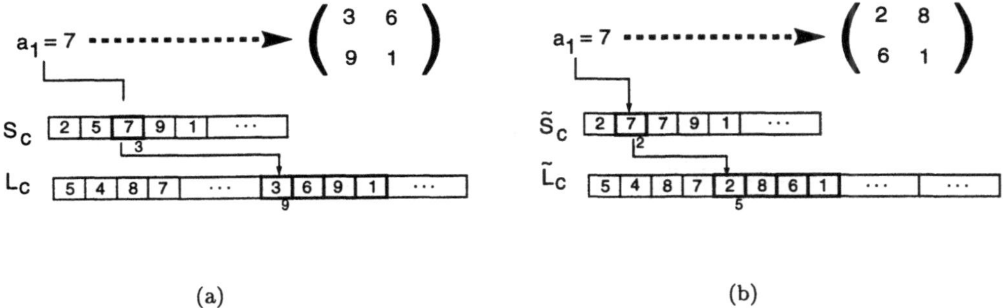

Figure 1: One element is replaced by four elements in the recursion step via the small chromosome $S_C \rightarrow$ large chromosome L_C mapping. (b) The same step as in (a), but using the small chromosome $\tilde{S}_C$, which has been subject to mutations.

2 The Recursive Encoding Method for Neural Networks

Kitano [4] introduced a method which is based on Lindenmayer grammar systems in order to build the connection matrix of a *bulk* neural network; i.e. one without a predetermined layered structure. We maintained the idea of employing two strings of symbols and using a recursive mapping between them to construct the connection matrix. However, we simplified the original procedure by using two integer strings without any different rule or operator sets and a straightforward mapping between these strings.

The recursive encoding of the connection matrix is the replacement of each element in the connection matrix by a 2×2 matrix of new elements in each step. The new elements are specified by a mapping from a first string of integers, the small *chromosome* S_C to a second string of integers, the large *chromosome* L_C. The length of the small chromosome N_{S_C} is variable, the length of the large one is fixed by the condition $N_{L_C} = 4 \cdot N_{S_C}$. At each step i the *first* place $N(a^i_{\mu\nu})$ of each connection matrix element $a^i_{\mu\nu}$ in S_C is determined; for example position $N(a_1 = 7) = 3$ in Figure 1(a). The element is then replaced by the four elements at the positions

$$\begin{aligned}(4 \cdot (N(a^i_{\mu\nu}) - 1) + 1,\ 4 \cdot (N(a^i_{\mu\nu}) - 1) + 2, \\ 4 \cdot (N(a^i_{\mu\nu}) - 1) + 3,\ 4 \cdot (N(a^i_{\mu\nu}) - 1) + 4\ )\end{aligned} \quad (1)$$

in the large chromosome L_C. If the first occurrence of $a^i_{\mu\nu}$ is on the last position in S_C, $N(a^i_{\mu\nu}) = N_{S_C}$, the last element of the new 2×2 matrix is given by (Equation 1) $4 \cdot (N_{S_C} - 1) + 4 = 4 \cdot N_{S_C}$. Therefore, the mapping fixes the length of L_C to $4 \cdot N_{S_C}$. Figure 1(a) shows the replacement of an element $a_1 = 7$ by the four elements $(3, 6, 9, 1)$ at the positions $(9, 10, 11, 12)$). In case $a^i_{\mu\nu}$ is not in S_C, it is replaced by four so called terminal symbols (in the notation of integer strings, the most convenient choice is zero). A terminal symbol is in turn always replaced by another four terminal symbols in a recursion step. Finally, the connection matrix is simplified by deleting all neurons which do not have any output. Figure 2 shows the evolution of a 8×8 connection matrix M_{con} following the introduced rules.

This network connection matrix is a function of the mutation and crossover probabilities, the chromosome length, the number of iteration steps N_{steps} and of the size of the integer interval $[1, N_{sym}]$ of allowed values for both strings. Although the transition from genotype (the two integer strings) to phenotype (the connection matrix of the neural network) makes the effect of the evolutionary operators more intrinsic, it also opens up the possibility to influence and guide the optimization process and to favour special network structures like modularity. Since we are also interested in the analysis of the evolutionary process itself, we chose an encoding method where theoretical analysis of the relation between operators and mapping is still possible.

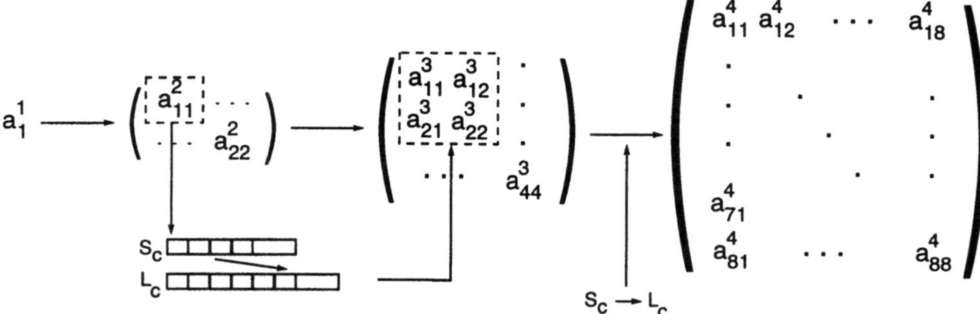

Figure 2: Scheme of the recursive development of the connection matrix up to a size of 8 × 8. Each element in each step is replaced by a 2 × 2 matrix via the mapping $S_C \rightarrow L_C$.

Figure 1(a) and (b) show the effect that a mutation of the small string S_C can have on the mapping. At the same time we notice that both biological paradigms, hidden mutation and redundant coding, can be found in our scheme. The encoding is redundant: If any element occurs more than once in the small *chromosome* it has no effect on the building process of the connection matrix. The mapping is initiated at the first occurence of the element. By the same token, hidden mutations can occur; if any of the redundant numbers mutate it does not have any effect. However, if the first symbol mutates in the following generation, the earlier mutation additionally comes into effect and the overall result might be a large step in the phenotype space.

3 Time Series Prediction

As an application we chose the prediction of a chaotic time series, both one step and three steps iterated. The time series has been generated by the Lorenz [6] differential equation ($a = 0.25$, $b = 4.0$, $F = 8.0$, $G = 1.0$) with an integration step size $\Delta t = 0.05$).

$$\begin{aligned}\frac{dx(t)}{dt} &= -y^2 - z^2 - a(x - F) \\ \frac{dy(t)}{dt} &= xy - bxz - y + G \\ \frac{dz(t)}{dt} &= bxy + xz - z\end{aligned} \tag{2}$$

We used the state $\underline{q}(t) = (x(t), y(t), z(t))$ as the input to predict the state one time step ahead $\underline{q}(t + \Delta t)$. Furthermore, in a second experiment we used the output of the network $\underline{\tilde{q}}(t + \Delta t)$ as its input for the next prediction step. The network used its own output for the three steps, before it was given a new input value without noise.

$$\begin{aligned}\underline{q}(t) \rightarrow \underline{\tilde{q}}(t+\Delta t) &\rightarrow \underline{\tilde{q}}(t + 2\Delta t) \\ \underline{\tilde{q}}(t+2\Delta t) \rightarrow \underline{\tilde{q}}(t+3\Delta t) &\rightarrow \underline{\tilde{q}}(t+4\Delta t) \ \ldots\end{aligned} \tag{3}$$

Feedforward neural networks, although generally rather robust against noise, have severe difficulties with these kind of prediction systems [7]. Successful modelling, which includes iteration, of chaotic time series have been reported for iterated chaotic systems [2], but not for differential equation systems. It seems logical that neural networks will have less difficulty learning the iteration function than solving the differential equations (2).

We used the encoding procedure described above to determine the number of neurons (the network size), the connection between these neurons and the initial setting of the weights. Each network was then trained for 50 cycles using a standard back propagation algorithm with a learning rate $l = 0.01$ and a momentum term $m = 0.3$. The Lorenz data were split up into three data sets, each consisting of 500 three dimensional input-output pairs. The test set was used to terminate the learning on the training

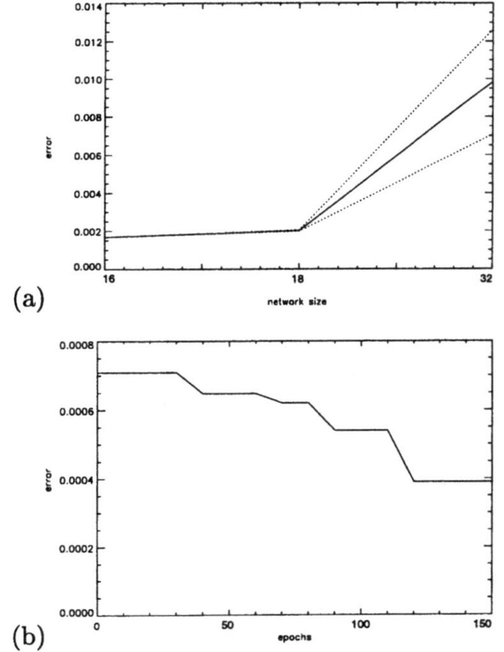

Figure 3: One step prediction of the Lorenz-84 chaotic times series: (a) Squared prediction error of neural networks with given size averaged over 100 runs with the variation as dotted line (50 learning cycles). (b) Squared prediction error of structural optimized networks with initial weight settings (50 learning cycles).

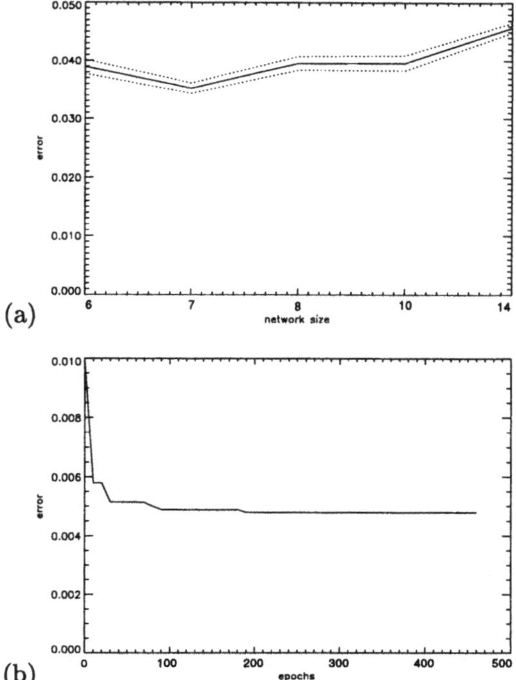

Figure 4: Three step prediction of the Lorenz-84 chaotic times series: (a) Squared prediction error of neural networks with given size averaged over 100 runs with the variation as dotted line (50 learning cycles). (b) Squared prediction error of structural optimized networks with initial weight settings (50 learning cycles).

set before the 50 cycles were finished, when the error on the test set was increasing. The third data set, the validation set, then determines the fitness of the network used in the evolutionary algorithm.

The triple $(N_{S_C}, N_{sym}, N_{steps})$ parameterizes the coding and is stored in a separate chromosome for each individuum. They are optimized in the evolutionary process on a different *time scale* which is realized through a reduced mutation and crossover probability on this chromosome. The next two chromsomes are S_C and L_C which are the integer vectors for the recursive *building* procedure of the connection matrix, Figure 2. A fourth chromosome contains integer values in the same range $[1, N_{sym}]$ for the initial weight setting of the input weights. Finally the initial weights of the neurons in the connection matrix are determined by mapping the interval $[1, N_{sym}]$ onto the interval $[-0.8, 0.8]$ for all matrix elements and all elements from the fourth chromosome. Thus, besides determining the network structure, the encoding is also used to initialize the neuron setting, so that the following backpropagation algorithm only has to fine tune them.

We see from Figures 3 and 4 that the evolutionary algorithm with the proposed encoding outperforms the neural nets with various numbers of neurons, full connection and random initialization of the weights (averages over 100 instances). More interesting than the performance of the networks, is the fact that the evolutionary process for the one step prediction problem chose a coding with large values for N_{S_C} and N_{steps} and a relatively small number of different integers N_{sym}. Large chromosomes and small values for N_{sym} guarantee fully connected networks since the probability that all symbols from the in-

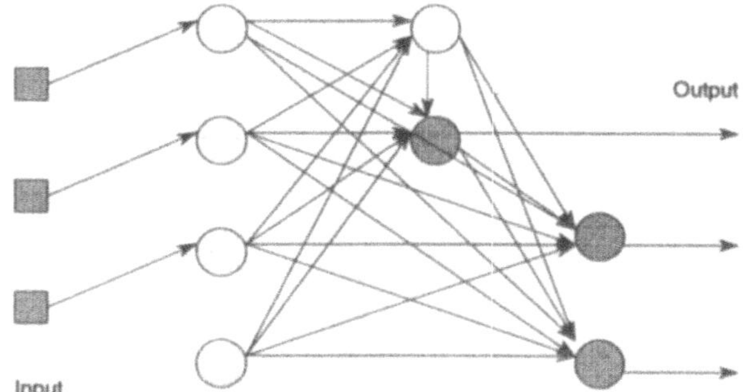

Figure 5: The network structure after 490 optimization generations for the 3 step prediction problem.

terval $[1, N_{sym}]$ are contained in S_C is very high and therefore the probability for a terminal symbol is very low. Thus, the best networks for the one step prediction problem are relatively large 32×32 networks and are fully connected.

The more difficult 3 step problem is solved using a small network of 8 units (see Figure 5) combined with a relatively small chromosome size and a large number of allowed integers. This combination guarantees large variability in the initialization of the weights and at the same time the networks tend to be sparsely connected. One could speculate that larger networks tend towards overfitting which is especially dangerous if the task is more complex, however more research has to be carried out.

4 Conclusion

The proposed *recursive encoding* procedure has proven useful for the structure optimization of neural networks. The chosen problem of predicting and modeling chaotic time series seems to be appropriate for the examinations concerning both the understanding of the chosen coding in the evolutionary algorithm and the structure of the network, as we pointed out in section 3. We believe that the extension of the network to recurrent structures will prove useful in the determination of models of chaotic time series. At the same time the structure of recurrent networks is even more complex which highlights the need for an effective opimization process, like the one proposed here. Additionally, it seems necessary to further understand the evolutionary process itself especially when complicated genotype → phenotype mappings are used, which make it difficult to predetermine the effect of the mutation and crossover operators on the search process.

References

[1] P.J. Angeline, G.M Saunders, and J.B. Pollack. An evolutionary algorithm that constructs recurrent neural networks. *IEEE Trans. Neural Networks*, 5(1):54–65, 1994.

[2] G. Deco and B. Schürmann. Neural learning of chaotic dynamics. *Neural Processing Letters*, 2(2):23 - 26, 1995.

[3] F. Gruau. Genetic synthesis of modular neural networks. In S. Forrest, editor, *Proc. 5th Int. Conf. on Genetic Algorithms*, pages 318–325, San Mateo, CA, 1993. Morgan Kaufmann Publishers.

[4] H. Kitano. Designing neural networks using genetic algorithms with graph generation system. *Complex Systems*, 4:461 - 476, 1990.

[5] H. Kitano. Neurogenetic learning:an integrated method of designing and training neural networks using genetic algorithms. *Physica D*, 75:225 - 238, 1994.

[6] E.N. Lorenz. Irregularity: A fundamental property of the atmosphere. *Tellus*, A(36):98, 1984.

[7] J. Principe, A. Rathie, and J. Kuo. Prediction of chaotic time series with neural networks and the issue of dynamic modeling. *International Journal of Bifurcation and Chaos*, 2(4):989, 1992.

[8] B. Sendhoff, M. Kreutz, and W. von Seelen. Causality and the analysis of evolutionary algorithms. 1997. Submitted to *IEEE Transactions on Evolutionary Computation*.

Using Genetic Engineering To Find Modular Structures for Architectures of Artificial Neural Networks

C. M. Friedrich
Institute for Technology Development and Systems Analysis, University of Witten/Herdecke,
Alfred-Herrhausen Str. 50; 58448 Witten, Germany
Email: chris@uni-wh.de

Abstract

Starting with an evolutionary algorithm to optimize the architecture of an artificial neural network (ANN), it will be shown that it is possible, with the help of a graph database and genetic engineering, to find modular structures for these networks. A new graph rewriting is used to construct families of architectures from these modular structures. Simulation results for two problems are given. This technique can be useful as an alternative to automatic defined functions for computing intensive structure optimization problems, where modularity is needed.

1 Introduction

One of the major problems using ANNs is the design of their architecture. The architecture of an ANN greatly influences its performance. If the architecture is too small, the net is not able to learn the desired input/output mapping. On the other hand, if the architecture is too large, the net generalizes poorly on unseen data. In addition to constructive and pruning techniques, evolutionary algorithms have been suggested by many scientists to find good architectures for ANNs. Much work has been done in this area. For a survey of recent work the paper of Branke [2] is suggested. Most of the difficulties in this field arise from the problem of choosing the right representation, to encode a network graph for the evolutionary algorithm. The second problem is the scalability of the chosen encoding technique, in this work the main interest is therefore on modularity, to obtain a scalable method for this problem.

2 Cellular Encoding

A method to optimize the architecture and weights of boolean neural networks, where the weights are restricted to the values -1 and 1, was suggested by F. Gruau [3] who named it cellular encoding. In this encoding technique, the information to develop an architecture is obtained by interpreting the information from a grammar tree. The nodes of this tree encode information about graph rewriting operations. The development of a neural network starts with a graph consisting of one node (cell), having ingoing connections from the input area and outgoing connections to the output area. Every cell has a reading head pointing to one node of the grammar-tree (cellular code). The nodes of the cellular code contain symbols defining the graph rewriting operation. This technique is comparable to a Turing machine. Instead of writing to a tape in cellular encoding the cells are changed. The operation #par: for example symbolizes a parallel cell division, both cells inherit the input and output connections from the mother cell. The reading heads of the new created cells are moved to the left and right subtree of the grammar tree. For biological plausibility, the development of the cells should be parallel. This can only be achieved by using a FIFO Queue. The development of a network ends, if all cells evaluate the #end: operation, located as terminal at the leaves of the grammar-tree.

As mentioned above, the cellular code is given as a grammar tree. In this case it is possible to use genetic programming to optimize the cellular code. Genetic programming is a genetic algorithm that uses trees instead of bitstrings as representation. The recombination operator crossover is realized by exchanging the subtrees of two parent individuals. The mutation operator brings some variations to the genome and is realized by inclusion of randomly initialized trees.

Gruau proved some properties of his encoding technique, including completeness, compactness, closure, modularity and scalability. He achieved these properties using a recursion operator making it possible to repeat parts of his cellular code.

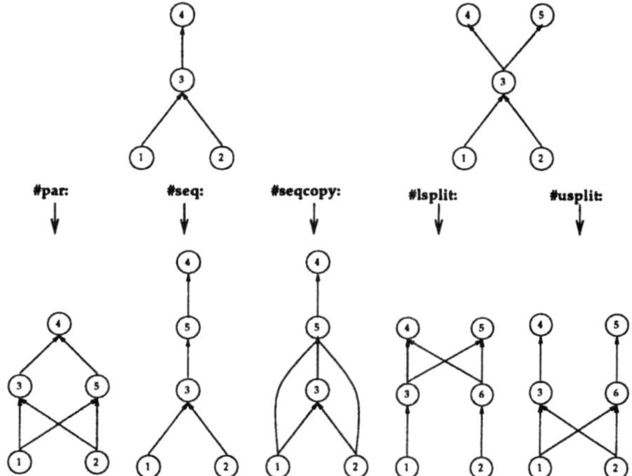

Figure 1: The used rewriting operations for the modified Cellular Encoding.

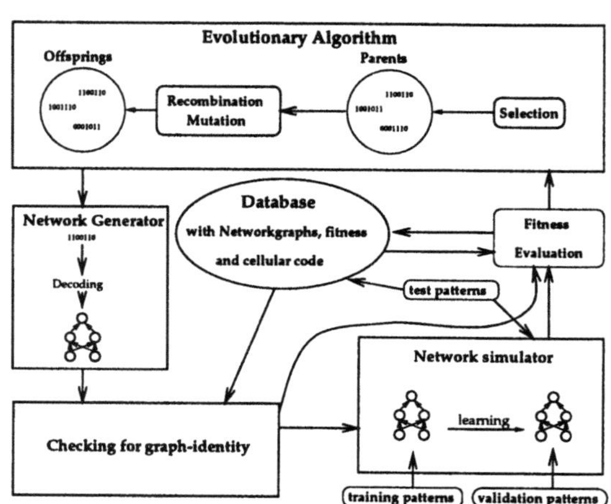

Figure 2: Modified cellular encoding using a graph-database.

3 Modified Cellular Encoding

For the optimization of the architecture of ANNs with real-valued weights, the cellular encoding method was modified. Some new development operators were developed and tested. The set of operators {#par:, #seq:, #seqcopy:, #usplit:, #lsplit:, #end:} was used to develop the architectures. Figure 1 shows the effects of rewriting the hidden node of a network with the development operators.

One property of the created networks should be their correctness. A created graph is a correct architecture for a feedforward neural network, if it contains only feedforward connections and if there are no isolated nodes, which means that all hidden nodes are on a path from an input node to an output node. All operators were designed to hold this property, so only correct network graphs could be created. Different to Gruau's work, it is possible with the operator #seqcopy:, to create networks with shortcut connections.

The second modification to the evolutionary development cycle was the use of a graph-database. This database contains the cellular code, network graphs and fitness parameters like the number of necessary epochs for learning, classification error on the learning, validation and test set and other parameters of the networks. Figure 2 shows the working structure of the evolutionary cycle. Before the fitness calculation of a developed network, it is checked, whether the fitness of this network was calculated five times before The number five was used as a compromise between statistical evaluation possibility and necessary computing time. It would be better to use the fitness distribution as information for this number.

If not, the fitness is calculated again and saved in the database. The resulting fitness of a network is the mean of all fitness evaluations of this network. This method allows it to minimize the fitness distortion resulting from the random initialization of the weights. This increases the validity and robustness of the obtained results. Other authors document only solutions which are results of a single fitness evaluation. In this work, only results are given for architectures tested five times. Another advantage of the database is the possibility to save computing-expensive fitness evaluations, if the network was tested five times before.

3.1 Genetic Engineering

Optimizing the architecture of an ANN using evolutionary algorithms leads to a proper performing architecture for a problem. But it gives no insight into the problem, how good architectures are built. It would be much more interesting to find building principles for good architectures to create modular architectures. In his work on genetic programming, Koza [4] suggested automatic defined functions as a possible solution for this problem. Unfortunately for the task of architecture optimization the fitness evaluation is very computing intensive, so it is not possible to use big population sizes and many epochs. Especially for genetic programming, where the representation of a problem is a grammar tree, Altenberg [1] suggested a method to find good modular structures that are subtrees of the representation. He called this technique *genetic engineering*. It depends on the assumption, that some parts of the genetic program have a higher impact on the complete fitness of the genetic program than others. It is a problem to attach a fitness value to sub-programs, because it is only possible to obtain the fitness of the complete phenotype.

In this work the fitness of modular structures of the genetic programs were found by analyzing the cellular code in the graph-database, which was built during the formerly described optimization process. This makes it possible to obtain information about many optimization runs that start with different populations. The fitness of a subtree is equivalent to the frequency of this subtree in the database. The found modular structures are called modules and noted as a preorder traversal of the subtree.

3.2 Graph-Rewriting

Having found good modular structures for neural network architectures, it would be interesting to see how architectures built of these modules perform. One possibility is to include these subtrees as encapsulated functions in the next evolutionary optimization process. In contrast to this, in this work a special graph-rewriting method was used. The (i, k)-rewriting of a network graph with a module starts with a fully connected $n - i - m$ network, where n, m are determined by the problem. Then all hidden nodes are rewritten k times with the rewriting operations defined by the module. For $k = 0$ this results in standard architectures with one hidden layer. With this method it is possible to create families of architectures consisting of good performing modules. Figure 3 shows an example of a network created by $(1, 1)$-rewriting of a 3-1-2 network with the module #usplit:#par:#lsplit:. The hidden node will be first rewritten with a #usplit: operation and the resulting nodes will be rewritten with the operation #par: or #lsplit:.

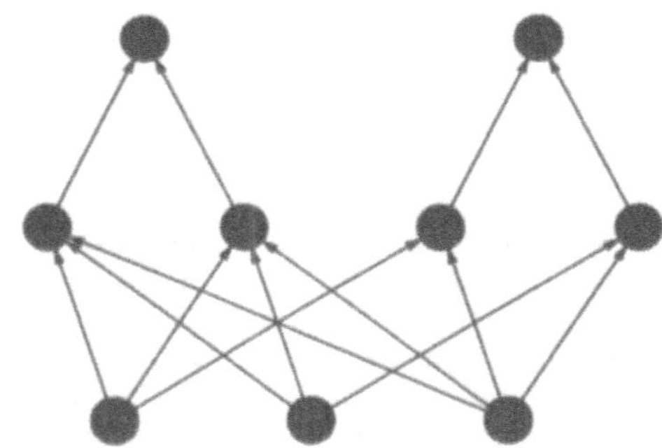

Figure 3: Resulting network of $(1, 1)$-rewriting of a 3-1-2 network with the module #usplit:#par:#lsplit:.

4 Experimental Results

The data for the first experiment are taken from the benchmark suite from Prechelt [7]. The task for the diabetes problem is to classify from some diagnostic data (e.g. blood pressure, result of glucose tolerance test etc.), whether a female Pima indian was diabetes positive or not. The patterns sets were divided into three datasets with the size (576/96/96). There are 8 input parameters and 2 output classes. The classification method is WTA (winner takes all). The problem is difficult, because some data is not available for all patterns and therefore set to zero. Prechelt [7] and Michie et al. [6] found networks having a classification error of 24.8 % on the test set. The activation function of the network was the $tanh(x)$. The learning set was presented up to 500 times, with RPROP as learning algorithm but cross validation usually stopped the run before. The fitness function for this problem was a linear combination of the classification error on the test set (factor 1) and the learning set (factor 0.3). Several optimization runs with this problem were made. About 100,000 architectures were tested, this task needed about 16 days of UltraSparc computing time. The best architecture found by the evolutionary algorithm is a network with 33 nodes (23 hidden nodes) and 148 weights. This sparsely connected network shows a classification error in the mean of five runs of 23.125% on the test set and 15% on the learning set.

The application of genetic engineering on the data in the graph-database found the module #par:#usplit:#usplit: with the highest frequency. The elements of the family of architectures that can be constructed through $(i, 1)$-rewriting perform very well on this problem. The architecture created by $(5, 1)$-rewriting, a net with 26 nodes (16 hidden nodes) and 160 weights shows the same classification error of 23.125% on the test set, but the classification error on the learning set was about 5% better with the evolutionary optimized architecture.

The second tested problem was the approximation of a mexican-hat function with an ANN. Mandischer [5] used it to show the effectiveness of his technique to optimize the architecture of an ANN. The problem has 2 inputs, the x-coordinate and y-coordinate of the function and one output, the z-coordinate. The network was learned with 841 patterns describing the function in the range $[-2.1, 2.1]$. The fitness-function was a linear combination of the classification error on the learning set (factor 1), remaining sum of squared errors (factor 10) and number of used epochs (factor 1) to obtain a mean squared error of 0.01. The learning set was presented up to 1000 times, but occasionally less epochs were needed. All other parameters were set as in the diabetes problem.

100 generations with a population size of 50 were tested. This results in 5000 fitness evaluations. For these evaluations approximately 12 hours of SparcStation 10 computing time was needed. The best found network had 35 nodes (32 hidden) and 214 weights. This network needed 58 epochs in the mean of five runs to approximate the mexican-hat function. All standard architectures up to 200 hidden nodes needed approximately 150 epochs. Mandischer found in his work a network with 100 nodes and 590 weights, that needed 64 epochs in one test. The works are not directly comparable, because Mandischer uses standard backpropagation and in this work RPROP was used. As a conclusion it can be seen that the evolutionary optimization process finds better performing architectures.

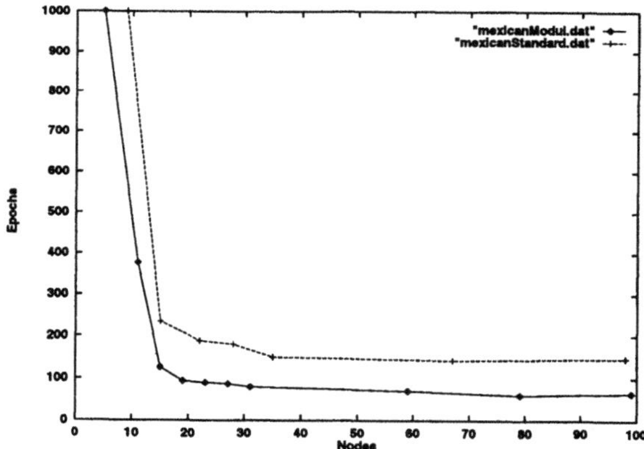

Figure 4: Comparison of needed Epochs for standard architectures and architectures built by graph-rewriting for the mexican-hat problem.

The application of genetic engineering found the module #seqcopy:#seqcopy:#usplit:. The family of architectures built by $(i, 1)$-rewriting with this module shows a better fitness than standard architectures on this problem. Figure 4 gives a comparison of this result, it should be noted, that the number of weights for standard architectures and architectures built by the rewriting operations are the same, if they have the same number of nodes.

5 Conclusion

It was shown that it is possible to find good performing architectures for non-boolean ANNs, with a modified cellular encoding method. The functionality of the method was demonstrated on two problems. The results were compared with results available from literature. All evolved architectures show better performance for the optimized criteria than comparable standard architectures. With the help of a graph-database it was furthermore possible to find modular structures for the architecture of ANNs. Architectures created with a new graph-rewriting method from this modular structures show better performance than standard architectures. This technique can be used as an alternative to automatic defined functions for problems that restrict the number of possible fitness evaluations. Using this method, it is possible to create a library of good modular structures for problem-specific modular network architectures. Further investigations on these modular structures and graph-rewritings may give some new insights into the working principles of ANNs and their learning algorithms.

References

[1] L. Altenberg. The evolution of evolvability in genetic programming. In K. E. Kinnear, editor, *Advances in Genetic Programming*. MIT Press, 1994. http://pueo.mhpcc.edu/ altenber/PAPERS/Papers2.html.

[2] J. Branke. Evolutionary algorithms for neural network design and training. In *Proceedings of the 1st Nordic Workshop on Genetic Algorithms and its Applications*, 1995. ftp://ftp.aifb.uni-karlsruhe.de/pub/jbr/Vaasa.ps.

[3] F. Gruau. Cellular encoding of genetic neural networks. Technical Report 92-21, Ecole Normale Superieure de Lyon, Institut IMAG, 1992. ftp://lip.ens-lyon.fr/pub/Rapports/RR/RR92/-RR92-21.ps.Z.

[4] J. R. Koza. *Genetic programming, on the programming of Computers by means of natural selection.* MIT Press, 1992.

[5] M. Mandischer. Representation and evolution of neural networks. In R.F. Albrecht, C.R. Reeves, and N.C. Steele, editors, *Artificial Neural Networks and Genetic Algorithms*, pages 643–649, Wien New York, 1993. Springer.

[6] D. Michie, D. J. Spiegelhalter, and C. C. Taylor. *Machine Learning, neural and statistical classification.* Ellis Horwood Ltd., 1994.

[7] L. Prechelt. PROBEN1 — A set of benchmarks and benchmarking rules for neural network training algorithms. Technical Report 21/94, Fakultät für Informatik, Universität Karlsruhe, D-76128 Karlsruhe, Germany, September 1994. ftp://ftp.ira.uka.de/pub/papers/techreports/1994/1994-21.ps.Z.

Evolutionary Learning of Recurrent Networks by Successive Orthogonal Inverse Approximations

C. Gégout
Laboratoire de l'Informatique du Parallèlisme, École Normale Supèrieure de Lyon,
Centre de Mathèmatiques Appliquèes, École Polytechnique, France
Email: gegout@email.enst.fr

Abstract

Recurrent networks have proved to be more powerful than feedforward neural networks in terms of classes of functions they can compute. But, because training of recurrent networks is a difficult task, it is not clear that these networks provide an advantage over feedforward networks for learning from examples. This communication proposes a general computation model that lays the foundations for characterizing the classes of functions computed by feedforward nets and convergent recurrent nets. Then a mathematical statement proves that convergent nets outperform feedforward nets on data fitting problems. It provides the basis to devise a new learning procedure that constraints the attractor set of a recurrent net and assures a convergent dynamic by using orthogonal inverse tools. The learning algorithm is based on an evolutionary selection mechanism. Using the previous procedure as evaluation function, it has been shown to be robust and well adapted to train convergent recurrent nets when feedforward nets cannot approximate a real parameter mapping.

1 Introduction

Previous works have proved that recurrent networks having convergent activation dynamics can be designed to accomplish useful tasks like pattern recognition, classification and combinatorial optimization. Due to the presence of feedback connections which give the net an additional computing power, they have proved to be more powerful than feedforward neural networks in terms of classes of functions they can compute. But because the activation dynamic may not provide convergent and stable orbits for neuron states, devising effective learning procedures is a very challenging task. Hence, it is not clear that these networks provide an advantage over feedforward networks for learning a complex mapping from examples.

The learning abilities of recurrent networks have been extensively studied essentially in the framework of connection matrices and architectures. Mathematical results guaranteeing the convergence and the stability of the activation dynamic often concern continuous time nets [7] or specific designed nets [9]. However the recent theory of difference equations [1] allows the study of functionalities and asymptotic behaviors of discrete-time activation dynamics for all kinds of recurrent networks.

2 The Computation Model

We restrict our attention to synchronous dynamics — all the nodes are updated simultaneously — but it is known that any deterministic asynchronous dynamics can be simulated on synchronous dynamics by adding nodes in the net. In our model of network computation, the states of neurons are updated until the network converges to an equilibrium (if it exists); the output vector Y is then read from a designated set of neurons. The neurons which have no entries are considered as inputs. Their states stay constant and form what we define the input vector of the net X. We identify two types of nodes: input nodes and others also called *units*. Because the input nodes do not compute anything, the input vector X is considered as a parameter of the net. Let $\mathcal{N}$ be a neural net with N units and L input nodes. Let $a_i(t)$, G_i and θ_i be respectively the state of the ith unit at time t, its transfer function and its bias. We use the vector notations $a(t) = (a_i(t))_i$,

$G(z) = (G_i(z_i))_i$ and $\Theta = (\theta_i)_i$. Let $W = (w_{ij})$ and Ω be respectively the weight matrix of the connections between the units, and the weight matrix of the connections leaving the input nodes. By convention, if there is no connection between the unit i and the unit j then $w_{ij} = 0$. At time t, the ith unit receives the signal

$$R_i(t) = \sum_{l=1}^{L} \Omega_{il} X_l + \sum_{j=1}^{N} w_{ij} a_j(t) + \theta_i.$$

Because X stays constant on an orbit, it is incorporated into a vector $T_X^{\Omega\Theta}$ defined by:

$$T_X^{\Omega\Theta} = \Omega X + \Theta.$$

$T_X^{\Omega\Theta}$ is written T when no ambiguity is possible. If the orbit $\{a(t)\}_t$ converges to an equilibrium a_∞ then we have the nonlinear equation:

$$\lim_{t\to\infty} a(t) = \lim_{t\to\infty} G(R(t)) \Leftrightarrow a_\infty = G(W a_\infty + T_X^{\Omega\Theta}).$$

The equilibrium state is often called a fixed point. Let $\mathbb{FP}(\mathcal{N})$ be the set which contains all the fixed points of $\mathcal{N}$. In our model, feedforward nets are just specific nets verifying: let $\mathcal{N}$ be a feedforward net and let $M(\mathcal{N})$ be its number of layers, the state $a(t)$ stays constant and reaches its equilibrium if $t > M(\mathcal{N})$.

3 Computational Power of a Convergent Network

In running a net to obtain its fixed points, we need to specify its parameters (W, Ω, Θ, X), the initial state $a(0)$ and the activation dynamic governing the update rules. An activation dynamic is considered as an operator K verifying the property:

$$\forall a_\infty \in \mathbb{R}^N,\ a_\infty \in \mathbb{FP}(\mathcal{N}) \Leftrightarrow K(a_\infty) = a_\infty.$$

At time t, the next activation states are assumed to be computed by the following equation: $a(t+1) = K(a(t))$. As pointed out in [7], relevant dynamics are not sensitive to noise and round-off errors occurred when reading the initial state of the parameters. In this paper, the most studied dynamic, $K(a) = G(Wa + T)$ is used. With the help of difference equation theory [1], relevances of some dynamics were analyzed in [6] and it has been shown that K has a non-chaotic asymptotic behavior for recurrent networks only if:

$$\forall a_\infty \in \mathbb{FP}(\mathcal{N}), \rho(DK(a_\infty)), < 1,$$

where $\rho(M)$ is the spectral radius of the matrix M (i.e., the maximum of the modules of the eigenvalues of M) and DK is the differential of K.

We propose to evaluate convergent recurrent networks according to their set of fixed points. The following definitions are needed:

Definition 1: *Let G be a function from $\mathbb{R}^N$ to $\mathbb{R}^N$ verifying the following statement: $\forall x = (x_i)_i \in \mathbb{R}^N, G(x) = (G_i(x_i))_i$. $DG(x)$ designates the differential of G computed at x.*

$$\begin{cases} \mathcal{Z}_G(W) = & \{T,\ \mathbb{FP}(W,T) \neq \emptyset\} \\ \mathcal{Z}_G^*(W) = & \{T \in \mathcal{Z}_G(W),\ \forall a \in \mathbb{R}^N, \\ & \rho(DG(Wa+T)W) < 1\} \\ \mathcal{Z}_G^+(W) = & \{T \in \mathcal{Z}_G(W),\ \forall a_\infty \in \mathbb{FP}(W,T), \\ & \rho(DG(Wa_\infty + T)W) < 1\} \\ \mathcal{Z}_G^\circ(W) = & \{T \in \mathcal{Z}_G(W),\ \exists k \in \mathbb{N},\ \forall \{a_i\}_{1\le i \le k}, \\ & \forall i,\ a_i = K(a_{i-1}), \\ & \prod_{i=1}^{k}(DG(Wa_i + T)W) = 0\} \end{cases}$$

Feedforward nets are characterized by the set $\mathcal{Z}_G^\circ(W)$ because when $t > M(\mathcal{N})$ the state of the net stays constant and does not depend on the initial state a_1. It is obvious that $\mathcal{Z}_G^\circ(W) \subset \mathcal{Z}_G^*(W) \subset \mathcal{Z}_G^+(W) \subset \mathcal{Z}_G(W)$. If F is a function from $\mathbb{R}^N$ to $\mathbb{R}^P$ which can be computed by a convergent net with the following parameters (W, Ω, Θ), F must verify:

$$\exists W_o,\ \forall X \in \mathbb{R}^N,\ T_X^{\Omega\Theta} \in \mathcal{Z}_G(W)$$
$$F(X) \in W_o \mathbb{FP}(W, T_X^{\Omega\Theta}),$$

where W_o is the projection on the subset of output units in our computation model.

But the activation dynamic K may not provide convergent behavior, some fixed points may not be reached [6, 7]. Indeed we can easily prove that if $\mathcal{R}_G^c$ is the set of functions computable by convergent recurrent nets using K, we have $\mathcal{R}_G^* \subset \mathcal{R}_G^c \subset \mathcal{R}_G^+$

where:

$$\begin{cases} \mathcal{R}_G^* = \{F, \exists(W,\Omega,\Theta,W_o),\ \forall X \in \mathbb{R}^N, \\ T_X^{\Omega\Theta} \in \mathcal{Z}_G^*(W) \text{ and } F(X) \in W_o\mathbb{F}\mathbb{P}(W,T_X^{\Omega\Theta})\} \\ \mathcal{R}_G^+ = \{F, \exists(W,\Omega,\Theta,W_o),\ \forall X \in \mathbb{R}^N, \\ T_X^{\Omega\Theta} \in \mathcal{Z}_G^+(W) \text{ and } F(X) \in W_o\mathbb{F}\mathbb{P}(W,T_X^{\Omega\Theta})\} \\ \mathcal{R}_G^\circ = \{F, \exists(W,\Omega,\Theta,W_o),\ \forall X \in \mathbb{R}^N, \\ T_X^{\Omega\Theta} \in \mathcal{Z}_G^\circ(W) \text{ and } F(X) \in W_o\mathbb{F}\mathbb{P}(W,T_X^{\Omega\Theta})\} \end{cases}$$

$\mathcal{R}_G^+$, $\mathcal{R}_G^*$ and $\mathcal{R}_G^\circ$ characterize the sets of functions computable by respectively non-chaotic recurrent networks, some convergent recurrent nets and feedforward nets. In addition, we have $\mathcal{Z}_G^\circ(W) \subset \mathcal{Z}^+(W) \Rightarrow \mathcal{R}_G^\circ \subset \mathcal{R}_G^c$. Hence feedforward nets are less powerful in this sense than convergent recurrent nets.

4 A Learning Procedure for Data Fitting Problems

We are interested in training recurrent networks to map input sequences to output sequences for applications in complex mapping approximation. Considering Atiya's and Pineda's works (both in [2]) and more recent studies [6, 10], a learning algorithm based on gradient descents can be set up when the net has a differential. Learning algorithms used for convergent recurrent networks are usually based on computing the gradient of a cost function with respect to the parameters of the net. But as pointed out in [3] and shown in Section 5, gradient descent algorithms are inefficient for learning of long-term dependencies in the input/output sequence and more certainly for learning convergent nets. We propose another learning algorithm directly computing the fixed points from a set of D examples $\{(x^k, y^k)\}_k$. Let $\{a^k\}_k$ be a set of fixed points. The net has to fit the examples by minimizing the square error $\sum_{k\leq D} \left|\left|y^k - F(x^k)\right|\right|^2$ on the set of mappings F. Actually, we search the closest function of $\mathcal{R}_G^c$ to F. Let us define the following sets: $S_{W,\Omega,\Theta} = \{(a^k)_k,\ \forall k,\ G(Wa^k + \Omega x^k + \Theta) = a^k \text{ and } \rho(DG(Wa^k + \Omega x^k + \Theta)W) < 1\}$ and $\mathcal{S} = \bigcup_{W,\Omega,\Theta} S_{W,\Omega,\Theta}$.

To have the closest function of $\mathcal{R}_G^+$ we need to find the fixed points $A_+ = (a_+^k)_k$ verifying:

$$A_+ = \underset{(a^k)_k \in \mathcal{S}}{\text{argmin}} \left\{ \min_{W_o} \sum_{1\leq k\leq D} \left|\left|W_o a^k - y^k\right|\right|^2 \right\} \tag{1}$$

Obviously, Equation (1) cannot be solved analytically. We propose a numerical method based on successive minimizations. First, suppose the inverses $\{G_i^{-1}\}_i$ exist. Let A, X and Y be matrices containing respectively the vectors (a^k), (x^k) and (y^k). Let L and $G^{-1}(A)$ be respectively a $(1 \times D)$ matrix with coordinates equal to 1 and the $(N \times D)$ matrix $\left(G_i^{-1}(a_i^k)\right)_{ik}$. Assume an approximation A of $A_+ = (a_+^k)_k$ is given. A finer approximation can be obtained by the following approximating process:

1. Start with a matrix A_1, $t = 1$.
2. Compute $(\tilde{W}, \tilde{\Omega}, \tilde{\Theta})$ which is the solution of $\underset{(W,\Omega,\Theta)}{\text{argmin}} \left|\left|WA_t + \Omega X + \Theta L - G^{-1}(A_t)\right|\right|^2$.
3. Do $\tilde{W} = \alpha\dfrac{\tilde{W}}{\rho(\tilde{W})}$ if $\rho(\tilde{W}) > \alpha = \sup_b \dfrac{1}{||DG(b)||}$ and compute the new fixed points A_t.
4. If the net is not convergent and $t < \tau$ do $t = t + 1$ and go to step 2 else evaluate $\min_{W_o} ||W_o A_t - Y||^2$.

For steps (2) and (4) we use the formalism of orthogonal inverses (see [4] for a rigorous description). If [] and <> designate respectively the horizontal concatenation of matrices and the vertical concatenation, results of orthogonal inverse theory prove that the following equation can be deduced:

$$\left[\tilde{W}, \tilde{\Omega}, \tilde{\Theta}\right] = G^{-1}(A)\,\langle A, X, L\rangle^\diamond \text{ and } W_o = YA^\diamond.$$

The notation $B^\diamond$ means the orthogonal right inverse of B. The orthogonal right inverse of a matrix B is defined by the following property: $(Id - B^\diamond B)$ is the orthogonal projector onto the kernel of B.

The procedure provides a convergent sequence $\{A_1, \ldots, A_\tau\}$ where A_{t+1} is a finer approximation of A_+ than A_t. It consists of successive projections on convex sets, and it converges toward a set of fixed

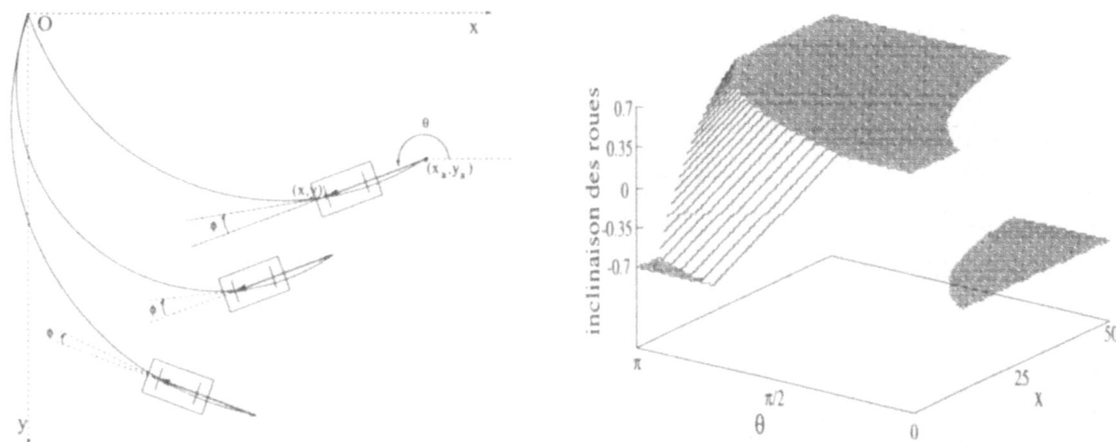

Figure 1: Optimal trajectories and response of the best controller (cross-section $y = 12$).

points computing an element of $\mathcal{R}_G^*$. But, to obtain an element of $\mathcal{R}_G^c$ we only have to stop the procedure when a convergent net is reached. The computations of an orthogonal inverse and a spectral radius need both N^3 operations, so if the maximum number of iterations is bounded then the number of operations required for the approximation procedure is proportionate to N^3. The quality of A_τ is evaluated by $\min_{W_o} ||W_o A_\tau - Y||$. It strongly depends on the initialization A_1. So, to obtain a good approximation of the solution of Equation (1), we must search the matrix A_1, the best initialization for the approximation procedure.

5 Simulations

The search for the best initialization $A = (a^k)_k$ cannot be accomplished using the gradient descent algorithm because no direction of descent can be computed. Stochastic algorithms are more adapted to the search. Evolutionary algorithms are global stochastic approaches which minimize an arbitrary fitness function by mimicking natural selection. They have proved efficient for minimizing real parameter functions (e.g. [5]). Our evolutionary algorithm is initiated by choosing randomly a population of 100 matrices A. Each matrix designs a genotype of an individual. The genetic operators work directly on the real coordinates of the matrix. An individual is evaluated by using the iterative procedure. The evolutionary algorithm then proceeds as a repeated three-step process. First the best elements of the population are identified (selection mechanism) as those which produce the lowest values of $\min_{W_o} ||W_o A_\tau - Y||$. These are then allowed to cross-breed (linear combination) and to mutate (vector translation). In the third step it forms a new population in which the best matrix is kept. The algorithm is stopped after 200 generations if no better matrices are found during the last 10 generations. A similar evolutionary algorithm is given in [5]. The learning procedure for feedforward nets is an efficient conjugate gradient descent: the Polak-Ribire gradient method computes the gradient at each iteration and makes a one-dimensional minimization requiring twenty or so computations of the error function. We have trained 9 feedforward architectures: from 3 to 9 units with 3 layers and from 5 to 9 units with 4 layers. For each architecture we have run 40 or so gradient descents with different initializations. The learning times for both feedforward nets and recurrent nets are then similar. We have kept the performances of the best initialization.

Our approach has been tested on a control problem. A neural network drives a car, the goal is to park the car at the origin facing the front of the wall $y = 0$. Let L be respectively the distance between the rear axle and the front axle and (x, y) the position of center of the front axle. V is the speed

Table 1: Square errors for the different architectures.

hidden units		3	5	7	9
1 hidden layer	train	0.82	0.46	0.425	0.295
	test	1.24	1.07	0.835	0.790
2 hidden layers	train		0.67	0.5	0.51
	test		0.975	0.91	0.74
recurrent nets	train	0.54	0.49		
	test	0.70	0.63		

of the car and θ designates the angle between the direction of the car and the $y-$axis. The controller has 3 inputs (x, y, θ) and gives the steer orientation ϕ as output. Therefore the motion of the car is governed by the following equations:

$$\begin{cases} x(t+1) = x(t) + V cos(\phi + \theta) \\ y(t+1) = y(t) + V sin(\phi + \theta) \\ \theta(t+1) = \theta(t) + \arcsin\left(\frac{V}{L}\sin(\phi)\right) \end{cases}$$

Using simple geometrical arguments and previous work [8], a controller has been designed. With the help of this controller we build a training set of 70 examples and a testing set of 140 examples. Figure 1 shows good parking trajectories performed by the controller and a cross-section of the controller response. The response function has proved to be discontinuous and very irregular.

Simulation results show that feedforward nets are not adapted to approximate such a function. The following table shows the square errors for the different architectures:

No learning algorithms for recurrent nets based on gradient descent were able to give a correct net: local optima were encountered too early and the car controllers are chaotic. But the evolutionary method has provided recurrent nets that outperform feedforward nets: with 3 units the square errors on the training set and the testing set are respectively 0.54 and 0.70, with 5 units they are 0.49 and 0.63. Practically, the behavior of the feedforward controller is not robust: it is very sensitive to the speed of the car (the parameter V) and it does not work well if the initial position of the car is not in the training set. The recurrent nets show robust behavior: dividing by 2 or multiplying by 2 the speed does not change the performance and the controllers work well when the initial position is not in the training set (this explains why the square error on the testing set are low).

6 Conclusion

$\mathcal{R}_G^c$ has been defined as the set of functions computable by convergent recurrent nets. A new learning algorithm searching the closest element $\mathcal{R}_G^c$ to an arbitrary function F has been devised. It constrains the set of fixed points of a recurrent network and assures a convergent activation dynamic. Based on an evolutionary selection mechanism, it breeds populations of sets of fixed points $A = (a^k)_k$. Each set A is evaluated by computing the closest set of fixed points corresponding to a convergent net. This algorithm allows us to handle all the convergent recurrent nets. Simulation results show that these nets are more advantageous tools than feedforward nets, and illustrate the mathematical results concerning their computational power. Additional works are needed to estimate the computational cost of the learning processes. The promising results point out that it seems an important direction for future research.

References

[1] R. P. Agarwal. *Difference Equations and Inequalities : Theory, Methods and Applications*. Pure and Applied Mathematics. Springer, E. J. Taft and Z. Nashed ed. edition, 1991.

[2] American Institute of Physics. *Neural Information Processing Systems Conference*, volume 1, 1987.

[3] Y. Bengio, P. Simard, and P. Frasconi. Learning Long-Term Dependencies with gradient is difficult. *IEEE Trans. on Neural Networks*, 1993. Special issue on Recurrent Networks.

[4] O. Christensen. Frames and pseudo-inverses. *J.Math.Anal.Appl.*, 2, 1995.

[5] C. Gégout. Improvement of Multilayer Perceptron Trainings with an Evolutionary Initialization. In *ICANN'95*, volume 2, pages 153–158. EC2, 1995.

[6] C. Gégout. Stable and Convergent Dynamics for Discrete-time Recurrent Networks. In *World Congress of Nonlinear Analysts*. IFNA, July 1996. to be published in 1997.

[7] M. W. Hirsch. Convergent Activation Dynamics in Continuous Time Networks. *Neural Networks*, 2:331–349, 1989.

[8] M. Schoenauer, E. Ronald, and S. Damour. Evolving Networks for Control. In *Neuronîmes 93*, Paris, 1993. EC2.

[9] H. T. Siegelmann, B. G. Horne, and C. L. Giles. Computational capabilities of recurrent narx neural networks. Technical Report 95-78, University of Maryland, College Park, Md, 1995. Accepted also IEEE Trans. on Syst., Man and Cybern.

[10] P. Y. Simard. *Learning State Space Dynamics in Recurrent Networks.* PhD thesis, University of Rochester, New York, mars 1991. 383.

Evolutionary Optimization of Neural Networks for Reinforcement Learning Algorithms

H. Braun and T. Ragg
Institute of Logic, Complexity and Deduction Systems
University of Karlsruhe, 76128 Karlsruhe, Germany
email: {braun,ragg}@ira.uka.de

Abstract

In this paper we study the combination of two powerful approaches, evolutionary topology optimization (ENZO) and temporal difference learning (TD(λ)) which is up to our knowledge the first time. Temporal difference learning was proven to be a well suited technique for learning strategies for solving reinforcement problems based on neural network models, whereas evolutionary topology optimization is concurrently the most efficient network optimization technique. On two benchmarks, a labyrinth problem and the game Nine Men's Morris, the power of the approach is demonstrated. We conclude that this combination of evolution and reinforcement learning algorithms is a suitable framework that uses the advantages of both methods leading to small and high performing networks for reinforcement problems.

1 Introduction

In the last decade many approaches for reinforcement learning problems have been proposed. Dynamic programming and temporal difference learning [1, 8] were shown to be powerful techniques to explore and learn about the environment by experience, e.g., for control tasks [6] or learning a winning strategy in a game [4, 9]. Often, neural networks are used as learning agents, which learn to assign each state a scoring value.

When using neural networks for these tasks one is facing several problems: Firstly, finding an appropriate input coding for the network is a nontrivial task. Secondly, determining the optimal network complexity, i.e., size of the network and degrees of freedom, is important for the generalization behaviour, but not addressed by standard training methods. The combination of evolution and learning forms a good framework to overcome these difficulties. A suitable topology can be evolved automatically and thus avoids the time consuming search by trial end error methods. In the last years we developed ENZO [2, 7], an evolutionary neural network optimizer which surpasses other optimizing algorithms with regard to performance and scalability [5]. ENZO implements an evolutionary algorithm as meta-heuristic that uses gradient descent to optimize networks locally.

In this paper we propose the combination of these two powerful techniques, evolutionary topology optimization and temporal difference learning which is up to our knowledge the first time. The combination with our evolutionary algorithm ENZO provides the following advantages:

1. *Elimination of both redundant and irrelevant features:* By the removal of unnecessary input units the coding is reduced to the relevant features. By that and the removal of hidden units the analysis of the neural network is simplified.

2. *Improved performance:* Smaller topologies have in general a better generalization behaviour. In particular, the elimination of irrelevant input units improved the generalization capabilities in many of our examined benchmark problems. Moreover, the generalization performance can easily be integrated into the fitness function of the evolutionary algorithm.

3. *Smaller networks:* Cost for hardware realizations are minimized and the network analysis is simplified.

4. *Faster networks:* Smaller networks can be computed faster on standard (sequential) hardware. A speedup of the computation is especially important for real time applications. Moreover it allows for further improvement by depth search.

We examine our approach using two benchmarks: Planning strategy for a two dimensional labyrinth problem (Figure 2), which allows for easy visualization, and learning a winning strategy in a game. We chose the endgame of Nine Men's Morris (Figure 4), which is a non-trivial game, since it has about 56000 essentially different board positions, but still solvable by exhaustive search [3]. That is, results can be evaluated using a database, thus receiving an exact measure of performance.

2 Reinforcement Learning

Reinforcement problems can be characterized as a search for a policy that optimizes the reinforcement signals while interacting with a given environment. In contrast to supervised learning the target actions are not given by a teacher, but have to be learned by experience. Formally, they can be described by: *a set of states* S, e.g., the fields of the labyrinth problem in Figure 2, *a set of actions* A, by which a successor state can be chosen in each state, e.g., a movement to one of the four neighbour fields in Figure 2, and *a reinforcement signal* $r(s,s')$ which is caused by selecting the successor state s' in state s.

A policy $\pi(s)$ of choosing an action is to select in state s_t a favourable successor state s_{t+1}. The task is now to optimize π, such that

$$V^{\pi}(s_0) = \sum_{n=0}^{\infty} \gamma^n r(s_t, \pi(s_t))$$

is minimized for all states s_0, where V^{π} is called the value function for a given policy π and $\gamma \in [0,1]$ is a decay parameter for weighting future reinforcement signals. The value function V^{π} can be computed by the following iteration formula:

$$\begin{aligned} V^{\pi}(s_t) &= r(s_t, \pi(s_t)) + \gamma V^{\pi}(s_{t+1}) \\ &= r(s_t, \pi(s_t)) + \gamma V^{\pi}(\pi(s_t)) \qquad (1) \end{aligned}$$

The dynamic programming approach for solving this task is to construct iteratively the value function V^{π^*} for the optimal policy π^*. To each value function V we can construct a policy π by choosing always the successor with the lowest value:

$$\begin{aligned} &r(s,\pi(s)) + \gamma V(\pi(s)) = \\ &\min\{r(s,s') + \gamma V(s') | s' \text{ successor of } s\} \qquad (2) \end{aligned}$$

If V is the value function of a given policy π', i.e. $V = V^{\pi'}$, it is easy to prove that the policy π induced by $V =$ trough Equation (2) is better than π', whenever there is a state s such that Equation (2) is not fulfilled. Moreover, it can be proven that policy π' was already optimal, when for all states s Equation (2) is fulfilled [1]. For a finite set of states S there are only finitely many different policies and we may conclude, that the iteration of this policy meliorization converges to an optimal strategy. This approach is called policy iteration (Figure 1). Combining Equations (1) and (2) we get for an optimal policy π^*:

$$\begin{aligned} V^{\pi^*}(s) =& r(s_t, \pi(s_t)) + \gamma V^{\pi^*}(\pi(s_t) \\ =& \min\{r(s,s') + \\ & \gamma V^{\pi^*}(s') | s' \text{ successor of } s\} \qquad (3) \end{aligned}$$

Vice versa, it can be proven that every value function satisfying Equation (3) is already an optimal value function. The method of value iteration constructs an optimal value function V by iterating Equation(3) for all states until a fix-point is reached (fulfilling Equation(3)).

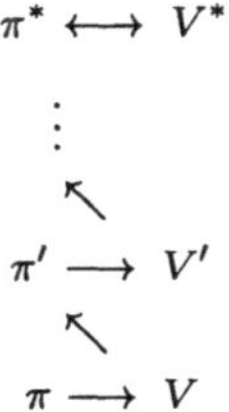

Figure 1: The method of policy iteration. The transition from π to V is based on exploring the state space, while the value function V induces in turn a policy π'. If no more changes take place, an optimal policy π^* is found.

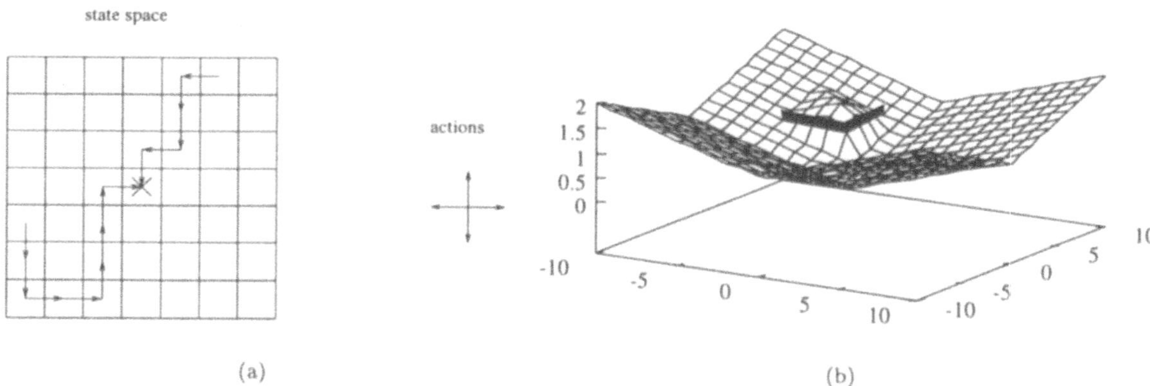

Figure 2: a) Part of the state space of the two-dimensional labyrinth problem and possible actions in each state. b) Graph of a perfect value function for the two dimensional labyrinth problem (20x20). The two black bars are an obstacle. The agent has to find his way from an arbitrary point to the point in the middle, which has the lowest value.

Temporal difference learning is an efficient method using both approaches (value iteration and policy iteration), where the parameter λ is weighting both: we get value iteration for $\lambda = 0$ and policy iteration for $\lambda \to 1$. The updating rule is given by

$$V^{k+1}(s_0) = \sum_{t=0}^{\infty}(1-\lambda)\lambda^{-t}\, V_t^{\pi(V^k)}(s_0) \qquad (4)$$

where $V_t^{\pi(V^k)}$ is a t-step approximation of the value function $V^{\pi(V^k)}$ for the policy induced by V^k:

$$V_t^{\pi(V^k)}(s_0) = \sum_{i=0}^{t}\gamma^i r(s_i, s_{i+1}) + \gamma^t V^k(s_t)$$

Note that the weighting factors $(1-\lambda)\lambda^{-t}$ sum up to 1, i.e., $\sum_{t=0}^{\infty}(1-\lambda)\lambda^{-t} = 1$. Moreover, for the convergence to the optimal value function it is not necessary to update V^{k+1} in Equation (4) for all states in each iteration (e.g. one is already sufficient), but every state has to be considered repeatedly.

All three methods based on dynamic programming are computable in polynomial time (relative to the number of states). Nevertheless, this is, in typical application problems intractable, because the number of states is too large. The idea of using neural networks for computing the value function is to speed up the learning process of the value function by the generalization capability of neural networks, i.e. updating the value for a state influences all states in its surrounding (assuming similar states should have similar values). The weight updating rule is given by minimizing the quadratic difference between the left and right side of the update rule of TD(λ) through gradient descent for each state s_0. If we perform the weight update for all states s_t following a randomly chosen starting state s_0 by using the actual policy $\pi(V^N)$, then the associated equations can be transformed in the more efficient computable equation (known as TD(λ) update rule):

$$\Delta w_t = \alpha \left(r(s_t, s_{t+1}) + \gamma V^N(s_{t+1}) - V^N(s_t)\right) \cdot \sum_{k=0}^{t}\lambda^{t-k}\nabla_w V^N(s_k), \qquad (5)$$

This approach has been proven to be a powerful learning method for sophisticated reinforcement problems. However, the success depends strongly on the chosen neural network topology: Under the constraint of the limited learning time we have to balance the tradeoff between increasing network size for approximating fair enough the optimal value function and decreasing network size for speeding up learning by coarser generalization. Moreover, each TD(λ) learning step is proportional to the network size, such that smaller network size allows more learning steps in a given time limit. By the intertwining of temporal difference learning and network topology optimization we may expect more robust and better performing solutions.

3 Evolutionary Topology Optimization

Derived from the biological example the basic principles of the evolutionary optimization may be summarized as follows [5]:

Initially, a population of $\mu \geq 1$ networks is created. Each network is evaluated according to a fitness function which in turn serves as optimization criterion (evaluation). For each generation $\lambda \geq 1$ offsprings are generated by first selecting their parents preferring the fitter ones, and then applying a mutation operator. Based on a strategy to rank the weights, respectively units, with respect to their importance, e.g., size, increase in error, the mutation removes weights, respectively units, by chance preferring the ones with higher ranking. The resulting offspring networks are trained using temporal difference learning and also evaluated according to the fitness function. Several parent networks might then be replaced by fitter offsprings. The loop is repeated until a user-defined stop criterion, in general a fixed number of generations, is fulfilled. The local optimization based on temporal difference learning is performed by first initializing the offspring network with the weights of the parent, that is the parent's knowledge is transferred to the offspring thus by exploiting Lamarckism a more explorative search is possible (speedup of 10). Than we alternately create a sequence of states using the current (fixed) policy and train the sequence. After each cycle the weights are updated according to Equation (5). By that the policy is changed.

4 Experimental Investigations

The labyrinth problem with obstacles (Figure 2) is solvable by a neural network with two hidden layers each with 8 units, using temporal difference learning. We trained 100 randomly initialized networks for 20000 epochs, where the patterns trained in each epoch are the states which were visited starting from an

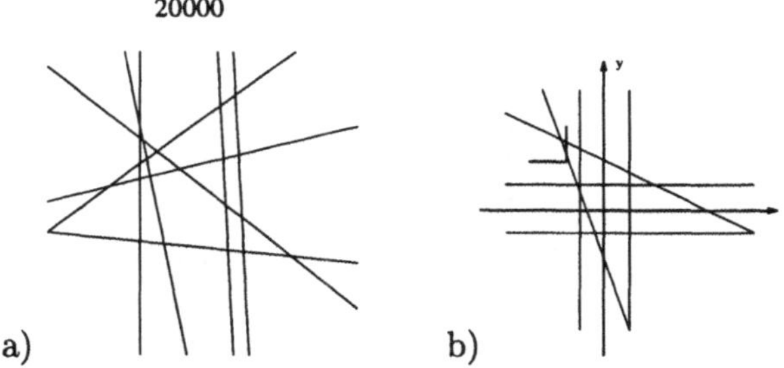

Figure 3: Separating lines of the first hidden layer of networks with optimal performance: a) Standard trained network; after 20000 training epochs no more essential changes take place. b) Evolved network; the lines separate the input space in a sensible way.

arbitrary state consecutively until a termination state or the maximal number of states was reached. Only 12 networks learned an optimal value function (Figure 2b), 23 networks chose some suboptimal moves, but always reached the target point. The other 65 networks got stuck in local minima. Figure 3a shows the hyper planes computed by the hidden units.

Using the combination of evolution and temporal difference learning led to the impressive results shown in Figure 3b. An optimal value function is found using the *minimal* network size, where the role of each hidden unit is explicit and easily analyzable: Four hidden units are used (and necessary) to solve the labyrinth problem without obstacle, while the two other units form the separating lines around the obstacle. In other words, four hidden units form the coarse structure of the value function, whereas the remaining two hidden units handle the exception at the obstacle. Furthermore, the evolution always converged to this optimal solution.

Nine Men's Morris is a typical instance of the class of deterministic two-player games. The game is played on a board as shown in Figure 4.

The endgame has about 96 million possible board positions which can be reduced by various symmetries to 56922 essentially different configurations. Winning the game may take up to 26 half-moves, if optimal play is assumed. The board positions were encoded in 51 features, whereas the output of the network is interpreted as the associated value of the board position. The network topology was

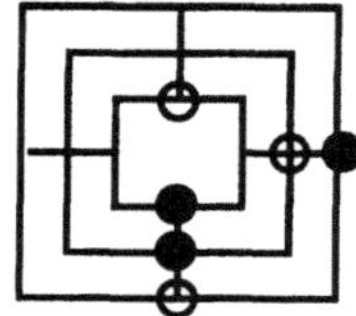

Figure 4: An endgame position of the game Nine Men's Morris. The player is allowed to move a piece to any other free position. A player wins if he has three pieces in a row on one line (called "a mill"). For the shown board position white is to move and wins in 25 moves, if both players play optimal.

51-30-10-1 with tangens hyperbolicus as activation function for the hidden units and a linear activation function for the output unit.

Using an improved version of Sutton's $TD(\lambda)$ algorithm [8] an almost optimal winning strategy for the endgame could be learned by a neural network [4]. With the combination with ENZO, the input dimension could be reduced to a third of the original input space. Furthermore, the size of the evolved network was decreased to about 10% of the original size without loss of performance. This speedup by a factor of 10 is important for game playing programs as a real time application. If the time constraint allows for additional evaluations of the value function, the playing performance can be improved by depth search.

5 Conclusion

The approach we presented combines evolutionary topology optimization (ENZO) and temporal difference learning. On the labyrinth benchmark we could demonstrate that the combination of both methods always leads to a sensible network architecture of the hidden units reflecting the properties of the problem. On the other hand, we could reduce the input dimension for the Nine Men's Morris benchmark from 51 to 18. This is a non-trivial task, since it is whether obvious which features are important for an optimal strategy nor easily computable with statistical methods since the target values are a priori unknown. Furthermore, by exploiting Lamarckism, the approach allows for a more explorative search. Other techniques of knowledge transfer are currently under investigation. We may conclude that our approach exploits the advantages of both temporal difference learning and evolutionary optimization. This combination of evolution and reinforcement learning is a suitable framework that leads to small and high performing networks, which moreover allow in case of real time tasks for further improvements of the policy by applying depth search methods.

References

[1] A. G. Barto, S. J. Bradtke, and S. P. Singh. Learning to act using real-time dynamic programming. *Artificial Intelligence*, (72):81–138, 1995.

[2] Heinrich Braun and Joachim Weisbrod. Evolving feedforward neural networks. In R. F. Albrecht, C. R. Reeves, N. C. Steele, (editors), *Artificial Neural Networks and Genetic Algorithms*, Wien New York, 1993. Springer-Verlag.

[3] R. Gasser and J. Nievergelt. Es ist entschieden: Das Mühlespiel ist unentschieden. *Informatik Spektrum*, 5(17):314–317, 1994.

[4] T. Ragg, H. Braun, and J. Feulner. Improving temporal differnce learning for deterministic sequential decision problems. In *Proceedings of the ICANN '95, Paris*, 1995.

[5] T. Ragg, H. Braun, and H. Landsberg. A Comparative Study of Neural Network Optimization Techniques. In this volume, pages 341–345.

[6] M. Riedmiller. Learning to Control Dynamic Systems. In *Proceedings of the Europeean Meeting on Cybernetics and System Resarch EMCSR, Vienna*, 1996.

[7] J. Schäfer and H. Braun. Optimizing classifiers for handwritten digits by genetic algorithms. In D. W. Pearson, N. C. Steele, R. F. Albrecht (editors), *Artificial Neural Networks and Genetic Algorithms*, pages 10–13, Wien New York, 1995. Springer-Verlag.

[8] R.S. Sutton. Learning to predict by the method of temporal differences. *Machine Learning*, 3:9–44, 1988.

[9] G. Tesauro. Practical issues in temporal difference learning. *Machine Learning*, 8:257–277, 1992.

Generalising Experience in Reinforcement Learning: Performance in Partially Observable Processes

C. H. C. Ribeiro
Neural Systems Engineering Group, Imperial College,
Exhibition Road, London SW7 2AZ, England
Email: c.ribeiro@ic.ac.uk

Abstract

Reinforcement learning algorithms have been used as reasonably efficient model-free techniques for solving small, perfectly observable Markov decision processes. When perfect state determination is impossible, performance is expected to degrade as a result of incorrect updates carried out in wrong regions of the state space. It is shown here that in this case a modified spreading version of Q-learning which takes into account its own uncertainty about the visited states is advantageous if the spreading mechanism fits a measure of similarity on the action-state space. In particular, an agent with an active perception capacity can use an expectation of similar past histories leading to similar results as a justification for this spreading mechanism.

1 Introduction

In the last few years, reinforcement learning (RL) techniques have been put forward as convenient model-free methods for programming agents to solve discrete Markov decision processes. This problem can be formally described by a tuple $\langle \mathcal{X}, \mathcal{A}, \mathcal{P}, \mathcal{R} \rangle$, where $\mathcal{X}$ is the set of process states, $\mathcal{A}$ is the set of possible actions to be taken by the controlling agent, $\mathcal{P} : \mathcal{X} \times \mathcal{A} \rightarrow \mathcal{X}$ is the *state transition function*— which gives for every visited state $x_t \in \mathcal{X}$ and agent action $a_t \in \mathcal{A}$ a probability distribution $P(x_{t+1} \mid x_t, a_t)$ over all the possible x_{t+1}, and $\mathcal{R} : \mathcal{X} \rightarrow \Re$ is a *reward function* that gives a real-valued immediate reward $r(x)$ for visiting state x.

The agent's goal is to find a policy of actions that optimises some performance criterion (usually maximisation of expected discounted reward). Our concern here is Q-learning [6], a well-known recursive RL algorithm. It assumes the existence of a tabular representation where estimates of action values $Q(x, a)$ are stored. Each action value represents the expected discounted reward to be gained by choosing action a in state x and following an optimal policy thereafter.

Q-learning — and, by extension other RL techniques — suffer from a drawback: One of the conditions for convergence is an infinite number of visits to every possible state-action pair. This burden can be relieved by considering a) that nearly optimal action policies can many times be found long before the action values actually converge and b) that generalisation strategies (neural networks or adaptive encoding schemes) provide quicker ways of "filling-up" the table of action values. Nevertheless, the whole process is usually not controlled: generalisation methods that work well in certain tasks do not perform satisfactorily in others, and there is no way to find out an optimal stopping condition that guarantees nearly optimal policies.

When information about states is incomplete, the theoretically sound approach of accessing all the relevant past information and considering this resulting *information vector* as an extended state of the process may be unfeasible due to the size of this new state space. Not only memory requirements may be excessive, but more importantly convergence can get very slow. In practice, the controller has always limited information resources to discriminate among the states it visits, and if the information for perfect discrimination is not available, performance degrades because of updates carried out in wrong regions of the state-space (the so-called perceptual aliasing phenomenon [7]). Nevertheless, the controller may somehow compensate this weakness if it is equipped with an *attentional* capacity that directs its observation focus towards relevant parts of the sensory field at different instants of time. It is shown here that an agent with this capacity can benefit from using a modified spreading version of Q-learning if an expectation of similar past histories leading to similar results is satisfied. This idea is implemented by defining an attentional mechanism coupled to an agent with a limited memory capacity and by spreading learned information along similar regions of an extended state space corresponding to the *approximate* information vectors defined by the memory.

In the next section, we review a modified Q-learning algorithm that generalises along the state-action space and that is guaranteed to converge. We will then present an extension of this technique for use with perceptually active agents.

2 The QS Algorithm

Consider the following variant of Q-learning. At time t, the agent:

• visits state x_t;
• selects action a_t and observes the next state x_{t+1};
• for every $x \in \mathcal{X}$ and action $a \in \mathcal{A}$ updates $Q_t(x,a)$ according to:

$$\Delta Q_t(x,a) = \alpha_t \sigma_t(x,a)[r_t + \gamma V_t(x_t) - Q_t(x,a)] \quad (1)$$

• repeats steps above until convergence;

where $V_t(x_t) = \max_a Q_t(x_{t+1}, a)$ and $\sigma_t(x,a)$ is the *spreading function* $(0 \leq \sigma_t(x,a) \leq 1)$. The standard Q-learning algorithm corresponds to Equation 1 with $\sigma_t(x,a) = \delta(x,x_t)\delta(a,a_t)$, where $\delta(.,.)$ is the Kronecker delta function. A spreading mechanism based on a similarity function s on the state space can be defined through $\sigma_t(x,a) = s(x,x_t)\delta(a,a_t)$.

The convenience of using a particular spreading function depends on how well it fits the optimal action values. Unfortunately, it is hard to deduce *a priori* a function perfectly suited to the problem at hand. Thus, using spreading involves the risk of choosing an inappropriate $\sigma_t(x,a)$. However, if the intensity of spreading is reduced along time at a certain rate both the advantages of using the spreading mechanism and the convergence properties of standard Q-learning are kept. We call this algorithm the *QS algorithm*.

A proof of convergence and results for QS in perfectly observable processes can be found in [2]. Briefly, the results showed that it leads to better final policies than standard Q when the spreading function fits the world to a certain extent. This superior performance is achieved after an initial phase when QS may actually perform worse than standard Q. We interpreted this result in the following way: this initial phase consists of a smoothing of the action value mapping that can imply a destructive interference with previous updates based on reliable temporal reinforcement. Once this smoothing process is over, usual Q-learning takes place in an environment where there is some indication — more or less valuable depending on the degree of smoothness of the optimal action values — about the possible shape of the action value mapping. Convergence to better policies is then achieved through the use of this additional global information.

3 QS in Partially Observable Processes

Consider now the interaction agent↔process in the more general case of partial observability. A possible model for an agent with an attentional capability is illustrated in Figure 1. The memory summarises all the information available to the agent at time t: the past observations $\{o_{t-1}, o_{t-2}, \ldots, o_{t-W}\}$ and the previously taken actions $\{a_{t-1}, a_{t-2}, \ldots, a_{t-W}\}$, where W is the size of the memory window. This information corresponds to an *approximate information vector* $\tilde{I}_t$, possibly insufficient for perfect state discrimination. The agent somehow compensates its memory limitation by using an active attentional system that defines how it is going to observe the next process state. This attentional system is simply a function associating available information and attentional settings e, but as it defines the attention for the *next* state (*i.e.*, the state at time $t+1$), it can include the action a_t as an additional piece of available information. Any observation o_{t-k} thus depends both on the real state x_{t-k} and on the attentional setting e_{t-k-1}.

Once the approximate information vector is de-

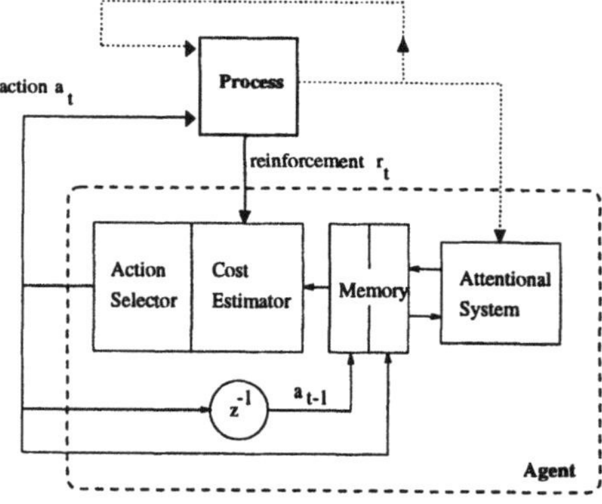

Figure 1: A model for the interaction between an attentional agent and a process.

fined as an extended state seen by the agent, the situation is rather similar to the one for Q-learning on perfectly observable processes, but with the real state replaced by the information vector. The problem, however, is the exponential increase in the size of the state space: for instance, a two-action agent with memory size $W = 3$ acting on a 25 states process on two state variables with attentional settings corresponding to the observation of each of these variables will have 80,000 possible information vectors (instead of just 25). The increase in the number of possible action policies can make a search — even when it is directed as in RL — absolutely hopeless [3, 5]. Traditional reinforcement learning is actually too conservative to deal with this situation, and a capacity to generalise experience through the use of some spreading mechanism is essential.

Consider then a modification to the QS algorithm that spreads learned information among extended states (information vectors) that have a common characteristic. For this case, similarity can be defined in terms of past history instead of state space closeness, as in Section 2. At time t, the agent:

- makes an observation o_t associated with current state x_t;
- selects action a_t and defines attentional setting e_{t+1};
- makes new observation o_{t+1} associated with the consequent state x_{t+1};
- for every $\tilde{I}$ and action $a \in \mathcal{A}$ updates $Q_t(\tilde{I}, a)$ according to:

$$\Delta Q_t(\tilde{I}, a) = \alpha_t \sigma_t(\tilde{I}, a)[r_t + \gamma V_t(\tilde{I}_t) - Q_t(\tilde{I}, a)] \quad (2)$$

- repeats steps above until convergence;

The similarity in terms of past history is represented by a spreading function of the form:

$$\sigma_t(\tilde{I}, a) = f_t(\bar{o}_t) \prod_k \delta(o_{t-k}, o_{-k})\delta(a_{t-k}, a_{-k}) \quad (3)$$

where $\bar{o}_t$ is any observation derived from an attentional setting different from the one adopted for generating the observation o_t. The product of Kronecker deltas means that spreading occurs only among observations with similar past histories, and the function f_t defines the intensity and temporal behaviour of the spreading mechanism.

3.1 Results

Consider an agent that moves in a stochastic world represented as a $X^{c1} \times X^{c2}$ states edged grid, trying to avoid trapping states from which transitions to other states are impossible. At each time step t, one out of two possible actions -1 or $+1$ must be chosen by the agent and, as a consequence, it moves to a neighbouring state x_{t+1} with a probability $P(x_{t+1} \mid x_t, a_t)$. The reinforcement received by the agent when in a trapping state is -20, otherwise it is zero. A temporal discounting factor $\gamma = 0.95$ was used in all experiments.

The agent is not able to observe x_t: instead, it sees only one of its coordinates x_t^{c1} or x_t^{c2} (chosen according to its attentional policy) and tries to control the process based on the new set of observations (the remotest observation is removed so that the memory contents can be shifted along the window to accommodate the new observation defined by the new attentional setting).

Experiments were carried out in a 3×3 world with one trapping state, randomly selected transition probabilities and a $W = 2$ memory window (that corresponds to a set of 864 possible information vectors). Three possible attentional policies E_1, E_2 and E_3 were selected in advance and assessed by the average (with respect to the possible action policies) estimation error they led to in a previously

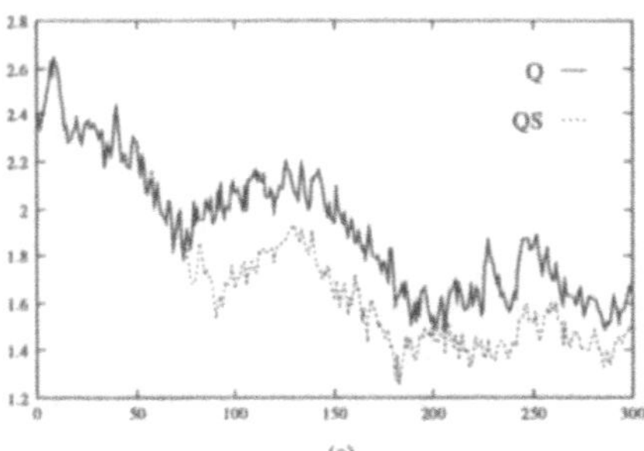

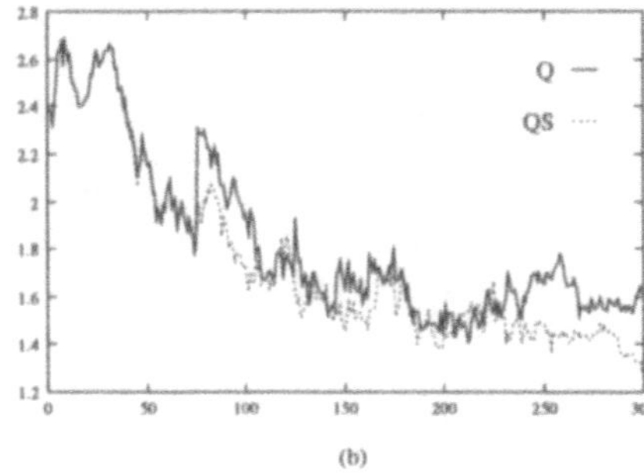

Figure 2: Learning curves for Q and QS. In (a), the sequence of attentional policies adopted is {E1,E2,E3}. In (b), the sequence is {E3,E2,E1}.

designed maximum likelihood estimator. This assessment permitted the ordering of policies according to state estimation quality. It is expected that, on average, the best attentional policies will lead to better agents: an agent with superior estimation capabilities must be able to learn comparatively better action policies. Figure 2 shows the learning curves for standard Q and QS averaged over 50 learning courses. The attentional policies were changed at steps 100 and 200. In both cases the learning rate α decreases linearly from 0.7 to 0.01. For the QS variant, the additional free parameter is the spreading function f_t. As the agent cannot define what should be a 'most likely' observation $\bar{o}_t$ (it sees only o_t), the function f_t must be constant in the observation space, for a given t. A linearly decreasing function with $f_0 = 0.5$ and $f_{150} = 0.0$ was chosen.

The results show that both when the agent gets worse attentional policies along time (sequence $E1$, $E2$, $E3$) or in the opposite case, performance of the QS version tends to be superior.

3.2 Discussion

An agent that uses QS pays a price for taking this risk: many information vectors may be incorrectly updated as a result of the spreading process. However, if update information is somehow reduced along time, a situation similar to the one described in Section 2 is likely to occur (case (b), Figure 2). On the other hand, if the agent uses worse attentional policies at later stages (case (a), Figure 2), the correct indications learnt from the initial stages — when better attentional policies were used — are carried over to facilitate the action policy learning for the new attentional policies. Spreading effectively reduces the demands on the learning of attentional policies by propagating information from one attentional policy to the other. In general, the conditions for efficient use of a spreading mechanism are twofold: first, there must be a similarity concept representable as a spreading function, which in this case is the expectation of similar results for similar past histories. Second, the spreading mechanism must decrease along time to guarantee convergence of the QS algorithm.

4 Conclusion

Using QS in processes where the action value mapping does not have strong discontinuities seems to be advantageous. It was shown here that a particular case of similarity that provides a smooth action value mapping is the expectation of similar future consequences for similar past histories in an active agent on a partially observable system. The idea of using spreading provides greater flexibility when compared to connectionist approaches, and such flexibility allows for convenient ways of deciding *how* to spread information and *when* to stop it. Naturally, spreading does not solve the storage problem (a look-up table is still needed).

A possible enhancement for the algorithm would be a finer control over the interference process carried out by QS. It might be exerted through a selective spreading mechanism based on frequency of state visitation. Although the choice of accurate spreading parameters is usually not critical due to the implicit capacity of the QS algorithm to recover from interference errors, there may be situations where choosing adequate spreading is difficult.

Finally, it is important to compare the performance of this algorithm with others that have been proposed to generalise experience, such as [1, 4].

References

[1] S. Mahadevan and J. Connell. Automatic programming of behavior-based robots using reinforcement learning. *Artificial Intelligence*, 55:311–365, 1992.

[2] C. H. C. Ribeiro and C. Szepesvári. Q-Learning combined with spreading: Convergence and results. In *Procs. of the ISRF-IEE International Conf. on Intelligent and Cognitive Systems (Neural Networks Symposium)*, pages 32–36, 1996.

[3] S. J. Russell and P. Norvig. *Artificial Intelligence: a modern approach*. Prentice-Hall, 1995.

[4] R. S. Sutton. Generalization in reinforcement learning: Succesful examples using sparse coarse coding. In Daniel S. Touretzky, Michael C. Mozer, and Michael E. Hasselmo, editors, *Advances in Neural Information Processing Systems 8*, pages 1038–1044. MIT Press, 1996.

[5] C. K. Tham. *Modular On-Line Function Approximation for Scaling Up Reinforcement Learning*. PhD thesis, University of Cambridge, 1994.

[6] C. J. C. H. Watkins. *Learning from Delayed Rewards*. PhD thesis, University of Cambridge, 1989.

[7] S. D. Whitehead and D. H. Ballard. Active perception and reinforcement learning. *Neural Computation*, 2:409–419, 1990.

Optimal Control of an Inverted Pendulum by Genetic Programming: Practical Aspects

F. Gordillo and A. Bernal
Escuela Superior de Ingenieros Industriales, Universidad de Sevilla,
Avda. Reina Mercedes s/n. 41012 Sevilla, Spain
E-mail: gordillo@esi.us.es

Abstract

During the past several years, numerous papers and applications designing control systems with genetic algorithms (GAs) have been written. Many of these studies end when the simulated behaviour of the system with the controller is satisfactory. They suppose that the final stage of application of the controller to the real system will be similar to the one from the traditional design.

This paper explains the conclusions of the real application of one such publication: the control of an inverted pendulum using a well-known variant of GAs: genetic programming (GP). The aim of this paper is to study the existence of possible special problems in the application stage of genetic-designed controllers. As will be seen, the application stage is more difficult for GAs than for traditional methods, and more knowledge is needed about the system.

1 Introduction

In order to apply GAs to the resolution of a particular optimization problem, several stages must be carried out. The procedure appears in Figure 1. These stages are:

1. Definition of the fitness function. When designing a control system, the controllers are usually simulated by a model of the system to assign a goodness measurement to each individual. Therefore, at this stage, three different elements must be defined:
   (a) The model of the system.
   (b) The performance index to be evaluated during each simulation.
   (c) The conditions under which the simulation is performed.
2. Specification of the parameters which define the GA, such as population size, crossover probability, etc.
3. Execution of the algorithm.
4. Verification of the results in order to know if the algorithm has converged correctly. If it does not converge correctly, another iteration is needed and changes must be made in stages 1 and/or 2.

In this process, two nested loops exist in which iterations are carried out:

- The inner loop is not represented in Figure 1 and corresponds to the iterations characteristic of GAs. They are performed automatically by means of a computer program.
- The outer loop appears in Figure 1. It is directed toward the improvement of the convergence of the GA. That is, the purpose of this

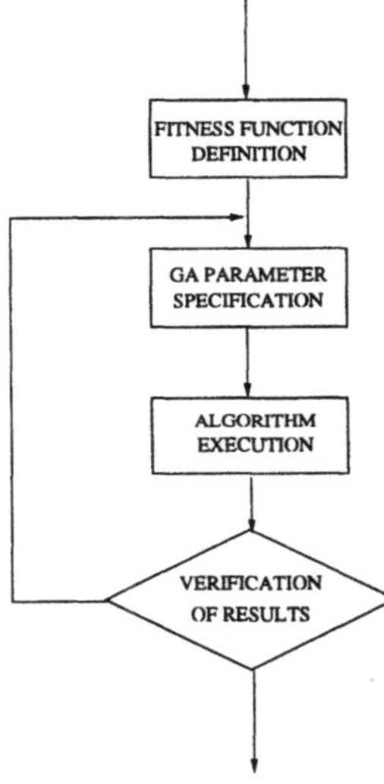

Figure 1: Stages of the resolution process of an optimization problem with GAs.

loop is to define a GA which, once solved, obtains individuals with better fitness. Usually, this loop is carried out by hand. It must be pointed out that this loop is not directed toward the improvement of the practical performance of the controller but the simulated one.

If this procedure is successful, it gives a controller with an optimal (or near optimal), simulated behaviour. But this simulation does not mean a correct real behaviour, above all, if the experiments are the same which have been carried out during the run of the GA. To complete the design, a new stage is necessary: verification of the real performance of the controller with the physical system. Usually, at this stage, another loop appears as is depicted in Figure 2. In this loop, iterations are performed so the real system will behave correctly. As this stage is also necessary in traditional control system designs, it is often supposed that this phase would be similar to the traditional one and that its features are outside the characteristic issues of a genetic-based design. The main conclusion of the study presented in this paper is that, at least in the problem considered, this final stage has its own features when GAs are used, making the design more difficult.

There exist several reasons which confirm the necessity of studying the particular characteristic of the implementation stage when using GAs:

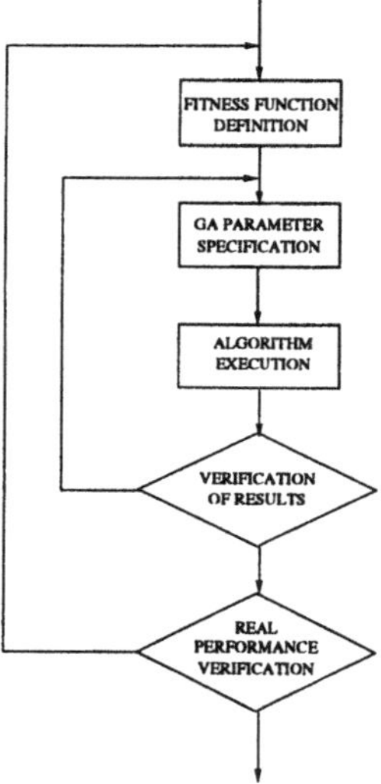

Figure 2: Stages of the resolution process of an optimization problem with GAs with physical implementation.

- Usually, when a controller that has been tested by means of simulations is faced with the real world, some of the limitations of the model used appear and the design must be revised. Since a genetic-based design intensively exploits the model, its faults may be more harmful than when using other design methods.

- The high degree of freedom that is normally associated with a genetic design, and especially when using GP, may lead to unrobust controllers if this characteristic is not taken into account during the design [1]. Robustness is a property needed in the implementation phase.

- As can be seen in Figure 2 the implementation stage loop obliges the fitness function (that is, the model, the performance index, and the experiments) to be redesigned. These elements are intrinsic parts of genetic design.

2 The Inverted Pendulum

The system for which a controller must be found is an inverted pendulum (also called cart-pole balancer). The system appears schematically in Figure 3.

The system is a wheeled cart which can move linearly along a track and an inverted pendulum that rotates with the motion of the cart. The objective is to keep the pendulum vertical and to locate the cart at a desired position. The action variable is the force which is applied to the cart by means of an electric motor. Two sensors are available to measure the linear position of the cart and the angular position of the pendulum.

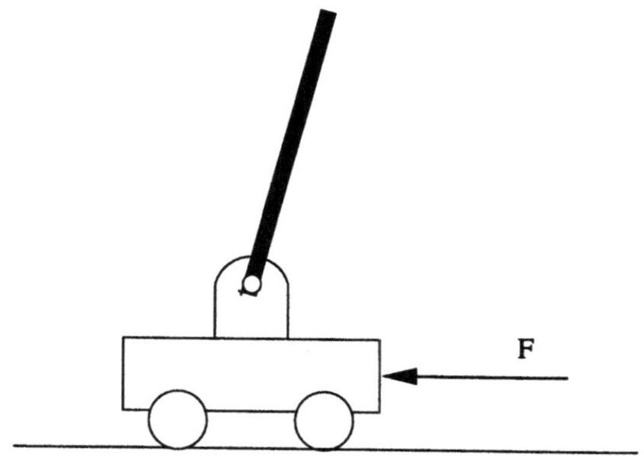

Figure 3: Inverted pendulum.

To control the pendulum, a computer is used with two A/D and D/A converters. The controller is implemented as a program in the computer.

3 Controller Design

3.1 LQR Design

Before designing the controller by means of GP, a traditional design was performed in order to compare the procedures. The traditional method adopted is a well-known, optimal control technique: the *linear quadratic regulator* (LQR) [2].

This method is based on a linear model of the system and a quadratic performance index which penalizes, on one hand, divergences of the output of the system with respect to the desired values and, on the other, the control effort. The result is a linear feedback of the state variable of the system multiplied by a set of gains. Once a model is obtained and a performance index is selected, the resolution of the LQR problem gives the value of these gains.

In this case, the state variables are: the linear position of the cart (x_1), the linear speed of the cart (x_2), the angular position of the pole (x_3), and the angular speed of the pole (x_4). A linear model can be obtained by approximating $\sin(\alpha) \approx \alpha$ and $\cos(\alpha) \approx 1$ when $\alpha \approx 0$. Notice that this is an approximation.

The selection of the performance index has been made by trial and error using simulations so the cart would not exceed the limits of the track.

Once the controller was obtained, it was tested with the real system. Its behaviour was found to be satisfactory with no need for any redesign in spite of the simplifications introduced:

- A continuous time study was carried out, while the introduction of the computer leads to a discrete-time system.
- The model is linear while the system is not.
- Two state variables (the speeds of the track and the pole) are not accessible and are computed from the positions.
- No noise study was carried out.

The correct behaviour of this controller is explained by the good robustness properties intrinsic to every LQR design [6].

3.2 GP-Based Design

In spite of the fact that the LQR design was satisfactory, a new controller was designed using GP. Of course, if a more specific method is possible, GP (or GAs) must not be used. The reason GP is applied to this simple problem is to study the problems associated with this design and to compare them with traditional methods. Another additional result will likely be the improvement of the design. Notice that the design performed in [5] must be redrawn because the available inverted pendulum is not exactly the same.

Initially, a simple model of the pendulum was used [5]. Nevertheless, this model was not as simple as the one used in the LQR design, since linearity is not a necessary characteristic of the model when using GAs. Therefore, the simplifications commented before were not introduced in this model. This fact must give an advantage to the GA design over the LQR design. The initial performance index was chosen to be equal to the LQR index. A complete set of experiments with different initial conditions was used to test each one of the GP algorithms.

The result of this initial design was very unsatisfactory because a large steady-state position error appeared. The cause was the finite length of the experiments which prevented the performance index from reflecting this error. This anomaly was detected even without testing the controllers in the real system. A possible solution would be to increase the length of the simulations but this would result in extending the time of the design. Therefore, an increase in the position-error penalty was selected.

Nevertheless, the results continued to be unsatisfactory. The fitness associated with each individual did not properly reflect the real performance of the associated controller with the real system. This fact was due to defects in the model and in the fitness function as a measurement of the real performance of the controllers. Notice that these defects were not essential in the LQR design, even with a simpler model. Therefore, the model had to be improved to reflect the reality better, so the following phenomena were incorporated to the model: the saturation of the signals, the A/D and D/A conversions, a filter included in the acqusition stage, an observer to estimate the speeds, the dead-zone of the electrical motor, a friction model and a noise model.

With these modifications of the model, the simulated behaviour reflects the real one better but the control obtained with GP showed big position error and big oscillations in the control signal (most industrial processes do not permit large increments in the input variables). This oscillation was also presented in the LQR design but it was increased in most of the controllers obtained with GP. In order to avoid such problems the fitness function was redesigned by hand in a trial and error procedure: the penalty of position errors was increased and increments of the control signal were also included.

In the end, satisfactory solutions were obtained. One interesting result is that the linear feedback of the state variables always appears in the best controllers, at least as a part of the control law (the set of primitive functions was: $\{+,-,\times,\div,\sqrt[3]{\ },\square^3\}$). For example, the best controller obtained is:

$$u(x_1,x_2,x_3,x_4) = 20x_1 + 30x_2 + 100x_3 + 18.35x_4 + 0.068\sqrt[9]{x_2}$$

This controller presents a lot less oscillations than the LQR with similar (but better) behaviour of the angle of the pendulum and the position of the track. Therefore, it can be said that this controller improves the LQR design.

4 Conclusions

Genetic programming was used to obtain the control law of an inverted pendulum with the real application of the controller in mind. The main conclusion of the study is that, at least in the problem considered, the final stage of the design of the controller, involving the verification of the design with real tests, has its own features when applying genetic techniques. If traditional control system design is not completed until real tests are performed, this final stage is more critical in the case of genetic designs. Designs which end with the same simulations which have been carried out during the application of the GA are especially incomplete.

It should not be thought, as a conclusion of this study, that genetic techniques are not useful for designing control systems (in fact a very good controller was obtained), but that a good model is necessary (as it could be thought in advance) and that the design must be completed with real tests. Of course, the usefulness of GP (or GA) designs is manifested when faced with problems which are difficult to solve with traditional methods.

5 Acknowledgements

Supported by DGICYT from the Spanish Ministerio de Educación under grant TAP–94–0491.

References

[1] T. Álamo, F. Gordillo, and J. Aracil. Robust fuzzy control using genetic algorithms. In *Proc. Third European Congress on Intelligent Computing, EUFIT'95*, pages 781–785. Aachen, Germany, 1995.

[2] B. D. O. Anderson. *Optimal Control. Linear Quadratic Methods.* Prentice-Hall, New Jersey, CA, 1990.

[3] M. G. Cooper and J. J. Vidal. Genetic design of fuzzy controllers. In *Proc. 2nd Int. Conference on Fuzzy Theory and Technology.* Durham, NC, 1993.

[4] P. J. Fleming and C. M. Fonseca. Genetic algorithms in control systems engineering. In *Proceedings of the IFAC 1993 World Congress.* Sydney, Australia, 1993.

[5] J. R. Koza. *The Genetic Programming Paradigm: Genetically Breeding Populations of Computer Programs to Solve Problems*, pages 203–321. John Wiley & Sons, 1992.

[6] M. G. Safonov and M. Athans. Gain and phase margin of multiloop lqg regulators. *IEEE Trans. on Automatic Control*, Apr. 1977.

Evolutionary Artificial Neural Networks and Genetic Programming: A Comparative Study Based on Financial Data

S.-H. Chen and C.-C. Ni
Department of Economics, National Chengchi University,
Taipei, Taiwan 11623
E-mail: chchen@cc.nccu.edu.tw, g2258503@grad.cc.nccu.edu.tw

Abstract

In this paper, the stock index *S&P* 500 is used to test the predicting performance of genetic programming (GP) and genetic programming neural networks (GPNN). While both GP and GPNN are considered *universal approximators*, in this practical financial application, they perform differently. GPNN seemed to suffer the *overlearning* problem more seriously than GP; the latter outdid the former in all the simulations.

1 Introduction and Motivation

In this paper, we compare the prediction performance between ***evolutionary artificial neural networks*** (EANNs) and ***genetic programming*** (GP). EANNs can be regarded as a subset of the function space defined by GP, i.e., $Space_{EANN} \subseteq Space_{GP}$. To exemplify this *set relation*, an artificial neural network (ANN) and its corresponding LISP tree representation are depicted in Figures 1 and 2.

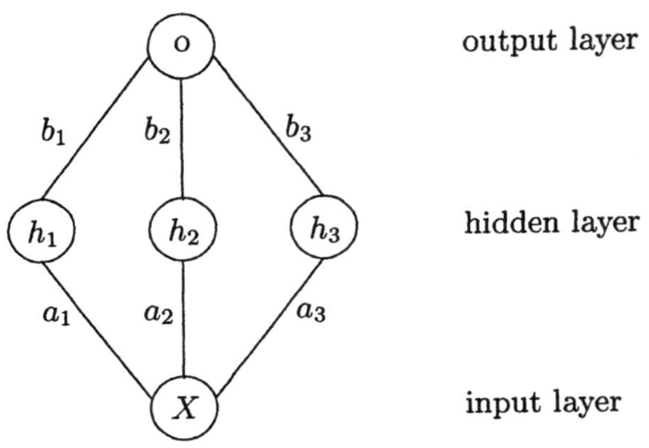

Figure 1: A 1-3-1 ANN Architecture.

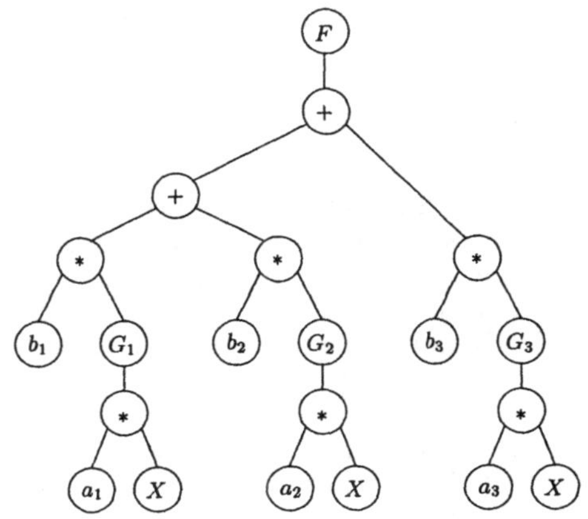

Figure 2: The LISP Tree of ANN(1-3-1).

While this paper only considers *feedforward* ANNs, this illustration can be easily extended to the case of *recurrent* ANNs. The interested reader is referred to [3].

The ANN in Figure 1 is a 1-3-1 architecture, i.e., a single input-output node and one hidden layer with three hidden nodes. The input for the ANN is denoted by X. The transfer function of the hidden nodes and output node are given in Equations (1) and (2),

$$h_i = G_i(a_i \times X), \quad i = 1, 2, 3 \tag{1}$$

$$o = F(\sum_{j=1}^{3}(b_j \times h_j)), \tag{2}$$

where a_i and b_j are weights. In a typical application, G_i can be the identity function and F the *sigmoid function*, the *Gaussian basis function* or the *Gaussian kernel function*. Generally speaking, in the EANN literature, choice of these functions should be *automatically* determined [3].

Since each ANN can be coded as a LISP tree, EANNs can also be implemented with GP. In fact, there has been a growing interest in using GP to evolve the ANN architecture over the last two years [3]. The difference between GPNN and GP lies in the *search space*. The search space defined by GPNN is a network architecture space with additional syntactic restrictions imposed upon the original GP search space. From the perspective of *approximation theory*, these two search spaces are *asymptotically equivalent*, i.e.,

$$\lim_{d\to\infty} Space_{GP}(d) \approx \lim_{d\to\infty} Space_{GPNN}(d), \quad (3)$$

where d denotes the maximum depth of the LISP tree. In other words, if there is no limit set upon *the depth of the tree*, then these two search spaces are approximately equivalent in the sense that both GP and GPNN are *universal approximators*. However, in practice, the upper limit of d always exists. Take Koza [4] as an example; d was set to be 17 in most of the applications. Therefore, in application, it is very likely that $Space_{GP}(d)$ is not the same as $Space_{GPNN}(d)$, and neither is a subset of the other. Even if they are the same, the difference in representation (syntax) may cause these two learning schemes to have such different dynamics that their performance will differ. Therefore, while GPNN provides EANN research with a promising technique, many genetic programmers are still quite reserved about GPNN.

Since the GP-GPNN equivalence issue is difficult to solve analytically, and since ANNs have been extensively used in *financial engineering*, it is desirable to have an empirical exploration of this issue based on financial applications. Motivated by this equivalence question, we conducted two series of experiments of predicting stock returns. Each series is composed of ten simulations. The first series of experiments were carried out by using GP, and the second by GPNNs.

2 Simulation Design

The data used in this study is *inherently difficult to predict*. Such 'inherent difficulty' is made precise via the *minimum description length (MDL) principle*. The MDL principle is applied to daily returns of the *S&P* 500 index to identify highly unpredictable subsets of samples with size 200. Details of this procedure are well documented in Chen and Tan [1]. Daily observations of the S&P 500 index from 1/2/1953 to 9/9/1994 are used to create percentage returns. The MDL methodology is applied to this whole dataset of percentage returns and the subset period 1/3/92 to 10/16/92 is chosen. This subset is further decomposed into the in-sample set and post-sample set in the ratio of 10 to 1.

To implement genetic programming, the program GP-Pascal is written in Pascal 4.0 by following the instructions given in [3]. The chosen parameters to run GP-Pascal are given in Table 1. % and *RLOG* appearing in the function set are the protected division function and the protected natural logarithm function respectively [3]. In this paper, all simulations conducted are based on the terminal set, which includes the ephemeral random floating-point constant R ranging over the interval [-9.99, 9.99] and the rate of return lagging up 10 periods, i.e., $R_{t-1},...R_{t-10}$. To escape local optima, the mutation rate is set to be 0.2. In addition, *elite operator* is "on" and is to keep the best-so-far program to the next generation.

The fitness criterion *Mean Absolute Percentage*

Table 1: Tableau for GP Parameters.

Population size	500
Number of trees created by complete growth	50
Number of trees created by partial growth	50
Function set	$\{+, -, \times, \%, EXP, RLOG, Sin, Cos\}$
Terminal set	$\{R_{t-1}, R_{t-2}, ..., R_{t-10}, R\}$
Number of trees generated by reproduction	50
Number of immigrants	50
Number of trees created by crossover	300
Number of trees created by mutation	100
Elite Operator	on
Probability of mutation	0.2
Maximum depth of the tree	17
Probability of leaf selection under crossover	0.5
Number of generations	200
Maximum number in the domain of Exp	1700
Criterion of fitness	MAPE

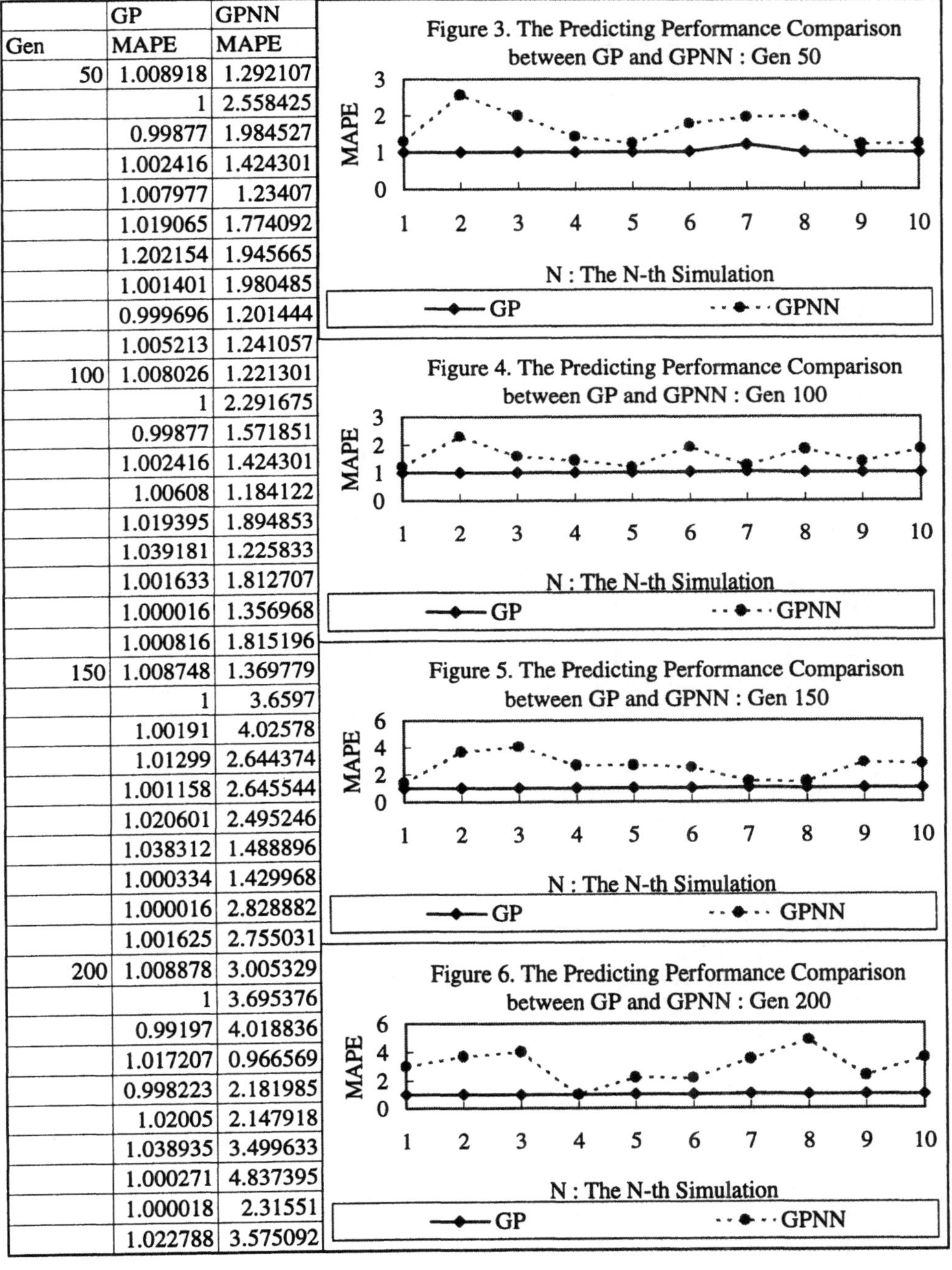

Gen	GP MAPE	GPNN MAPE
50	1.008918	1.292107
	1	2.558425
	0.99877	1.984527
	1.002416	1.424301
	1.007977	1.23407
	1.019065	1.774092
	1.202154	1.945665
	1.001401	1.980485
	0.999696	1.201444
	1.005213	1.241057
100	1.008026	1.221301
	1	2.291675
	0.99877	1.571851
	1.002416	1.424301
	1.00608	1.184122
	1.019395	1.894853
	1.039181	1.225833
	1.001633	1.812707
	1.000016	1.356968
	1.000816	1.815196
150	1.008748	1.369779
	1	3.6597
	1.00191	4.02578
	1.01299	2.644374
	1.001158	2.645544
	1.020601	2.495246
	1.038312	1.488896
	1.000334	1.429968
	1.000016	2.828882
	1.001625	2.755031
200	1.008878	3.005329
	1	3.695376
	0.99197	4.018836
	1.017207	0.966569
	0.998223	2.181985
	1.02005	2.147918
	1.038935	3.499633
	1.000271	4.837395
	1.000018	2.31551
	1.022788	3.575092

Figure 3. The Predicting Performance Comparison between GP and GPNN : Gen 50

Figure 4. The Predicting Performance Comparison between GP and GPNN : Gen 100

Figure 5. The Predicting Performance Comparison between GP and GPNN : Gen 150

Figure 6. The Predicting Performance Comparison between GP and GPNN : Gen 200

Table 2: Tableau for EANN Parameters.

Population Size	196
Number of input nodes	11
Window size	3
Learning rate	self adapting
Error tolerance	self adapting
Momentum rate	self adapting
Output processing	non-linear but fixed
Network model	genetica

Error (MAPE) is defined as follows:

$$\text{MAPE} = \sum_{i=1}^{m} \frac{\mid \hat{R}_i - R_i \mid}{m \mid R_i \mid}, \tag{4}$$

where m is the sample size and $\hat{R}_i$ is the prediction value of R_i. The choice of the *mean absolute percentage error* as the fitness function is attributed to [5], who suggested a modified form of MAPEs as the most appropriate measure satisfying both theoretical and practical concerns while allowing meaningful relative comparisons.

Based on those control parameters, multiple runs of simulations were executed. For each of the simulations, the MAPE is calculated for the in-sample period and the post-sample period. The results of the post-sample MAPEs of generations 50, 100, 150 and 200 under each simulation are exhibited in Figures 3-6.

As to the implementation of EANNs, there are lots of different encoding strategies, including genetic algorithms, evolutionary programming, and genetic programming. In this paper, we follow Wong [6] to encode ANNs and use *NeuroForecaster 4.2* to conduct the experiments. The controlled parameters to run EANNs are given in Table 2. Notice that the node transfer function is exogenously given and is fixed. As a comparison, the results of the post-sample MAPEs of generations 50, 100, 150 and 200 under each simulation are also depicted in Figures 3-6.

3 Simulation Results and Conclusions

From Figures 3-6, we can see that GP's performance is *uniformly superior* to GPNN's for all generations. Moreover, by comparing Figures 3-4 with Figures 5-6, it is interesting to note that GPNN suffers the *overfitting* problem more seriously than GP. These results indicate that presence or absence of the ANN architecture can make a difference in implementing GP driven search. In this typical financial engineering application, imposing the ANN architecture did not bring anything good. It remains to be investigated whether it is helpful to make automatic the determination of *node transfer functions.*

4 Acknowledgements

Research support from NSC grant No.85-2415-H-004-001 is gratefully acknowledged. The authors are grateful to two anonymous referees for helpful comments. This is a short version of the full paper with the same title.

References

[1] S.-H. Chen and C.-W. Tan. *Measuring Randomness by Rissanen's Stochastic Complexity: Applications to the Financial Data*, pages 200–211. World Scientific, 1996.

[2] S.-H. Chen and C.-H. Yeh. Bridging the gap between nonlinearity tests and the efficient market hypothesis by genetic programming. In *Proceedings of the IEEE/IAFE 1996 Conference on Computational Intelligence for Financial Engineering*, pages 34–39. IEEE Press, 1996.

[3] A. I. Esparcia-Alcazar and K. C. Sharman. Evolving recurrent neural network architectures by genetic programming. In *Proc. Genetic Programming 1996 Conference.* Stanford, CA, U.S.A., July 28-31 1996.

[4] J. Koza. *Genetic Programming: On the Programming of Computers by Means of Natural Selection.* The MIT Press, Cambridge, MA, 1992.

[5] S. Makridakis. Accuracy measure: Theoretical and practical concerns. *International Journal of Forecasting*, 9:527–529, 1993.

[6] F. Wong. Neurogenetic computing technology. *NeuroVe$t Journal*, 2(4):12–15, 1996.

A Canonical Genetic Algorithm Based Approach to Genetic Programming

F. Oppacher and M. Wineberg
Intelligent Systems Lab, School of Computer Science,
Carleton University, Ottawa K1S 5B6 Canada,
Email: {oppacher,wineberg}@scs.carleton.ca

Abstract

This paper studies genetic programming (GP) and its relation to the genetic algorithm (GA). Since the programs used as chromosomes by GP are non-homologous, GP uses a different crossover operator than GA. Thus, by modifying the GA, GP loses the theoretical foundations which have been developed for GA. This paper describes an algorithm (called EPI for evolutionary program induction) that stays within the canonical GA paradigm yet breeds programs in a similar manner to GP. EPI has been tested on three problems whose behavior under GP is known; EPI performed identically to GP over this test suite. The success of the implementation shows that the special crossover used in GP is not necessary to solve program induction using a GA.

1 Introduction

In this paper we study the behavior of genetic algorithms on the problem of program induction. The genetic algorithm (GA) is an adaptive search heuristic which searches a solution space [4, 5]. There have been many attempts to implement the task of program induction using a GA [1, 3]. The most recent attempt is genetic programming (GP) [6]. GP uses a modification of the GA to breed successive populations of (initially arbitrary) programs until a program that solves the problem emerges.

GP, however, does not follow the GA model exactly. It has a very different 'crossover' and uses a variable structured, non-linear chromosome [6]. Consequently, the behavior of GP is not well understood, and the theoretical underpinnings of the GA model cannot be directly applied to GP [7]. In order to understand which of the innovations utilized in GP are required to evolve a population of programs towards the correct general solution of a given task, we develop a new GA representation scheme called EPI. EPI encodes programs, which GP uses as variable length chromosomes, into linear chromosomes with fixed internal structures. EPI's underlying canonical GA is then applied to the new chromosomes, and a test suite of program induction problems are solved. The success of the implementation shows that the special crossover used in GP is not necessary to solve program induction using a GA.

2 Genetic Programming

At the heart of GP lie some basic insights: that it is desirable and feasible to directly manipulate parse trees that are treated as variable length chromosomes; that different problem specific primitives may have to be used for different tasks; and that the primitives should satisfy the property of closure (here closure is a design principle that any operator used in a parse tree should be able to take as input any given atomic expression (i.e., terminals), or the output of any operator). Using these insights helps alleviate the problems of both context sensitive interpretation, and order dependency (these problems are described in [2]: order dependency occurs when the position of program structures is important; context sensitive interpretation occurs when the meaning of an expression is affected by changes in another part of the program) genetic programming begins with a set of domain dependent, user-defined operators and terminals, and an initial population of randomly generated solutions. Once the initial population has been created, each program is evaluated, to see how well it operates in comparison with known output. Using the results of the fitness function, programs are selected by a standard GA selection method. Reproduction of the selected programs is done by either copying a program, or mating two programs together using a special crossover operator for parse trees (GP crossover). Reproduction continues until the new population is complete. This cycle of evaluation and reproduction continues until an acceptable program is created, or until a maximum number of generations have passed.

GP crossover exchanges subtrees of two parent parse trees. The crossover points (the location of the subtrees) in the parents are chosen independently of each other. Because of this random choice of the crossover points, the heights of the subtrees are variable. Therefore the heights of the offspring may differ from those of the parents. Since it would be computationally infeasible to allow unrestricted growth in height, a maximum height restriction is imposed on the offspring of GP crossover. If, as a result of a crossover, an offspring has a tree height that exceeds the maximum, it is rejected.

3 Evolutionary Program Induction: EPI

While GP is based on GA methods, there are differences. GA has fixed length chromosomes of fixed structure, and GP has non-linear, tree structured, variable length chromosomes. Accordingly GA and GP employ different operators, in particular, GA uses a mutation operator while GP does not, and the GP crossover differs markedly from its GA counterpart.

The differences between GP, and a canonical GA, should be studied, to see which of the new, or modified, techniques, introduced by GP, are fundamental to the process of program induction through evolution, and which are merely expedient. In particular, is there anything 'magical' about the use of GP crossover that accounts for GP's demonstrated success at program induction?

There are a few other advantages of couching program induction in a canonical GA framework. Two of these advantages are: the extant GA theory becomes applicable to the program induction domain; and techniques thoroughly studied in other GA domains become available to solve problems in program induction.

The reason that GP uses a non-homologous crossover is to accommodate the non-linear structures of the parse trees. The traditional crossovers in GA, such as the 1-point crossover, are defined only for linear chromosomes. A naive application of the traditional crossovers would render the resulting offspring programs completely meaningless. In order to be compatible with the genetic operators used in a canonical GA, a chromosome must, therefore, have the following two properties:

1. the chromosome must be linear;
2. after any genetic operator is applied, the resulting chromosome must produce a syntactically correct computer program.

The second property will be called Syntactic Closure for the rest of the paper. The proposed approach to program induction, i.e., evolutionary program induction, EPI for short, encodes the parse tree in a way that conforms to both of the above requirements.

3.1 Fixing the Tree Height

To simplify the description of EPI, we shall for the moment restrict ourselves to the case of binary operators (this restriction shall be lifted in Section 3.3 below). The property of syntactic closure requires the meaning of each locus to be fixed in the chromosome. This allows the crossover operator to exchange genes with their alleles, and not with incompatible genes.

As pointed out previously, GP uses a variable structured genotype. It should be noted, however, that the parse trees produced by the creation and reproduction routines have a maximum tree height. If a parse tree is produced with a greater height than the maximum it is rejected.

This suggests a way of using parse trees as the genotype and yet still keep a fixed structure for the chromosome. If one views the genome as a full binary tree, expanded out to its maximum depth, with the original parse tree as a subtree, then the length of the genotype is actually unchanging. Therefore the length of the chromosome is now fixed.

3.2 Full Tree Embedding of the Parse Tree

While the above observation allows one to fix the size of the chromosome, it does not determine a fixed structure within the chromosome.

In the proposed embedding of the parse tree the terminals 'sink' in the expanded tree, becoming leaves. Therefore when crossover is applied, each matched pair of loci will be drawn from the same set. If the pair consists of leaves of the expanded tree, both genes are from the terminal set, otherwise, both genes are from the operator set. The internal structure of the chromosome is now static.

The embedding of the parse tree into the expanded parse tree as defined so far is obviously incomplete: there will usually be more operator nodes in the expanded parse tree than operators from the parse tree. This also holds for terminals. To solve

Depth First Encoded Genome:

ghRR14Lii*+i2R02/LL44R7jRfjjLki

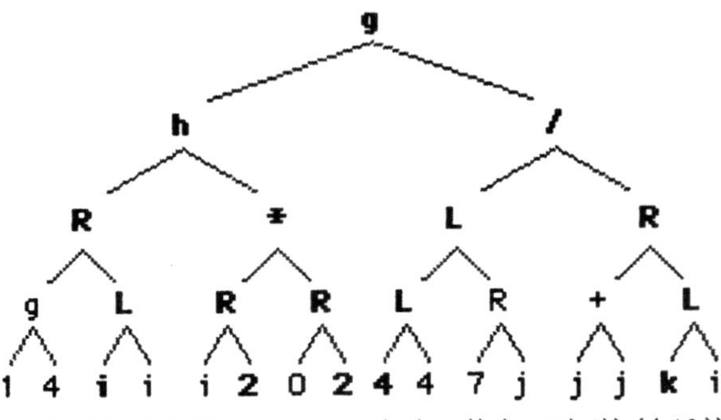

Embedded S-Expression: (g (h i (* (+ i 2) 2)) (/ 4 k))

Figure 1: An embedded parse tree.

this problem, a small set of related operators, called selector operators, are added to the operator set.

A selector operator is a simple function that selects one outgoing edge for traversal. In a binary tree there are only two outgoing edges, so there are only two selector functions: selector-0 which selects the left edge, and selector-1 which selects the right edge (in case of n-ary trees, we will designate the set of selectors as 'selector-k' , $k = 2, \ldots, n$). These operators can now be used to properly embed the parse tree into the expanded tree.

The basic idea for the embedding is to use the selector operators to route the traversal of the expanded tree 'around' areas that do not correspond to any part of the parse tree. The selector operators form a chain hooking up the various code fragments. The elements that are not part of the code, nor in a selector chain, can hold any value, as long as the value is from the correct set: leaves must have genes from the terminal set, and internal nodes must have genes from the operator set. See Figure 1 for an example.

3.3 EPI Genetic Operators

Now that the internal structure of the chromosome is fixed by EPI, the canonical GA operator can be applied. Just as with GP's special purpose operators, any manipulation of the chromosome with any canonical genetic operators will lead to a syntactically correct program under EPI. Since we can now use any traditional GA operator, EPI opens the doors for using any operators established in other contexts for the purpose of program induction.

EPI with n-ary operators can be easily defined. The first step is to set the fan-out of each node of the tree to equal the maximum number of parameters needed by any operator in the operator set. Once this fan-out is found from the operator set, and the tree-height is known, a full tree can be created, which can be used as the template for a fixed length chromosome.

While setting the fan-out equal to the maximum number of parameters that any operator can use, fixes the structure of the expanded parse tree, there are still many operators that will take fewer parameters than the maximum. Therefore, the operators at many nodes will be given too many operands. To handle this problem, the operator can be applied to a subset of the values returned by the node's children.

4 Experimentation with EPI

4.1 Experimentation Design

Three applications are used to test EPI. All are taken from [6]: symbolic regression ($x^4+x^3+x^2+x$); programming a 6 bit multiplexor; and inducing the recursive formula for the Fibonacci series.

The traditional one-point crossover was used in all experiments. The selection technique used was rank selection with elitism, however some experiments were done with fitness proportional selection alone which obtained similar results (although with a slightly lower convergence rate). Crossover was performed 80% of the time with 20% of the population copied directly into the next generation as elite. The probability of mutation used was 0.05. The population size used was 500. The maximum number of generations permitted in a run was 51. A tree height of 4 was used in all cases.

The probability of convergence, which is the ratio of successful runs (where a solution to the problem is found) to total number of runs, for each of the three problems is compared with the published result for GP.

Since each run is independent, the true probability of converence has a binomial distribution, and so the estimated probability (the experimentally derived ratio used for the comparison) will lie within a plus or minus error bound which can be calculated by the formula

$$\pm Z_{\frac{\alpha}{2}} \sqrt{\frac{\widehat{P}(1-\widehat{P})}{n}}$$

Table 1: Results of the experiments.

	GP	EPI
Symb. Regr.	0.35 ± .09	0.47 ± .18
6 Bit Multiplexor	0.66 ± .12	0.60 ± .16
Fibonacci	0.01 ± .01	0.08 ± .04

where $\widehat{P}$ is the estimated probability of convergence, n is the number of experiments and Z is the z-score. There were 30 runs done for each test to obtain the probability of convergence except for the Fibonacci test which was run 200 times since the probability of convergence was small. The error bound was calculated using $z = 1.960$ which corresponds to a 95% confidence level ($\alpha = 0.025$).

4.2 Results

EPI obtained comparable results to GP on the same problems (see Table 1). Neither GP nor EPI have been optimally tuned for the various applications, so a direct comparison is impossible. For the Fibonacci experiment, we ran GP to obtain the results since they are not published in [6].

5 Conclusion

Overall, the EPI system, using the canonical GA approach, solves program induction with comparable performance to GP. This shows that there is nothing 'magical' about the otherwise very different GP crossover. That the two systems behave so similarly when using the same primitives, leads to the possible conclusion that there is some underlying property to program design, which affects the ability of a GA system to evolve the programs.

References

[1] N. L. Cramer. A representation for the adaptive generation of simple sequential programs. In *Proceedings of an International Conference on Genetic Algorithms and their Applications*, Hillsdale, New Jersey, 1985. Lawrence Erlbaum Associates.

[2] K. A. De Jong. On using genetic algorithms to search program spaces. In *Proceedings of the Third International Conference on Genetic Algorithms*, Hillsdale, San Mateo, California, 1989. Morgan Kaufmann.

[3] C. Fujiki and J. Dickinson. Using the genetic algorithm to generate lisp source code to solve the prisoner's dilemma. In *Proceedings of the Second International Conference on Genetic Algorithms*, Hillsdale, New Jersey, 1985. Lawrence Erlbaum Associates.

[4] D. E. Goldberg. *Genetic Algorithms in Search, Optimization and Machine Learning.* Addison Wesley, 1989.

[5] J. H. Holland. *Adaptation in Natural and Artificial Systems.* MIT Press, 1992. first published by University of Michigan, 1975.

[6] J. R. Koza. *Genetic Programming: On the Programming of Computers by means of Natural Selection.* MIT Press, Cambridge, MA, 1992.

[7] U.-M. O'Reilly and F. Oppacher. On using genetic algorithms to search program spaces. In *Proceedings of the Third International Conference on Genetic Algorithms*, Hillsdale, San Mateo, California, 1989. Morgan Kaufmann.

Is Genetic Programming Dependent on High-level Primitives?

D. Heiss-Czedik
Institute for Information Systems, Vienna University of Technology
Auhofstraße 160/11, A-1130 Vienna, Austria.
Email: dorothea@tbi.univie.ac.at

Abstract

The aim of this paper is to refute the claim that the success of genetic programming depends on problem-specific high-level primitives. We therefore apply genetic programming to the λ-calculus, a Turing complete formalism with only two (very low-level) primitives.

Genetic programming is suited to find the predecessor function in the space of λ-definable functions without *a priori* knowledge. The predecessor function is historically important and documented to be 'a challenge' and 'difficult' to find.

1 Introduction

Genetic programming [9] is an evolutionary method of program induction. Its goal is to find a program in a given language and with a given behavior. Genetic programming starts with a population of randomly generated programs. A program is chosen for reproduction with a probability proportional to its 'fitness' (i.e. how well the program meets the problem-specific behavior requirements). Reproduction yields either an exact copy, a mutant, or a recombinant with another randomly chosen program. This scheme is repeated until a solution is found or some resources are exhausted.

The success of genetic programming was demonstrated in such diverse applications as solving symbolic regression problems [9], inducing decision trees [9], generating controllers (e.g. for robots) [6, 9, 11, 12], discovering game-playing strategies [2, 9], cracking and evolving randomizers [7, 9], and many others.

Nevertheless it is often criticized [1, 13] that in most successful applications the *primitives*, which genetic programming combines to programs, are customized carefully to fit the problem. The user who wants a program, solving a particular problem, needs enough insight into this problem to provide these primitives, which are often sophisticated high-level functions. It may be argued that in this case genetic programming plays only a minor role in finding the solution compared with the contribution of the user.

To examine this allegation we want to induce programs in a very general representation which is not tuned for the given problem. General programming languages unfortunately have so many different constructs that they are tedious to work with. A simple yet general framework is the λ-calculus, which was invented by Alonzo Church [3, 4] to study the properties of functions. Turning to this abstract universe of functions has several advantages: its definition is transparent, a large body of mathematical theory is available, and it captures a programming language paradigm. In fact, it is the syntactically unsugared core of functional programming languages.

In λ-calculus functions are represented as so-called λ-expressions, *expr*, whose syntax (according to [10]) is:

$$expr ::= x \mid \lambda x.expr \mid (expr)expr \qquad (1)$$

where x is a variable. Thus, the simplest λ-expression is a variable (without type or sort, infinite in supply). To build more complex λ-expressions there are only two operators: Abstraction, $\lambda x.expr$, introduces a formal parameter x to make an unary function (the usual notation of this function would be $f(x) = expr$) from the given *expr*. Inside *expr* all occurrences of x are called 'bound'. Application, $(expr)expr$, is the application of one λ-expression (the operator - here in parentheses) to another λ-expression (the operand, which represents also a function).

The semantic of a λ-expression is its reduced form - or more precisely its *normal form*, which cannot be reduced any further. There are several reduction rules, the most important is:

$$(\lambda x.P)Q \Rightarrow [Q/x]P \qquad (2)$$

where P and Q are λ-expressions and x is a variable. $[Q/x]P$ means the substitution of Q for all occurrences of x in P. The situation corresponds to the substitution of the actual parameter Q for the formal parameter x of a function P. The reduced form corresponds to the result of the evaluated function. E.g. $(\lambda x.x)y \Rightarrow [y/x]x = y$.

2 Evolving the Predecessor Function

As an example we try to evolve the predecessor function. This may not seem a very challenging task, but Church himself had just about convinced himself that there is no λ-definition of the predecessor function, when Stephen C. Kleene found a representation for it [8, 10]. This was an important result, because otherwise the predecessor function would be a counter example to the notion of λ-definability, i.e. everything computable can be defined with just the syntax (1). In other words, λ-definability would not be equivalent with the Turing notion of 'computability' and the Herbrand-Gödel notion of 'general recursiveness' [8].

A λ-implementation of the predecessor function only makes sense in the context of some representation for the natural numbers. Like Kleene, we will use the Church numerals here: The numeral representing the number n iterates its first argument to its second argument n times.

$$n \cong \lambda f.\lambda x.\underbrace{(f)\ldots(f)}_{n \text{ times}}x \cong \lambda f.\lambda x.(f)^n x. \qquad (3)$$

The behavior of the predecessor function is defined as follows: When applied to a numeral representing $n > 0$, its result is the numeral representing $n-1$; when applied to 0, the result is 0.

A simplified version of Kleene's representation for the predecessor function can be found in [10]:

$$\begin{aligned} pred_K \cong\ & \lambda n_1.(((n_1)\lambda p.\lambda z_1.((z_1) \\ & \underbrace{(\lambda n_2.\lambda f_1.\lambda x_1.(f_1)((n_2)f_1)x_1}_{successor})(p) \\ & \underbrace{\lambda v_1.\lambda w_1.v_1}_{true})(p)\underbrace{\lambda v_2.\lambda w_2.v_2}_{true})\lambda z_2.((z_2) \\ & \underbrace{\lambda f_2.\lambda x_2.x_2}_{zero})\underbrace{\lambda f_3.\lambda x_3.x_3}_{zero})\underbrace{\lambda v_3.\lambda w_3.w_3}_{false} \end{aligned}$$

This representation of the predecessor function uses ordered pairs $[a, b]$ represented as $\lambda z.((z)a)b)$ whose result is a when applied to $true$ and b when applied to $ffalse$. If $pred_K$ is applied to a numeral n, then $n+1$ ordered pairs are iteratively generated, where the zeroth pair is $[0, 0]$ and the i-th pair ($0 < i \leq n$) is $[i, i-1]$. It is easy to obtain $[i+1, i]$ from$[i, i-1]$. Finally, the second element of the pair $[n, n-1]$ is the result of the predecessor function.

The usual approach in genetic programming to evolve a representation for the predecessor function would start with designing useful primitives, like 'ordered pair', successor function, true, false and zero. The goal would be that genetic programming combines these primitives to a correct predecessor function. But this approach requires that the problem is solved in advance, because Kleene's solution is the only rationale for the choice of these primitives.

We want to tackle this problem as if it were still unsolved. We cannot foresee what high-level primitives may be useful and consequently provide none. Instead plain λ-calculus is used as the representation. There is only one peculiarity: A function is usually meant to act independently from the context it is in. In λ-calculus this behavior is achieved, when expressions contain only bound variables. Such expressions are termed *closed expressions*. Thus, in the present case genetic programming is not operating on closed λ-expressions. We introduce new genetic operators to preserve syntactical legality and closure. Mutation is the deletion or insertion of the following underlined constructs: $\underline{\lambda x.}\bullet$, where $\bullet$ stands for an expression, and x is a variable (which never occurs in $\bullet$), $\underline{(simple)}\bullet$, and $\underline{(}\bullet\underline{)simple}$, where $simple$ is $\lambda x.x$, or a bound variable. Crossover is the exchange of closed sub-expressions.

By using pure λ-calculus, we comply with the idea of genetic programming as a totally task-independent problem solver: The representation must not favour any area of the search space. The claim of genetic programming is that 'because no explicit knowledge exists in the evolutionary algorithm, this knowledge must emerge from the interaction of the simple problem solver and the task environment' [2, p.10].

The task environment is modeled by the *fitness function* whose result is needed as a feedback for the problem solver. Nevertheless, the fitness function contains no knowledge which areas of the search space are promising, either. Instead it grades the actual behavior of a program with respect to the desired target behavior. The behavior of a λ-expression $expr$ when applied to the representation of the number i is its resulting normal form $result_i$. The $result_i$ is compared to the desired result $target_i$ for a small number of *fitness cases*. But $result_i$ will typically not be a correct numeral let alone the desired $target_i$. Therefore, a λ-expression is rewarded to the extent that its actual $result_i$ is 'similar' to some numeral num_i. Similarity is determined by using ***regular expressions*** simply to find the biggest numeral num_i that is ***contained*** in the $result_i$.

$$\text{dist}(\text{result}_i, \text{target}_i) =$$

$$\underbrace{\frac{\text{primitives}(\text{num}_i)}{\text{primitives}(\text{result}_i)}}_{\text{syntactical distance}} \cdot \underbrace{\frac{1}{|\text{num}_i - \text{target}_i| + \epsilon}}_{\text{arithmetic distance}} \tag{4}$$

where $primitives(expr)$ denotes the sum of the number of applications, abstractions, and variable occurrences in $expr$. Total fitness is the sum of the distances over the different fitness cases.

With this setting (9 fitness cases (numerals 0 to 8), $\epsilon = 0.1$ (maximum fitness is therefore 90); population size of 1000, 100 to 200 generations, steady-state) genetic programming was able to find predecessor functions. All the ones we examined are based on the same principle, which is different than Kleene's. We will demonstrate this principle on the shortest predecessor function found (most of them were much longer).

The expression reads:

$$\lambda x_1.((x_1)\underbrace{\lambda x_2.((x_2)\lambda x_3.x_3)\lambda x_4.\lambda x_5.((x_2)x_4)(x_4)x_5)}_{S}$$
$$\underbrace{\lambda x_6.(x_1)\lambda x_7.\lambda x_8.\lambda x_9.x_9}_{A},$$

Applying this expression to a numeral n, see (3), gives, after some reductions,

$$(pred)n = (S)^n A_n.$$

with $A_n = \lambda x_6.\lambda x.(\lambda x_7.\lambda x_8.\lambda x_9.x_9)^n x$. When pred is applied to the numeral $n = 0$, the result is $A_0 = \lambda x_6.\lambda x.x$ which is the Church numeral for 0 (modulo names of bound variables). Therefore, $(pred)0 = 0$.

In the case of $n > 0$, $A_{n>0} = A' = \lambda x_6.\lambda x.\lambda x_8.\lambda x_9.x_9$ independently of $n > 0$. When S is applied to A', the result is the numeral representing 0. When S is applied to a Church numeral, it acts exactly like a successor function! The mechanism of this predecessor function is quite elegant: The first application of S to A' yields zero. The next $n-1$ applications of S increment zero to $n-1$:

$$(pred)n = (S)^n A' = \underbrace{(S)...(S)}_{n\ times} A'$$

$$= \underbrace{(S)...(S)}_{n-1\ times}(S)A' = \underbrace{(S)...(S)}_{n-1\ times} 0 = n-1.$$

3 Discussion and Conclusion

Among almost 300 runs, four of them were successful. One of the successful runs generated 805 different predecessor functions (in normal form) until it was stopped. In 56% of the runs no expression was found that returned the correct value for at least three out of the nine fitness cases. In only 6% of the runs an expression solved at least four fitness cases. What are the reasons for this extraordinary poor success rate?

1. Fitness proportional selection without scaling is known to be problematic. [5]. This may be counteracted by playing with scaled fitness or tournament selection [9]. However, our point was not to fine-tune the system.
2. The demand of Kleene that $pred(0) = 0$ is irregular. There is no logical reason why the predecessor of zero should be zero except to assure closure. We included this demand in our experiments to ensure equivalence of our work with Kleene's. Our success rate would have been more impressive without this requirement. Dropping it, resulted in 11% of the runs finding a 'predecessor' function, compared to the meager 1.3% that ended with a genuine predecessor complying with $pred(0) = 0$.

The reason for this behavior is not that the 'predecessor' functionality without the zero case is so much simpler to realize. As a matter of fact, the distance between a predecessor expression (with $pred(0) = 0$) and a 'predecessor' (with $pred(0) = something$) was at times just one mutation. The point is that when $pred(0)$ had to be zero, the constant function $f(x) = 0$ was a trap (returning always a numeral and satisfying two fitness cases). The population converged to this local maximum. Given the syntactic simplicity of its realizations, such as $\lambda x_1.\lambda x_2.\lambda x_3.x_3$, the production of more complicated and better behaving functions proved difficult.

We could use our knowledge of the attractiveness of the constant function $f(x) = 0$ when searching for the predecessor function. To avoid this trap the fitness function could somehow 'punish' λ-implementations of $f(x) = 0$. But that's exactly what we didn't want: to introduce knowledge which areas of the search space are unfavourable. This kind of knowledge should emerge, while the search progresses: The population explores areas of the search space which are more and more likely to contain the solution.

Genetic programming did find solutions, indeed. But is this a proof that genetic programming needs no high-level primitives or *a priori* knowledge? We will investigate some counter arguments:

Surely, we did not use any high-level functions as primitive functions, but plain λ-calculus, instead. Nevertheless, we restricted the search space to closed expressions. It would be interesting to analyze the consequences of dropping this constraint.

The fitness function also contains no problem-specific knowledge as it is applicable to any arithmetic target function. The only aspect in which the fitness function does more than that, is in keeping expressions viable that are not functions mapping numerals to numerals, while inducing a selection pressure towards arithmetic functions in general. Obviously this 'smoothed' the fitness landscape. When expressions were assigned zero fitness, if their results were not numerals, no solution was found.

The predecessor function may seem like a disappointingly simple function, but first, Kleene called it 'a challenge' and 'difficult' to find [8], and secondly, in λ-calculus our intuition about intricacy of functions is wrong. For example the power function was almost always in the randomly generated initial population. Multiplication was also very easy to find with genetic programming. Addition posed a harder problem, but genetic programming was successful again.

The unsatisfactory success rate may indicate that genetic programming is handicapped by the low-level representation. On the other hand, the 'predecessor' function failing on zero was found much easier, although the same representation was used. A single example is not enough to explain why and when genetic programming is successful or not. Nevertheless, the ability of genetic programming to evolve a historically challenging λ-expressions demonstrates its usefulness even if the primitives contain no task-dependent knowledge.

4 Acknowledgments

The author is grateful to Dr. Walter Fontana for his contributions to this work.

References

[1] R. Abbott, 1993. mailing list genetic-programming@cs.stanford.edu.

[2] P. J. Angeline. *Evolutionary Algorithms and Emergent Intelligence*. PhD thesis, The Ohio State University, 1993.

[3] A. Church. A set of postulates for the foundation of logic. *Annals of Math.*, 33:346–366, 1932.

[4] A. Church. A set of postulates for the foundation of logic (second paper). *Annals of Math.*, 34:839–864, 1933.

[5] D. E. Goldberg. *Genetic Algorithms in Search, Optimization, and Machine Learning*. 1989.

[6] S. G. Handley. *The automatic generation of plans for a mobile robot via genetic programming with automatically defined functions*. MIT Press, Cambridge, MA, 1994.

[7] J. Jannink. *Cracking and co-evolving randomizers*. MIT Press, Cambridge, MA, 1994.

[8] S. C. Kleene. Origins of recursive function theory. *Annals of the History of Computing*, 3(1), 1981.

[9] J. R. Koza. *Genetic Programming*. MIT Press, Cambridge, MA, 1992.

[10] G. E. Revesz. *Lambda-Calculus, Combinators, and Functional Programming*. Cambridge University Press, 1988.

[11] C. W. Reynolds. *Evolution of obstacle avoidance behavior: Using noise to promote robust solutions*. MIT Press, Cambridge, MA, 1994.

[12] G. Spencer. *Automatic generation of programs for crawling and walking*. MIT Press, Cambridge, MA, 1994.

[13] S. Taylor, 1993. mailing list genetic-programming@cs.stanford.edu Sep 9, 1993.

DGP: How To Improve Genetic Programming with Duals

J.-L. Segapeli, C. Escazut and P. Collard
Laboratory I3S-CNRS UNSA, Bât. 4, 250 av. A. Einstein,
Sophia-Antipolis, 06560 Valbonne, FRANCE
Email: {segapeli,escazut,pc}@unice.fr

Abstract

In this paper, we present a new approach, improving the performances of a genetic algorithm (GA). Such algorithms are iterative search procedures based on natural genetics. We use an original genetic algorithm that manipulates pairs of twins in its population: DGA, dual-based genetic algorithm. We show that this approach is relevant for genetic programming (GP), which manipulates populations of trees. In particular, we show that duals can transform a deceptive problem into a convergent one. We also prove that using pairs of dual functions in the primitive function set, is more efficient in the problem of learning boolean functions. Here, in order to prove the theoretical interest of our approach (DGP: dual-based genetic programming), we perform a numerical simulation.

1 Introduction

GAs find their origins in biology. GAs handle populations of fixed-length character strings, that evolve according to Darwin's Natural Selection principle. GP (proposed by John R. Koza [4]) is based on GAs. Nevertheless, instead of character strings, GP handles programs. Koza chose to use LISP programs (also called *S*-expressions).

2 Dual-Based Genetic Programming

A new approach of GAs, DGA, was proposed by P. Collard and J. P. Aurand [1]: duals are introduced in the population.

Experiments have brought to the fore the interest of the approach on deceptive problems. Simulations showed that duals are selected so that they constitute stable or semi-stable monotonous predicates [6]. It was also shown that the introduction of duals, in GAs, is a source of diversity that can conceive robust systems.

In GP, individuals consist of terminals and functions in various numbers. So we introduced duals in a different way from the one proposed for GAs: we need a *dual function* for each function (in the internal points set); a dual function will evaluate its arguments in a different way from the original function [2].

The dual functions presented here are based on symmetry. We can present the idea of 'symmetry' in the following way. If a function f is defined in the range $[M-D, M+D]$, we can define its dual function $\overline{f}$ by: $\overline{f}(X) = f(2M - X)$. This allows us to obtain different trees, but with the same evaluation. An example of dual function is *sine*, that can be considered as a dual function of *cosine*; indeed: $sin(X) = cos(\pi/2 - X)$.

Let us take an example. We are given the function '+' with two arguments. We assume that the set of terminals that can be used is {2,23}, as in the arithmetical problem presented below. Then we define the function $\oplus$, dual of the function +, in the following way. The function $\oplus$ has two arguments, like +, but evaluates them in a different way. The function $\oplus$ makes the addition (+) of its two arguments after having transformed them: if the argument is 2 (resp. 23), it becomes 23 (resp. 2) ; if the argument is an S-expression beginning with +, we replace this + by a $\oplus$, and vice versa. Each S-expression is thus evaluated by using recursively this process (c.f. Figure 1).

In order to evaluate the leftmost tree, we start from the root and go down to the leaves, giving successively the three other trees. The rightmost tree is easily evaluated, since the only function it contains is '+'. Since these four different trees are evaluated with the same value (i.e. 75), we can say that they are duals.

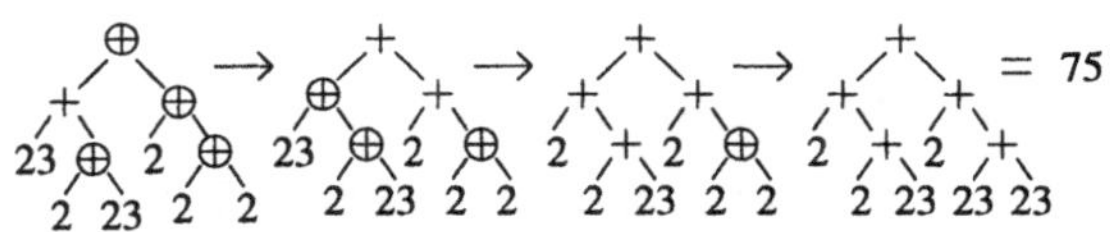

Figure 1: Evaluation of a tree containing function $\oplus$.

Thus, each population contains trees which are the duals of other ones. The interest of introducing duals in GP is that some programs can be expressed in an easier way with some particular forms of duals. Schemata, in GP, are programs containing, as sub-programs, one or several specified sub-programs [4, p 117–118]. Duals enable us to handle schemata that are duals of standard schemata; they can sometimes be more interesting. We can think that introducing duals leads to an increase of the combinatory; indeed, the size of the genotypes' space is doubled. However, we must notice that the phenotypes' space and the population size remain unchanged: the number of individuals in the population is not doubled. Thus, we could, *a priori*, say that we increase the possibilities of finding an optimal value.

3 Experimental Results

In this section, we present two important results that we obtained with duals in [2].

3.1 A Deceptive Problem in Arithmetic

We first studied a deceptive problem and compared duals to mutations. This problem was derived from a deceptive problem dealing with two-dimensional vectors, proposed by David J. Montana [5]. We observed that the algorithm looped in 50% of the runs performed, that means the population got stuck in a local minimum. Then we introduce duals in this problem, through the function $\oplus$, dual of function $+$, presented above. We only introduced the dual function $\oplus$, because in our original function set, the only function was $+$. The only terminals were 2 and 23. We did not have to use knowledge of the solution to build this dual function. This dual-based version never looped and was able, in all the runs, to get out of the local minima. Would a non zero mutation rate enable us to improve the convergence? We performed trials with a version without duals, but where mutations were allowed. These tests with mutations found the solution on the average four times slower than the version with duals. Duals have thus made it possible to transform a deceptive problem into a non-deceptive one.

3.2 Boolean Functions

Then we studied the usefulness of duals in the search of boolean functions. For a given boolean target function f, we tried, using GP, to generate a tree which implements f. Each tree of the initial population consisted of terminals and internal functions given by the programmer. We wanted to determine whether, by choosing the internal functions in a set formed by pairs of duals, the algorithm's performances are better than with a set containing no pairs of dual functions. We performed the tests using four sets of internal functions, the third and the fourth consisting of two pairs of dual functions. The two sets formed with pairs of duals obtained results far better than the two sets without duals. Thus, we showed that duals enable us to form an efficient set of functions for the search of boolean functions.

4 Numerical Simulation

4.1 Equating the Minimum Deceptive Problem for GP

In this section, we shall perform numerical simulations in order to bring to the fore the fact that a DGA is able to transform a *minimum deceptive problem* (MDP) into a convergent one. The MDP studied here, derives from the one proposed by David E. Goldberg [3]. In this problem, a too high important proportion of deceptive individuals in the initial population, prevents the algorithm from converging towards the solution. Thus, in order to consider each individual of the search space, our objective is to write equations expressing the evolution of the proportion of each individual. These equations will represent the gains and the losses, due to fitness proportionate reproduction and crossover (we suppose, as Koza did, that no mutation is performed).

We propose to adjust the Type II MDP, initially concerning binary strings, to GP: the 'linear' chromosomes are replaced by trees. Darrel Whitley [7] focused on the study and the equating of this same problem, concerning GAs. Here, we shall consider populations of four trees: $f(0,0)$, $f(0,1)$, $f(1,0)$, $f(1,1)$. In other words, the node of each tree is the function f with two arguments, and the terminals are 0 and 1. In the following, we note these four individuals: $f_{00}, f_{01}, f_{10}, f_{11}$. Let us note that these notations represent trees. Let us also assume that their fitness are $f_{00} \to 1.0$, $f_{01} \to 0.9$, $f_{10} \to 0.5$ and $f_{11} \to 1.1$. As a standard GA is not able to optimize such a problem, we can hope that applying DGP, the population converges towards the optimum f_{11}.

Our aim is to exhibit the MDP in GP, we thus restrict our study to trees of depth 1, and the crossovers will only consist of the exchange of simple leaves.

Let us now write the equations. Each individual corresponds to an equation composed of a term representing the reproduction's impact, of terms representing the gains obtained by crossover, and of terms representing the losses caused by this same crossover. Let us first describe the term concerning reproduction. Let $F(x)$ be the strength of the individual x, related to the population (i.e. its fitness divided by the average fitness of the entire population). If we denote by $P_t(x)$ the proportion of the individual x in the population at generation t, the reproduction's effect can be written: $P_{t+1}(x) = P_t(x) \times F(x)$. Considering the crossover, the evolution of an individual x from generation t to generation $t+1$ is: $P_{t+1}(x) = P_t(x) \times F(x) \times (1 - P_c \times losses(x)) + P_c \times gains(x)$, where $losses(x)$ represents the disappearance of x by crossover, and $gains(x)$ the appearance of x by crossover. If we denote by $N_t(x)$ the product $P_t(x) \times F(x)$, the previous equation becomes: $P_{t+1}(x) = N_t(x) \times (1 - P_c \times losses(x)) + P_c \times gains(x)$.

Let us now detail the terms '*losses*' and '*gains*'. As we only consider trees of depth 1 representing a function with 2 arguments, there are 4 possible crossovers. Indeed, let us consider $f(a,b)$ and $f(c,d)$; there are 4 pairs of possible crossing sites: $\{(1,1),(1,2),(2,1),(2,2)\}$. For example, the pair $(1,2)$ indicates that the first argument of f in $f(a,b)$ is swapped with the second argument of f in $f(c,d)$. The two offspring thus created are $f(d,b)$ and $f(c,a)$. The term $losses(x)$ represents the crossovers in which the individual x is involved and disappears. For instance, f_{11} disappears by crossing with f_{00}. So, the term $losses(f_{11})$ will contain the term $N_t(f_{00})$. There may be some cases where an individual participates in a crossover and is one of the two offspring created. Such a crossover is not taken into account in the losses. As for the term $gains(x)$, it represents the fact that the individual x can be obtained from the crossover of two individuals. For example, f_{10} can be created by crossing f_{00} and f_{11}. Thus the term $gains(f_{10})$ will contain the product $N_t(f_{00}) \times N_t(f_{11})$.

The terms '*losses*' and '*gains*' are thus sums of products, themselves multiplied by a coefficient: 1/4, 1/2 or 1. Let us consider the appearance or disappearance of an individual x. If we cross 2 individuals, different from x, we may obtain x. For instance, crossing f_{01} and f_{11} on the pair of site $(1,2)$ produces an individual f_{10}. The 3 other crossovers (i.e. on the sites $(1,1)$, $(2,1)$ and $(2,2)$) give no individual f_{10}. In other words, in 4 possible crossovers, only one individual f_{10} can be obtained. So, in the term $gains(f_{10})$, there will be the product: $1/4 \times N_t(f_{01}) \times N_t(f_{11})$. Four equations can thus describe the evolution of the 4 individuals f_{00}, f_{01}, f_{10} and f_{11}. For instance, let us write this equation for f_{01}:

$$\begin{aligned}
P_{t+1}(f_{01}) =& N_t(f_{01}) \times (1 - P_c \times losses(f_{01})) \\
&+ P_c \times gains(f_{01}) \\
losses(f_{01}) =& 0.25 \times N_t(f_{00}) + N_t(f_{01}) \\
&+ 0.5 \times N_t(f_{10}) + 0.25 \times N_t(f_{11}) \\
gains(f_{01}) =& N_t(f_{00}) \times N_t(f_{11}) \\
&+ 0.25 \times N_t(f_{10}) \times N_t(f_{11}) \\
&+ 0.25 \times N_t(f_{10}) \times N_t(f_{00})
\end{aligned}$$

If we introduce duals in the search space, by using the function g, dual of f, we have now 4 pairs of dual trees. The function g transforms a 1 into a 0, and a 0 into a 1, thus creating 8 individuals such as: $f_{00} = g_{11} = 1.0$, $f_{01} = g_{10} = 0.9$, $f_{10} = g_{01} = 0.5$ and $f_{11} = g_{00} = 1.1$. There are 8 equations relating the evolution of the individuals. For instance, the one concerning f_{01} is:

$$\begin{aligned}
P_{t+1}(f_{01}) =& N_t(f_{01}) \times (1 - P_c \times losses(f_{01})) \\
&+ P_c \times gains(f_{01}) \\
losses(f_{01}) =& N_t(f_{01}) + 0.5 \times N_t(f_{10}) \\
&+ 0.25 \times N_t(f_{11}) + 0.25 \times N_t(f_{00}) \\
&+ 0.5 \times N_t(g_{01}) + 0.5 \times N_t(g_{10}) \\
&+ 0.5 \times N_t(g_{11}) + 0.5 \times N_t(g_{00}) \\
gains(f_{01}) =& N_t(f_{11}) \times N_t(f_{00}) \\
&+ 0.25 \times N_t(f_{10}) \times N_t(f_{11}) \\
&+ 0.25 \times N_t(f_{10}) \times N_t(f_{00}) \\
&+ 0.5 \times N_t(f_{11}) \times N_t(g_{00}) \\
&+ 0.5 \times N_t(g_{11}) \times N_t(f_{00}) \\
&+ 0.25 \times N_t(g_{10}) \times N_t(f_{11}) \\
&+ 0.25 \times N_t(g_{10}) \times N_t(f_{00}) \\
&+ 0.25 \times N_t(f_{11}) \times N_t(g_{01}) \\
&+ 0.25 \times N_t(f_{00}) \times N_t(g_{01})
\end{aligned}$$

4.2 Results of the Simulation

We studied the evolution of the proportions of the individuals in the search space with the standard approach and the dual-based one, using the equations described above. Two categories of tests (simulations) have been performed depending on the crossover probability P_c and on the initial proportions of the individuals.

In the first category of experiments, we set P_c to the value 0.9, as Koza did in GP. With proportions in the initial population unfavourable to the optimum, the algorithm is deceptive (that means it converges to the deceptive attractor f_{00}), if we do not use duals (Figure 2).

On the other hand, if we introduce duals, the algorithm always converges to the optimum, even with initial conditions very unfavourable to the optimum (Figure 3).

So this first category of experiments has shown that introducing duals may transform a deceptive problem into a convergent one.

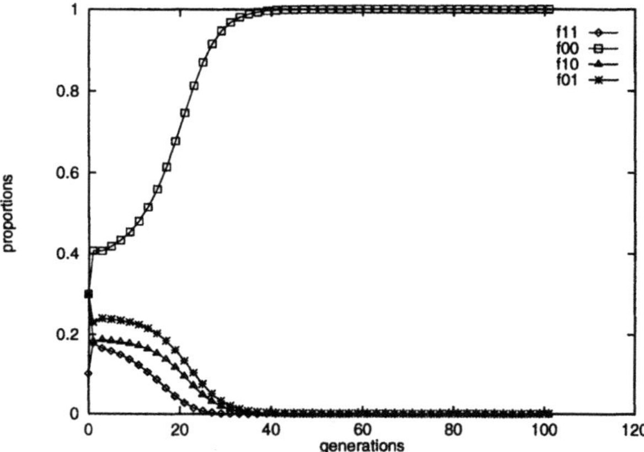

Figure 2: Minimum deceptive problem.

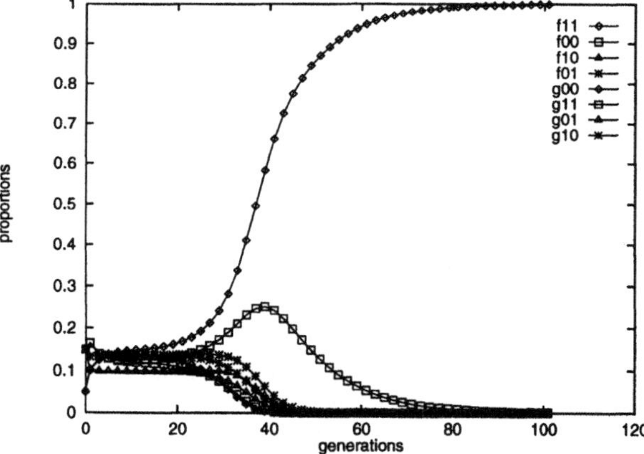

Figure 3: Duals lead to convergence.

Then we performed a second category of experiments in order to bring to the fore the influence of crossovers. Therefore, we computed the final proportions of individuals (after 100 generations) according to various crossover probabilities. Using initial conditions unfavourable to the optimum, if we do not use duals in the population, the problem may be deceptive or convergent, depending on the crossover probality used during the run. On the other hand, with the same initial proportions, if we use duals in the population, the results show that whatever the crossover probality is, the final population is only composed of the optimum. So this second category of experiments also showed minimum deceptive problems transformed by duals into convergent ones.

The simulations have shown the interest of introducing duals in such deceptive problems in GP. They also confirm the results obtained on 'linear' chromosomes [1].

5 Conclusion

In this paper, we explored the power of an original GA in the GP paradigm. The dual-based genetic programming paradigm described here provides a useful extension of GP standard search procedure.

We have shown that the introduction of duals in the population has advantages for GP. It enabled to transform an arithmetical deceptive problem into a convergent one, and we showed that it was interesting to use a set of initial functions composed of pairs of duals, in the search of boolean functions. The theoretical simulation presented here has reinforced these experimental results.

The simulations presented, are meant to stimulate further debate on the interest of DGA approach. The underscored behaviours present a theoretical interest, but other tests will be necessary in order to confirm their practical interest.

References

[1] P. Collard and J.-P. Aurand. *DGA: An Efficient Genetic Algorithm.* ECAI'94. 1994.

[2] P. Collard and J.-L. Segapeli. Using a Double-based Genetic Algorithm on a Population of Computer Programs. In *ICTAI'94: Proceedings of the 6th IEEE International Conference on Tools with Artificial Intelligence.* New Orleans. USA. 1994.

[3] D. E. Goldberg. Simple Genetic Algorithms and the Minimal Deceptive Problem. *Genetic Algorithms and Simulated Annealing.* L. Davis ed. 1987.

[4] J. R. Koza. *Genetic Programming.* MIT Press, Cambridge, MA. 1992.

[5] D. J. Montana. *Strongly Typed Genetic Programming.* BBN Technical Report #7866. 1993.

[6] M. D. Vose. Generalizing the notion of schema in genetic algorithms. *Artificial Intelligence*, 50: 385–396. 1991.

[7] D. Whitley. An Executable Model of a Simple Genetic Algorithm. *Foundations of Genetic Algorithms 2*, Morgan Kaufmann. 1993.

Fitness Landscapes and Inductive Genetic Programming

V. Slavov[1] and N. I. Nikolaev[2]
[1] Information Technologies Lab, New Bulgarian University, Sofia 1113, Bulgaria,
[2] Department of Computer Science, American University in Bulgaria, Blagoevgrad 2700, Bulgaria
Email: vslavov@inf.nbu.acad.bg, nikolaev@nws.aubg.bg

Abstract

This paper proposes a study of the performance of inductive genetic programming with decision trees. The investigation concerns the influence of the fitness function, the genetic mutation operator and the categorical distribution of the examples in inductive tasks on the search process. The approach uses statistical correlations in order to clarify two aspects: the global and the local search characteristics of the structure of the fitness landscape. The work is motivated by the fact that the structure of the fitness landscape is the only information which helps to navigate in the search space of the inductive task. It was found that the analysis of the landscape structure allows tuning the landscape and increasing the exploratory power of the operator on this landscape.

1 Introduction

Evolutionary algorithms [1, 4] are random search methods that could be employed for inductive concept learning. They learn by reformulating a population of concept descriptions with reproduction, recombination and mutation operators. An advantage of the evolutionary algorithms is the possibility to control the population diversity and so avoid sticking in local optima [2]. In addition, reliable statistical means have been developed [3, 6] for examining the performance of evolutionary algorithms.

This paper presents an empirical study of inductive genetic programming [4] with decision trees [7] (GPDT). The influence of the fitness function, the mutation operator, and the categorical distribution of the examples on the search process is investigated with statistical measures. The search is viewed as navigation, with the fitness function and the mutation operator, and structure, determined by the fitness landscape of the same operator. The global and local characteristics of the landscape provide the only information for search navigation. The global search depends on the correlation between the fitness of a point and its distance to a global optimum. This is evaluated with the fitness distance correlation measure [3]. The local search depends on the relation between the fitness of a point and its neighbors. This is estimated with the autocorrelation function and the correlation length [6].

This research reveals how the examination of the landscape structure can help to improve the performance of GPDT. Two ways for mitigating the search difficulties are proposed: tuning the fitness landscape, and increasing the exploratory power of the genetic operator on this landscape. The first idea, concerning the global landscape structure, is to repair the fitness function when the fitness and the distance to the optima are uncorrelated. It is demonstrated that a careful design of the fitness function can increase the fitness distance correlation of a landscape and thus enable efficient search. The second idea, concerning the local landscape structure, is to adjust the parameters of the genetic operator if the correlation between the neighboring points is low. The claim of this paper is that the most useful practical ranges of the mutation operator parameters could be determined by landscape structure analysis.

2 Genetic Programming with Decision Trees

The genetic programming paradigm may solve inductive learning tasks [4] by breeding a population of decision trees. The modifications of the shape and complexity of the decision trees, are considered search for concept descriptions. An important feature of GP that makes it especially suitable for inductive learning is that they evolve the size of the concepts.

2.1 Decision Tree-Like Programs

GPDT manipulates variable length genotypes in the form of decision tree-like programs [7]. A decision tree is a set of concept descriptions. In terms of the chosen concept description language, the nodes of the decision tree are attributes of the concept features, and the leaves denote the class of the concept. Since a decision tree is an abstract representation of a composition of functions, it follows that the concept attributes should be functional genes; and the concept classes should be terminal genes.

2.2 Reproduction and Mutation

The GPDT approach evolves a population of decision trees with two genetic operators: steady state reproduction and mutation.

A uniform replacement mutation operator has been developed especially for inductive GPDT. This operator traverses a decision tree in depth-first manner and changes each visited node or leaf with probability $Pm = x/length(DT)$, where: x is a parameter. When a functional node is encountered, it is replaced with another randomly chosen functional node or with a randomly chosen terminal leaf with equal probability. If a terminal is reached, it is replaced with a randomly chosen functional node or a randomly chosen other terminal. Again, the possibility for choosing a node or a leaf is 50%.

2.3 The Fitness Function

The purpose of inductive learners is to identify concepts that best model given examples. This is estimated precisely with stochastic complexity measures. They provide criteria for isolation of accurate and parsimonious trees. We employ here the measure of Quinlan [8] and modify it to a new stochastic complexity $SC(DT)$ estimate which uses conditional probabilities, $P(e|tp)$ and $P(e|tn)$:

$$SC(DT) = min\{I(DT) + I(e|DT)\}$$

$$I(DT) = n_f + n_l + n_f \times log_2(f) + n_l \times log_2(l)$$

$$\begin{aligned} I(e|DT) = & \; tp \times (-log_2(P(e|tp)))+ \\ & fp \times (-log_2(1 - P(e|tp)))+ \\ & tn \times (-log_2(P(e|tn)))+ \\ & fn \times (-log_2(1 - P(e|tn))) \end{aligned}$$

$$P(e|tp) = P(e|tp_{l1} + tp_{l2} + ... + tp_{l_{N_p}})$$
$$P(e|tn) = P(e|tn_{l1} + tn_{l2} + ... + tn_{l_{N_n}})$$

where: n_f-nodes in DT; n_l-leaves in DT; f-possible functions; l-possible leaf classes; tp-true positive examples; fp-false positive examples; tn-true negative examples; fn-false negative examples.

3 Studies of Inductive GPDT

3.1 The Correlation Measures

The fitness distance correlation (FDC) [3] is calculated in GPDT with fitness-distance pairs recorded during runs with one tree. The distance is the smallest number of one-point mutations needed to produce the optimal tree from a particular tree. Since a fitness function with minimizing effect is used in GPDT, the FDC is 1 when the correlation is maximal.

The autocorrelation function (AC) [6] of the fitness landscape in GPDT is measured with fitnesses taken during a random walk from a randomly chosen decision tree through a set of trees generated by uniform replacement mutation. The correlation length (CL) [6] is the number of steps where $AC = 0.5$.

3.2 The Mutation Landscape

The mutation landscape relates decision trees with their fitnesses calculated with the stochastic complexity function. Such a view enables us to examine reliably the decision trees because their landscape could be analyzed with statistical measures. The differences between decision trees could be precisely identified as their fitnesses are points on different hills, valleys, and slopes of the landscape. From a global perspective, the FDC summarizes whether some global optima is accessible from every point on the landscape. The fitness should improve with decreasing the distance to the global optima. From a local perspective, a high AC during a long random walk, is an indication for easy searchable landscape [2]. If the mutation landscape is rugged often the search is difficult.

3.3 Experimental Results

Study of Global Landscapes

The analysis of the global landscape with FDC started using the measure of Quinlan [7]. The experiments were made with 2500 sampled points on the landscape and parameter $x = 0.05$.

Example sets for 6 decision trees with up to 6 attributes have been generated. The trees were of two kinds: balanced and unbalanced, and each tree had three different depths: short, medium, and long. Series of 10 tests have been conducted with each set and the results averaged. It has been noted that the tasks have low *FDC* in [0.31, 0.023],hence they are difficult for GPDT. This, however, was against the common sense as the examples have been deliberately distributed around globally optimal trees. Thus, the global landscape analysis with *FDC* revealed that there is something inconvenient in the measure of Quinlan [7] as a fitness function for GPDT.

The formula was modified to use conditional probabilities, which means that the fewer the branches the better the examples modelling. The tests with the modified function showed that the landscape is more correlated with the global optimum as the *FDC* increased to [0.58, 0.61]. This indicates a straightforward landscape [3]. Thus, the modified fitness function helped to tune the global landscape from misleading to straightforward. Having a straightforward global landscape, however, does not imply that an evolutionary learner can search efficiently on it.

Study of Local Landscapes

A group of 5 distributions of examples into positive and negative have been generated. For each of them 10 subsubgroups of examples for trees with up to 6 attributes have been produced. The parameters were: *PopulationSize*=50, *Generations*=200, and selected values of x: 0.001, 0.003, 0.005, 0.007, 0.01, 0.05, 0.07, 0.1, 0.15, 0.2, 0.25, 0.3, 0.5, 1, 2, 3, 5, 10, and 15. Average fitnesses from each of the 10 subgroups for every distribution were derived in cycles. The results from distribution 25% positives are plotted in Figure 1.

Figure 1 revealed that the curves are divided in two bunches: the lower bunch from values of x in [0.001, 1.0], and the upper bunch from x in [2.5, 15].

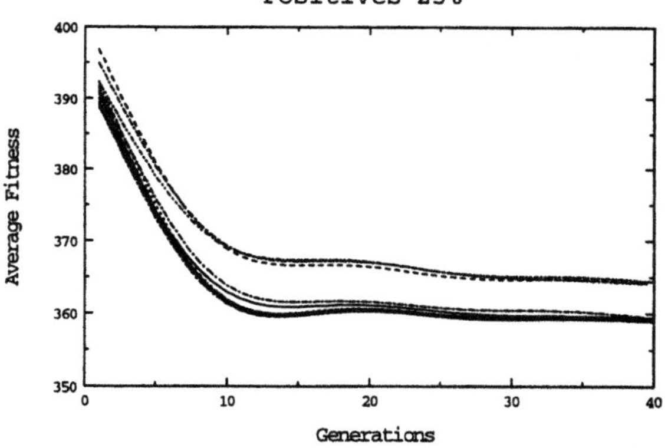

Figure 1: Results from distribution 25% positives.

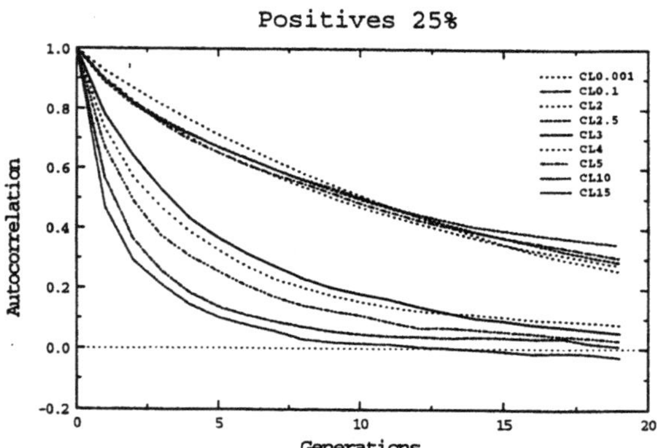

Figure 2: AC curves emphasizing that the performance is different for different values of x

The obvious separation of the curves prompted that maybe there is a relation between x and the degradation of the performance. That is why we studied the change of the local landscape structure with different x using the *AC* measure. The *AC* curves on Figure 2 emphasize that the performance is different for different values of x: the search is more efficient when x is in [0.001, 2] (i.e., the upper bunch of curves), and less efficient for values of x in [3, 15] (i.e., the lower bunch of curves). This means that the *AC* measure reveals the sharp transition in the performance of inductive GPDT.

When a transition exists, then it will be of benefit if the influence of x on the local fitness landscape structure is investigated. If the precise transition values of x can be determined, then they could be used to define the boundaries for efficient work. When the searcher is adjusted with respect to the critical transition values of x, then the behavior would not degrade. The answer of this question could be to take another look at the local fitness landscape structure with the *CL* metric computed over the *AC* values [6].

The *CL* from the 5 examples distributions are plotted on Figure 3. It shows that the behavior of GPDT when induced from 50% positives suddenly degrades, exactly as it suddenly degrades when induced from 25% and 75% positives. The sharp transition in the performance when learning from 50% positives resembles that found when learning from 25% and 75% positives. The transitions occur after the following critical values of x: when learning from 50% positives approximately $x = 1$, and when learning from 25% and 75% positives approx-

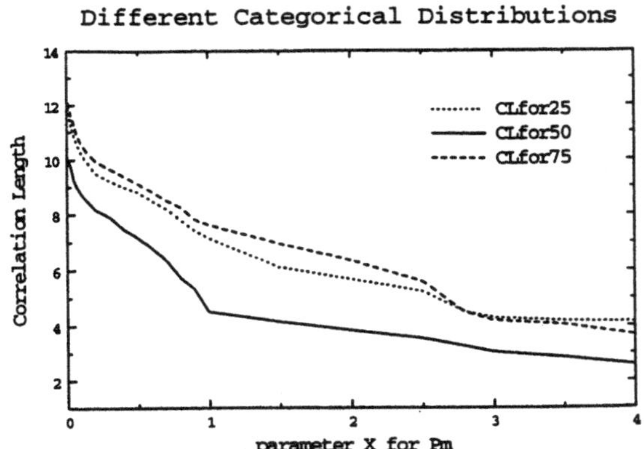

Figure 3: Correlation length from the 5 examples distributions.

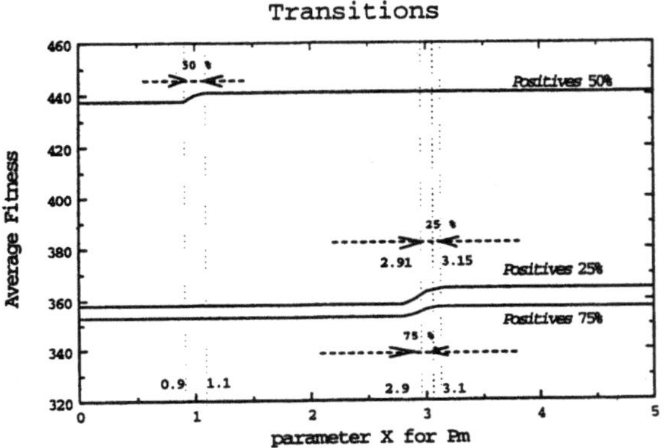

Figure 4: Critical values of x for different categorical examples distributions.

imately $x = 2.9$. In Figure 4 the critical values of x for different categorical examples distributions are plotted.

The implication of this phenomena is that it helps to determine the practical ranges of x. Figure 5 shows the values of x, above which the behavior of GPDT sharply degrades. When less than 25% of the given examples are positive, the range of practical values of x is relatively large and varies from 0.001 to 2.91. For tasks where there are provided between 25% and 50% positive examples, the upper practical value of x quickly decreases from 2.91 to 1.0. The third case again allows greater values for x from 1.0 to 2.9.

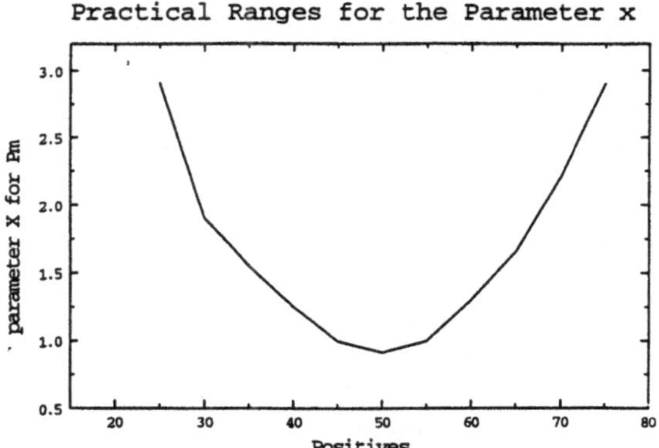

Figure 5: Practical ranges of x, which the GPDR sharply degrades.

4 Discussion

This study investigated the influence of the fitness function, the mutation operator and the categorical distribution of the examples on the search performance during evolutionary inductive learning. One conclusion is that the fitness function should fairly assign different fitnesses to the different decision trees and should account for the global optimum. It was demonstrated that the stochastic complexity fitness functions are very appropriate for evolving accurate and parsimonious decision trees.

The results point out the importance of the careful selection of values for the mutation operator parameter x. The choice of values for x critically influences the evolutionary behavior. First, the search with the mutation operator, tuned with special values of x, can be really efficient. Second, the search abruptly degrades after certain transition value of x.

It was found that the influence of the categorical distribution of the examples could be of three kinds: first, if there are given only positive or only negative examples the search is straightforward; second, if the positives and negatives are mixed the search is more difficult; and, third, if the positives and negatives are equally distributed the search is extremely difficult. This coincides with the known results derived by other learning paradigms.

5 Conclusion

This paper initiated the study of the navigation and the structure of the search process in evolutionary inductive learning. It has been demonstrated how the correlation measures could help to analyze the global and local landscape structure. Such an approach prompts for possible adjustments of the operator parameters. Therefore, it could serve for clever development of evolutionary inductive learners.

References

[1] J. Holland. *Adaptation in Natural and Artificial Systems.* MIT Press, Cambridge, MA, 1992.

[2] J. Horn and D. E. Goldberg. *Genetic Algorithm Difficulty and the Modality of Fitness Landscapes*, pages 243–269. Morgan Kaufmann Publishers, CA, 1995.

[3] T. C. Jones and S. Forrest. Fitness distance correlation as a measure of search difficulty for genetic algorithms. In *Proc. Sixth Int. Conference on Genetic Algorithms*, pages 184–192, 1995.

[4] J. R. Koza. *Genetic Programming: On the Programming of Computers by Means of Natural Selection.* MIT Press, Cambridge, MA, 1992.

[5] W. G. Macready, A. G. Siapas, and S. A. Kauffman. Criticality and parallelism in combinatorial optimization. Working Paper 95-06-054, Santa Fe Institute, Santa Fe, NM, 1995.

[6] B. Manderick, M. de Weger, and P. Spiessens. The genetic algorithm and the structure of the fitness landscape. In *Proc. Fourth Int. Conference on Genetic Algorithms*, pages 143–150, 1991.

[7] J. R. Quinlan. Induction of decision trees. *Machine Learning*, 1(1):81–106, 1986.

[8] J. R. Quinlan. MDL and categorical theories (continued). In *Proc. Int. Conference on Machine Learning, ICML-95*, Tahoe City, CA, 1995.

Discovery of Symbolic, Neuro-Symbolic and Neural Networks with Parallel Distributed Genetic Programming

R. Poli
School of Computer Science, The University of Birmingham,
Birmingham B15 2TT, UK
E-mail: R.Poli@cs.bham.ac.uk

Abstract

Parallel Distributed Genetic Programming (PDGP) is a new form of genetic programming suitable for the development of parallel programs in which symbolic and neural processing elements can be combined in a free and natural way. This paper describes the representation for programs and the genetic operators on which PDGP is based. Experimental results on the XOR problem are also reported.

1 Introduction

Genetic Programming (GP) is a method of program discovery consisting of a special kind of genetic algorithm (GA) capable of handling programs and an interpreter to run them [3].

Programs are expressed in GP as parse trees. For example, the expression `max(x*y,3+x*y)` would be represented as shown in Figure 1(a). The set of possible internal nodes used in GP parse trees is called *function set*, $\mathcal{F}$. $\mathcal{F}$ can include arithmetic operators, mathematical and Boolean functions, etc. The set of terminal nodes in the parse trees is called *terminal set* $\mathcal{T}$. $\mathcal{T}$ can include variables, constants, etc. The basic search algorithm used in GP is a GA with mutation and crossover specifically designed to operate on parse trees.

GP has been applied successfully to induce *sequential programs* and solve a large number of difficult problems (see [5] for an extensive bibliography). However, only a very small number of results have been reported where GP, appropriately modified, has gone beyond the production of sequential tree-like programs. For example, using cellular encoding GP has been used to develop neural nets [2] and electronic analogue circuits [4], while using interpreters implementing parallel virtual machines it has been used to develop special kinds of parallel programs [1, 7].

This paper describes Parallel Distributed Genetic Programming (PDGP), a new form of GP which is specialised in the development of parallel programs in which symbolic, numeric and neural processing elements can be combined in a totally free and natural way. PDGP is based on a graph-like representation for parallel programs and genetic operators which guarantee the syntactic correctness of the offspring. In the following sections PDGP is described and some results on the XOR problem are reported.

2 Representation

In PDGP we represent (parallel) programs as graphs with labelled nodes and oriented links. The nodes are the functions and terminals used in the program while the links determine which arguments are used by each function-node.

Figure 1(b) shows how `max(x*y,3+x*y)` can be represented in a parallel distributed form as a graph. The execution of the program should be imagined as a "wave of computations" starting from the terminals and propagating upwards along the graph, like the updating of the activations of the neurons in a multi-layer perceptron.

Graph-like representations of programs can be more compact (in term of number of nodes) and more efficient than tree-like representations (e.g. in Figure 1(b) the sub-expression `x*y` is computed only once). However, the direct handling of graphs within a GA presents some problems.

PDGP uses a direct representation of graphs which allows the definition of crossover operators which always produces valid offspring in a very efficient way. The representation is based on the idea of assigning each node in the graph to a physical location in a grid of pre-fixed shape and limiting the connections between nodes to be forward and between adjacent layers only.

By adding the identity function (i.e. a pass-

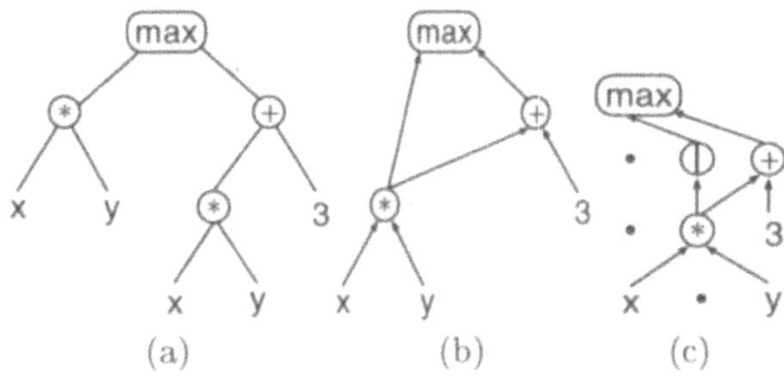

Figure 1: Parse-tree representation of the expression `max(x*y,3+x*y)` (a), the corresponding graph-like representation (b), and its PDGP grid-based representation (c).

through node) to the function set, any parallel distributed program (i.e. any directed acyclic graph) can be rearranged so that it can be described with this grid-like graph representation. For example, the program in Figure 1(b) can be transformed into the network in Figure 1(c).

In order to study all the possibilities offered by our representation, we decided to expand it to explicitly include introns ("unexpressed" parts of code) by associating a function or a terminal to *every* node in the grid, i.e. also to the nodes not directly or indirectly connected with the output. In some experiments we also added numeric labels to links to allow the direct development of neural networks.

3 Genetic Operators

Several kinds of crossover and mutation can be defined in PDGP. The crossover operator most similar to the one used in GP is called *Sub-graph Active-Active Node (SAAN) crossover.* It works as follows:

1. a random active node is selected in the first parent,
2. the sub-graph including all the active nodes which are used to compute the output of the selected node is extracted and its height and width is determined,
3. an active node in the second parent is selected such that its vertical position is compatible with the height of the sub-graph,
4. the sub-graph is inserted in the second parent to generate the offspring. In this last phase, if the horizontal position of the insertion node in the second parent is not compatible with the width of the sub-graph, the sub-graph is wrapped around.

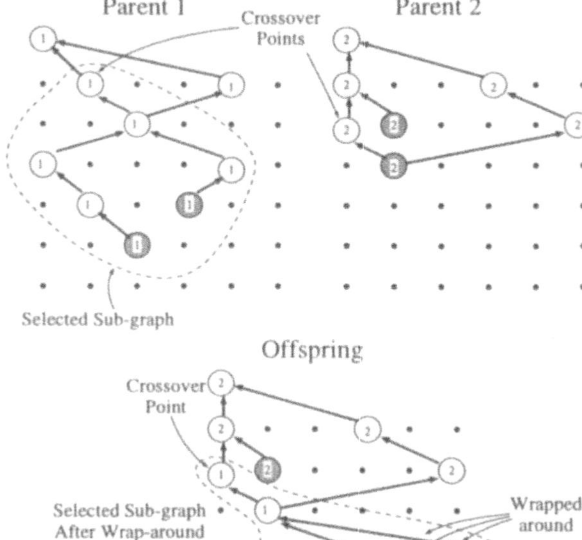

Figure 2: Sub-graph active-active node (SAAN) crossover.

An example of SAAN crossover is shown in Figure 2. The idea behind the SAAN crossover is that connected sub-graphs are functional units whose output is used by other functional units.

Therefore, by replacing a sub-graph with another sub-graph, we tend to explore different ways of combining the functional units discovered during evolution.

Other forms of crossover can be defined by modifying SAAN crossover. In this paper we have adopted the ***Sub-graph Inactive-Active Node (SIAN) crossover*** in which the crossover point in the second parent is randomly selected among the active nodes, while the crossover point in the first parent is randomly chosen among all nodes in the grid.

In standard GP, mutation consists of swapping a random sub-tree in an individual with a new randomly generated tree. This technique, which we call *global mutation*, can easily be extended to PDGP. We also use another form of mutation, *link mutation*, which makes local modifications to the connection topology of the graph (see [6] for more details). Mutation operators are applied after recombination and cloning.

4 Experimental Results

In this section we report on some experimental results obtained by applying PDGP to the problem of finding parallel distributed programs implementing the XOR function.

In the experiments, the population included 200 individuals, the maximum number of generation was 20, the crossover probability was 0.7, the global mutation probability was 0.25 and the link mutation probability was 0.25. The GA used tournament selection with tournament size 7. The other parameters were: "grow" initialisation method and SIAN crossover. The fitness of a solution was the number of entries in the XOR truth-table it correctly predicted.

Logic solutions In these experiments we used the function set $\mathcal{F}$={AND, OR, NAND, NOR, I} (I is the identity function) and the terminal set $\mathcal{T}$={x1, x2}.

Figure 3 shows two typical solutions obtained by PDGP. In the figure the active nodes and the active links have been drawn with thick lines and the output node has been centred horizontally for displaying purposes. In order to study the behaviour of PDGP on this problem we performed 20 runs (with different seeds for the random number generator) with three different grid sizes: 2×2, 2×3 and 3×4.

To assess the performance of PDGP we used two criteria: the computational effort E used in the GP literature (E is the number of fitness evaluations necessary to get a correct program, in multiple runs, with probability 99%) and the total number of nodes N to be evaluated in order to get a solution with 99% probability. We (over)estimated N by multiplying E by the number of nodes in the grid. The results are summarised in the first column of Table 1.

These results indicate that increasing the size of the grid reduces considerably the number of fitness evaluations necessary to get a solution. This seems reasonable considering that smaller grids impose harder constraints on the search. However, if we look at the values of N the advantage of larger grids is not so clear. In fact, the slightly greater number of node evaluations required by smaller grids is balanced by the fact that they produce better solution in terms of size, execution speed and generalisation.

Algebraic solutions In these experiments we used $\mathcal{F}$={+, -, *, PDIV, I} (PDIV is the protected division, which returns its first argument if the second is 0) and $\mathcal{T}$={x1, x2}. In these and the following experiments the output is considered to be 1 if it is greater than 0.5, 0 otherwise.

Figure 4(a) shows a typical solution to the XOR problem obtained using algebraic operators.

In 20 runs with three different grid sizes we obtained the results in the second column of Table 1 which indicate that algebraic operators make the search easier and that larger grids reduce the number of fitness evaluations but not the number of node evaluations.

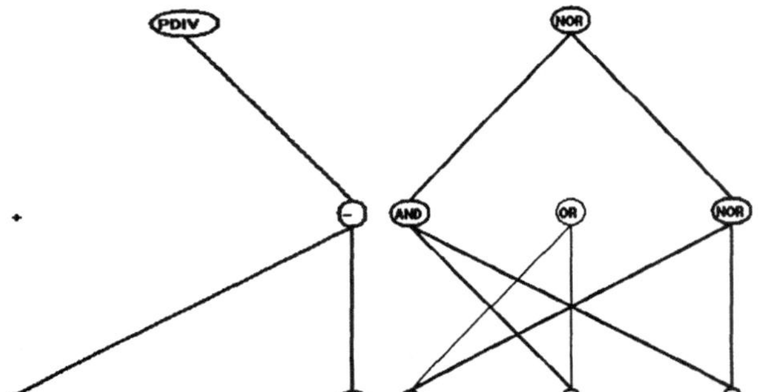

Figure 3: Symbolic networks implementing the exclusive-or function with Boolean processing elements.

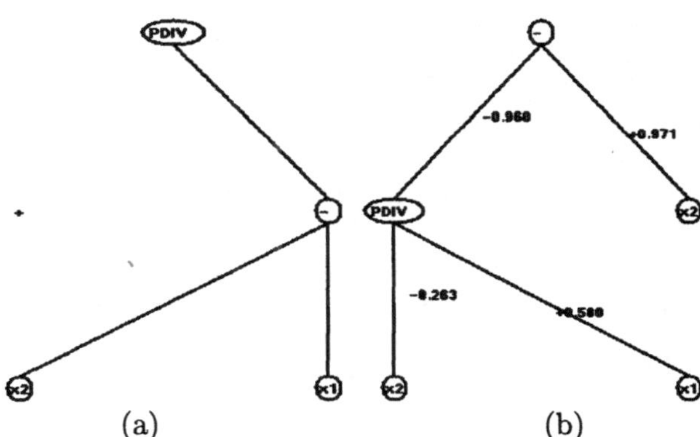

Figure 4: Algebraic (a) and neuro-algebraic (b) realisations of the XOR function.

Table 1: Computational effort required to solve the XOR problem with PDGP using different representations.

	Logic		Algebraic		Neuro-algebraic		Weight-less neural		Neural	
Grid size	E	N	E	N	E	N	E	N	E	N
2×2	5,200	26,000	2,400	12,000	9,600	48,000	10,200	54,000	342,000	1,710,000
2×3	4,200	29,400	2,800	19,600	12,000	84,000	6,800	47,600	378,000	2,646,000
3×4	1,600	20,800	1,600	20,800	7,000	91,000	6,000	78,000	46,200	600,600

Neuro-Algebraic solutions In these experiments we used the same function and terminal sets as in the previous section, but we added random weights in the range $[-1, 1]$ to the links. The weights act as pre-multipliers for the arguments of the functions in $\mathcal{F}$.

Figure 4(b) shows a typical solution to the XOR problem obtained using neuro-algebraic operators.

In 60 runs with this setting we obtained the results in the third column of Table 1 which indicate that the use of random weights makes the search much harder, at least without specialised weight-altering operators. Also, again increasing the size of the grid reduces the number of fitness evaluations but not the number of node evaluations.

Weight-less Neural Solutions In these experiments we used the function set $\mathcal{F}$={+, −, S2, S3, P2, P3, I} where + and − are introduced to simulate linear neurons, S2 and S3 are neurons whose inputs are added and then passed through a sigmoid activation function, and P2 and P3 are Π neurons which compute the product of their inputs. The terminal set included also a random constant generator, to create biases in the range [-1.0,+1.0]. The links had no weights.

Figure 5 shows two XOR implementation obtained by PDGP, while the fourth column of Table 1 reports the results obtained in 60 runs.

These results suggest that the use of neurons instead of Boolean or algebraic nodes makes the search harder. The reason for this might be the limited expressive power of neurons with respect to other classes of functions, at least for Boolean classification problems.

Neural Solutions In these experiments we used the same function and terminal sets as in the previous section but we added weights to the links.

Figure 6 shows two typical solutions to the XOR problem obtained with this setting. The results obtained in 60 runs are summarised in the last column of Table 1.

As expected, the combination of the negative effects of weights and neural processing elements produces a considerable degradation of the performance of PDGP. However, in this case a large grid is a big relative advantage.

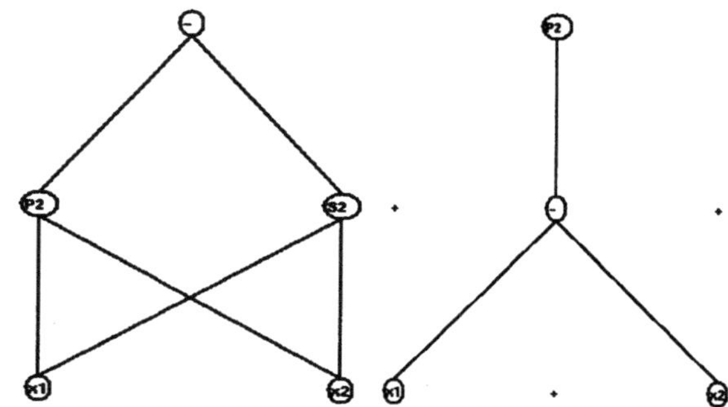

Figure 5: Weight-less neural networks implementing the XOR function.

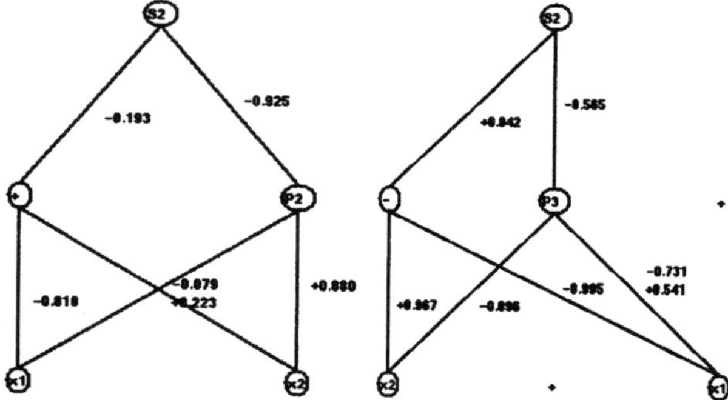

Figure 6: Neural XOR realisations.

5 Conclusions

PDGP is a new form of GP which is suitable for the automatic discovery of parallel network-like programs in which symbolic and sub-symbolic primitives can be combined in a free and natural way. In this paper we have presented PDGP and studied its representational capabilities using a very simple problem: learning the XOR function.

The results described here, along with recent results obtained on a variety of more complex problems [6], are very promising as they clearly show how PDGP can explore entirely new spaces of programs in which neural nets and classical tree-like programs are just special cases.

6 Acknowledgements

The author thanks the members of the EEBIC (Evolutionary and Emergent Behaviour Intelligence and Computation) group for useful discussions and comments. This research is partially supported by a grant under the British Council MURST CRUI agreement.

References

[1] F. H. Bennett III. Automatic creation of an efficient multi-agent architecture using genetic programming with architecture-altering operations. In *Proc. of Genetic Programming'96*, pages 30–38, Stanford University, July 1996. MIT Press.

[2] F. Gruau. Genetic micro programming of neural networks. In K. E. Kinnear, Jr., editor, *Advances in Genetic Programming*, chapter 24, pages 495–518. MIT Press, 1994.

[3] J. R. Koza. *Genetic Programming: On the Programming of Computers by Means of Natural Selection.* MIT Press, 1992.

[4] J. R. Koza, F. H. Bennett III, D. Andre, and M. A. Keane. Automated WYWIWYG design of both the topology and component values of electrical circuits using genetic programming. In *Proc. of Genetic Programming'96*, pages 123–131, Stanford University, July 1996. MIT Press.

[5] W. B. Langdon. A bibliography for genetic programming. In P. J. Angeline and K. E. Kinnear, Jr., editors, *Advances in Genetic Programming 2*, chapter B, pages 507–532. MIT Press, Cambridge, MA, USA, 1996.

[6] R. Poli. Parallel distributed genetic programming. Technical Report CSRP-96-15, School of Computer Science, The University of Birmingham, September 1996.

[7] A. Teller and M. Veloso. Neural programming and an internal reinforcement policy. In J. R. Koza, editor, *Late Breaking Papers at the Genetic Programming 1996 Conference*, pages 186–192, Stanford University, July 1996. Stanford Bookstore.

[8] P. Walsh and C. Ryan. Paragen: A novel technique for the autoparallelisation of sequential programs using genetic programming. In *Proc. of Genetic Programming'96*, pages 406–409, Stanford University, July 1996. MIT Press.

A Neural Network Technique for Detecting and Modelling Residential Property Sub-Markets

O.M. Lewis[1,2], J.A. Ware[1] and D. Jenkins[2]
[1]School of Accounting and Mathematics, University of Glamorgan,Trefforest, Mid Glamorgan, UK
[2]Centre for Research in the Built Environment. University of Glamorgan, Trefforest, Mid Glamorgan, UK

Abstract

A number of published studies have investigated the application of neural network technology to residential property appraisal. The majority of these studies have concentrated on homogeneous areas (that is areas where properties are subject to the same environmental and locational factors). This is generally done to restrict the data set to one local sub-market. However, the models created are specialised and not locationally portable. This paper presents a methodology, which builds on research in which a Kohonen map is used to uncover sub-markets within a large data set that are subsequently independently used to train a series of multi layer perceptron (MLP) networks. The study concludes that by modelling possible sub-markets an acceptable accuracy over a heterogeneous area can be achieved. The work presented in this paper is funded via a Realising Our Potential Award under the auspices of the ESRC.

1 Introduction

In the UK, and indeed in many countries, the valuation of residential property is based on the method of direct capital comparison. However, the principal weakness of this method of valuation is the problem of obtaining suitable comparable properties as evidence [8]. Given this weakness, research has been carried out on directly calculating the value of a property from its locational and physical attributes [2, 4, 5, 8]. Many of these research studies have considered the application of neural networks to residential property appraisal [4, 5], with the majority of studies using data from a homogenous area (i.e. an area where all properties are subject to the same environmental and locational forces). This approach is taken as the valuation function can become very elaborate when spread across a heterogeneous area. This however leads to neural network models which are not locationally portable. Nevertheless, the studies have reported a high level of success, with average absolute percentage error levels of between 5 and 7.5% not being uncommon [3, 5].

Adair [1] hypothesises that sub-markets can be identified by stratifying the market into increasingly homogeneous subsets. Using the hypothesis that a heterogeneous area consists of many homogeneous areas, the authors postulate that a heterogeneous area can be modelled indirectly using many MLP, trained on subsets of the parent data set. To use such a system to predict the value of a previously unseen property, requires a method of selecting the appropriate neural network model. This paper describes such a method, in which a Kohonen feature map is used both to identify groupings within the parent data set and to act as a panel judge to decide which neural network model to select when asked to give a valuation.

2 Overview of Methodology

The methodology involves training a Kohonen network on historical data, collected from approved mortgage transactions over a number of years. The data set from each significant grouping — formed by the Kohonen network — become the training set for a MLP. Each MLP is trained using a save best approach. This involves periodical testing of the network with a validation set and, if the current state produces better results than the previous best, the current state becomes the new best state. At the end of training the current best state is adopted as the network to be used during day-to-day operations. Figure 1 provides an overview of the methodology.

The advantage of using the Kohonen self organising map for this application is that it can identify clusters within the parent data set that are difficult to achieve using simple sort procedures. However, it is sometimes difficult to identify class boundaries

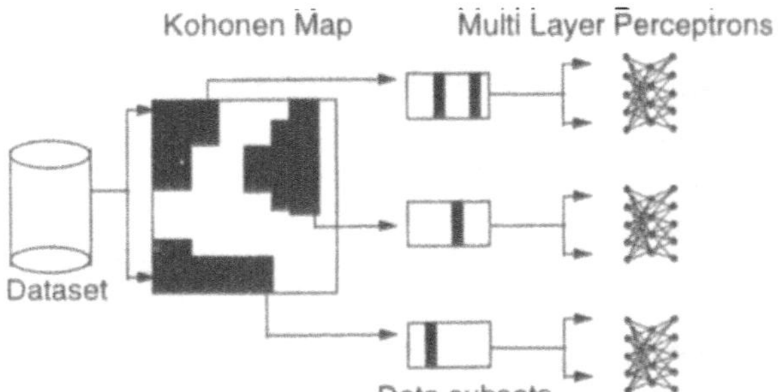

Figure 1: An Overview of the Methodology.

within a trained Kohonen map [6], and this in turn leads to problems in generating training sets for the MLP. To overcome this problem a simple method of identifying class boundaries or discriminants can be used, Zurada [9] explains:

> A pattern added to the cluster has to be closer to the centre of the cluster than to the center of any other cluster.

Using this rule, each node can be examined and the distance from the surrounding centroids can be calculated. The subject node can then be added to the nearest cluster [6].

However, in this application, which aims to generate useful training data sets, the formation of a class boundary using this may dramatically increase the variance of the training data. This increase will in turn reduce the potential accuracy of the MLP model.

In addition to identifying boundaries around input clusters, it is also important for input clusters to be matched by corresponding output clusters. If, for example, the Kohonen map has clustered residential properties from two different locational areas, it is reasonable to expect similar types of houses from each area to have a similar property value. Faced with this problem, the authors decided to investigate a recently published variance estimation routine known as the gamma test [7] with the expectation that this would address the modelling requirements.

The gamma test is a data analysis routine which aims to estimate the best mean square error (MSE) which can be achieved by any smooth data modelling technique using the data. In essence, the test estimates the variance in the output mapping by considering the cumulative average distances in Euclidean space for each record and its nearest neighbours. This is achieved by performing least squares over the difference in input and output distances from each vector and a specified number of its nearest neighbours. Knowing the variance within a data set allows prediction beforehand of the MSE of the best possible neural network trained on that data [7].

An algorithm was developed by the authors which attempts to identify useful clusters by selecting a centroid and adding neighbouring nodes — where the addition of a node increases the variance significantly it is subsequently removed. This process iterates until the cluster size is maximised within a specified variance threshold.

As the gamma test estimates the variance within a data set, prior to boundary detection the variance threshold can be set, and therefore the level of performance of a neural network trained on the sub-clusters can be predicted.

3 Testing the Methodology

A database containing information on residential property transactions during the period January 1993 to December 1995 was selected to test the methodology. The input features used for this study were: Street, District, Type (Mid-Terraced etc.), Floor Area and Valuation Date. The target variable was value.

A 10 by 10 Kohonen map was used to find the groupings in a historical data set containing 990 records. Value, the output feature, was omitted from the Kohonen training set. The records from each cluster identified using the boundary technique over the Kohonen map were examined and common features within the group removed. Obviously, as the data set is partitioned into classes, the classes contain only a portion of the original 990 records. However, this is accompanied by a decrease in the number of dimensions - as constant columns were removed.

4 The Impact of the Methodology on Prediction Accuracy

In order to provide a bench mark for analysing the methodology, a single MLP network was trained on the whole data set. After training the ability of the network to appraise residential properties with known values was tested. The results of this test are shown in Figure 2; the graph shows the actual and predicted values for 117 test properties (for ease of interpretation the properties have been ordered according to actual value). Figure 3 shows the re-

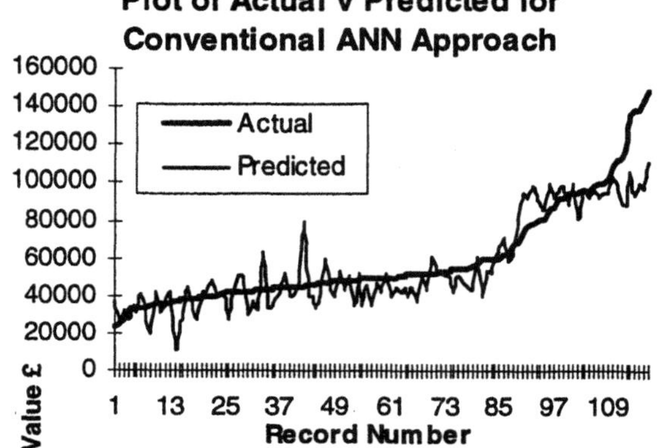

Figure 2: A graph of actual and predicted value gained using a conventional neural network approach.

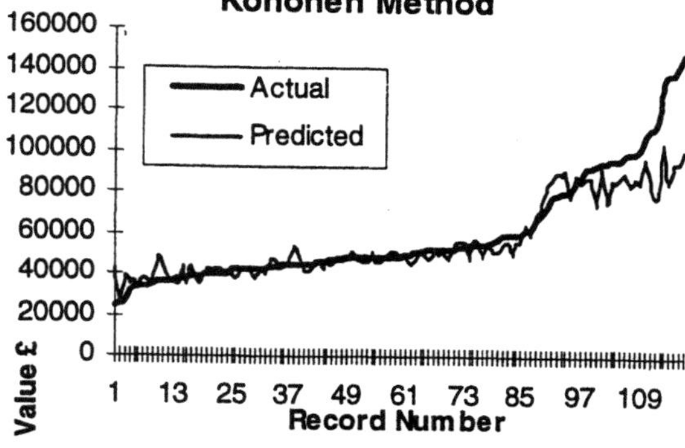

Figure 3: A graph of actual and predicted value gained using the hybrid approach.

Table 1: Prediction Accuracy of Models.

	Conventional ANN Method	Kohonen Method
Mean absolute % difference	18%	8%
% of Records with an error > 10%	74%	22%
Maximum absolute % error	310%	49%

sults achieved using the described methodology on the same 117 properties in the test set. The differences between the value returned by the valuer and with those predicted by the neural network models are shown in Table 1.

5 Conclusion

It is evident, from the results obtained, that the methodology proposed in this paper compares very favourably with the more conventional neural network approach. An average increase in prediction accuracy of 10% was achieved using the new method over the conventional approach. This implies that the original data set either contained more than one underlying function (pattern) [6] or the function was too elaborate to be modelled using a single MLP. Moreover the Kohonen network can discern different classes within the data, which when independently modelled yield a greater predictive accuracy than those computed for the original data set [6].

From the work outlined in this paper, the authors have concluded that the techniques described may usefully be applied to other data sets.

The authors are currently in the process of analysing data from the 1991 census, using the methods described, with the aim of grouping together enumeration districts that contain the same valuation functions. This will allow the modelling process to move from specialised models to locationally portable systems.

References

[1] A. S. Adair, J. N. Berry, and W. S. McGreal. Hedonic modelling, housing submarkets and residential valuation. *Journal of Property Research*, 13:67–83, 1996.

[2] A. S. Adair and S. McGreal. The application of multiple regression analysis in property valuation. *Journal of Valuation*, 6:57–67, 1987.

[3] R. A. Borst. Artificial neural networks: The next modelling/calibration technology for the assessment community? *Journal of Property Tax*, 10(1):69–94, 1991.

[4] Q. Do and G. Grudnitski. A neural network approach to residential property appraisal. *The Real Estate Appraiser*, pages 38–45, 1992.

[5] A. Evans, H. James, and A. Collins. Artificial neural networks: An application to residential valuation in the UK. *Journal of Property Valuation and Investment*, 11:195–204, 1992.

[6] H. James. An automatic pilot for surveyors. In *Proc. RICS Cutting Edge Conference*, 1994.

[7] N. Koncar. *Optimisation Methodologies for Direct Inverse Neurocontrol.* PhD thesis, Department of Computing, 180 Queens Gate London SW7 2XZ, UK, 1997.

[8] E. Worzola, M. Lenk, and A. Silva. An exploration of neural networks and its application to real estate valuation. *Journal of Real Estate Research*, pages 185–201, 1995.

[9] J. M. Zurada. *Introduction to Artificial Neural Systems.* West Publishing Company, 1992.

Versatile Graph Planarisation via an Artificial Neural Network

T. Tambouratzis
Institute of Nuclear Technology—Radiation Protection, NCSR 'Demokritos',
Aghia Paraskevi, Athens 153 10, Greece.
Email: tatiana@zeus.int-rpnet.ariadne-t.gr

Abstract

An artificial neural network which is based on harmony theory has been applied to the graph planarisation problem. The presented artificial neural network is capable of solving both aspects of graph planarisation (creation of the optimally planarised as well as of the maximally planar forms of any graph). Optimal solutions are always produced independent of the complexity or planarity of the original graph; if more than one equally good solutions exist they are produced with the same probability.

1 Introduction

An artificial neural network (ANN) which is based on the principles of harmony theory (HT) [1] is proposed for solving the graph planarisation problem. Graph planarisation constitutes an NP-complete problem with important applications in the area of VLSI circuit design. It consists of determining an *optimally planarised graph*, i.e. drawing the edges between the connected vertices of the original graph such that the least possible number of (edge) crossings are necessary; the single-row routing representation for the vertices of the graph is generally followed. If no crossings occur the graph is planar, else it is characterised as non-planar. An important problem for a non-planar graph is to find a *maximal planar subgraph*, i.e. to determine the maximum number of edges of the original graph which can be accommodated such that no crossings are necessary. Clearly, the two aspects of the planarisation problem coincide for planar graphs.

For the HT ANN implementation of graph planarisation, construction is easy to perform, the solution is given directly by the activation values of the nodes, and an optimal solution is always produced. Additionally, both aspects of the problem are solved with a single ANN; the solution depends on the direction of the activation flow within the ANN which is—in turn—accomplished by simply interchanging the two layers of nodes of the HT ANN.

2 Harmony Theory

HT supports an ANN architecture which falls in the category of consensus-function ANNs. The particular consensus function is called harmony (H); H is calculated over the activation values of all the nodes in the ANN and denotes the internal consistency of the state, i.e. the optimality of the corresponding solution. An optimal solution (HT ANN state of highest H) is reached after repeated updates of the activation values of the nodes (simulated annealing).

Owing to the temperature parameter T, which is progressively lowered during simulated annealing, the updates are not purely deterministic. Simulated annealing begins from an adequately high T, whereby the node activation updates are completely probabilistic and the entire problem-space has equal likelihood of being sampled. T is gradually decremented, causing the node activation updates to become more deterministic and the HT ANN to focus upon specific neighbourhoods of states corresponding to solutions of high internal consistency. At the end of simulated annealing T approaches zero: the node activation updates are completely deterministic and a HT ANN state of highest H is reached with a probability of $+1$; this constitutes an optimal solution of the problem.

The HT ANN does not require training since the nodes and connections can be set directly by the programmer. The nodes are binary and are arranged in exactly two layers (lower and upper). Connectivity is symmetric and is only allowed between nodes of different layers. The following differences between the nodes of the two layers can be defined.

2.1 Semantic Distinction

The nodes of the lower layer represent the basic elements of the problem; they assume binary activation values $\{+1, -1\}$ which denote whether the encoded element appears in the solution of the problem. The collection of activation values over all the nodes of the lower layer expresses the solution of the problem.

The nodes of the upper layer encode the constraints which apply between the elements of the problem; they assume binary activation values $\{+1, 0\}$ which denote whether the particular constraint is enforced. The collection of the active nodes of the upper layer expresses the set of constraints which are satisfied by the solution of the problem.

2.2 Functional Distinction

Connectivity is defined by the upper layer. The connections are logically assigned from each node of the upper layer to all the nodes of the lower layer and are:

- *Positive* if the constraint encoded by the node of the upper layer supports the presence of the element represented by the node of the lower layer.
- *Negative* if the constraint supports the absence of the element.
- *Zero* if the constraint has no relation to the element.

The weights of the HT ANN are determined by the connectivity of the nodes of the upper layer. In the simplest case, the absolute value of the weight of each connection is inversely proportional to the number of connections emanating from the node of the upper layer.

The two layers affect H in a different manner. Only active nodes of the upper layer contribute to H, their positive/negative contribution depending on the agreement in sign between the connections from that node and the activation values of the connected nodes of the lower layer. Conversely, all the nodes of the lower layer influence H.

3 Graph Planarisation

Let us assume that graph G is to be planarised, where $G = (V, E)$, $V = \{v_1, v_2, \ldots, v_n\}$ denotes the set of vertices and $E = \{(i,j) : v_i$ and v_j are connected$\}$ denotes the set of edges of G; each $v_i \in V$ is assumed to be connected to at least one $v_i \in V$, $i \neq j$. Provided that the single-row routing representation of the vertices of V has been applied so that v_i appears before v_j in the row of vertices if and only if $i < j$, every $(i,j) \in E$ can be drawn as a curved line which appears above or below the row of vertices and connects vertices v_i and v_j.

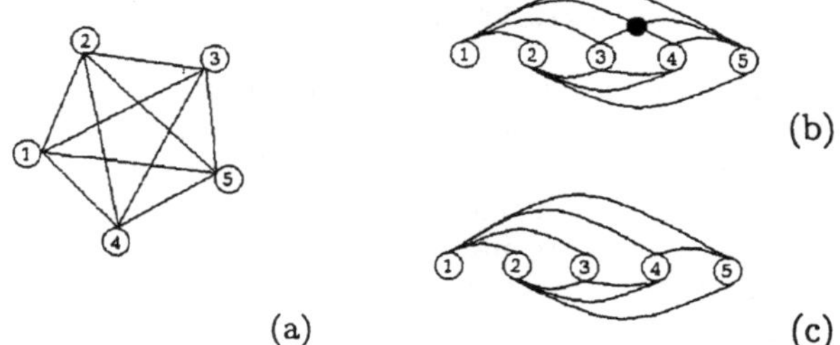

Figure 1: A graph (a) and an example of its optimally planarised (b) and its maximally planar (c) form.

The constraints that apply to the planarisation problem are:

- *Uniqueness constraint.* The edge between two connected vertices — if it appears in the planarised graph — must be drawn only once, either above or below the row of vertices.
- *Local planarity constraint.* For any pair of edges (i,j) and (k,l) such that $i < k < j < l$, if (i,j) is drawn above/below the row of vertices (k,l) must be drawn below/above for a crossing to be avoided.

If it is possible for all the edges in E to be drawn in the single-row routing representation of the vertices in V such that none of the aforementioned constraints are violated, G is planar and a solution of the graph planarisation problem is produced. Else (for a non-planar graph), it is important to find a maximally planar subgraph (such that the greatest possible number of edges of E be drawn, once each and in an appropriate configuration of above/below directions that no crossings exist). An optimally planarised graph as well as a maximal planar subgraph of a simple non-planar graph are illustrated in Figure 1. While the original graph (a) and the optimally planarised graph (b) contain 10 edges, the maximal planar subgraph (c) contains only 9; on the other hand, one local planarity constraint is violated in (b) (crossing between edges (1,4) and (3,5)), but no constraints are violated in (c).

4 ANN Construction

Special emphasis is placed on the functional distinction between the two layers (see Section 2.2) of the HT ANN and this is utilised for solving the two aspects of the graph planarisation problem. The two layers of nodes are interchanged, whereas the activation values, the connectivity and H are reconfigured to provide the different solutions. One layer encodes the edges of E, while the other layer encodes the constraints described in Section 3. More specifically:

- A pair of nodes is employed for each edge: if the first node of the pair is active the related edge is supported as being drawn above the row of vertices, while if the second node of the pair is active the related edge is supported as being drawn below.

- A pair of nodes is employed for each constraint. Not all of the constraints need to be encoded in the HT ANN, as it suffices for an edge to be involved either in uniqueness or in local planarity constraints. When an edge (i, j) is involved in at least one local planarity constraint (e.g. with edge (k, l)), a pair a nodes is encoded for each such constraint: if the first node of the pair is active and $i < k < j < l$ (i, j) is supported as being drawn above the row of vertices and (k, l) below, while if the second node of the pair is active (i, j) is supported as being drawn below the row of vertices and (k, l) above; the reverse is true if $k < i < l < j$. When an edge is not involved in any local planarity constraints, a pair of nodes is used to encode the uniqueness constraint: if the first node of the pair is active (i, j) is supported as being drawn above the row of vertices, while if the second node of the pair is active (i, j) is supported as being drawn below.

This manner of encoding the graph planarisation problem renders the HT ANN transparent. The graph that corresponds to a particular HT ANN state is given directly by the activation values of the nodes. The active nodes specify:

- Which edges have been drawn in the graph and whether they have been drawn above or below the row of vertices.

- Which constraints have been satisfied.

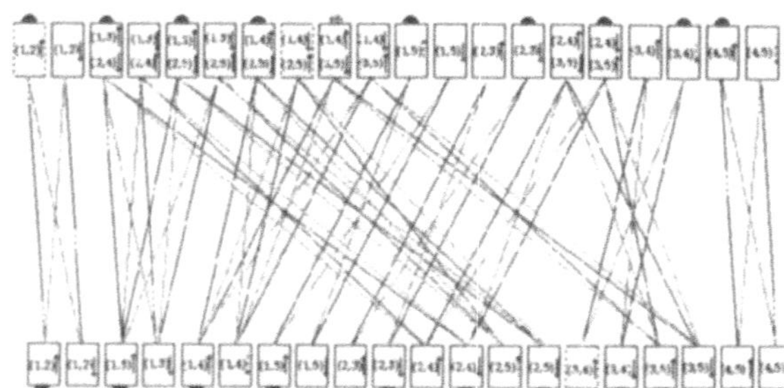

Figure 2: HT ANN for the creation of the optimally planarised forms of the graph of Figure 1(a).

4.1 Creation of the Optimally Planarised Graph

The HT ANN employed for optimally planarising the graph of Figure 1(a) is depicted in Figure 2. The nodes encoding the edges appear in the lower layer, while the nodes encoding the constraints appear in the upper layer. Black lines represent positive connections and grey lines negative connections (zero connections have been omitted). Pairs of nodes of the upper layer have connections of opposite sign to pairs of nodes of the lower layer, ensuring that exactly one node of the lower layer from every pair is active in a HT ANN state of high H.

If the HT ANN contains L nodes in the lower layer and U nodes in the upper layer, exactly $\frac{L}{2}$ are active after completion of simulated annealing and these represent a solution of the problem of optimal planarisation. If the graph is planar, exactly $\frac{U}{2}$ nodes of the upper layer remain active. If the graph is non-planar and U^* equals the number of active nodes of the upper layer, the number of crossings is given directly from $\frac{U}{2} - U^*$. This is demonstrated in Figure 2, where the active nodes of both layers have been marked by a black semi-circle for the solution corresponding to the optimally planarised graph of Figure 1(b). The grey semi-circle marks the inactive node of the upper layer whose corresponding edge is responsible for the crossing. $\frac{U}{2} = 10$ and $U^* = 9$, which implies that one crossing exists in the optimally planarised graph. An HT ANN state with 10 active nodes in the lower layer and 9 in the upper constitutes an optimal solution.

4.2 Creation of the Maximal Planar Subgraph

Figure 3 illustrates the HT ANN that produces a maximal planar subgraph for the graph of Figure

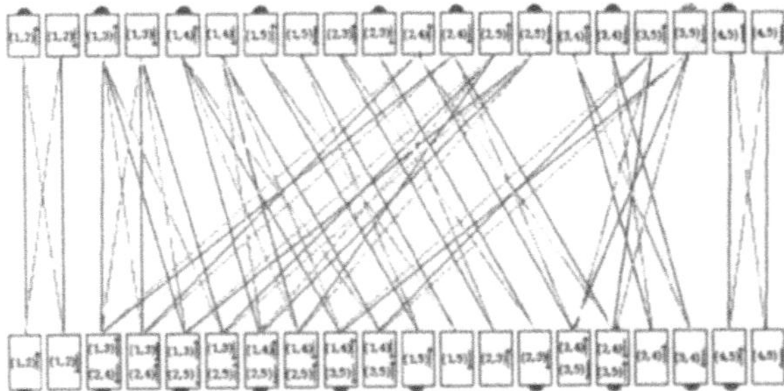

Figure 3: HT ANN for the creation of the maximally planar forms of the graph of Figure 1(a).

1(a). Here, the reverse arrangement of the two layers is followed, i.e. the nodes encoding the edges appear in the upper layer and the nodes encoding the constraints appear in the lower layer.

If L nodes appear in the lower layer and U in the upper, exactly $\frac{L}{2}$ nodes from the lower layer are active in a HT ANN state of highest H — one from each pair — denoting that every uniqueness and local planarity constraint is correctly enforced. For nodes encoding uniqueness constraints such a configuration of activation values is straightforward and exactly one of the connected nodes of the upper layer becomes active while the other remains inactive. The same is true of all the nodes encoding local planarity constraints for a planar graph; consequently, half of the nodes of the upper layer ($\frac{U}{2}$) — one from each pair — are active in an optimal solution representing a maximal planar subgraph which also constitutes an optimally planarised graph of the planar graph. For a non-planar graph $U^* < \frac{U}{2}$, and $\frac{U}{2} - U^*$ provides the number of edges which must be omitted from the graph in order for a maximal planar sugbraph to be produced.

5 Results

The decrement of the temperature parameter T has been scheduled as:

$$T_i = T_{i-1} T_{decr}, \ i = 1, 2, \ldots, Min$$

where T_o denotes the initial and T_{Min} the final temperature, while T_{decr} denotes the scaling factor $(0 < T_{decr} < 1)$ that causes simulated annealing.

It has been found that, as the number of nodes in the HT ANN increases (which corresponds to planarising larger graphs), the number of node activation updates rises until a solution can be consistently produced. This implies that different values of T_o, T_{decr}, and T_{Min} are required for efficient and accurate simulated annealing of different graphs.

Performance has been found not to be affected by the following:

- The complexity of the original graph, i.e. the number of nodes and edges in the graph.
- The planarity of the original graph, i.e. whether it is planar and — if not — by its number of crossings.
- The number of optimal solutions that can be produced for the original graph.

The HT ANN always produces an optimal solution of the desired aspect of the planarisation problem and, if more than one optimal solutions exist, these are produced with equal probabilities.

6 Conclusions

A HT ANN has been proposed for solving the two aspects of the graph planarisation problem, i.e. determining an optimally planarised graph and a maximal planar subgraph. The HT ANN is:

- *Transparent.*

  1. By utilising its two layers to separately encode the edges of the graph and the constraints of the problem, the solution corresponding to the ANN state is easy to glean.

  2. The activation values of the nodes specify which edges have been drawn in the graph as well as which constraints have been satisfied by the solution.

- *Versatile.* The aspect of the solution (optimally planarised graph or maximal planar subgraph) depends solely on the relative arrangement of its two layers of nodes, which affects the flow of activation within the ANN, its connectivity and the optimality of the HT ANN state.
- *Accurate.* For appropriate values of its parameters, a correct solution is always produced.

References

[1] P. Smolensky. *Information processing in dynamical systems: foundations of harmony theory*, volume 1, pages 184– 281. MIT Press, Cambridge MA, 1986.

Artificial Neural Networks for Generic Predictive Maintenance

C. Kirkham and T. Harris
The Centre for Neural Computing Applications, Brunel University
Egham, Surrey, UK
Email: {Christopher.Kirkham, Thomas.Harris}@brunel.ac.uk

Abstract

This paper outlines a research project to develop artificial neural networks as a diagnostic tool for the automatic identification of rotating machine faults. This work was instigated by the DTI Neural Computing Learning Solutions Campaign's AXON Neural Projects Club. Industrial sponsors are Entek/IRD, Diagnostic Instruments Ltd., and Arjo Wiggins Paper Mills. The biggest problem encountered by developers of SMART software systems is that examples of all conditions to be identified are required. In practice this is not possible due to the routine method of plant machinery data collection, and due to the individual behaviour of the machinery. A method is required which is capable of diagnosing a previously unseen fault upon any bearing. This paper proposes a hybrid neural network approach which first determines a novel condition, then a knowledge base categorises the condition. The system is currently working off-line in support of the maintenance technician at the sponsors paper mill plant.

1 The Problem

The development of non-intrusive monitoring techniques, such as vibration and acoustic analysis, thermal imaging, and oil analysis, coupled with the ability to record these measurements, has enabled industry's maintenance strategies to move forward from reactive and preventative methods towards more predictive approaches such as Hanna [1].

These original approaches relied primarily upon the skills of a manual operator to determine whether the machine was operating abnormally, but they were only able to gauge the overall condition of the machine. They had difficulty though in determining small changes, and this is particularly important for detecting changes in single components, such as bearings, fan blades, etc.

This inability to identify such small defects or changes which could indicate the cause of the machines condition often lead to a constant failure being accepted as normal, when corrective action could be taken to prevent such events occurring. Howieson [4] explains how such small changes can also lead to large failures developing within the machine.

By implementing predictive maintenance strategies, it is possible for industry to achieve large financial savings (emphasised by Kershaw [5]), offsetting the initial investment in monitoring equipment and staff training. Predictive maintenance (PM) utilises non-intrusive data collection methods to develop a picture of the machines condition. Trends are identified in this data from which it is then possible to extrapolate the serviceable life before failure of the machine.

These savings arise as a result of reduced downtime, often less catastrophic failures (consequently cheaper repair work), and the efficient use of stores by extending the lifetime of components before replacement, when compared to the results of reactive and preventative approaches.

1.1 Off the Shelf Predictive Maintenance

This project is concerned with PM machine condition monitoring strategies. The focus is upon walk-around maintenance using vibration analysis, implemented by portable data collectors/analysers downloading data to a desktop PC.

Walk-around maintenance is not an on-line method; as its name implies, data is collected by a human visiting each machine on a tour. Since only very little dedicated hardware (a desktop PC, analyser) and software (database, e.g. Emonitor) are required, it is an economic solution and is used widely in manufacturing plants.

The method requires a technician to visit in turn each machine that is scheduled to be monitored (typically weekly or monthly) and collect the vibration spectra with the portable analyser and transducer, downloading the data to PC based software

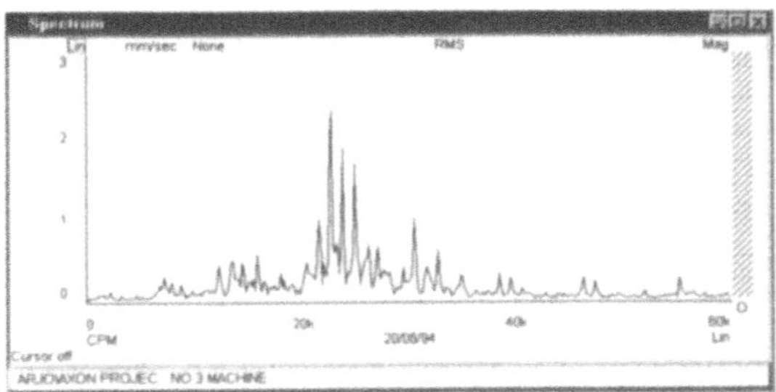

Figure 1: Typical collected vibration spectrum data.

for analysis. This provides an image or 'snapshot' view of the machine's condition at that moment in time (see Figure 1). Diagnosis of the current condition and a prediction of the expected time to component failure is then performed by the technician.

Desktop PC packages now provide the ability to incorporate various non-intrusive collection methods e.g. vibration, acoustic, thermal, oil, and motor current analysis. These are normally used individually, the most widely used being vibration analysis, but by combining these methods they can provide complementary analysis data. This can improve the accuracy of the diagnosis procedure, but is only economic for expensive or critical plant.

Despite the power of PC software, providing tools such as component alarm settings and trend plots, diagnosis and analysis is a time consuming process requiring training and expertise, and one which currently can only be performed away from the production floor.

Thus the disadvantages with the current approaches are that they are primarily reliant upon human knowledge, experience and expertise to perform a diagnosis; plus this process has to be performed upon large amounts of data, off-line and away from the machines, which can be a very time consuming process, particularly in large plants. Ideally a system is required that is capable of providing immediate diagnosis at the point of collection as described and introduced by Harris [2].

1.2 Data

Most artificial intelligence technologies and certainly all neural computing systems are critically dependant upon a good quantity of quality example data—the garbage in, garbage out situation, Serridge [6], applies where poor development data will produce a poor system. To perform a complete diagnosis, a neural network (and any SMART software system, whether an expert system, fuzzy based system, etc.) will require examples of all the conditions it is expected to diagnose. Herein lies the problem.

One difficulty with diagnosis of roller element bearings is due to the nature of PM data collection, normally scheduled for once per week or month. Therefore very little data will normally be collected as the fault develops. Consequently databases are deficient in fault data available to train a neural network. This problem has plagued any system development in the field.

Difficulties in collecting fault data arise because:

1. A company will not let you run their machine close to failure just to collect data, due to either the expense of the machinery and/or the cost of loss of production due to consequent shutdown and repair time;

2. All types of fault do not normally occur in a machines lifetime, if maintained properly;

3. A company will not let faults be induced to allow data collection, due to (1) above.

This has the effect of making it very difficult to develop a system capable of diagnosing the condition of a particular machine, and practically impossible to produce a single network approach to provide a generic diagnosis system for any machine.

Collection of test rig data is one approach to solving this, but it lacks the reality of in service use and wear. Serridge [6] discusses the implications of identical bearings operating in separate machines which behave differently. This occurs to the extent that one machine will operate in good condition with higher vibration amplitudes than the other, or will require much higher vibration levels before failure occurs, or will take longer to fail due to a certain fault.

Despite this variation in magnitude of vibration levels, there are certain modes of bearing failure or a combination of these, which all are subject to, which are illustrated and discussed in detail by Herraty [3]. Therefore another approach is required which can utilise these generic bearing defect characteristics to develop a system capable of diagnosing bearing faults without requiring examples of all conceivable conditions.

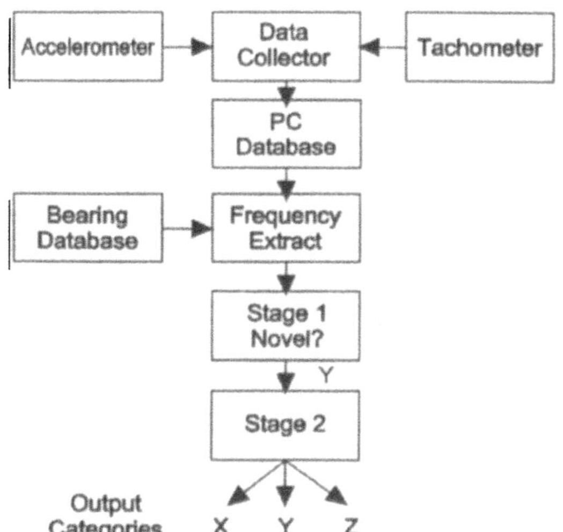

Figure 2: Block diagram of AXON system.

2 Solution

The approach developed by this project and illustrated by Figure 2, utilises a hybrid neural network system in which a first stage network determines the novelty of the input data, and a second network provides generic diagnosis.

The first stage networks use a Kohonen architecture and are machine specific. These are trained upon examples covering their entire good condition operating envelope (e.g. varying load and speed). From this model of 'normality', any novel or abnormal changes to the machine can be identified.

The data is deemed as novel automatically by the first stage networks. This has been achieved by assigning a boundary value to the nodes in the network, which is determined by the interneuron distances which result from training. If the input data exceeds the boundary of the winning node, then the data is considered novel (see Figure 3).

The second stage uses the error distances between the input and the winning node in the first network as inputs for training. The knowledge base is developed using known modes of failure conditions from various machines, and represents the generic relationships of the fault data.

Consequently, data from a machine is passed through a first stage network which is specific to that machine. The outputs, if deemed novel, are then passed to the second stage for diagnosis. Outputs from the whole system classify the data into shaft and bearing related faults (see Figure 2). Using this approach, the system is capable of diagnosing a fault condition upon any machine without the need for example fault data from that machine.

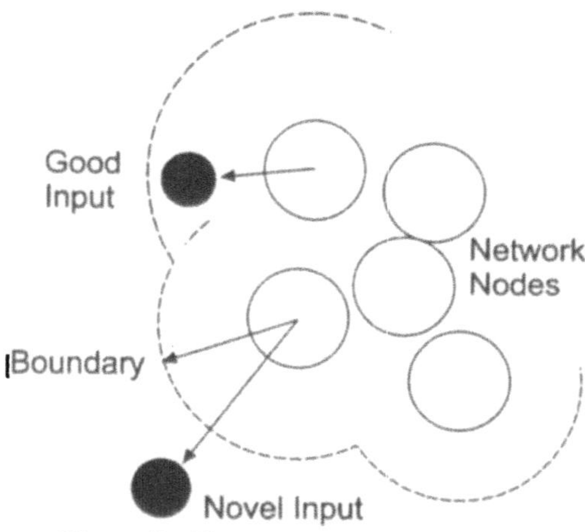

Figure 3: First stage novelty boundary.

3 Results to Date

This complete system has been developed in software, with consideration for implementation in hardware.

A knowledge base has been developed with the limited data currently available, and this has been operated in parallel with the maintenance technician at the paper mill. The results are based upon the performance of the system compared to the technician's diagnosis upon the same data.

No data from the mill machinery has been used to develop the Knowledge Base stage of the system. Therefore all results are for unseen fault data.

Stage 1: Sensitivity = 83% — The number correctly considered as good or novel. False negatives = 3%. False positives = 14%.

Stage 2: Specificity = 100% — Of the correct novelties, the total diagnosed into the correct fault class.

The stage 1 values do not represent as yet a satisfactory level of categorisation for successful implementation into hardware. The sensitivity level may provide a better screening level than a novice maintenance technician, but the number of misdiagnosed measurements is the most important factor of the system's first stage.

False positives — 14% of the applied measurements are incorrectly considered as novel and subsequently categorised into a fault class. In practical terms, this means that a false alarm is triggered for these measurements and would require further investigation.

False negatives — 3% are incorrectly considered normal when in fact they are novel. This means in practice that the measurements are not presented as a problem to the technician, which is an unacceptable outcome.

For this application, a low rate of false positives is acceptable, whereas the occurrence of false negatives has to be as minimal as possible.

This all stems from the accuracy of the first stage networks and their boundarys, which determines which measurements are diagnosed by the second stage.

The second stage is providing an acceptable level of categorisation with only a low number of output states. Work will continue to expand the number of output classifications.

4 Conclusions

The availability of suitable quality data has hindered the development of this project since it began. As such, to date, the system has been created with very limited data, no doubt having a great influence upon the accuracy and inhibiting the development of the system.

To expand the system to diagnose and categorise further typical machine faults, a much larger and diverse database than already exists for the project is required. This is an ongoing task.

Further development in particular of the first stage networks is required to increase their acceptance level, as is verification of the whole system. This is being performed through continued trials off-line at the sponsors paper mill plant.

This work and its results indicate that it is possible for a silicon based generic classification system to diagnose previously unseen data into typical machine faults. It is intended that the final system will be implemented into the Entek and DI product lines.

References

[1] A.J. Hanna. *Predictive Maintenance via Vibration Analysis.* TAPPI.

[2] T. J. Harris. A Kohonen SOM based, machine health monitoring system which enables diagnosis of faults not seen in the training set. In *IJCNN*, 1993.

[3] A. G. Herraty. Bearing vibration — failures and diagnosis. *Mining Technology*, pages 51–53, February 1993.

[4] D. D. Howieson. *A Practical Introduction to Condition Monitoring of Roller Element Bearings Using Envelope Signal Processing.* Diagnostic Instruments Ltd., Scotland, UK.

[5] R. J. Kershaw and B. Robertson. Condition-based maintenance program increases production. *Paper Trade Journal*, pages 34–36, February 1995.

[6] M. Serridge. What makes vibration condition monitoring reliable? *Noise and Vibration Worldwide*, pages 17–24, September 1991.

The Effect of Recurrent Networks on Policy Improvement in Polling Systems

H. Sato[1], Y. Matsumoto[2] and N. Okino[1]
[1] Division of Applied Systems Science, Faculty of Engineering, Kyoto University
[2] I.T.S., Inc., Japan

Abstract

This paper considers polling policies represented by recurrent neural networks and investigates the effect of feedback weights on the optimality. The polling system consists of a single server and multiple stations and whenever the server finishes serving one of the stations, it determines the next station to visit according to the output of the neural network for the current system state. By using the simulated annealing method, we improve the polling policy in such a way that the mean delay of customers is to be minimized in the steady state. The benefit of applying recurrent networks is in that they can represent a broader class of policies than feedforward networks. Numerical results show that recurrent networks can substantially reduce the (sub-)optimal mean delay in comparison with feedforward networks.

1 Introduction

The basic polling system consists of a single server and multiple stations, where it takes non-zero time for the server to switch over between stations. Customers arrive at each station according to a certain stochastic process and they queue up waiting for service upon arrival, where the service time of a customer at each station obeys a probability distribution function. Just when the server finishes serving customers at a station, it must determine which station it should visit next. This instant is called the ***polling instant*** and the decision is made (i.e., one of the stations is polled) according to the server's ***polling policy***.

Polling systems are regarded as mathematical models for many practical systems such as elevator systems, intersections with traffic signals or telecommunication networks where the common bandwidth is shared by multiple users on a demand basis (e.g., token-ring network). A variety of applications can be found in [9]. For these applications, optimization of the server's polling policy is the central issue. The objective function to be optimized depends on the application and usually it is the mean delay (which is defined as the elapsed time from the arrival instant of a customer to the end of its service) of customers or system throughput. Some papers have dealt with theoretical derivation of optimal polling policies [1, 2, 3, 4], but they are based on the assumption that all stations are more or less stochastically identical in terms of arrival and service processes as well as switch-over time. As real systems are rarely symmetric, this assumption is too restrictive to apply the results directly to practical situations. On the other hand, it is still an open question to treat asymmetric models in theory.

Recently, one of the authors has presented a new computational approach to policy optimization of asymmetric polling systems [7]. The polling policy is represented by the learning vector quantization (LVQ) [5]. Namely, at a polling instant the server gets the information on each queue length as well as the index of the station at which the server stays. Then the server inputs the vector of such information to the LVQ and leaves for the station into which the input vector is classified. To optimize the polling policy represented by the LVQ, a weight-perturbation method [6, 8] is applied in [7].

This study extends the approach in [7] by adopting recurrent networks for policy representation and investigates the impact of feedback weights on policy improvement. The benefit of applying recurrent networks is in that they can represent a broader class of policies than feedforward networks. Because of the feedback weights, the output is affected by all previous inputs; that is, the polling decision is influenced by not only the current system state but also system states at previous polling instants. This feature seems to play an important role in the optimality especially for the case that the arrival process is not i.i.d.(independent and identically distributed). We also note that policies represented by recurrent networks entirely include policies represented by feedforward networks if all feedback weights are set equal to zero.

2 The Model

We consider a single-server polling system consisting of N stations with infinite waiting room. Customers are assumed to arrive at station i ($i = 1, 2, \cdots, N$) according to an independent general process with rate λ_i. The service time of a customer at station i is assumed to be an independent random variable, whose first moment is denoted by b_i. For convenience, we denote the traffic load of station i by ρ_i, where $\rho_i \stackrel{\text{def}}{=} \lambda_i b_i$. It takes non-zero time for the server to switch over service to another station. The switch-over time (or walking time) from station j to station k ($j, k = 1, 2, \cdots, N$) can be a random variable, whose first moment is denoted by s_{jk}.

The server is assumed to be instantaneously and correctly informed of the queue length of every station at each polling instant. Suppose that according to a predetermined service policy, the server finishes serving station j at time t. Let $q_i(t)$ ($i = 1, 2, \cdots, N$) denote the queue length of station i at time t. Then the server determines the next station it should visit in the following manner. The server presents the current system state $[j, q_1(t), \cdots, q_N(t)]$ to a neural network as an input and it moves to the station that corresponds to the largest output. Here we assume that the time required for neural computation is negligible. If the server decides to stay at an empty station, it makes a decision again when a new customer arrives to any of the stations.

Unlike [7, 8], this study adopts recurrent networks for policy representation. The recurrent network consists of three layers (input, hidden and output) and there are full connections among neurons in the hidden layer. The input, hidden and output layers have $N + 1$, H and N neurons, respectively, so that the total number of weights equals $(N + 1)H + H^2 + HN = H^2 + 2HN + H$. Let $w_{kl}^{(IH)}$, $w_{lm}^{(HH)}$ and $w_{ln}^{(HO)}$ denote weights from neuron k in the input layer to neuron l in the hidden layer, weights from neuron l in the hidden layer to neuron m in the hidden layer and weights from neuron l in the hidden layer to neuron n in the output layer, respectively, for $k = 1, \cdots, N + 1; l, m = 1, \cdots, H; n = 1, \cdots, N$. Let $i_k(t)$ denote an input to neuron k in the input layer at time t. Then the output of neuron l in the hidden layer at time $t + 1$, which is denoted by $h_l(t + 1)$, is given by $f(\sum_{k=1}^{N+1} i_k(t)w_{kl}^{(IH)} + \sum_{m=1}^{H} h_m(t)w_{ml}^{(HH)})$ and the output of neuron n in the output layer at time $t + 2$, which is denoted by $o_n(t + 2)$, is given by $f(\sum_{l=1}^{H} h_l(t + 1)w_{ln}^{(HO)})$, where $f(x)$ is a sigmoid function defined by $1/(1 + \exp(-x))$.

3 Optimization

The goal is to obtain a set of weights representing a polling policy that minimizes the mean delay of customers for given traffic conditions. Following the approach in [7, 8], we apply the simulated annealing method based on weight perturbation [6] to optimize the polling policy.

To reduce computation time, it is very important to start with an appropriate initial policy [7, 8]. As for the initial policy we train the utilization maximization (UMA) policy [7] to the recurrent network by backpropagation (BP). In the UMA policy, the server moves to station n, such that $n = \arg\max_i \frac{\rho_i s_{ji} + b_i q_i(t)}{s_{ji} + b_i q_i(t)}$, given that it finishes serving station j at time t. We note that the UMA policy is an individual policy in the sense that the utilization is maximized between two polling instants for Poisson arrival and exhaustive service [7], but does not always lead to global optimality [10]. We also note that previous decisions do not influence the current decision under the UMA policy, so that it can be taught to the neural network not by recurrent BP but simply by BP.

Let $\pi(n)$ ($n = 0, 1, \cdots$) and π^* denote the polling policy represented by the recurrent network at the n-th iteration and the current solution, respectively. In addition, let $\bar{D}(n)$ and $\bar{D}^*$ denote the mean delays of customers under $\pi(n)$ and π^*, respectively. The optimization procedure can be summarized as follows:

Step 1) Initialization: Randomize weights $w_{kl}^{(IH)}$, $w_{ln}^{(HO)}$ in the interval $[W_{\min}, W_{\max}]$ and set $w_{lm}^{(HH)}$ equal to zero. Teach the UMA policy to the recurrent network by BP to obtain $\pi(0)$. Evaluate $\bar{D}(0)$ under $\pi(0)$. Let $n \Leftarrow 0$, $\pi^* \Leftarrow \pi(0)$, and $\bar{D}^* \Leftarrow \bar{D}(0)$.

Step 2) Perturbation: Let $n \Leftarrow n + 1$. Choose one weight randomly out of the $H^2 + 2HN + H$ weights and add ξ to it to obtain a new policy $\pi(n)$, where ξ obeys a Gaussian distribution with mean 0 and variance σ^2. Evaluate $\bar{D}(n)$ under $\pi(n)$.

Step 3) Acceptance: Accept $\pi(n)$ as the current solution with probability P_{ac} which is given by

$$P_{ac} = \begin{cases} 1, & \text{if } \bar{D}(n) \leq \bar{D}^*, \\ \exp\left(\frac{-\bar{D}(n)-\bar{D}^*}{T(n)}\right), & \text{if } \bar{D}(n) > \bar{D}^*, \end{cases}$$

where $T(n)$ is the *temperature* at the n-th iteration. If $\pi(n)$ is accepted, then let $\pi^* \Leftarrow \pi(n)$ and $\bar{D}^* \Leftarrow \bar{D}(n)$. Otherwise, no change.

Step 4) Termination: If $n > \nu$, then stop, where ν is a stopping criterion. Otherwise, go to Step 2.

4 Numerical Results

In numerical examples we consider asymmetric polling systems with exhaustive (i.e., all customers are served until the queue is emptied) and first-in first-out service. We consider the case of $N = 4$ stations, where the arrival rate is given by $\lambda_i = 0.05 \times i$, the service time obeys the exponential distribution whose mean b_i is determined to satisfy $\rho_i = 0.1$, and the switch-over time is constant with $s_{ij} = j - i + 4 \pmod 4$ for $i, j = 1, \cdots, 4$.

The number of neurons in the hidden layer of the recurrent network is $H = 3$, so that the total number of weights sums up to 36. At Step 1 in the optimization procedure, initial weights are chosen randomly between $[W_{\min}, W_{\max}] = [-5, 5]$. To teach the UMA policy to the recurrent network, we generate 100,000 training samples as follows: pick up one of the four stations with equal probability, set the station's queue length equal to zero because of the exhaustive service, and choose integers between 0 and 10 with equal probability for each queue length of the other stations. The learning rate at the t-th iteration of BP is empirically determined as $0.85/\log(1.15t)$. After teaching the UMA policy, we checked the average percentage of correct answers for 25 test sets, each of which has 100,000 test samples and the result was 87.3%.

The variance of the Gaussian distribution for weight perturbation is given by $\sigma^2 = 3.5$. Since it is impossible to analytically derive the mean delay for the polling policy $\pi(n)$, we evaluate $\bar{D}(n)$ by simulation, where the capacity of the waiting rooms is assumed to be up to 1,000 customers and to take account of the steady state in simulation, the mean delay is evaluated for 4,000 customers after the server has finished serving 1,000 customers from the beginning. In case of overflow, we discard the polling policy with probability one at Step 3. The temperature is reduced by $T(n) = 4/\log(12n)$

Table 1: Result for Poisson arrival & exponential service.

i	λ_i / b_i	$E[D_i]$ & 95% Confidence Interval			
		Re-NN	FFNN	UMA	LVQ [8]
1	0.05 / 2.0	3.685 ±0.086	5.304 ±0.030	4.121 ±0.013	5.776 ±0.026
2	0.1 / 1.0	4.183 ±0.084	5.812 ±0.027	4.390 ±0.011	5.609 ±0.032
3	0.15 / 0.667	4.014 ±0.086	4.388 ±0.011	4.537 ±0.013	4.468 ±0.025
4	0.2 / 0.5	4.171 ±0.051	3.658 ±0.018	4.910 ±0.018	4.101 ±0.022
Overall		4.077 ±0.056	4.473 ±0.016	4.615 ±0.012	4.680 ±0.017

and the annealing process is stopped at $\nu = 1,000$ iterations.

For the case that the arrival process of customers is Poisson and the service time distribution is exponential at each station, Table 1 compares the mean delays under three different policies: the recurrent network (Re-NN), the three-layered feedforward network (FFNN) which has the same number of weights (36) as the Re-NN but has no feedback weights, and the UMA. For comparison, we applied the same optimization procedure to the FFNN as to the Re-NN. The result for the LVQ [8] is also shown for reference though the operation of the server in [8] is a little bit different from the present system in that it keeps staying at the last station when it has emptied all queues. In Table 1 the 95% confidence intervals of the mean delays are also given. We observe that the recurrent network can reduce the mean delay about 12% in comparison with the UMA and 9% in comparison with the feedforward network. We notice that the orders of priorities given to the stations are different among the neural networks as the goal is to minimize the overall mean delay.

In Figure 1 we checked how the mean delays increase if we apply the policies obtained in Table 1 to more congested situations. Starting from $\sum_{i=1}^{4} \rho_i = 0.4$, we increased the arrival rates of the stations by 20%, 40%, 60%, 80%, 100%, 120% to obtain the total loads $\sum_{i=1}^{4} \rho_i$=0.48, 0.56, 0.64, 0.72, 0.80, 0.88. For $\sum_{i=1}^{4} \rho_i$=0.56 and 0.72, we also plot the optimal solutions by the Re-NN, which are recalculated by the optimization procedure. We observe that once an optimal solution is obtained by the Re-NN, the optimality is rather insensitive to

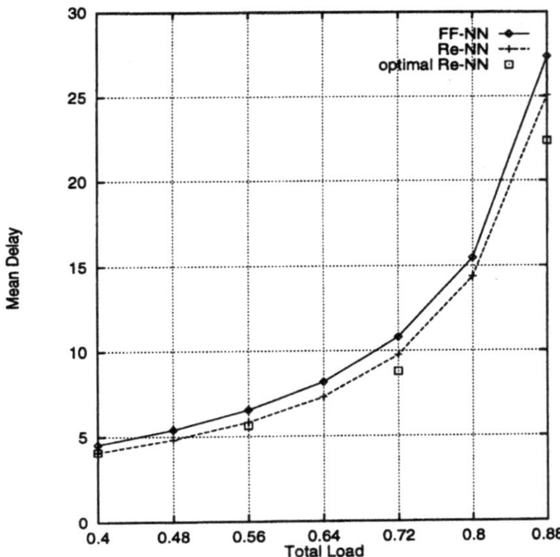

Figure 1: Robustness to the increase of arrival rates.

the proportional increase of arrival rates at the stations.

5 Conclusion

In this paper we applied the recurrent network to policy representation in polling systems and optimized the policy by a weight perturbation method. From numerical results, we found that the policy represented by the recurrent network has better performance than the one represented by the feedforward network. This is an expected result as structurally feed-forward networks are nothing but a subset of recurrent networks. For future research, application to non-i.i.d. arrival processes and rule extraction from the optimal weights remain.

References

[1] O. J. Boxma, H. Levy, and J. A. Weststrate. Optimization of polling systems. In *Proc. of PERFORMANCE'90*, pages 349–361. Elsevier Science Publishers B.V., 1990.

[2] S. Browne and U. Yechiali. Dynamic priority rules for cyclic-type queues. *Adv. Appli. Prob.*, 21:432–450, 1989.

[3] O. Fabian and H. Levy. Polling system optimization through dynamic routing policies. In *Proc. of IEEE INFOCOM'93*, pages 2b.3.1–2b.3.7, 1993.

[4] M. Hofri and K. W. Ross. On the optimal control of two queues with server set-up times and its analysis. *SIAM J. on Computing*, 16:399–419, 1987.

[5] T. Kohonen. The self-organizing map. *Proc. of the IEEE*, 78(9):1464–1480, 1990.

[6] M. Markon, H. Kita, and Y. Nishikawa. Reinforcement learning for stochastic system control by using a feature extraction with bp neural networks. *Tech. Rep. of IEICE*, NC91-126:209–214, 1991.

[7] Y. Matsumoto. On optimization of polling policy represented by neural network. In *Proc. of ACM SIGCOMM'94*, pages 181–190, 1994.

[8] H. Sato, Y. Matsumoto, and N. Okino. Policy optimization by neural network and its application to queuing allocation problem. In D. W. Pearson, N. C. Steele, R. F. Albrecht (editors), *Artificial Neural Networks and Genetic Algorithms*, pages 344–347, Wien New York, 1995. Springer-Verlag.

[9] H. Takagi. *Analysis of polling systems*. MIT Press, 1986.

[10] J. Walrand. *An introduction to queueing networks*. Prentice Hall, NJ, 1988.

EXPRESS - A Strategic Software System for Equity Valuation

M.P. Foscolos[1] and S. Nilchan[2]
[1] School of Information Systems, Faculty of Commerce, University of New South Wales
Sydney 2033, Australia.
[2] Centre for Process Systems Engineering, Imperial College
London SW7 2BY, United Kingdom.

Abstract

This paper examines the problems associated with equity valuation and exposes the weaknesses of current computer based modelling systems. The paper identifies the necessary requirements of a strategic software system for equity valuation and proposes a software system with an integrated architecture, which combines both artificial intelligence technologies with conventional software. The paper then describes a powerful strategic software system, EXPRESS, developed by the authors to significantly de-skill and improve stock valuation. The EXPRESS system is a state-of the-art windows based application capable of performing the three accepted methods of stock valuation namely, quantitative, technical and fundamental analysis. The paper outlines the integrated architecture of EXPRESS and describes the function of each of the three integrated software systems, the EXPRESS decision support generator system, the EXPRESS artificial neural network system and the EXPRESS extension model system.

1 Introduction

In the financial market, stock valuation has remained an extremely difficult task. Practitioners of the three accepted method of equity valuation, fundamental, technical and quantitative analysis require high levels of experience and expertise. Financial analysts are not only responsible for constructing financial models to accommodate the three methods of equity evaluation, but in the face of increasing investor scrutiny must also ensure that these models, despite information restrictions, perform adequately.

To deal with the increased vigilance of investors, analysts employing computer based models endeavour to construct models in manner that will furnish a decision support environment. Such an environment is crucial to formulate and support sound investment decisions. Unfortunately, the models analysts apply to justify stock recommendations often fall short of market expectations, frequently represent no more than specialised computer based information systems (CBIS).

Until recently, various branches of artificial intelligence (AI) such as expert systems and artificial neural networks, have remained independent. The process of integration involves the development of a system architecture that provides links between AI technologies. Integration allows different AI technologies to interact with each other and/or with other computer based information systems. The aim is to combine the advantages of each technology and thus reduce system limitations. While recent research advocates integration as means to develop high performance systems for complex problem areas [1, 2, 8, 11, 12, 13, 14, 15], there has been no attempt to develop an integrated system specifically for equity valuation.

2 System Architectures for Integration

The primary requirement in system integration is cohesion between hardware, traditional software and artificial intelligence software. According to [14], there are two main architectures that support integration, the access architecture and embedded architecture. In access architectures software components are loosely coupled, with two or more independent applications interacting to provide integration. In an embedded architecture, software systems are tightly coupled within a single executable application. Though embedded architectures are preferable, they are significantly more difficult to produce, as they require a significant amount of programming [6].

The EXPRESS system is a powerful integrated software system, developed specifically to address the problems faced in equity valuation. The system is designed for Window 3.1 and 95 platforms, and employ an embedded architecture to achieve integration. Artificial intelligence technologies, in the form of neural networks and expert systems and conventional software are integrated within the single executable EXPRESS application.

More specifically the EXPRESS system consists of three integrated software systems. A conventional computer based information system, or decision support generator system and two systems supporting artificial intelligence, namely, the artificial neural network system and the extension model system. These three systems provide the functionality of a modern computer based information system and offers additional facilities to cater for the process of equities valuation via quantitative, technical and fundamental analysis [3].

In order to provide seamless integration, the EXPRESS system has been coded in Visual Basic for Windows, with lower level C++, assembler and Pascal routines, in the form of dynamic link libraries, ensuring efficiency.

Visual Basic was selected as the primary development environment, in preference to more traditional environments, such as C++, due to the object orientated event driven programming style. While only supporting one of the three concepts of object orientation, encapsulation, the Visual Basic environment provided a far superior paradigm for financial system development [3, 7]. The only noticeable limitation, resulting from the inability for the language to support inheritance and polymorphism, was that new objects could not be created in run time. However, this short fall was adequately compensated by supplementing standard Visual Basic objects, with the required new objects, referred to a custom controls, at design time. Any number of new objects can be created for the Visual Basic environment using lower languages such as C++, or, turbo Pascal.

3 The Decision Support Generator System

The EXPRESS decision support generator system, shown in Figure 1, followed the designed proposed by Sprague and Watson [10] and advocated by Ramirez *et al.* [9] and Turban [13]. The primary aim in the development of the EXPRESS decision support generator system was to improve efficiency and reliability relative to available decision support generators, such as Microsoft Excel.

The EXPRESS decision support generator pro-

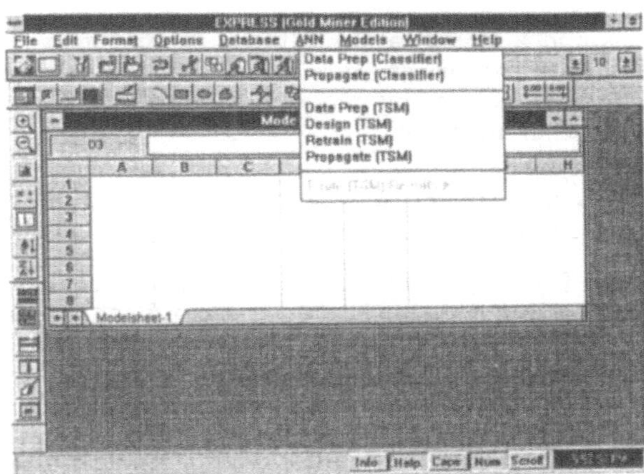

Figure 1: The EXPRESS decision support generator.

vides all necessary data processing facilities and features of commercial spreadsheet applications and, while consuming fewer system resources, additionally provides a comprehensive database and scientific graphic facility. The significant saving in resources (20% less than Microsoft Excel), provided by highly efficient spreadsheet engine encoded in C++, not only stabilises the software, but ensures adequate resources for the operations of the two systems supporting artificial intelligence. To further insure system integrity, a system management component, encoded in Visual Basic with assembler routines, provides significant error trapping and monitors both hardware and warn of major system faults. This component conservatively avoids up to 25% of the major system failures, experienced by windows based applications.

4 The Artificial Neural Network System

The EXPRESS system furnishes both technical and quantitative analysis via the artificial neural network system. Technical analysis, primarily rests on the application of time series forecasting techniques to identify and predict the trend in stock price over time. Quantitative analysis, employs either statistical decision making or casual time series forecasting to estimate a stocks value an/or investment potential.

The artificial neural network system consists of a time series forecasting software component, which permits the design, development and application of either trend or casual time series forecasting models and a classification software component which permits the design, development and application of decision making models, employing advanced Bayesian inference.

4.1 The Time Series Forecasting Software Component

EXPRESS furnishes an accurate general purpose forecasting system via the C++ Spirit of Progress neural network engine, developed by the authors. Unlike most neural network forecasting systems, which incorrectly employ a standard backpropagation neural network paradigms that are only capable of modelling spatial patterns, the EXPRESS time series forecasting component employs a recurrent neural network paradigm. The architecture of the recurrent neural network, while similar to the standard feedforward neural network, incorporates feedback loops, or connections, which provide a time lag, thus, a neuron's previous outputs will influence its subsequent outputs and therefore provides temporal pattern recognition. The EXPRESS recurrent neural network supports four separate learning laws, Momentum, Example by Example Steepest Descent, Batch Steepest Descent and Quick Prop. In addition, seven separate transfer functions are provided.

The EXPRESS system provides an interface (see Figure 2) to automate data preparation and training set development. The data preparation interface, test each variable in the training set for outliers using z score analysis, tests for normality using kurtosis, skewness and a frequency histogram and via a test of hypothesis, based on the t distribution, tests the correlations between variables. A rule base

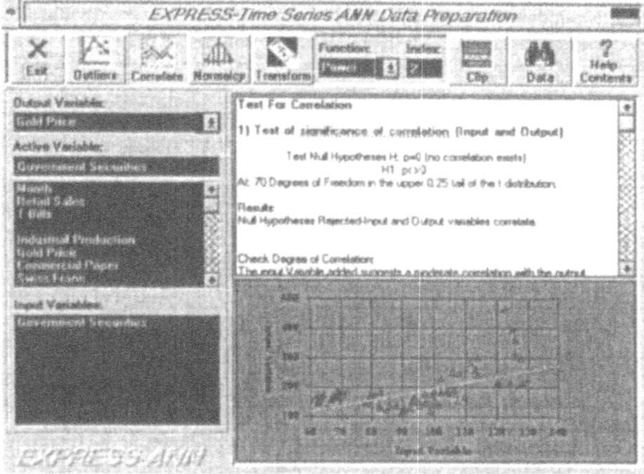

Figure 2: Data preparation interface.

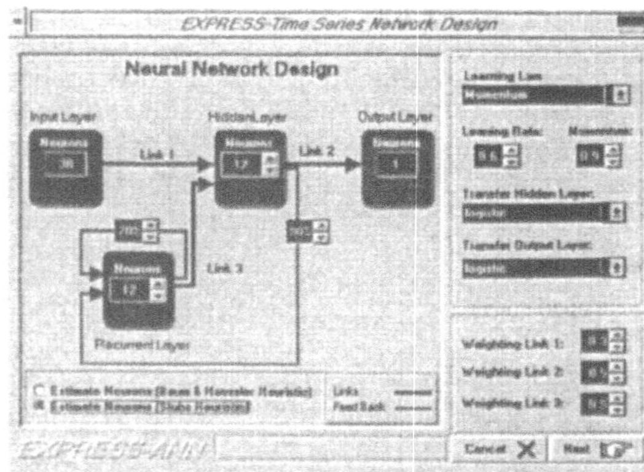

Figure 3: Neural network design interface.

expert system, on the basis of the degree of correlation, comments on the relevance of each input variable. Facilities are provided to modify the training set, transform variables to ensure normalcy and map variables to ensure representative data.

The EXPRESS system provides an interface (see Figure 2) to automate data Once data is prepared, the design interface is used to set network parameters (see Figure 3). When the design interface is loaded, a rule base expert system inspects the user specified training set, select optimum network parameters and initialises the networks weights via the random initialisation method. Once loaded, the user can manually change network parameters from this graphic user interface.

From the design interface, the user can enter the training interface (see Figure 4) which when loading, automatically constructs the neural network and via an integrated expert system determines the correct mapping of the training set required for the selected transfer functions set in the hidden and output layers. One of the unique features of this inter-

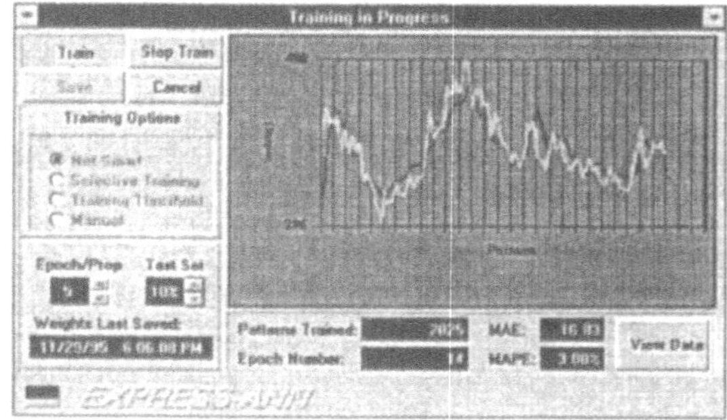

Figure 4: Neural network training interface.

face is the Net Smart training algorithm, developed by the authors, which ensures optimal generalisation of the network by automatically stopping the training process prior to the onset of over training. Furthermore, during training all other express systems remain available to the user. From the training interface, the networks design and weights can be caved as a binary file.

Once trained, the binary file containing the neural network model can be opened and retrained, via the training interface, or, can be opened to a propagation interface. The propagation interface requests the entry of appropriate input variables, which are passed through the network model to determine forecasts. The forecasts from the propagation interface can be pasted directly into the database or spreadsheet of the EXPRESS decision support generator.

4.2 The Classification Software Component

Bayesian decision theory, offers a unified approach to decision making. It can be implemented by replacing the transfer function of backpropagation neural network with a statistically derived one. The decision boundary of such a network, is able to asymptotically approach the Bayes optimal decision surface. Such a network is referred to as a probabilistic network. EXPRESS implements such a network via the Spirit of Progress neural network engine.

A design interface (Figure 5) allows the EXPRESS user to select input variables and the output classes. This interface also automatically constructs and stores the training set as a binary file. The interface provides the user with the ability to conduct a test of hypothesis, based on the t distribution, to determine whether the correlation between input variables is significant. A rule base expert system, on the basis of the Pearson correlation coefficient, comments on the relevance of each input variable. The propagation interface provides a mechanism for the user to apply the model. The network is trained as propagation interface is loaded, as the probabilistic neural network trains in real time. The user enters the requested input variables into the interface, on the basis of which, the network determines an appropriate membership class. In addition the prior probability of membership in each of the specified class is provided. This is particularly useful in an equity selection and asset allocation, when a decision, as well as a measure of the certainty of the decision. In the case of the gold stock selection model, shown in Figure 6, the decision is to sell highlands gold (HGL), the result tab showed the decision to be associated with a 80% posterior probability of occurance, in other words, a strong sell.

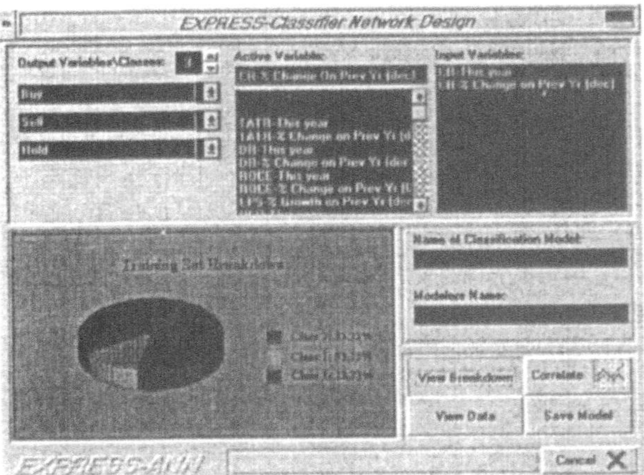

Figure 5: Classifier design interface.

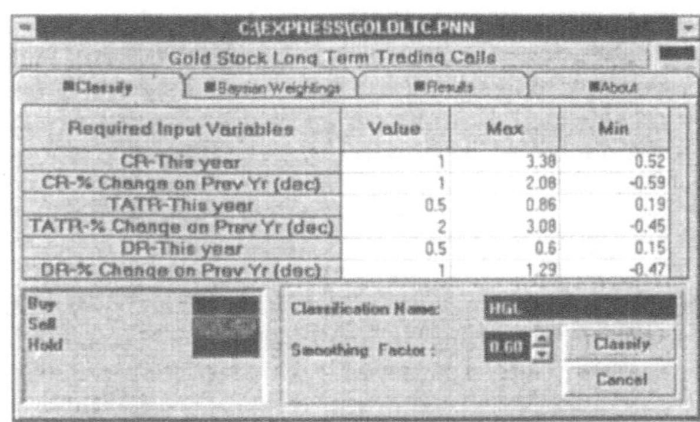

Figure 6: Classifier propragation interface.

5 The Extension Model System

The EXPRESS extension model system is aimed at fundamental analysis. At present this system consists of a dividend pricing model, which estimates intrinsic stock price on the basis of dividend payout and estimates on the basis of market price whether a stock is over or under valued.

EXPRESS is able to construct and manage commodity specific mineral project valuation models. A corporate consolidation model permits these project models to be consolidated at the corporate level. Both the mineral project valuation models and the corporate models contain an integrated executive information system (EIS), which provides the user with expert valuation reports. Furthermore, both these models contain an integrated neural network forecasting system, which allow users to directly apply models developed via the forecasting component of the neural network system.

6 Conclusion

Extensive testing conducted with financial institutions in both the UK and Australia, found EXPRESS provided superior results in comparison to the current stock valuation methods. The forecasting system, on average, provides a mean absolute percentage error of 3.15% compared with analyst forecasts which, on average, produce a mean absolute percentage error of 11.25%, [5]. The classifier, on average, provides 90% correct classification, compared with analysts, which on average, provide 30% correct classification [4]. The EXPRESS extension model system automates the construction of mineral project valuation models and the corporate models, providing expensive amount of relevant financial fundamental information. The EIS, expedites and de-skills the production of expert reports. In addition the dividend valuation model, provides a more general measure of fundamental stock value. While designed for equity valuation, the potential exists to apply EXPRESS to problems in other disciplines.

References

[1] D. R. Dolk. An Introduction to Model Integration & Integrated Modelling Environments. *Decision Support Systems*, 10, 1993.

[2] J. J. Elam and B. Konsynski. Using AI Techniques to Enhance the Capabilities of Model Management Systems. *Decision Science*, 1987.

[3] M. P. Foscolos. *The Development of a Strategic Software System for Resource Equity Valuation.* PhD thesis, Imperial College, University of London, 1996.

[4] M. P. Foscolos and S. Nilchan. Improving Decision-Making in the Financial Market via the Probablistic Neural Network Paradigm. *Neurove$t Journal*, March 1997.

[5] M. P. Foscolos, S. Nilchan, and P. E. Bell. Improving Financial Market Forcasts via the Recurrent Neural Network Paradigm. In *AIC Modelling Techniques in Portfolio Management Conference*, Inter-Continental Hotel, Sydney, Australia, August 1996.

[6] M. P. Foscolos and C.T. Shaw. Fundamental Resource Equity Evaluation & Modelling. In *IASTED International Conference of Applied Modelling and Simulation*, Lugano, Switzerland, 1994.

[7] D. King. Intelligent Decision Support: Strategies for Integrating Decision Support, Database Management and Expert System Technology. *Expert Systems with Applications*, 1, 1990.

[8] T. J. Martin. Integration With Conventional Information Systems. In Watkins & Eliot, editor, *Expert Systems in Business and Finance.* Wiley & Son, 1993.

[9] R. G. Ramirez, C. Ching, and R.D. Louise. Independence and Mappings in Model Decision Support Systems. *Decision Support Systems*, 10, 1993.

[10] R. H. Sprague and J.H. Watson. *Decision Support Systems.* Prentice Hall, 1986.

[11] G. S. Swales and Y. Yoon. Applying Neural Networks to Investment Analysis. *Financial Analysts Journal*, Sept/Oct 1992.

[12] J. T. C. Tseng. A Unified Architecture for Intelligent DSS. In *21st HICSS*, 1988.

[13] E. Turban. *Decision Support and Expert Systems-Management Support Systems.* McMillan, 1988.

[14] E. Turban. Expert Systems Integration with Computer Based Information Systems. In Watkins & Eliot, editor, *Expert Systems in Business and Finance.* Wiley & Son, New York, 1993.

[15] W. Wu. An Integrated System Based on the Synergy Between Systems. In *International Conference of the Systems Dynamic Society*, June 1988.

Virtual Table Tennis and the Design of Neural Network Players

D. d'Aulignac, A. Moschovinos and S. Lucas
Department of Electronic Systems Engineering, University of Essex,
Colchester CO4 3SQ, UK
Email: sml@essex.ac.uk

Abstract

This paper discusses the design of a virtual table-tennis environment, and the design of neural network based controllers to play in that environment. The motivation behind the work is to provide an interesting and entertaining forum in which to carry out research on adaptive control and planning problems that stretch the limits of current neural network paradigms.

1 Introduction

Currently there is much interest in neural networks and genetic programming methods for control of real robots. However, to experiment with such things requires expensive hardware that can be time consuming to set up and maintain. An alternative to experimenting with real robots is to experiment with virtual robots. The complexity of designing controllers for such virtual robots depends on the 'physics' of the virtual environment, and the task at hand. This paper describes the design of a virtual table-tennis environment and some initial work on the design of algorithmic and neural network based bat controllers.

Neural networks have been designed or evolved for solution for control problems (e.g. [2, 3]) and separately, for strategy games such as tic-tac-toe[1]. Table tennis (or for that matter, any racquet sport) provides an interesting mix of control problems and strategy problems, making the design of good players extremely challenging. In a recent tournament of table tennis for real robots, rallies were extremely few and far between. By concentrating on the design within a virtual environment, however, we have control over exactly how difficult to make the game, and at least for the initial stage, we can ignore completely problems of visually tracking the ball and accurate actuation of the robot.

2 The Simulated Environment

Here we choose a 2-dimensional simulated table tennis game as our environment. This is implemented in an object-oriented style. The system was initially implemented in C++ for both Unix and Windows platforms, but is currently being ported to Java, to exploit Java's platform independence and superior networking capabilities.

The implementation model has three aspects, the logical model, the physical model and the graphical model.

2.1 Logical Model

This defines the rules of the game. The responsibilities of the logical model are to detect the end of the game, detect the end of a point, and the winner of the point, and decide which player currently has service. For detecting the end of a point, and the winner of that point, it interacts with the physical model. The logical model contains a state machine for this purpose. The states correspond to the logically distinct states of the game, such as *left_serve, right_serve, left_bat, right_bat, left_table, right_table, left_wins, right_wins etc.* The actual state table is a little more complex than this, since the rule governing a serve is different to the normal run of the game. Each time the physical model detects a collision between the ball and the table (either left or right side), the net, or the left or right bat, a state transition is made depending on the object that the ball collided with.

2.2 Physical Model

The physical model is responsible for applying the laws of physics to the objects in the game. To do this it must update the dynamic (ball and bats) objects and monitor their collisions with each other and with the static objects (table, net, floor and ceiling). After each collision the physical model informs the logical model in order to keep the state machine updated.

The Ball

The ball is a dynamic yet passive object, which gets hit around according to the laws of physics. The physical model accurately describes most of the features of the real game (except for the missing third dimension) including gravity, air resistance, the effects of spin on ball trajectory and collisions, and the coefficients of friction between bat and ball and between table and ball. These effects can also be switched off in order to provide a simpler game environment if necessary.

The Bats

The bats are active dynamic objects. Each bat has an associated bat controller. The bat controller must implement a method called `getForce` that takes as parameters the current bat position vector, the position of the opponent's bat, the position of the ball and a boolean variable to indicate whether or not it is this players turn. Of course, velocity and acceleration information is also useful to the bat controller — but this can be derived from successive values of the position.

2.3 Graphical Model

This is used to display an animated view of a game. For the machine based controllers it is entirely unnecessary, and games can be played much faster without one. However, it is useful to observe the traits of various machine-based players in order to better understand their strengths and weaknesses. Also, it is essential if it is required to allow human players to play against machine-based opponents.

3 An Algorithmic Controller

Given the above simulated environment, it is possible to make accurate predictions of ball trajectory, make decisions on where to intercept the ball, and which shot to play when there, and then make a perfect execution of the chosen shot. It may seem that such a controller should never lose a point, but this is not the case. We limit the force that can be applied to the bat at each time instant, and hence, not all shots are possible, and if a bat can be caught out of position it may even be unable to make contact with the ball.

We have implemented an algorithmic controller based on the above ideas, and as expected, it plays a good game of table tennis — it is difficult for human players to win a point against it.

The main reason for implementing an algorithmic controller was to provide training data for supervised neural networks. To make it more interesting, and to provide more varied training data for the networks, the algorithmic controller makes pseudorandom choices regarding the point at which to intercept the ball, and the shot to execute when there.

4 Neural Network Based Controllers

The main aim was to design neural networks and train them to play a 'good' game. Two approaches were tested. Firstly, a single network was used and trained by the algorithmic controller. For this approach both multilayer perceptron (MLP) and radial basis function (RBF) architectures were tested. For the second approach the task was divided into a small number of neural networks. Each network was trained to do a particular task. Then, they were all combined to integrate a player. For this case an MLP was used.

4.1 Inputs and Outputs

A realistic input vector for a neural network would be the position and velocity of the ball, and the position and velocity of the bat it handles. An output vector would consist of the forces (in x and y coordinates) and the torque that the controller applies to its bat. However, for simplicity at this stage the torque is ignored. We are just interested in moving the bat, while leaving at a fixed angle. Using the algorithmic controller a set of training pattern pairs can be produced. This should include as many representative cases as possible.

4.2 Single Module Network

The results of the single module networks at first appeared to be strange. Both the RBF and MLP networks were trained on the training data, and repeatedly tested on the test data until the test-set error reached a minimum. In the case of each network, this was a reasonably small error (of the order of 0.001 mean square error). These learned weights were then hard-wired into a bat controller to play in an actual game. The RBF marginally outperformed the MLP, but both networks (many different configurations of each one were experimented with) generally performed poorly compared to the algorithmic controller — frequently missing the ball or hitting it way off the table, and on some occasions, even appearing to actively avoid the ball.

The most probable cause of this is that while the network behaves well in the regions of input space which the algorithmic controller inhabits, it has no reason to behave well outside of these regions. As soon as the neural network begins to stray from what the algorithmic controller would have done in a given situation, the problem then accumulates — and the bat is rapidly sent into regions of input space where the algorithmic controller has never explored.

Perhaps a further problem is the pseudo-random behaviour of the bat controller. The neural network models are capable of approximating functional mappings, but the data given to them is not of this nature if we include a random element in the algorithmic controller, since, given identical input conditions, the algorithmic controller can produce different outputs

However, the neural networks still performed significantly better when trained on the random algorithmic controller than when trained on a non-random version. Perhaps the best possibility is to simply generate a large number of random input training vectors together with what the algorithmic controller would output given those inputs, but we have not yet done this.

4.3 Modular Neural Networks

The second approach to implement a neural network player is to decompose the task into a number of smaller tasks. Each smaller task can then be handled by an independent specialist neural network. With this modular approach, it is of course possible to have different modules based on different paradigms — there is no need for all modules to be neural networks. During development of the system, it is sensible to begin with an algorithmic module for each task. Having checked that this functions well, each module can then be replaced in turn by its neural network alternative. In this way, it is possible to identify which neural network modules are performing well and which ones are performing poorly.

We decompose the problem into three parts: prediction of the intercept point, calculation of intercept vector and movement of bat to achieve the desired intercept.

Calculation of Desired Intercept Point

The first step is to have a network which can predict some point of interception. This involves predicting the position of the intercept and the time at which the ball will be at that position. This is chosen to be the highest point of the trajectory in which the player is allowed to hit the ball. An MLP is used for this stage and actually predicts the point of interception very accurately. Inputs to the network are the position and velocity of the ball when leaving from the opponents bat. The outputs are the x, y coordinates and time of the predicted intercept. The training set was produced from the outputs of the algorithmic controller, designed to estimate the highest point of the trajectory. An MLP with a single hidden layer of 22 neurons proved to be good enough for this task.

Calculation of Desired Intercept Velocity

Based again on the outputs of the algorithmic controller, targets can be derived to train an MLP to output what velocity the bat should have at the predicted point. This velocity must be such so as to return a good shot. For this task, neural networks have also been trained successfully. Thus, if we cheat and warp the bat to the desired point with the desired velocity at the correct time, we have a combination of two neural networks which can play as good as the robots (i.e. a game lasting for more than 20 hits!).

Moving the Bat

The third network module has the task of moving the bat over successive time intervals in order that at the time of intercept, the bat has the desired position and velocity.

However, another network must be trained so as to apply legal force to move the bat to arrive at the desired point at the correct time with the correct velocity. A neural network for this has been designed but not yet tested. It seems likely that by employing a modular decomposition of the problem we shall be able to develop a highly proficient neural network bat controller — one that plays as well as the original algorithmic controller. This leads on to the next step — evolving superior players.

5 Discussion

We are almost at the stage where a multi-module neural network can play a good game of table tennis within the current environment. The next stage in the work is to take the most successful neural network individuals and apply tournament based evolutionary methods to evolving successively better individuals.

The current implementation of the game has been designed to be an accurate simulation of a real, but 2-dimensional, table tennis environment. There are several ways the set up can be varied. The game can be made simpler by eliminating the effects of air resistance and spin. Alternatively, the parameters which control these can be adjusted to increase their effect, hence making the game more difficult.

Other ways in which the game can be made more difficult are: extend the simulation to a three dimensional environment; make the robot controllers act on some multi-segment robot arm in order to control the bat, rather than applying forces directly to the bat as they do now; limit the amount of computation allowed for each controller at each timestep. This would have the effect of favouring controllers who could not only play a good game, but do it within some bounded amount of computation. Before these are explored, however, there is plenty of scope for improving the performance of the current robot players.

There has been a good deal of interest around the world in our virtual table tennis project, and it is planned to hold an internet-based virtual table tennis tournament. The idea is that people wishing to enter a competitor would submit the code for their controller (having already developed it and tested it on their own machine) — the newly submitted controller would then be pitted against a league of all the best controllers so far submitted, and if sufficiently successful, earn its own place in the league or otherwise be discarded. Over time it would be interesting to see the type of architectures that dominate the tournament, and the kind of games they play. To facilitate this, we are currently porting the simulator to JAVA, and also working out details of a GUI-based neural network controller design system.

6 Conclusions

This paper has described a framework that allows the development and evaluation of robot controllers for a simulated table-tennis environment. The current status of the project is that algorithmic controllers have been designed that play a good game of virtual table tennis. The initial experiment to train a single feedforward neural network to play virtual table tennis was largely unsuccessful, with the single neural controller struggling to maintain a rally of more than about 2 shots, hence proving no match for the algorithmic controller.

The modular neural networks are far more promising, and a successful bat controller has been constructed using a neural network for prediction of the intercept point, a neural network for the calculation of the intercept velocity, and an algorithmic module for seeing that the bat achieves the desired intercept velocity at the chosen place and time.

Already the project has generated a good deal of interest on the internet. This will hopefully increase when the Java version of the simulator becomes available, which will include a system for the interactive design of robot controllers, and an easy means for people to participate in an internet-based tournament.

Finally, this kind of work provides a natural bridge into the design of real robot game players (i.e. to play table tennis against each other, and/or against humans on a real table tennis table). By developing the details of the robot controller in a virtual environment, much of the design work can be done much more quickly than if having to deal with real robots. Although not a feature of the current implementation, it is of course possible to implement models of real robots within our simulated environment.

7 Related WWW Sites

For more information on the project and related links, or to download our table tennis simulator, visit our project home-page: `http://giwww.essex.ac.uk/`

References

[1] D. Fogel. Using evolutionary programming to create networks that are capable of playing tic-tac-toe. In *Proceedings of IEEE International Conference on Neural Networks*, pages 875–880, San Francisco, 1993. IEEE.

[2] F. Gruau, D. Whitley, and L. Pyeatt. A comparison between cellular encoding and direct encoding for genetic neural networks. Technical Report NeuroColt series NC-TR-96-048, 1996.

[3] A. Wieland. Evolving controls for unstable systems. In *Proceedings of the 1990 Connectionist Models Summer School*, pages 91–102, San Francisco, 1990. Morgan Kaufman. (D. Touretzky, J. Elman, T. Sejnowski, and G. Hinton, eds.).

Investigating Arbitration Strategies in an Animat Navigation System

N.R. Ball
Engineering Design Centre, Department of Engineering, University of Cambridge,
Trumpington Street, Cambridge, CB2 1PZ U.K.
Email: nrb@eng.cam.ac.uk

Abstract

This paper reports on recent experiments applying classifier systems to the problem of supporting both local and global navigation in a simulated animat. The basis of this research is a hybrid learning system that extends the classifier representation to enable environmental feedback to impinge directly upon the classifier population. The system applies a connectionist representation to the condition sets of classifiers which enables the direct encoding of classifier condition/fitness values onto network nodes. The goal of the system is to achieve domain objectives by calibrating classifier behaviour during the exploration of the domain and evolving new classifiers to exploit the domain by discovering goal states.

1 Introduction

Animat artificial intelligence (AAI) emphasizes the role of an animat's continuous interaction with its environment in driving the selection and performance of behaviour [8,10]. A typical AAI system utilizes a reactive control architecture which directly couples perceptual activity with action without the use of intervening symbolic representations [1,5]. A key issue in such systems is the problem of tailoring perception to meet motor demands. Perceptual representations and/or algorithms need only provide the information that is required to achieve a particular motor activity i.e. that of local navigation. Animats that are tasked with finding and following paths to target locations in large scale space need additional global navigation behaviour. This requires the incremental development of internal structures that can encode at least some aspects of the animat's past experience of its environment [9]. In this case multiple subsystems are used to control of the animat with each system maintaining a representation of those aspects of the domain that it finds important.

2 Feature-Based Representation

The basis of this research is a hybrid learning system (HLS) that extends the classifier representation to enable environmental feedback to impinge directly upon the classifier population [2]. The HLS applies a sub-symbolic, distributed data representation — the Kohonen self-organizing feature map (SOFM) [7] — to the condition sets of a classifier system which enables the direct encoding of multiple classifier condition/fitness values onto the network nodes (Figure 1). The goal of the system is to achieve (fixed) problem domain objectives by calibrating classifier behaviour during the exploration of the domain and evolving new classifiers to exploit the domain by discovering goal states. The key concept is to use SOFMs to represent the behaviour of each classifier within the population in terms of its effects on the external environment in a way similar to Arkin's motor schemas [1]. These effects are encoded within each SOFM as state vectors with each vector representing the domain state before and after classifier activation. The self-organizing process calibrates classifiers by adapting their pre/post activation state vectors and eliminating redundant state data unchanged by classifier activation. Elimination of this redundancy is equivalent to the induction of general rules that describe a classifier's action from a set of specific examples. The problem domain is represented as a set of 'f' features in HLS. Each feature may be either ecocentric (defining the domain space) or egocentric (defining the animat state). Each classifier is allocated 'f' SOFMs to encode its condition set. The resulting multi-layer network self-organizes to represent the relative effect that the classifier has across the feature space when it is activated. Activation causes changes to the feature set that encodes the domain/animat state. These changes then are applied directly to the SOFMs of the activated classifier as post activation values. Self organization of a SOFM changes the conditions under which a classifier operates and effects the probability of its selection (for activation) given a current (pre-activation) state and goal (post activation) state. Divergence from goals triggers competition between classifier feature maps and enables the system to focus on sequences of behaviour that lead to goal convergence. The classifier selection process described above is essentially a noisy hill climber working towards a fixed goal using strictly local information. The noise in this context is introduced by inter-classifier competition which is a measure of how efficient the system is in achieving its goals. High levels of competition degrade each classifier's fitness ranking and reduce the effectiveness of the genetic algorithm based adaptive strategy. Accurate calibration of classifiers through stable self-organization is therefore highly desirable if the system is to converge on goal states [3]. Classifier SOFMs provide a reactive learning capacity which enables the system to calibrate classifier behaviour independently of global feedback. An additional SOFM, the feature correlation network, is utilized as an associative memory to correlate current domain states with predefined landmark targets. Interrogation of this network produces the goal definitions that correspond to the desired post classifer activation state.

3 Problem Domain

The specific problem domain is a simulation of a typical autonomous guided vehicle (AGV) produced by second year undergraduates at the Cambridge University Engineering Department. The AGV is designed to navigate a course (defined by white lines and constrained by obstacles) and perform various pallet handling tasks. The AGV is controlled by a Pascal program which interacts with the vehicle via a fixed set of commands that can retrieve course information from infra-red sensors/microswitches and control motion via two direct drive wheels. Local behaviour in the simulation is defined as obstacle avoidance whilst global behaviour is defined in terms of following white lines denoting specific landmarks. The simulation decomposes the problem domain into a feature set defined as:

- microswitch status;
- infrared status;
- x-coordinate;
- y-coordinate;

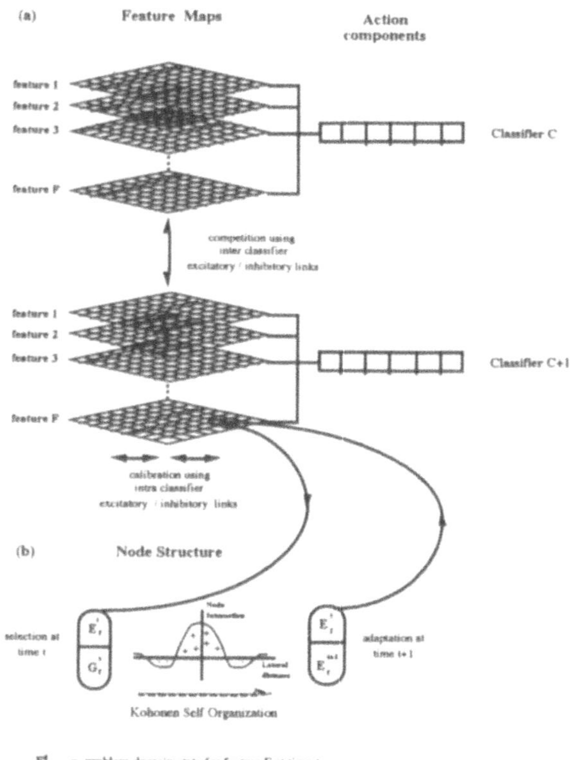

Figure 1: HLS classifier architecture.

- motion;
- rotation;
- heading.

The first four features are ecocentric and require an associative internal map. The remaining features are egocentric and impinge directly upon the SOFMs linked to each classifier. The behaviour of the AGV is simulated by a classifier set defined as:

- motor forward;
- motor backward;
- motor stop — for each motor.

Previous experiments have considered minimal representations to encode obstacles and support 'wall following' behaviour by building associative links between microswitch states and $x - y$ coordinates in the feature correlation network [4]. Current work is attempting to define effective arbitration strategies that can select an optimal domain feature to be focussed on at each time step.

4 Arbitration Experiments

The basic behaviour selection mechanism in HLS (Figure 2) is:

1. detect changes in the domain and map onto the ecocentric / egocentric feature set;
2. update the Feature Correlation Network with feature set data;
3. perform an associative lookup of next target given ecocentric goals and egocentric animat state;
4. select a set of candidate classifiers to achieve each feature target;
5. effect changes to the domain through activation of the 'best' classifier;
6. perform self-organization of classifier feature maps based on domain state before and after (5).

A single feature target is required because HLS activates a single classifier at (5). This activation may produce multiple changes in the domain since a classifier may have evolved multiple action components. An arbitration strategy in this context has to define 'best'. Two sets of experiments have been conducted in this research based on different goals being set for the animat. The first experiment sets a fixed goal G specified in terms of $x - y$ coordinates. Only local behaviour is required from the animat to achieve this goal since coordinate feedback is available continuously from the starting state S. Four arbitration strategies were investigated and ranked according to animat performance in achieving the goal. The strategies were — random selection, heading only, rotation only and $x - y$ coordinates only. Results from this first experiment show that selecting feature targets that enable stable self organization of the classifiers' SOFMs are more effective in optimizing behaviour than random selection (Figure 3). During the early stages of the task when the animat has to to follow a fairly stable heading, a heading-only strategy worked best (Figure 4). Later on, when the animat has to maintain its position at the goal G, a rotation-only strategy worked best (Figure 5). Essentially the calibration behaviour of the system varied depending on

HYBRID LEARNING SYSTEM (HLS)

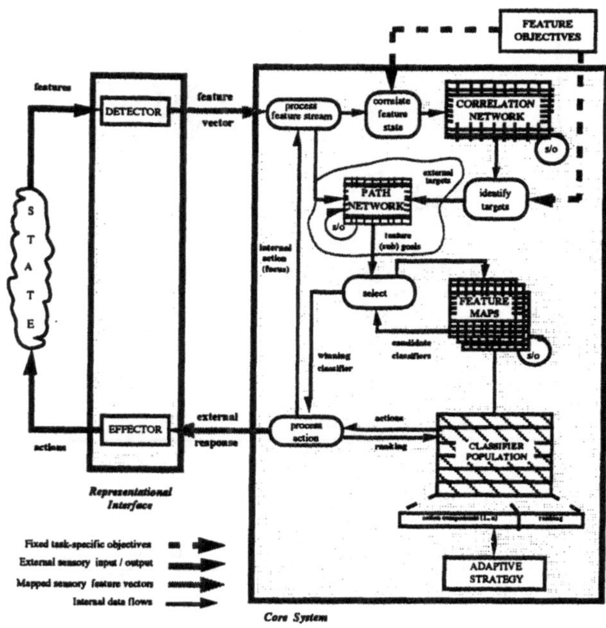

ADAPTIVE STRATEGY - application of G.A operators (based on classifier ranking)
S/O - self organization of neural networks

Figure 2: HLS system architecture.

the feature being targetted e.g. classifiers such as 'right motor forward' cannot be calibrated against coordinate feature state since their behaviour is unpredictable (Figure 6). However in the event of a desired 'left motor forward-right motor forward' behaviour, the calibration process for these classifiers will work against heading feature state. The conclusion from this experiment is that an optimal arbitration strategy would select the feature whose candidate classifier had the most stable SOFM for that feature. The second experiment set a fixed goal in terms of infrared state and microswitch state — to find a white line and avoid walls. This experiment required the acquisition of global behaviour since feedback was only available intermittently from the domain whenever the animat hit an obstacle or ran over a white line. Initial results using the same arbitration strategies produced uniformly poor results suggesting that development and representation of derived features (not directly ecocentric or egocentric) will be required as an emergent behaviour of the animat.

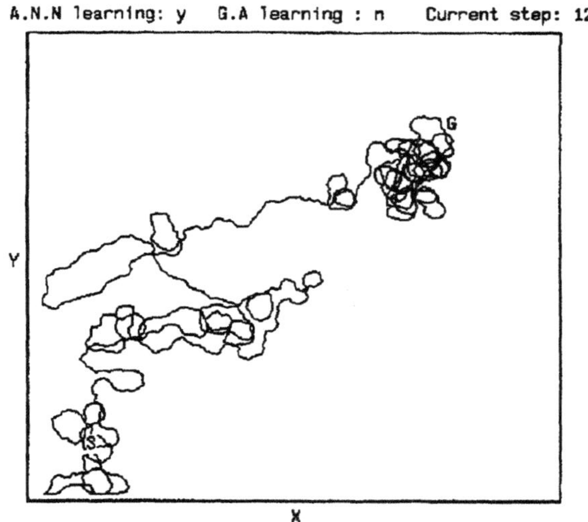

Figure 3: Random feature selection.

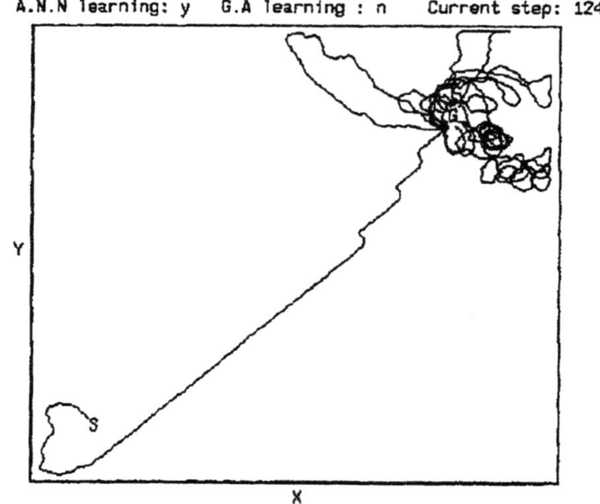

Figure 4: Heading feature selected.

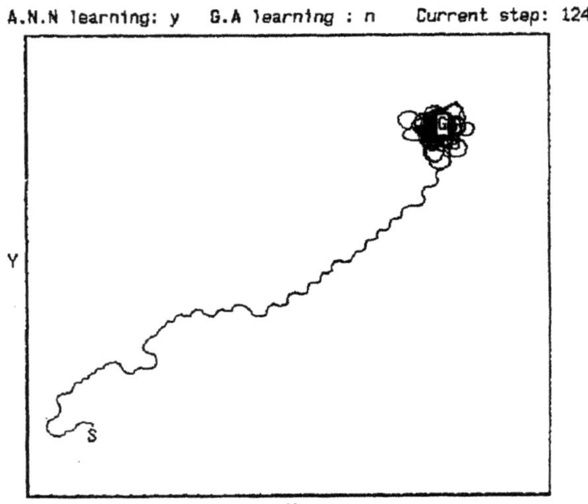

Figure 5: Rotation feature selected.

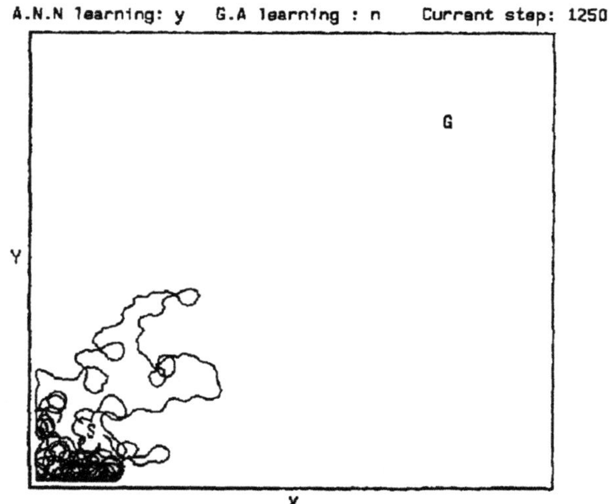

Figure 6: $X - Y$ coordinate feature selected.

5 Conclusion

The use of low level feedback within HLS makes it applicable in domains where fitness is specified over multiple, discrete features that can impinge directly upon the classifier population via SOFMs. Reactive control of simulated AGVs is one such domain. The application of an associative SOFM (the feature correlation network) within HLS enables the system to maintain landmark-based as well as behaviour-based perspectives. Current research suggests that the trajectory representation used to enable obstacle avoidance (presented in [4]) will need to be evolved to support new types of behaviour such as line following.

References

[1] R. Arkin. Behaviour-based robot navigation for extended domains. *Adaptive Behaviour*, 1(2):201–

225, 1992.

[2] N. Ball. *Cognitive Maps in Learning Classifier Systems.* PhD thesis, University of Reading, Reading, UK., 1991.

[3] N. Ball. *Organizing an Animat's behavioural repertoires using Kohonen Feature Maps.* MIT Press, 1994.

[4] N. Ball. *Application of a neural network based classifier system to AGV obstacle avoidance*, volume 1237, pages 1–12. Elsevier, 1996.

[5] R. Brooks. A robust layered control system for a mobile robot. *IEEE Journal of Robotics and Automation*, 2(1):14–23, 1986.

[6] J. Holland, K. Holyoak, R. Nisbett, and P. Thagard. *Induction: Processes of Inference, Learning and Discovery.* MIT Press, 1986.

[7] T. Kohonen. *Self Organization and Associative Memory.* Springer, 1984.

[8] P. Maes. *Behaviour-based artificial intelligence.* MIT Press, 1992.

[9] T. Prescott. Spatial representation for navigation in animats. *Adaptive Behaviour*, 4(2):85–123, 1996.

[10] S. Wilson. *The Animat Path to AI.* MIT press, 1990.

Sequence Clustering by Time Delay Networks

N. Allott, P. Halstead and P. Fazackerley
Computing Department, The Nottingham Trent University,
Burton St, Nottingham, NG1 4BU
Email: nma@doc.ntu.ac.uk

Abstract

This paper outlines the form and structure of an activation passing context rich network ideally suited to tasks such as; recovering from noisy or damaged data, recognition or spelling correction. Such a network has been shown to be in many ways similar to N-Gram analysis or the transition matrix of Markov Models. However it is superior in that the scope of context (N) to be considered is dynamically and locally identified for each node. A pair of algorithms are presented which can be used to produce such networks. The first uses global working memory to deal with the time dependent nature of the input data, the second uses time delay nodes within the network. The latter, however, is considered superior as it reduces algorithmic complexity and increases computational efficiency.

1 Introduction

One of the keys to successfully recovering from noisy, damaged or incomplete data within the fields of spelling correction or recognition is to maintain and apply an accurate model of the orthographic regularity of sub-word components and using this to modify the probability estimates from a user error module. N-gram analysis and Markov models have both been used successfully to perform this task [3, 5, 8], statistically deriving their information from a corpus. However both are limited in that the depth of context (the value of N) must be specified before the corpus analysis commences: low values of N mean corpora need only be small however transition estimates are of limited value due to the narrow context; high values require ridiculously large corpora for reliable probability estimates [6]. In this paper we look at several algorithms that produce context rich networks from corpus analysis. The network production processes differ from above in that the value for N is locally and dynamically determined for each node.

For convenience, here on, we use the term composition-decomposition (CDC) network to identify a particular type of connectionist network originally developed in [1]. This network was developed initially in order to represent high level semantic information for application to an automated assessment task. However, it was the intention of the authors as stated in [1] that the same representational and problem solving scheme could be applied to more primitive natural language processing tasks. Such tasks as spelling correction or error detection, where a sophisticated models of orthographic or phonetic regularity could improve performance.

We shall in this paper consider firstly the format and specification of the proposed network. Secondly, we shall consider two alternative algorithms for the production of these networks. Finally shall consider a fuzzy element retrieval from the network using the second of these algorithms.

2 Formal Specification of Data Structure and Heuristics

All the algorithms considered below generate a network of the same general form and structure [2] and are modifiable by two configurable heuristics. Each network initially consists of a set of non-connected primitive nodes. In the case of spelling correction these primitive nodes are the 26 letters of the alphabet. When a string is presented to the network the appropriate primitive nodes are activated. A linkage heuristic must be defined which specifies a relationship that must hold between two activated nodes. If the linkage heuristic holds for any two nodes, a context history is recorded for those nodes. These context histories are constantly updated as new strings are processed. When the context history exceeds a threshold defined by the clustering heuristic the two nodes found within each other's context are made into a new parent node and child links are created and maintained. If in the future, the two child nodes become active in the correct order, the parent node becomes active also.

Formally this is specified as: N the nodelist is a set of s nodes,

$$N = \{n_1, n_2, \ldots, n_s\}.$$

Here each component n is a composite with the following attributes:

$$n = (WordID, Frequency, Children, Linkage).$$

$WordID$ is a string to identify the node, Frequency is an integer in which we record the number of times the node has been found. Children, parents and linkage are all lists and defined as follows:

$$Children_x = \{c_{1x}, c_{2x}, c_{3x}, \ldots, c_{qx}\},$$

where each member is of type c comprising the single attribute: $c = (WordID)$. Similarly

$$Parents_x = \{p_{1x}, p_{2x}, p_{3x}, \ldots, p_{rx}\},$$

and: $p = (WordID)$. Finally

$$Linkage_x = \{l_{1x}, l_{2x}, l_{3x}, \ldots, l_{rx}\}$$

where l is a composite type with two attributes

$$l = (WordID, Frequency).$$

If we use the syntax where a function of the same name as an attribute performed on an appropriate composite extracts the value of that attribute, it follows that:

$$Freq(n_x) = \sum_{1}^{s} frequency(l_{ix}).$$

In other words the frequency of a node is equal to the sum of the frequency of the distinct contexts it has been found in. A useful variable is the total number of primitive nodes encountered

$$Global = \sum_{1}^{s} freqprims(i)$$

where:

$$freqprims(i) = \begin{cases} freq(n_i) & \mid word(n_i) \mid = 1 \\ 0 & \mid word(n_i) \mid \neq 1 \end{cases}$$

and

$$Word(n_i) = \text{length of string}.$$

The input string we may interpret mathematically as a set of ordered pairs,

$$String = \{k_1, k_2, k_3, \ldots, k_z\}.$$

Each element is a composite with the first attribute the $letterID$ the second is the position in the string, i.e. $k = (letterID, position)$.

Given this the linkage heuristic may be defined as a boolean returning function. Simple right adjacency heuristic could be defined as:

$$Link(x,y) = \begin{cases} 1, & position(k_x) = position(k_y) + 1 \\ 0, & position(k_x) \neq position(k_y) + 1 \end{cases}$$

Clustering heuristics are also boolean returning functions. A simple heuristic relying on absolute frequency could be defined as

$$Cluster(x,y) = \begin{cases} 1, & \frac{freq(l_{xy})}{Global} \geq 0.01 \\ 0, & \frac{freq(l_{xy})}{Global} < 0.01 \end{cases}$$

As a general note the linkage heuristic defines the form or nature of the network, i.e. what the network actually represents. In the simplest case this is right adjacency. The clustering heuristic determines the breadth and depth of the network, and is usually a statistical threshold. A low threshold will produce a deep tree that will have nodes that correspond to full lexical items.

3 Time Series through Global Working Memory

A standard statistical procedure attempts to identify a relationship between N variables. The algorithm outlined does this with a set of primitive nodes, however when a strong relationship is found between two or more nodes they are concatenated into a new node and the procedure will then look for relationships with this node also. The number of variables is therefore constantly growing as the data is recursively applied to itself.

If we are to implement this as an adaptive network there are two problems that have to be overcome:

1. With strings we are dealing with time dependent data. Within a standard connectionist network it is difficult to come up within an encoding where a nodes position within the string (the time dependant data) is preserved.

2. Within the learning phase we have the problem of making the distinction between the type and token of a node. This is best exemplified in the word banana. Here there are two instances of the cluster 'ana' that occupy non unique positions within the string. If there is a single node to represent the type 'ana' we must at least temporarily be able to discriminate two distinct instances of the token 'ana' at different but overlapping positions.

One solution to this problem is to couple a working memory onto the connectionist network. Working memory becomes the blackboard upon which all instances of nodes are recorded as they are activated. A node instance is created within working memory as soon as it receives input from its first child node and the time at which it was first activated is bound to this instance. When the node receives input from its final child node the node itself is activated and this time is also recorded.

In outline, a new unconnected primitive node is instantiated for each primitive unit encountered in a string. Whenever this node is subsequently encountered within a string this node is instanced in working memory and remains there for a specified lifespan. The linkage heuristic is constantly applied to working memory and contextual information is recorded for any node instances satisfying the criteria.

When the context history combining two (or more) nodes exceeds a value specified by the clustering heuristic a new compound unit is instantiated which describes the conjunction of the children nodes.

Certain metrics are attached to the various entities in the tree. It is with respect to these metrics that the clustering and linkage heuristic are defined. Of significance is the measure of frequency (attached to nodes, links and global measures). An absolute measure of frequency would be inadequate for inter-node comparisons as new nodes are being created all the time and the new nodes are unaware of how often they occurred before they were created. More suitable would be a measure of acquisition velocity, that is to take an estimate of the differential of frequency over time. It was found that the regular resetting of all frequency values served as a satisfactory approximation to this.

The algorithm as discussed may be defined in pseudo code as follows:

```
LOOP (for all strings)
{
   LOOP (for all units)
   {
      IF (node not previously seen)
         DO Instantiate_Node
      ELSE (node already seen)
      {
         DO Activate_This_Node
         DO Pass_Activation_Parents
         DO Instance_Node_In_Memory
         LOOP (all instanced nodes)
         {
            IF (LINKAGE HEURISTIC)
            {
               DO Add_Link_To_History
               IF (THRESHOLD HEURISTIC)
               {
                  DO Make_Nodes_Compounds
                  DO Link_Children+Parent
               }
            }
         }
      }
      DO Clear_Working_Memory
   }
}
```

Although there are several inefficiencies in this procedure it does work and rapidly produces networks of considerable complexity embodying deep contextual information from the problem domain. Training time is roughly proportional to the square of the current number of identified nodes and therefore assuming a constant rate of node acquisition increases exponentially over time. The rate of node acquisition is however completely determined by the clustering heuristic and therefore further generalisation is difficult.

In summary global working memory is being used to perform two distinct activities.

1. Sequence Processing: to allow time dependent information to propagate up the network so that the 'th' node only becomes active when the 't' node and the 'h' node are activated immediately after one another.

2. Learning: to provide a blackboard on which activated nodes can be recorded and upon which the linkage heuristic can be applied.

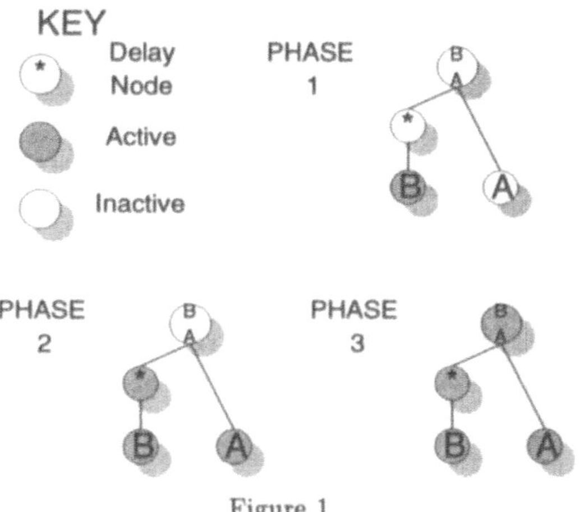

Figure 1

4 Time Series through the Synchronous Activation Update of Gateway Nodes

With a more sophisticated activation model it is possible to remove the need for working memory to model the propagation of time dependent data. A stricter activation model must be applied where each single unit evidence node is activated at distinct phases. Consider the example of 'banana': 'b' must be activated on phase 1, 'a' activated on phase 2, 'n' activated on phase 3 etc.

A composite when identified must be gatewayed with activation delay nodes to offset the activation latency of primary and secondary child nodes. For example if 'ba' was identified as a composite node 'b' and 'a' would both be recorded as child nodes, but a delay node would be inserted between 'b' and its parent 'ba'. This would give an activation schedule as follows:

Phase 1: 'b' becomes active.
Phase 2: 'b' passes activation to delay node, 'a' becomes active.
Phase 3: delay node passes activation to 'ba', a passes activation to 'ba'.

This way 'ba' receives its activation from both its child nodes at the same time even though the respective child nodes themselves were active at different times.

Not only is this computationally more efficient but it is architecturally more consistent with the connectionist model. Working memory is now only necessary for the application of the learning phase, i.e. the application of the linkage heuristic.

Further, as we shall see below, such a model makes it easier to introduce a notion of variable activation which will be necessary for the recovery of damaged data.

5 Application of Network

As discussed such context rich trees are ideally applicable to recognition type tasks. In their simplest they can be shown to produce identical information to that produced by N-Gram analysis or the transition matrix of a Markov model. However, we do not have to specify our grain of analysis (bigram - trigram etc.) before processing takes place nor must the grain be unique throughout the analysis. If the heuristics are set up correctly the algorithm dynamically and locally determines these for each node. We can therefore use the trees in the same way as we would with N-gram or Markov models, i.e. produce estimates of word probability from supplied evidence, as a multiple of transition probabilities between primitives. But using the context tree we may augment these estimates by describing our units in higher level terms and using transition probabilities between these to modify our original figure. Similarly we could use the identified tokens as the principal resource for N-gram [4, 7] indexing as used for spelling correction, with parallel benefits.

But, by extending the synchronous activation model discussed in the final algorithm we could introduce the notion of partial activation into the network. By applying a time decay function to each node we can see how two perfectly timed child nodes will lead to full activation of a parent node, whereas if the child nodes are slightly misplaced the parent node will only be partially activated. If the clustering heuristic was adjusted such that the tree was built to the deepest level ie full lexical items, the lexical items that most closely matched the supplied evidence would become most active. Further by implementing mutually inhibitory nodes between all full lexical items through interactive activation and competition [9] the network itself could resolve the best fitting match.

6 Further Work

The learning algorithms presented above consider only the composition element of CDC networks. No algorithm has yet been developed which attempts to automatically identify the decomposition element and hence the various subtypes of nodes. This is an area within the problem domain ripe for exploitation. For example, constants and vowels are the two crudest subtypes within the domain each of which has a completely distinct contextual distribution which could be used to great effect in modifying probability estimates for data recovery. This is seen as the next phase for development.

References

[1] N. Allott, P. Halstead, and P. Fazackerley. A knowledge driven aid to the automated assessment of free text. *AISBQ*, 88, 1994.

[2] N. Allott, P. Halstead, and P. Fazackerley. Clustering algorithm to produce context rich networks. In *Proceedings of Applied Decision Technologies*, pages 265–269, 1995.

[3] P. Brown, H. Lee, and Spohrer. Bayesian adaptation in speech recognition. In *Proceedings of the ICCASP*, pages 761–764, Boston, USA.

[4] M.W. Du and S.C. Chang. An approach to designing very fast approximate string matching algorithms. *IEEE Transactions on Knowledge and Data Engineering*, 6(4):620–633, 1994.

[5] F. Jelinek, R. Mercer, and Bahl. Continuous speech recognition: Statistical methods. *IEEE Transactions on Pattern Analysis and Machine Intelligence*, PAM-5, 1983.

[6] F. Keenan. *Large Vocabulary Syntactic Analysis for Text Recognition.* PhD thesis, The Nottingham Trent University, 1992.

[7] K. Kukich. Techniques for automatically correcting words in text. *ACM Computing Surveys*, 24(4):377–439, 1992.

[8] E. Riseman and A. Hansen. A contextual postprocessing system for error correction using binary n-grams. *IEEE Transactions on Computers*, C-23:490–493, 1974.

[9] D. E. Rumelhart and J. McClelland. *Parallel Distributed Processing.* MIT, Cambridge,MA, 1968.

Modeling Complex Symbolic Sequences with Neural Based Systems

P. Tiňo and V. Vojtek
Department of Computer Science and Engineering,
Slovak University of Technology,
Ilkovicova 3, 812 19 Bratislava, Slovakia
Email: {tino,vojtek}@decef.elf.stuba.sk

Abstract

We study the problem of modeling long, complex symbolic sequences with recurrent neural networks (RNNs) and stochastic machines (SMs). RNNs are trained to predict the next symbol and the training process is monitored with information theory based performance measures. SMs are constructed using Kohonen self-organizing map quantizing RNN state space. We compare generative models through entropy spectra computed from sequences, or directly from the machines.

1 Introduction

Given a long sequence S with positive entropy, i.e. is difficult to predict, we wish to construct a stochastic model whose information theoretic properties are similar to those of S. In this contribution, we confine ourselves to model classes of recurrent neural networks (RNNs) with probabilistically interpretable outputs and finite state stochastic machines (SMs).

We train RNN on S to predict the next symbol and monitor the training process with information theory based performance measures. Trained RNNs are then used as sequence generators and compared with SMs constructed as ***hybrid neural-neural models*** consisting of trained RNNs and Kohonen self-organizing maps.

We use statistical mechanical metaphor to study and compare different generative models. Entropy spectra computed from sequences, or directly from machines, serve as a means to understand statistical sequence structure. In particular, the long term behavior of a SM is captured by the sequence- and block length-independent machine entropy spectrum. Different temperatures accentuate different probability levels of subsequences.

2 Statistics on Symbolic Sequences

We consider sequences $S = s_0 s_1 s_2 ...$ over a finite alphabet $\mathcal{A}$ generated by stationary information sources. To study the statistical structure of S one can use a 'sliding window' $w = w_1 ... w_n$ of length n and determine the (empirical) probabilities $P_n(w)$ of finding a particular window w in S, if a block of n symbols (an n-block) is randomly chosen. A measure of uncertainty of n-blocks is given by the block entropy

$$H_n = H(P_n) = - \sum_{w \in \mathcal{A}^n} P_n(w) \log P_n(w). \qquad (1)$$

A measure of predictability of an added symbol independent of block length is then

$$h = \lim_{n \to \infty} h_n. \qquad (2)$$

Entropy provides only a partial information concerning the sequence distribution P. A more fulfilling description is obtained through a spectrum of entropy measures describing P. The spectrum is constructed using a formal parameter β that can be thought of as the inverse temperature in the statistical mechanics of spin systems [1]. The original distribution of n-blocks, $P_n(w)$, is transformed to the 'twisted' distribution [6]

$$Q_{\beta,n}(w) = \frac{P_n^\beta(w)}{\sum_{w \in \mathcal{A}^n} P_n^\beta(w)}. \qquad (3)$$

The most probable and the least probable n-blocks of the original distribution $P_n(w)$ become dominant in the positive zero and the negative zero temperature regimes, $Q_{\infty,n}(w)$ and $Q_{-\infty,n}(w)$ respectively. Varying β from 0 to ∞ amounts to a shift from all allowed n-blocks to the most probable ones by accentuating still more and more probable subsequences. Varying β from 0 to $-\infty$ accentuates less and less probable n-blocks with the extreme of the least probable ones.

Thermodynamic entropy density, is approximated from the distribution over n-blocks by

$$h_{\beta,n} = \frac{-\sum_{w\in\mathcal{A}^n} Q_{\beta,n}(w)\log Q_{\beta,n}(w)}{n} \quad (4)$$

and is given asymptotically by

$$h_\beta = \lim_{n\to\infty} h_{\beta,n}. \quad (5)$$

3 Stochastic Machines

Stochastic machines (SMs) are much like non-deterministic finite state machines except that the state transitions take place with probabilities prescribed by a distribution T. To start the process, the machine M chooses the initial state according to the 'initial' distribution π, and then, at any given time step after that, the machine is in some state $i \in Q$, and at the next time step moves to another state $j \in Q$ outputting some symbol $s \in \mathcal{A}$, with the transition probability $T_{i,j,s}$.

It is convenient to denote the transition matrix associated with a symbol s and the stochastic state transition matrix by $T(s)$ and $\mathcal{T}$ respectively

$$T(s)_{i,j} = T_{i,j,s} \quad \text{for all } i,j \in Q \text{ and } s \in \mathcal{A}, \quad (6)$$

$$\mathcal{T} = \sum_{s\in\mathcal{A}} T(s). \quad (7)$$

Ignoring the state transition labels, $\mathcal{T}$ describes a Markov chain over the machine states Q.

As in the previous section, we introduce parameterized transition probabilities $T^{\beta}_{i,j,s}$ and think of each setting of the formal parameter β as emphasizing a different set of sequences generated by M. The state transition matrix becomes ($\mathcal{T}_\beta$ is no longer a stochastic matrix)

$$\mathcal{T}_\beta = \sum_{s\in\mathcal{A}} (T(s))^\beta. \quad (8)$$

Denote the left and right eigenvectors of $\mathcal{T}_\beta$ associated with the maximum eigenvalue λ_β by v^L_β and v^R_β respectively. The equivalent stochastic process with transition probabilities weighted according to $\mathcal{T}_\beta$ is given by the stochasticized version

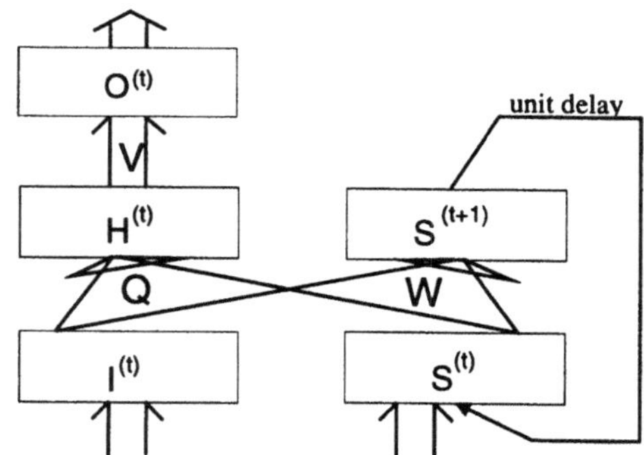

Figure 1: RNN architecture.

$$(\mathcal{R}_\beta)_{ij} = \frac{(\mathcal{T}_\beta)_{ij}\left(v^R_\beta\right)_j}{\lambda_\beta\left(v^R_\beta\right)_i}. \quad (9)$$

The metric entropy of the parameterized stochastic machine M_β,

$$h_\mu(\beta) = -\sum_{i,j\in Q} p_{\beta,i}\,(\mathcal{R}_\beta)_{ij}\log(\mathcal{R}_\beta)_{ij} \quad (10)$$

is an average of transition uncertainty over all machine states. In the deterministic case, when the skeletal structure of M is deterministic, i.e. for every state, each symbol s uniquely determines the next state, $h_\mu(\beta)$ is also the thermodynamic entropy density h_β (eq. (5)) of sequences generated by M [6].

4 Neural models

The RNN presented in Figure 1 was shown to be able to learn mappings that can be described by finite state machines We ask the reader to consult the architecture details in [5].

The network is trained with RTRL [2] on a single, long symbolic sequence $S = s_0s_1s_2...$, to predict, at each point in time, the next symbol.

To start the training, the initial network state $R^{(0)}$ (activations of recurrent neurons at time 0) is randomly generated. The network is reset with $R^{(0)}$ at the beginning of each training epoch.

After the training, the network is seeded with the initial state $R^{(0)}$ and the first symbol s_0. For the next T_1 'pre-test' steps, for the current input and state, the next network state is computed, and it comes into play, together with the next symbol from S, at the next time step. This way, the network is given a right 'momentum' in the state path starting in the initial 'reset' state $R^{(0)}$.

After T_1 pre-test steps, the network generates a symbol sequence by itself. In each of T_2 test steps, the network output is interpreted as a new symbol that will appear at the net input at the next time step. The network state sequence is generated as before. The output activations $O^{(t)} \in (0,1)$ are transformed into 'probabilities' $P_i^{(t)}$,

$$P_i^{(t)} = \frac{O_i^{(t)}}{\sum_{j=1}^{A} O_j^{(t)}}, \quad i = 1, 2, ..., A, \tag{11}$$

and the new symbol $\hat{s}^{(t)} \in \mathcal{A}$ is generated with respect to the distribution $P_i^{(t)}$ (the number of output neurons is equal to the number of symbols in $\mathcal{A}$).

We assume that the reader is familiar with the standard unsupervised training procedure for SOM [3]. SOM places a fixed number of codevectors w_i (codevectors) into the map input space Y, subject to a minimum distortion constraint. w_i represents a part of the input space,

$$V(i) = \left\{ y \in Y |\ d(y, w_i) = \min_j \{d(y, w_j)\} \right\}, \tag{12}$$

where d is the Euclidean distance. $V(i)$ is referred to as the Voronoi compartment of the codevector i.

5 Stochastic Machines Induced by Symbolic Sequences

The stochastic machine $M_{RNN} = (Q, \mathcal{A}, T, \pi)$ is extracted from the RNN trained on a sequence $S = s_0 s_1 s_2 ...$ using the following algorithm:

1. Quantize the RNN State space by running a Kohonen SOM on RNN states recorded during the RNN testing.
2. The initial state is a pair (s_0, i_0), where i_0 is the index of the Kohonen unit defining the Voronoi compartment $V(i_0)$ containing the network 'reset' state $R^{(0)}$, i.e. $R^{(0)} \in V(i_0)$. Set $Q = \{(s_0, i_0)\}$.
3. For T_1 pre-test steps $1 \leq t \leq T_1$
   - $Q := Q \cup (s_t, i_t)$, where $R^{(t)} \in V(i_t)$
   - add the edge (from the state (s_{t-1}, i_{t-1}) to the state (s_t, i_t), labeled with s_t) $(s_{t-1}, i_{t-1}) \to^{s_t} (s_t, i_t)$ to the topological skeletal state-transition structure of M_{RNN}.
4. For T_2 test steps $T_1 < t \leq T_1 + T_2$
   - $Q := Q \cup (\hat{s}_t, i_t)$, where $R^{(t)} \in V(i_t)$ and $\hat{s}_t$ is the symbol generated at the RNN output.
   - add the edge $(\hat{s}_{t-1}, i_{t-1}) \to^{\hat{s}_t} (\hat{s}_t, i_t)$ (noting that $\hat{s}_{T_1} = s_{T_1}$) to the set of allowed state-transitions in M_{RNN}.

The probabilistic structure is added to the topological structure of M_{RNN} by counting, for all state pairs $(p, q) \in Q^2$ and each symbol $s \in \mathcal{A}$, the number $N(p, q, s)$ of times the edge $p \to^s q$ was invoked while performing steps 3 and 4. The state-transition probabilities are then computed as

$$T_{p,q,s} = \frac{N(p, q, s)}{\sum_{r \in Q, a \in \mathcal{A}} N(p, r, a)}. \tag{13}$$

The philosophy of the extraction procedure is to let the RNN act as in the testing mode, and interpret the activity of RNN, whose states have been factorized into a finite set of clusters, as a stochastic machine M_{RNN}.

6 Experiments

For a stationary ergodic process that has generated a sequence S of length N, the Lempel-Ziv codeword length for S, divided by N, is computationally efficient and reliable estimate $h_L Z(S)$ of the source entropy [7].

The notion of 'distance' between distributions used in this paper is a well-known measure in information theory, called Kullback-Leibler divergence. It is also known as the relative, or cross entropy. Let P and Q be two Markov probability measures, each of some (unknown) finite order. The divergence between P and Q is defined by

$$d^{KL}(Q|P) = \limsup_{n \to \infty} \frac{1}{n} \sum_{w \in A^n} Q_n(w) \log \frac{Q_n(w)}{P_n(w)}.$$

d^{KL} measures the expected additional code length required when using the ideal code for P instead of the ideal code for the 'right' distribution Q.

Suppose we have only length-N realizations S_P and S_Q of P and Q respectively. Analogically to Lempel-Ziv entropy estimation, there is an estimation procedure for determining $d^{KL}(Q|P)$ from S_P and S_Q [7]. The procedure is based on Lempel-Ziv sequential parsing of S_Q with respect to S_P.

In the experiment, we used Santa Fe competition data recorded from a laser in a chaotic state available on the internet at (`http://www.cs.colorado.edu/~andreas/Time-Series/SantaFe.html`).

Time series of approximately 10,000 points was transformed into a symbolic sequence over $\{a, b, c, d\}$ by partitioning the signal range into 4 regions $[0, 50)$, $[50, 200)$, $[-64, 0)$ and $[-200, 64)$. The regions were determined by close inspection of the data and correspond to clusters of low and high positive/negative laser activity.

We trained two RNNs with 2 and 5 recurrent (state) neurons. The two RNNs are referred to as RNN_2 and RNN_5 respectively. The training process consisted of 10 runs. There were respectively, 150 and 100 passes through S in each run of RNN_2 and RNN_5 training.

After certain runs, we let the RNN generate a sequence $S(RNN)$ of length equal to the length of the training sequence S and computed the entropy and cross entropy estimates $h_{LZ}(S(RNN))$ and $h_{LZ}^{KL}(S|S(RNN))$ respectively. Figures 2 and 3 show the summary of entropic measures for the training process.

On average, the 5 state neuron network, RNN_5, did better than its 2 state counterpart RNN_2, because RNN_5 developed more sophisticated dynamical representations of the temporal structure in S, than RNN_2. A consequence of a more powerful potential for developing dynamical scenarios in RNN_5 is a greater liability to bad local minima solutions and systematic underestimation of the training sequence entropy. On the other hand, relatively simple dynamical regimes in RNN_2 resulted in oversimplifying dynamical patterns of allowed n-blocks, and RNN_2 systematically overestimated the entropy of S.

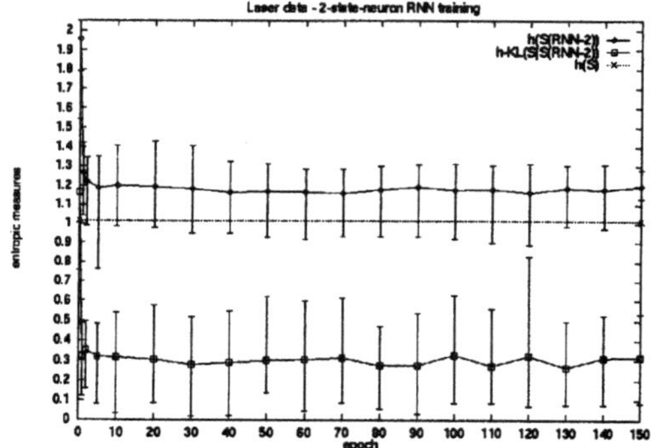

Figure 2: Training of 2-state-neuron RNN on laser data S.

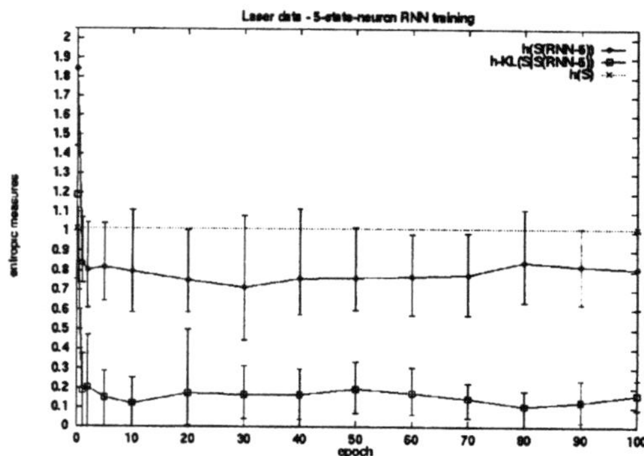

Figure 3: Training of 5-state-neuron RNNs on laser data S.

We observed that the complexity of computational structure in training sequences was reflected by dynamical state representations of trained RNNs, and hence by complexity of extracted machines M_{RNN}.

From the modeling point of view, the most important model characteristics in entropy spectra are metric ($\beta = 1$) and topological ($\beta = 0$) entropies. We found that the training sequence, trained RNNs, as well as extracted SMs M_{RNN}, shared very similar metric and topological characteristics. Measures at other high positive/negative inverse temperatures reflect details in model construction procedures and can be of importance only when generating much longer sequences than the training one. Finite length sequences do not contain sufficient information for determining extreme inverse temperature statistics. A detailed account of work presented in this contribution can be found in [4].

References

[1] J.P. Crutchfield and K. Young. Computation at the onset of chaos. In W.H. Zurek, editor, *Complexity, Entropy, and the physics of Information, SFI*

Studies in the Sciences of Complexity, vol 8, pages 223–269. Addison-Wesley, 1990.

[2] J. Hertz, A. Krogh, and R.G. Palmer. *Introduction to the Theory of Neural Computation*. Addison–Wesley, Redwood City, CA, 1991.

[3] T. Kohonen. The self–organizing map. *Proceedings of the IEEE*, 78(9):1464–1479, 1990.

[4] P. Tiňo and M. Koteles. Modeling complex sequences with neural and hybrid neural based approaches. Technical Report STUFEI-DCSTR-96-49, Slovak University of Technology, Bratislava, Slovakia, September 1996.

[5] P. Tiňo and J. Sajda. Learning and extracting initial mealy machines with a modular neural network model. *Neural Computation*, 7(4):822–844, 1995.

[6] K. Young and J.P. Crutchfield. Fluctuation spectroscopy. In W. Ebeling, editor, *Chaos, Solitons, and Fractals, special issue on Complexity*, 1993.

[7] J. Ziv and N. Merhav. A measure of relative entropy between individual sequences with application to universal classification. *IEEE Transactions on Information Theory*, 39(4):1270–1279, 1993.

An Unsupervised Neural Method for Time Series Analysis, Characterisation and Prediction

C. Fyfe
Department of Computing and Information Systems,
The University of Paisley, UK.
Email:fyfe0ci@paisley.ac.uk

Abstract

We present a novel neural network method for extraction of the embedding function of a time series. We give results on two sets of computer-generated data which are known to show exponentially increasing divergence from nearby initial conditions. We use the network to predict the future evolution of these artificial mappings.

1 Introduction

Time series prediction is based on the assumption that an observable feature of a system is determined by an underlying deterministic system. If the evolution of the system can be described by a set of n ordinary differential equations in n variables, there exists a unique trajectory through every point a in R^n. In order to make a prediction we would like to know both the underlying rules of the deterministic system and the current state of the system. For example, consider a system in which we can observe a single scalar quantity, x_t, which is the value of x at time t and which is determined by the state at time t, a_t. Now if the underlying system is based on the n dimensional set of differential equations described above, then the state of the system at time $t+1$, a_{t+1}, can be described by the set of equations a_{t+1}=F(a_t), while the observable at time $t+1$ is given by $x_{t+1} = g(a_{t+1}) = g(F(a_t))$. It can be shown that there exists a value d such that the vector $x = (x_1, x_2, ..., x_d)^T$ consisting of d consecutive observations of x, fully characterises the system, i.e. we may absolutely specify the evolution of the system using the function $F()$ or equally by using the function $H()$ where $x_{t+1} = H(x_t) = H(x_t, x_{t-1}, ..., x_{t-d+1})$. Therefore merely by finding a sufficiently long set of consecutive observations of x (d is known as the embedding dimension of the system) we have a complete specification of the future values of the future observations of the system. It is well known, of course, that, for non-linear systems, the underlying dynamics are often such that there will be divergence of trajectories from nearby initial conditions. Thus since we can never measure observables to infinite precision, we can only predict such systems a finite length into the future.

Most neural methods used in time series prediction are based on supervised learning (see e.g. [3] for a full discussion). This paper will discuss a self organising neural network method of finding the embedding function $H()$ for two simple non-linear systems and use it to provide an estimate of future observable events in the systems. The theory is based on that developed by Deco and Obradovic [1] who have used a 'triangular volume preserving architecture' to make the output of their network independent of the inputs (where independence implies that knowledge of one conveys no information about the other) and so the weights of the network converge to reveal the underlying mapping of the time series. The actual network used is novel and extremely simple.

2 The FIR Hebbian Model

We begin by preprocessing the input data from the time series at time t, $data_t$, by calculating a set of functions of this data where such functions are drawn from a basis of function space. Thus if the functions are $f_1(), f_2(), ...$, then we calculate

$$y_1(t) = x_{10} = f_1(data_t)$$
$$y_2(t) = x_{20} = f_2(data_t) \text{ etc.}$$

The last subscript on the x values is a time parameter so that

$$x_{10} = f_1(data_t)$$
$$x_{11} = f_1(data_{t-1})$$
$$x_{12} = f_1(data_{t-2}) \text{ etc.}$$

This can be modelled by a neural network with input data being preprocessed by the functions spanning the function space and then each x-neuron in turn passing its activation to subsequent x-neurons at each discrete time step. Before learning we have negative feedback of activation to the y-neurons:

$$y_i(t) \leftarrow y_i(t) - \sum_{j=1, j\neq i}^{m} \sum_{k=1}^{d} w_{ijk} x_{jk} \quad (1)$$

where x_{jk} is the value of the jth input at time (t-k), and w_{ijk} is the weight from the ith neuron to this input. Thus x_{i0} denotes the initial value of the ith input at time t *before* any feedback from other neurons. The d parameter measures the length of the embedding dimension. Learning uses simple Hebbian learning: $\Delta w_{ijk} = \eta y_i x_{jk}$ where η is a learning rate which may be decreased during the course of the simulation. The negative feedback in the network ensures that the network does not suffer from the usual Hebbian problem of weights growing without bound. Now since the expected value of $\Delta w_{ijk} = 0$ only when y_i and x_{jk} are decorrelated, we are consecutively decorrelating y_i and x_{j0} and then y_i and x_{j1} etc. and then similarly with respect to the time-delayed values of the other inputs. At convergence then we have an approximation to independence between the neuron's outputs and the input data stream.

In our initial simulations we use simple polynomials as the basis of function space, so that

$$y_1(t) = x_{10} = 1$$
$$y_2(t) = x_{20} = data_t$$
$$y_3(t) = x_{30} = data_t^2$$
$$y_4(t) = x_{40} = data_t^3$$

and so on. Thus the value of $x_{0k} = 1, \forall k$ while $x_{1k} = data_{t-k}, \forall k$ and $x_{2k} = data_{t-k}^2, \forall k$ etc.

We see that the network is a finite impulse response (FIR) network — each neuron sees a weighted sum of the inputs but this is not fed back to the other neurons and our aim is to extract the embedding function, $H()$, from the network's weights.

3 Results

We review the results of applying the network to artificial data. All simulations reported herein were carried out over 1000 iterations and with a learning rate of 0.1 which is annealed to 0 during the simulation.

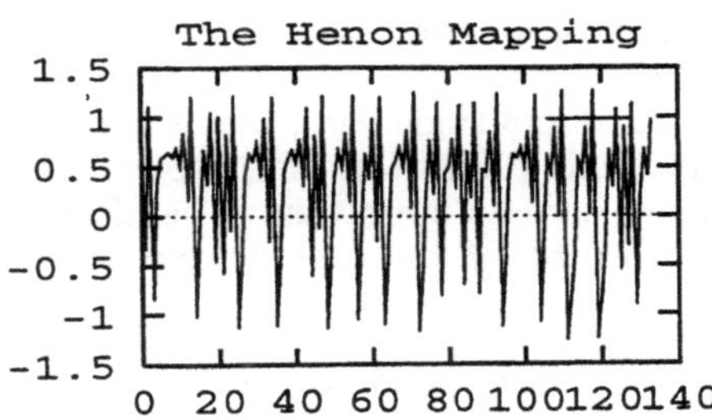

Figure 1: Samples from the Henon series.

Table 1: The converged weights when the network was trained on the Henon data.

Neuron	2 $= (data_t)$	3 $= (data_t)^2$	4 $= (data_t)^3$
time t	1.00	0.00	-0.01
t-1	0.01	1.39	0.00
t-2	-0.29	0.01	0.00

3.1 The Henon Map

The Henon Map is an iterative map given by the equations $q_t = 1 - 1.4q_{t-1}^2 + 0.3q_{t-2}$ where the values of the parameters have been chosen to be close to those which give an exponential divergence from initially close starting points. (This may in fact not be chaotic since it has been shown [2] that nearby parameters give an infinity of periodic orbits with very small basins of attraction. However this still gives a difficult to predict function.) We show in Figure 1 the time series data for one set of examples of the Henon Map.

It is clearly not an easy series to predict but we can show that there is structure in the data by mapping q_t against q_{t+1} from which we can see that, while there is not a mapping from q_t to q_{t+1} there is some relationship between q_t and q_{t+1} .

We run the Henon map from a random initial value in the range [0,1] and perform 2000 iterations of the mapping before collecting 1000 samples of the mapping which are our data points. We use a 4 input function set $(1, data_t, (data_t)^2, (data_t)^3)$. The weights *into* the first y-neuron converged to the values shown in Table 1.

Now each x-neuron is attempting to turn the other y-neurons off. When we look at the output of the first neuron we see that the other neurons have been largely successful—its output is approximately zero for all inputs suggesting that the mapping has been captured. Recall that the above values of the weights are used as inhibition. Therefore since the initial activation of the first neuron is 1, its final value is

$$1 - 1.00x_t - 1.39x_{t-1}^2 + 0.29x_{t-2}. \quad (2)$$

By letting this equal zero, the value the network is attempting to achieve, we can approximately recover the original equation

$$x_t = 1 - 1.39x_{t-1}^2 + 0.29x_{t-2}$$

which we can use to make subsequent predictions.

3.2 Estimating Noise

A noisy quadratic map can be created with the iteration

$$q_t = 1 - 4q_{t-1}(1 - q_{t-1}) + \gamma_t \quad (3)$$

where zero mean Gaussian noise of standard deviation 0.05 has been added to the map. Notice that since the noise appears in the iteration, it appears as coloured noise in the data sequence.

Again the same type of network was run with this data and the weights into the first neuron recorded. Again it was extremely simple to extract the underlying equations which created the data. To illustrate the independence of the outputs we have shown in Figure 2 the residual at the first neuron and the noise at time t. We can see that the first residual output is extemely close to the noise though not exact since the actual noise in the data contains noise from previous iterations. This tends to corroborate the assertian that this output has been made independent from the mapping which created the data model. Similar results can be achieved with noise added to the Henon map using the formula

$$q_t = 1 - 1.4q_{t-1}^2 + 0.3q_{t-2} + \gamma_t$$

where γ_t is the noise at time t.

It is of interest though that the fact that the noise is coloured is essential to the accuracy of the above graph: if we perform the Henon mapping and then add white Gaussian noise to the final values, the output of the first neuron does not accurately reflect the magnitude of the noise, i.e. if we calculate

$$q_t = 1 - 1.4q_{t-1}^2 + 0.3q_{t-2} \quad (4)$$

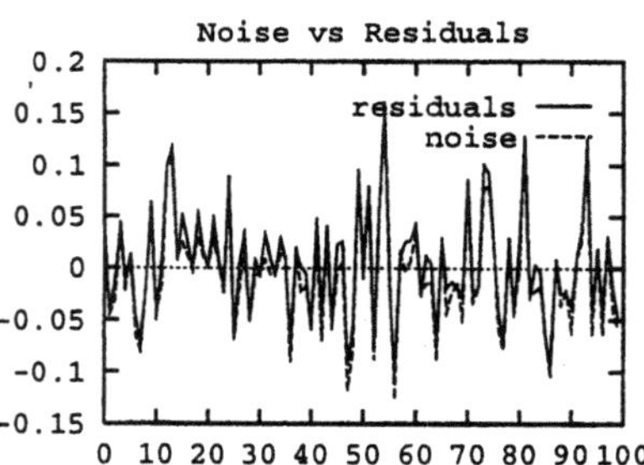

Figure 2: Graphing the residual output of the first neuron at time t against the Gaussian noise at time t.

Table 2: The embedding for this network is one unit too long.

Neuron	2 (= $data_t$)	3 (= $data_t$)^2
time t	0.50	0.00
time $t-1$	0.50	0.70
time $t-2$	-0.15	0.70
time $t-3$	-0.15	0.00

for all values of the Henon map and then add noise to all calculated values by $q_t \leftarrow q_t + \gamma_t$ where γ_t is the noise at time t, the network's output after training is not an accurate reflection of the magnitude of the noise. Thus this method can be used to estimate noise inherent in the data series (equivalent to modelling error) but not noise due to, for example, measurement inaccuracies. The underlying map, though, is recovered in both cases.

3.3 Finding the Length of the Embedding

If we have an embedding which is too short, we will find that the magnitude of no output neuron decreases to zero. This suggests that we start with a short embedding and increase the length till the magnitude of the output of the first neuron is zero. However we have a simple test to find if we have gone too far, i.e. have too long an embedding. Table 2 shows the weights learned on the Henon mapping when the embedding is only a single time unit too long.

This is effectively stating that the network has minimised

$$1 - (0.5x_t - 0.7x_{t-1}^2 + 0.15x_{t-2}) \\ - (0.5x_{t-1} - 0.7x_{t-2}^2 + 0.15x_{t-3})$$

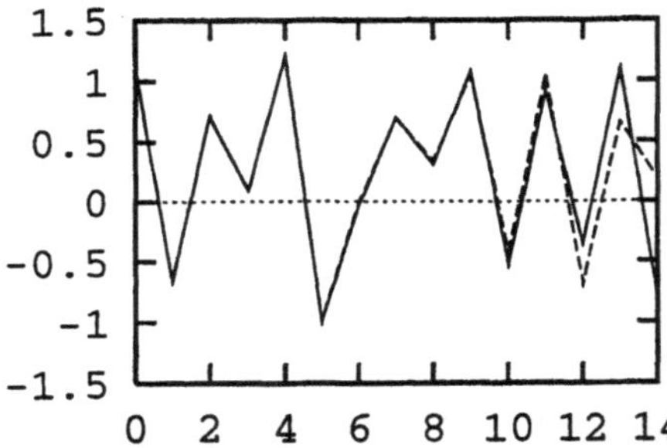

Figure 3: Accuracy of forecast on Henon mapping. Dotted line is forecast, solid line is actual.

i.e. the network is giving two equivalent mappings for the data.

3.4 Other Bases

Finally we note that a basis of function space of polynomials may not be appropriate for some mappings. However the network described above need not be based on polynomials. Using a network whose inputs are $1, \sin(x), \sin^2(x)$ etc. on the Henon mapping we find that we can achieve equivalent accuracy (using the residuals as a measure of accuracy). However it is no longer the case that we can extract a simple function from the learned weights.

We have similar results when using a function basis of $1, \sin(x), \sin(2x), \sin(3x)$. There are in fact an infinite number of bases which can be used in the network. Some of these may be more appropriate for the extraction of information from time series data from real data.

3.5 Forecasting using the Network

The network was trained on 1000 samples of the noisy Henon series. The last data points were then moved through the embedding sequence in the same manner as used in the learning phase and the network's output at the first output neuron used as its estimate of the next value from the sequence. The results are shown in Figure 3.

We are in this figure comparing the continuation of the Henon series without noise with the network's predictions. We use a noiseless Henon series for the comparison (though we used a noisy series during training) since we wish to estimate the accuracy of the network's mapping. It is not possible for any system to estimate the noise a priori and so the degree of correspondence between these is a better measure of the network's accuracy.

The method used for these results is repeated one-step lookahead prediction: the results for time $(t+1)$ are used to predict the value of the observable at time $(t+2)$ etc.

It should be emphasised that the network did not see the continuation data during training: we have different training data from test data. The network uses the data structure it has learned during training to forecast the future, i.e. this is not a look-up table method.

4 Conclusion

We have discussed an extremely simple neural network which has been shown to be capable of extracting the structure of a time series from instances of its visible variable. Future work will investigate the capabilities of the network on data from real data such as that from financial or traffic distributions.

References

[1] G. Deco and D. Obradovic. *An Information Theoretic Approach to Neural Computing.* Springer, 1996.

[2] C. Robinson. Bifurcation to infinitely many sinks. *Communications in Mathematical Physics*, pages 433–459, 1990.

[3] A. Weigend and N. Gershenfeld. *Time Series Prediction, Forecasting the Future and Understanding the Past.* Addison Wesley, 1996.

Time-Series Prediction with Neural Networks: Combinatorial versus Sequential Approach

A. Dobnikar[1], M. Trebar[1] and B. Petelin[2]
[1] Faculty of Computer and Information Science, University of Ljubljana, Trzaska 25, 1000 Ljubljana, Slovenia
[2] Tomos - Informatika, Koper, Slovenia
Email: Andrej.Dobnikar@fri.uni-lj.si

Abstract

In this paper, two different approaches of time-series prediction with neural networks are presented. The first is called combinatorial because it deals with a finite set of classes, obtained from the differences between several consequent function values. It is implemented through a modular neural network. The second describes time-series with interval functions or sequences of successive function values and is therefore a sequential approach, employing Kalman neural gas networks. In the first case the future value (prediction) of an input vector depends on the classes (from input vectors, possibly together with the next values) obtained from learning the history of a time-series. In the second, based on the sequence of last input vector(s), the closest covering neuron (interval function) is defined, and is responsible for a future value calculation. A linear autoregressive method (AR) and multilayer perceptron (MLP) are used as references, and with the help of three different time-series, the efficiency of the suggested methods is given.

1 Introduction

Recently neural networks have often been used for time-series prediction, [4, 5, 6], in which different approaches are detailed. The following models of neural networks are most frequently used for the purpose of forecasting future values of different time functions: multilayer perceptron, probabilistic neural networks, Box-Jenkins ARMA/ARIMA model, GMDH, RBF neural networks, and recurrent neural networks.

The purpose of this paper is to present a different way of looking at the solution of a time-series prediction problem based on neural networks. The aim is to attach some meaning to the neurons of the network, and to not treat it as a magic black box that is supposed to learn a history of a function and be able somehow to predict its future values. Two ways are suggested for assigning some role to the neurons.

The first approach uses the differences between the successive components of an observation vector (input vector and possibly a next value) as a prediction class. Based on the input vector (if observation vector is equal to the input vector) or the probabilities of the finite number of possible prediction classes taken from a time-series history, the future value of an input vector can be obtained from the modules realised by the relevant classes. This approach is called combinatorial in accordance with the nature of the formation of prediction classes. For the purpose of the first approach a modular neural network (MNN) is used [4, 9], where each module covers one possible prediction class, which is taught by corresponding pairs of the form: (input vector, next value). Different types of neural networks can be used to realise the modules. For the purpose of this paper a multilayer perceptron (MLP) with backpropagation (BP) learning algorithm was used, as no significant improvements with modular recurrent neural networks have been found [9].

In the second, sequential approach, where the history of a time-series is seen as a set of successive interval functions or sequences of observations, the so-called Kalman neural gas network (KNGN) [1, 2, 7] is used. Each neuron in the gas learns an individual interval function from the set of input/output pairs, where the input is the weighted sum of input vectors and the output is the corresponding future value. A special unit (sometimes called the voter or selection unit) is responsible for selection of the neuron which is used to predict the future value from the last input vector. A local principle is the simplest, whereby the last input vector, shifted backwards one step, is tested on every neuron in the gas and the one that gives the closest prediction value

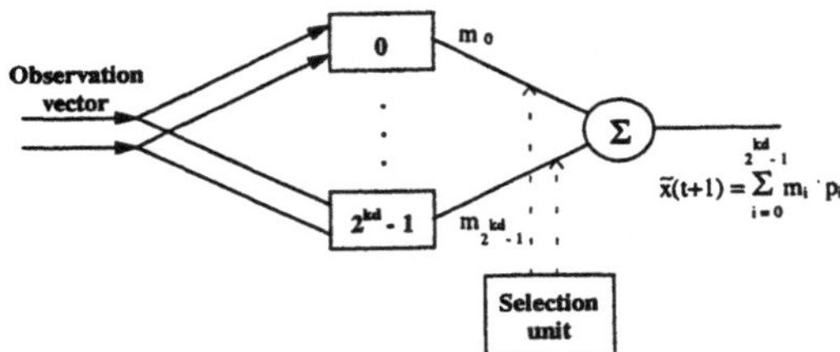

Figure 1: Modular neural network.

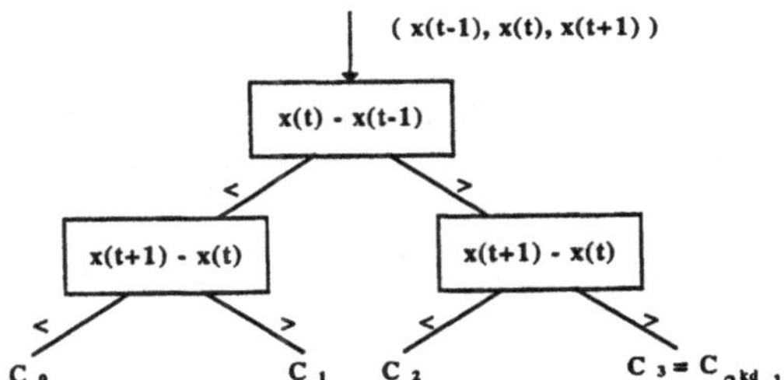

Figure 2: Example of class recognition

to the very last known function value is selected, but other methods take into account more than the most recent vector.

2 Modular Neural Networks

For the purpose of the prediction problems a definition of modular neural networks that differs slightly from the one given in [4] was employed. Namely, it was suggested that each module belongs to only one class of possible outcome from the differences of the components of an observation vector. The length of the observation vector defines the number of possible classes (modules) according to the expression:

$$N = 2^{kd} \tag{1}$$

where k is the resolution (number of bits) of the difference between the two successive function values in time-series and d is the observation vector length. There are two possibilities for class formation. In the first case the observation vector is equal to the input vector ($\vec{x}$), and in the second case it is equal to the input vector together with the next value ($\vec{x}, x(t+1)$). In the simplest case, where $k = 1$, classes are defined with binary notation. In this case the difference 1 means 'greater or equal', and 0 'less' or all other cases. One should note that there are many possible ways to describe a time-series according to this definition: the higher the value of k, the more accurate the description of the time-series but also the larger the number of classes. There are many different observation vectors that lead to the same class and therefore pertain to the same training set for the suitable module. Figure 1 shows the modular neural network described.

The selection unit makes a difference vector Δ of length $d-1$ from an observation vector $\vec{x}$ of length d or difference vector Δ of length d from vector ($\vec{x}, x(t+1)$) by successively comparing its adjoining values, for example:

$$\begin{aligned}(\vec{x}, x(t+1)) &= (x(t-d+1), \\ &\quad \ldots, x(t-1), x(t), x(t+1)) \\ \vec{\Delta} &= (\Delta x(t-d+1), \ldots, \Delta x(t)) \\ \Delta x(t-i) &= x(t-i+1) - x(t-i).\end{aligned} \tag{2}$$

Based on the number (of base k) constructed from the components of the difference vector Δ, the class index is determined by converting it to a decimal number, which in turn addresses a corresponding module in MNN.

The transition diagram of class recognition from the observation vector is equivalent to a decision tree with k branches per level, k being the resolution of the differences and d the number of levels. The procedure stops at the bottom of the tree where the destination leaf addresses the actual module. Figure 2 shows the example of the decision tree that follows the above description.

Based on the input vector the selection unit determines module(s) responsible for a prediction calculation. There are two different methods of selection. In the first, with observation vector ($\vec{x}$), the module is simply addressed according to the class obtained from the input vector, and in the second, with observation vector ($\vec{x}, x(t+1)$), all modules are selected but its output values are weighted with the corresponding class probabilities (calculated during the learning phase) to obtain the prediction.

In the combinatorial approach there are three parameters that influence the result of the prediction. These are the length of the observation vector, the resolution of the difference between successive function values, and the type of class formation. It is therefore necessary to search for their optimal values in order to get the best possible RMS error in the test part of a time-series.

3 Kalman Neural Gas Network

The neural gas network (NGN) was first introduced by Martinetz *et al.* [3] for vector quantisation problems in which neurons act as attractors that map the problem domain onto Voronoi polyhedra. In [2] the authors presented a Kalman neural gas network (KNGN) in which neurons cover the functional space similar to the manner in which the neurons of the NGN cover Voronoi polyhedra. A KNGN was used for two different types of prediction: for real-time and non-real-time problems. In the first case [2], a functional neural gas was suggested based on the family of time-series expected. The corresponding neurons were pre-taught and its weights were only updated during real-time processing. Real-time tracking of moving objects is a typical example that fits well to this method of prediction. In the second case [1], a KNGN was used for a non-real-time application, where there was enough time to complete Kalman learning. Time series found in economics belong to this type of prediction, e.g. exchange rates, stock prices.

Learning by neurons in the gas follows the Kalman filter algorithm. An exellent review of this subject can be found in [8]. For the sake of clarity and brevity only the basic expressions are given here.

The standard Kalman filter is defined by observation and dynamic equations. In the case of neuron learning these are of the form:

$$\begin{aligned} y &= x^T\omega + n \\ \omega(t) &= \omega(t-1), \end{aligned} \tag{3}$$

where x is the input vector, ω an unknown weight vector, n is measurement noise with variance R and zero mean, and x^T the transpose of x. The observation equation describes the method of calculating the output function from the input vector, allowing for deviations. The dynamic equation describes the target situation in which no further learning is necessary, i.e. the weight vector has achieved its final values. This means in this case that the neuron has been taught the interval function of a time-series.

The following equations describe a single step in updating the weight vector:

$$\begin{aligned} k &= Px/[x^TPx + R] \\ P &= [I - k\omega^T]P \\ \omega(t+1) &= \omega(t) + k(y - x^T\omega), \end{aligned} \tag{4}$$

where P is the covariance matrix of the weight vector estimation, which has initial value of αI, I being the unit matrix and α being a large number describing the situation in which the initial estimation is optional, and k is the Kalman gain vector. In each step of the learning procedure all three equations are processed. Firstly Kalman gain is calculated from the current covariance matrix and input vector, then the covariance matrix is updated, and finally the weight vector is updated. It should be noted that the covariance matrix decreases (reduction of uncertainty) with the number of steps, because $k\omega^T$ is always between 0 and 1, and the weight vector approaches the optimum value of MSE. The equations also illustrate that the convergence of learning is better if the variance of the noise R is smaller.

In the sequential approach the following parameters were optimised: the length of the input vector and N the number of input vectors used by the voter. They all affect the quality of prediction.

4 Experimental Work

The two approaches were tested using three time-series: the first (sunspots) is frequently used for comparing different prediction methods, while the second and the third describe actual sales of bicycle and scooter tyres on a weekly basis by a domestic producer.

For the purpose of evaluation, a comparison with traditional methods of AR and MLP is given, although these are not described, as they can be found elsewhere, for example in [4, 6].

In all three cases the time-series is divided into two parts. The first 80% is for learning with validation, and the last 20% for testing. The efficiency of prediction is calculated on the basis of RMS error in the test part of a time-series. Figure 3 shows all three time-series used in the experimental work. Table 1 gives the best results of the prediction for all three time-series obtained with the two methods, and the results using the reference methods. The optimal values of the parameters are also given in the table. d in AR and MLP were obtained with the help of partial autocorrelation [6]. To illustrate the quality of prediction Figure 4 presents the test part of the second time-series (sales of scutter tyres), together with predictions based on combinatorial and sequential approaches.

It is well known that pre-processing plays an important role in searching for the best possible prediction result. Centring, detrending and filter-

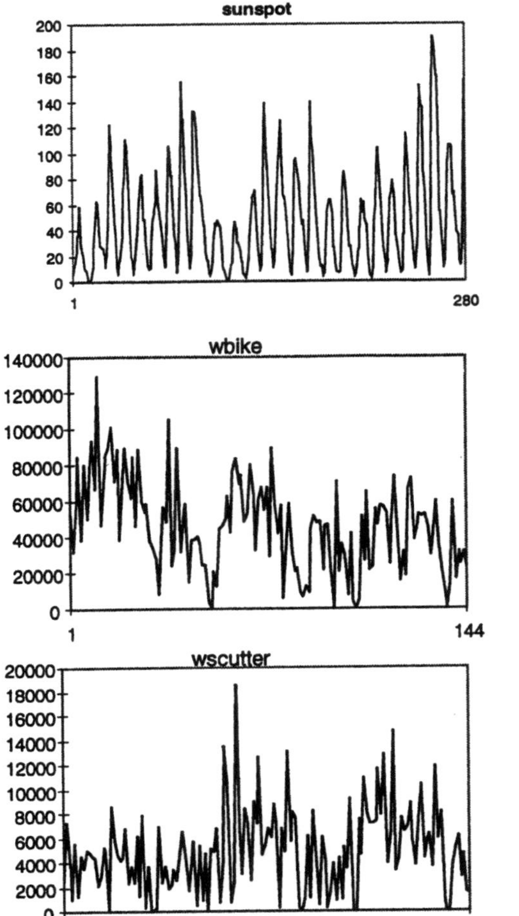

Figure 3: The three time-series used in experimental work.

ing are the most frequent functions used for that purpose. A further experiment was therefore performed. Instead of applying the prediction method directly on the time-series, it was first filtered with the help of low-, band- and high-pass filters [6]. The sum of the three filtered time-series gives exactly the original time-series. Then the two suggested prediction methods were applied to all three filtered time-series and the results of the predictions summed. Table 2 shows the results obtained for scutter time-series. They may even be too good, due to the characteristics of the filters used, which suppose that all data is available at any moment, implying that the future values are also known to some extent. However, more realistic filtering is suggested.

Table 1: The results of prediction for all time-series and methods.

MNN:

Time series	RMS train	RMS test	k	d	class
sunspot	10.1	15	1	5	x
wbike	12638.5	19225	1	3	$x, x(t+1)$
wscutter	2253	3304	1	3	$x, x(t+1)$

KNGN:

Time series	RMS train	RMS test	d	N
sunspot	12.43	17.1	8	1
wbike	20032.4	19443	8	1
wscutter	3281.0	3769.8	3	1

AR:

Time series	RMS train	RMS test	d
sunspot	14.4	19.1	8
wbike	21077	19850	6
wscutter	3346	3473	6

MLP:

Time series	RMS train	RMS test	d
sunspot	10.8	20.3	8
wbike	15419	24873	6
wscutter	1917	4569	6

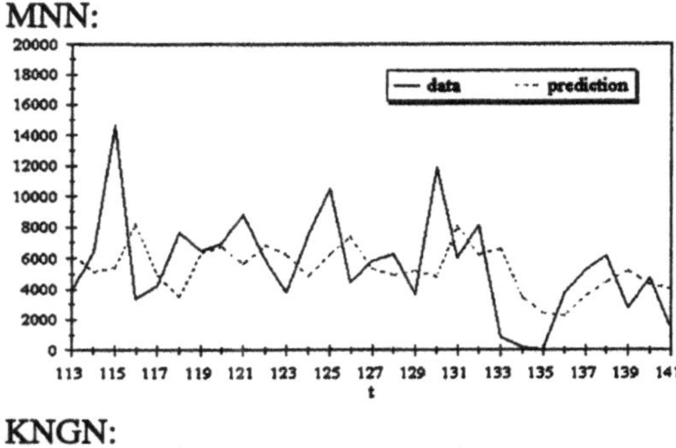

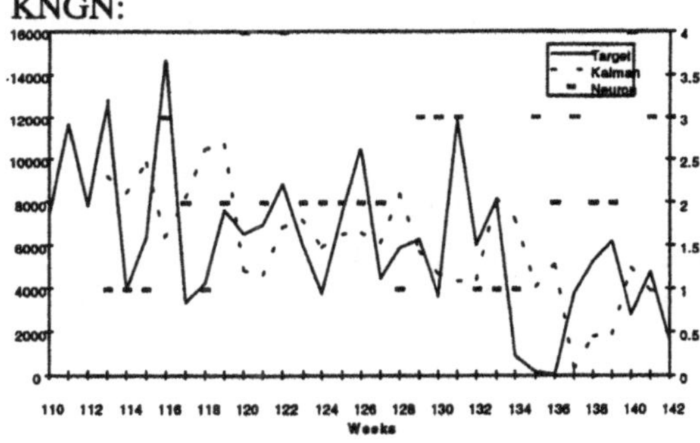

Figure 4: Illustration of the two methods on test part of the scutter time-series.

5 Conclusion

Two methods for time-series prediction with neural networks are outlined. In both cases a role was assigned to the neurons in the networks. The results obtained for three actual time-series indicate that both methods outperform AR and MLP methods significiantly, and seem to be accurate enough for real life applications. The combinatorial approach is slightly better than the sequential, as the time-series used appear somewhat noisy. We believe that with the proper pre-processing by some realistic filtering the results might be even improved. This will be the topics of our future investigations.

References

[1] A. Dobnikar. Kalman neural gas network for time series prediction. In *Brain Processes, Theories and Models*. MIT Press, 1995.

[2] A. Dobnikar, A. Likar, and M. Trebar. *Time-series Prediction with Online Correction of Kalman Gain - A Connectionist Approach*. Elsevier, 1995.

[3] T.M. Martinetz et al. Neural-gas network for vector quantization and its application to time-series prediction. *IEEE Trans. on Neural Networks*, 4(4):558–569, 1993.

[4] S. Haykin. *Neural Networks*. MacMillan College Publishing Company, 1994.

[5] G. Janacek and L. Swift. *Time-Series Forecasting, Simulation, Applications*. Ellis Horwood Limited, 1993.

[6] T. Masters. *Neural, Novel and Hybrid Algorithms for Time Series Prediction*. John Wiley and Sons, 1995.

[7] D. W. Pearson, A. Dobnikar, B. Petelin, and G. Dray. Estimating neural network based predictors for production processes. In *Proc. CESA'96*. Lille, France, 1996.

[8] S. Shar, F. Palmieri, and F. Datum. Optimal filtering algorithms for fast learning in feedforward neural networks. *Neural Networks*, 5:779–787, 1992.

[9] M. Trebar and A. Dobnikar. Time series prediction using modular recurrent neural networks. In *International Conference CESA96*, Lille, France, 1996.

A New Method for Defining Parameters to SETAR(2;k_1,k_2)-models

J. Kyngäs
Department of Computer Science, University of Joensuu,
P.O.Box 111, FIN-80101 Joensuu, Finland.
E-mail:Jari.Kyngas@cs.joensuu.fi

Abstract

This paper describes a new numerical method for defining the *threshold* and *delay* parameters of k-order self-exciting threshold autoregressive (SETAR) models. The idea of the method is to divide a time series into ascending and descending parts. This division can especially be exploited in building prediction models: it is a considerably easier task to predict the ascending and descending parts separately than to try to predict both of them at the same time.

Another issue of this paper is to build the prediction model with only the most significant predictor variables. This is achieved with the help of evolutionary optimizing. In the first step we generate a number of models (networks) with different predictor variables, then we train each of the networks and the fittest ones are used for reproduction. In the beginning we also set the possible number of predictor variables to some constant. In the next step we reduce the number of predictor variables by leaving out the ones not chosen by the algorithm in the first step. Then we repeat the training and dropping procedure until the genetic algorithm does not drop any predictors anymore.

I have shown in this paper that the division procedure works very well, that building separate prediction models increases the accuracy of the predictions noticeably, and that using evolutionary optimization is succesful in dropping out the irrelevant predictors.

1 Introduction

Threshold models were introduced by H. Tong in a long series of papers in the late 1970's and the beginning of 1980's. A detailed account of the theory is given in the monograph by Tong [6]. Here I will only briefly introduce the k-order SETAR(2)-model.

A k-order SETAR(2)-model is a combination of 2 linear models of the form

$$X_t=\begin{cases} a_0 + a_1X_{t-1}+\ldots + a_{k1}X_{t-k1}, & \text{if } X_{t-r} < d \\ b_0 + b_1X_{t-1}+\ldots + b_{k2}X_{t-k2}, & \text{if } X_{t-r} \geq d \end{cases}$$

where $a_0,\ldots,a_{k1}$,$b_0,\ldots,b_{k2}$ are constants, r is the delay parameter and d is the threshold parameter.

To use this kind of prediction function, we need to know the delay parameter, which is located r lags backwards in the time series and the threshold parameter, which determines which of the equations to use. The technique, introduced by Tong, to define these parameters requires a great deal of mathematical understanding. Due to that fact, I had to find a simpler way of defining them.

The selection of relevant predictor variables is a difficult task. The problem can be solved with the help of evolutionary algorithms [5, 8]. We have programmed an evolutionary neural network simulator [3], which idea is to find these relevant inputs and the optimal size of the hidden layer. The simulator uses feedforward neural networks with one hidden layer. A detailed description of the simulator can be found in [3, 4].

2 Methods

The method proposed for defining parameters to SETAR(2,k_1,k_2) consists of two quite simple numerical searches: the delay value search and the threshold value search. The threshold value is a value which divides the observations into ascending and descending groups. So the task is to find the threshold value which is located some lags backward in the time series. Algorithm 1 describes this procedure in detail.

The calculation of percentual correctness is done by examining successive values one period at a time. For example, if the division algorithm has defined 7 successive values to be in one of the groups, there are 6 pairs of values to be examined.

Often there are a few division results that are equally good if measured only with percentual correctness. In such cases the division which divides the time series closest to equally sized subsets is preferred.

ALGORITHM 1:

```
If data is normally distributed then
  use the mean as an average value
else use the median as an average value
delay value = 1
max delay value = user defined
repeat
  threshold value = average value - 10% of the
   datas range
  repeat
    observation to check = value of
    time series [1+delay value]
    observation at lag = value of time series [1]
    repeat
      if observation at lag ≥ threshold value then
          include observation to check to group 1
      else
          include observation to check to group 2
      observation at lag = observation at lag+1
      observation to check = observation to
      check+1
    until whole time series scanned
    if group1 consists mostly of ascending values
    then
      mark group 1 as the ascending and group 2
      as the descending
    else
      mark group 1 as the descending and group 2
      as the ascending
    calculate percentual correctness of the
    divisions
    threshold value = threshold value+[0.01|0.1|1]
  until threshold value > average value + 10% of
  the datas range
  delay value = delay value+1
until delay value > max delay value
sort the division results according to percentual
correctness of the division
```

3 Results

I have tested the division method with three time series: average temperatures (measured in Fahrenheits) [2], the log_{10} transformation of the lynx trappings [7] and the annual sunspot numbers and square-root transformed annual sunspot numbers [1, 7]. The average temperatures series has been chosen because it has a stable cycle and is therefore an excellent meter of the goodness of the division method. The two other series have been chosen because they have an irregular cycle and they are taken to be very hard to model and to predict. Table 1 shows the results of the division.

Table 1: Division results of time series of the average temperatures (column A), log_{10} transformation of the lynx trappings (column B), annual sunspots (column C), and square-root transformed annual sunspots (column D).

	A	B	C	D
Ascending	100.00	96.00	82.02	82.02
Descending	100.00	92.11	93.26	93.26
Total	100.00	94.32	87.64	87.64
Delay	3	2	3	3
Threshold	46.1	3.01	36.8	10.29

As can be seen from the table the percentual correctnesses are very good. The relatively low correctness in the ascending part of the sunspot series is due to the fact that the ascending parts are much steeper than the descending. This in turn leads to more errors in the division.

If only measured with percentual correctness of the division, lynx trappings data can be divided equally well with threshold values 3.01–3.11. Tong's method ends up with a threshold value of 3.116. The same thing applies to the sunspot series: threshold values 36.8–37.6 and 35.1–35.4 divide the data equally well. In this case, Tong's method defines the threshold value to be 36.6. However, it has to be noticed that Tong uses different criteria for his selections.

The biggest difference is in the square-root transformed sunspot numbers. Tong's method ends up with the delay value as 8 and the threshold value as 11.93. But as I stated before, Tong uses different criterions. In fact, he stated that the best value for the delay parameter is 3, but he rejected that model for other reasons.

4 Application

To really test if this division is of any use, I built a neural network model to predict the future sunspot numbers. This model consists of two neural networks of which the first has been trained to predict the ascending period and the second has been trained to predict the descending period of the sunspot series. As training data I used the sunspot numbers of years 1700–1920, as testing data the sunspot numbers of years 1921–1955 and as validation data the sunspot numbers of years 1956–1995.

The networks were trained using our evolutionary neural network simulator [3, 4]. I started with a population of 250 networks. In each generation 50 children were reproduced for the 200 best performing networks. The size of the hidden layer was allowed to vary between 2–5 hidden neurons. Because the sunspot cycle is taken to be about 11 years, I chose to let the simulator choose the relevant predictor variables from the last twelve lags.

For the descending part of the time series the simulator dropped out lag 12 three times in five runs. Lags 3, 5, 6 and 10 were dropped out once in five runs. For the ascending part of the time series the simulator dropped out lag 4 four times, lags 3, 10 and 12 three times and lags 7, 9 and 11 one time.

I then executed five more runs so that I let the simulator choose the relevant predictor variables only from lags once dropped out in the previous runs. The rest were locked as on or off depending on the drop outs.

The remarkable thing in doing these two sets of runs was that in the first five runs all the networks finally chosen by the algorithm included 4 or 5 hidden neurons. However, after the second set of runs the chosen networks included only 2 or 3 hidden neurons. This probably happens because the search space is much smaller in the second set of runs. The search space is smaller because most of the predictor variables are locked. In addition, the prediction accuracy is more stable in the second set of runs (i.e the gap between the worst pair of networks and the best pair of networks is smaller).

If we compare the results of choosing predictor variables by the Tong method and the evolutionary optimizing method we find that for the descending part of the series both methods picked lags 1–11 to be relevant. For the ascending part, Tong's method picked lags 1–3 whereas the evolutionary optimizing method picked lags 1, 2, 5–9 and 11 to be relevant. These results were achieved with the untransformed series. Here, however I compare my results to the Tong's model built to the transformed data. I chose to build my model to the raw data, because I wanted to keep the model building as simple as possible. In addition, I think that the neural network model would do even better if I could build the model for the square-root transformed data.

In Table 2, I have collected the results of all the runs. There is maybe no idea in choosing the worst pair of networks, but I have done this because I want to make it perfectly clear that by building the prediction model in the way described in this paper, one can always get much better results than doing it some other way.

Table 2: Column A, results of the worst pair of networks; column B, results of the best pair of networks. The worst and the best networks were defined according to their ability to predict the validation set. Column C, results of one network trained for both ascending and descending periods [4]. Column D, results of the best statistical model [6] found in the literature.

RMSE lag 1	A	B	C	D
training set	12.39	12.97	11.38	12.26
test set	8.93	9.94	9.78	10.76
validation set	16.03	13.56	19.39	19.89

I also performed five runs in which I replaced the test data with the training data. In that case the prediction accuracy of the validation set with the worst pair of networks did decrease to 17.68 (RMSE) and with the best pair of networks to 16.48 (RMSE). Despite the decrease, the results are still better than those of the whole model (Table 2 column C) and the statistical model (Table 2 column D).

5 Conclusion

As the results shows, division of data into ascending and descending parts is very successful. When the time series has a regular cycle the division can be made with an 100% accuracy. Even when the cycle is of irregular length, the division is successful with an about 90% accuracy.

The advantages of the division can be exploited in forecasting. It is obvious that building two separate models to predict the ascending and descending parts of the time series is easier than building one model to handle both situations. It can be thought of as dividing the forecasting problem into two smaller problems and solving them separately.

The problem of choosing the relevant predictor variables (lags) has also been a big problem in the field of time series analysis. Most statistical methods are able to find the biggest lag that is a good predictor variable. However, in most cases finding this biggest lag means that the model also has to include all the preceding lags. When using evolutionary optimizing this is not the case — the genetic algorithm picks only the relevant predictor variables of a given set.

References

[1] National Geophysical Data Center. Sunspot numbers of years 1989–1995. Internet www-page, at URL: ftp://ftp.ngdc.noaa.gov /STP /SOLAR_DATA /SUNSPOT_NUMBERS /yearly.html (version current at 4 September 1996).

[2] J. D. Cryer. *Time Series Analysis.* R.R. Donnelley & Sons Company, USA, 1986.

[3] J. Hakkarainen, A. Jumppanen, J. Kyngäs, and J. Kyyrö. An evolutionary approach to neural network design applied to sunspot prediction. Technical Report A-1996-3, Department of Computer Science, University of Joensuu, 1996. Available via ftp from cs.joensuu.fi/pub/Reports as file A-1996-3.ps.

[4] J. Kyngäs and J. Hakkarainen. Predicting sunspot numbers with evolutionary optimized neural networks. In *Proceedings of the Second Nordic Workshop on Genetic Algorithms and their Applications.* Vaasa University, 1996.

[5] G. F. Miller, P. M. Todd, and S. U. Hegde. Designing neural networks using genetic algorithms. In *Proceedings of the Third International Conference on Genetic Algorithms and their Applications.* Morgan Kaufman, 1989.

[6] H. Tong. *Threshold Models in Non-linear Time Series Analysis*, volume 21. Springer-Verlag, New York, 1983.

[7] H. Tong. *Non-linear Time Series: A Dynamical System Approach.* Oxford University Press Inc, New York, 1995.

[8] Y. Xin. A review of evolutionary neural networks. *International Journal of Intelligent Systems*, 8, 1993.

Predicting Conditional Probability Densities with the Gaussian Mixture – RVFL Network

D. Husmeier and J. G. Taylor
Department of Mathematics,
King's College London, UK

Abstract

The incorporation of the random vector functional link (RVFL) concept into mixture models for predicting conditional probability densities achieves a considerable speed-up of the training process. This allows the creation of a large ensemble of predictors, which results in an improvement in the generalization performance.

1 Introduction

Several approaches to the problem of predicting noisy time series with neural networks have been developed in the last few years ([1, 2, 4, 5, 8, 10]), all of which try to model the conditional probability density of the target $y = x(t+1)$ conditioned by the lag-vector of previous observations $x(t) := (x(t), x(t-1), \ldots, x(t-m+1))$ with a mixture model

$$P(y|x,q) = \sum_{k=1}^{K} P(k|x,q)P(y|x,q,k). \qquad (1)$$

Here the $k = 1, \ldots, K$ are labels for the different subprocesses of the mixture, q is a vector of network parameters, $P(k|x,q)$ a discrete probability for the occurrence of the kth subprocess, and $P(y|x,q,k)$ a conditional probability density chosen to be of a simple unimodal form (henceforth referred to as the *kernel function*). In this text we shall focus on the model introduced in [4], [5] and define

Definition 1 (GMCD model).
Let $a_1, \ldots, a_K \in I\!R^+$ *with* $\sum_{k=1}^{K} a_k = 1$; *let* $\beta_1, \ldots, \beta_K \in I\!R^+$; $w \in I\!R^N$;

$$q := (a_1, \ldots, a_{K-1}, \beta_1, \ldots, \beta_K, w^\dagger)^\dagger \in (I\!R^+)^{2K-1} \otimes I\!R^N.$$

Let $\mu_k : I\!R^m \to I\!R$, $x \to \mu_k(x|w)$ *be the output of a standard feedforward network with weights* w *and sigmoidal units in a single hidden layer. A probability model of the form of Equation (1) with*

$$P(k|x,q) := a_k$$

and

$$P(y|x,q,k) := \sqrt{\frac{\beta_k}{2\pi}} \exp\left(-\frac{\beta_k}{2}[y(t) - \mu_k(x(t)|w)]^2\right)$$

is called a Gaussian mixture conditional density model (GMCD).

This model can be interpreted as a two-hidden-layer network with sigmoidal units in the first layer, radial basis units in the second layer, and positive normalized output weights. The following universal approximation theorem can be shown to hold [4, 5]:

Proposition 1. *A GMCD network is a universal approximator for conditional probability densities.*

Given a time series $D = \{x(t)\}_{t=-m+2}^{T+1}$ as a training set, the objective of a training process is to adapt the parameters q so as to maximize the likelihood

$$P(D|q) = \left(\prod_{t=1}^{T} P\left(x(t+1)|x(t),q\right)\right) P(x(1))$$

or, equivalently, minimize the 'error' function

$$E(q) := -\ln P(D|q) \simeq -\sum_{t=1}^{T} \ln P(x(t+1)|x(t),q) \qquad (2)$$

(where we have assumed that the time series can be modeled by an mth order Markov process, and have dropped the term $P(x(1))$ in Equation (2) since it does not depend on the network parameters).

The standard training method of *maximum likelihood* adapts the network parameters q according to a steepest descent scheme, $q \sim -\nabla E(q)$. However, since the network has a two-hidden-layer structure, the convergence of such an approach is extremely slow [5].

2 Expectation Maximization (EM) Algorithm

A faster training scheme is the Expectation Maximization (EM) algorithm [3, 7]. Consider the posterior probability for 'class' k,

$$\pi_k(t) := P(k|y(t), x(t), q) = \frac{a_k P(y(t)|x(t), q, k)}{\sum_i a_i P(y(t)|x(t), q, i)}, \tag{3}$$

where the expression on the right follows from Bayes' rule and Definition 1, and the vector q denotes the *current* network parameters. Then the *new* network parameters $\tilde{q} = (\tilde{a}_1, \dots, \tilde{a}_{K-1}, \tilde{\beta}_1, \dots, \tilde{\beta}_K, \tilde{w}^\dagger)^\dagger$ are given by

$$\tilde{\beta}_k^{-1} \sum_{t=1}^{T} \pi_k(t) = \sum_{t=1}^{T} \pi_k(t) \Big(y(t) - \mu_k(x(t)|w)\Big)^2 \tag{4}$$

$$\tilde{a}_k = \frac{1}{T} \sum_{t=1}^{T} \pi_k(t) \tag{5}$$

$$\tilde{w} = \text{argmin}_w \{U(w)\} \tag{6}$$

$$U(w) := \frac{1}{2} \sum_{k=1}^{K} \sum_{t=1}^{T} \pi_k(t) \beta_k \Big[y(t) - \mu_k(x(t)|w)\Big]^2 \tag{7}$$

The adaptation of the a_k and β_k can be performed immediately. However, since the $\mu_k(x(t)|w)$ are nonlinear functions of the weights w, U is a complex *non-convex* function of w. Its minimum therefore cannot be found in a single adaptation step, which prevents any improvement in training speed over a standard *maximum likelihood* scheme.

3 Combining GMCD and RVFL

A solution seems to be offered by the random-vector functional link net approach (RVFL). Let us start with the following definition:

Definition 2 (RVFL). *Let $v_i \epsilon \mathbb{R}^m$; $v_{io} \epsilon \mathbb{R}$; $w := (w_0, w_1, \dots, w_M)^\dagger \epsilon \mathbb{R}^{M+1}$; $i \epsilon \{1, \dots, M\}$; and let $\Phi : \mathbb{R} \to \mathbb{R}$ be a sigmoidal, i.e. nonconstant, monotonically increasing and bounded continuous function. A one-hidden-layer feedforward network $\mu : \mathbb{R}^m \to \mathbb{R}$, $x \to \sum_{i=1}^{M} w_i \Phi(v_i^\dagger x + v_{io}) + w_0$ with v_i and v_{io} constrained to randomly and independently selected values is called a random vector functional link network (RVFL).*

In this way U becomes quadratic in w, so its minimum can easily be found in a single step by matrix inversion techniques. The justification of this approach is given by the following proposition:

Proposition 2. (i) The RVFL is a universal approximator for continuous functions on bounded finite dimensional sets, and (ii) the rate of approximation error convergence to zero is of order $\mathcal{O}(1/\sqrt{M})$, where M is the number of basis functions.

The second statement implies that the rate of error convergence for the RVFL is of the same order as for a standard feedforward network. A proof was given by Igelnik and Pao [6]. An empirical corroboration can be found in [9]. This encourages the definition and, somewhat loosely formulated, proposition:

Definition 3 (GMCD-RVFL). *A GMCD probability model (Definition 1) with $\mu_k : \mathbb{R}^m \to \mathbb{R}$ given by Definition 1 is called a GMCD-RVFL model.*

Proposition 3. *a GMCD-RVFL model is a universal approximator for conditional probability densities.*

The proof follows that given in [5] for the GMCD model, making use of the fact that in the deterministic limit, $a_i \to \delta_{i,i^*}$, $\beta_{i^*} \to \infty$, the universal approximation capability is preserved due to Proposition 2.

In practice, we add adaptable direct connections between the input and second hidden layer to the GMCD-RVFL in order to extract the linearly predictable part of the function to be learned and free up the nonlinear resources, i.e. the random part of the network, to be employed only where really needed. Let us denote the weights feeding into the kth second-hidden-layer node by $\mathrm{w}_k \epsilon \mathbb{R}^{M+1}$, and let $g[\mathrm{x}(t)] \epsilon \mathbb{R}^{M+1}$ contain the activities of all the nodes in the input, bias and first hidden layers upon presenting pattern x(t). Then

$$\mu_k\left(x(t)|w\right) = \mu\left(x(t)|w_k\right) := w_k^T g[x(t)], \tag{8}$$

and U becomes quadratic in $w = (w_1^\dagger, \ldots, w_K^\dagger)^\dagger \in \mathbb{R}^{(M+1)K}$. Let us introduce the definitions

$$y := (y(1), \ldots, y(T))^\dagger \in \mathbb{R}^T, \tag{9}$$

$$G := (g[x(1)], \ldots, g[x(T)]) \in \mathbb{R}^{(M+1)} \otimes \mathbb{R}^T, \tag{10}$$

$$\Pi_k \in \mathbb{R}^T \otimes \mathbb{R}^T, \quad (\Pi_k)_{tt'} := \pi_k(t)\delta_{tt'}, \tag{11}$$

$$I \in \mathbb{R}^{(M+1)} \otimes \mathbb{R}^{(M+1)}, \quad I_{ij} := \delta_{ij} \tag{12}$$

where we assume $T > (M+1)$. From Equation (7) and $\nabla_w U = (\nabla_{w_1}^\dagger U \ldots \nabla_{w_K}^\dagger U)^\dagger = 0$ we obtain

$$\left(G\Pi_k G^\dagger\right) w_k = G\Pi_k y. \tag{13}$$

By inverting the matrix on the left, the M-step with respect to w can in principle be carried out. In practice, however, $G\Pi_k G^\dagger$ is likely to be ill-conditioned, in which case there is no unique solution for w. We therefore adopt the standard pseudo-inverse technique and solve

$$\left(G\Pi_k G^\dagger + \lambda^2 I\right) w_k = G\Pi_k y \tag{14}$$

for some small value λ. Since the matrix on the left is now strictly positive definite, this equation does have a unique solution, choosing amongst the degenerate solutions of (13) the one with the smallest norm $||w||$.

4 Empirical Study

We applied the GMCD-RVFL network to the prediction of two stochastic time series with bimodal conditional probability distributions. The first time series was generated by randomly switching between two stochastic dynamical systems, giving the first-order Markov process $\{x(t)\epsilon[0,1]\forall t\}$ with

$$\begin{aligned} x(t+1) = &\Theta\left(\xi_t - \vartheta\right) \alpha_t x(t)[1 - x(t)] + \\ &[1 - \Theta\left(\xi_t - \vartheta\right)]\left[1 - x(t)^{\kappa_t}\right], \end{aligned} \tag{15}$$

where $\xi_t\epsilon[0,1]$, $\alpha_t\epsilon[3,4]$, $\kappa_t\epsilon[0.5, 1.25]$ are random variables uniformly distributed in the respective intervals, $\vartheta := \frac{1}{3}$, and Θ is the Heaviside function. Three sets of 1000 data points for training, cross-validation, and testing the generalization performance were generated from Equation (15). The task was to predict $P(x(t+1)|x(t))$.

As a second problem, we chose the time series studied in [8]. A particle moves in a double-well potential $V(x) = 0.5x^4 - x^2 + 1$ subject to the Brownian dynamics

$$\frac{d^2x}{dt^2} = -\frac{dV}{dx} - \alpha\frac{dx}{dt} + R(t), \tag{16}$$

where $R(t)$ is a Gaussian stochastic variable with zero mean and intensity $\langle R(t)R(t')\rangle = 2\alpha\delta(t-t')$. We chose the same parameter as in [8], $m = 1$, and integrated Equation (16) numerically with the Leapfrog algorithm (using a stepsize of $\Delta t = 0.1$) to generate three sets of 10,000 data points for training, cross-validation, and testing the generalization performance. In order to enable a comparison with the results in [8], these data were normalized. The task was to predict the position of the particle 25 time steps ahead, i.e. the conditional probability density $P\left(x(t+25) = y|x(t)\right)$.

In the first example, one network architecture was employed, containing one unit in the input and ten nodes in each of the hidden layers. The parameters were initialized as follows: All output weights were chosen to be equal and satisfy the positivity and normalization constraints, $a_k \geq 0$, $\sum_k a_k = 1$; all kernel widths were set to $\sigma_k = 1.0 \Leftrightarrow \beta_k = 1.0$. The adaptable weights were drawn from a Gaussian distribution of standard deviation 0.1, and the fixed weights from a Gaussian distribution of standard deviation σ_{rw}. Six different values for σ_{rw} were chosen ($\ln\sigma_{rw} = -2, -1, \ldots, 3$), and the simulations were repeated with 20 different initializations (starting from different random number generator seeds), yielding a total ensemble of 120 predictors.

The treatment of the second example was similar, except that we applied four different network architectures ((i) 3 inputs, 10 tanh-units, 9 kernels, (ii) 5 inputs, 10 tanh-units, 9 kernels, (iii) 5 inputs, 10 tanh-units, 5 kernels, (iv) 10 inputs, 15 tanh-units, 9 kernels,) and combined them with three different values ($\ln\sigma_{rw}\epsilon\{-1, 1, 3\}$) for σ_{rw}. For each combination, the training simulation was repeated with three different initializations, yielding an overall ensemble of 36 networks.

5 Results

In order to assess the prediction performance of the model, we compared the results with those ob-

Table 1: Comparison of typical training times. N denotes the number of adaptation steps, T the CPU time on a 75 MHz Pentium PC.

Prediction Problem	Training Time	
	GMCD	GMCD-RVFL
Time series 1 Equation (15)	N = 5000-7000 T = 6-8 h	N = 30-50 T = 3-5 min.
Time series 2 Equation (16)	N = 2000 T = 48 h	N = 10 T = 50 min.

Table 2: Comparison between the generalization performance of the GMCD and the GMCD-RVFL model (time series 1, Equation (15)). The last column shows the generalization 'error' ε_{gen}, as defined above. Smaller values correspond to a better generalization performance.

Network	Prediction	ε_{gen}
GMCD	single model	-1.020
GMCD-RVFL	single model (average over committee)	-1.035
GMCD-RVFL	committee	-1.047

tained in [5] and [8]. Table 1 shows typical training times for the GMCD and GMCD-RVFL networks, and suggests that a considerable acceleration of the training process can be achieved. In this way a whole ensemble of GMCD-RVFL models (using different random number generator seeds, and different values for σ_{rw}) can be trained at the same computational cost as required for training a single GMCD model. From this ensemble, we chose a committee of 'best predictors', whose cross-validation 'error' $\varepsilon_{cross} := -\ln P(D_{cross}|q)$ (where D_{cross} denotes the cross-validation set) was smaller than a certain threshold θ, $\varepsilon_{cross} < \theta$. With the somewhat arbitrary values of $\theta = -1.02$ in the first and $\theta = 0.32$ in the second study, we obtained committees of 13 and 11 predictors, respectively.

In order to estimate the prediction performance, the generalization 'error' ε_{gen} was measured on an independent test set D_{gen}, $\varepsilon_{gen} := -\ln P(D_{gen}|q)$. The results are listed in Tables 2 and 3. It can be seen that the typical performance of a single GMCD-RVFL is comparable to the performance of a fully-adaptable conventional model, and that in both examples the best predictions were obtained with a committee of GMCD-RVFL networks (last rows in the tables). This can be understood in-

Table 3: Comparison between the generalization performance of the GMCD-RVFL network and alternative models (time series 2, Equation (16)). The performance increases with decreasing generalization 'error' ε_{gen}.

Network	Prediction	ε_{gen}
Best model in [8]	single model	0.334
GMCD	single model	0.336
GMCD-RVFL	single model	0.337 ± 0.027
GMCD-RVFL	committee	0.276 ± 0.009

tuitively: given the same computational resources, the fast training process for the GMCD-RVFL networks allows a wider exploration of the model and configuration space, so that it will most probably eventually come across predictors that are better than those obtained from the slow training simulations of the fully-adaptable conventional models. Combining these best models in a committee will further decrease the generalization 'error'.

6 Conclusions

Predicting conditional probability densities of noisy time series with neural networks requires architectures with at least two hidden layers, which renders standard maximum likelihood training schemes extremely slow. A considerable improvement in this respect can be achieved by adopting the random vector functional link net approach (GMCD-RVFL). The idea is to constrain a subset of the parameters to randomly chosen initial values, and adapt the remaining ones by the faster EM algorithm. In this way a speed-up of the training process by about two orders of magnitude can be achieved, allowing training of a whole ensemble of models at the same computational cost as required for *one* fully adaptable model. Since a larger range of the model and configuration space can be explored, one will most likely come across predictors that are better than those obtained from the conventional approach. Combining the best GMCD-RVFL predictors in a committee can therefore expect to significantly improve the generalization performance.

7 Acknowledgements

Dirk Husmeier is supported by a Postgraduate Trust Studentship from the University of London. We would like to thank Dr. R. Dale for proofreading the manuscript.

References

[1] D. Allen and J. G. Taylor. Learning time series by neural networks. In *Proc. ICANN 94*, pages 529–532, 1994.

[2] C. M. Bishop. *Neural Networks for Pattern Recognition.* Oxford University Press, New York, 1995.

[3] A. P. Dempster, N. M. Laird, and D. B. Rubin. Maximum likelihood from incomplete data via the em algorithm. *Journal of the Royal Statistical Society*, B39(1):1–38, 1977.

[4] D. Husmeier, D. Allen, and J. G. Taylor. *A Universal Approximator Network for Learning Conditional Probability Densities.* Kluwer, 1997. to appear.

[5] D. Husmeier and J. G. Taylor. Predicting conditional probability densities of stationary stochastic time series. *Neural Networks*, 10, 1997. to appear.

[6] B. Igelnik and Y. H. Pao. Stochastic choice of basis functions in adaptive functional approximation and the functional-link net. *IEEE Transactions on Neural Networks*, 6:1320–1329, 1995.

[7] M. I. Jordan and R. A. Jacobs. Hierarchical mixtures of experts and the em algorithm. *Neural Computation*, 6:181–214, 1994.

[8] R. Neuneier, F. Hergert, W. Finnoff, and D. Ormoneit. Estimation of conditional densities: A comparison of neural network approaches. In *Proc. ICANN 94*, pages 689–692. Springer-Verlag, 1994.

[9] Y. H. Pao, G. H. Park, and D. J. Sobajic. Learning and generalization characteristics of the random vector functional-link net. *Neurocomputing*, 6:163–180, 1994.

[10] A. N. Srivastava and A. S. Weigend. Computing the probability density in connectionist regression. In *Proc. ICANN '94*, pages 685–688. Springer-Verlag, 1994.

An Artificial Neuron with Quantum Mechanical Properties

D. Ventura and T. Martinez
Neural Networks and Machine Learning Laboratory
Department of Computer Science
Brigham Young University, Provo, Utah 84602 USA
dan@axon.cs.byu.edu, martinez@cs.byu.edu

Abstract

Quantum computation uses microscopic quantum level effects to perform computational tasks and has produced results that in some cases are exponentially faster than their classical counterparts. Choosing the best weights for a neural network is a time consuming problem that makes the harnessing of this 'quantum parallelism' appealing. This paper briefly covers necessary high-level quantum theory and introduces a model for a quantum neuron.

1 Introduction

The field of artificial neural networks has at least two important goals: (A) creation of powerful artificial problem solving systems and (B) furthering understanding of biological neural networks including the human brain. Much effort has been made in both areas and some progress has been realized. The field of quantum computation [2], which has been completely unrelated to that of neural networks until very recently, applies ideas from quantum mechanics to the study of computation and has made interesting progress. Most notably, quantum algorithms for prime factorization and discrete logarithms have recently been discovered that provide exponential improvement over the best known classical methods [5]. Recently some work has been done in the area of combining classical artificial neural networks with ideas from the field of quantum mechanics in pursuit of goal (A).

It is the purpose of this paper to further this pursuit of goal (A), that is, to show the usefulness of some ideas from the field of quantum mechanics (in particular those of linear superposition and coherence/decoherence) to that of artificial neural networks in order to improve neural networks' abilities as problem solving systems. Our approach is to introduce a mathematical model of an artificial neuron with quantum mechanical properties that allow it to discriminate linearly inseparable problems using only linear thresholding as its activation function.

2 Some Quantum Ideas

Quantum mechanics is in many ways extremely counterintuitive and yet it has provided us with perhaps the most accurate theory (in terms of predicting experimental results) ever devised by science. The theory is well established and is covered in its basic form by many textbooks (see for example [6]). Several ideas from this theory that are necessary for the following presentation must be briefly mentioned.

Linear superposition is closely related to the familiar mathematical principle of linear combination. For example, in a vector space with bases $\vec{x}$ and $\vec{y}$, any vector $\vec{v}$ can be defined as $\vec{v} = a\vec{x} + b\vec{y}$. In some sense $\vec{v}$ can be thought of as being both $\vec{x}$ and $\vec{y}$ at the same time. In quantum mechanics, this principle actually applies to physical variables in physical systems. The vector space is generalized to a Hilbert space whose bases are the classical values normally associated with the system. For example, the position of an electron orbiting a nucleus is usually a superposition of all possible positions in 3-d space, where each possible position is a basis state for the Hilbert space and has a finite probability of being the actual position. One of the most counterintuitive aspects of quantum theory is this — at the quantum or microscopic level, the electron is not in any one position in an orbit, but it is in a superposition of all of them at once. In some sense it is in all positions at the same time. However, at the macroscopic or classical level, the location of the electron is a single definite position in 3-d space. This apparent contradiction is still not fully understood, but it is explained as follows. A quantum mechanical system remains in a superposition of its basis states until it interacts with its environment.

Coherence/decoherence is related to the idea of linear superposition. A quantum system is said to be *coherent* if it is in a linear superposition of its basis states. As mentioned above, a result of quantum mechanics is that if a system that is in a linear superposition of states is observed or interacts in any way with its environment, it must instantaneously choose one of those states and 'collapse' into that state and that state only. This collapse is called *decoherence* and is governed by the wave function Ψ. Just as the basis states of the Hilbert space have a physical interpretation (as the classical values associated with a system), so too does the amplitude of the wave function Ψ describing the system. This wave function actually represents the probability amplitudes (in the complex domain) of all bases (possible positions, etc.) such that the probability for a given basis (position, etc.) is given by $|\Psi|^2$.

Operators are a mathematical formalism used to describe how one wave function is changed into another. They are analogical to matrices operating on vectors. Using operators, an eigenvalue equation can be written $\hat{A}\phi_i = a\phi_i$. The solutions ϕ_i to such an equation are called eigenstates and are the basis states of a Hilbert space. In the quantum formalism, all properties (position, momentum, spin, etc.) are represented as operators whose eigenstates are the classical values normally associated with that property.

3 Related Work

To date, very little has been done in combining the fields of quantum mechanics and neural networks. However, a few notable exceptions do exist. Perus has published an interesting set of mathematical analogies between the quantum formalism and neural network theory [4]. Menneer and Narayanan have proposed a model weakly inspired (their term) by the many worlds interpretation of quantum mechanics in which a network is trained for each instance in the training set and the final network is a superposition of these [3]. Finally, Behrman *et al.* have developed a novel approach to implementing a quantum neural network using quantum dots [1]. It uses a quantum dot for each input and the system is allowed to evolve quantum mechanically through time while being observed (and thus forced to decohere) at fixed time intervals. Interestingly, the different time slices act as the neurons in a hidden layer of a neural network — the more time slices, the more hidden layer neurons. Perhaps most notably, Behrman *et al.* have actually implemented this quantum dot neural network and give results for its learning several two input boolean functions.

4 Quantum Neuron

The simplest classical artificial neuron is the perceptron that takes as input n bipolar (or binary) values, $\{i_j\}$. It is defined as a weight vector $\vec{w} = (w_1, w_2, ..., w_n)^T$, a threshold θ and an output function f where

$$f = \begin{cases} 1 & \text{if} \sum_{j=1}^{n} w_j i_j > \theta \\ \text{-1} & \text{otherwise} \end{cases} . \tag{1}$$

This neuron model is well understood, but it is mentioned for several reasons. First, it cannot solve problems that are not linearly separable, and second, though this classical perceptron is extremely simple and well understood, this will not be the case for its quantum counterpart and thus it is important to start with the very simplest concepts as quantum ideas are incorporated.

We now define a simple quantum analog to the perceptron, which also takes as input $\{i_j\}$. It is similar in all respects to the classic perceptron except that the single weight vector $\vec{w}$, is replaced by a wave function, $\Psi(\vec{w}, t)$ in a Hilbert space whose basis states are the classical weight vectors. This wave function represents the probability amplitude (in general this is a complex wave as opposed to a real one) for all possible weight vectors in weight space together with the *normalization condition* that for any time t

$$\int_{-\infty}^{\infty} |\Psi|^2 d\vec{w} = 1. \tag{2}$$

Thus, the weight vector of the perceptron is replaced with a quantum superposition of many weight vectors which on interaction with its environment will decohere into one classical weight vector, according to the probabilities given by $|\Psi|^2$.

For example, consider the one-input, one-output bipolar function that inverts its input (NOT). For convenience the weights will be bounded such that

$$-\pi \leq w_j \leq \pi. \tag{3}$$

In order for a quantum neuron to learn this function, $\Psi(\vec{w}, t)$ and θ must be found. We assume that Ψ is

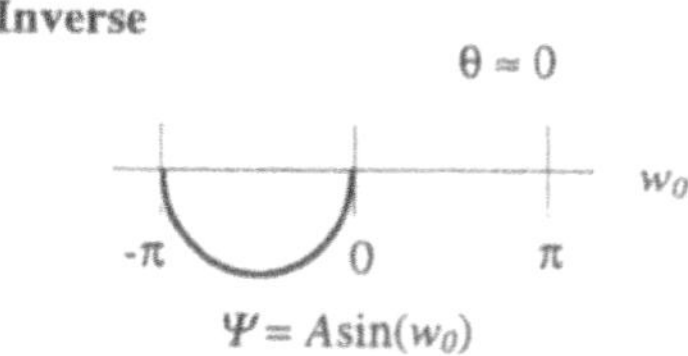

Figure 1: One solution for the 1-d NOT function.

time-invariant and concentrate only on $\vec{w}$, which is in this case a one-dimensional vector. Because of the strict bounds given in (3), finding Ψ is equivalent to solving the one-dimensional rigid box problem common to most elementary treatments of quantum mechanics, whose solutions are of the form

$$\Psi(w_0) = A\sin(\frac{n\pi}{a}w_0). \tag{4}$$

Here A is a normalization constant that can be found using (2), $n = 1, 2, 3, \ldots$, w_0 is the single element of $\vec{w}$, and a is the width of the box (in general this is 2π, but the width may be altered to be smaller).

Figure 1 shows a graph of one Ψ for the one-input bipolar NOT function, and for comparison Figure 2 shows a solution for another one-input bipolar function, TRUE (the extra π term shifts the function, which is necessary since the general solution (4) assumes for convenience that the left hand side of the box occurs at 0). To understand what these graphs represent, it must be realized that they are graphs in weight space and that Ψ is the probability amplitude for a given weight vector $\vec{w}$. In a quantum neuron the weight vectors exist in coherent superposition of all possible classical weight vectors in weight space with non-zero probability amplitude.

When the superposition of weight vectors interacts with its environment (for example when it encounters an input) it must decohere into one basis state — a classical weight vector within the bounds enforced in (3) — and this decoherence occurs with probability governed by $|\Psi|^2$. Note that in the case

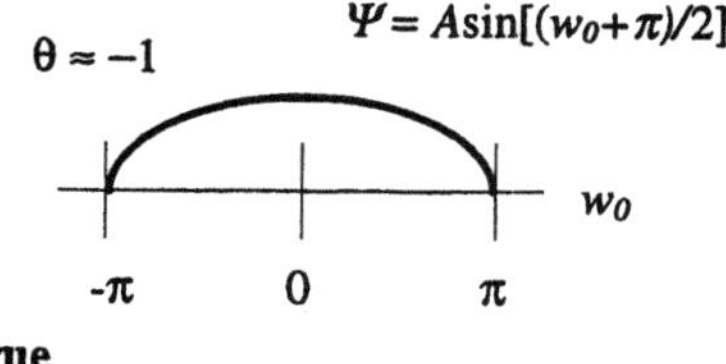

Figure 2: One solution for the 1-d TRUE function.

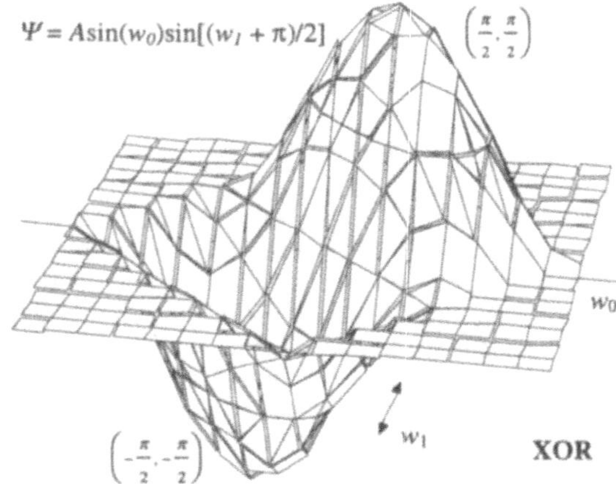

Figure 3: Plot of Ψ for the XOR problem.

of these simple functions, no matter what weight vector is chosen at decoherence, it will result in the correct output.

Now consider the more complicated two-input XOR problem, which is of course, not linearly separable. Using a similar argument for that of the one-dimensional case, it is not surprising to find that the solutions to problems with two-dimensional weight spaces are equivalent to those of the two dimensional rigid box

$$\Psi(w_0, w_1) = A\sin(\frac{n_{w0}\pi}{a}w_0)\sin(\frac{n_{w1}\pi}{a}w_1), \tag{5}$$

where the variables and constants have the same meanings as in (4) with the added note that now there are two different n's, one for each weight.

Figure 3 shows a contour plot of one such solution, rotated by 45°. Note again that this is a graph of Ψ, the probability amplitude, whereas the probability, $|\Psi|^2$, is what is really important. Also notice that its contour plot (Figure 4) gives two solutions with maximal probability, at $(\pi/2, \pi/2)$ and at $(-\pi/2, -\pi/2)$. It is also important to note that the nodal line that exists at $w_0 = -w_1$ in Figure 3 ensures that no solutions exist on this line which must be the case in order to learn the function. Finally note that since the quantum neuron maintains a coherent superposition of weights, it can solve non-linearly separable problems.

Of course, these examples are simple and their solutions are simple. Even in the general case however, the solutions will always be equivalent to those of an n dimensional rigid box so that Ψ will always be of the general form introduced above. This need

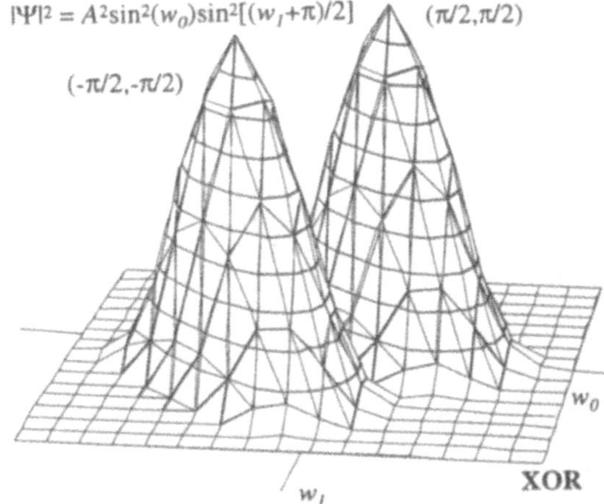

Figure 4: Plot of $|\Psi|^2$ for the XOR problem.

not be true, however, if other physical models, such as the non-rigid box, simple harmonic oscillator, hydrogen atom, etc. are considered. Further, the time variable has not been incorporated into the equations. Another important topic is training of the neuron, which entails changing the wave function Ψ. Since Ψ is governed by the (time-independent for now) Schrödinger equation,

$$\nabla^2\Psi = \frac{2m}{\hbar^2}[U - E]\Psi, \tag{6}$$

note that the only variable that may be changed is the potential, U. Therefore Ψ may be changed by changing U. The n's introduced in (4) and (5) are termed quantum numbers and play an important role in quantum mechanics. Further investigation of their role with regards to quantum neurons is needed. A series of classification operators that ensures appropriate wave function decoherence is also required. For example, in the XOR problem, two of the possible input patterns, $(-1,1)$ and $(1,-1)$, are correctly classified using any of the superposed weight vectors; however, for the other two input patterns, $(-1,-1)$ and $(1,1)$, this is not the case. The necessary operators for the quantum neuron will be analogical to a matrix form of the input vector. The operator associated with the input vector $(-1,-1)$ must greatly decrease the probability amplitude (and thus the probability) for all the negative weight vectors and correspondingly increase the amplitudes for the positive weight vectors. The operator associated with the input vector $(1,1)$ must do the opposite.

5 Concluding Remarks

Though many of these topics are beyond the scope of this paper, it has been demonstrated that the application of quantum mechanical ideas to the field of neural networks is a fertile area for further research. This includes development of a learning algorithm for the quantum neuron, further investigation of the classification operators, and theoretical analysis of the quantum neuron's capabilities. Further, the idea of linear superposition may be applied not only to the weight vector of a neuron, but also to its inputs, its output, and its activation function, among other things. See [7] for an application to the problem of choosing useful features from the exponentially large set of possibilities. Also, other quantum mechanical concepts such as the quantum nature of energy, spin, momentum, etc. and EPR phenomenon may find application as this field is explored further.

References

[1] E. C. Behrman, J. E. Steck, and S. R. Skinner. A quantum dot neural network. preprint submitted to Physical Review Letters, 1997.

[2] G. Brassard. *New Trends in Quantum Computation, 13th Symposium on Theoretical Aspects of Computer Science.* 1996.

[3] T. Menneer and A. Narayanan. Quantum-inspired neural networks. Technical Report R329, Department of Computer Science, University of Exeter, Exeter, UK, 1995.

[4] M. Perus. Neuro-quantum parallelism in brain-mind and computers. *Informatica*, 20:173–183, 1996.

[5] P. W. Shor. Polynomial-time algorithms for prime factorization and discrete logarithms on a quantum computer. *SIAM Journal of Computing*, 1996. to appear.

[6] J. R. Taylor and C. D. Zafiratos. *Modern Physics for Scientists and Engineers.* Prentice Hall, Englewood Cliffs, New Jersey, 1991.

[7] D. Ventura and T. Martinez. Application of quantum mechanical properties to machine learning. In *International Conference on Machine Learning*, 1997.

Computation of Weighted Sum by Physical Wave Properties — Coding Problems by Unit Positions

I. Kumazawa and Y. Kure,
Department of Computer Science, Tokyo Institute of Technology,
Tokyo 152 Japan
email: kumazawa@cs.titech.ac.jp

Abstract

An architecture for neural computation is proposed. The costliest parts of neural computation: inter-unit-communication, weight representation, and weighted sum execution are all implemented by natures of wave propagation and interaction. As a result, unit interaction is realized without wires, weighted sum is executed without adders or multipliers, and the weight values are coded as unit positions. Some experimental results, which are based on software models of the wave natures, show that the approach provides a promising computation potential regardless of its limited ability to represent connection weights.

1 Introduction

This paper proposes a computation system which consists of units with a weighted sum mechanism embodied by physical wave properties. Problems to be solved are expressed by the positions of the units. The inter-unit-communication is wireless and realized with an electromagnetic wave, a sound wave or, for a two dimensinal system, a surface acoustic wave.

Each unit computes the weighted sum of signals from other units using the nature of wave propagation mechanisms, that is, the waves from different sources are physically summed with different weights depending on their phase differences at the observation point. In this way, the signal is already the weighted sum when it is detected by a sensor. There is no need for any additional hardware for summation or weighting.

No hardware for representing or storing weight values is needed either. Weights are represented by relative positions of sensors. The problems to be solved are represented by the pattern of the sensor position arrangement.

This framework is effective to simplify the hardware and reduce the cost as the most part of neural computation hardware is devoted to communication, weight value representation, and weighted sum computation. In ordinary hardware, communication wires, memories to store weights and multipliers and adders for computing weighted sums are all required with the order of n^2 to implement a network of n units.

In return for the simplicity, a restriction is imposed on the computational ability of the framework, that is the limited freedom in the weight representation. In contrast to an ordinary neural computing unit, which has n weight values independently determined for each of its n inputs, the framework offers only three independent variables for the weights, which are the coordinates to specify the position of the unit in the three dimensional space. The range of communication is also restricted due to the decaying nature of the waves. There may be ways to avoid these problems, but the purpose of this paper is to show experimentally that, even with the limited freedom in the weight representation, by compensating for it with the increased number of units, the framework has a practical computation ability for some problems.

The experiments were performed not by a physical hardware but by computer software which simulates the physical properties of wave propagation. Promising results were obtained for several tasks including a multiplication problem for a set of single digit integers with a binary expression.

2 Details of the Computation Framework

The network units are located in a three dimensional space, each with a wave emitting device, serving as a wave source. The wave emitting devices are all synchronized and emit sine waves with the same frequency and the same phase. Each unit also has a wave detector which receives wave signals from other units. The waves arrive at the detector with some delays depending on the distances between the detector and the wave sources. The delays cause differences in the phases of waves at the detecting point. The signal detected is actually the sum of these waves. And when the waves are detected as a form of sum, the phase difference works as the weight difference. For example, waves with the phase difference $\pi/2$ cancel with each other. This is regarded as the cancelation between excitatory and inhibitory interactions. As the phase difference becomes smaller, the waves come to interact more cooperatively, just like the cooperative interactions among excitatory, or inhibitory synapses. With a standard sine wave, the kind of interaction (excitatory or inhibitory) can be determined by the phase difference from the standard wave.

The details of the weighting mechanism are as follows. Assume N units: $\{ U_1, \cdots, U_N \}$ positioned in a three dimensional space. The unit U_i, where i is the index to identify the unit, can change its position within the sphere of a radius $\lambda/2$. Depending on the position in the sphere, phases of the waves from other units vary. However, they can not vary independently. The waves from the same direction change their phases by the same amount when the receiving unit moves. This dependency results from the fact that the position can change only in the three dimensional space and, as the result, the weight vector space has the limited dimension of three. In addition to this, The decaying feature of the waves should also be considered. The wave amplitude gets smaller as the wave propagates. In case of the electromagnetic wave, the amplitude changes in inverse proportion to the distance.

These wave properties are formalized by the following equation:

$$f(t) = \sum_{j=1}^{N} \frac{A_j}{D_{ij}} sin(2\pi(ft - D_{ij}/\lambda)). \tag{1}$$

where $f(t)$ denotes the signal detected by the unit U_i and D_{ij} denotes the distance between the unit U_i and U_j. The frequency and the wave length are denoted by f and λ respectively. Units U_j, $j = 1, 2, \cdots N$ are assumed to emit the sine waves: $A_j sin(2\pi ft)$, $j = 1, 2, \cdots N$.

With this natural weighting mechanism, signals from other units are integrated and used to decide the output signal for the unit U_i to emit. The output signal has a sine wave form with amplitude A_i:

$$A_i sin(2\pi ft). \tag{2}$$

The amplitude A_i is determined according to the weighted sum of inputs to the unit. The frequency f and the phase of the output are the same for all the units. In this way, the output is expressed by a kind of amplitude modulation. Other kinds of modulation techniques might also be considerable as a means of representing the output information but these are not discussed in this paper.

There are several ways for computing A_i from the signal detected by the unit $f(t)$. In any event, it is important to find a way to prevent one's own output from being used as one of its inputs. Otherwise, the unit's own output occupies the major part of its inputs burying inputs from other units. One of the ways is to use sampled values of the signal at $t = n/f$ where $n = 0, 1, 2, \cdots$. The unit output: $A_i sin(2\pi ft)$ at these sampling times are zero. So the sampled signals only have inputs from other units as their components. In this case, the output A_i is computed by the following procedure.

$$\frac{ds}{dt} = -s + f(n/f), \tag{3}$$

$$A_i = sigmoid(s) \tag{4}$$

where $f(t)$ denote the signal detected by U_i and

$$sigmoid(s) = \frac{1}{1 + e^{-\alpha s}}. \tag{5}$$

The time evolution nature of the internal state s, which is described by Equation (3) is substituted by

$$\frac{ds}{dt} = -s + \int_{t}^{t+T} f(t) cos(2\pi ft) dt. \tag{6}$$

As the signals $cos(2\pi ft)$ and $sin(2\pi ft)$ are orthogonal under the inner product used above, the unit's own output, which is the $sin(2\pi ft)$ component in $f(t)$, is eliminated.

These procedures also eliminate signals from other units if they have the same phase as the unit's output. However, if we have a sufficient number of units, and the learning procedure finds suitable unit positions to avoid this problem, the system still has a promising computational ability.

A set of units, each of which operates with the mechanism mentioned above, constitutes a recurrent system and converges to one of the local minima of an energy function. As the interaction among units is symmetric, in this case, the existance of the energy function and the convergence are guaranteed.

3 Assignment of Unit Positions

Our question is if there is an arrangement of unit positions with which any practical problems can be solved as a convergent state of the proposed system. As the framework is based on a kind of recurrent network, we applied the generalized (recurrent) backpropagation developed by Pineda *et al.* [7] (see also [7]) and its modified version by Kumazawa [3] to simplify computation. Another modification is needed to account for the limited freedom of the values the weights can take.

The ordinary learning methods [1, 3, 7] provide gradients of an error function with regard to connection weights, which are denoted by:

$$\frac{\partial Error}{\partial w_{ij}} \tag{7}$$

where $Error$ is an error evaluating function and w_{ij} is a connection weight between the unit U_i and the unit U_j. One of the coordinates to specify the position of the unit U_i such as x_i of (x_i, y_i, z_i) affects the values of w_{ij}, $j = 1, 2, \cdots N$. So the gradient of the error function with regard to x_i is given by:

$$\frac{\partial Error}{\partial x_i} = \sum_{j=1}^{N} \frac{\partial Error}{\partial w_{ij}} \frac{\partial w_{ij}}{\partial x_i}. \tag{8}$$

When Equation (3) is used for the time evolution, w_{ij} has the form:

$$w_{ij} = \frac{1}{D_{ij}} sin(2\pi(-D_{ij}/\lambda)). \tag{9}$$

By differentiating this, Equation (8) can be computed.

Using this gradient, the modification procedure for x_i is given by

$$x_i = x_i + \varepsilon \frac{\partial Error}{\partial x_i} \tag{10}$$

where ε is a small positive number. Modifications of y_i and z_i can be done by similar procedures.

4 Computer Simulations

As a first attempt, we used a two dimensional system for the sake of simplicity. This means each unit has only two independent parameters to decide its weight values. In this case, the two parameters are the coordinates which specify the position of each unit in a plane. The task was to execute multiplications among the integer numbers {1, 2, 3 } which are represented in a binary fashion. In total, 25 units were used to constitute the system. Four of them were input units, divided into two groups each with two units, used to represent a pair of integers. The two units of each group were used for 2-bit-binary-representation of an integer. Output is a single integer and represented by using four units for its 4-bit-binary-representation. The positions of these units on the two dimensional plane were decided by using the recurrent backpropagation algorithm and the procedure described in Section 3.

The actual procedures for computing unit output were slightly modified from the originals described in Section 2 and the parameters were tuned so that the better results were obtained. We show the actual procedures and exact values of parameters for those who would like to reproduce our results.

[Computation procedure (detailed version)]

$$\Delta u_i = 0.05(-u_i + \sum_{j=1}^{25} w_{ij}x_j + \theta_i) + 0.8\Delta u_i \tag{11}$$

where θ_i is a threshold and the term $0.8\Delta U_i$ is introduced as a momentum factor.

$$u_i = u_i + \Delta u_i. \tag{12}$$

$$x_i = \frac{1}{1 + e^{-0.1u_i}} - 0.5. \tag{13}$$

$$w_{ij} = \frac{500}{D_{ij}} sin(2\pi D_{ij}). \tag{14}$$

Note that, according to Equation (13), the unit output x_i takes values between -0.5 and 0.5 instead of those between 0 and 1 produced when Equation (5) is applied. So the value 0 for the binary representation is represented by -0.5 and 1 is represented by 0.5 using the x_i value.

Table 1: Input and output units and the binary representation of integers.

	Input 1 Unit No.		Input 2 Unit No.		Output Unit No.			
	9	12	14	18	1	5	21	25
$1 \times 1 = 1$	0	1	0	1	0	0	0	1
$1 \times 2 = 2$	0	1	1	0	0	0	1	0
$1 \times 3 = 3$	0	1	1	1	0	0	1	1
$2 \times 1 = 2$	1	0	0	1	0	0	1	0
$2 \times 2 = 4$	1	0	1	0	0	1	0	0
$2 \times 3 = 6$	1	0	1	1	0	1	1	0
$3 \times 1 = 3$	1	1	0	1	0	0	1	1
$3 \times 2 = 6$	1	1	1	0	0	1	1	0
$3 \times 3 = 9$	1	1	1	1	1	0	0	1

Table 2: Unit positions and thresholds for the multiplication task.

Unit No.	Position	Threshold
1	(0.007970628620,0.004255573458)	-10.825931381643
2	(9.995055626932,0.030542275502)	-10.335042922914
3	(19.958548305400,0.012646481011)	-9.961512683236
4	(30.084554473132,-0.086384786902)	-10.122858749392
5	(40.047926787712,-0.105880681069)	-9.914247798641
6	(4.991292734862,8.652346085399)	-9.382742471627
7	(14.969058662590,8.681443823645)	-10.209607388229
8	(24.938292449168,8.647860981801)	-10.693555673904
9	(35.037285356564,8.631137145760)	-10.000000000000
10	(44.951121585977,8.715712437771)	-10.158573112297
11	(9.932232338411,17.376415311939)	-10.352768343046
12	(20.175805655988,17.469296844932)	-10.000000000000
13	(30.118194365472,17.250691878320)	-10.358598688497
14	(39.854711346031,17.251097086843)	-10.000000000000
15	(49.994754728607,17.339868911601)	-10.005820901464
16	(15.001709546748,25.930718318703)	-9.520146464486
17	(25.020866530470,26.065915693132)	-10.611202167573
18	(34.998186170387,26.074992182403)	-10.000000000000
19	(45.125144288189,25.908607389403)	-10.116844161024
20	(54.948759902181,26.035760588905)	-9.071763110205
21	(19.895923153528,34.551808482619)	-10.435221302085
22	(30.032188724272,34.596329006162)	-7.889145167194
23	(39.950477362734,34.732820320518)	-10.721451704381
24	(49.971444165347,34.665915261937)	-10.534720998924
25	(59.997353832163,34.582900103775)	-11.922181267709

We considered the set of u_i, $i = 1, \cdots, 25$ converged when $\Delta u_i < 0.001$ for $i = 1, \cdots, 25$. The initial states for u_i's were all set to 0.

The positions were successfully found and with which the multiplication among the three integer numbers was correctly executed. The positions and thresholds obtained for the 25 units were shown in Table 2. The position is expressed in the form of (x, y) and scaled so that the wave length corresponds to one.

Table 3: Outputs of the system with units positioned acording to Table 1. Note that 0 is represented by -0.5 and 1 by 0.5 in the system.

	Input 1 Unit No.		Input 2 Unit No.		Output Unit No.			
	9	12	14	18	1	5	21	25
$1 \times 1 = 1$	-0.5	0.5	-0.5	0.5	-0.37	-0.49	-0.37	0.46
$1 \times 2 = 2$	-0.5	0.5	0.5	-0.5	-0.43	-0.39	0.42	-0.39
$1 \times 3 = 3$	-0.5	0.5	0.5	0.5	-0.39	-0.49	0.38	0.49
$2 \times 1 = 2$	0.5	-0.5	-0.5	0.5	-0.45	-0.39	0.39	-0.34
$2 \times 2 = 4$	0.5	-0.5	0.5	-0.5	-0.35	0.46	-0.39	-0.49
$2 \times 3 = 6$	0.5	-0.5	0.5	0.5	-0.46	0.38	0.43	-0.44
$3 \times 1 = 3$	0.5	0.5	-0.5	0.5	-0.37	-0.49	0.36	0.35
$3 \times 2 = 6$	0.5	0.5	0.5	-0.5	-0.41	0.39	0.40	-0.46
$3 \times 3 = 9$	0.5	0.5	0.5	0.5	0.44	-0.39	-0.40	0.46

5 Conclusion

The framework proposed in this paper significantly decreases the complexity and the cost of hardware required for implementing connections of neural computation. One of the disadvantages of the framework is its limited capability in weight representation. However, even under this limitation, a difficult task such as the multiplication of integers was successfully realized. The result demonstrates the potential of this framework despite its limited freedom in weight representation.

6 Acknowledgments

The authors would like to thank Mr. S. L. Funk for his helpful discussions while writing this article.

References

[1] L.B. Almeida. Backpropagation in perceptrons with feedback. *Neural Computers*, 1988.

[2] J.J. Hopefield and D.W. Tank. Neural computation in optimization problems. *Biol. Cybernetics*, 52, 1985.

[3] I. Kumazawa. A signal sharing scheme for unifying computation and learning phases of recurrent networks. In *Artificial Neural Network II*. Elsevier, 1992.

[4] I. Kumazawa. Neural computation by interaction of waves: Coding problems by unit positions. Technical report, Neural Computation Group, March 1997. Just submitted.

[5] I. Kumazawa and Y. Kure. Computation by interaction waves: Realisation of recurrent neural networks. In *IEICE, Spring National Conference*, 1997. Just submitted.

[6] Y. Kure and I. Kumazawa. Neural computation by interaction of waves: Weight representation by differences in wave phases. Technical report, Neural Computation Group, March 1997. Just submitted.

[7] F.J. Pineda. Generalization of back-propagation to recurrent neural networks. *Physical Review Letters*, 59, 1987.

Some Analytical Results for a Recurrent Neural Network Producing Oscillations

T. P. Fredman and H. Saxén
Heat Engineering Laboratory, Åbo Akademi University,
Biskopsgatan 8, FIN-20500 Åbo, Finland
Email: {tfredman,hsaxen}@abo.fi

Abstract

A two-node fully recurrent neural network producing oscillations is analysed. The network has no true inputs and the outputs from the network exhibit a circular phase portrait. The weight configuration of the network is investigated, resulting in analytical weight expressions. The theoretical predictions are compared with numerical weight estimates obtained by training the network on the desired trajectory. The analysis shows that the analytical expressions agree well with the findings of the numerical study.

1 Introduction

The ability of recurrent neural networks to approximate and produce oscillating sequences has been demonstrated by many authors in the literature [2, 4, 5, 7, 9], and more theoretical aspects of recurrent networks have also attracted recent interest [1, 3, 6, 8]. A typical benchmark task used in studies of recurrent networks is the problem of producing two trigonometric oscillations, which form a circle in the phase plane of the output units. However, not much attention has been paid to the way in which the trained networks solve the task. This paper analyses a fully recurrent two-node network, which arises if both outputs of a feedforward network, without true inputs, are fed back to the input layer (Figure 1). Analytical conditions for the weights are derived for the task where the outputs produce a circular phase portrait, and the results are compared with findings from a numerical study. The results show that the values of the weights can be derived theoretically.

2 Analysis

Consider the discrete-time recurrent neural network depicted in Figure 1, with feedback weights fixed to equal the backward shift operator, i.e., outputs

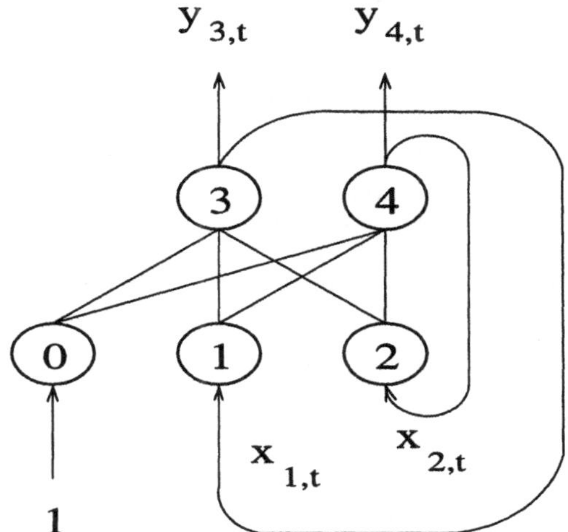

Figure 1: The recurrent network studied.

are fed into the network lagged by one time step. The network is trained to approximate a circular trajectory (radius r) on the set of $K+1$ observations

$$\begin{aligned} z_1(i) &= r\cos(i\Delta\alpha) \\ z_2(i) &= r\sin(i\Delta\alpha) \end{aligned}, i=0,\dots,K, \quad (1)$$

where $\Delta\alpha$ is a constant angle increment. By noting that the equation can be expressed as

$$\begin{aligned} z_1(i) &= r\cos[(i-1)\Delta\alpha+\Delta\alpha] \\ z_2(i) &= r\sin[(i-1)\Delta\alpha+\Delta\alpha], \end{aligned} \; i=0,\dots,K, \quad (2)$$

we may apply the formulae for sums of arguments in the trigonometric functions, yielding

$$\begin{aligned} z_1(i) &= r\cos[(i-1)\Delta\alpha]\cos\Delta\alpha \\ &- r\sin[(i-1)\Delta\alpha]\sin\Delta\alpha \\ z_2(i) &= r\sin[(i-1)\Delta\alpha]\cos\Delta\alpha \\ &+ r\cos[(i-1)\Delta\alpha]\sin\Delta\alpha. \end{aligned} \; i=0,\dots,K. \quad (3)$$

This is seen to be equivalent to

$$\begin{bmatrix} z_1 \\ z_2 \end{bmatrix}(i) = \begin{bmatrix} \cos\Delta\alpha & -\sin\Delta\alpha \\ \sin\Delta\alpha & \cos\Delta\alpha \end{bmatrix} \begin{bmatrix} z_1 \\ z_2 \end{bmatrix}(i-1). \quad (4)$$

For arbitrary activating functions, f, in the output neurons, the network will produce a mapping according to

$$\begin{bmatrix} y_3 \\ y_4 \end{bmatrix}(i) = f\left(\begin{bmatrix} w_{31} & w_{32} \\ w_{41} & w_{42} \end{bmatrix} \begin{bmatrix} y_3 \\ y_4 \end{bmatrix}(i-1)\right), \quad (5)$$

where w_{ij} denotes the weight on the connection from node j to node i. Due to the symmetry of the training instances generated by Equation (1), the bias terms have been disregarded. The objective is now to find expressions for the weights in Equation (5), as functions of the discrete time instant, by utilizing Equation (4) and the requirement that the net be able to reproduce the observations of Equation (1). We proceed by formally inverting the activating function in Equation (5) as well as the relation in Equation (4), with the goal that the network outputs should approximate z_1 and z_2. Thus, Equation (5) can be written as

$$f^{-1}\left(\begin{bmatrix} y_3 \\ y_4 \end{bmatrix}\right)(i) = \begin{bmatrix} w_{31} & w_{32} \\ w_{41} & w_{42} \end{bmatrix} \cdot \begin{bmatrix} \cos\Delta\alpha & \sin\Delta\alpha \\ -\sin\Delta\alpha & \cos\Delta\alpha \end{bmatrix} \begin{bmatrix} y_3 \\ y_4 \end{bmatrix}(i). \quad (6)$$

With four unknowns and only two equations, two further constraints on the weight matrix are needed in order to find a unique solution to the problem. Comparing Equations (4) and (6), it may be natural to require the weights to exhibit the same symmetry properties as the training patterns, i.e.

$$w_{31} = w_{42} \equiv w_I, -w_{32} = w_{41} \equiv w_{II} \quad (7)$$

Equation (6) now takes the form

$$f^{-1}\left(\begin{bmatrix} y_3 \\ y_4 \end{bmatrix}\right) = \begin{bmatrix} y_3\cos\Delta\alpha + y_4\sin\Delta\alpha & y_3\sin\Delta\alpha - y_4\cos\Delta\alpha \\ -(y_3\sin\Delta\alpha - y_4\cos\Delta\alpha) & y_3\cos\Delta\alpha + y_4\sin\Delta\alpha \end{bmatrix} \begin{bmatrix} w_I \\ w_{II} \end{bmatrix}, \quad (8)$$

where the time indices (i) on z, y and w have been dropped for convenience of notation. The matrix on the right-hand side has the same symmetry properties as the weight matrix. Furthermore, its determinant equals r^2, which is seen upon equating $y_3 = z_1$ and $y_4 = z_2$, and inserting z_i from Equation (1). Consequently, the solution for the two weights is

$$\begin{bmatrix} w_I \\ w_{II} \end{bmatrix} = \frac{1}{r^2} \begin{bmatrix} y_3\cos\Delta\alpha + y_4\sin\Delta\alpha & y_4\cos\Delta\alpha - y_3\sin\Delta\alpha \\ y_3\sin\Delta\alpha - y_4\cos\Delta\alpha & -(y_3\sin\Delta\alpha - y_4\cos\Delta\alpha) \end{bmatrix} f^{-1}\left(\begin{bmatrix} y_3 \\ y_4 \end{bmatrix}\right). \quad (9)$$

Inserting (1) into the coefficient matrix and the activating function now yields

$$\begin{aligned} \begin{bmatrix} w_I \\ w_{II} \end{bmatrix}(i) &= \frac{1}{r} \begin{bmatrix} \cos i\Delta\alpha\cos\Delta\alpha + \sin i\Delta\alpha\sin\Delta\alpha & \sin i\Delta\alpha\cos\Delta\alpha - \cos i\Delta\alpha\sin\Delta\alpha \\ \cos i\Delta\alpha\sin\Delta\alpha - \sin i\Delta\alpha\cos\Delta\alpha & \cos i\Delta\alpha\cos\Delta\alpha + \sin i\Delta\alpha\sin\Delta\alpha \end{bmatrix} \cdot f^{-1}\left(\begin{bmatrix} z_1 \\ z_2 \end{bmatrix}\right)^T(i) \\ &= \frac{1}{r} \begin{bmatrix} \cos(i-1)\Delta\alpha & \sin(i-1)\Delta\alpha \\ \sin(1-i)\Delta\alpha & \cos(i-1)\Delta\alpha \end{bmatrix} \cdot f^{-1}\left(\begin{bmatrix} z_1 \\ z_2 \end{bmatrix}\right)^T(i), \end{aligned} \quad (10)$$

where, in the last equality, the addition formulae for trigonometric functions were used.

If the activation function for the output units is the identity function, $f(x) = x$, we should, obviously, get the weight set of the coefficient matrix of Equation (4). With $f^{-1}(x) = x$ in Equation (10), we correctly obtain

$$\begin{bmatrix} w_I \\ w_{II} \end{bmatrix}(i) = \frac{1}{r} \begin{bmatrix} \cos(i-1)\Delta\alpha & \sin(i-1)\Delta\alpha \\ \sin(1-i)\Delta\alpha & \cos(i-1)\Delta\alpha \end{bmatrix} \cdot f^{-1}\left(\begin{bmatrix} r\cos i\Delta\alpha \\ r\sin i\Delta\alpha \end{bmatrix}\right)(i) = \begin{bmatrix} \cos\Delta\alpha \\ \sin\Delta\alpha \end{bmatrix}. \quad (11)$$

In the case of the $f(x) = \tanh(\frac{x}{2})$ activating function we have

$$\begin{bmatrix} y_3 \\ y_4 \end{bmatrix}(i) = \tanh\left(\frac{1}{2}\begin{bmatrix} w_{31} & w_{32} \\ w_{41} & w_{42} \end{bmatrix} \begin{bmatrix} y_3 \\ y_4 \end{bmatrix}(i-1)\right), \quad (12)$$

where the function exercises elementvise action on the variables. In this case, Equation (10) becomes

$$\begin{bmatrix} w_I \\ w_{II} \end{bmatrix}(i) = \frac{2}{r} \begin{bmatrix} \cos(i-1)\Delta\alpha & \sin(i-1)\Delta\alpha \\ \sin(1-i)\Delta\alpha & \cos(i-1)\Delta\alpha \end{bmatrix} \cdot \tanh^{-1}\left(\begin{bmatrix} r\cos i\Delta\alpha \\ r\sin i\Delta\alpha \end{bmatrix}\right). \quad (13)$$

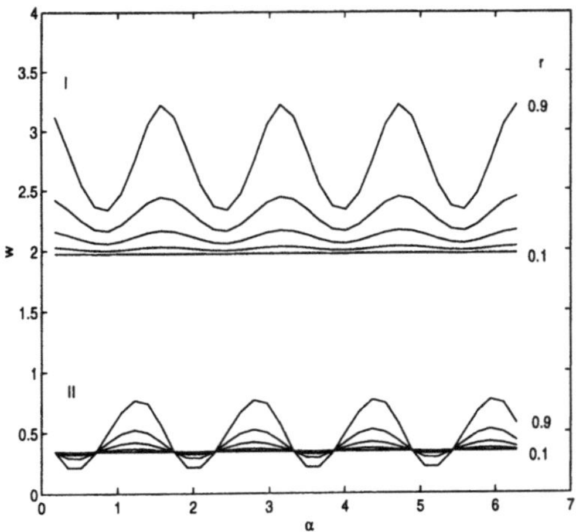

Figure 2: The two weights, w_I and w_{II} of Equation (13) for different values of the radius r, as functions of the angle α, with $\Delta\alpha = \frac{\pi}{18}$ and $K = 36$.

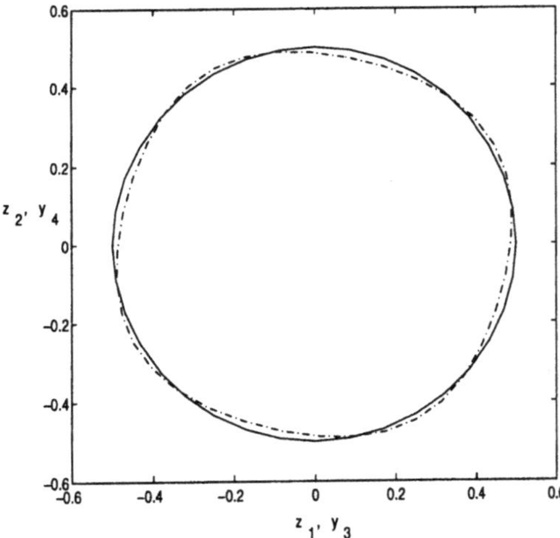

Figure 3: The target (solid line) generated by Equation (1) with $r = 0.5$, $\Delta\alpha = \frac{\pi}{18}$ and $K = 36$ and the approximation (dashed-dotted line) by the recurrent network.

3 Conclusions

Figure 2 shows the weights of Equation (13) for different values of the radius, $r = 0.1, 0.3, \ldots, 0.9$, with $\Delta\alpha = \frac{\pi}{18}$ and $K = 36$ (producing one full revolution for each circle). For small r, the weights are practically constant, which is understandable, since

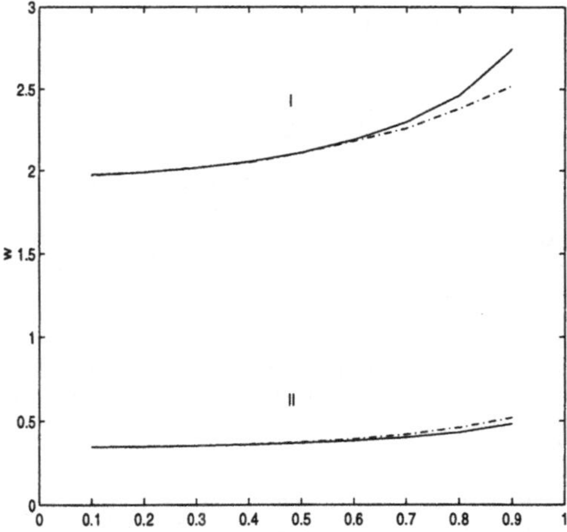

Figure 4: Arithmetic average of the analytical weights (solid line) and the weights obtained by training the recurrent network (dashed-dotted line) for different values of r.

the part of $\tanh(\frac{x}{2})$ close to the origin can easily approximate a linear model. For higher values of the radius, the oscillations of the weights are amplified. The reason for this behavior is understood if the approximation of the circle in the phase portrait is studied. Figure 3 depicts the numerical approximation [2] provided by a network trained on observations with the radius $r = 0.5$. Clearly, the approximating trajectory (dashed-dotted line) oscillates from one side of the target to the other. Finally, it is interesting to compare the mean values of the oscillating weights computed from Equation (13) with the findings from the numerical study. Figure 4 depicts the weights $w_{31} \approx w_I$ and $w_{41} \approx w_{II}$ for some values of r. These results show an excellent agreement between the numerical and analytical results for $r < 0.6$.

References

[1] *IEEE Transactions on Neural Networks*, volume 5(2), 1994. Special issue on dynamic recurrent neural networks.

[2] A. B. Bulsari and H. Saxén. A partially recurrent connectionist model. In B. Neumann, editor, *Proceedings of the 10th European Conference on Artificial Intellligence*, pages 198–202, Chichester, August 1992. Vienna, Austria, John Wiley & Sons.

[3] A. B. Bulsari and H. Saxén. A recurrent network for

modeling noisy temporal sequences. *Neurocomputing*, 7:29–40, 1995.

[4] J. L. Elman. Finding structure in time. *Cognitive Science*, 14:179–211, 1990.

[5] M. I. Jordan. Attractor dynamics and parallelism in a connectionist sequential machine. In *Proceedings of the Eight Annual Conference of the Cognitive Science Society*, pages 531–546, Amherst, 1989. Hillsdale, Erlbaum.

[6] O. Nerrand, P. Roussel-Ragot, L. Personnaz, G. Dreyfus, and S. Marcos. Neural networks and nonlinear adaptive filtering: Unifying concepts and new algorithms. *Neural Computation*, 5:165–199, 1993.

[7] B. Pearlmutter. Learning state-space trajectories in recurrent neural networks. *Neural Computation*, 1:263–269, 1989.

[8] F.-S. Tsung. *Modeling Dynamical Systems with Recurrent Neural Networks*. PhD thesis, University of California, San Diego, 1994.

[9] R. J. Williams and D. Zipser. Experimental analysis of the real-time recurrent learning algorithm. *Connection Science*, 1:87–111, 1989.

Upper Bounds on the Approximation Rates of Real-valued Boolean Functions by Neural Networks

K. Hlaváčková, V. Kůrková and P. Savický
Institute of Computer Science, Academy of Sciences of the Czech Republic,
Pod Vodárenskou věží 2, 182 07 Prague 8, Czech Republic
Email: katka@uivt.cas.cz

Abstract

Real-valued functions with multiple boolean variables are represented by one-hidden-layer Heaviside perceptron networks with an exponential number of hidden units. We derive upper bounds on approximation error using a given number n of hidden units. The bounds on error are of the form $\frac{c}{\sqrt{n}}$ where c depends on certain norms of the function being approximated and n is the number of hidden units. We show examples of functions for which these norms grow polynomially and exponentially with increasing input dimension.

1 Introduction

The rate of approximation of real functions by feedforward neural networks has been studied recently. Jones [5] introduced a recursive construction of approximant with 'dimension-independent' rates of convergence. Together with Barron he proposed to apply it to functions computable by one-hidden-layer neural networks. Several authors (e.g., Barron [1], Kůrková *et al.* [7], Mhaskar and Micchelli [9]) have characterized sets of functions with d real variables which can be approximated by networks with n hidden units of various types within an error $\mathcal{O}\left(\frac{1}{\sqrt{n}}\right)$.

In some applications, input data are only binary values. For example, Sejnowski and Rosenberg's NETtalk [11] sufficiently approximates a real function having 200 input Boolean variables with one output by a neural network with 80 hidden units. Motivated by similar experiments, we investigated representation and approximation of real functions with several Boolean variables using one-hidden-layer perceptron type networks. Ito's results [4] indicate that all *real* functions of d boolean variables can be computed *exactly* by perceptron networks with any sigmoidal activation function. We took two exact representations of functions obtained from two standard bases as functions computable by one-hidden-layer perceptron networks. An approximation of a function is obtainable, only if a limited number of elements of the basis are used in the representation.

In Section 2, we derive a bound on the approximation error obtained in this way improving Mhaskar and Micchelli's result. If the basic functions create an orthonormal basis, the bound is also slightly stronger than it would be when directly applying Jones-Barron's theorem. The achieved upper bound is of the form $\frac{c}{\sqrt{n}}$, where n is the number of hidden units and c depends on l_1 and l_2 norms of the function to be approximated. If only these two norms are known, the bound differs at most by a constant factor from the best possible bound.

Section 3 presents corollaries of the above bound for approximation rates of Boolean real-valued functions by Heaviside perceptron networks. Section 4 gives examples of functions where the upper bounds grow polynomially and exponentially. We also discuss the relation of the bounds gained from Euclidean and Fourier bases as well as the approximation error obtainable by a limited number of general perceptrons.

2 Upper Bounds on Rates of Approximation in Finite Dimensional Vector Spaces

We derive our results based on two theorems, Barron's extension [1] of Jones' result [5] and our extension of the result by Mhaskar and Michelli [9].

Let $\mathcal{X}$ be a real vector space with a norm $\|.\|_2$ which is generated by an inner product $f \cdot g$ for any two functions $f, g \in \mathcal{X}$. Let *cl conv* $\mathcal{G}$ be the closure of the convex hull of $\mathcal{G}$, where $\mathcal{G}$ is a subset of $\mathcal{X}$. The closure is taken with respect to the topology generated by the norm $\|.\|_2$ ($\|.\|_2 = \sqrt{f \cdot f}$). $\mathcal{N}$ denotes the set of positive integers.

Theorem 1 (Jones-Barron). *Let $\mathcal{X}$ be a real vector space with a norm $\|.\|_2$ generated by an inner product on $\mathcal{X}$, B be a positive real number and $\mathcal{G}$ be a subset of $\mathcal{X}$ such that for every $g \in \mathcal{G}$ $\|g\|_2 \leq B$. Then for every $f \in cl\ conv\ \mathcal{G}$, for every real number c such that $c > B^2 - \|f\|_2^2$ and for every $n \in \mathcal{N}$ there exists f_n that is a convex combination of n elements of $\mathcal{G}$ such that $\|f - f_n\|_2 \leq \sqrt{\frac{c}{n}}$.*

If f is a linear but not convex combination of functions from $\mathcal{G}$, $\mathcal{G}$ can be replaced by real multiples of functions from $\mathcal{G}$ bounded by a constant. This leads to the term of variation, first introduced by Barron for a set of characteristic functions of half-spaces. For a normed vector space $(\mathcal{X}, \|.\|)$ consisting of real functions on $J \subset \mathcal{R}^d$ for an integer d, let the *variation of a function $f \in \mathcal{X}$ with respect to a subset* $\mathcal{G}$ of $\mathcal{X}$ be $V(f, \mathcal{G}) = \inf\{B \geq 0; f \in cl\ conv\ \mathcal{G}(B)\}$, where the closure is taken with respect to the topology generated by the norm $\|.\|$ and $\mathcal{G}(B) = \{wg; g \in \mathcal{G}, w \in \mathcal{R}, |w| \leq B\}$. Kůrková [6] investigated the concept of variation with respect to a general family of functions. It is claimed that the infimum in the definition of variation can be replaced by minimum.

The following theorem is a corollary of the Jones-Barron theorem formulated by means of variation. Since in our applications the set $\mathcal{G}$ is finite, we use a stronger formulation of the theorem for compact sets $\mathcal{G}$.

Theorem 2. *Let $(\mathcal{X}, \|.\|)$ be a real vector space with the norm $\|.\|$ generated by an inner product and $\mathcal{G}$ be a compact subset of $\mathcal{X}$. Then for every $f \in \mathcal{X}$ such that $V(f, \mathcal{G}) < \infty$ and for every $n = 1, \dots, card\ \mathcal{G}$ there exists f_n that is a linear combination of n elements of $\mathcal{G}$ such that $\|f - f_n\|_2 \leq \sqrt{\frac{B^2 - \|f\|_2^2}{n}}$, where $B = V(f, \mathcal{G}) \sup_{g \in \mathcal{G}} \|g\|_2$.*

If $\mathcal{G}$ is an orthonormal basis, we can prove a stronger estimate. We present a result that improves a result of Mhaskar and Micchelli in [9] by a factor of two. The bound is formulated for finite dimensional spaces. For any orthonormal basis let A of $\mathcal{X}$ denote by $\|.\|_{1,A}$ the l_1-norm with respect to A, i.e. for every $f \in \mathcal{X}$ $\|f\|_{1,A} = \sum_{g \in A} |f \cdot g|$.

Theorem 3. *Let $\mathcal{X}$ be a finite dimensional real vector space with a norm $\|.\|_2$ generated by an inner product and let A be its orthonormal basis. Then for every $f \in \mathcal{X}$ and for every $n = 1, \dots, \dim\ \mathcal{X}$ there exists f_n which is a linear combination of n elements of A such that $\|f - f_n\|_2 \leq \frac{\|f\|_{1,A}}{2\sqrt{n}}$.*

If also $\|f\|_2$ is known, the bound from Theorem 3 can be improved.

Theorem 4 *Let $\mathcal{X}$ be a finite dimensional real vector space with an inner product, let A be its orthonormal basis, let $f \in \mathcal{X}$ and let $1 \leq n \leq \dim \mathcal{X}$. Then, there exists a function g expressible as a linear combination of at most n functions from A satisfying*

$$\|f - g\|_2 \leq \frac{\|f\|_{1,A}^2 - \|f\|_2^2}{2\|f\|_{1,A}\sqrt{n-1}}.$$

If both $\|f\|_{1,A}$ and $\|f\|_2$ are known, then Theorem 4 yields a good bound only if $4n \geq \|f\|_{1,A}^2 / \|f\|_2^2$. Otherwise, the trivial bound $\|f\|_2$ for the error of the approximation by the zero function is better. In fact, these two bounds together, i.e. the minimum of $\|f\|_2$ and the bound from Theorem 4, yield a bound that differs from the best possible bound based only on $\|f\|_{1,A}$ and $\|f\|_2$ by a constant factor.

3 Representations of Real-Valued Boolean Functions

The linear space of all real functions of d Boolean variables (where d is a positive integer) is denoted by $\mathcal{F}(\{0,1\}^d)$. It is easy to prove that this space is isomorphic to $\mathcal{R}^{2^d}$. For any $f, g \in \mathcal{F}(\{0,1\}^d)$, the standard Euclidean inner product is

$$f \cdot g = \sum_{x \in \{0,1\}^d} f(x) g(x).$$

In this section, we study representations and approximations of functions in $\mathcal{F}(\{0,1\}^d)$ by functions computable by networks with one linear output unit and one hidden layer with the *Heaviside function* ϑ defined by $\vartheta(t) = 0$ for $t < 0$ and $\vartheta(t) = 1$ for $t \geq 0$. The set of functions expressible by such networks with a bounded number of hidden units can be denoted by: $\mathcal{P}_d(n) = \{f \in \mathcal{F}(\{0,1\}^d); f(x)\}$, where $f(x) = \sum_{i=1}^n w_i \vartheta(v_i \cdot x + b_i)$; $w_i, b_i \in \mathcal{R}, v_i \in \mathcal{R}^d$.

We apply Theorem 4 to an appropriate orthonormal basis of $\mathcal{F}(\{0,1\}^d)$, namely, the Euclidean and the Fourier basis. Denote by $E = \{e_u; u \in \{0,1\}^d\}$ the *Euclidean orthonormal basis* of $\mathcal{F}(\{0,1\}^d)$, i.e. $e_u(u) = 1$ and $e_u(x) = 0$ for $x \neq u$. For any $u \in \{0,1\}^d$, let u^* be defined by $u_i^* = 2u_i - 1$.

It is easy to verify that for every $u, x \in \{0,1\}^d$, we have $x \cdot u^* \geq u \cdot u^*$, if and only if $x = u$. Thus, e_u can be computed by one Heaviside perceptron, i.e. $e_u \in \mathcal{P}_d(1)$. Together with the representation of any function $f \in \mathcal{F}(\{0,1\}^d)$ as $f(x) = \sum_{u\in\{0,1\}^d} f(u)e_u$, this yields that $\mathcal{F}(\{0,1\}^d) = \mathcal{P}_d(2^d)$.

A representation of a different type can be obtained from the *orthonormal Fourier basis* $F = \{\frac{1}{\sqrt{2^d}}\cos(\pi u \cdot x); u \in \{0,1\}^d\}$ of $\mathcal{F}(\{0,1\}^d)$ (as in [12]). Since in our context both x and u are Boolean vectors, we have $\cos(\pi u \cdot x) = (-1)^{u\cdot x}$. Thus every function $f \in \mathcal{F}(\{0,1\}^d)$ can be represented as

$$f(x) = \frac{1}{\sqrt{2^d}} \sum_{u\in\{0,1\}^d} \tilde{f}(u)(-1)^{u\cdot x},$$

where the Fourier coefficients $\tilde{f}(u)$ are given by the formula

$$\tilde{f}(u) = \frac{1}{\sqrt{2^d}} \sum_{x\in\{0,1\}^d} f(x)(-1)^{u\cdot x}.$$

Note that for any $f \in \mathcal{F}(\{0,1\}^d)$,

$$\|f\|_{1,F} = \|\tilde{f}\|_1 = \sum_{u\in\{0,1\}^d} |\tilde{f}(u)|.$$

Furthermore, all functions from the Fourier basis are computable by Heaviside perceptron networks. In contrast to the Euclidean basis, where one hidden unit was sufficient for one basis function, $d+1$ hidden units are needed for the members in the Fourier basis. Indeed, the function

$$\hat{\vartheta}(t) = 1 + 2\sum_{j=0}^{d-1}(-1)^{j+1}\vartheta\left(t - j - \frac{1}{2}\right)$$

satisfies the condition $\hat{\vartheta}(t) = (-1)^t$ for every $t = 0, \ldots d$. Hence for every $x, u \in \{0,1\}^d$ we have $(-1)^{u\cdot x} = \hat{\vartheta}(u \cdot x)$. Thus we have a representation of any $f \in \mathcal{F}(\{0,1\}^d)$ as an element of $\mathcal{P}_d((d+1)2^d)$ if we replace $(-1)^{u\cdot x}$ by $\hat{\vartheta}$ in the Fourier representation.

Note that all norms on $\mathcal{R}^{2^d}$ are topologically equivalent, in particular for every $f \in \mathcal{F}(\{0,1\}^d)$,

$$\|f\|_2 \leq \|f\|_1 \leq \sqrt{2^d}\|f\|_2$$

and

$$\|f\|_2 \leq \|\tilde{f}\|_1 \leq \sqrt{2^d}\|f\|_2.$$

Since each of these inequalities is tight, the differences between the norms may be exponential in the d dimension.

Theorem 5. *Let d be a positive integer and $f \in \mathcal{F}(\{0,1\}^d)$ and $n \geq 2$. Then, we have*
(i) there exists a function $f_n \in \mathcal{P}_d(n)$ such that

$$\|f - f_n\|_2 \leq \frac{\|f\|_1^2 - \|f\|_2^2}{2\|f\|_1\sqrt{n-1}};$$

(ii) there exists a function $f_n \in \mathcal{P}_d((d+1)n)$ such that

$$\|f - f_n\|_2 \leq \frac{\|\tilde{f}\|_1^2 - \|f\|_2^2}{2\|\tilde{f}\|_1\sqrt{n-1}}.$$

4 Tightness of the Bounds

This section presents two examples of functions for which the upper bounds on the approximation error from Theorem 5 yield a feasible approximation. The bounds are compared with the approximation error by general half-spaces.

We can easily show that Theorem 5 implies $\|f - f_n\|_2 \ll \|f\|_2$ for a feasible n only if $\min\{\|f\|_1, \|\tilde{f}\|_1\}$ is not much larger than $\|f\|_2$. In fact, if the equality holds, the bound implies an exact representation. For every orthonormal basis A, $\|f\|_{1,A} = \|f\|_2$ is satisfied if and only if f is a multiple of just one of the elements of the basis. The functions for which this situation occurs for the Fourier basis are functions represented by $f(x) = (-1)^{u\cdot x}$, $u \in \{0,1\}^d$. These functions correspond to the Boolean functions called parity functions, since the value of $(-1)^{u\cdot x}$ depends on the parity of the sum $\sum_{i\in I} x_i$, where $I = \{i; u_i = 1\}$.

Let f be a function represented by a decision tree of polynomial size and let the ratio $(\max_x |f(x)|)/(\min_x |f(x)|)$ be defined and polynomially bounded. Using the method of [8], it can be proven that $\|\tilde{f}\|_1/\|f\|_2$ is polynomially bounded. Using Theorem 5, this implies that f can be approximated by a polynomial number of hidden units.

We now turn to the functions for which our two bases do not yield a good approximation. A function from $\mathcal{F}(\{0,1\}^d)$ is called *bent*, if for every $x, u \in \{0,1\}^d$ $|f(x)| = 1$ and $|\tilde{f}(u)| = 1$.

Bent functions were introduced by Rothaus [10]. Recall that a bent function of d variables exists if and only if d is even. For every bent function, $\|f\|_1 = \|\tilde{f}\|_1 = \sqrt{2^d}\|f\|_2$. Thus, Theorem 5 does not imply a good approximation error. Moreover, it is possible to prove that for any bent function, the approximation error cannot be small if we take only approximations in the two bases. For any bent function f and any function f_n, which is a linear combination of at most n elements of the Euclidean basis or a linear combination of at most n elements of the Fourier basis, $\|f - f_n\|_2 \geq \sqrt{2^d - n}$ holds. For every even d, let the function $\phi_d : \{0,1\}^d \rightarrow \{-1,1\}$ be defined by

$$\phi_d(x) = \begin{cases} -1 \text{ if } |x| \equiv 0 \pmod 4 \text{ or } |x| \equiv 1 \pmod 4 \\ 1 \text{ otherwise} \end{cases}$$

where $|x|$ denotes the number of ones in a vector $x \in \{0,1\}^d$. This function is symmetric, i.e. it does not depend on the order of input variables. In other words, it depends only on the number of ones in the input vector x. We can easily show that every such function is a linear combination of functions $g_j(x) = \vartheta(\sum_{i=1}^d x_i - j)$ for $j = 0, 1, \ldots, d$. Hence, ϕ_d is easily expressible by perceptrons, in particular $\phi_d \in \mathcal{P}_d(d+1)$.

For an even positive integer d, represent a vector $x \in \{0,1\}^d$ as a pair of two vectors of length $d/2$. Let x_l and x_r denote the leftmost and the rightmost $d/2$ coordinates of x, respectively. Define a function $\beta_d : \{0,1\}^d \rightarrow \{-1,1\}$ by a $\beta_d(x) = (-1)^{x_l \cdot x_r}$. It was shown by Rothaus [10] that for every even positive integer d, β_d is a bent function. It is not known whether β_d may be approximated with a small error by function from $\mathcal{P}_d(n)$ for n bounded by a polynomial in d. The next two statements at least imply that such an approximation cannot be derived from Theorem 2. Using the result by Hajnal *et al.* [2], we establish a lower bound on $V(\beta_d, \mathcal{H}_d)$, where $\mathcal{H}_d$ is the set of functions of the form $\vartheta(v \cdot x + b)$, where $b \in \mathcal{R}$ and $v \in \mathcal{R}^d$.

Lemma 1 (Hajnal *et al.*) *For every even integer d and for every $g \in \mathcal{H}_d$ we have $|\beta_d \cdot g| \leq 2^{5d/6}$.*

Hajnal *et al.*[2] used this lemma to prove an exponential lower bound on the number of hidden units needed to compute β_d in any neural network of the following type: the network has one layer of hidden units. Both the hidden units and the output unit are Heaviside perceptrons and all output weights of the hidden layer (i.e. input weights of the output perceptron) are integers bounded by a polynomial in d. It is not known whether allowing networks with arbitrary real input weights in the output perceptron would considerably decrease this exponential lower bound.

Using Lemma 1, it is possible to prove the following lower bound on the variation with respect to half-spaces of β_d.

Theorem 6. *For every even positive integer d we have $V(\beta_d, \mathcal{H}_d) \geq 2^{d/6}$.*

Since $\|\beta_d\|_2 = \max_{g \in \mathcal{H}_d} \|g\|_2$, Theorem 2 implies the existence of $f_n \in \mathcal{P}_d(n)$ satisfying $\|\beta_d - f_n\|_2 \ll \|\beta_d\|_2$ only if n is exponentially large with respect to d.

5 Acknowledgements

This work was partially supported by GA AV grants A2030602, A2075606 and GA ČR 201/95/0976.

References

[1] A. R. Barron. Universal approximation bounds for superpositions of a sigmoidal function. *IEEE Transactions on Information Theory*, 39:930–945, 1993.

[2] A. Hajnal, W. Maass, P. Pudlák, M. Szegedy, and G. Turán. Threshold circuits of bounded depth. In *Proceedings of the 28th Annual Symposium on Foundations of Computer Science*, pages 99–110. IEEE Computer Society Press, 1987.

[3] K. Hlaváčková and V. Kůrková. Rates of approximation of real-valued Boolean functions by neural networks. In *Proceedings of ESANN'96*, pages 167–172. Bruges 1996, Belgium, 1996.

[4] Y. Ito. Finite mapping by neural networks and truth functions. *Math. Scientist*, 17:69–77, 1992.

[5] L. K. Jones. A simple lemma on greedy approximation in hilbert space and convergence rates for projection pursuit regression and neural network training. *Annals of Statistics*, 20:601–613, 1992.

[6] V. Kůrková. *Dimension–independent rates of approximation by neural networks.* Birkhauser, 1997. in press.

[7] V. Kůrková, P. C. Kainen, and V. Kreinovich. Estimates of the number of hidden units and variation with respect to half–spaces. *Neural Networks*, 1997. in press.

[8] E. Kushilevitz and Y. Mansour. Learning decision trees using the fourier spectrum. In *Proceedings of 23rd STOC*, pages 455–464, 1991.

[9] H. N. Mhaskar and C. A. Micchelli. Dimension-independent bounds on the degree of approximation by neural networks. *IBM Journal of Research and Development*, 38(3), May 1994.

[10] O. S. Rothaus. On "bent" functions. *J. Combin. Theory, Ser. A*, 20:300–305, 1976.

[11] T. J. Sejnowski and C. Rosenberg. Parallel networks that learn to pronounce english text. *Complex Systems*, 1:145–168, 1987.

[12] H. J. Weaver. *Applications of discrete and continuous Fourier analysis.* John Wiley, New York, 1983.

A Method for Task Allocation in Modular Neural Network with an Information Criterion

H.-H. Kim and Y. Anzai
Dept. of Computer Science, Keio University,
3-14-1 Hiyoshi, Kohoku-ku, Yokohama 223, Japan
Email : {kim,anzai}@aa.cs.keio.ac.jp

Abstract

It is well known that large-scale neural networks suffer from serious problems such as the scale problem and the local minima problems. Modular architecture neural network is an approach to alleviate these problems. It is important that the construction of modular neural network is the selection or construction of a network that can converge and has the good generalization ability for a task, and the Akaike Information Criterion (AIC) is a criterion of evaluation of estimated model from observed parameters is a very useful tool for selection of network. This paper proposes a method for task allocation in a modular architecture neural network. The method allocates a best fit network that has a good generalization ability from multiple neural networks for a task with (AIC) and the state of convergence of a network, simply and certainly. The performance of proposed method is evaluated with the Fisher' Iris data and the What and Where vision tasks.

1 Introduction

There are many different types of artificial neural networks, each of which has different strengths particular to their application. The abilities of different networks can be related to their structure, dynamics, and learning methods. It is well known that, however, the large-scale neural networks suffer from serious problems such as the scale problem and the local minima problems. Modular architecture neural network is an approach to alleviate these problems. In addition, previous researchers have shown that the modular neural network has many advantages in terms of speed and generalization capabilities [1,2,3,4,5].

Generally, construction of a modular neural network is achieved by three steps. First of all, the task has to be decomposed into sub-tasks using some *a priori* knowledge on the task, then the neural modules (we call them sub-networks) have to be properly organized considering the sub-tasks, and, finally a way of interaction between modules has to be integrated in the whole architecture. Jacobs *et al.* [2] proposed a model that consists of some expert networks and a gating network. The expert networks compete to learn the training patterns, and the gating network mediates this competition. After training, each expert network computes different task. Namely, through the competitive learning using a gating network, a sub-task is allocated to an expert network.

In this paper, we propose a method for task allocation in modular neural network with AIC and the state of convergence of the networks to given task. The idea is that if a network cannot converge a task then the network is excepted from the candidate networks for the task through the competition on the gating network. The dynamics of the proposed method is that firstly, some networks are prepared for given task, next all the networks start learning for the given task at the same time and the AIC of all networks are calculated on each epoch of learning, then the desired values of gating network is defined by these AIC of all networks. Finally, when the gating network finishes the learning, a network will be selected. The property of the proposed method can be explained as follows. The proposed method selects a best fit network using AIC and the state of convergence that were approximately observed and computed through learning course. The method does not need pre-learning to calculate AIC on the maximum log likelihood of all sub-networks and its learning dynamics are very simple.

2 Akaike Information Criterion

AIC was provided to evaluate an estimated model from observed parameters as a statistical quantity by Akaike [6]. The goodness of value of parameters of a specific network can be measured by the expected log likelihood. The larger the expected log likelihood the better the values of parameters. AIC is defined by

$$AIC = -2(MLL) + 2F, \quad (1)$$

where MLL is the maximum log likelihood, and F is the number of free parameters of a network. A network that minimizes AIC is considered to be the most appropriate parameters. Equation (1) implies that when there are several networks whose values of MLL are about the same level, a network that has the smallest number of free parameters is selected by the second term of right side.

3 Modular Neural Network and Task Allocation

In this paper, the modular architecture that was introduced by Jacobs *et al.* [2] is used, and back propagation learning algorithm is used for learning of all sub-networks and a gating network.

3.1 Computing AIC of Sub-Network

AIC of each sub-network is obtained by calculating the log likelihood of the sub-network for given data. When a sub-network has I input nodes, H hidden nodes, and O output nodes, the dynamics of the neural network can be described by

$$\Psi_h = \sum_{i=1}^{I} W_{hi}x_i - \Theta_h, \quad (2)$$

$$y_h = f(\Psi_h) = \frac{1}{1+exp(-\Psi_h)}, \quad (3)$$

$$\Phi_o = \sum_{h=1}^{H} W_{oh}y_h - \Theta_o, \quad (4)$$

$$z_o = f(\Phi_o) = \frac{1}{1+exp(-\Phi_o)}, \quad (5)$$

where y_h is the output values of hidden units, z_o is the output values of output units, Ψ_h is internal states of hidden units, Φ_o is internal states of output units, W_{hi} is connection weights between the hidden units and input units, W_{oh} is connection weights between the hidden units and output units, Θ_h and Θ_o are threshold values of the hidden units and output units. Also this network has free parameters that consist of $I \times H + H \times O$ weights and $H + O$ threshold values. Suppose that the pth input data of the training data N is described as x_{pn}, $(p = 1,..,P, n = 1,..,N)$, its output vector is defined as z_{pk}, $(k = 1,..,K)$, and the desired values of each input data is defined as t_{pk}, where P is the number of training data, N is the degree of each data and K is the number of output nodes. Then the likelihood, L, of this network that the output is a desired response for each training data can be computed by

$$L = \sum_{p=1}^{P}\sum_{k=1}^{K} z_{pk}^{t_{pk}}(1-z_{pk})^{(1-t_{pk})}. \quad (6)$$

The log likelihood of L is

$$l = -\sum_{p=1}^{P}\sum_{k=1}^{K}\{(1-t_{pk})\Phi_{pk} + log(1+exp(-\Phi_{pk}))\}. \quad (7)$$

l is maximized through updating the network parameters on training process. Next, in order to judge whether a sub-network is fitted to given task or not, the convergent state, E_p, of the sub-network is added to Equation (7).

$$E_p = 1 - \sum_{k=1}^{K}(t_{pk} - Z_{pk})^2. \quad (8)$$

Then, the log likelihood of a sub-network is changed as follows.

$$\bar{l} = -\sum_{p=1}^{P}\sum_{k=1}^{K}\{(1-t_{pk})\Phi_{pk} + log(1+exp(-\Phi_{pk})) - E_p\}. \quad (9)$$

Finally, AIC of the sub-network can be computed by

$$AIC = -2\bar{l} + 2(H \times I + H \times O + H + O). \quad (10)$$

3.2 Task Allocation

The gating network allocates tasks to sub-networks as follows. Let's define the adaptability of the ith sub-network for a task as $AIC_i(t)$ at the time t. The adaptability of all sub-networks are computed

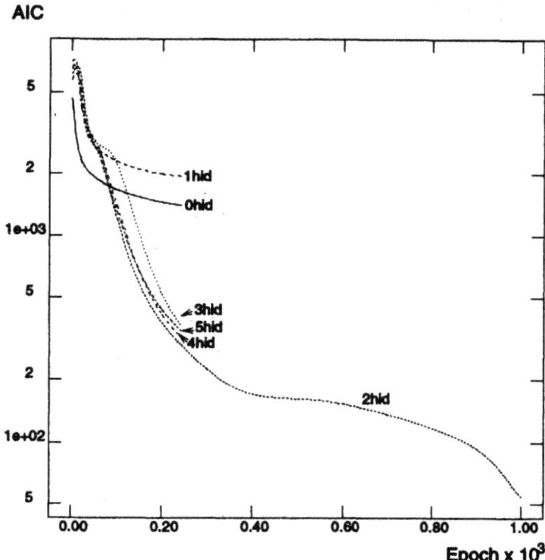

Figure 1: Network selection for Iris data classification. This shows that the variation of AIC of all sub-networks through the training course. At epoch 235, the network that has 2 hidden nodes is selected for the task and the other networks stop the learning.

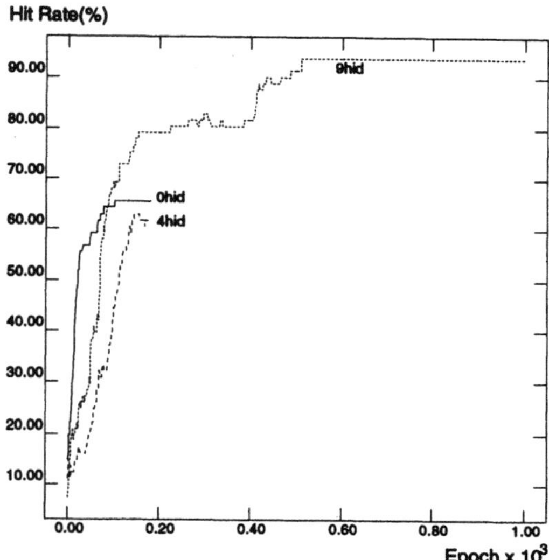

Figure 2: The generalization ability of all sub-networks for what task, the network that has 9 hidden nodes is the best network.

as in Equation (10). The desired values, $g_i^*(t)$, of the output units of gating network are decided as follows,

$$g_i^*(t) = \begin{cases} 1, \min \{AIC_i(t) | (i = 1, ...N)\} \\ 0, \text{otherwise.} \end{cases} \quad (11)$$

And the gating network uses G to learn tasks as an evaluation function.

$$G = \frac{1}{N} \sum_{i=1}^{N} (g_i^* - g_i)^2, \quad (12)$$

where N is the number of sub-networks. When the evaluation function G is satisfied with the convergent condition, a task is allocated to a best fit sub-network, and only the sub-network is training by satisfying the convergence condition of itself. And the others are renewed to learning the another task.

4 Simulation

Simulations of the proposed method were applied to Fisher' Iris data and the What and Where vision tasks used in [2]. We used backpropagation learning algorithm for learning of all sub-tasks and gating network. The weights of all networks were updated at each time step. Desired output values of 0 and 1 were used. The weights of the all networks were initialized with values randomly selected from a uniform distribution over the interval [-0.3, 0.3]. The learning momentum of 0.1 and learning factor of 0.45 were used. Selecting an optimal network for a data classification task was simulated using Fisher' Iris data. The data set contains 3 classes of 50 instances each, where each class refers to a type of iris plant. One class is linearly separable from the other two, the others are not linearly separable from each other. In order to test the selection of the optimal network for the data classification task, we prepared some networks that have 4 input nodes, 3 output nodes, and 0 to 5 hidden nodes each. The number of training data is 75 instances that consist of 25 instances from each class. Figure 1 shows the network selected for this task that has 2 hidden nodes.

The network that has no hidden nodes or 1 hidden node did not converge, but the other networks did. The generalization ability of the networks that have more than 2 hidden nodes was about the same. In this case, the network having 2 hidden nodes was selected, because it has the smallest free parameters.

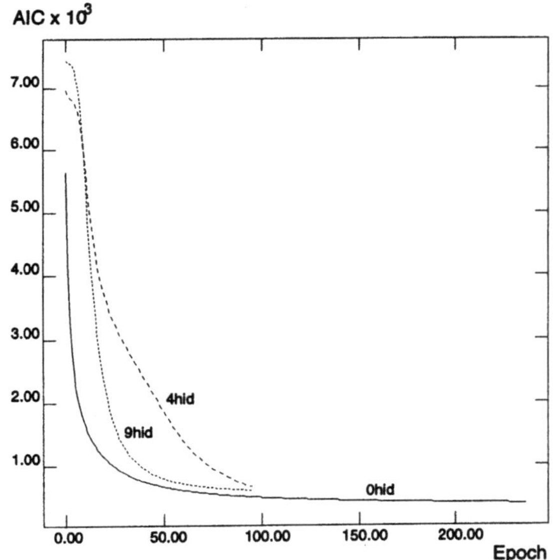

Figure 3: The variation of AIC of all sub-networks for where task. At epoch 95, the network that has no hidden layer is allocated to where task. The network is a winner whose AIC is the smallest from the start of the training.

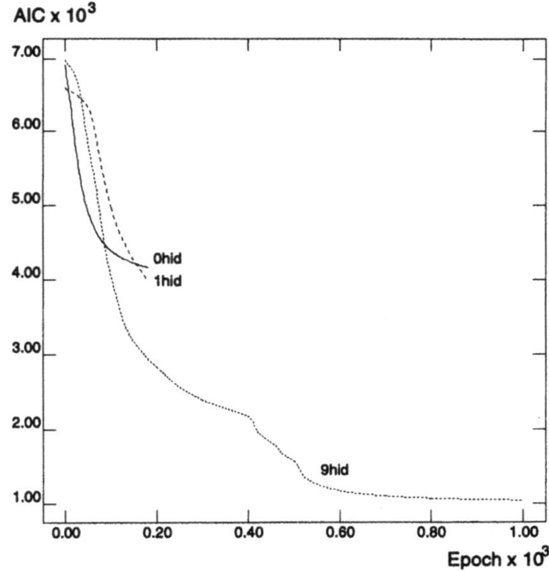

Figure 4: The variation of AIC of all sub-networks for what task. At epochs 7 and 88, the winner network is changed. The gating network satisfied the convergent condition at epochs 181, and the network that has 9 hidden nodes is selected for what task.

Next, the task decomposition and allocation was applied to the What and Where vision tasks. The What sub-task is to recognize the object and the Where sub-task is to localize it. The input space corresponds to a retina of matrix of 5×5. Each object is a shape which can be defined in 3×3. There are nine positions and shapes possible. At each time step of the simulation, one of the nine objects is located at one of the nine positions on an input space. The two sub-tasks are allocated to the modular network that is composed by three sub-networks. One of them has no hidden layer and another networks each has 4 hidden nodes and 9 hidden nodes. All sub-networks have 27 (5×5 data +2 task identifier values) input nodes and 9 output nodes. The gating network has 2 input nodes, 3 output nodes and no hidden layer. We defined the task identifier as 0 1 (what task) and 1 0 (where task). The result of the simulation shows that the Where sub-task was allocated a network that has no hidden layer, and What sub-task was allocated to a network that has 9 hidden nodes. Because the Where sub-task can be linearly separated and then does not need a hidden layer, whereas What sub-task is a non-linear problem, therefore implies a more complex network structure than the Where sub-task.

The learning adaptability of all sub-networks for both tasks is shown in Figure 3 and 4. They show that the learning of the network, that is best fit to a sub-task, is continued till satisfies the convergent condition, and other sub-networks stop the learning at the point that a sub-task is allocated to a sub-network. Figure 2 shows the generalization ability of all sub-networks for What task.

The learning curve of the gating network for What task is shown in Figure 5. In this case there are two changing points of the winner network at epochs 7 and 88.

5 Conclusion

The use of modular neural network leads to so many advantages, and we think that the usefulness of modular neural network will be expanded more and more. In this paper, we proposed a method for task allocation in modular neural network using Akaike information criterion and the state of convergence of network. The method allocates a task through

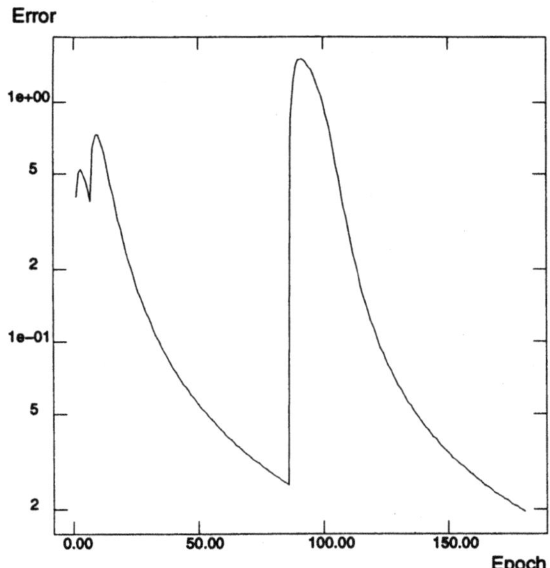

Figure 5: Task learning curve of the gating network for what task. This shows the change of the winner network at epochs 7 and 88. It means that the change of the winner network is the change of the desired values of output units of gating network.

competitive learning of all sub-networks with the gating network, to a best fit network. The method has some merits such that it can allocate a task to a best fit network in the given environment quickly and the training of the gating network is very simple.

However, some important matters remain to be solved to enable a systematic use of modular neural network. It is that the task decomposition method and interaction between the modules (subnetworks) lack formalization. And one of the important problems is that if there is not a network that can train a sub-task, the goal task cannot be run. These matters have to be solved to apply modular neural network to practical application more easily and systematically. Those are left as our future work.

References

[1] B. L. M. Happel and J. M. J. Murre, 'Design and Evolution of Modular Neural Network Architectures', Neural Networks, Volume 7, pages 985—1004, 1994.

[2] R. A. Jacobs, M. I. Jordan and A. G. Barto, 'Task Decomposition Through Competition in a Modular Connectionist Architecture: The What and Where Vision Tasks', Cognitive Science 15, pages 219—250, 1991.

[3] R. A. Jacobs and M. I. Jordan, 'A Competitive Modular Connectionist Architecture', Neural Information Processing Systems, Volume 3, pages 767—773, 1991.

[4] S. J. Nowlan and G. E. Hinton, 'Evaluation of Adaptive Mixtures of Competing Experts', Neural Information Processing Systems, Volume 3, pages 774—780, 1991.

[5] E. Ronco and P. Gawthrop, 'Modular Neural Networks: A State of the Art', CSC-95026, University of Glasgow, UK, 1995.

[6] Y. Sakamoto, M. Ishiguro and G. Kitagawa, 'Akaike Information Criterion Statistics', D. Reidel Publishing Company, 1986.

A Meta Neural Network Polling System for the RPROP Learning Rule

C. McCormack,
Department of Computer Science, University College Cork,
Cork, Ireland.
E-mail: colin@odyssey.ucc.ie

Abstract

This paper proposes an application independent method of automating learning rule parameter selection using a group of supervisor neural networks, known as meta neural networks, to alter the value of a learning rule parameter during training. Each meta neural network is trained using data generated by observing the training of a neural network and recording the effects of the selection of various parameter values. A group of meta neural networks is then polled to obtain a parameter value for a learning rule. Experiments are undertaken to see how this method performs by using it to adapt a global parameter of RPROP.

1 Introduction

Despite the development of more efficient learning rules it remains necessary to manually select appropriate learning rule parameter values in order to achieve an acceptable solution. The learning rule parameters which yield a performance of highest quality (where quality can be defined as the speed of convergence and the accuracy of the resultant network) are usually specific for each problem with no effective method of judging what parameter value is suitable for which problem. This is a significant shortcoming in the area of learning rule usage as selection of an inappropriate parameter can have a marked effect on the performance of most learning rules [1]. Even learning rules which have been shown to be tolerant to arbitrary selection of initial learning rule parameters, such as RPROP [5], are inhibited by inappropriate parameter selection [1, 2, 5]. In most learning rules parameters are initialised with general values which are suggested by the rules authors. These general values are those which have been found to deliver the best results on average and may not be the ideal parameter values for every problem domain they are used for. A learning rule is therefore constrained in its performance by a non-deterministic process of selecting a parameter value.

This paper investigates a method of parameter adaptation which involves the use of a group of separate neural networks (called meta neural networks) to select appropriate values for the η^- parameter of the RPROP learning rule. We look at the results obtained when a standard RPROP rule and a group of Meta Neural Networks are applied to three benchmark problems.

1.1 RPROP

Resilient backpropagation (RPROP) [5] is a local adaptive learning scheme. In it the sign of the derivative is taken to indicate the direction of the weight update. The weight update values (unique for each weight) are adjusted as training is carried out. The parameters η^- and η^+ are used to adjust the weight update value.

2 Meta Neural Network Description

The experiments performed investigate the effectiveness of using a group of meta neural networks to adapt the η^- parameter of RPROP. A meta neural network (MNN) is a form of supervisor neural network which makes suggestions to a conventional learning rule on the values of various parameters. Earlier work using individual MNNs [1, 2] to adapt a learning rules parameter value indicated that the approach was successful but not always guaranteed to produce a consistent result. This paper proposes a polling system which uses a group of three different MNN's in an attempt to improve the operation of a MNN based system.

2.1 MNN Methodology

There are three stages involved in training a MNN.

1. A backtracking system is set up which allows a learning algorithm to see the results of the

selection of the next learning rule parameter value (i.e. the value of η^- for RPROP). This method is known as backtracking since it allows the learning rule to backtrack from a parameter value choice that does not lead to a short term decrease in error value. At each epoch a potential parameter value is evaluated with the value leading to the greatest reduction in the training set error being retained and the training process continued. The value of the parameter is limited to six values, in the range 0.3 to 0.8. This range of parameter values has been determined from prior experimentation as the set of values from which the 'optimal' (or best available) parameter for a problem is usually found. At each epoch the parameter value is allowed to increase or decrease by one step (where a step is a factor of 0.1) in its range or remain at the current value, these three values are then evaluated. After the evaluation the results are used to augment a set D with the current network slope, the previous network slope and the action (increment/maintain/decrement) which produced the best value of the parameter (i.e. the value of the parameter which caused the largest reduction in error).

2. The set D is used as a training set for a MNN, where the inputs are: current network slope, previous network slope and the output is a single value which indicates whether the value of the parameter increased, decreased or remained the same.

3. At each epoch the learning rule passes the value of the current network slope and the previous network slope to the MNN which suggests an increase/decrease or no change in the value of the parameter.

2.2 Polling the Meta Neural Networks

Individual MNNs can produce outstanding performances [1, 2], however they are not always consistent when applied to problems outside the domain they were trained on. In an effort to produce a system which would make the best possible suggestion for a parameter alteration a scheme in which a set of MNNs is combined is evaluated. Each MNN in the scheme receives information about the current and past slopes for the network ($S(t)$ and $S(t-1)$) and each MNN then contributes an opinion as to the action that should be performed (i.e. raise, lower or maintain the value of η^-). The autonomous nature of a MNN means that the individual MNNs can operate in a parallel environment. A polling program receives each MNN's suggestion for the next value of η^- and returns the value of η^- which received the most 'votes' to the RPROP learning rule. It is possible to combine any number of MNNs in this scheme. For the experiments conducted the three MNNs trained (Building, Thyroid and Heart) were polled. The method differs from that of other parallel methods of parameter selection by network duplication (the Spy system [3] for example) in that the MNNs are pretrained and thus incur minimum processing overhead. The system is also flexible as regards the number of MNNs that are polled.

2.3 Benchmark Problem Description

Three benchmark problems, the Thyroid problem, the Building problem and the Heart problem are used to evaluate the effectiveness of the proposed system. Three of the problems are also used to produce data for the MNN's which are in turn trained and evaluated. The problems are taken from a comprehensive study of neural network benchmarks [4] and are referred to in [4] as 'Thyroid 1', 'Building 1' and 'Heart 1'. All initial learning rule and network architecture parameters were fixed apart from the initial weight set which was random for each network.

3 Experiments

In the set of experiments undertaken groups of thirty networks were trained using the standard RPROP method to gauge the average performance. Additional standard RPROP learning algorithms were trained using 'optimal' parameter values derived from previous experimentation. Each MNN is trained using a set D derived from backtracking on a particular problem.

3.1 Result Evaluation

The set of available examples is divided into three sets: a training set is used to train the network, a validation set is used to evaluate the quality of the network during training and to measure overfitting, finally a test set is used at the end of training to evaluate the resultant network. In the series of experiments undertaken 50% of the problems total available examples are allocated for the training set, 25% for the validation set and 25% for the test set.

The error measure, E, used was the squared error percentage [4], this was derived from the normalisation of the mean squared error to reduce its dependence on the number of coefficients in the problem representation and on the range of output values used.

Training progress P [4] is measured after a training strip of length k, which is a sequence of k epochs numbered $n+1, \ldots, n+k$ where n is divisible by k:

$$P_k(t) = 1000 \cdot \left(\frac{\sum_{t' \in t-k+1 \ldots t} E_{tr}(t')}{k \cdot \min_{t' \in t-k+1 \ldots t} E_{tr}(t')} - 1 \right),$$

where E_{tr} is the training set error. In the experiments detailed in this paper $k = 5$. The training progress gives the extent of the difference between the average training set error in the strip and the minimum training set error and is used to determine when the network has reached a point where no further training is effectively taking place. In the experiments performed in this paper training is halted when the progress P drops below 0.1.

4 Results

The networks trained using the MNN to suggest parameter values are known as 'Thyroid MNN' (i.e. the MNN was trained using results obtained from training on the Thyroid problem), 'Building MNN' and 'Heart MNN'. The suggestions of each MNN are passed to a polling program which decides which value of η^- to use. The results for individual MNN are included in this paper to illustrate their effectiveness and the effectiveness of combination of groups of MNN.

Networks were trained for the normal RPROP parameter ($\eta^- = 0.5$), the parameter for η^- which was found to be optimal from trial and error experiments, three MNNs and a polled set of MNN. The 'optimal' η^- values which resulted in the most accurate results obtained by trial and error in previous experiments were: $\eta^- = 0.3$ for the Thyroid problem, 0.5 for the Building problem and 0.5 for the Heart problem, these are listed in the results under the key 'Optimal RPROP'. No 'optimal' results are shown for the Building or Heart problems since these problems 'optimal' η^- values were equal to the normal η^- value.

The average of the errors at the cessation of training for the training set (Train), validation set (Validation) and the test set (Test) are presented in Figures 1-3. The standard deviation for the validation set is illustrated in the form of an error bar. The validation set standard deviation was chosen because it was usually the most significant and gave a good idea of the consistency of the learning rule used. The average number of epochs taken to reach cessation of the training process is included in the figures as a column. The keys used are 'Normal' for RPROP using the normal parameter, 'Optimal' for RPROP using the 'optimal' parameter, 'Thyroid MNN', 'Building MNN' and 'Heart MNN' for the MNN using the individual training sets and 'Poll MNN' for the MNN using a polling system.

4.1 Results for the Building problem

For the building problem (Figure 1) the results in general for the MNNs are quite satisfactory. A network aided by a MNN will produce significantly lower validation set errors. The standard deviations for the MNNs are also lower. One point of interest however is the fact that the MNNs are taking longer to reach termination (significantly so in the case of the Thyroid and Heart MNN). Using a polled MNN system results in an averaging of performance of the MNNs.

4.2 Results for the Thyroid problem

For the thyroid problem (Figure 2) we can see that using an optimal value for the parameter does not have a significant impact on the results. Using a MNN aided scheme however improves the results significantly. For a polled MNN scheme the results are also considerably better than a normal or optimal parameter value and represent an average of the other MNNs performance. The standard deviations for the polled MNN system are as low as the other MNNs indicating it is a consistent performer.

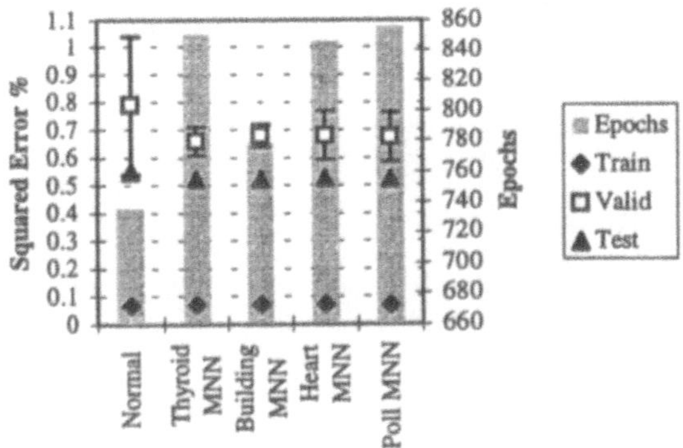

Figure 1: Results for the building problem.

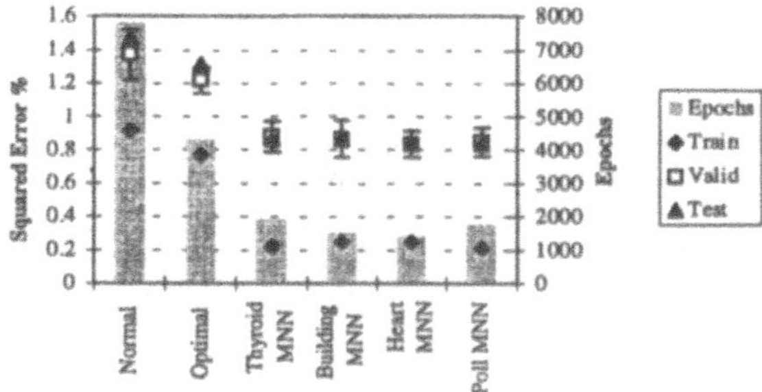

Figure 2: Results for the thyroid problem.

4.3 Results for the Heart problem

Results for the heart problem (Figure 3) show that the performance of a MNN aided network is superior to that of a normal parameter value. The polled MNN scheme performs well although not as well as some of the MNNs. The standard deviations are as low or lower for the polled MNN as the other MNN.

4.4 Summary of Results

The aim of a polled MNN scheme was to build a parameter adaptation scheme which contained all of the advantages of a MNN based scheme without being encumbered by any of the disadvantages of that scheme. We saw that while a MNN based scheme can be used to select a parameter as an alternative to tuning the learning rule the training strategies used by the three MNNs trained seem to be suitable for specific problems. This leads to the situation where instead of having to select the best learning rule parameter by trial and error we must now select the most appropriate MNN. The use of a system of polling MNNs overcomes that problem and shows that MNNs can now be implemented without the need to intervene in the choice of MNN.

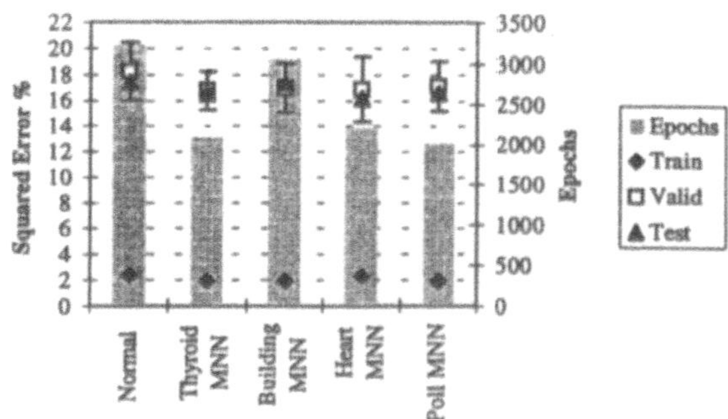

Figure 3: Results for the heart problem.

5 Conclusion

A meta neural network is a means of acquiring and using information about the learning mechanism of a neural network. It is a method of combining neural networks to improve the quality of the solution by using a neural network as an extension to a learning rule. A polled MNN learning scheme uses a number of meta neural networks to produce a consistently better training scheme. Using a scheme whereby the suggestions for the new parameter made by a MNN are polled has been shown to improve the consistency of a MNN based learning method without reducing its performance.

References

[1] C. McCormack. A study of the adaptation of learning rule parameters using a meta neural network. In *13th European Meeting on Systems and Cybernetic Research*, volume 2, pages 1043–1048, 1996.

[2] C. McCormack. Using a meta neural network for RPROP parameter adaptation. In *Proc. European Symposium on Artificial Neural Networks*, pages 7–12, 1996.

[3] I. Pitas. *Parallel Algorithms for Digital Image Processing, Vision and Neural Networks*. John Wiley and Sons, Chichester, 1993.

[4] L. Prechelt. PROBEN1: A set of neural network benchmarking rules. Technical Report 21/94, Dept. of Informatics, University of Karlsruhe, Germany, 1994. ftp://ftp.ira.uka.de/pub/neuron/proben1.tar.gz.

[5] M. Riedmiller. Advanced supervised learning in multilayered perceptrons: From backpropagation to adaptive learning algorithms. *Computer Standards and Interfaces*, 16:265–278, 1994.

Designing Development Rules for Artificial Evolution

A. G. Rust[1,2], R. Adams[1], S. George[1] and H. Bolouri[2,3]
[1] Division of Computer Science, University of Hertfordshire, UK
[2] Engineering Research and Development Centre, University of Hertfordshire, UK
[3] Biology 216-76, California Institute of Technology, USA
E-mail : a.g.rust@herts.ac.uk

Abstract

Using artificial evolution to successfully create neural networks requires appropriate developmental algorithms. The aim is to determine the least complex set of rules that allow a range of networks to evolve. This paper presents a set of generic growth rules that abstractly model the biological processes associated with the development of neuron-to-neuron connections. Substantially different 3D artificial neural structures can be grown by changing parameter values associated with the rules. A genetic algorithm has been successfully employed in determining parameter values that lead to specific neural structures.

1 Artificial Development and Evolution

In biological systems, the creation of precise neural circuitry is governed by the expression of the development programmes contained within genes. The programmes consist of rules that are believed to be simple [12], encompassing the creation of structure through chemically-driven events and the modification of structure by neural activity. The interactions of these simple rules represents developmental self-organisation, which is robust, modular and scalable. Evolutionary changes of neural systems are due to genetic modifications of these interactions between the developmental rules.

Efficient evolution of complex artificial neural systems requires the design of robust and effective developmental rules. Artificial evolution may then be used to optimise the interactions of these rules. The subject of our research is to determine a set of generic developmental rules capable of generating a variety of network architectures.

2 Modelling Development

Harnessing the self-organising principles of biological development to design artificial neural networks has received recent attention [7]. Developmental artificial neural networks (DANNs) encode and express developmental rules using a wide-spectrum of algorithmic methods. This ranges from models which are simply *ad hoc*, having minimal reference to biological processes [1], through to biologically plausible models with large numbers of rules and parameters [4].

This paper reports a DANN which has been designed to incorporate a sufficient but abstract model of biological development without explicitly modelling detailed molecular and chemical processes. Our aim is to find the least complex, computational rule sets which create large 3D networks that mimic biological structure. This is a two stage process, firstly to identify the rules and secondly to determine the optimum parameter settings which govern their actions.

3 Structure versus Function

It is unknown which developmental rules produce the best building blocks for creating artificial neural systems since function and structure are interrelated. The functionality of neurons is greatly determined by the pattern of interconnections [6] and the locations of synapses on dendritic trees [9]. Different structures may be functionally similar and can be produced using many different developmental routines: the *competing conventions problem* [2].

We aim to search for developmental rules which specifically produce structures that mimic biological neural networks. We believe that function will follow from having the 'right' architecture. By examining biological neural systems, actual neural structures can be used to directly assess grown networks.

4 The DANN Environment and Growth Rules

A more detailed analysis of the DANN can be found in [10].

4.1 Chemical Environment

Networks develop within a virtual chemical environment, which can be modelled in 2 or 3 dimensions. Axons and dendrites (neurites) grow in response to chemical stimulation in the form of gradients. The model represents a simplified version of the environment in which biological development occurs. An underlying gradient is imposed to encourage neurites to move in pre-defined directions.

4.2 Neuron Model

Neurons are modelled as discrete elements which locally interact with each other and the environment. Each neuron remains in a fixed position but a set of internal genetic rules enable them to extend neurites. The same rules are used for both axon and dendrite processes.

Neurites emit chemicals from every point along their length, which produce local chemical gradients determined by the parameters: strength, range and a diffusion law (e.g. $1/\text{distance}^n$). Axon growing neurons emit negative chemicals whilst dendritic neurons emit positive chemicals.

The tips of growing neurites sense their local chemical environment and modify their state based on the internal genetic rules. The response can be to extend, split/branch or make a connection. Movement is restricted to 8 directions in 2D and 26 in 3D.

4.3 Growth Rules

Neurite growth is governed by the following simple developmental rules:

1. *Follow the path of the steepest gradient of the opposite polarity to that emitted.* Therefore, axons perform hill climbing and dendrites gradient-descent.

2. *Maintain the same path unless attracted by a larger gradient.* Neurites are initially guided by the underlying substrate gradient before being attracted to target neurites.

3. *Split if genetically programmed or if there are two strong local gradients.* Environmental splitting is a stochastic process regulated by the local chemical gradients at the tip of a neurite. Intrinsic splitting occurs at pre-determined time intervals regulated by the neuron itself and is deterministic.

Parameters are associated with the rules which govern their individual and collective interactions.

Neurons are arranged into layers. Those in the same layer possess the same set of rules but the range and strength of the chemical emitted by each neuron may vary. Connections form when the tips of growing trails collide with either a trail, tip or cell body of another neuron.

4.4 Results

Figure 1 gives an example of two structurally dissimilar 3D neural networks grown from the same starting conditions but using different parameter sets. (Neurons are represented as spheres). This demonstrates that the rules can produce classes of developmental programmes that generate different structures; this is required in neural network design, since generally no two applications have the same network architecture.

5 Developmental Parameter Search

5.1 Genetic Algorithm Roles

The initial networks were developed using a manual search of the parameters and visual selection of desirable structures. To automate the search for optimal growth parameters the genetic algorithm (GA) is being used. The use of the GA is not however just confined to the role of modifying the interactions between development rules. Once rules describing the building blocks of development have been defined and encoded, the GA can then be used to explore the mechanisms of artificial evolution, i.e. how the rules themselves can be optimised.

Identifying optimal parameter values in DANNs requires large problem spaces to be searched. This is computationally expensive [4] and for those models which are more biologically defensible, this may ultimately be intractable [3].

Currently, we are aiming to determine the best growth parameter values as opposed to examining the evolution of developmental programmes. The format of our rules is fixed and it is the parameters of the rules which are encoded in the genome of the GA.

5.2 Fitness Function Design

The traditional testbed application for DANNs has been the design of animat controllers [7]. Animat

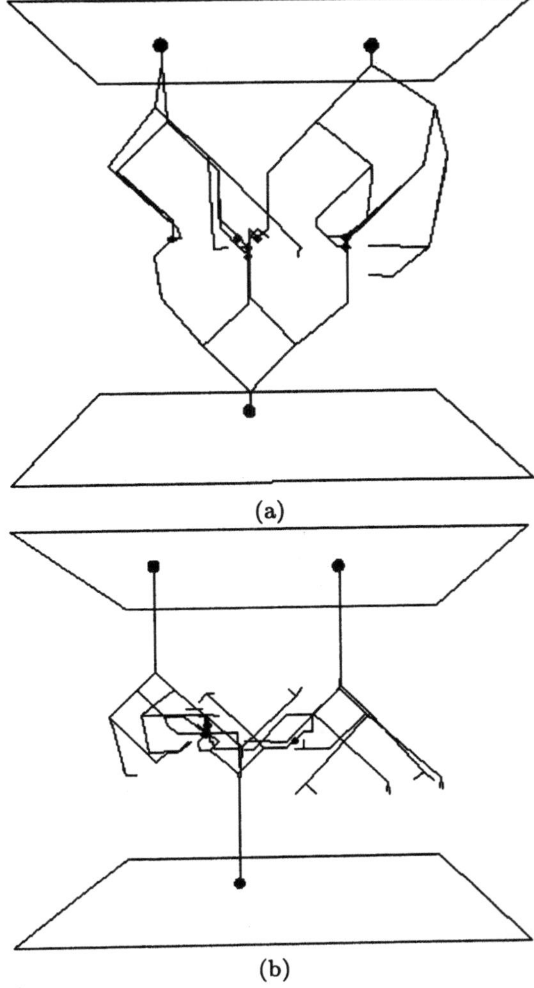

Figure 1: Examples of 3D neural growth.

fitness function design is however a complex task because of such requirements as feedback mechanisms, neural activity phase-locking and non-deterministic interactions of animats with their environment [8]. We suggest that some of these problems can be overcome by using geometric and topographic properties of biological neural structures to measure the fitness of grown networks.

In the current work the primate retina has been chosen as the 'target' neural structure used to assess the development rules of the DANN [10]. In this study triad junctions, which are found in the outer plexiform layer of the retina, are the chosen test structures. A triad junction is formed by a cone cell connecting with single dendrites from a bipolar cell and two horizontal cells.

In the DANN triads are formed using a two stage growth process. Initially, cone cells are grown to bipolar cells where any junctions formed become 'targets' for subsequent horizontal cell growth. Triad development was chosen for investigation since trying to visually assess and hand-tune parameter values of these structures is a non-trivial, inaccurate process.

The overriding aim in designing a fitness function was to maximise the formation of triad connections. This critically depends on the number and distribution of the initial cone-to-bipolar junctions. An arbitrary number of these junctions is defined to provide both a target and restraint for cone-to-bipolar cell growth. The value is not a maximum limit and is based on the size of the network. To encourage a general spread of triads across the network, the fitness function also aims to maximise the number different cones to which a bipolar cell connects. The theoretical limit for an individual bipolar cell is the entire layer of cone cells.

The fitness value is determined by three criteria: the number of triads, the sum of all the different cone cells contacted by individual bipolar cells and the closeness of the total number of initial cone-to-bipolar connections to the arbitrary, target figure. These criteria were respectively assigned the relative weightings of 4, 2 and 1 within the fitness function.

5.3 Genetic Algorithm Specification

Preliminary work on the optimisation of parameter values used the GENESIS GA package from John J. Grefenstette [5].

The genome was 80 bits in length, containing 34 variable genes that encoded parameters such as timing of intrinsic splits, the relative values of chemical gradient and the direction of environmental splits. The GA had a population of 50 sets of neurons, arranged in 3 layers representing 36 cone, 25 horizontal and 13 bipolar cells. The genomes that allowed the initial population to grow were randomly generated. A network's fitness value was assessed after all 3 layers had been fully grown.

The target number of initial cone-to-bipolar connections was 800. The crossover and mutation rates were set to 0.6 and 0.001 respectively. Recombination selection was directly proportional to the structure's fitness value. The replacement strategy was generational, in that the entire population was replaced.

5.4 Results

The results for a single run of the GA, over 57 generations is presented in Table 1. It summarises data for the percentage of triads formed from the initial cone-to-bipolar junctions, and for the average number of cones to which a bipolar cell connects.

There is a significant improvement in the chosen structural measurements, where the percentage of triads formed increases by an order of magnitude on average within the population and the result for the best individual more than quadruples. The number of different cone cells which a bipolar cell contacts also improves, more than doubling for the average case.

The fitness function had to balance the need to maximise the formation of triads against preventing the number of cone-to-bipolar connections from becoming too large. In this way, a greater percentage of triads would be formed from initial cone-to-bipolar junctions. The average figure of nearly 70% went beyond expectations. Hand-tuning parameters to gain similar performance would be non-trivial.

Analysis of the early generations of the networks showed that a range of developmental programmes were examined: from very sparse to highly branching. The speed with which the GA made structural improvements, however, suggests that there may have been premature convergence, albeit to a better than expected result. Further experimentation will consider rank based selection rather than proportional selection.

Table 1: Summary of GA results.

Criteria	Average		Best	
	Initial	Final	Initial	Final
% triads formed	7.0	66.8	18.9	76.9
Cone connections	4.1	9.0	7.8	9.2

6 Future Work

In parallel with designing developmental rules for neuron connectivity growth, we are currently examining rules that model spontaneous neural activity in biological structures. Spontaneous activity, in conjunction with other biological processes, is known to refine pre-natal neural structure during development before real world stimuli are present [11]. A similar method of rule design is being employed, such that candidate rules are specified and parameter values are subsequently optimised using the GA.

7 Conclusion

This paper shows that a set of developmental rules abstractly modelled on biological neuron interconnection growth could evolve complex, but significantly different neural-like structures. A GA found a good set of parameter values for the rules that solved a non-trivial problem. We are continuing to use the GA to model the retina, a much more complex task.

References

[1] E. J. W. Boers, H. Kuiper, B. L. M. Happel, and I. G. Sprinkhuizen-Kuyper. Designing modular artificial neural networks. Technical Report 93-24, Dept. of Computer Science, Leiden University, Netherlands, September 1993.

[2] J. Branke. Evolutionary algorithms for neural network design and training. In *Proceedings of the 1st Nordic Workshop on Genetic Algorithms and its Applications*, 1995.

[3] F. Dellaert and R. D. Beer. *A Developmental Model for the Evolution of Complete Autonomous Agents.* MIT Press / Bradford Books, 1996.

[4] K. Fleischer. *A Multiple-Mechanism Developmental Model for Defining Self-Organizing Geometric Structures.* PhD thesis, California Institute of Technology, May 1995.

[5] J. J. Grefenstette. Genesis 5.0. ftp://www.aic.nrl.navy.mil /pub/galist/src/.

[6] J. Hertz, A. Krogh, and R. P. Palmer. *Introduction to the Theory of Neural Computation.* Addison Wesley, Redwood, CA, 1991.

[7] J. Kodjabachian and J-A. Meyer. Evolution and development of control architectures in animats. *Robotics and Autonomous Systems*, 16:161–182, 1995.

[8] M. Mataric and D. Cliff. Challenges in evolving controllers for physical robots. Technical Report CS-95-184, Brandeis University, USA, November 1995.

[9] B. W. Mel. Information-processing in dendritic trees. *Neural Computation*, 6(6):1031–1085, 1994.

[10] A. G. Rust, S. George, H. Bolouri, and R. Adams. Developmental artificial neural networks for shape recognition: A model of the retina. Technical Report Technical Memorandum ERDC/1996/0011, ERDC, University of Hertfordshire, UK, May 1996.

[11] C. J. Shatz. Role for spontaneous neural activity in the patterning ofconnections between retina and LGN during visual system development. *International Journal of Developmental Neuroscience*, 12(6):531–546, 1994.

[12] M. P. Stryker. Precise development from imprecise rules. *Science*, 263:1244–1245, 4 March 1994.

Improved Center Point Selection for Probabilistic Neural Networks

D. R. Wilson and T. R. Martinez
Neural Networks and Machine Learning Laboratory, Computer Science Department,
Brigham Young University, Provo, Utah 84602, USA
Email: randy@axon.cs.byu.edu, martinez@cs.byu.edu

Abstract

Probabilistic neural networks (PNN) typically learn more quickly than many neural network models and have had success on a variety of applications. However, in their basic form, they tend to have a large number of hidden nodes. One common solution to this problem is to keep only a randomly selected subset of the original training data in building the network. This paper presents an algorithm called the reduced probabilistic neural network (RPNN) that seeks to choose a better than random subset of the available instances to use as center points of nodes in the network. The algorithm tends to retain non-noisy border points while removing nodes with instances in regions of the input space that are highly homogeneous. In experiments on 22 datasets, the RPNN had better average generalization accuracy than two other PNN models, while requiring an average of less than one-third the number of nodes.

1 Introduction

Probabilistic neural networks (PNN) [6] often learn more quickly than many neural network models such as backpropagation networks [5], and have had success on a variety of applications. PNNs are a special form of radial basis function (RBF) network [8] used for classification.

The network learns from a *training set T*, which is a collection of examples called *instances*. Each instance i has an input vector $\vec{y_i}$, and an output class, denoted as $class_i$. During execution, the network receives additional input vectors, denoted as x, and outputs the class that x seems most likely to belong to.

The probabilistic neural network used in this paper is shown in Figure 1. The first (leftmost) layer contains one input node for each input attribute in an application. All connections in the network have a weight of 1, which means that the input vector is passed directly to each hidden node.

There is one hidden node for each training in-

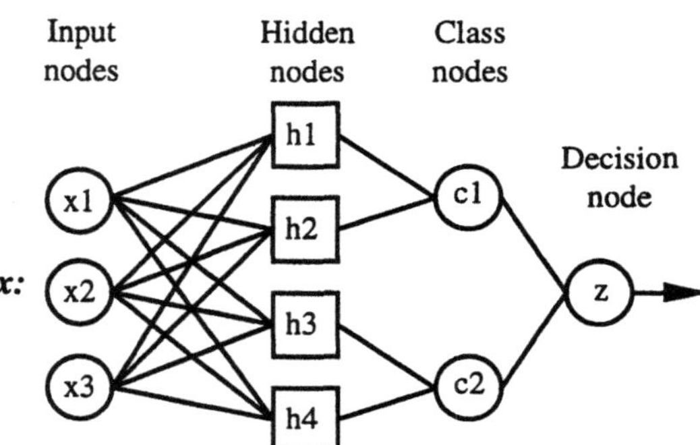

Figure 1: Probabilistic neural network.

stance i in the training set. Each hidden node h_i has a center point y_i associated with it, which is the input vector of instance i. A hidden node also has a spread factor, σ_i, which determines the size of its receptive field. There are a variety of ways to set this parameter. In this paper, we set σ_i equal to a fraction f of the distance to the nearest neighbor of each instance i. The value of f begins at 0.5 and a binary search is performed to fine-tune this value. At each of five steps the value of f that results in the highest average confidence of classification is chosen.

A hidden node receives an input vector $\vec{x}$ and outputs an activation given by the Gaussian function g, which returns a value of 1 if $\vec{x}$ and $\vec{y_i}$ are equal, and drops to an insignificant value as the distance grows:

$$g(\vec{x}, \vec{y_i}, \sigma_i) = exp[-D^2(\vec{x}, \vec{y_i})/2\sigma_i^2] \quad (1)$$

The distance function D determines how far apart the two vectors are. By far the most common distance function used in PNNs is Euclidean distance. However, in order to appropriately handle applications that have both linear and nominal attributes, we use a heterogeneous distance function HVDM [9,10] that uses normalized Euclidean distance for linear attributes and the value difference metric (VDM) [7] for nominal attributes. It is defined as follows:

$$\mathrm{HVDM}(\vec{x},\vec{y}) = \sqrt{\sum_{a=1}^{m} d_a^2(x_a, y_a)} \qquad (2)$$

where m is the number of attributes. The function $d_a(x,y)$ returns a distance between the two values x and y for attribute a and is defined as:

$$d_a(x,y) = \begin{cases} 1, & \text{if } x \text{ or } y \text{ is unknown} \\ \mathrm{vdm}_a(x,y), & \text{if } a \text{ is nominal} \\ \mathrm{diff}_a(x,y), & \text{if } a \text{ is linear} \end{cases} \qquad (3)$$

The function $d_a(x,y)$ uses the following function, based on the Value Difference Metric (VDM) [7] for nominal (discrete, unordered) attributes:

$$\mathrm{vdm}_a(x,y) = \sqrt{\sum_{c=1}^{C} \left| \frac{N_{a,x,c}}{N_{a,x}} - \frac{N_{a,y,c}}{N_{a,y}} \right|^2} \qquad (4)$$

where $N_{a,x}$ is the number of times attribute a had value x; $N_{a,x,c}$ is the number of times attribute a had value x and the output class was c; and C is the number of output classes.

For linear attributes the following function is used:

$$\mathrm{diff}_a(x,y) = \frac{|x-y|}{4s_a} \qquad (5)$$

where s_a is the sample standard deviation of the values occurring for attribute a in the training set.

Each hidden node h_i in the network is connected to a single class node. If the output class of instance i is j, then h_i is connected to class node c_j. Each *class node* c_j computes the sum of the activations of the hidden nodes that are connected to it (i.e., all the hidden nodes for a particular class) and passes this sum to a decision node. The ***decision node*** outputs the class with the highest summed activation.

One of the greatest advantages of this network is that it does not require any iterative training, and thus can learn quite quickly. However, one of the main disadvantages of this network is that it has one hidden node for each training instance and thus requires more computational resources (storage and time) during execution than many other models. When simulated on a serial machine, $O(n)$ time is required to classify a single input vector. On a parallel system, only $O(\log n)$ time is required, but n nodes and nm connections are still required (where n is the number of instances in the training set, and m is the number of input attributes).

The most direct way to reduce storage requirements and speed up execution is to reduce the number of nodes in the network. One common solution to this problem is to keep only a randomly selected subset of the original training data in building the network. However, arbitrarily removing instances can reduce generalization accuracy. In addition, it is difficult to know how many nodes can be safely removed without a reasonable stopping criterion.

Other subset selection algorithms exist in linear regression theory [4], including ***forward selection***, in which the network starts with no nodes and nodes are added one at a time to the network. Another method that has been used [2] is k-means clustering [1].

This paper presents an algorithm called the reduced probabilistic neural network (RPNN) that begins with all of the available training instances as node centers and selectively removes them one at a time until classification accuracy suffers. The algorithm tends to retain only non-noisy border points while removing nodes with instances in regions of the input space that are highly homogeneous. The next section gives details of this algorithm.

2 Reduction Algorithm

The reduced probabilistic neural network (RPNN) begins with one node per training instance, as does the original PNN, and then uses the following basic rule to determine which nodes are removed from the network:

- Remove a node if it does not cause more instances in the original training set to be classified incorrectly by the nodes remaining in the network.

In other words, if the removal of a node does not hurt classification, remove it. When applying this rule to the network, the order of removal is important. In particular, it may be desirable to remove instances far from decision boundaries first, since they have the least effect on decisions. RPNN does this by finding the distance of every instance from its nearest *enemy*, which is the nearest neighbor of a different class, and then sorting the instances by that distance. The above rule is then applied beginning with the node furthest from its nearest enemy and proceeding to that which is closest to its nearest enemy.

In order to decide if the removal of a node degrades classification accuracy, each instance in the original training set is queried to see if its classification would be altered by the removal of the instance in question.

Specifically, in our serial implementation each instance maintains a vector of activations with one activation level for each class. The removal of a particular node would subtract some amount of activation (dependent on the distance) from its own class if removed. In addition, if the removed instance I is the nearest neighbor of some other instance A, then A must find a new nearest neighbor, and update its σ accordingly, which in turn changes what effect A has on all other instances.

The change in activation due to both the removal of I and the possible change in σ of other nodes may be enough to cause the classification of some instances to change. The change can cause an instance that used to be correctly classified to be misclassified, or cause an instance that was misclassified to be correctly classified. Such changes are counted, and if the number of newly misclassified instances is less than or equal to the number of new correctly classified instances, then the removal is performed, and the changes in activation values and σ parameters are made permanent. Otherwise they are restored to their previous values.

In order to reduce the effect of noisy instances on the network, the instance corresponding to the node that is being considered for removal is not included in the tabulation. This means that a node can be removed even if its instance is itself no longer classified correctly, as long as other instances are not hurt.

To further reduce the effect of noise, a noise-reduction pass through the network is done first, beginning with the instance closest to its nearest enemy, since noisy instances are often close to instances of another class. During the noise reduction step, the criteria for removal is more strict. In order to be removed, an instance must not hurt classification, as explained above, and it must also strictly increase the average *confidence* of classification. The confidence for each node is defined as the activation of the correct class divided by the sum of activations for all of the output classes.

Noisy instances are often located near instances of another class but far from instances of their own class, so their removal will increase confidence of nearby instances classification while having a much smaller effect on instances of their own class. Other instances, however, will typically lower confidence of nearby neighbors, which are largely of the same class, while having a smaller effect on instances of different classes. Therefore, the test of confidence is appropriate during the noise reduction pass, but would prevent almost any pruning from taking place if used in the remainder of the algorithm.

3 Empirical Results

The RPNN algorithm was implemented and tested on 22 applications from the Machine Learning Database Repository at the University of California, Irvine [3].

Each test consisted of ten trials. Each trial consisted of learning from 90% of the training instances, and then seeing how many of the remaining 10% of the instances were classified correctly.

The RPNN was compared to EPNN, a standard probabilistic neural network (PNN) that retains 100% of the instances in the training set and uses a normalized Euclidean distance metric with σ set to the distance of a nodes instance to its nearest neighbor. The RPNN was also compared to HPNN, a PNN that uses the same heterogeneous distance function HVDM as RPNN , but retains 100% of the instances. HPNN and RPNN both used a dynamically-adjusted spreading factor, as explained in Section 1, and RPNN used only a subset of the available instances for generalization.

Table 1 summarizes the empirical results. For each database the table shows the average accuracy for the EPNN and HPNN using all of the instances, and for the RPNN, using the percentage of instances shown.

The last line of Table 1 shows that RPNN had the highest average accuracy over all 22 datasets of the three algorithms, while using less than one-third of the instances (on average) for generalization. RPNN 's average accuracy was slightly higher than HPNN, and both of these were substantially higher than EPNN, due in part to the use of the HVDM distance function.

Using a dynamically-adjusted spreading factor had very little effect on the accuracy of HPNN (less than 1% on average), but resulted in a large improvement on RPNN (75.5% accuracy instead of 71.5%) as well as improved size reduction.

The success of RPNN varies depending upon the application. For example, on the *Vowel* dataset, it retained almost two-thirds of the instances while

Table 1: Generalization accuracy of PNN and RPNN.

Dataset	EPNN	HPNN	RPNN	(size)
Anneal	76.2	76.2	94.9	38.3
Audiology	36.0	57.0	54.0	38.9
Australian	80.1	83.8	79.7	27.5
Breast Cancer (WI)	97.0	94.7	92.9	20.0
Bridges	52.4	55.2	57.3	21.4
Crx	75.4	84.4	82.3	27.2
Echocardiogram	78.0	76.8	90.9	9.0
Flag	45.7	53.6	47.0	35.7
Heart (Hungarian)	64.0	66.7	80.3	25.1
Heart (More)	46.0	68.8	71.8	21.3
Heart	80.7	81.5	73.3	24.8
Heart (Swiss)	38.9	93.5	78.3	1.4
Hepatitis	79.3	80.6	77.3	15.3
Horse-Colic	67.1	67.1	68.7	17.8
Iris	94.0	91.3	94.7	44.2
Liver-Bupa	62.5	66.6	57.6	29.2
Pima-Indians-Diabetes	76.3	74.1	67.2	30.3
Promoters	54.3	84.5	88.7	52.5
Soybean-Large	13.0	13.0	49.9	51.5
Vowel	92.0	92.4	84.7	65.2
Wine	94.4	97.2	92.2	40.7
Zoo	78.9	71.1	82.2	26.1
Average	67.4	74.1	75.7	30.2

suffering a large drop in accuracy compared to the other two models. However, in the *Echocardiogram* dataset, the RPNN used only 9% of the data while improving generalization accuracy by over 12%. Future research will focus on identifying characteristics of applications that help determine whether the RPNN model is appropriate.

It should be noted that these datasets are not especially large (only a few hundred instances in most cases), and that the reduction in size can be even more dramatic when there are more instances available. This is especially true when the number of instances is large compared to the complexity of the decision surface.

4 Conclusion

The RPNN reduces the size and execution time of a PNN by removing nodes from the network that are estimated to be least needed for proper generalization. It tends to retain non-noisy border points in the input space while removing nodes that are either noisy or have centers that are far from the decision boundaries. By so doing, it can fairly quickly find a reasonable subset of nodes to include in the PNN, thus reducing network complexity and execution time, as well as reducing sensitivity to noise.

The RPNN requires $O(n^2)$ time for learning on a serial machine, but only $O(n \log n)$ time in a parallel network, and in our experiments on 22 datasets reduced storage by over two-thirds on average.

It is possible that the RPNN could achieve even better size reduction as well as more robust accuracy by employing search techniques such as genetic algorithms after initial pruning. Such search techniques could find additional nodes to remove, fine tune the spreading factor of individual nodes, and even adjust the nodes center points. Future research will address this question, and continue to seek improved size reduction techniques. The results of this study are encouraging and show the potential for substantial reduction without sacrificing generalization ability.

References

[1] J.A. Leonard, M.A. Kramer, and L.H. Ungar. Using radial basis functions to approximate a function and its error bounds. *IEEE Transactions on Neural Networks*, 3(4):624–627, 1992.

[2] J. MacQueen. Some methods for classification and analysis of multivariate observations. In *Proceedings of the Fifth Berkeley Symposium on Mathematics, Stistics and Probability*, pages 281–297, Berkeley, CA, 1967.

[3] C.J. Merz and P.M. Murphy. UCI machine learning databases. Technical report, Irvine, CA: University of California, Department of Information and Computer Science, 1996. Internet: http://www.ics.uci.edu/-mlearn/MLRepository.html.

[4] O.J. Rawlings. *Applied Regression Analysis*. Wadsworth and Brook/Cole, Pacific Grove, CA, 1988.

[5] D.E. Rumelhart and J.L. McClelland. *Parallel Distributed Processing*. MIT Press, 1986.

[6] D.F. Specht. Enhancements to probabilistic neural networks. In *Proceedings of the International Joint Conference on Neural Networks*, volume 1, pages 761–786, 1992.

[7] C. Stanfill and D. Waltz. Toward memory-based reasoning. *Communications of the ACM*, 29, 1986.

[8] P.D. Wasserman. *Advanced Methods in Neural Computing*. Van Nostrand Reinhold, New York, NY, 1996.

[9] D.R. Wilson and T.R. Martinez. Heterogeneous radial basis functions. In *Proceedings of the International Conference on Neural Networks*, volume 2, pages 1263–1267, 1996.

[10] D.R. Wilson and T.R. Martinez. Improved heterogeneous distance functions. *Journal of Artificial Intelligence*, 6(1):1–34, 1997.

The Evolution of a Feedforward Neural Network trained under Backpropagation

D. McLean, Z. Bandar and J. D. O'Shea
The Intelligent Systems Group, The Manchester Metropolitan University
Manchester, UK.
Email: z.bandar@doc.mmu.ac.uk

Abstract

This paper presents a theoretical and empirical analysis of the evolution of a feedforward neural network (FFNN) trained using backpropagation (BP). The results of two sets of experiments are presented which illustrate the nature of BP's search through weight space as the network learns to classify the training data. The search is shown to be driven by the initial values of the weights in the output layer of neurons.

1 Introduction

The problem of training FFNN involves finding a mapping that approximately transforms all the input vectors in the training set in to their associated class. This is broken down in to a set of transformation subproblems that must be solved at each layer of neurons. If the network is pre-constructed prior to training (as in BP) then each of the transformation subproblems is individually incomplete, i.e. either the desired domain, range or both is unknown. This is known as the credit assignment problem [3].

2 Classification by Function Mapping in FFNN

Firstly an FFNN's ability to approximate a function mapping of arbitrary complexity through consecutive spatial transformations is defined. These transformations are implemented using decision regions constructed from the hyperplanes of neuronal layers.

A set of data T consists of a number p of tuples of arity 2, $t(v,c)$ where v is a set of n attributes $\{I_1, I_2, .., I_n\}$ from the domain, from which T is taken, that describe a single object or occurrence and c is the associated class or set of outcomes $\{\zeta_1, \zeta_2, .., \zeta_m\}$.

$T = \{t_1, t_2 ..., t_p\}$

$t_\alpha \in T \mid t_\alpha = (v_\alpha = \{I_1^\alpha, I_2^\alpha, ..., I_n^\alpha\}, c_\alpha = \{\zeta_1^\alpha, \zeta_2^\alpha, ..., \zeta_m^\alpha\})$

$\alpha = 1..p,$

$I_n^\alpha \in D$, where $D = \{0,1\}$ or $D = \Re$, the set of real numbers

$\zeta_m^\alpha \in \{0,1\}$.

The task of learning from this data set can be considered as finding an approximation F', of a mapping function F which transforms any set of attributes from the domain v, to its corresponding set of outcomes c. For an example α from the training set:

$$F'(T) \approx F(T) : v_\alpha \longrightarrow c_\alpha \quad \text{for } \alpha = 1, \ldots, p.$$

The function F may be linearly inseparable and thus require multiple layers of neurons to approximate it. In this case the transformation is broken down into simpler mapping functions where each layer of neurons provides a further abstraction until the overall mapping is accomplished.

$$F'(t_\alpha) : v_\alpha \xrightarrow{f_1} h_\alpha^1 \xrightarrow{f_2} h_\alpha^2 ... h_\propto^\Omega - 1 \xrightarrow{f_\Omega} C_\alpha$$

For $\alpha = 1, \ldots, p, \quad F' = f_1 \circ f_2 \circ \ldots \circ f_\Omega$

where $h_\propto^\omega$ are the pattern vectors presented to the layer ω neurons for any training example α and $\omega = 1..\Omega$ the number of neuronal layers.

Each layer of the network, in Figure 1, implements a mapping function f_ω which performs a

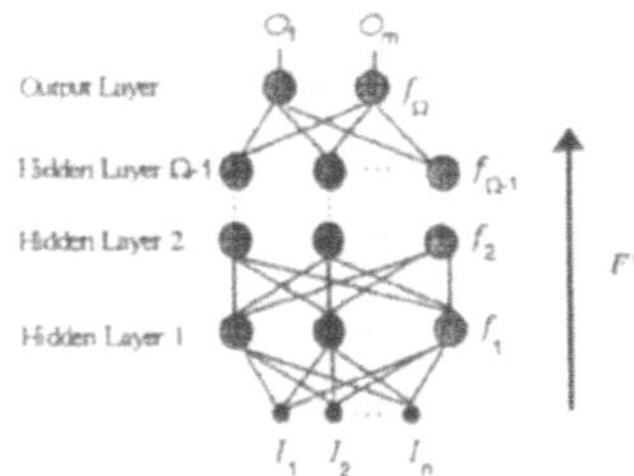

Figure 1: Mapping functions in a layered FFNN.

transformation on the vectors fed to its neurons. The composition of these functions F', approximately maps all the subsets of input vectors $V_\alpha = \{I_1^\alpha, I_2^\alpha, ..., I_n^\alpha\}$, in the training set, to the corresponding output vectors $O_\alpha = \{O_1^\alpha, O_2^\alpha, ..., O_m^\alpha\}$ where each $O_i^\alpha \approx \zeta_i^\alpha$.

At each iteration BP attempts to improve on each of the transformations f_ω to f_1 (see Figure 1) individually and in that order. Each error correcting step is based on the error at the outputs of the current layer ω, of neurons and the outputs of the previous layer $\omega - 1$. The weights are adjusted so as to improve the transformation f_ω of the previous layer's outputs (the domain D) to the desired outputs at the current layer (the range R) and thus reduce the error term for each training pattern.

In Figure 2, each set of weight updates at a layer ω improves upon the transformation f_ω of the domain D (the previous layer's outputs calculated at the last forward propagation of an input vector) to the range R (the desired outputs at the current layer governed by the last backwards propagation of error terms).

The desired neuronal outputs, for a layer of neurons ω, are governed by equation (1) below [3]:

$$\delta_j = y_j(1 - y_j) \times err \tag{1}$$

where

$$err = \begin{cases} [y_j - \zeta_j] & \text{for the output layer,} \\ \sum \delta_k W_{kj} & \text{otherwise.} \end{cases}$$

which, with the exception of the output layer, is affected by the weights in the later layers ($\omega + 1, \omega + 2, ..$). Each backwards propagative step involves the error terms being multiplied by the weights on the connections and then summed together. So the error terms at layer ω are controlled by the weights at layer $\omega + 1$.

In Figure 3, the domain of each transformation at each hidden layer is set by forward propagations of input patterns through the network. In the early stages of training each layer of neurons ω is likely to have badly placed hyperplanes due to the random weight initialisation. Therefore with the exception of the first layer, as it is preceded by the input units, the transformation corrections at each layer $\omega + 1$ will be attempting to map an ill-defined domain to some range specified by the back propagated error terms.

Thus we have an evolving chain of cause and effect which governs the evolution of a single transformation related to a hidden layer of neurons ω:

- The domain - the outputs of the neurons in layer $\omega - 1$, affected by transformations from previous layers $\omega - 2$, $\omega - 3$ etc.
- The range - the desired outputs for the neurons in layer ω governed by the error terms, affected by the current weights in later layers $\omega+1$, $\omega+2$ etc.
- As the transformation at layer ω changes, the new weights will affect the domain for layer $\omega+1$ and the range for layer $\omega - 1$.

With these considerations in mind it would be difficult to ascertain the nature of the weight space

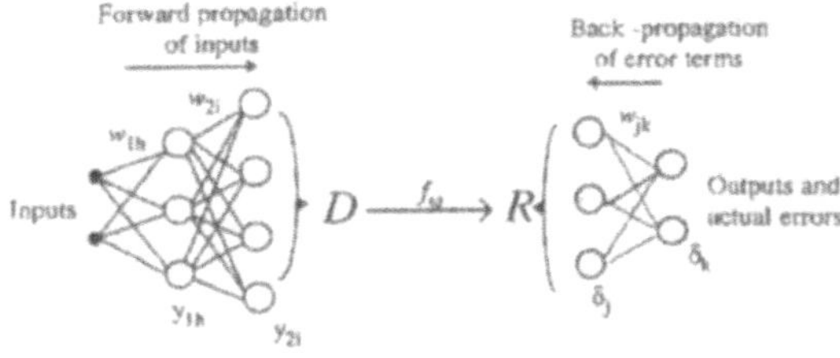

y - neuron output
w - weights on the connections
δ - error term for a neuron

Figure 2: A single layer in a layered FFNN.

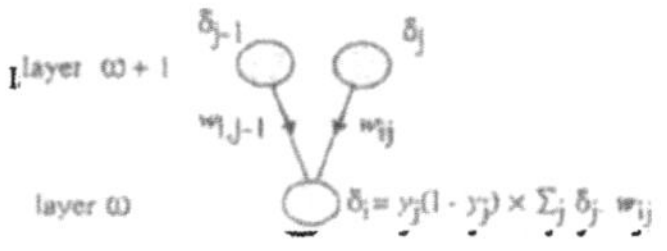

Figure 3: Backpropagation of error terms.

search which might occur in a training run. The BP paradigm will always attempt to minimise the overall network error function by gradient descent through iterative corrections but the corresponding evolution of the weights and inter-layer transformations is extremely complex. As error terms are backpropagated they are corrupted by the current inter-layer weights, as are the network input vectors by the preceding transformations. Neuronal layers 1 & 2 which form the initial input space partitionings will be updated using error terms which have been backpropagated through the rest of the network. The transformation corrections made at these layers will thus be heavily dependant on the conditions of the rest of the network. The current weights in the later layers of neurons $(\Omega, \Omega - 1, ..)$ will be controlling, or driving, the transformation changes which take place in the earlier layers of neurons (1,2,..).

This complexity in training also illustrates the problems with the scaling up of these networks, with greater numbers of layers and neurons making the development of the inter-layer transformations even more complex.

3 Empirical Analysis

Two sets of experiments were undertaken to illustrate the evolutionary nature of a FFNN trained under BP. We are attempting to establish which layers of weights are subject to the most drastic changes, how frequently hyperplanes change direction and how the interdependencies in the layer transformations affect each other as training progresses. Due to the large number of weights contained in an FFNN for a non-trivial problem it would be difficult to monitor every hyperplane throughout training, so a simpler scheme was devised.

4 Experiment 1

A set of four networks were trained, using BP, to monitor swings in hyperplane orientation as the network developed and record which layers of neuronal hyperplanes changed direction with the greatest frequency. This monitoring was conducted by observing changes of sign in weights throughout the training schedule. If a weight does not change sign then the weight vector has stayed within a volume of pattern space defined by a swing of 90^0 in one axis.

The tests were conducted using a real world continuous data set consisting of 270 instances. Each instance is a patient's record, made up of 13 attributes and is classed as either positive or negative for heart disease. A learning rate of 0.05 and a momentum parameter [3] of 0.3 were used during these experiments with a network topology as defined in Table 1.

Table 1

	No of Neurons in Layer	No. of weights
4th - Out Layer	1	5
3rd Layer	4	32
2nd Layer	7	77
1st Layer	10	140
Input units	13	Total : 254

Table 2

Layer	Exp. Swings	Ave
Out - 4th	2.0	0.95
3rd	12.6	2.87
2nd	30.3	13.15
1st	55.1	83.03

4.1 Results

Table 2 displays the results for the four experiments. The rows are labelled with each of the 4 layers of weights in the network. The column labelled Exp. Swings contains the number of weights in that layer as a percentage of the total number of weights in the network. These would be the expected results if all the weights in the network changed sign with equal frequency. Again each value is given as a percentage of the total number of sign shifts throughout the training schedule. The last column displays the averaged results.

These experiments showed that initially all layers of neurons made frequent swings in hyperplane orientation. After this initial stage, the first hidden layer of neurons accounted for between 70 and 90% of the sign changes recorded for all the weights in the network. The second layer of weights accounted for most of the rest of the changes and occasionally other layers' weights displayed sign changes.

4.2 Conclusion

Most of the search for a solution through weight space was conducted along the axes corresponding to the first layer's weights. These neurons form the hyperplanes which partition the input space directly. The second layer neurons which associate the partitions in to decision volumes accounted for most of the other weight sign changes. This infers that the search is primarily governed by the initial starting conditions of the later layers of neurons (the driving weights) and that large numbers of decision volumes of differing size, orientation and complexity are tried until the resulting transformations fit in with the requirements set by the weights in the later layers. This illustrates the importance of the first two layers of neurons with respect to the others. These hyperplanes form the initial decision region boundaries which partition the input vectors [4]. As most of the network's function is handled by these layers it makes sense that a suitable weight space search should concentrate on these hyperplanes.

5 Experiment 2

To verify the conclusions reached from the first set of experiments a second set was performed which involved training two sets of four networks of the same topology as above, from the same initial conditions with the exception of a single weight value. The weights in all eight networks were randomly initialised in the range $[0.5.. - 0.5]$ to the same values. At this stage we have eight identical networks with identical weights.
(i) For each of the first set of four networks, a single randomly, though exclusively, selected weight value from the output layer was incremented by a small amount (0.01).
(ii) The second set of four networks were dealt with in exactly the same way though the altered weight was selected from the input layer of connections.

5.1 Results

(i) The first set of four networks were trained and tested. They were all found to give entirely different solutions to the mapping problem. Each network correctly classified and missclassified different subsets of the test data and thus had found alternatively oriented sets of decision regions to partition the training data.

(ii) Once training was completed all four networks were found to generalise in the same way, though some small negligible differences in outputs were apparent. Each network was implementing the same set of decision regions and thus had found approximately the same set of weights to learn the training data.

5.2 Conclusion

Due to the sigmoidal squashing function used in BP a small change in a first layer weight should make little difference to the output of the neuron when forward propagating input vectors. The same is true for weights in the output layer, though once the backpropagation of errors begins, the altered weight will have a dramatic effect on the error terms which pass through the associated connection. This was clearly shown in Experiment 2 and illustrates the networks evolutionary dependence on the starting conditions of the *driving weights* in the later layers of neurons. A small change in the initial conditions of the later layers of weights, can cause a large difference in the final network state. This implies that FFNN trained under BP may be exhibiting chaotic behaviour. Chaotic behaviour in neural networks has been previously reported for networks with random asymmetric connections trained using a steadily increasing learning rate [5]. It was also used as a technique to increase the speed of BP training [6] again by altering the learning rate.

6 Summary

As a BP trained network becomes more complex the associated training times will increase dramatically. Many variations on BP training have been suggested [1,2,6]. This study validates the use of a separate learning rate for each layer in the network. Each of these should be set so as to reflect the proportion of the search which takes place at that particular layer. As the search for a solution is concentrated on the first layer of weights a large learning rate would enable the frequent changes in hyperplane orientation to occur more quickly. The later layers of weights have been shown to drive this search and undergo far fewer changes in hyperplane orientation. Accordingly the corresponding learning rates should be small relative to those used for the other layers. These proposals could be used in isolation or in conjunction with an adaptive learning rate scheme, as a technique for initialising the various step sizes.

References

[1] S.E. Fahlman. Faster-learning variations on BP: An empirical study. In D. Touretszky, G. Hinton, and T. Sejinowski, editors, *Proceedings of the 1988 Connectionist Models Summer School*, pages 38–51, 1989.

[2] R. A. Jacobs. Increased rates of convergence through learning rate adaptation. *Neural Networks*, 1:295–307, 1988.

[3] D. Rumelhart, G. Hinton, and R. Williams. Learning representations by BP errors. *Letters to Nature*, 323:533–535, 1996.

[4] I. K. Sethi. Entropy nets: From decision trees to neural networks. *Proceedings of the IEEE*, 78(10):1605–1613, 1990.

[5] H. Sompolinsky and A. Crisanti. Chaos in random neural networks. *Physical Review Letters*, 61(3):259–262, 1988.

[6] P.F.M.J. Verschure. Chaos-based learning. *Complex Systems*, 5:359–370, 1991.

Fuzzy Vector Bundles for Classification via Neural Networks

D. W. Pearson, G. Dray and N. Peton
Nonlinear and Uncertain Systems Group
Laboratoire de Génie Informatique et d'Ingénierie de Production
EMA-EERIE, Parc Scientifique Georges Besse, 30000 Nîmes, France.
Email: {pearson,dray,peton}@eerie.fr

Abstract

In this paper we propose a method of classification based on standard feedforward neural networks. The novelty of the approach is that we calculate local approximations of Lie algebras which generate the leaves of a foliation, each leaf corresponds to a class. From these linear approximations we pass to the case where a point on a leaf is not known with precision but can be specified using fuzzy set theory. Integrating the approximating linear equations then provides us with 'fuzzy leaves' or fuzzy classes.

1 Introduction

In a recent paper [8] we introduced the idea of constructing continuous classes from a neural network trained on discrete sample pairs. We make use of a standard two-layer backpropagation neural network, where the second hidden layer also acts as the output layer. Also we assume that it has been trained as a classifier using data pairs in the form (x_k, y_k), where $x_k \in M$ corresponds to an input vector and $y_k \in N$ the associated output vector. M and N are, in general, differentiable manifolds of dimension m and n respectively.

We treat the neural network purely as a nonlinear mapping from the input space to the output space which is written as

$$\begin{aligned} \pi &: M \to N \\ \pi(x) &= \sigma(W_2\sigma(W_1 x + b_1) + b_2) \end{aligned} \tag{1}$$

where $W_1 : \mathbb{R}^m \to \mathbb{R}^p$ and $W_2 : \mathbb{R}^p \to \mathbb{R}^n$ are the synaptic weight matrices, $b_1 \in \mathbb{R}^p$ and $b_2 \in \mathbb{R}^n$ are the bias vectors, σ is the neuron transfer function which we assume to be any of the habitually used functions and impose only the condition that it has to be sufficiently many times differentiable. We also place the blanket hypothesis on the mapping π that it is regular [2].

The point set $F_{y_0} = \pi^{-1}(y_0)$ for some fixed output value y_0 is called the *fibre* above the point y_0 [1] and obviously corresponds to all the input vectors which belong to the class corresponding to the point y_0. In [8] we presented a method whereby a local approximation can be obtained for $\pi^{-1}(y_0)$, hence the notion of constructing a continuous class out of discrete data samples.

2 Vertical Vector Fields

Let π_* denote the differential of the mapping π, then if a tangent vector $\dot{x}$ can be found such that

$$\pi_* \dot{x} = 0$$

then $\dot{x}$ is said to be a vertical vector and a vertical vector field v is such that $v : M \to \pi_*^{-1}(0)$. Because of the regularity hypothesis placed on the mapping π there will be $m - n$ independent vertical fields. A trajectory of a vector field passing through a point x_0 is called the flow of the vector field and is calculated from the following differential equation

$$\begin{aligned} \dot{x} &= v(x) \\ x(0) &= x_0 \end{aligned}$$

Such a flow is usually written as $x(t) = exp(tv)x_0$ and the act of solving the above differential equation is called exponentiation of the vector field. It will be assumed, without stating explicitly each time, that only valid solutions of this equation are considered, ie $x(t)$ for $\mid t \mid \leq \epsilon$ so that the vector field is defined.

If, for a point x_0 which satisfies $\pi(x_0) = y_0$, there exists a vertical vector field defined in some neighbourhood of x_0 then [1]

$$\pi(x)(t) := \pi(x(t)) = \pi(\exp(tv)x_0) = y_0.$$

In other words if $x_0 \in \pi^{-1}(y_0)$ and v is vertical then $x(t) \in \pi^{-1}(y_0)$ and so $x(t) \in F_{y_0}$. The exponentiation of the $m - n$ vertical vector fields provides a foliation called the fibre foliation of π, the leaves of this foliation are the connected components of the fibres of π.

It would be difficult to calculate a vertical vector field satisfying $v(x) \in \pi_*^{-1}(0)$ for all $x \in M$. For this reason we work locally and in [4] we illustrated how a local linear approximation can be found such that $\dot{x} = Ax$ and $\pi_* Ax = 0$ for x in some neighbourhood of a point x_0 on which the approximation is based, where $A : \mathbb{R}^m \to \mathbb{R}^m$.

From the $m - n$ approximations to the vertical vector fields we deduce the equation for an arbitrary trajectory in M

$$\begin{aligned} \dot{x} &= \textstyle\sum_{k=1}^{m-n} (u^k A_k) x \\ x(0) &= x_0 \end{aligned} \tag{2}$$

where the u^k, $k = 1, \ldots, m-n$ in Equation (2) are functions of t which are usually taken to be piecewise constant.

3 Fuzzy Differential Equations

The linear fields correspond to linear differential equations. In fact they are slightly more structured than that because they are elements of a lie algebra at least in a neighbourhood of the point on which they were calculated. Integration of linear equations is a fairly well understood technique. However we introduce a fuzzy initial state. That is to say we assume that the initial state is not known exactly but can be modelled by a vector of fuzzy numbers [3, 9].

A fuzzy number, a, can be determined from its α-level sets [3, 9]

$$[a]_\alpha := \{s : \mu_a(s) \geq \alpha\}, 0 < \alpha \leq 1,$$

where μ_a denotes the membership function for the fuzzy number a. Certain constraints are placed on the membership function μ_a, in particular we require that μ_a be normal

$$\exists s_0 : \mu_a(s_0) = 1$$

and convex

$$\begin{aligned} &\alpha_1 \geq \alpha_2 \\ &\Rightarrow [a]_{\alpha_1} \subseteq [a]_{\alpha_2}. \end{aligned}$$

An α-level set therefore corresponds to an interval for each given value of α

$$[a]_\alpha = [\underline{a}, \overline{a}]$$

With such a system each element of the vector x in Equation (2) at the time instant t is a fuzzy number where [6]

$$x_\alpha^k(t) = \left[\underline{x}_\alpha^k(t), \overline{x}_\alpha^k(t)\right], k = 1, \ldots, n \tag{3}$$

Then, by the method introduced in [6] and [7], integration of the linear fields provides a state vector parameterised as $x(t)$ where for each instant t the vector $x(t)$ is a vector of fuzzy numbers.

Each of these fuzzy trajectories lies on a connected component of a fibre of the mapping π in this way we are able to build up a picture of all the fibres, ie classes, in a fuzzy set theoretical framework.

4 Numerical Example

In this section we present a numerical example in order to illustrate the theoretical developments. Data was generated to satisfy the following equation:

$$x^3 = 2x^2(x^1)^3 - 3x^1(x^2)^2 + \epsilon,$$

where ϵ is a random variable, normally distributed with mean 0 and variance 0.04 and x^1, x^2, x^3 are the components of the input vector x, thus corresponding to a 2-surface in 3-space with some random noise added to make things interesting.

A 3-2-1 neural network like Equation (1) was trained, with neuron transfer function $\sigma = \tanh$, to produce an output of 0.9 for all the input data lying on the surface defined above. We did not include other constraints and outputs simply because we needed a straightforward example in order to illustrate the procedure. The resulting synaptic weight matrices and bias vectors were:

$$W_1 = \begin{bmatrix} 1.8569985 & 0.70268784 & -0.042531887 \\ 0.13555284 & 0.22024515 & 1.2998821 \end{bmatrix}$$

$$W_2 = [\, 1.6008923e-05 \;\; 1.9462921e-05 \,]$$

$$b_1 = \begin{bmatrix} 1.3535401 \\ -0.13520780 \end{bmatrix}, \; b_2 = 1.4722034.$$

For the calculations the point

$$x_0 = \begin{bmatrix} -1.0000000e-01 \\ 0.0000000e+00 \\ 1.3499409e-02 \end{bmatrix}$$

was chosen, $m - n = 2$ and the following two matrices were calculated for Equation (2) using the method illustrated in [4]

$$A_1 = \begin{bmatrix} 2.7114406 & 0.37350079 & -0.22245481 \\ -9.7104838 & 0.49034488 & -1.7605167 \\ 1.8287380 & -0.35202544 & 1.2626121 \end{bmatrix}$$

$$A_2 = \begin{bmatrix} 8.3374343 & -1.0357271 & -2.1199826 \\ 1.5699999 & 5.3379702 & -0.65405057 \\ -4.6581633 & -1.0265828 & 0.92952923 \end{bmatrix}$$

A typical trajectory is illustrated in Figure 1, where the classification surface is plotted and a trajectory for the field $-A_1 + 0.5A_2$ is indicated by the $\star$'s. Using the method introduced in [5] for calculating the intervals in Equation (3) we define

$$\underline{x}_\alpha(t) = \begin{bmatrix} \underline{x}_\alpha^1(t) \\ \underline{x}_\alpha^2(t) \\ \underline{x}_\alpha^3(t) \end{bmatrix}$$

and

$$\overline{x}_\alpha(t) = \begin{bmatrix} \overline{x}_\alpha^1(t) \\ \overline{x}_\alpha^2(t) \\ \overline{x}_\alpha^3(t) \end{bmatrix}.$$

Then each matrix A_k is split into positive and negative parts as follows:

$$A_k = A_k^+ + A_k^-,$$

where the elements of A_k^+ are all positive and those of A_k^- are all negative. A composite matrix is then constructed as

$$A_\alpha = \begin{bmatrix} A^+ & A^- \\ A^- & A^+ \end{bmatrix},$$

where A^+ and A^- are linear combinations of the A_k^+ and A_k^- for $k = 1, 2$. For each α-level set defining the initial state $[x_0^i]_\alpha = [\underline{x}_{\alpha 0}^i, \overline{x}_{\alpha 0}^i]$ for $i = 1, 2, 3$ the dynamic evolution of these values is determined from

$$\begin{bmatrix} \underline{x}_\alpha(t) \\ \overline{x}_\alpha(t) \end{bmatrix} = \exp(tA_\alpha) \begin{bmatrix} \underline{x}_{\alpha 0} \\ \overline{x}_{\alpha 0} \end{bmatrix}.$$

We make use of a particularly simple form for a fuzzy number [3] which is a triangular shaped function. For example in Figure 2 the fuzzy number

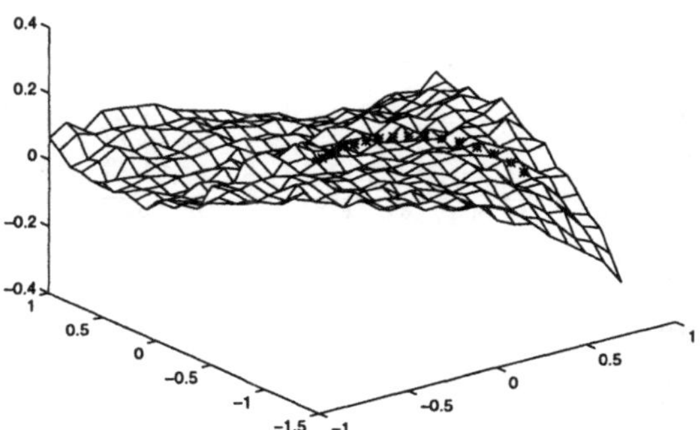

Figure 1: Classification surface and trajectory.

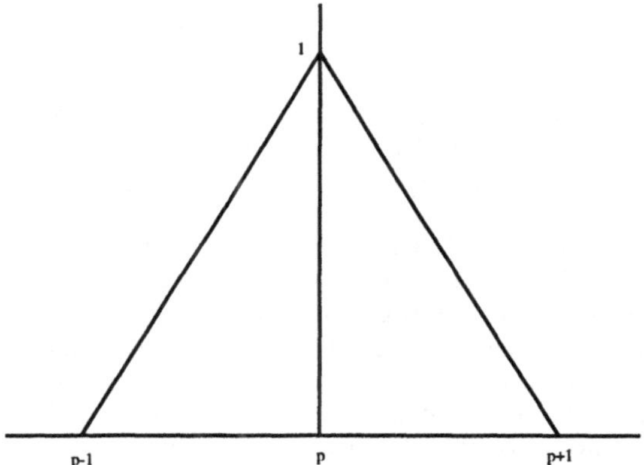

Figure 2: Fuzzy number 'about p'.

'about p' is depicted. Note that the base here is fairly large (p±1) and so if p is small then this represents an extremely fuzzy fuzzy number.

The membership function for the fuzzy number in Figure 2 is

$$\mu_p(s) = 1 - p + s \text{ for } p - 1 \leq s < p$$
$$\mu_p(s) = 1 + p - s \text{ for } p \leq s \leq p + 1$$
$$\mu_p(s) = 0 \text{ otherwise.}$$

With this formula the α-level set for a number 'about p' becomes

$$[p]_\alpha = [\alpha + p - 1, 1 + p - \alpha]$$

As explained above, the support of the fuzzy number used is very large in relation to the elements of

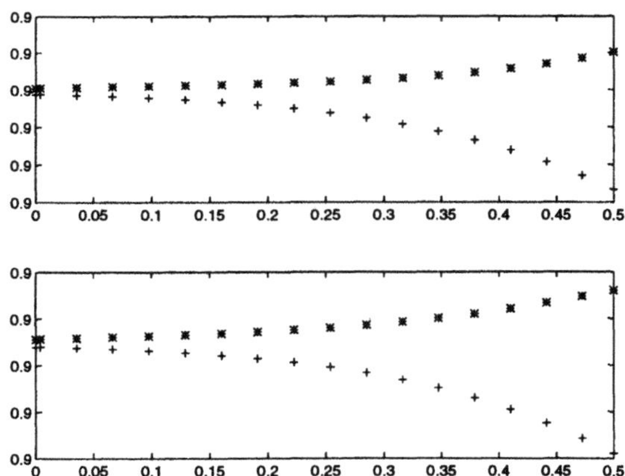

Figure 3: Output of neural network.

x_0 and so we use fairly high values of α in our simulations. It would be interesting to investigate the effect of modifying the support as a function of the value p.

As an example we set $\alpha = 0.97$ and calculate the two trajectories corresponding to the two endpoints of the α-level set interval, then we repeat the procedure with $\alpha = 0.95$. As a test, the output of the neural network is calculated for all the points along the trajectories of the interval endpoints. The output should be 0.9 of course and one can see in Figure 3 that this is indeed so, the top graph corresponds to $\alpha = 0.97$ and the bottom to $\alpha = 0.95$.

Finally, in Figure 4 we see the original 'crisp trajectory' indicated by $\star$'s along with the 'fuzzy trajectories' indicated by $+$'s.

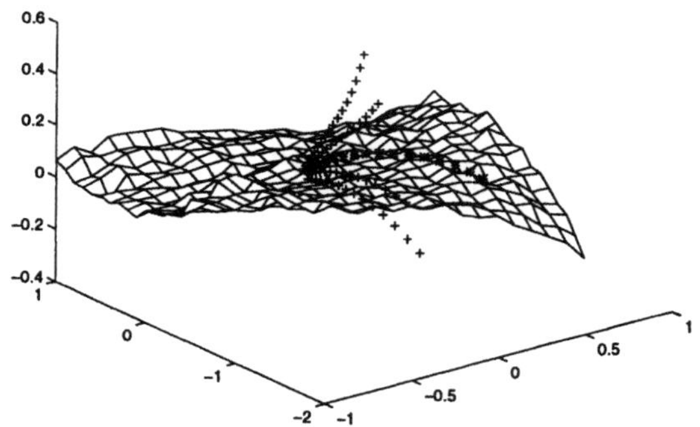

Figure 4: Fuzzy trajectories.

5 Conclusion

Neural networks have proved themselves to be good numerical tools, with useful capabilities of achieving a good balance between precision and generalisation. A frequent task required of neural networks is to classify data. However, data can be unreliable and unaccurate and so it is important to address these problems by incorporating fuzzy set based notions into a neural network framework.

Our research in vector bundle calculations for feedforward neural networks has led us to combine neural and fuzzy via the relatively new field of research known as fuzzy dynamical systems. Our work is continuing along these lines.

References

[1] C. Ehresmann. Les connexions infinitésimales dans un espace fibré. In *Proceedings Colloque de Topologie*, pages 29–55. Bruxelles, 1950.

[2] J. G. Hocking and G. S. Young. *Topology*. 1988.

[3] M. Mizamoto and K. Tanaka. Algebraic properties of fuzzy numbers. In *Proceedings International Conference on Cybernetics and Society*, pages 559–563. Washington DC, USA, 1976.

[4] D. W. Pearson. Approximating vertical vector fields for feedforward neural networks. *Applied Mathematics Letters*, 9(2):61–64, 1996.

[5] D. W. Pearson. On linear fuzzy dynamical systems. In *International Symposium on Soft Computing*. Nîmes, France, 1997. to appear.

[6] D. W. Pearson. A property of linear fuzzy differential equations. *Applied Mathematics Letters*, 1997. to appear.

[7] D. W. Pearson. Some structural properties of fuzzy linear dynamical systems. In *Proceedings International Symposium on Fuzzy Logic*. Zurich, Switzerland, February 1997.

[8] D. W. Pearson and G. Dray. Construction of continuous classes using vertical vector fields. In *Proceedings Engineering Systems Design and Analysis*. Montpellier, France, July 1996.

[9] S. Seikkala. On the fuzzy initial value problem. *Fuzzy Sets and Systems*, 24:319–330, 1987.

A Constructive Algorithm for Real Valued Multi-category Classification Problems

H. Poulard[1,2] and N. Hernandez[2]
[1] ACTIA, 25 Chemin de Pouvourville - 31432 Toulouse FRANCE.
[2] Laboratoire d'Analyse et d'Architecture des Systèmes - CNRS,
7 avenue du Colonel Roche - 31077 Toulouse FRANCE.
Email: poulard,hdez@laas.fr

Abstract

In this paper, an overview of a new constructive algorithm is proposed. Most common binary-unit based constructive algorithms are faced with 3 major drawbacks: binary inputs, only one output and a complex connectivity. The proposed algorithm aims to overcome these problems. An extension to multiple outputs of the sequential learning algorithm in combination with the Barycentric correction procedure is used. The performance of this new algorithm is evaluated in comparison with cascade correlation, on the *Vowel* classification benchmark.

1 Introduction

Recently, constructive algorithms have been proposed to overcome a number of backpropagation drawbacks such as the computational time and the *a priori* choice of the network structure. The rationale of these approaches to the neural network design relies on an incremental construction of the network. Units are added when required throughout the process and training, i.e. parameter adjustment is done separately for each unit. Many strategies have been proposed. A number of them involve a continuous unit, such us the well-known cascade correlation algorithm [2], but constructive algorithms furnish a network made up of binary units. Threshold units are particularly interesting for VLSI implementation. Compared with continuous units, the activation function simplicity allows a potential integration of a higher number of neurons.

However, binary-unit based constructive algorithms often encounter three major problems. Firstly, a limited number of algorithms are capable of training real-valued training sets without preprocessing (analog-digital conversion or projection on a convex body). Secondly, they are usually limited to one output and, therefore, restricted to classification tasks involving only two categories. Recently, extensions to multiple output have been proposed for a number of constructive strategies [4]. Thirdly, the connection topology obtained from these algorithms is often complex and not suited to a potential VLSI implementation. To our knowledge, among all the constructive algorithms , none of them addresses these three problems.

In this paper, we present an overview of a new constructive method clarifying up these limitations. The method is based on the sequential learning [3] technique associated with a new threshold unit learning algorithm, known as the Barycentric correction procedure (BCP) [6]. A special version of the BCP (referred to as BCP Max) has been devised for use with the sequential learning. The association of BCP and sequential learning (called BCPSL) allows learning of any mapping for two-category classification problems and leads to a simple architecture consisting of one hidden layer. In this paper, the multiple output version proposed is a natural extension of the BCPSL.

After a brief presentation of the BCP and sequential learning in the first section, the multiple output extension is presented. Then, a few simulation results are also given before the conclusion.

2 The BCPSL algorithm

Above all, a binary unit can be viewed as a hyperplane separating the input space into two open halfspaces: positive and negative. The former contains the points x for which

$$w.x + \theta > 0$$

where w is the weight vector of the unit and θ its bias, and the latter the points x such that

$$w.x + \theta < 0.$$

The sequential learning strategy appears as an incremental modelling of the boundary between the two classes. One hidden layer is constructed as follows. Each unit is built up so as to have a *pattern exclusion location*, i.e. all the patterns located in one of the halfspaces belonging to the same class. These patterns are referred to as *excluded patterns.* Moreover, they are correctly classified: if the target is 1, they are contained in the positive halfspace, if not, in the negative one. When such a unit is built, the excluded patterns are removed from the training set and the construction process is continued until the training set consists of patterns belonging to the same class. With this type of construction, Marchand [3] has shown that the resulting mapping for learning the output unit, obtained from the internal representations on the hidden layer, is linearly separable, and easily solved. Compared with the other, the main benefit of this algorithm lies in that there is theoretically no difficulty in constructing a network with real-valued inputs. The sufficient condition for the convergence of the algorithm is to find an exclusion position for any mappings, with at least one excluded pattern. However in order to minimize the number of units constructed in the hidden layer and, therefore, the generalization ability, one has to maximize the number of excluded patterns when a unit is built up. To do this, the authors of the sequential learning [3] proposed a method using the perceptron on several different training sets. Theoretically, it can lead to the optimal number of excluded patterns. But the combinatorial explosion limits the application of this method to small training sets.

Recently, a new generic threshold unit training method has been developed. The BCP [6] stands for an efficient set of methods for binary unit learning, which is particularly well-suited for various constructive algorithms including sequential learning. The heuristic version presented in [6] appears to be a powerful algorithm for finding a solution for linearly separable mappings. In terms of number of iterations needed to achieve convergence, a comparison with the perceptron has shown that it greatly speeds up computation by several orders of magnitude. Two extensions have been associated with this method. The first allows minimization of the number of misclassified patterns and the second maximizes the number of excluded patterns. The latter is referred to as BCP Max. The association of the sequential learning and the BCP Max represents an efficient algorithm capable of learning complex classification tasks in a short amount of time.

We will briefly tackle the heuristic BCP algorithm. A few details about the pattern exclusion procedure, needed in the sequel of the paper, will equally be given. Assume that the training set consists of a set of patterns C_1 of target 1 and a set C_0 with target 0. Throughout the learning phase, the heuristic BCP computes a hyperplane

$$\mathcal{H} : w.x + \theta = 0$$

with the weight vector w defined as follows:

$$w = b_1 - b_0.$$

The point b_1 belongs to the convex hull of C_1 and b_0 belongs to the convex hull of C_0. These points are computed as two convex combination of patterns. Learning is based on an iterative modification of the position of these points (see [6] for details).

The pattern exclusion extension concurrently computes another hyperplane

$$\mathcal{H}_c : \varepsilon\, w.x + \theta_c = 0,$$

with $\varepsilon \in \{-1, 1\}$ for maximizing the number of excluded patterns at each iteration. The principle is straightforward. Consider the function

$$\vartheta : \mathbb{R}^n \longrightarrow \mathbb{R}$$

Step a : Computation of ext_1, ext_2, ext_3, ext_4, t_- and t_+
Step b : If $t_- \neq t_f$ and $t_+ \neq t_f$ then $Ex_c = 0$ and **Stop**
Step c : If $(t_- = t_f)$ then computation of $\max(P_-)$, N_- and $\mathcal{G}^- = (ext_2 - \max(P_-))/||w||$
Step d : If $(t_+ = t_f)$ then computation of $\min(P_+)$, N_+ and $\mathcal{G}^+ = (\min(P_+) - ext_3)/||w||$
Step e : If $[\ t_- = t_f\ ]$ and $[\ (N_- > N_+)$ or $(N_- = N_+ > 0$ and $\mathcal{G}^- \geq \mathcal{G}^+)\ ]$ then
$\theta_c = (ext_2 + \max(P_-))/2, \quad \mathcal{G}_c = \mathcal{G}^-$ and $Ex_c = N_-$
If $(t_- = 1)$ $\varepsilon = 1$
If $(t_- = 0)$ $\varepsilon = -1$ and $\theta_c \longleftarrow -\theta_c$
Step f : If $[\ t_+ = t_f\ ]$ and $[\ (N_- > N_+)$ or $(N_- = N_+ > 0$ and $\mathcal{G}^- < \mathcal{G}^+)\ ]$ then
$\theta_c = (\min(P_+) + ext_3)/2, \quad \mathcal{G}_c = \mathcal{G}^+$ and $Ex_c = N_+$
If $(t_+ = 0)$ $\varepsilon = 1$
If $(t_+ = 1)$ $\varepsilon = -1$ and $\theta_c \longleftarrow -\theta_c$

Figure 1: Fixed target t_f excluded patterns maximization

computing the bias of a hyperplane normal to w including the point $p : \vartheta(p) = -w.p$. For a point p, $\vartheta(p)$ will be called its *projection*. Now consider the sets

$$\vartheta_1 = \{\vartheta(p_i)/p_i \in C_1\},$$

$$\vartheta_0 = \{\vartheta(m_i)/m_i \in C_0\}$$

and

$$\vartheta = \vartheta_1 \cup \vartheta_0.$$

At each iteration, the following extrema are computed: $\min \vartheta_1$, $\max \vartheta_1$, $\min \vartheta_0$ and $\max \vartheta_0$. Then, one can simply test whether the hyperplane $\mathcal{H}$ can separate C_1 from C_0. Indeed, if $\max \vartheta_1 < \min \vartheta_0$ the training set is linearly separable and takes the bias θ equal to $(\max \vartheta_1 + \min \vartheta_0)/2$ yields a correct hyperplane. In this case, the algorithm is stopped. If the training set is not linearly separable, the condition $\max \vartheta_1 < \min \vartheta_0$ is never fulfilled and the sets ϑ_1 and ϑ_0 are always overlapping.

The aim is to find an exclusion position. To do so, one renames the previous extrema as ext_1, ext_2, ext_3 and ext_4 such that $ext_1 \leq ext_2 \leq ext_3 \leq ext_4$. Therefore, one defines the sets

$$P_- = [ext_1, ext_2) \cap \vartheta,$$

$$P_+ = (ext_3, ext_4] \cap \vartheta$$

and

$$P_{ov} = [ext_2, ext_3] \cap \vartheta.$$

The set P_{ov} is called the *overlapping zone* whereas the sets P_- and P_+ are referred to as the exclusion zones. The exclusion zones contain the projection of patterns belonging to the same class. For each of the two exclusion zones, the pattern targets depend on the relative positioning of the extrema. Indeed, there exist four different cases (see [6] for details). Two feasible exclusion positions are simply found at the boundary of these zones, the one retained at each iteration excluding the highest number of patterns. Throughout the training, the best position occupied by hyperplane $\mathcal{H}_c$ is stored in a third hyperplane $\mathcal{H}_{poc}$.

To increase the robustness of the hyperplane retained at the end of the training process, the quantity referred to as *gap* is equally optimized throughout learning. The gap is the width of the *dead zone* around the hyperplane. If the minimum distance between the hyperplane and the patterns located in the positive (resp. negative) halfspace is denoted d^+ (resp. d^-), then the gap is given by $\mathcal{G} = d^+ + d^-$. The higher the gap, the more robust the hyperplane becomes, i.e. less sensitive to noise added to the patterns. Increasing the robustness of each hyperplane in a network increases the overall generalization ability.

3 Multiple outputs BCPSL

For a multi-category classification problem, the training set consists of N couples of vectors (ξ^μ, t^μ), where ξ^μ are the patterns in $\mathbb{R}^n$ and t^μ the target vectors in $\{0,1\}^p$. Then, the network has p output neurons. The patterns are also divided into several classes defined by the different target vectors. The way in which the classes are coded is not significant. With p output neurons, 2^p different classes can be coded.

The strategy is also based on building up one hidden layer such that each unit excludes a set of patterns of the same class. This set is referred to as the excluded cluster. But the number of classes being greater than 2, one can no longer associate a halfspace with a class. Actually, the excluded patterns will always be located in the positive halfspace. Then the negative halfspace contains all the patterns of $q-1$ classes. As in the one output version, the excluded patterns are removed from the training set and the construction process goes on until there are only patterns of the same class in the working training set. To find an excluded cluster, the technique is more complex than with the BCPSL. Note first that patterns of a given class cannot always be excluded. By way of example, if you consider the mapping involving a target 1 pattern x at the center of a circle mainly made up of target 0 patterns, x cannot be excluded. Thus to ensure the exclusion of a cluster, for each class c_i a training set is composed such that a target 1 is assigned to all the patterns belonging to c_i and a target 0 otherwise. A particular version of the BCP Max is also run on these training sets to exclude target 1 patterns. The position finally kept will be the one excluding the highest number of patterns among the positions found for the different classes. If two positions yield the same number of excluded patterns, the one exhibiting the highest gap is chosen.

The BCP Max must also be modified in order to exclude only those patterns with a given target. This modified version is proposed in Figure 1. The algorithm allows the exclusion of fixed target t_f patterns (t_f equal to 0 or 1). For the exclusion positions, the number of excluded patterns is evaluated $N_- = \text{Card}\ (P_-)$ and $N_- = \text{Card}\ (P_-)$ as well as the gaps $\mathcal{G}^-$ and $\mathcal{G}^+$. The target of the patterns in the set P_- (resp. t_-) is denoted t_- (resp. t_+). The number of excluded patterns thus obtained is Ex_c. Note that in certain cases, the normal vector of the hyperplane $\mathcal{H}_c$ has to be inverted. Due to its definition, the hyperplane $\mathcal{H}$ is always oriented in a direction corresponding to a decrease of $\vartheta(\xi)$, i.e. from P_+ to P_-. Then, if the set P_+ provides more excluded patterns than the set P_-, the normal vector of the hyperplane $\mathcal{H}_c$ must be the opposite of the vector w. This is the rationale for the introduction of the coefficient ε. One can see that for the two exclusion positions, the gaps are computed at steps c and d. These formulas can easily be proven, because by definition there is no pattern projection in the open ranges $(\max(P_-), ext_2)$ and $(ext_3, \min(P_+))$. Finally, one can notice that when a position is retained, the bias is always taken as the middle of two consecutive pattern projections, either

$$\theta_c = (\max(P_-) + ext_2)/2$$

or

$$\theta_c = (ext_3 + \min(P_+))/2,$$

in order to increase the robustness of the position. In the multiple output sequential learning this algorithm is also run for each class, as described above, with $t_f = 1$.

Once the hidden layer is constructed, the internal representations are computed to constitute the training sets for learning the output neurons. The patterns yielding the same internal representation have necessarily been excluded from the same cluster. Note that an excluded cluster can provide several different internal representations, except the last cluster obtained at the end of the algorithm. To learn a given output neuron k, the targets assigned to the internal representations are easily defined with the target vectors of the patterns. If an internal representation is obtained with a set of patterns having a target vector equal to $t = (t_1, \ldots, t_p)$, one will assign the target t_k. It can be shown that these new training sets are linearly separable. The proof is similar to the one proposed by Marchand et al. [3] for the one output sequential learning. It also provides also a direct solution, but due to the exponential weight growth, this solution cannot be employed. Rather than, setting the output neuron weight vectors to this solution, the heuristic BCP is used to learn these mappings.

4 Simulation results

The performance of the algorithm proposed, referred to as MOBCPSL (Multiple Outputs BCPSL), is compared to the well-known cascade correlation [2] on the *Vowel* classification benchmark. This

Table 1: Simulation results on the *Vowel* data set.

	MOBCPSL			CasCor
N_i	40	100	200	200
$\bar{g}$	83,8	83,5	83,3	82,6
σ_g	1,86	1,97	1,96	1,70
$\bar{t}$	7,38	14,8	25,5	29,0
σ_t	1,79	3,89	6,91	7,62
$\overline{n_u}$	27,8	24,6	22,5	12,4
σ_{n_u}	4,47	4,55	4,25	2,24

database is made up of 528 patterns in a 10-dimensional input space. These data correspond to different vowel signature pronunciations. With the MOBCPSL, the simulations were made using three different numbers of iteration (40, 100 and 200) to learn each unit. cascade correlation was run with 200 iterations and the sigmoïd activation function. Both algorithms were run on the same training sets and evaluated on the same test sets. These sets were obtained by randomly dividing the database into two parts: 35% (185 vectors) in the training sets and 65% in the test sets. The results listed in Table 1 were averaged on 500 trials. This table contains the mean value and the standard deviation for 3 quantities: generalization ability (percentage of correct network outputs on the test set), learning-time (in seconds) and number of neurons constructed.

It is worth pointing out that the generalization ability obtained of the MOBCPSL is slightly better than that of the cascade correlation. Moreover, a high number of iterations is not necessary to reach good generalization. The number of units constructed by the MOBCPSL is twice greater than with the cascade correlation but this is not a significant factor since the units used are simpler. Moreover the connectivity of the networks provided by the MOBCPSL is much less complex than the one obtained with the cascade correlation networks.

5 Concluding remarks

In this paper, a new way of constructing neural networks for general classification tasks has been briefly presented. Sequential learning algorithm which is at the core of this method has several advantages compared with the other constructive algorithms. But this technique was not very successful due to the abscence of an efficient algorithm dedicated to pattern exclusion. With the BCP and its derivatives, sequential learning and the multiple output version presented here now offers a promising approach to neural network design. However, for complex mappings (involving numerous neurons in the first hidden layer), a drawback appears for computing the internal representations : this step becomes time consuming and requires a large memory storage. Another version has been proposed in [5] to overcome this. Moreover, the extended version is well-suited for VLSI implementation, and a hardware design is proposed in [1].

References

[1] A. Bermak and H. Poulard. On VLSI implementation of multiple output sequential learning networks. In this volume, pages 93–97.

[2] S. Fahlman and C. Lebiere. *The cascade-correlation learning architecture*, pages 524–532. Morgan Kaufmann, San Mateo, CA, 1990. (D. Touretzky, ed.).

[3] M. Marchand, M. Golea, and P. Rújan. A convergence theorem for sequential learning in two-layer perceptron. *Europhysics Lett.*, 11:487–492, 1990.

[4] R. Parekh, J. Yang, and V. Honavar. Constructive neural network learning algorithms for multi-category pattern classification. Technical Report ISU-CS-TR 95-15a, Department of Computer Science, Iowa State University, 1995.

[5] H. Poulard and N. Hernandez. Two efficient constructive algorithms. Working paper, 1996.

[6] H. Poulard and S. Labrèche. A new algorithm for learning threshold unit. Technical report, LAAS-CNRS, 1996. Submitted paper, available at URL http://www.laas.fr/~ poulard/papers/bcp.ps.gz.

Classification of Thermal Profiles in Blast Furnace Walls by Neural Networks

H. Saxén, L. Lassus and A. Bulsari
Heat Engineering Laboratory, Åbo Akademi University
Biskopsgatan 8, FIN-20500 Åbo, Finland
Email: hsaxen@abo.fi

Abstract

The wall or cooling water temperatures in the ironmaking blast furnace are important sources of information about the internal conditions of the process, which cannot be measured directly because of high temperatures, mechanic wear and hostile environment. Two alternative methods of classification of wall thermal load profiles are studied: a feedforward neural network trained on manually classified temperature profiles, and an unsupervised classification using a Kohonen network. Both approaches were found to yield interesting results. In the former approach, the results can be easily explained since the characteristics of the profile classes are known, while in the latter approach the generalization properties and the ease of retraining the networks were considered advantageous. The classifiers are being implemented in the automation system of a Finnish blast furnace.

1 Introduction

The blast furnace is the major process for ore-based production of iron for steelmaking. The furnace, which acts mainly as a huge exchanger of oxygen and energy, is very complicated. Because of high temperatures, attrition, and presence of soot and dust, it is extremely difficult to carry out measurements of the internal conditions of the furnace. However, the operators need information about the inner state in order to take proper control actions.

Because of the complexity of the blast furnace, knowledge-based techniques and neural networks have been applied successfully to aid the operators in interpreting and controlling the state of the process. Neural networks have been applied to, for instance, pattern recognition and classification of data from probes [1, 2, 6], burden distribution [3], and heat level control problems [5].

2 The Approach

Data from a Finnish blast furnace was used in the study. Temperatures measured with thermocouples in the lining were averaged for each vertical level, from the tuyere level (where the hot blast is injected) to the furnace shaft. By this arrangement, a profile consisting of six observations was obtained. Furthermore, it was decided that daily average values of the profiles be used in order to filter out short-term fluctuations, the effects of which on the overall operation of the process are minor. About 200 observations, available from the start of the campaign of the furnace were used in the study. Two alternative approaches were made, one based on supervised classification using feedforward neural networks, and the other based on Kohonen's feature map [4].

3 Results and Discussion

In the first approach, different classes were selected a priori on the basis of visual inspection of the material. It was found that about 10 classes would be required: A possible set of classes has been depicted in Figure 1. For this set, 'rules' on (likely) feasible transitions between the classes could also be deduced. By training feedforward networks, with the six averaged temperatures as inputs and boolean signals indicating the (crisp) temperature profile classes as outputs, it was observed that the best choice seemed to be to apply two different networks to carry out the classification. This decision served to make the classifier robust and also to keep the dimension of the networks (i.e., number of hidden nodes) moderate. Classes $A_1 \dots A_5$ and B_1 were classified by the first network, and the remaining classes by the second one.

It is interesting to note that the errors in the classifications (residuals) by the trained networks in al-

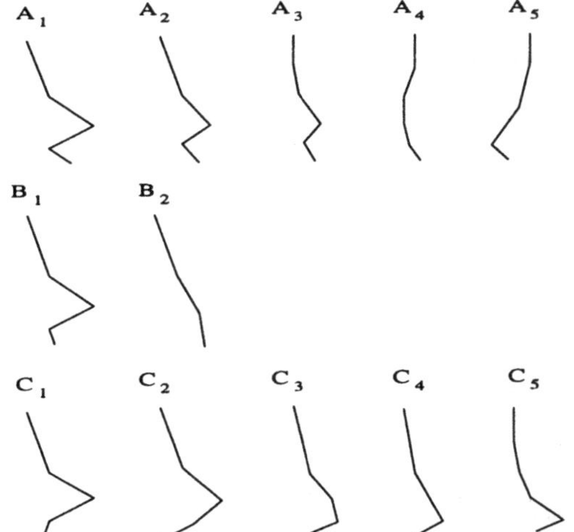

Figure 1: Classes used for training the feedforward neural networks.

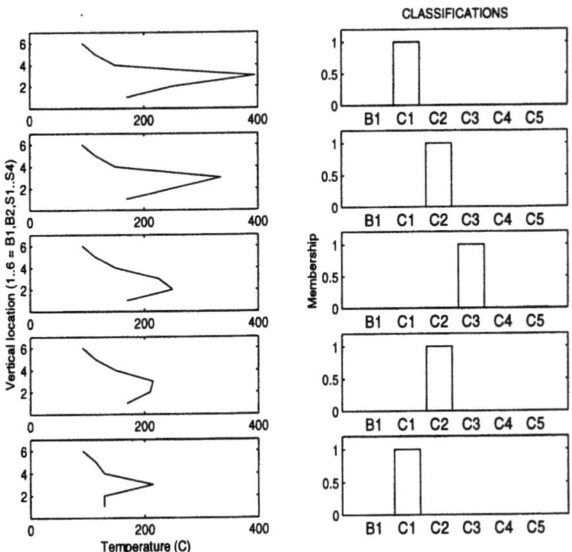

Figure 2: An example of a classification provided by the feedforward neural network.

most all cases turned out to be due to inconsistencies in the 'manual' classification, or occurred for patterns that were on, or close to, the classification boundaries. Similar findings have been reported for other classification problems from the same field [1]. Figure 2 shows an example of the classification on a set of five patterns not included in the training set.

As an alternative, the temperature patterns were used as inputs to Kohonen networks. Using a reasonable number of nodes (e.g., 5×5), a classification was obtained that — despite some differences — showed considerable agreement with the results of the classification method described above. Figure 3 shows the temperature profiles on the two-dimensional feature map. The network has clearly been able to separate the most important profile classes. Referring to the classification by the feedforward networks (cf. Figure 1), the classes $A_2 \ldots A_5$ are, for instance, found on the lower left edge of the map, B_2 appears on the top, and C_1 and C_2 on the right edge.

Figure 4 shows the evolution of the classification during a 20-day period: during the first 10 days of the period, the furnace exhibited relatively flat profiles, corresponding to operation with considerable amounts of accretions on the walls. After some control actions, the temperatures in the lower region started to increase, forming a profile with a pronounced peak at the furnace belly. This indicates that an efficient gas flow has developed in the region, which also led to better performance indices of the process (in terms of productivity, fuel rate, etc.). The evolution on the feature map thus illustrates the shift from one point of operation to another, which is a useful feature for the operator of the furnace.

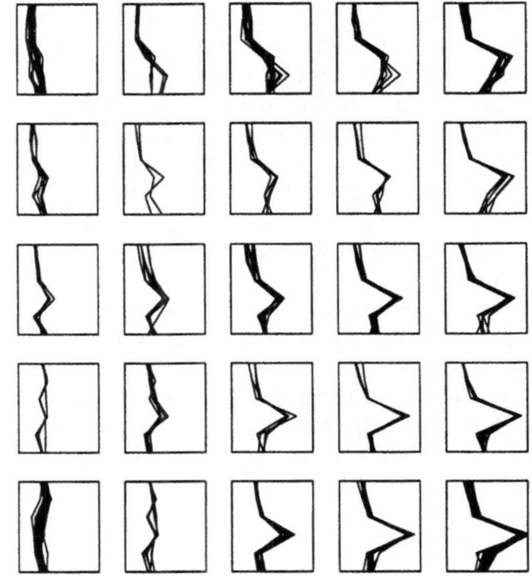

Figure 3: The location of the temperature profiles of the training set in the Kohonen network.

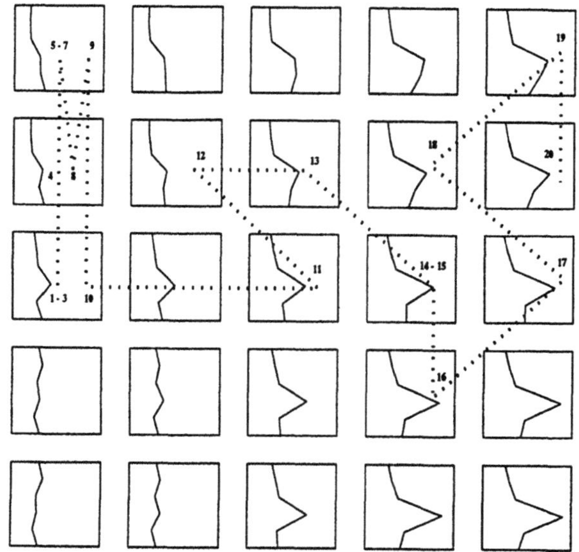

Figure 4: Example of a transition between the classes on the feature map during a 20-day period.

In general it was found that the most frequently occurring classes corresponded to appropriate points of operation, as expected, while abnormal profiles clearly corresponded to days of low productivity and/or high fuel rates. Occasionally, the classification by the Kohonen network showed some problems in that profiles with clearly different performance indices were lumped together in the same class or region of the feature space. On the other hand, the feature map has been found to be robust, and usually yields a reasonable classification of completely novel profiles; such profiles usually create problems for the feedforward networks, which do not classify the patterns or do not suggest any classification at all (i.e., with all outputs equal to zero).

However, the two different schemes are merely considered complementary than rivalling techniques, since the characteristics of the information they provide are different. The two methods are at present being implemented in the supervision system of a Finnish iron works, where the practical use of the tools will be evaluated.

References

[1] A.B. Bulsari and H. Saxén. Classification of blast furnace probe temperatures using neural networks. *Steel Research*, 66:231–236, 1995.

[2] T. Hirata et al. Blast furnace operationsystem using neural networks and knowledge-base. In *Proceedings of the 6th International Iron and Steel Congress*, pages 23-27. Nagoya, Japan, ISIJ, 1990.

[3] S. Hirose et al. Application of AI systems to Mizushima no.3 BF. In *Proceedings of the 51st Ironmaking Conference*, pages 163–170. Toronto, Canada, 1993.

[4] T. Kohonen. *Self-Organization and Associative Memory.* Springer-Verlag, Berlin, 3rd edition, 1989.

[5] Y. Otsuka et al. Application of neural network systems to pattern recognition of blast furnace operation data. *Kobelco Technology Review*, 15:12–16, 1992.

[6] H. Takada et al. A design environment for neural networks and its applications. *Control Engineering Practice*, 2:123–128, 1994.

Geometrical Selection of Important Inputs with Feedforward Neural Networks

F. Rossi[1,2]
[1] THOMSON-CSF/AIRSYS/RD/RDTE, 7-9, rue des Mathurins, 92221 BAGNEUX, France
[2] Universit Paris-IX Dauphine, UFR MD, Place du Marchal de Lattre de Tassigny, 75016 PARIS, France
Email: rossi@ceremade.dauphine.fr

Abstract

In this paper, we introduce a method that allows to evaluate efficiently the 'importance' of each coordinate of the input vector of a neural network. This measurement can be used to obtain information about the studied data. It can also be used to suppress irrelevant inputs in order to speed up the classification process conducted by the network.

1 Introduction

Variable selection is one of the key issues of classification tasks (and more generally of model estimation). In order to solve a classification problem, we start in general with a lot of measures coming from the real world. If these measurements are well chosen and if the problem is simple enough, we can reasonably assume that a classifier could be designed from the data. Unfortunately, it will in general perform its classification as a black box, without giving any information about the underlying task. One of the direct applications of variable selection is to suppress the useless input variables of the classifier: this suppression helps to understand the classification and give important informations about the task itself. Moreover, the suppression allows to simplify the classifier itself and therefore to speed up the classification process. It is also very important to notice that variable suppression speeds up the acquisition process (i.e., the measurement of the real world in order to produce the input variables).

A great deal of work has been done by statisticians in order to study variable selection when the regression tool is linear (e.g., [5]). The purpose of this article is not to review these methods but to work on problems that cannot be correctly solved with linear tools. When the classifier is non-linear, the problem is more complex and several neural based method have been proposed to solve it (e.g., [2, 7, 8]). In this paper, we propose an extension of a previously introduced method [8] and compare it with another neural approach, OCD [2], and with a statistical method which was proposed in the neural network community [1].

2 Geometrical Variable Selection (GVS)

2.1 The proposed method

The method presented in this section allows us to choose which attributes to suppress in an efficient and reliable way. The key idea is to analyze an already trained classifier in order to measure how much its calculation depends on each attribute: the evaluation is global. Moreover, the method can be applied to non-linear classifiers and therefore solve the linearity limitation of traditional methods.

Let $F(x,w)$ be a parametric classifying function (also called a non-linear regression model, e.g. a MLP): x is an input vector, w is a weight vector that allows to modify the computation performed by F (e.g. the connection weights in a MLP) and $F(x,w)$ is the output of the function which belongs to $[0,1]^c$, where c is the number of classes of the problem. Let C_k be the studied k-th class and let $\chi_{C_k}(x)$ be the membership function corresponding to C_k, i.e. $\chi_{C_k}(x) = 1 \Leftrightarrow x \in C_k$. Let $C(x)$ be the perfect classifying function, i.e. $C(x) = (\chi_{C_1}(x), \chi_{C_2}(x), \ldots, \chi_{C_c}(x))$. The goal of the learning phase is to find w such that $F(x,w) \simeq C(x)$. In general w is chosen in order to minimize some distance criterion between $F(x,w)$ and $C(x)$ (for instance, the total quadratic error, $E(w) = \sum_{x \in T} \|F(x,w) - C(x)\|^2$, where T is the training set of the classification task). In fact, minimizing this quadratic distance is equivalent to minimizing a quadratic probabilistic distance between $F(x,w)_k$ (the k-th output of F) and $P(C_k \mid x)$, the *a posteriori* probability of class C_k, given x (see [9]).

Let us assume that we can compute the differential of F with respect to its first variable (i.e., the input vector), called $\frac{\partial F}{\partial x}(x,w)$. The optimal Bayes decision rule [3], if F_k is a good approximation of $P(C_k \mid x)$, is to assume that x belongs to class j if $F_j(x,w)$ is strictly greater than $F_i(x,w)$ for all $i \neq j$. The boundary between C_k and C_l is the set of points $x \in \partial C^{k,l}$ such that $F_k(x,w) = F_l(x,w)$ and for all $i \neq k$ and $i \neq l$, $F_i(x,w) < F_l(x,w)$. Therefore, the boundary is locally described by the equation $F_k(x,w) = F_l(x,w)$. Then, the unitary normal to the boundary at point x is $n^{k,l}(x)$, is given by:

$$n^{k,l}(x) = \frac{\frac{\partial F_k(x,w)}{\partial x} - \frac{\partial F_l(x,w)}{\partial x}}{\left\| \frac{\partial F_k(x,w)}{\partial x} - \frac{\partial F_l(x,w)}{\partial x} \right\|} \tag{1}$$

A low value for $\left|n_i^{k,l}(x)\right|$ (the i-th coordinate of the normal unitary vector) shows that the boundary normal is perpendicular to the i-th coordinate axis and therefore that the local separation hyperplane (which approximate the boundary) contains this axis: the i-th coordinate of the input vector is locally useless (this is true only if the tangent hyperplane does not run through the boundary, a case which is quite unlikely to happen) for separating elements from C_k from elements of C_l. Therefore, $\left|n_i^{k,l}(x)\right|$ is a good measure of the local importance of the i-th coordinate axis.

The remaining problem is to combine the $\left|n_i^{k,l}(x)\right|$ when x belongs to $\partial C^{k,l}$ in order to obtain a global understanding of the relative importance of the different coordinate axis. On a theoretical point of view, it will be interesting to compute the mean normal vector of boundary between C_k and other classes. But C_k may contain several separated clusters for which the boundaries are parallel but with opposite normal vectors: therefore, computing the integral of $n^{k,l}$ on $\partial C^{k,l}$ is not really meaningful.

In order to obtain a simple criterion, we define the following matrix:

$$S_{k,i} = \sum_{l \neq k} \int_{x \in \partial C^{k,l}} n_i^{k,l}(x)dx, \tag{2}$$

The main problem is now to obtain points belonging to $\partial C^{k,l}$. As this boundary has at most a dimension of $n-1$ (if n is the dimension of the input space of the classifier, i.e., the number of variables), we cannot obtain points in $\partial C^{k,l}$ with a random selection.

In order to find a point belonging to $\partial C^{k,l}$, we start by randomly choosing two points (in the training or validation set), x_1 and x_2 such that $x_1$is classified by F in C_k and x_2 is classified in C_l. Then, we study the function

$$g(\lambda) = F_k\left(\lambda x_1 + (1-\lambda)x_2\right) - F_l\left(\lambda x_1 + (1-\lambda)x_2\right) \tag{3}$$

The definition of g and assumptions on x_1 and x_2 implies that $g(0) < 0$ and $g(1) > 0$. Moreover, as F is assume to be differentiable with respect to x, g is obviously continuous. Therefore, for a specific $\lambda \in]0,1[$ (which can be easily found with simple zero finding algorithms [6] such as dichotomy), we have $g(\lambda) = 0$. Of course, there is no guarantee that the corresponding $x = \lambda x_1 + (1-\lambda)x_2$ belongs to $\partial C^{k,l}$ (because we can have for a specific j, $F_j(x) > F_k(x)$) but this is quite likely to happen. This method can be used to build a first description of $\partial C^{k,l}$. Of course, this description is far from being perfect, but it is easy to test if the obtained points really belongs to the boundary and therefore, we will end up with $\partial \tilde{C}^{k,l}$, a finite subset of $\partial C^{k,l}$.

The obtained value $S_{k,i}$ is therefore the global score associated to the i-th coordinate axis as a classifying axis for class C_k. Finally, we can define a mean score, the vector S as:

$$S_i = \frac{1}{c}\sum_{k=1}^{c} S_{k,i} \tag{4}$$

S_i is the mean global score associated to the i-th coordinate axis as a classifying axis.

Each coordinate axis of the input space is associated to a variable. Therefore, a high axis score is equivalent to an important variable: in order to suppress variables, we just have to discard the one with the lowest score.

2.2 Links with Previous Works

The method explained in the previous section is closely related to an algorithm introduced in [7]. In this article, the authors introduce an attribute ranking method based on first order differentials. There are two important differences between the method presented here and their algorithm:

- Priddy *et al.* combine the individual differential $\left|\frac{\partial F_k}{\partial x_i}(x,w)\right|$ without normalization;

- they take into account every example (and even additional points which are not examples) without focusing on boundary examples.

In fact, the main justification of Priddy *et al.* is a statistical one, whereas we are working on geometrical arguments, which are in our opinion more suited to the attribute suppression goal.

In a previous paper [8], we have introduced a first version of the method presented in the previous section. This paper shows that this method was more efficient than the one introduced in [7]. The main difference between the old version and the improved one is the use in the current method of an exact determination of boundary examples and an exact value for the normal vector. The current method has therefore stronger justification. On the real data given in the following section, there is no important differences between results obtained by the previous method and the current one, but such differences were observed for artificial problems (for which GVS performs better).

2.3 Feedforward Neural Network Case

We have demonstrated in a previous paper [4] that an extended backpropagation algorithm can be defined for arbitrary feedforward neural networks (including in the same framework, RBF networks and wavelet networks [10] for instance). This algorithm allows to compute efficiently the differential of $F(x,w)$ with respect to its input x, $\frac{\partial F}{\partial x}(x,w)$, if F is the output of a neural network (with w as generalized weight vector). Therefore, our algorithm can be applied to any feedforward neural network.

3 Experiment on Real Data

Some experiments were conducted on real world data: we have chosen to work with radar data. In this case, we have 32 inputs corresponding to different physical measurements. The goal is to decide whether a given input vector represents a target or some clutter. We have a big database, with about 40 000 points. This database contains only 5 000 target points, therefore, we will characterize the performances with two numbers: the average classification rate on the whole test set and the mean of the recognition rate of both classes. In order to applied a stopped training method, the databse was split into three parts : around 20 000 training examples, 10 000 validation examples and 10 000 test examples.

The goal of our simulation is to keep as few variables as possible. We tried several different methods:

3.1 Neural Methods

We trained a simple MLP with 32 inputs, 16 hidden neurons and 2 output neurons (i.e., a 32-16-2 MLP). After 200 iterations of Polak Ribiere Conjugate Gradient [6] (PRCG), the best MLP (selected with the help of the validation set) obtain 96.43 % as classification rate on the test set (and 89.93 % as mean classification rate).

We applied on this best MLP two neural based variable saliency computation methods: OCD [2] and GVS (as described earlier). The saliencies obtained by these algorithms allow to rank the different variables. The order for the 7 best variables is given in the following table:

OCD	2	21	22	18	20	15	6
GVS	2	21	20	22	13	6	15

In fact, there is only one difference : OCD chooses 18 attributes, whereas GVS chooses 13.

3.2 Statistical Method

We also used Battiti's algorithm, as described in [1]. This method combines two measures: the mutual information between a feature and class information, $MI(f,C)$, and the mutual information between a feature and the already selected features, $MI(f_1, f_2, \ldots, f_p, f)$. The algorithm uses a parameter λ which measures the importance of the between features mutual information. It has to be heuristically chosen. Battiti states that values ranging from 0.5 to 1.0 give good results. We have chosen to compare three values : 0.5, 0.75 and 1.0. These choices give the following attribute orders:

λ =0.5	2	22	21	15	14	20	18
λ =0.75	2	32	19	26	6	8	29
λ =1.0	2	32	19	26	6	8	19

The first set ($\lambda = 0.5$) is called B1 and the second one (valid for $\lambda = 0.75$ and $\lambda = 1$) is called B2.

Comparison

In order to compare the different methods (and in fact the selected attribute sets), we have trained a 7-16-2 MLP on the data. For neural based methods, we have pruned useless inputs in the best MLP and we have retrained the obtained MLP from this starting point. For Battiti's method, we have chosen randomly a starting point for a 7-16-2 MLP. In order to allow a fair comparison, we have increased the learning time for this MLP (400 iterations instead of 200). The performance for one starting point is the classification rate on the test set for the best MLP obtained after the training (the MLP is selected with the help of its mean square error on the validation set). The following table shows the performances:

attribute set	rate	mean rate
GVS	95.81%	87.64%
OCD	95.42 %	87.79%
B1	94.00%	81.70%
B2	93.41%	80.98%

This table shows that our neural network based selection obtains good performance. It overcomes limitations of the mutual information based method [1] which does not select attributes needed to maintain a satisfactory recognition rate of the target examples. Our method obtains results very similar to OCD [2]. The main advantage of our method is that it can be applied to any neural network, even one for which a zero weight does not mean that this weight can be suppressed (for instance RBF networks). Moreover, OCD is quite difficult to use because it cannot be applied if the network is not at a minimum of the error function. This is not the case for GVS.

4 Conclusion

In this paper we have introduced a new method that allows to suppress data attributes in a classification task. The goal of this suppression is to reduce the preprocessing and classification times. It is based on an analysis of the calculation performed by a parametric classifier such as a multi-layer perceptron. With the help of previous results, we have shown that this method was easy to use for arbitrary feedforward neural networks. An experiment conducted on real data shows that this method is efficient and can therefore be used for real world applications. Additional work is needed to prove the consistency of this method and to demonstrate its performances on other real world data.

5 Acknowledgments

This work was performed on Mrs Kim K. PHAM's responsibility, at THOMSON-CSF/AIRSYS.

References

[1] R. Battiti. Using Mutual Information for Selecting Features in Supervised Neural Net Learning. *IEEE Trans. On Neural Networks*, 5(4):537–550, July 1994.

[2] T. Cibas, F. Fogelman-Soulié, P. Gallinari, and S. Raudys. Variable selection with neural networks. *Neurocomputing*, 8(12):223–248, 1996.

[3] R. O. Duda and P. E. Hart. *Pattern Classification and Scene Analysis.* Wiley, New York, 1973.

[4] C. Gégout, B. Girau and F. Rossi. Generic Back-Propagation in Arbitrary Feedforward Neural Networks. In D. W. Pearson, N. C. Steele, and R. F. Albrecht, editors, *Int. Conf. on Artificial Neural Networks and Genetic Algorithms*, pages 168–171, Als, April 1995. Springer Verlag.

[5] A. J. Miller. *Subset Selection in Regression.* Chapman and Hall, 1990.

[6] W. H. Press, S. A. Teukolsky, W. T. Vetterling, and B. P. Flannery. *Numerical Recipes in C.* Cambridge University Press, second edition, 1992.

[7] K. L. Priddy, S. K. Rogers, D. W. Ruck, G. L. Tarr, and M. Kabrisky. Bayesian selection of important features for feedforward neural networks. *Neurocomputing*, 5:91–103, 1993.

[8] F. Rossi. Attribute suppression with multi-layer perceptron. In *CESA Multiconference*, Symposium on Robotics and Cybernetics, pages 542–547, Lille-France, July 1996. IMACS.

[9] H. White. Learning in Artificial Neural Networks: A Statistical Perspective. *Neural Computation*, 1(4):425–464, 1989.

[10] Q. Zhang and A. Benveniste. Wavelet networks. *IEEE Trans. On Neural Networks*, 3(6):889–898, November 1992.

Classifier Systems Based on Possibility Distributions: A Comparative Study

S. Singh[1], E. L. Hines[2] and J. W. Gardner[2]
[1] School of Computing, University of Plymouth, Plymouth PL4 8AA, UK
[2] Department of Engineering, University of Warwick, Coventry CV4 7AL, UK

Abstract

The main aim of this paper is three fold: a) to understand the working of a classifier system based on possibility distribution functions, b) to evaluate its performance against other superior methods such as fuzzy and non-fuzzy neural networks on real data, c) and finally to recommend changes for enhancing its performance. The paper explains how to construct a possibility based classifier system which is used with conventional error-estimation techniques such as cross-validation and bootstrapping. The results were obtained on a set of electronic nose data and this performance was compared with earlier published results on the same data using fuzzy and non-fuzzy neural networks. The results show that the possibility approach is superior to the non-fuzzy approach, however, further work needs to be done.

1 Introduction

Classifier systems for most pattern recognition problems need to satisfy various criteria to be considered reliable for decision making. These criteria relate not only to the classification results in terms of false negative and false positive classifications and their related costs, but also to their ability to work with imprecise and uncertain data, to produce reliable results with small samples, to reduce the cost and time of learning, to work with error-estimation techniques, and finally to avoid over- and under-generalization. All these qualities are not readily possible to model when defining the constructional parameters of a traditional classifier, and often poor results are obtained as a result. Hence we usually work with more than one classifier and to evaluate their performances, we use different error estimation techniques depending on the test sample size [7], and establish key parameters in the developmental stages of proposing a system that may be changed later to improve performance in different pattern classification environments, for example as in neural networks.

In this paper, we present a classification system based on possibility distributions. The original concept of possibility distributions and fuzzy theory dates back to the work done by Zadeh and followers [1, 2, 8, 9, 10]. Possibility distribution functions depend on the descriptive statistics of data and converge to normal distributions for large samples. The possibility of the mean value of a sample to occur is the highest, i.e. 1, and decreases quadratically as we move towards the minimum and maximum values. A set of formulae used for calculating possibility distributions are available in Zadeh [8], Mamdani *et al.*[2] and Singh [3, 5]. Possibility measurements are themselves supposed to be resistant to the effects of noise in data. The argument behind this belief is as follows: in a data set A of size n $\{x_1, x_2, \ldots, x_n\}$, if all measurements are uniformly affected by noise, and the new set B is $\{x_1+N, x_2+N, \ldots, x_n+N\}$ where N is the uniform noise, then the possibility of a measurement x_1 belonging to set A is the same or very similar to that of x_1+N belonging to B since possibilities depend on the descriptive statistics of the sets (minimum, mean and maximum) than individual values in them. This characteristic of possibility distribution is therefore attractive for using them in classifier systems which need to be robust to noisy and imprecise data. The above description however should be taken with caution. The performance of possibility based systems is not clear with varying levels of noise or when two different data sets may have the same descriptive statistics (excluding variance and standard deviation) but different underlying distributions.

2 Possibility Based Classifier

A possibility based classifier is shown in Figure 1. This scheme includes a re-sampling step. In Figure 1, the data is initially preprocessed. The presence of outliers can significantly harm the classifier by showing false estimates of distribution statistics, and they need to be eliminated. The data is then re-sampled for true error estimation using either cross-validation (for larger sets) or using bootstrapping or leave-one out method for particularly small sets. In this manner, the data is trained and tested with different sets (folds) and the true error rate is taken to be the average for these folds. Once the descriptive statistics are calculated on the training set, the possibilities are next computed. Let us assume that the test set is $K = \{K_1, K_2, \ldots, K_r, \ldots K_n\}$ where K_r, $r = 1$ to n, are patterns. Each pattern in turn will contain measurements for the features selected, for example $K_r = \{k_1, k_2, \ldots, k_i, \ldots, k_m\}$ for m features. For any test pattern K_r, we estimate its possibility of belonging to different classes. For instance, for a particular class j whose statistics is available $(s_{j1}, s_{j2}, \ldots, s_{jm})$ where s_{1j} represents the availability of mean, min. and maximum values for the first feature and class j using all samples in j, we can now find the possibility that K_r belongs to class j. This is done by: calculating the possibility that k_1 belongs to j using s_{j1} statistics, k_2 belongs to j using s_{j2} ...and k_m belongs to j using s_{jm} statistics. These individual features in a pattern may be called nodes. Once we have possibility of occurrence of each node to belong to class j, these are combined together to give us the possibility that the overall pattern K_r belongs to class j. The process iterates for all patterns K_1 to K_n in the test set and for all different classes available in the training set. The manner in which possibilities of the nodes are combined is important. Conventionally, for fuzzy intersection

$$p(k_1) \text{ AND } p(k_2) \ldots \text{ AND } p(k_m),$$

where $p(k_i)$ is the possibility of occurrence of k_i, is taken equal to $\min(p(k_1), p(k_2), \ldots, p(k_m))$. This may seem unreasonable for the proposed classifier system where valuable information in the form of all possibility values which are greater than their minimum is being wasted. Also, higher possibility values should be given more weight, i.e. if $p(k_i)$ increases from 0.8 to 0.85, then it should contribute more significantly to the decision making than if it changed from 0.5 to 0.55. After experimentation with different statistical functions, an exponential function has been found best for adding individual node possibilities for our problem. Hence, for each test pattern K_r, we calculate a final index of possibility that it belongs to say a given class j using, individual node possibilities. This index is therefore given by:

$$I_{K_{r,j}} = e^{p(k_1,C_j)} + \ldots + e^{p(k_2,C_j)} + \ldots + e^{M(k_m,C_j)},$$

where $I_{K_{r,j}}$ is the overall possibility that pattern K_r belongs to class j, and $p(k_i, C_j)$ is the possibility that node k_i in pattern K_r belongs to class j (same as C_j). The inference engine in Figure 1 compares the possibility index of all classes for K_r ($r = 1$ to n) and assigns it to the class with the highest index. It is in this way that test cases are predicted to belong to different classes.

3 Performance evaluation

The proposed classifier system has been tested previously on simulated data and encouraging results have been obtained [5]. Singh and Steinl [5] note that possibility distributions can be reliably used for large data, in their case automated assembly data containing over 12,000 patterns. These authors report two main advantages: most calculations are performed on a smaller file size containing statistical descriptions, and therefore it is computationally cheap, and that good classification rates can be achieved in real-time. In the present study, two powerful systems have been chosen for comparing possibility based classifiers performance: a non fuzzy backpropagation neural network, and a fuzzy neural network. Details on the working of an ordinary backpropagation network are widely available in literature, for example [6]. The fuzzy network is a relatively new approach, and the model used here and in related work by authors [3, 4] is based on using a conventional network with fixed starting weights. These weights are derived using possibility distributions, and it has been reported by

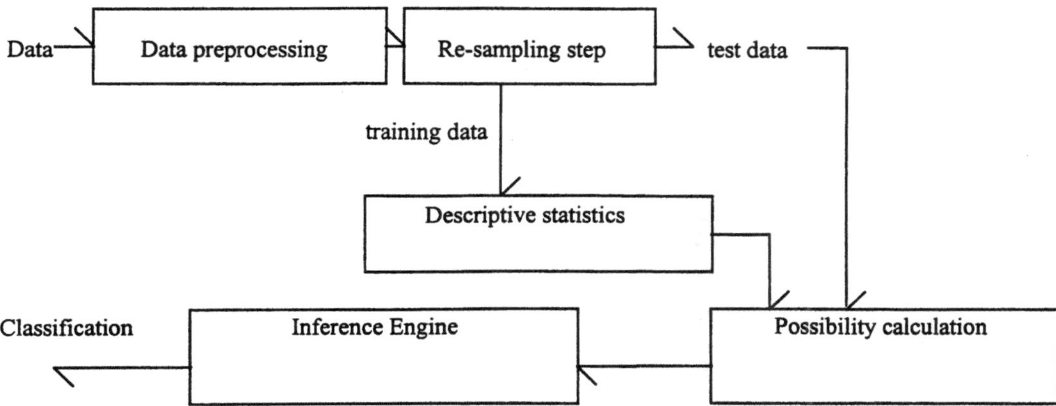

Figure 1: Possibility based classification system.

Singh *et al.* [4] that this method performs far superior compared to a network starting with random weights (non-fuzzy). The performance in this paper will be evaluated on coffee data which was also used in [4]. The coffee data has a total of 89 patterns for three different types of coffee (different blends). For each coffee pattern, there are a total of twelve input sensor measurements. The fuzzy and the non-fuzzy network used for classifying data both have 12×3×3 architecture. Singh *et al.* [4] have commented in detail on their relative performances on coffee and other data. The task here is to compare the possibility classifier with these two systems on coffee data.

The data was trained and tested for all three systems using a ten-fold cross-validation method. This technique has been recommended to obtain correct estimates of the true error rate [7]. The performance of the system was gauged on the basis of the number of misclassifications on test data (error rate obtained as an average of ten folds): false positives and false negatives were not separately considered. The data was normalized for neural network training, however, for possibility classifier the possibilities themselves were within the [0,1] range.

4 Results

The first set of results obtained using the possibility classifier were pessimistic with a poor 63% recognition rate. This didn't come altogether as a surprise because the coffee data was highly non-linear and noisy, and most possibility indexes were highly competitive during decision making. The evaluation of final possibility index was now modified taking into account the number of nodes in a pattern whose possibility of occurrence was zero. Test patterns where more nodes had a zero possibility were given less weight in their final index. This meant that possibility indexes computed earlier were modified by a standardized factor which took into account the number of zero possibilities of nodes in that pattern. The results now obtained are presented in Figure 2.

In Figure 2, the number of misclassified patterns in each fold are summarized. The fuzzy neural network and the possibility classifier in general produce the same result with different trials using the same test and train data. The performance of a non-fuzzy backpropagation network however varies with different random starting weights. In Figure 2, only the best results of the non-fuzzy network over a number of trials have been compared using a ten fold cross-validation. It should also be noted here that the fuzzy network was also trained using backpropagation (see [3] for details). The final summary of classification rates for the coffee data therefore is: fuzzy neural network (92.14%), non-fuzzy network (82.03%) and possibility classifier (84.27%).

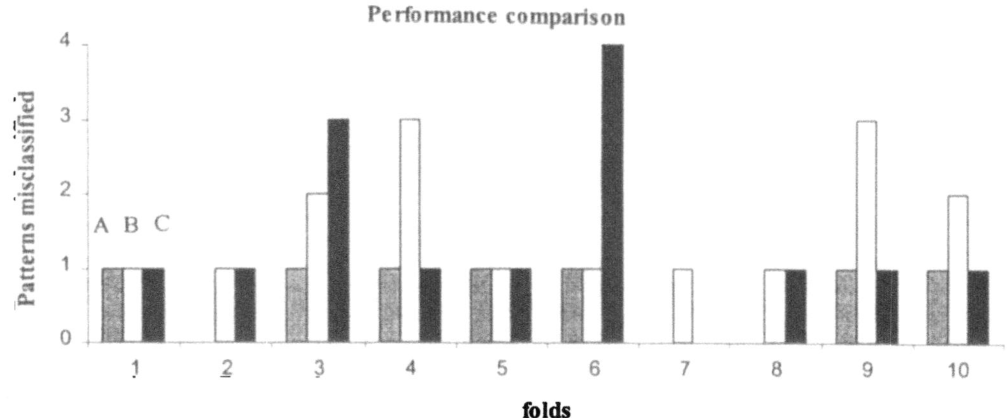

A = Fuzzy neural net, B = Non-fuzzy net, C = Possibility classifier

Figure 2: Misclassified patterns for the three techniques using cross-validation.

5 Conclusion

The above results are encouraging. The most important result presented is that the possibility classifier performs better than a non-fuzzy network using backpropagation. However, the performance is not the best of the three, as the fuzzy network is still superior. During the course of the experiment, a number of important observations were made:

1. the possibility classifier was quick;
2. possibility index calculation needs further revision and it may be worth exploring the concept of conditional possibility;
3. the performance of the system is as good as the data itself and especially on small data sets, possibility indexes of a pattern that belong to different classes are sometimes very similar — the classification decisions are highly competitive;
4. the classifier is not computationally expensive when new data is added to the training set.

Only the descriptive statistics need recalculation, no iterative learning process takes place. In conclusion, this paper suggests the utility of a possibility based classifier which upon further revisions and validation in other fields, promises to be an important pattern recognition system in areas where data is limited, noisy and uncertain.

References

[1] B. Kosko. *Neural Networks and Fuzzy Systems — A Dynamical Systems Approach to Machine Intelligence.* Prentice Hall, 1992.

[2] E. H. Mamdani and B.R. Gaines, editors. *Fuzzy Reasoning and its Applications.* Academic Press, 1981.

[3] S. Singh. Fuzzy neural networks for managing uncertainty. Master's thesis, University of Warwick, UK, 1993.

[4] S. Singh, E. L. Hines, and J. W. Gardner. Fuzzy neural computing of coffee and tainted water data on electronic noise. *Sensors and Actuators B*, 30(3):190–195, 1996.

[5] S. Singh and M. Steinl. Fuzzy search techniques in knowledge-based systems. In *Proc. 5th Intl Conference on Data on Knowledge Systems for Manufacturing and Engineering.* Reno, 1996.

[6] P. D. Wasserman. *Neural Computing : Theory and Practice.* Van Nostrand Reinhold, NY, 1989.

[7] S. M. Weiss and C. A. Kulikowski. *Computer Systems that Learn.* Morgan Kauffman, CA, 1991.

[8] L. A. Zadeh. *Fuzzy Logic and Its Applications.* Academic Press, New York, 1965.

[9] L. A. Zadeh. *A Fuzzy-Algorithm Approach to the Definition of Complex or Imprecise Concepts*, pages 147–192. John Wiley, 1987.

[10] L. A. Zadeh. *Fuzzy Sets as a Basis for a Theory of Possibility*, pages 193–218. John Wiley, 1987.

Learning by Co-operation: Combining Multiple Computationally Intelligent Programs into a Computational Network

H. L. Viktor[1,2] and I. Cloete[1]
[1]Computer Science Department, University of Stellenbosch, Stellenbosch 7600, South Africa
[2]Department of Informatics, University of Pretoria, Pretoria 0002, South Africa
Email: hlviktor@econ.up.ac.za; ian@cs.sun.ac.za

Abstract

This paper introduces a computational network which combines computational intelligent programs into a general framework for intelligent data analysis. Integrating more than one program may potentially lead to more powerful and versatile results.

1 Introduction

Computational intelligent algorithms, including decision trees, set covering algorithms and artificial neural networks, learn classification rules from data. These programs are capable of discovering interesting relationships in the data and usually require little technical knowledge of the programs. Computational intelligent methods can be very effective if the learning problem is sufficiently narrowly defined and the distribution of the attributes contained in the data set favours the particular program. Many complex real-world problems, however, pose learning problems which cannot effectively be solved by a single program. The data sets yielded by these problems contain uncertain and/or incomplete data, are generally very large and complex and contains dynamically changing data. In addition, the different computational intelligent programs offer complimentary advantages which may lead to complimentary results. In view of this, the development of a computational network which integrate two or more computational intelligent programs should prove worthwhile.

This paper gives an overview of a computational network for executing multiple computational intelligent programs. In Section 2 the computational intelligent programs which we use in the computational network, are introduced. The computational network is presented in Section 3. Section 4 includes a brief discussion of some experimental results which were obtained when considering two medical data sets. Finally, in Section 5, some conclusions are reached and future extensions are discussed.

2 Computational Intelligent Programs

There is a large amount of literature on computational intelligent approaches and methods, including [1], [2] and [6]. These methods learn concept descriptions from training data sets. The concept descriptions are subsequently tested on previously unseen test examples to give an estimated accuracy of the concepts learned.

Symbolic methods focus on producing discrete combinations of features, while ***subsymbolic*** methods adjust continuous, non-linear weighting of their inputs. We consider two symbolic methods, namely set covering algorithms and decision trees. The other two methods combine artificial neural networks, a subsymbolic method, with (a) symbolic rule insertion and (b) symbolic rule extraction.

2.1 Set Covering Algorithms

Set covering algorithms construct concept descriptions by repeatedly generating conjunctive expressions until all positive instances of a concept are covered or some threshold is reached. One class of covering algorithms, including BEXA [7], constructs conjunctions using a general-to-specific search. In this approach, the algorithm starts with a general concept description and specialises it in steps until some termination criterion is met. Each conjunction is evaluated according to an error estimate to select the best conjunct for further specialisation. The computational network includes the BEXA and CN2 [1] set covering algorithms.

2.2 Decision Trees

A decision tree generates a classifier by means of a structure that is either (a) a *leaf*, indicating a class, or (b) a *decision node* that specifies some test to be carried out on a single attribute value, with one branch and subtree for each possible outcome of the test.

A data row (case) is classified by starting at the root of the tree and moving through it until a leaf is encountered. The decision tree program contains heuristic methods for simplifying the tree; with the aim of producing comprehensible structures without compromising unseen cases. The C4.5 decision tree [6] is currently considered to be the state of the art and is used in our computational network.

2.3 Artificial Neural Networks

Artificial neural networks (ANNs) is a class of learning systems that model the human brain. The network consists of a number of weighted units, which can be one of three types: input, hidden or output. Units do only one thing, i.e. they compute a real-numbered output that is a function of real-numbered inputs. Inputs receive the initial numeric attributes from the environment. Hidden units act as links between the inputs and outputs. The outputs correspond to the expected outcome of a training set data row. Artificial neural networks learn the relationships between numeric inputs and outputs by minimising the difference between the expected and actual outputs via weight adaption. Training of the ANN is done by adapting weights via a gradient search. In essence, an ANN performs a non-linear regression.

Knowledge-Based ANNs

In this approach prior knowledge is inserted in the network and subsequently refined by ANN training. In the first step, a set of inference rules that describe the domain knowledge is gathered, usually from domain experts. In this way, the ANN is able to effectively make use of prior knowledge to perform well. Secondly, the knowledge is re-represented in an ANN, and subsequently refined using ANN learning as well as a training data set.

The knowledge defines the topology and weights of the network it creates. The way in which knowledge is re-represented in an ANN is to individually translate each rule into a subnetwork that accurately reproduces the behaviour of the rule. Additional nodes may be introduced to handle disjunctions in the rule set. See [2] for a description of the rule insertion process included in the computational network.

Rules From ANNs

The numerical representation of the attributes as well as the *"black box"* nature of the ANN makes it difficult to determine how a particular decision was reached. Domain experts may be sceptic about the decision reached. If the network produced an interesting discovery, it would be beneficial if this was made explicit. Rule extraction from the ANN attaches symbolic meaning to the learning process.

The rule extraction algorithm considers the sum of the weighted inputs to each hidden and output unit. It forms rules by taking the combination of inputs that exceed a threshold as the antecedents and the hidden/output unit as the consequence. The final rule set consists of a combination of these rules which have been reduced using propositional logic and applying sensitivity analysis. See [11] for a detailed discussion of our ANNSER algorithm.

3 Computational Network

A computational network is a general framework for parallel and/or distributed computation [4]. It is modelled on a directed graph together with certain attributes and specifications. The combination of computation intelligent programs is problem dependant. Therefore, the aims of the proposed computational network are to be versatile and dynamically changeable. Our computational network consists of four general layers, as depicted in Figure 1. The activation and execution of each of the nodes in the four layers of the computational network are synchronized by a central co-operation manager. The central co-operation manager is responsible for determining the timing of the activations of the various layers, the routing data and the termination of execution. Execution is terminated after a certain number of iterations.

3.1 Layer L1: Data Set Manipulation

The first task of this layer is to translate the initial data set to an equivalent representation without any loss of information. The input data representations of the four component programs vary considerably. Data types include discrete or continuous numeric values and symbolic data. This layer includes programs to convert these types of data to CN2, BEXA, C4.5 and ANN format. For an ANN, the symbolic

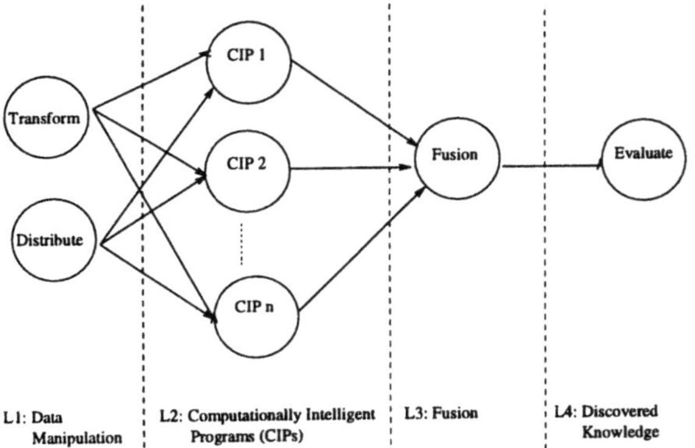

Figure 1: Computational network.

data are translated to a numeric representation by the use of a scaling algorithm.

Secondly, the selection of the various training data sets for insertion into the computational intelligent programs is addressed. The data set may be partitioned into non-overlapping sets, fully replicated or partially replicated as needed. Again, this selection process is problem dependent. Usually, if the data set is small, wide and the level of noise is high then the data are fully replicated. For example, a noisy tuberculosis data set consisting of 344 rows and containing 144 inputs yields the best results when fully replicated [9].

3.2 Layer L2: Individual Computational Intelligent Programs

This layer may consist of a number of sublayers, where the outputs of layer $L2_i$ act as input to layer $L2_{(i+1)}$. Each of the individual computational intelligent programs introduced in Section 2 represents a node in the directed graph. The programs execute autonomously by receiving data at its incoming edge, executing the specific algorithm, and computing and presenting output values. The programs have the ability to obtain and use the results of preceding programs. The number and type of individual computational intelligent programs are problem dependent. More than one program of a specific type may be executed in parallel. For example, a variety of ANNs, each using a different training function, may be utilized. In particular, optimal linear combinations of neural networks [3], which are constructed by forming weighted sums of the corresponding outputs of networks, may be incorporated into the layer.

3.3 Layer L3: Fusion Layer

The fusion layer consists of methods which interpret and combine the results obtained in the previous layer. It also contains procedures to test the accuracy and comprehensibility of the knowledge discovered by the individual components of layer L2. The fusion process is dependant on the fragmentation of the training set. If the training set has been partitioned, the fusion set usually combines the results of layer L2. Replication of data usually requires that some selection process is performed. Criteria such as the accuracy of individual rules, the number of examples covered and the comprehensibility of the rules are used to yield an optimum rule set.

3.4 Layer L4: Discovered Knowledge

The final layer contains a set of CN2-format rules which represents the knowledge discovered by the framework. Additional information, including the overall rule set accuracy, rule and attribute relevance, individual rule accuracy and coverage, rule set size and average number of attributes used, is also provided.

4 Discussion

Some results, which were obtained when applying the computational network to two medical data sets, are discussed elsewhere [9, 10]. For the first set of experiments, a tuberculosis (TB) data set was collected by a medical research team at the University of Stellenbosch, South Africa. The data concerned children under the age of five years who were diagnosed either confimed TB or not confirmed TB. The second data set concerned female diabetes patients of Pima Indian heritage living near Phoenix, Arizona in the United States of America.

In both cases, results indicate that, when combining the various computational intelligent programs, an increase in rule set accuracy and the overall comprehensibility are obtained.

5 Conclusion

The ability of a computational intelligent program to find the best solution to a given problem is partially determined by the data set representation. A number of factors, including training-set size and the ability of the program to discover interesting relationships, can mediate the effect of the data representation on the accuracy of the learned concept descriptions. By constructing a framework which takes advantage of each component program's strengths and data representation, the effect of the data set representation may be further minimized.

Current research includes the experimental evaluation and refinement of our framework. The development of a framework for executing computational intelligent programs is a promising approach to machine learning. Since it combines the strengths of various computational intelligent programs, it has the potential to address real-world problems which could not previously be solved by single methods.

References

[1] P. Clark and T. Niblett. The CN2 induction algorithm. *Machine Learning*, 3, 1989.

[2] L. M. Fu. *Neural Networks in Computational Intelligence.* McGraw-Hill, 1994.

[3] S. Hashem, B. Schemeiser, and Y. Yih. Optimal linear combinations of neural networks: An overview. In *Proc. IEEE ICNN'94.* Orlando, Florida, 1994.

[4] R. C. Lacher and K. D. Nguyen. *Hierarchical Architectures for Reasoning.* Kluwer Academic Publishers, 1995.

[5] C. J. Matheus, P. K. Chan, and G. Piatetsky-Shapiro. Systems for knowledge discovery in databases. *IEEE Transactions on Knowledge Data Engineering*, 5(6):904–913, 1993.

[6] J. R. Quinlan. *C4.5: Programs for Machine Learning.* Morgan Kaufmann, San Mateo, CA, 1994.

[7] H. Theron and I. Cloete. BEXA: A covering algorithm for learning propositional concept descriptions. *Machine Learning*, 14:321–331, 1996.

[8] G. G. Towell and J. W. Shavlik. *Refining Symbolic Knowledge using Neural Networks*, chapter 4. Morgan Kaufmann, San Mateo, CA, 1994.

[9] H. L. Viktor and I. Cloete. *Extracting Knowledge from Tuberculosis Data.* 1996.

[10] H. L. Viktor and I. Cloete. A computational network for diabetes diagnosis using intelligent data analysis. In *Proc. NEURAP'97.* Marselle, France, 1997.

[11] H. L. Viktor, A. P. Engelbrecht, and I. Cloete. Reduction of symbolic rules from artificial neural networks using sensitivity analysis. In *Proc. IEEE ICNN'95.* Perth, Australia, 1995.

Comparing a Variety of Evolutionary Algorithm Techniques on a Collection of Rule Induction Tasks

D. Corne
Parallel Emergent and Distributed Architectures Laboratory,
Department of Computer Science, University of Reading,
Reading, RG6 6AY, UK.
Email: D.W.Corne@reading.ac.uk

Abstract

Induction of useful rules from databases has been studied by several researchers. There remains need for systematic comparison of alternative such methods, especially considering the available variety of rule representation strategies, genetic operators, evolutionary algorithm designs, and so forth. Here, the performance of five commonly employed evolutionary algorithms are examined on a collection of 100 separate rule induction tasks on five freely available datasets. All tasks require the generation of rules in disjunctive normal form with either a fixed or free consequent maximising an accuracy/applicability tradeoff measure; tasks differ in terms of the dataset used, the identity of a fixed consequent (or no fixed consequent), and the maximum number of disjuncts allowed in the antecedent. Results generally indicate that *single-member based methods* (hill climbing, simulated annealing, tabu search) fare at least as well as population based techniques when rules are restricted to fairly low complexity, but this situation is reversed as rules are allowed to be more complex. These results are of import to data mining application developers and researchers wishing to find the appropriate search strategy for rule induction with respect to their particular needs.

1 Introduction

Expanding commercial and academic interest in data mining and knowledge discovery has led to much recent research activity in rule induction and other methods for finding useful information from large databases. A good tutorial and overview can be found in [5]. In particular, recent work suggests that evolutionary techniques compare favourably with traditional machine learning methods [2], while the main advantage of evolutionary techniques over traditional machine learning algorithms is the possibility for *undirected* data mining, in which the task is simply to generate good (in some specific sense) rules, rather than good rules with a fixed consequent, for example [7]. However, the broad range of potential evolutionary algorithm based techniques, and the differential performance of techniques with respect to different rule induction based tasks has been little explored as yet in the rule induction research literature (in contrast, much effort has been spent in looking at decision tree induction using methods such as the C4.5 learning algorithm [6]). For example, Iglesia *et al.* [2] show that simulated annealing and a genetic algorithm perform fairly similarly on *fixed-consequent* rule induction tasks on two datasets, but this result may not recur on an undirected rule induction task.

In place of comprehensive systematic examination of existing search schemes on the general rule induction problem, research has instead tended to focus on moves towards interesting or successful genetic algorithm based architectures for such problems [8], or inspired knowledge-intensive iterative or clustering techniques [3, 4]. Further, research in this area tends to be hampered by commercial sensitivity and consequent barriers to publishing results.

This work attempts to address some of these issues by reporting on the comparative rule induction performance of a wide range of evolutionary algorithm based techniques in the context of five freely available datasets. Each of six methods are compared on a total of 100 separate rule induction tasks (20 for each dataset). This ranges through different requirements on maximum rule complexity, and presence and identity of a fixed rule consequent.

Results indicate that single-member based methods (hill climbing, simulated annealing, tabu search) fare better than population based techniques when rules are restricted to fairly low complexity, but this situation is reversed as rules are allowed to be more complex (which in turn leads to fitter rules). This has been validated on experiments with commercially sensitive datasets associated with a major retailer. These results are of importance to data mining application developers and researchers wishing to find the appropriate search strategy for rule induction with respect to their particular needs.

2 Test Datasets

Understandably, applied data mining research is usually carried out with commercially sensitive datasets which are consequently unavailable for others to use for comparative and further work. There are a variety of freely available large datasets, however, which employ many of the characteristics appropriate for data mining research; essentially: they are relatively large, and hide a variety of interesting patterns. The datasets used in this paper are anonymised student/class enrolment records from various universities (available via `ftp://ftp.ie.utoronto.ca/pub/mwc/`), and used mainly in timetable optimisation [1].

Each such enrolment database consists of a number of records, one per student, listing the courses taken by that student. In this context, rule induction is used to find useful rules to predict if a student has enrolled for a course given information about other courses they have registered for. Such rules will then have an application in future course planning, future timetable construction, course design, course marketing, and so on.

A summary of the basic details of the problems used here is in Table 1. The problem-name abbreviations refer to the institution from which the enrollment records emerged.

This table provides the size of the pool of courses available to students, the number of students, and ('Choices') the number of courses taken by each student on average (i.e.: the mean number of fields in each record).

Table 1: Summary of enrolment datasets.

	Problem Dataset				
	EAR	HEC	STA	UTE	YOR
Courses	190	81	139	184	181
Records	1125	2823	611	2750	941
Choices	8	3	10	3	8

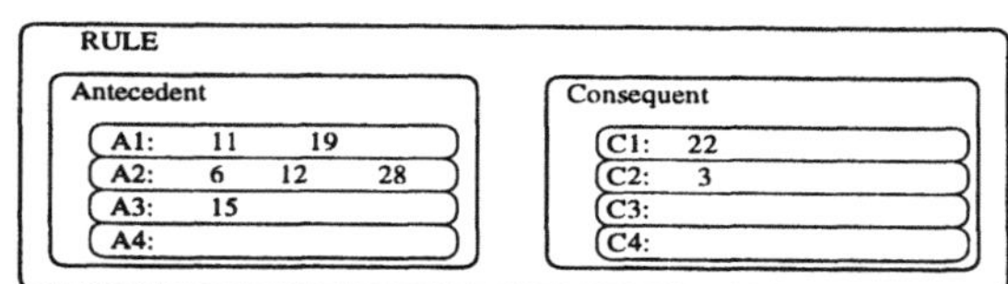

Figure 1: Rule representation.

3 Algorithms

Five algorithms are used in the comparative tests. It will be useful to first explain the simple rule representation method before detailing the algorithms employed. An explicit, structured rule representation is used, made of two main parts: an antecedent, and a consequent, with the obvious meanings. Each such part is made up of a variable number of *conjunct* structures, and each conjunct contains zero or more ***fields***. Figure 1 illustrates the notion.

In Figure 1, the antecedent substructure contains a number of conjuncts, labeled A1, A2, A3, etc ; similarly, the consequent substructure has conjuncts C1, C2, C3, and so on. The intended interpretation of the example in Figure 1 is:

If ((11 AND 19) OR (6 AND 12 AND 18) OR (15)) THEN ((22) OR (3))

Mutation operators on such rule structures operate directly on the fields, either altering the value of one or more fields in one or more conjuncts, or swapping fields between conjuncts. Crossover operators on such structures may combine the antecedent of one rule with the consequent of another, or construct a new antecedent by combining conjuncts from 'parent' antecedents, and so on.

In this report, all algorithms employed employ this representation, and are further detailed as follows:

stochastic hill climbing (SHC): This is the standard single-member hill climbing algorithm, employing three separate mutation operators: **M1**, altering the value of a single field in a conjunct; **M2** , remove a field from a conjunct; and **M3**, add a field to a conjunct.

simulated annealing (SA): Standard simulated annealing, employing the same operators as SHC.

tabu search (TS): Standard tabu search, again employing the same operators as SHC, and with a tabu list of size 7.

evolutionary algorithm 1 (EA1): A population-based version of the SHC algorithm, employing only mutation operators (same operators as SHC).

evolutionary algorithm 2 (EA2): An evolutionary algorithm employing the mutation operators **M1**, **M2** and **M3** along with one crossover operator: **X1**, in which the child antecedent is constructed from a combination of its parents antecedent conjuncts.

Both EA1 and EA2 used a steady-state [10] reproduction strategy, with a crossover rate of 0.7 in the case of EA2.

4 Experiments

Each individual trial ran for 10 000 evaluations, where each evaluation tested a candidate rule on the entire appropriate dataset. 20 trials each were run on each of 100 individual problem scenarios. Experiments took a long time. The 100 different problem scenarios are all combinations of the five problem datasets, four different sizes of antecedent (3, 6, 9, and 12 disjuncts allowed respectively), and five different kinds of consequent — 4 fixed and one variable.

Rule fitness was a weighted sum of accuracy (the proportion of cases in which both the antecedent and consequent were correct against a record) and applicability (the proportion of the entire dataset for which the antecedent was correct). The task was therefore to maximise this weighted sum.

5 Results

Space restrictions preclude a complete exposition of results. Instead, results are shown in Table 2 in full for one dataset, while Table 3 summarises the complete set of results in various ways.

First, Table 2 shows the average best over 20 trials for all experiments on the UTE enrollment records.

Trends emerging from Table 2 are:

1. A general distinction between the EA based methods and the others, with EAs generally performing at least as well as SHC, SA, and TS;
2. A general superiority of TS and SA over HC;
3. Increasing superiority of the EAs as the rules get more complex (ie: larger antecedents are allowed) and for the variable consequent cases.

Table 2: Full average results on UTE.

UTE Results: Fixed Consequent 1				
#Disjuncts:	3	6	9	12
SHC	1.05	0.95	0.87	0.84
SA	1.06	1.07	1.07	1.07
TS	1.06	1.07	1.07	1.06
EA1	1.07	1.09	1.09	1.1
EA2	1.07	1.09	1.09	1.1
UTE Results: Fixed Consequent 2				
SHC	1.015	0.83	0.78	0.72
SA	1.018	1.028	1.03	1.04
TS	1.017	1.025	1.03	1.04
EA1	1.025	1.038	1.05	1.05
EA2	1.023	1.04	1.05	1.05
UTE Results: Fixed Consequent 3				
SHC	1.020	0.94	0.78	0.73
SA	1.020	1.028	1.016	1.033
TS	1.020	1.029	1.010	1.032
EA1	1.022	1.039	1.054	1.059
EA2	1.020	1.039	1.053	1.058
UTE Results: Fixed Consequent 4				
SHC	1.022	0.9	0.76	0.89
SA	1.024	1.033	1.038	1.042
TS	1.024	1.033	1.030	1.055
EA1	1.028	1.045	1.06	1.065
EA2	1.028	1.043	1.06	1.065
UTE Results: Variable Consequent				
SHC	1.00	0.75	0.62	0.61
SA	1.01	0.98	0.92	0.96
TS	1.01	0.95	0.92	1.00
EA1	1.026	1.035	1.05	1.05
EA2	1.018	1.044	1.04	1.05

In Table 3, the results over all datasets are summarised and expressed in terms of rankings of the five algorithms tested. A rank of '1' indicates best, through to '5' indicating worst. Algorithms are distinguished by rank if a T-test reports a difference between them with 90% confidence or better.

The clearest trends are that rules of relatively high complexity tended to be better optimised by the EA methods. Also, as rules become large, and especially with variable consequents, differences start to show up between the single member methods, largely indicating a superiority of TS and SA over SHC. There generally seems to be little difference between the two EAs, but use of crossover (ie: EA2) seems to break through as a strong advantage on complex variable consequent problems.

6 Conclusions

It is important to be able to extract some general rules and guidelines from the current hive of activity in the rule induction arena. Several studies

Table 3: Ranking of algorithms over all problems.

Rankings on Fixed Consequent Problems					
#Disjuncts	SHC	SA	TS	EA1	EA2
3	3	3	3	1	2
6	3	3	3	1	1
9	5	3	3	1	1
12	5	3	3	1	1
Rankings on Variable Consequent Problems					
#Disjuncts	SHC	SA	TS	EA1	EA2
3	3	3	3	1	2
6	5	3	3	1	1
9	5	3	3	1	1
12	5	3	3	2	1
Overall Rankings by #Disjuncts					
#Disjuncts	SHC	SA	TS	EA1	EA2
3	2	2	2	1	2
6	3	3	3	1	1
9	5	3	3	1	1
12	5	3	3	2	1
Overall Rankings by Consequent					
Consequent	SHC	SA	TS	EA1	EA2
Fixed	5	3	3	1	2
Variable	5	3	3	1	1

report comparative results for a variety of methods on a variety of datasets, for example, but it can be difficult to tease out the underlying trends without careful control of independent factors. In this report, preliminary results find that control of a simple 'rule complexity' parameter, the maximum number of disjoined conjuncts allowed in a rule's antecedent, reveals differential performance along the 'simple hill climbing' through 'sophisticated evolutionary algorithm' continuum which may have been hidden in previous studies. In particular, the appropriate choice of rule induction technique appears to depend closely on maximum allowed rule complexity. This finding is important in an applications context, wherein rule complexity roughly equates to rule understandability; certain applications may require the induction of 'simple' rules, for example, while more complex rules may be allowable in other situations.

This paper has examined several standard search techniques on varied rule induction problems, and concludes that relatively high rule complexity and/or the need for a variable consequent warrant the use of a population-based method. Further work is under way to examine differential algorithm performance on more traditional machine learning datasets, using a more sophisticated rule representation, to see if these conclusions still hold. In particular, this will also consider the use of guided local search [9], a sophisticated hill climbing strategy recently found to work very well on certain combinatorial problems.

References

[1] M.W. Carter, G. Laporte, and S.Y. Lee. Examination timetabling: Algorithmic strategies and applications. *Operational Research Society*, 47(3):373–383, 1996.

[2] B. Iglesia, J.C.M. Debuse, and V.J. Rayward-Smith. Discovering knowledge in commercial database using modern heuristic techniques. Technical report, Department of Information Systems, University of East Anglia, 1996.

[3] M. Klemettinen, H Mannila, P. Roukainen, M. Toivonen, and I. Verkamo. Finding interesting rules from large sets of discovered association rules. In N. Adam, B. Bhargava, and Y. Tesha, editors, *Third International Conference on Information and Knowledge Management (CIKM94)*, pages 401–407. ACM Press, 1994.

[4] H. Mannila, M. Toivonen, and I. Verkamo. Efficient algorithms for discovering association rules. In Fayyad and Uthurusamy, editors, *Knowledge Discovery in Databases*, pages 181–192. AAAI Press, 1994.

[5] M. Hosheimer and A.P.J.M. Siebes. Data mining: the search for knowledge in databases. Technical Report Report CR-R9406, CWI, The Netherlands, 1994. (available via: ftp://ftp.cwi.nl/pub/CWIreports/AA/CS-R9406.ps.Z).

[6] R. Quinlan. *C4.5 Programs for Machine Learning.* Morgan Kauffman, San Mateo, CA, 1993.

[7] N.J. Radcliffe. GA-miner: Parallel data mining with hierarchical genetic algorithms. Technical Report GR/J99278, EPSRC AIKMS Grant, 1996.

[8] N.J. Radcliffe and P.D. Surry. Co-operation through hierarchical competition in genetic data mining. Technical Report EPCC-TR94-09, EPCC, 1994.

[9] C. Voudrais and E. Tsang. Partial constraint satisfaction problems and guided local search. Technical Report TR CSM-250, Dept. Computer Science, University of Essex, 1995.

[10] D. Whitley. The GENITOR algorithms and selection pressure. In J.D. Schaffer, editor, *The Proceedings of the Third International Conference on Genetic Algorithms*, San Mateo, 1989. Morgan Kaufmann.

An Investigation into the Performance and Representations of a Stochastic, Evolutionary Neural Tree

K. Butchart, N. Davey, R.G. Adams
School of Information Sciences, University of Hertfordshire
Hatfield, Herts., AL10 9AB, UK.
Email: {K. Butchart, N. Davey, R.G. Adams}@herts.ac.uk

Abstract

The stochastic competitive evolutionary neural tree (SCENT) is a new unsupervised neural net that dynamically evolves a representational structure in response to its training data. Uniquely SCENT requires no initial parameter setting as it autonomously creates appropriate parameterisation at runtime. Pruning and convergence are stochastically controlled using locally calculated heuristics. A thorough investigation into the performance of SCENT is presented. The network is compared to other dynamic tree based models and to a high quality flat clusterer over a variety of data sets and runs.

1 Introduction

This paper provides a thorough comparative analysis of the performance of a new unsupervised neural net architecture, SCENT , first introduced in [3]. This architecture contains two unique features: firstly in its combination of evolutionary growth with stochastic pruning and secondly in its dynamic, autonomous parameterisation; the user supplies no information to the net aside from the training data. Here we evaluate the network against the best networks with equivalent functionality using a collection of data sets chosen to provide a variety of clustering scenarios.

2 Dynamic Neural Trees and Stochastic Clustering

Unsupervised clustering using neural networks is a well established technique, with the simple competitive net (SCN) being the elementary model and Kohonen's self organising maps (SOMs) the most popular manifestation [5]. In a SOM the classifying units are arranged with some spacial topology, usually a grid. Recent research has investigated the possibility that the units could be arranged in a tree structure; such an arrangement has two potential benefits: searching a tree is fast and any hierarchical information in the data may be explicitly represented in the tree. Most clustering algorithms, neural net or otherwise, expect the number of classes used to be predefined; this is an obvious problem if no *a priori* knowledge about the data is available. Research has been done to investigate the possibility that a network can dynamically evolve the appropriate structure in response to the data. Such an approach can naturally lead to tree structures - a unit may produce child nodes if it feels it is not classifying its local data with sufficient granularity.

For the purposes of the results reported here we have compared our model with two other dynamic neural trees, the dynamic competitive learning technique of Racz and Klotz [8], and the neural tree of Li [6]. Apart from technical differences between the models SCENT has one fundamental difference: it requires no initial parameter setting. This is an important quality, as moving the need for the number of clusters to be specified, to one of specifying some other less transparent parameters is of doubtful benefit.

Most learning rules in neural networks perform gradient descent on an error function, and unsupervised clusterers are no exception, leading to the well documented problem of local minima in the error function. Several researchers have proposed the addition of stochasticity, often in conjunction with a simulated annealing technique, as a means of overcoming this difficulty for competitive networks. One of the best of these clusterers is the neural gas model of Martinetz et al. [7] which we use as another source of comparison. The idea of stochasticity and an associated temperature is also used in the SCENT model as described in the next section.

3 The SCENT Model

3.1 Classification and Learning

In the SCN model a search through all nodes is undertaken at each input presentation for the node that is closest, which is designated the winner. The winner then moves towards the input vector. In a tree structured network, the search is recursive and thereby restricted to the winning branch, by the following algorithm:

Find best-child of current, i.e. closest to the input
Move best-child towards input
Set current to best-child and recurse

Once learning is completed classification is provided by leaf nodes, whilst the rest of the tree provides a hierarchical classification. The search for a winner either in the learning or test phase is therefore $O(db)$, where d is the average depth and b is the average branching factor, which contrasts with an $O(b^{d-1})$ search of all leaf nodes.

3.2 Growth and Pruning

As already stated growth decisions in SCENT are devolved to units in the tree. A unit is allowed two modes of growth, it may produce children or it may produce a sibling. A node may only produce children once, and when it does so it creates a pair, offset from the parent by a small amount of noise. Further children are generated by an existing child producing a sibling, once again a noisy copy. The decision that a node makes to grow is determined in the SCENT network by *relative activity.* Broadly speaking this is the ratio of the number of times a node has won to the number of times its parent has won. In order for the performance of the network to be relatively invariant with respect to the order of presentation of the inputs, activity is calculated over a sliding window with a linearly decreasing weighting, as originally used in Li's neural tree.

Outwards or Downwards

Having decided to grow, a leaf node must decide whether it should spawn children or a sibling and to do this it makes use of its *tolerance* value. The tolerance of a node is the radius of its classificatory hypersphere. In order for finer grained classification to be performed in lower levels of the tree, tolerance is reduced from parent to child. The reduction is scaled by the success of the parent in placing those vectors it classifies within tolerance. If most activity is within tolerance the children are given tolerance values of significantly smaller size - reduction is up to 40%.

If a leaf node finds that the majority of the vectors it classifies are within tolerance then it is classifying a spatially compact cluster and should therefore produce children to subclassify this group. On the other hand, if most vectors are not within tolerance, there may be large scale structure that should be represented by another node at the same level - so a sibling is spawned.

Pruning

A leaf node does not have a guarantee of continued existence and it may die at birth or later in life. Short term pruning, or *growth rejection*, takes place if a new node does not reduce the error of its parent sufficiently; this decision is taken at the epoch end immediately following its creation. A leaf node that has established itself in the tree may still be removed if its long term performance is inadequate. The performance is measured by comparing the activity of the node against a threshold that decreases in lower levels of the tree; node removal is only performed at epoch boundaries.

It is in the pruning process that stochasticity is used. In early growth it is useful for the network to create much tentative new growth, and also for longer term pruning to be more common. To this end growth rejection, is initially made less likely — it becomes a probabilistic process, inversely proportional to a temperature value that is initially high and decreases exponentially over time. Similarly long term pruning is made more likely early in a nodes life - it is stochastic, but directly proportional to temperature. The addition of this mechanism leads to a more reliable performance by the model when compared to its non-stochastic predecessor, CENT [1]. See [2] for comparative results.

Parameterisation

As described earlier it is important for data exploration that a clusterer does not require the user to specify a set of opaque parameters that will define the final classification. In previous dynamic tree based clusterers two key values, tolerance and threshold had to be initialised. In the SCENT model the threshold value is subsumed in the node creation process and the tolerance value is calculated at run time. The root node has a tolerance such that 2/3 of the data lies within tolerance; this is simply accomplished by an initial pass through

Table 1

Set 1	2D single source Gaussian
Set 2	2D single source Gaussian
Set 3	2D Uniform within a square
Set 4	16 even clusters in 4 groups, 2D
Set 5	10 clusters with varying density and size
Set 6	Varied Clusters with hierarchical stucture
Set 7	Anderson's IRIS data [4]

the data set, which will also position the root node roughly at the mean of the data. Subsequent tolerance values for descendants of the root are calculated recursively from this initial value as described above.

4 Tests

The four neural net models identified earlier: NGas, neural tree (Li), dynamic learning (RK) and SCENT were used to cluster seven data sets summarised in Table 1. Each model was run over each data set for 15 complete runs. For all the networks except SCENT, experimentation was initially performed to find a satisfactory set of parameters. The number of nodes in the NGas model was set to be roughly that number of the leaf nodes that SCENT produced. The results produced are averages over these 15 runs.

It is very difficult to judge the performance of a hierarchical cluster, but relatively straightforward to measure the quality of a flat classification. The aim of a clusterer is to produce low Sum Squared Error (SSE) but in a dynamic network this is complicated by the ability of the net to use an unbounded number of nodes. A more reasonable network comparison can therefore be obtained by using the product of SSE and number of nodes. The results are summarised in Table 2, in which the column marked '*' is the product of SSE and Nodes. Data set 5 is shown in Figure 1.

The presentation of the results is to some extent unfair to the tree based networks as many nodes are not used as classifiers, and in this light the performance of SCENT is exceptionally strong. Its SSE by #nodes is roughly double that of NGas, which implies that the leaf nodes are almost certainly out-performing NGas. The performance gap to the other two dynamic trees is substantial.

When attempting to judge the quality of the hierarchies produced by the tree based networks the criteria to look at are more subjective. However for the data which is designed to be hierarchical

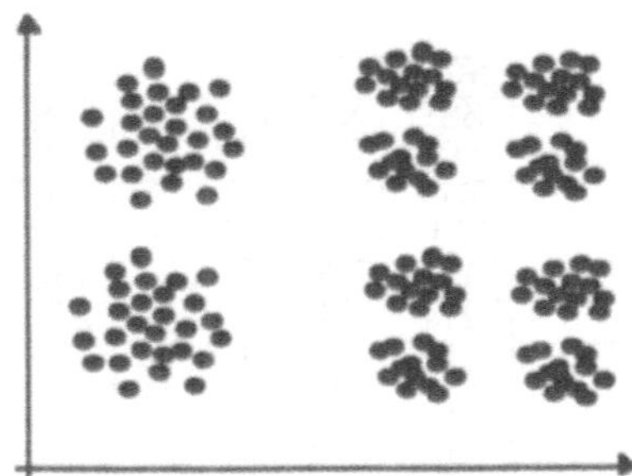

Figure 1: Data set 5, 10 clusters in 4 large groups.

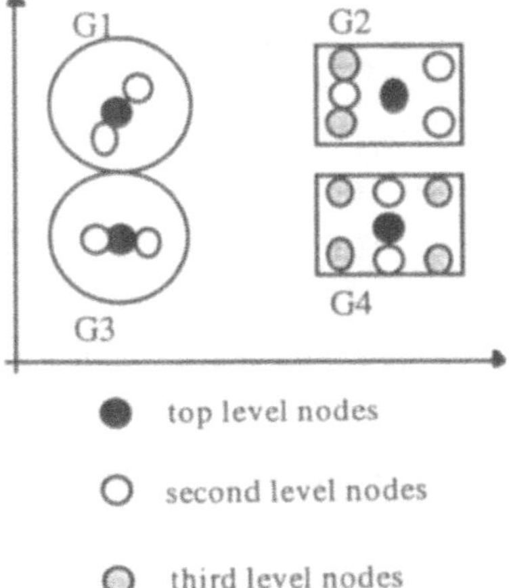

Figure 2: Position of nodes in a typical run of SCENT.

we would like the induced structure to reflect that of the data, so that, for example, the top level of the tree should contain representatives for each of the major groups in the data and subsequent layers should reflect subgroupings. Some behaviour is not desirable, specifically a tendency to produce long thin branches which do not add to the semantic content of the tree. It was found that both Li and RK had a tendency to do this which could only be avoided by very careful selection of parameters, although even this was not always possible. The high average node count for RK is explained by this phenomena. A typical example of the classication tree produced by SCENT for data set 5 is shown in Figure 3, with the position of the classifying nodes shown in Figure 2. It can be seen that the network has produced a near optimal classification and a reasonable tree structure. The results from SCENT were repeatable - tree stuctures produced varied across runs but were generally consistent.

Table 2

	NGas			Li			RK			SCENT		
Set	SSE	Nodes	*	SSE	Nodes	*	SSE	Nodes	*	SSE	Nodes	*
1	15	16	240	22	26	572	24	57	1368	20	28	560
2	679	16	10864	766	110	84260	346	685	237010	678	27	18306
3	11	25	275	11	43	473	48	20	960	21	29	609
4	125	32	4000	194	55	10670	602	42	25284	248	25	6200
5	272	16	4352	298	49	14602	455	45	20475	330	27	8910
6	132	16	2112	125	40	5000	79	56	4424	129	28	3612
7	1	16	16	6	8	48	3	32	96	2	36	72
Avg.	176	19.6	3123	203	47.3	16518	222	134	41374	204	28.6	5467

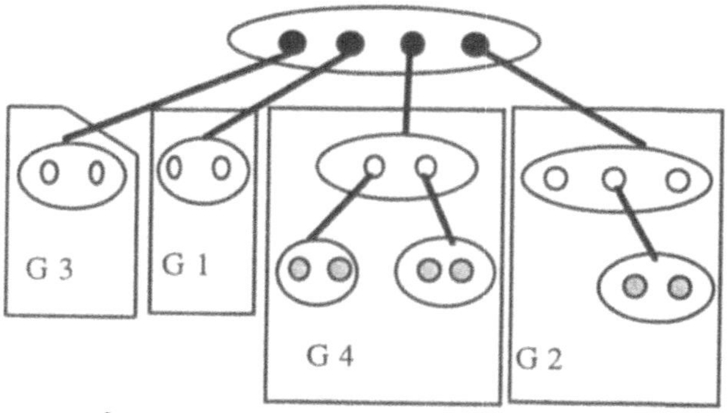

Figure 3: The organisation of the nodes from Figure 2 in the resulting tree.

5 Conclusions

Using neural nets to perform data exploration is difficult; most models require the preimposition of a maximum number of clusters, and will normally classify the data to utilise all classificatory units. Some recent architectures have attempted to dynamically create tree structures to overcome the need for prescribing cluster number and to give a hierarchical view of the data. Their use though has been problematic, with a tendency to display great sensitivity to initial parameter values. SCENT attempts to overcome this problem by being completely autonomous. In the tests reported here the model produced excellent results when viewed as a straightforward clusterer, giving comparable performance to the NGas network. It also produced useful, compact and repeatable tree structures to represent any hierarchical information in the data.

References

[1] K. Butchart, N. Davey, and R. Adams. Hierarchical classification with a competitive evolutionary neural tree. *Neural Networks*, 1995. to appear.

[2] K. Butchart, N. Davey, and R. Adams. *A Comparative Study of two Self Organising and Structually Adaptive Dynamic NeuralTree Networks*. John Wiley, Chichester, 1996.

[3] K. Butchart, N. Davey, and R. Adams. Hierarchical classification with a stochastic competitive evolutionary neural tree. In *Proceedings of ICNN96*, volume 2, pages 1372–1377, 1996.

[4] B. S. Everit. *Cluster Analysis*. Edward Arnold, London, 1993.

[5] J. Hertz, A. Krogh, and R. Palmer. *Introduction to the theory of Neural Computation*. Addison Wesley, New Jersey, 1991.

[6] T. Li, Y. Tan, S. Suen, and L. Fang. A structurally adaptive neural tree for recognition of a large character set. In *Proc. 11th IAPR International Joint Conference on Pattern Recognition*, volume II, pages 187–190, 1992.

[7] T. Martinetz, S. Berkovich, and K. Schulten. Neural-gas network for vector quantisation and its application to time-series prediction. *IEEE transactions on Neural Networks*, 44(4), 1993.

[8] J. Racz and T. Klotz. Knowledge representation by dynamic competitive learning techniques. *SPIE Applications of Artificial Neural Networks II*, 1469:778–783, 1991.

Experimental Results of a Michigan-like Evolution Strategy for Non-stationary Clustering

A. I. Gonzalez, M. Graña, J. A. Lozano and P. Larrañaga
Dpt. CCIA Univ. Pais Vasco/EHU, Aptdo 649, 20080 San Sebastián, España
e-mail: ccpgrrom@si.ehu.es

Abstract

Non-stationary clustering deals with the clustering of a sequence of data samples obtained at a diverse time instant. A paradigm case of non-stationary clustering is the color quantization of image sequences. We propose an efficient evolution strategy to compute adaptively the color representatives for each image in the sequence.

1 Introduction

Evolution strategies [2, 3, 15] have been developed mainly by Schwefel and Rechemberg since the 1960s. They belong to the broad class of algorithms inspired by natural selection. The features most widely accepted as characteristic of evolution strategies are: (1) vector real valued individuals, (2) the main genetic operator is mutation, (3) individuals contain local information for mutation so that adaptive strategies can be formulated to self-regulate the mutation operator. However, it is also widely recognized [15] that a lot of hybrid algorithms can be defined, so that it is generally difficult to assign a definitive "label" for a particular algorithm. Nevertheless, we classify the algorithm proposed here as an evolution strategy because it fits in the above characterization.

The application of evolution strategies to clustering and vector quantization problems [1], follows the general assumption of the stationarity of the statistical properties of the data. They also share the so-called (in the context of classifier systems) Pittsburg approach: each individual represents a feasible solution of the optimization problem.

The work reported in this paper differs from previous attempts to apply evolution strategies to clustering problems in several technical points and in a major philosophical point: we are looking for an adaptive strategy that can be applied in near real time. Our emphasis in adaptiveness implies that we look for near optimal results that can be computed in a bounded time frame. Also, non-stationary clustering is an unusual problem, in which the underlying statistical properties of the data are assumed to be time-varying, with an unknown time dependence model. Therefore, there is no guarantee that a solution to the clustering problem obtained in a given time instant will be optimal for the future data. New samples must be drawn and the solution to the clustering problem recomputed at each time step. Adaptive algorithms are desirable for this problem because they reduce computing requirements, profiting from past solutions, although there is no knowledge of a time dependence model.

We are working on a very simple and effective evolution strategy for the adaptive computation of the cluster centers for a non-stationary clustering problem. Salient features of the evolution strategy proposed here are:

1- The mapping of the problem into the population follows a Michigan-like approach : individuals correspond to individual cluster centres, at each time instant the population gives a feasible solution to the clustering of the sample.

2- To approach real-time response we impose one-generation adaptation for the sample that characterizes the problem at each time instant.

3- Only mutation operators are applied. These mutation operators are guided by the estimated covariance matrices of the clusters.

It is well known that vector quantization is a special case of clustering [5, 6, 7, 8, 10, 12]. Color quantization [11, 13, 14, 16, 17] is an instance of the more general problem of vector quantization [8] in the space of colors. Color quantization of image sequences [4, 9] is, thus, an instance of non-stationary clustering. Our approach is to state the problem as a time-varying clustering problem, and to propose an evolution strategy as an adaptive mechanism.

2 Non-stationary Clustering

A time variant formulation of the clustering problem must start with the explicit assumption of a time varying population described by an discrete time stochastic process $\{X_t t = 0, 1, ..\}$. In this framework, a working definition of the non-stationary clustering problem could read as follows: Given a sequence of sets of vectors $\aleph(t) = \{x_1(t), .., x_n(t)\}$ obtain a corresponding sequence of partitions of each of them into a sequence of sets of disjoint clusters $\{\aleph_1(t), .., \aleph_c(t)\}$ that minimizes a criterium function $C = \sum_{t\geq 0} C(t)$. The related non-stationary vector quantization can be stated as the search for a sequence of representatives $Y(t) = \{y_1(t), .., y_c(t)\}$ that minimizes the error function (distortion) $E = \sum_{t\geq 0} E(t)$. Functions C and E coincide when the criterium function is within the cluster variance and the error function is based in the Euclidean distance. The stochastic minimization problem that must be considered in order to derive adaptive algorithms can be stated as follows:

$$\min_{\{Y(t)\}} \sum_{t\geq 0} \sum_{j=1}^{n} \sum_{i=1}^{c} \| x_j(t) - y_i(t) \|^2 \delta_{ij}(t)$$

$$\delta_{ij}(t) = \begin{cases} 1 & i = \arg\min_{k=1,\dots,c} \| x_j(t) - y_k(t) \|^2 \\ 0 & \text{otherwise.} \end{cases}$$

3 The Evolution Strategy

A widely accepted pseudocode representation of the evolution strategy algorithm is as follows [3]:

```
t := 0
initialize P(t)
evaluate P(t)
while not terminate do
        P'(t) :=recombine P(t)
        P''(t) := mutate P'(t)
        evaluate P''(t)
        P(t+1) := select (P''(t) ∪ Q)
        t := t+1
end while
```

The population $P(t)$ is a set of solutions proposed at generation t. The algorithm iterates until some time or optimality condition is met. The evaluate operator computes the objective function for each individual. The recombine operator finds mates and recombines them producing a set of offsprings $P'(t)$, we have not defined it in our strategy. The mutate operator produces new individuals by the application of random perturbations. Usually evolution strategies define these random perturbations as samples of normally distributed random variables. The set Q can be either the set of parents or be empty, depending on the strategy.

We have defined each individual as a single cluster center:

$$P(t) = \{y_i(t); i = 1..c\}.$$

The local fitness of the individual is, then, its local distortion:

$$F_i(t) = \sum_{j=1}^{n} \|x_j(t) - y_i(t)\|^2 \delta_{ij}(t).$$

The solution proposed at time t is given by the entire population. The population as a whole can be evaluated to measure its fitness

$$F(t) = \sum_{i=1}^{c} F_i(t)$$

which corresponds to the objective function to be minimized.

The mutation operator is a random perturbation that follows a normal distribution. We have decided to perform a guided selection of the individuals to be subjected to mutation. The guide is to obtain mutations from the individuals with the highest distortion. We have considered two possibilities:

$$S^1(t) = \{i \,|F_i(t) \geq \bar{F}(t)\}$$
$$S^2(t) = \{k = \arg\max\{F_i(t)\}\}$$

By design we perform a fixed number of mutations in any case, so that the number of mutations per individual will depend on the selection strategy chosen. The mutation itself is performed adding to the selected individuals pseudorandom samples of a normal random variable:

$$P''(t) = \left\{\lambda = y_i + u; u \approx N\left(0, \hat{\Sigma}_i(t)\right) | y_i \in S^o(t)\right\}$$

where S^o denotes the mutation selection strategy employed (either S^1 or S^2).

The estimation of the covariance matrix is based on the actual cluster elements assigned to the mutated individual.

$$\hat{\Sigma}_i(t) = \frac{\sum_{j=1}^{n}(x_j(t)-y_i(t))(x_j(t)-y_i(t))^t \delta_{ij}(t)}{n-1}.$$

Finally, to define the selection of the next generation individuals we have followed the so called ($\mu+\lambda$)-strategy. We pool together parents and children:

$$P''(t) \cup Q = \{y_1, .., y_c, y_{c+1}, .., y_{c+\lambda}\}$$

where λ is the number of individuals generated by mutation.

The fitness function used for the final selection of an individual is the distortion when the sample is codified with the codebook given by $P''(t) \cup Q - \{y_k\}$, more formally:

$$F_k^s(t) = \sum_{i=1;i\neq k}^{c+\lambda} \sum_{j=1}^{n} \|x_j(t) - y_i(t)\|^2 \delta_{ij}(t)$$

The selection operator selects the c best individuals according to the above fitness. To define the selection operator formally, first consider the set:

$$P'''(t) = \left\{ y_{i_1}, .., y_{i_{c+\lambda}} \middle| i_j < i_k \Rightarrow F_{i_j}^s(t) > F_{i_k}^s(t) \right\}$$

Then the specification of the selection operator is: $P(t+1) = \{y_i \in P'''(t); i = 1..c\}$ This selection involves the fitness of the whole population with the addition of the mutations generated. This makes the algorithm sensitive to the number of mutations generated, forcing the above mentioned restriction to a fixed number of them.

The last critical decision in the design of the evolution strategy is the mapping of the generation number into the frame number of the image sequence. We have defined a one to one mapping. That is, for each image only one generation of the evolution strategy is computed.

4 Experimental Results

The sequence of images used for the experiment is a panning of the laboratory taken with an electronic Apple Quicktake camera. Original images have an spatial resolution of 480x640 pixels. Each two consecutive images overlap 50% of the scene.

The figures show the distortion results of the application of the evolution strategy to random samples of 1600 pixels of each image in the cases of $c = 16$ and $c = 256$. As a reference algorithm we have used a variation of the algorithm proposed by Heckbert [11] as implemented in MATLAB following [18]. This algorithm has been applied in two ways. First it has been applied independently to each image (Time Varying Min Var). Second, the set of color representatives obtained with the Heckbert algorithm for the first image has been used to color quantize the entire sequence (Time Invariant Min Var). The space left between both applications of the Heckbert algorithm is the range of behaviors that can be considered as adaptive in some way.

The results of the evolution strategy are shown with a 95% confidence interval based on 30 replications of its application to the image sequence. Overall the evolution strategy behaves adaptively in all the cases. As the data samples remain the same, the only source of variability is the mutation operator.

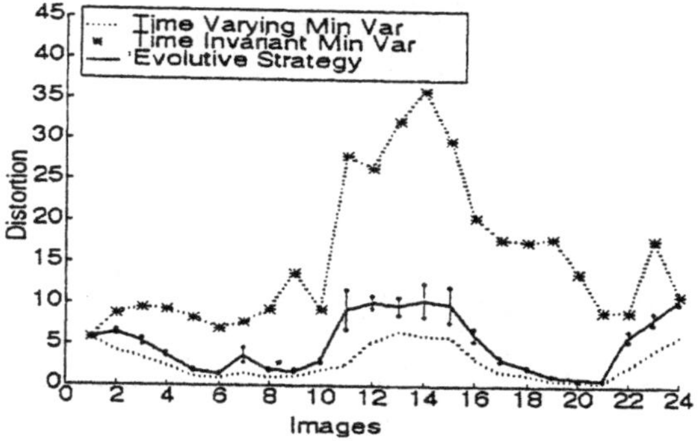

Figure 1: Results with c=16, $\lambda = 16$, and selection S[1] of the mutated individuals.

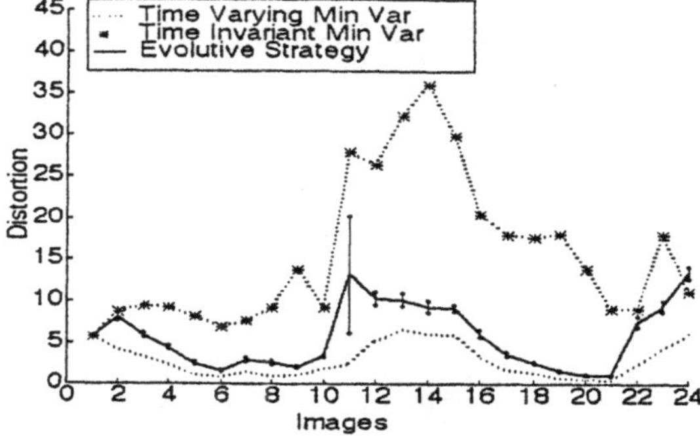

Figure 2: Results with c=16, $\lambda = 16$, and selection S[2] of the mutated individuals.

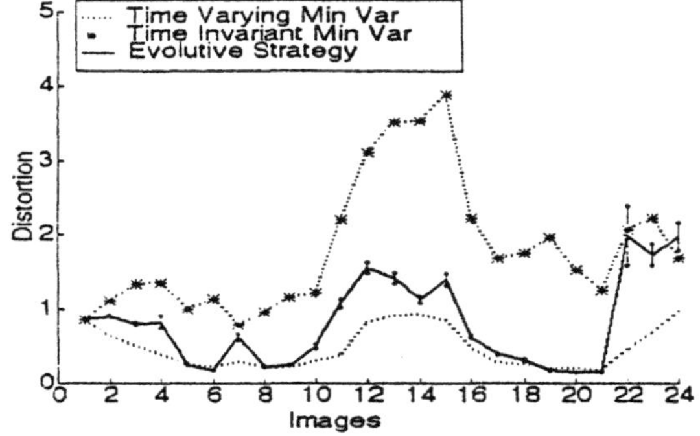

Figure 3: Results with c=256, λ = 128, and selection S^1 of the mutated individuals.

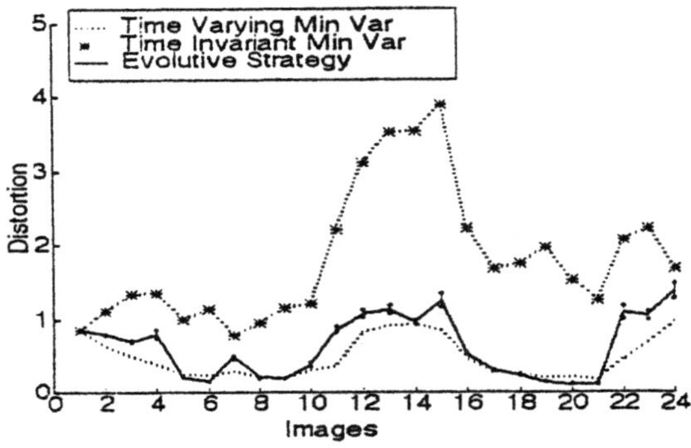

Figure 4: Results with c=256, λ = 256, and selection S^1 of the mutated individuals.

The results on quantization to 16 colors show that S^1 gives better results with lower variance. The results on quantization to 256 (assuming S^1) suggest that allowing $\lambda = c$ gives near optimal results.

5 Conclusions

We have proposed an evolution strategy for the adaptive computation of color representatives for color quantization that can be very efficiently implemented and reach almost real time performance for highly variable color populations. Preliminary experimental works show a good response for a realistic sequence of images. We are actually working to define deterministic mutation operators that could reduce the variance of the results.

6 Acknowledgements

This work is being supported by a research grant from the Dpto. de Economía of the Excma. Diputación de Guipuzcoa, and a predoctoral grant and project PI94-78 of the Dept. Educación, Univ. e Inv. of the Gobierno Vasco.

References

[1] G. P. Babu and N. M. Murty. Clustering with evolution strategies. *Pattern Recognition*, 27:321–329, 1994.

[2] T. Back and H. P. Schwefel. An overview of evolutionary algorithms for parameter optimization. *Evolutionary Computation*, 1:1–24, 1993.

[3] T. Back and H. P. Schwefel. Evolutionary computation: an overview. In *IEEE Int. Conf. Evolutive Computation*, pages 20–29, 1996.

[4] O. T. Chen, B. J. Chen, and Z. Zhang. An adaptive vector quantization based on the gold-washing method for image compression. *IEEE Trans Circuits & Systems for Video Techn.*, 4(2):143–156, 1994.

[5] E. Diday and J. C. Simon. *Clustering Analysis*, In K. S. Fu (editor), *Digital Pattern Recognition*, pages 47–94. Springer Verlag, 1980.

[6] R. D. Duda and P. E. Hart. *Pattern Classification and Scene Analysis.* John Wiley, Chichester, 1973.

[7] K. Fukunaga. *Statistical Pattern Recognition.* Academic Press, 1990.

[8] A. Gersho and R. M. Gray. *Vector Quantization and signal compression.* Kluwer Academic Publisher, 1992.

[9] Y. Gong, H. Zen, Y. Ohsawa, and M. Sakauchi. A color video image quantization method with stable and efficient color selection capability. In *Int. Conf. Pattern Recognition*, volume 3, pages 33–36, 1992.

[10] J. Hartigan. *Clustering Algorithms.* John Wiley, Chichester, 1975.

[11] P. Heckbert. Color image quantization for frame-buffer display. *Computer Graphics*, 16(3):297–307, 1980.

[12] A. K. Jain and R. C. Dubes. *Algorithms for clustering data.* Prentice Hall, 1988.

[13] M. S. Kankanhalli, B. M. Mehtre, and J. K. Wu. Cluster based color matching for image retrieval. *Pattern Recognition*, 29(4):701–708, 1996.

[14] T. S. Lin and L. W. Chang. Fast color image quantization with error diffusion and morphological operations. *Signal Processing*, 43:293–303, 1995.

[15] Z. Michalewicz. Evolutionary computation: practical issues. In *IEEE Int. Conf. Evolutive Computation*, pages 30–39, 1996.

[16] M. T. Orchard and C. A. Bouman. Color quantization of images. *IEEE Trans. Signal Processing*, 39(12):2677–2690, 1991.

[17] T. Uchiyama and M. A. Arbib. Color image segmentation using competitive learning. *IEEE Trans. Patt. Anal. and Machine Intelligence*, 16(12):1197–1206, 1994.

[18] X. Wu. *Efficient Statistical Computations for Optimal Color Quantization*, pages 126–133. Academic Press Professional, 1991.

Excursion Set Mediated Evolutionary Strategy

S. Baskaran[1,2] and D. Noever[2]
[1] Institut fuer Moleculare Biotechnologie, e.V.,
Beutenbergerstr. 11, DO-7745, Jena, Germany
[2] Biophysics Branch ES76, National Aeronautics and Space Administration
George C. Marshall Space Flight Center,
Huntsville, AL-35812 USA
Email: subbiah@darwin.msfc.nasa.gov

Abstract

By mediating excursion sets in the selection phase of the standard evolutionary strategies (ES), Plus and Comma, two possible new variants which we call, Semicolon, and Colon strategies have been devised and their optimization behavior on model landscapes reported. By an internal mechanism, these strategies systematically vary the excursion level, and achieve stronger self adaptive exploration of the fitness landscapes exhibiting emergent characteristics of both Comma and Plus to the advantage of them.

1 Introduction

Algorithms based on the Darwinian metaphor of evolution, like evolution strategies (ES) and genetic algorithms(GA) have largely revolutionized the field of combinatorial optimization and proved higher degree of success [4, 5, 7]. Although earlier models followed a stripped to essentials approach, later developments experimented many variations on the Darwinian central theme of reproduction, adaptation to external fitness measure and constant organization by selection. In this paper we present two new variants of evolutionary strategies which show mixed characteristics of the Comma and the Plus strategies for adaptive problem solving.

2 Excursion Sets, Fitness Landscapes and Global Optima

Let f(x) be a real valued and continuous function defined on a compact set $S \in R^N$. Let us assume that we are seeking the global minimum of $f(x)$ in S. Here x is the n - dimensional trial solution vector, and ξ is an arbitrary fitness level in $f(x)$.

Then the set $E(\xi) = \{x \in S : f(x) \leq \xi\}\xi \in R^1$ is called the excursion set for the landscape, $f(x)$ at excursion level, ξ.

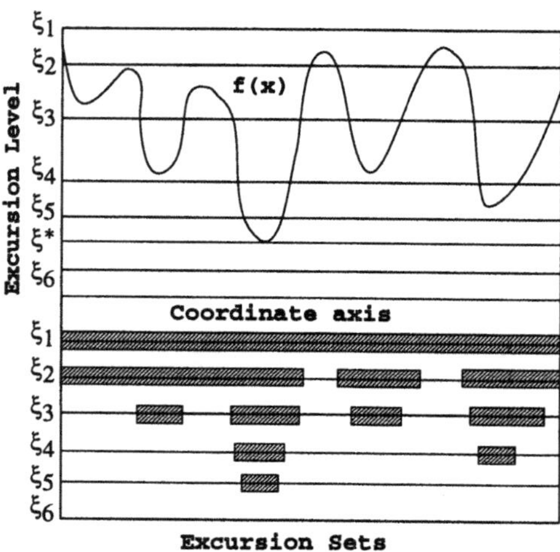

Figure 1: Excursion sets in a modal landscape. The hatched boxes are the excursion sets. The ξ's are excursion levels, ξ^* is the global optimum level.

Figure 1 shows a modal landscape (for brevity a one dimensional case is chosen), $f(x)$, marked with some hypothetical excursion levels ($\xi_1 - \xi_6$). At the bottom are shown the induced excursion sets at these levels. In general excursion sets will induce a fitness hierarchy resulting in a finite collection of excursion subsets (real interval boxes in the language of the interval arithmetic) whose union constitute the excursion set for that level. In the case of multimodal functions, they will be mostly disconnected subsets. These subsets will diminish in size as the level decreases (for minimization processes), and eventually become empty [1, 3].

We can easily see that $E(\xi_i)$ belong to the $\sigma -$ *algebra* of the Lebesgue sets in S. Let $m(\cdot)$ be a measure proportional to the Lebesgue measure on S, so that $\mu(S) = 1$. Since this measure is countably additive, we have then

$$\mu(\bigcup_{k=1}^{\infty})\mu(E_k) = \sum_{k=1}^{\infty} \mu(E_k)$$

Now let us define a function $\psi(\xi) = m(E(\xi))$ which maps R into R^+.

Suppose f^* be the global minimum of $f(x)$ in S (marked as ξ^* in Figure 1). Then for any $\xi < \xi^*$, $\psi(\xi) = 0$. Since $m(f^{-1}(f^*)) = 0$, this implies $\psi(\xi) = 0$ iff $\xi \leq f^*$. Also it follows that

$$m(f^{-1}(-\infty, \xi)) = \psi(\xi),$$

which shows such a function is bounded: $0 \leq \psi(\xi) \leq 1$.

From the above property, if a quantity β^* exists so that $\psi(\xi) = 0$ for $\xi \leq \beta^*$, and $\psi(\xi) > 0$ for $\xi > \beta^*$, then obtaining such β^* provides an efficient methodology for bracketing the global optimum by traveling through the excursion sets in a hierarchical fashion up to that level (see Figure 1).

In other words, assuming a minimization process, we define an excursion level, perform a sampling in the associated excursion set, and then adaptively update the level to further optimal values. By repeating these two steps, we will arrive at excursion sets with measure zero. And just before that happens, we confine with probability one, the excursion subset that contains the global optimum of $f(x)$ in S. This is exactly what is achieved by mediating the selection through excursion sets in evolutionary strategy.

This mathematically relates to obtaining an approximation for $\psi(\xi)$ in the interval $[f^*, \|f\|_\infty]$, where $\|f\|_\infty$ is the supremum of $f(x)$ in S.

3 Standard and Excursion Set Mediated ES

In the present study we consider only the standard strategies and two possible excursion variants. Correlated mutations are not considered. Figure 2 shows the algorithmic flow for a single generation in all the four strategies.

3.1 Comma and Plus Strategies

Figure 2(a) and (b) show the Comma and Plus strategies. These strategies start with a population of μ landscape points (called parents) and their strategy parameters. From them in a generation, λ new points are generated by applying the genetic operators of crossover and mutation. Each offspring is made by taking two parents at a time. The selection is called Comma if we select the new μ from the λ newly generated children without including the old parents. However the selection is called Plus if the μ new parents are selected from the combined set of $(\mu + \lambda)$ members.

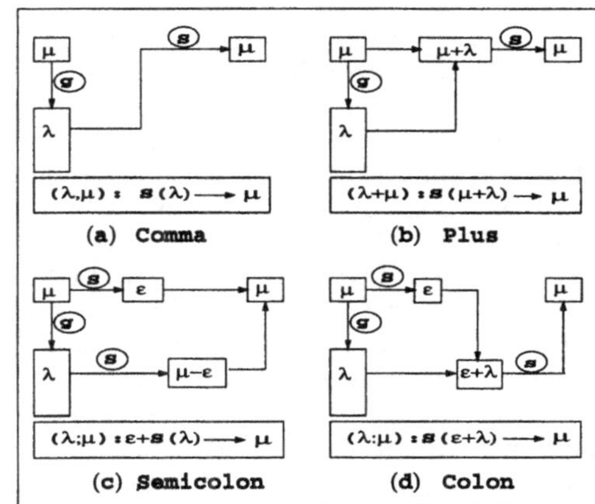

Figure 2: One generation algorithmic flow in standard and the proposed evolutionary strategies. μ=parents. λ=children. g is the reproduction operator. S is the selection operator. See text for further explanations.

In other words, in Comma strategy old parents are dropped at the end of each generation, after they pass their genetic material to children. Here age of any parent is strictly one generation only. In Plus, since they are allowed to compete with their children for selection into next generation, their age can be more than one generation.

Evolutionary strategies work because each generation improves the quality of the population through the selection phase. The selection phase does the differential pumping of points towards promising regions by selecting better performing members.

3.2 Semicolon and Colon Strategies

We introduce excursion sets as shown in Figure 2(c) and (d) to modify the flow and obtain two variants of the standard evolutionary strategies. For the new strategies to operate, we introduce an excursion level, ξ_i, and split the population into the set ϵ containing excursion members and its compliment.

Semicolon strategy is shown in Figure 2(c). In this during selection, the set ϵ is passed onto next generation, as a kind of elitism. The remaining $(\mu - \lambda)$ members are selected from the λ children generated. This constitutes one generation. No change is made to the excursion level. The new population is measured against the excursion level, and once again excursion partition made, the ϵ set selected and passed on to next generation.

Since selection favors best fitness members, ϵ value will increase towards μ. When it equals μ, we update the excursion level a better value (either current population best or average based on which one has been initially set for excursion level). This is called one excursion cycle. Here excursion members stay in the population for at least one excursion cycle.

Figure 2(d) illustrates the flow associated with the Colon evolutionary strategy. The operational steps are same as the Semicolon except that at each generation, the ϵ excursion members are mixed with the λ children, from this combined set, μ best ones are selected as parents for the next generation. This way it avoids longer life time for some excursion members. This is equivalent to Plus strategy, but instead of all old parents only excursion parents compete with children for selection into next generation.

In the beginning excursion cycle lengths will be small. Let $\tau_1, \tau_2, \tau_3, \ldots, \tau_n$ be the successive excursion cycle lengths in generations, then we have $\tau_i \propto 1/\psi(\xi)$. Thus at optimal levels, the waiting time to complete the cycles will approach ∞.

3.3 Choice of Excursion Level Parameter and Scaling

Any fitness in the range, $[f^*, ||f||_\infty]$ can be chosen as the excursion level for the Semicolon and the Colon strategies. Since we do not have a suitable means of adaptively updating when arbitrarily fixed, we have to somehow or other make the algorithm choose it. This will be readily achieved if we ask the algorithm to set either the best or the average population fitness. Since during evolution, both best and average fitnesses evolve to optimality, the adaptive excursion level update will be automatically done internal to the strategies.

We have four total combinations of strategies and excursion level choices: Semicolon with best fitness as excursion level, Semicolon with average fitness as excursion level; Colon with best fitness as excursion level, Colon with average fitness as excursion level. As an example, we illustrate in Figure 3 how in Colon strategy, best and average fitnesses

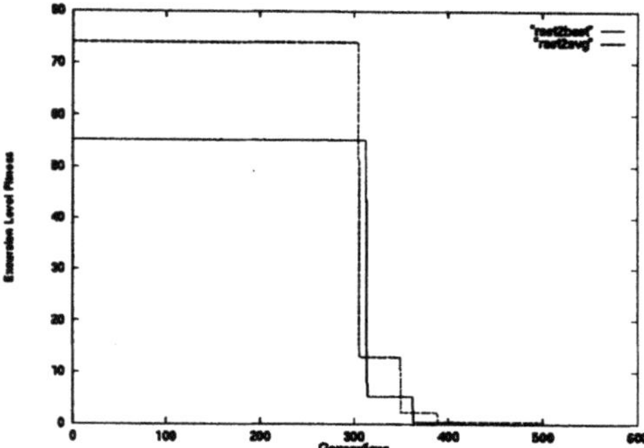

Figure 3: Excursion level fitness evolution for Rastrigin's generalized function set at population best fitness as well as population average fitness. Both have long stochastic cycle lengths but always be bound by a monotonically decreasing sequence.

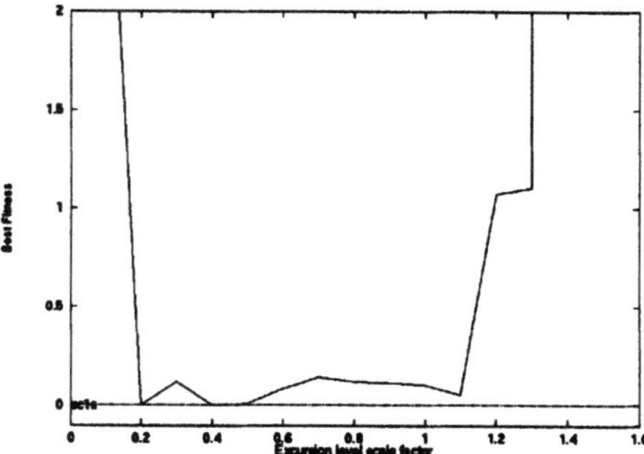

Figure 4: Best fitness attained at generation 500, vs. scale factor ρ for Schwefel's 6th problem. Semicolon strategy with average fitness for excursion level.

as excursion levels evolve in the case of Rastrigin's generalized function [6].

With all strategies, many excursion cycles are executed before we get a zero measure for $\psi(\xi)$ which signals obtaining the best approximation to the optimum.

In order to better control the level of pressure excursion levels exert on the selected size of ϵ and hence evolving population, we introduce a scale factor, $\rho > 0$. This multiplies the current excursion level, and hence either diminishes or boosts its effect. Figure 4 shows the effect of ρ on the best fitness attainable during evolution for Schwefel's 6th problem.

3.4 Emergent Mixed Strategy Characteristics

Within each excursion cycle, mixed Plus and Comma characters evolve and disappear in the new strategies. In the early phase of the excursion cycle when ϵ is small compared to μ, Colon strategy behaves like Comma strategy. However, towards the end of the cycle, ϵ will be closer to μ and so behaves like a Plus strategy. This mixed behavior will be recurring throughout the evolution. For Semicolon, the emergent behavior will be a convex combination of Comma and Plus.

When $\rho < 0$, the level will be lowered so more elitism and fractional Plus character will be imparted. However when $\rho > 1.0$, the level will be made high, and so more of Comma character will be imparted in the selection. Thus by tuning this parameter and choosing either Semicolon or Colon strategy, we will be able to impart Plus and Comma characters differentially into selection during evolution.

4 Tests with Model Landscapes

In this section we report the optimization characteristics of the Semicolon and Colon strategies and compare them with the standard ones. No attempt has been made to characterize these strategies in all parameter combinations, so the results are minimal but sufficient to bring about their salient features. The results of an extensive study will be reported elsewhere. However parameter setting in each experiment is done in a way that it made the problem hard for the standard ES.

4.1 Shekel's Foxholes

This problem [8] is known to be a hard problem for ES. To avoid some initial points that have already fallen in holes, the population is initialized far away from the holes. One sigma value was used for both variables.

$$\frac{1}{f_5(\vec{x})} = \frac{1}{K} + \sum_{j=1}^{25} \frac{1}{c_j + \sum_{i=1}^{2}(x_i - a_{ij})^6}$$

$$(a_{ij}) = \begin{pmatrix} -32 & -16 & 0 & 16 & 32 & -32 \cdots & 0 & 16 & 32 \\ -32 & -32 & -32 & -32 & -32 & -16 & \cdots & 32 & 32 & 32 \end{pmatrix}$$

$$K = 500 \quad ; \quad f_5(a_{1j}, a_{2j}) \approx c_j = j$$

$$-65.536 \leq x_i \leq 65.536$$

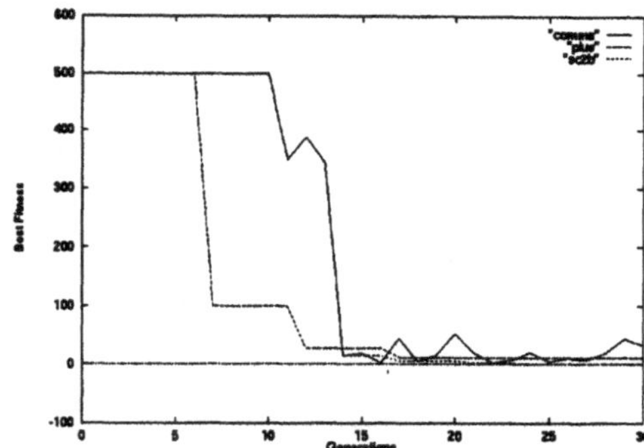

Figure 5: Evolution best fitness for strategies for foxhole landscape. The parameters are $\mu = 10$, $\lambda = 100$, x recombination = discrete, sigma recombination = arithmetic.

$$\min(f_5) = f_5(-32, -32) \approx 1$$

Figure 5 shows the evolution of the best fitness in the first 30 generations. Both the Colon and Semicolon strategies performed well in discovering the optimum, where as Comma, and Plus obtained premature convergence.

4.2 Rastrigin's Generalized Function

This function [6] is also a very difficult problem for standard ES as shown by Hoffmeister and Baeck [2]. we used the multiple sigma values which makes the problem even harder [9].

$$f_7(\vec{x}) = nA + \sum_{i=1}^{n} x_i^2 - A\cos(\omega x_i)$$

$$A = 10 \quad ; \quad \omega = 2\pi$$
$$-5.12 \leq x_i \leq 5.12$$

$$\min(f_7) = f_7(0, \ldots, 0) = 0$$

Figure 6 shows the computational results for this function. Although Comma closely followed the Colon and Semicolon, its convergence speed slowed after generation 300, and obtained premature convergence. Population best fitness was set as the excursion level.

4.3 Schwefel's 6th Function

Although this function is unimodal [7], searching along the coordinate axis only gets a poor rate of convergence as the gradient is not oriented along any axis. Other than correlated mutations, no strategies achieve the required accuracy.

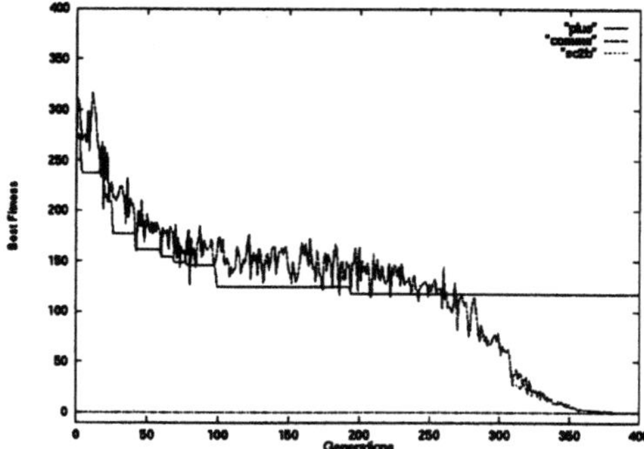

Figure 6: Best fitness evolution for Rastrigin's function. The parameters used are $\mu = 100$, $\lambda = 300$, x recombination = discrete, sigma recombination = arithmetic.

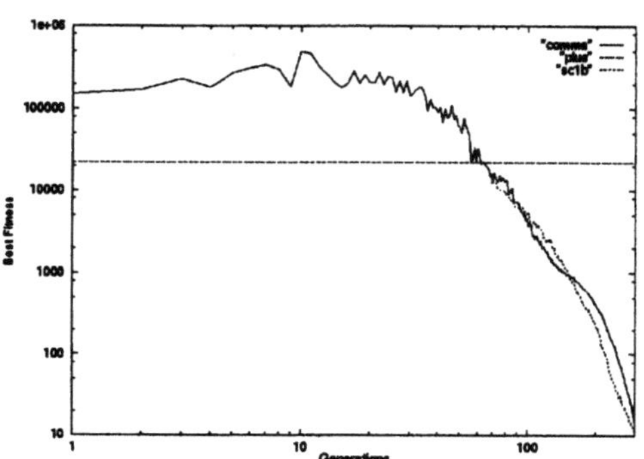

Figure 7: Best fitness evolution for the Schwefel's problem 1.2. The parameters: $\mu = 30$, $\lambda = 240$, n=20, x recombination = discrete, sigma recombination = geometric

$$f_6(\vec{x}) = \sum_{i=1}^{n} \left(\sum_{j=1}^{i} x_j \right)^2 = x^{\mathrm{T}} \mathbf{A} x + b^{\mathrm{T}} x$$

$$-65.536 \leq x_i \leq 65.536$$

$$\min(f_6) = f_6(0, \dots, 0) = 0$$

Figure 7 shows the results on this function. The Semicolon strategy with best fitness as excursion level performed well compared to Comma and achieved faster convergence.

5 Conclusions

Evolutionary adaptation on fitness landscapes can be improved with the introduction of excursion sets in the selection phase. Of the many possible variants, two strategies called Semicolon, and Colon were constructed and their characteristics studied. Early results show their potential of combining the characteristics of both Plus and Comma strategies for the optimal exploration of complex fitness landscapes during evolutionary computation.

6 Acknowledgments

SB thanks Professor Peter Schuster for the formative discussions while the author was with the Department of Theoretical Chemistry, University of Vienna, Austria.

References

[1] D. Adler. *The Geometry of Random Fields.* John Wiley & Sons, New York, 1981.

[2] T. Baeck and F. Hoffmeister. Genetic algorithm and evolution strategies - similarities and differences. In *PPSN I*, pages 455–469. Dortmund, FRG, October 1-3 1990.

[3] S. Baskaran, P. Stadler, and P. Schuster. Approximate scaling properties of RNA free energy landscale. *J. Theor. Biol.*, 181:299–310, 1996.

[4] J. H. Holland. *Adapdation in Natural and artificial systems.* University of Michigan Press, Ann Arbor MI, 1992.

[5] I. Rechenberg. *Evolution strategie: Optimierung technischer Systeme nach Prinzipien der biologischen Evolution.* Formmann-Holzboog, Stuttgart, 1973.

[6] G. Rudolph. *Globale Optimierung mit Parallelen Evolutionsstrategien.* PhD thesis, Department of Computer Science, University of Dortmund, Dortmund, FRG, 1990.

[7] H.-P. Schwefel. *Evolution and Optimum Seeking, Sixth Generation Computer Technology Series.* John Wiley & Sons, Chichester, 1995.

[8] J. Shekel. Test functions for multimodal search techniques. In *Fifth Annual Princeton Conference on Information Science and Systems*, 1971.

[9] J. Sprave. Linear neighborhood evolution strategy. In *Proceedings of the Third Annual Conference on Evolutionary Programming*, pages 42–51. World Scientific, 1994.

Use of Mutual Information to Extract Rules from Artificial Neural Networks

T. Nedjari
LIPN-CNRS URA 1507, Institut Galilée, Université Paris 13
Avenue J.-B. Clément 93430 Villetaneuse France
Email: nedjari@ura1507.univ-paris13.fr

Abstract

This paper investigates the application of the mutual information for the evaluation of neuron inputs and for the selection of the relevant ones. The rules extraction method is based on the notion of *weights templates*, parameterizing regions of weights space using the mutual information criteria. The simulation results obtained with this method are very satisfactory.

1 Introduction

Artificial neural networks (ANNs) have proved to be a powerful and general technique for machine learning [9]. However without some form of explanation capability, the full potential of trained ANNs may not be realized [2]. In this paper, a new method is proposed to remedy this problem.

This method is called *MITER* for mutual information and template for extracting rules. *MITER* searches a subset of relevant input weights for each unit within the trained ANNs. This is accomplished by using mutual information (MI) to measure arbitrary dependencies between random variables, and to compute the information content in each input during the learning process. In other words, MI evaluates the *information content* of each input connection with regard to the output unit. The approximate evaluation of MI for each connection is a pruning operation that selects a subset of all relevant input connections [3]. The latter are used to define a *weights template*. The symbolic rules generated from the template have a *N-of-M* form. This form has been used successfully on many experiments [6, 8].

In the following sections, we first define the MI criteria, then explain how it's used to measure input connections relevance, after that we describe how to generate a template. Finally, we present the *MITER* algorithm together with some simulation results.

2 Mutual Information Criteria

An algorithm of rule generation which can reflect the real ANNs behaviour can be considered as a system that reduces the initial uncertainty by consuming the information contained in the input data. This reduction is confronted with two problems: insufficient input information and suboptimal operation [5]. The second problem can be solved by selecting relevant features at a level of an ANNs and by selecting relevant connections at a level of each hidden and output neuron.

Mutual Information provides suitable formalism for realizing these two selections. We choose the MI to select relevant connections because it does not depend on a transformation of variables. It is a function of the joint probability distribution of two variables. The MI between two variables Inp_i (i^{th} input variable) and Out (output variable) is defined as follows:

$$I(Inp_i; Out) = H(Out) - H(Out|Inp_i) \quad (1)$$

where $H(Out)$ is the entropy that measures the initial uncertainty in the output variable;

$$H(Out) = -\sum_{Out} P(Out) \log P(Out), \quad (2)$$

where $P(Out)$ is the probability of the variable Out. $H(Out|Inp_i)$ is the conditional entropy that measures the average uncertainty after knowing the input vector;

$$H(Out|Inp_i) = -\sum_{Inp_i}\sum_{Out} P(Inp_i, Out) \log P(Out|Inp_i), \quad (3)$$

where $P(Inp_i, Out) = P(Out|Inp_i)P(Inp_i)$, and $P(Out|Inp_i)$ is the conditional probability of the output Out given the input Inp_i. Using the above equations

$$I(Inp_i; Out) = \sum_{Inp_i} \sum_{Out} P(Inp_i, Out) \log \frac{P(Inp_i, Out)}{P(Inp_i)P(Out)} . \quad (4)$$

The mutual information is therefore the amount by which the knowledge provided by the input vector decreases the uncertainty about the output.

3 Selection of Connections

In rule extraction from ANNs, we are often confronted with practical constraints on the number of neurons in ANNs and on the number of connections for each neuron. Since the information contained in some neurons and connections is sufficient to determine the correct ANNs outputs with low ambiguity, we may reduce the initial set of input connections to a small subset.

This *connections reduction* problem can be formulated as follows: for a given neuron i with n_i connections, find the subset with $m_i < n_i$ connections that is *maximally informative* about the neuron activation. Since the goal is to find the input connections subset that is most relevant to give the true output. This problem can be reformulated as follows: given a neuron i with the output Out and an initial set S_i with n_i input connections, find the subset $S'_i \subset S_i$ with m_i input connections that minimizes $H(Out|S'_i)$, i.e. that maximizes the mutual information $I(S'_i; Out)$.

The MI is used to identify and eliminate the irrelevant connections [4]. Since, each connection contributes to the neuron activation, when we select a subset of connections, we obtain an approximated solution. This solution is acceptable because first there is no guarantee that the optimal subset of connections will processed in the optimal way by the learning algorithm. Secondly, the availability of an *informative* input vector is not sufficient for the development of a correct results.

Before using Equation (4) to select relevant connections for each neuron, the data must be quantified by clustering, to have a fixed number of values. If the input values are in $\{0, +1\}$, the output value will be also limited between 0 and 1 because of the use of sigmoidal function which maps $I\!R$ into $[0, 1]$. All values are clustered in some spaces by dividing the interval $[0, +1]$ into k equally spaced segments. k is an experimental value which will be used in the next section to select the relevant inputs. If the input values are in $\{-1, +1\}$, in the same way, all values between -1 and $+1$ will be clustered in some intervals. Generally, all the data are clustered in three at five intervals.

For each neuron, the mutual information between each input and the output $I(Inp_j, Out_i)$ is computed. After that, the connections which maximizes the information about neuron activation is chosen. In order to be selected, a connection must be informative about the output of the neuron. The selection of the relevant inputs can be described by the following procedure called $SelInp$:

For each neuron i with n_i input connections:

1. $S_i = \{$ set of n_i connections$\}$, with $S'_i = \emptyset$.
2. For each connection Inp_j (j=1, n_i), compute MI with the output $I(Inp_j; Out)$.
3. Compute the average mutual information of all input connections by
$$AvgMI = \frac{\sum_j I(Inp_j, Out)}{n_i}.$$
4. Output the set S'_i containing all connections where $I(Inp_j; Out) \geq \dfrac{AvgMI}{k}$:
   - $S_i = S_i - \{Inp_j\}$,
   - $S'_i = S'_i \cup \{Inp_j\}$;

   where k is an experimental value defined during the clustering.

4 Rule Extraction

Unlike other decompositional methods such as *KT* [7] and *Subset* [9] which employ variations on search and test technique, our approach performs rule extraction by directly interpreting *weights templates* as rules. Consequently our approach avoids the computational problems of other decompositional techniques and the recourse to heuristics to control the search of the solution space.

We define a weight template as a parameterized region of weight space corresponding to a specific symbolic function [1]. To find a template, T, from neuron input weights, we associate with each weight one of the three values:

$$T_j \in \{-V, 0, +V\}$$

where V is a positive value which will be defined by the minimization of the Euclidean distance between the input weights and the template. The component T_i is equal to $+V$ (respectively $-V$) if Inp_j is belonging to the set S_i' of the selected connections and the input weight ω_{ij} is positive (respectively negative). The value of V, that minimizes the Euclidean distance $||T - W||^2$, is given by

$$V^* = \frac{\sum_j |\omega_{ij}| \, Val_j}{m} \qquad (5)$$

where m is the number of connections selected by the use of MI (m is the cardinality of S_i'). $Val_i = +1$ if $T_i \neq 0$, $Val_i = 0$ otherwise. The neuron i is active if

$$\sum_j \omega_{ij} x_j + \theta_i > 0 \qquad (6)$$

where θ_i is the bias of the neuron i. When we replace ω_{ij} by T_j and θ_i by b_i in the above equation, we obtain

$$\sum_j T_j x_j + b_i > 0 \qquad (7)$$

To find the formula that defines b_i, we distinguish two cases:

First case: The input values $x_i \in \{0, +1\}$.
To find a general value for b_i, we consider the worst case: $x_i = 1$ for all $T_i = -V$. In these conditions, we can rewrite Equation 7 as follows: $-m_n V + nV + b_i > 0 \Leftrightarrow (-m_n + n)V + b_i > 0$ where m_n represents the number of connections with the value $-V$ and n is the necessary number of connections with a value $+V$ to have an active neuron.
We search for a value b_i for which the latter equation is verified, i.e. $b_i > (m_n - n)V$. We choose for b_i a value limited by the two successive integer values $(m_n - n)$ and $(m_n - n + 1)$ multiplied by the real value V, we obtain: $b_i = (m_n - n + \frac{1}{2})V$.

Second case: The input values $x_i \in \{-1, +1\}$.
To find a general value for b_i, we consider the worst case: all $T_i = -V$ will be multiplied by $+1$ and all $T_i = +V$ will be multiplied by -1. We find $b_i = ((m_n + m_p) - 2n + 1)V$ where m_p represents the number of connections with the value $+V$.
Let $m = m_n + m_p$, we can rewrite b_i formula as follows: $b_i = (m - 2n + 1)V$.

To find the value of n, we look for the value of V' that minimizes the Euclidean distance $||T - W||^2$ which includes the biases θ_i and b_i. The optimal value of V', noted V'^*, is given by

$$V'^* = \frac{\sum_j |\omega_{ij}| \, Val_j + b\theta_i}{m + b^2}, \qquad (8)$$

$$\text{where } b = \begin{cases} m_n - n + \frac{1}{2} & \text{if } x_i \in \{0, +1\}, \\ m - 2n + 1 & \text{if } x_i \in \{-1, +1\}, \end{cases}$$

$$\text{and } Val_j = \begin{cases} 1 & \text{if } T_i \in \{-V, +V\}, \\ 0 & \text{if } T_i = 0. \end{cases}$$

The value of V'^* depends on the value of n and it is near to the value of V^* because they minimize nearly the same Euclidean distance. The value of n, for which $V'^* = V^*$, is

$$n = \begin{cases} \left\lceil \frac{(m_n + \frac{1}{2}) \sum_j |\omega_{ij}| \, Val_j - m\,\theta_i}{\sum_j |\omega_{ij}| \, Val_j} \right\rceil & \text{if } x_i \in \{0, +1\}, \\ \left\lceil \frac{(m+1) \sum_j |\omega_{ij}| \, Val_j - m\,\theta_i}{2 * \sum_j |\omega_{ij}| \, Val_j} \right\rceil & \text{if } x_i \in \{-1, +1\}. \end{cases} \qquad (9)$$

The following procedure, noted *TempRul*, is used to generate a template T, to determine the value of n, and to extract rule with *N-of-M* form:

For each sigmoidal unit i with n_i input connections and a bias θ_i:

1. For each input connection Inp_i:
   If $Inp_i \notin S_i'$ then $T_i = 0$
   else if $\omega_{ij} > 0$ then $T_i = +V$
   else $T_i = -V$,
   where S_i' is the set of selected input connections by the use of MI.
2. Compute the value of n using Equation (9),
3. The extracted rule from the neuron i is

   $$\text{If } n \text{ of } \{\bigcup_j x_j \text{ if } T_j = +V, \bar{x_j} \text{ if } T_j = -V\}$$

   then y_i,

   where each relevant input connection j is represented by the symbol x_j, and the output is described by the symbol y_i.

5 Algorithm

The basic steps in the algorithm *MITER*, which is used to extract *N-of-M* rules from each hidden and output unit using the mutual information and weights template, are:

1. learn the problem with ANNs using backpropagation algorithm;
2. for each input vector, compute the output of each neuron;
3. quantify the output values by clustering;
4. select the relevant input connections by the use of MI (*SelInp*, section 3);
5. define the Template and extract the rule of the *N-of-M* form (*TempRul*, section 4).

6 Simulation

As an example of logic expressions that can be processed by *MITER*, we describe a simple problem. This problem is defined by the following expression: $y = x_1x_2 \vee \bar{x_1}\bar{x_2}\bar{x_3}\bar{x_4}$ where x_1, x_2, x_3, $and x_4$ are boolean variables. To solve this problem, we define <4-2-1> neural networks architecture to learn a logic expression from the truth table using a backpropagation algorithm. The input values are in $\{-1, +1\}$, and $k = 5$. The intermediate results are:

First hidden unit				
input weights	0.791	0.792	0.765	0.765
bias	2.122			
mutual information	0.162	0.225	0.225	0.225
$\frac{AverageMI}{k}$	0.042			
extracted rule	If 1 of $\{x_1, x_2, x_3, x_4\}$ then x_5			
Second hidden unit				
input weights	0.730	1.730	0.298	0.266
bias	-1.304			
mutual information	0.311	0.311	0.000	0.000
$\frac{AverageMI}{k}$	0.031			
extracted rule	If 2 of $\{x_1, x_2\}$ then x_6			
Output unit				
input weights	-3.418		3.278	
bias	2.630			
mutual information	0.246		0.586	
$\frac{AverageMI}{k}$	0.083			
extracted rule	If 1 of $\{\bar{x_5}, x_6\}$ then x_7			

From the three extracted rules, we can obtain the following final rule: if $x_1x_2 \vee \bar{x_1}\bar{x_2}\bar{x_3}\bar{x_4}$ then x_7.

7 Conclusion

It is generally admitted that hybrid connectionist-symbolic models constitute a promising approach for developing more robust and more powerful system. The algorithm *MITER*, developed in this work, uses simples steps to extract symbolic rules from standard backpropagation ANNs which have not been specially constructed to facilitate rule extraction. The extracted rules, given by *MITER*, are both simple and efficient in the same time. Furthermore, this method can be applied easily to other problems. To compare our method with other techniques for extracting knowledge from trained ANNs as a set of symbolic rules, *MITER* will be tested on other applications like, for instance, the three monks problems. This work can be extended to the problem of finding an optimal neural networks architecture.

8 Acknowledgments

I wish to thank Y. Bennani for his general comments on this work, and the members of my laboratory for their helpful discussions. I would like also to acknowledge H. Merabet from Cambridge University for his interests and reading this paper.

References

[1] J. A. Alexander and M. C. Mozer. *Template-based algorithms for connectionnist rule extraction*, pages 609–616. MIT Press, Cambridge, MA, 1995.

[2] R. Andrews, J. Diederich, and A. B. Tickle. A survey and critique of techniques for extracting rules from trained artificial neural networks. Technical Report QUTNRC-95-01-02, Queensland University of Technology, Brisbane, Australia, 1995.

[3] R. Battiti. Using mutual information for selecting features in supervised neural net learning. *IEEE Transactions on Neural Networks*, 5(4):537–550, 1994.

[4] B. V. Bonnlander and A. S. Weigned. Selecting input variables using mutual information and non-parametric density estimation. In *Proceedings of the 1994 International Symposium on Artificial Neural Networks (ISANN'94)*. Tainan, Taiwan, 1996.

[5] K. J. Cherkauer and J. W. Shavlik. Rapid quality estimation of neural network input representations. *Neural Information Processing Systems*, 8, 1996.

[6] D. H. Fisher and K. B. McKusick. An empirical comparison of ID3 and back-propagation. In *Proceedings of the eleventh International Joint Conference on Artificial Intelligence*, pages 788–793, 1989.

[7] L. M. Fu. Rule generation from neural networks. *IEEE Transaction on Systems, Man, and Cybernetics*, 28(8):1114–1124, 1994.

[8] P. M. Murphy and M. J. Pazzani. ID2-of-3: Constructive induction of N-of-M concepts for discrimination in decision trees. In *Proceedings of the Eight International Machine Learning Workshop*, pages 183–187, 1991.

[9] J. W. Shavlik and G. G. Towell. Extracting refined rules from knowledge-based neural networks. *Machine learning*, pages 71–101, 1993.

Connectionism and Symbolism in Symbiosis

N. Allott, P. Fazackerley and P. Halstead
Computing Department, The Nottingham Trent University, Burton St,
Nottingham, NG1 4BU, England
Email: nma@doc.ntu.ac.uk

Abstract

In this paper we examine a previously published algorithm which addresses the problem of network growth by implementing a clustering algorithm to operate on time dependant data. The computational constraints of the problem forced the development of an architecture, which in retrospect can be analysed in terms of a computational and symbolic module operating symbiotically. Here we attempt to identify the computational constraints that necessitate the use of this architecture, and any further merits it has. Further, we analyse the nature of the interaction between the two modules and highlight the manner in which the behaviour the symbiotic modules correlates with what is known of human problem solving behaviour.

1 Introduction

This paper both develops and outlines the computational merits of a symbiotic symbolic and connectionist network used within previously published clustering algorithm [1]. Within this algorithm a connectionist network was used to embody the relationship between discrete observable elements, for example letters of the alphabet. In the simplest case the relationship modelled is the relative statistical distribution, or context, of the letters. As such the network can be shown to be very similar to N-gram analysis [3, 7, 10] or the transition network of a Markov model [6]. However it is superior in that the scope of analysis to be considered does not have to be specified globally, but is dynamically and locally determined for each node.

An algorithm was required to produce these connectionist *trees* from empirical data. It was the original intention that the algorithm be developed within the context of the connectionist paradigm. By this we mean capable of being implemented in a parallel manner such that, the functions used to compute the working parameters for each node have access only to those items to which the node is architecturally linked (such as the back propagation algorithm or simple Hebbian learning). However the fact that (a) the algorithm attempts to *grow* the network (b) time dependant data was being handled, made this design goal difficult to satisfy. In the next section we attempt to formalise the source of the difficulty.

2 Formal Definition of Problem

The network can be characterised as an n-tuple: $\langle P, N, L, a, \Sigma\rangle$. Where

P is the set of primitive nodes,
N is the set of all nodes, initially $N = P$,
L is the set of links between nodes, initially $L = \emptyset$, and each element of L is a 3-tuple $\langle p, c, s\rangle$, parent, child, strength, where $s = 1$.
a is the activation function, $a : N, t, \Sigma^* \rightarrow \{0, 1\}$
Σ is the set of possible primitve evidence.

Derived from this we have:-
Σ^* the set of sentences possible from Σ.
$P(N)$ the power set of all nodes,
A the set of abstract nodes, defined $A = N \cap P'$,
c is the set of children of a node, and can be defined in terms of L and N.

To simplify the problem, in the initial case the strength of all links is assumed to be 1, and the activation function is boolean returning {0,1}.

If $S \in \Sigma^*$ then $S[1] \in \Sigma$ and is the first element of S. It follows for simple sequence analysis (such as the text string discussed above) where there is a 1:1 mapping between P and Σ, the activation function for a node n, where $n \in P$,

$$a(n, t, S) = \begin{cases} 1, & S[t] = n \\ 0, & S[t] \neq n. \end{cases}$$

And for node n where $n \in A$, the activation is some function $f()$ of the activation of the children of n, $a(n, t, S) = f(c(n), S, t)$.

To illustrate, take the simple problem of clustering with the node immediately adjacent on the right. If $c(n)[x]$ is the xth child of node n, the activation function for a parent with two children becomes:

$$a(n,t,S) = a(c(n)[1],t,S) \wedge a(c(n)[2],t+1,S).$$

If it is our aim to grow the network, the process of growth is to identify a new node x such that

$$N = N \cup x$$

and to add links such that

$$L = L \cup \bigcup_{i \in c(x)} \{\langle x, i, 1\rangle\}.$$

It is hoped that each new identified node x should capture some abstract feature of the input domain thus giving the network greater depth of perception.

When it comes to implementing the above procedure on a connectionist network there are essentially two problems to be solved:

2.1 Type Token Distinction

In a network where there is localist representation (one node represents one feature in the problem domain) it is conceivable that a single feature occurs twice in the input pattern. This feature could occur in distinct positions within the input sting, in overlapping positions, or the most difficult: in recursively embedded positions. (The string 'ana' as it occurs in the input string 'banana' is an example of an overlapped position) The problem is how does a single node within the network simultaneously take on two distinct activations to reflect the two instances of the feature within the input string?

2.2 New Nodes, New Links

Most mainstream connectionist learning algorithms have a predefined set of nodes and a predefined set of links (often all nodes fully interconnected). The learning algorithm then adjusts the strength of the existing links to reflect the relationship between the nodes. This algorithm addresses the problem of network growth, all nodes are therefore initially unconnected and there are no links. The aim is to identify new nodes and new links that represent relationships between the existing nodes. However if we are to implement the algorithm to identify these relationships in a truly parallel, connectionist manner we have a problem, each node has no direct architectural link to any other node from which parameters could be computed which could lead to the instantiatation of a new node.

2.3 Solution

To solve these both problems a symbolic processing module was added to the connectionist network. Within this symbolic layer, nodes are instantiated when they become active within the network. It therefore identifies and records ***salient*** nodes. Each of these instances can be regarded as a token of the particular node type. Further, the symbolic layer provides a local area where the parameters of an instanced object may be compared against one another, in order to identify relationships.

3 Architecture

The architecture described in summary consists of a symbiotic connectionist and symbolic process. The connectionist process both provides a permanent store for the associations found between units and the perceptual framework for the overall process (i.e. identified units within data). The symbolic process provides a type of working memory for our network. Let us consider each of these processes in greater detail.

3.1 Symbolic Process

A symbolic process by definition operates on symbols. The question of what do these symbols represent is usually defined prior to the instigation of the process. However in the outlined architecture we circumvent the need to do this. We define only the lowest levels symbols - those at atomic level. The interaction between the connectionist and symbolic processes serves to identify new symbols that are hopefully more appropriate for the task in hand.

It is part of the function of the symbolic process to identify the relationship between active units. In the example outlined above this association is simple adjacency. There is no reason why this association may be considerably more complex than this.

3.2 Connectionist Process

The connectionist layer represents the relationship between identified clusters within the problem domain. This could well map out the hierarchical description of the problem domain, or by the use of excitatory and inhibitory links describe a causal link between nodes. However in the outlined design the exact nature of the relationship is embodied in the symbolic layer. By extracting this information from the network itself we allow for specialisation of the network.

4 Advantages

4.1 Type-Token Distinction

To maintain a type/token distinction between nodes within a connectionist architecture is a far from trivial task. Further we must ask ourselves the question when is it necessary to make the distinction. Using the letters example from above we need to distinguish between two instances of the node 'ana' only within working memory i.e. within the symbolic phase. Should we later need to create a more permanent distinction of the individual tokens of the 'ana' type we must ask ourselves what is going the be the discriminating aspect of the tokens. Necessarily this will have to be context, and a cluster contextually discriminated is a longer cluster, which would naturally be incorporated into the connectionist layer.

We have therefore made a distinction between node tokens which only need to be discriminated in the learning phase of the network and those which are to become part of the network structure itself. Symbolic memory, which is needed to make the token distinction, is therefore needed in the learning phase only.

4.2 Functional Normalisation

One of the prime differences between a symbolic and a connectionist system is the clarity of the distinction between data and process. Within a symbolic process the distinction is clear cut whilst one of the key reasons for the flexibility of the connectionist model is that there may be no such distinction. Data may be represented locally (the activation of one node represents one atom of data) or data may be represented in a distributed manner (there is no one-to-one relationship between the activation of any set of nodes and a piece of data). A process is modelled by the entire spread of activation through a set of nodes.

Take the situation where we wish to model a complex functional relationship between many pieces of data (this is the function $f()$ in the above formalisation). That functional relationship must be modelled itself by a set of nodes and its interconnecting links. If this is to hold between several pieces of data the nodes necessary to model the functional relationship must be repeated throughout the network many times, an obvious redundancy. In the design outlined above the functional relationship desired can be modelled outside of the connectionist network, eliminating this redundancy. This is analogous to the process of data normalisation as used in databases, for this reason we call the process functional normalisation.

4.3 Specialisation of Networks

To follow on from this point, by encapsulating the relationship that holds between items in a central place (in this case the symbolic layer) we allow for greater specialisation of networks. Where the connectionist layer models the hierarchical structure of, or the causal relationships that are to hold between, many items, and the symbolic layer models the relationship itself, we can use essentially the same type of network to model all types of things. For example, referring back to the clustering example, it is possible to use essentially the same network to represent left adjacency and right adjacency of clusters by just changing the function $f()$ the symbolic process.

4.4 New Data

If we are producing a network which is to reflect the structure of items between which a particular relationship holds, by maintaining this relationship outside of the network we have the means by which to test new, unseen and hence unrecorded items against each other.

5 Cognitive Correlations

There are certain aspects of the outlined model which correlate well with the what has been observed of our own human problem solving behaviour. We comment on the similarities in idle speculation only and do not consider the similarities to constitute any form of proof of the validity of the proposed model.

5.1 Memory Types

Psychologists have for some time maintained a distinction between short term memory (STM) and long term memory (LTM) [2]. The three major distinctions being: duration, capacity and coding [13]. Within the two modules discussed here there are similar distinction to be made. Items instantiated within the symbolic layer have a short life span (the length of the current input pattern) whilst the connectionist links are permanent. The coding of the nodes within the connectionist layer is contextual; all that is known about that node is inherent within the links attached to it. The tokens used within the symbolic layer are arbitrary representations. However, the capacity of the symbolic layer does not seem to be limited in the same way that STM seems to be. This could be due to their differing implementations (see later.)

Note, also, in overall function the symbolic layer is similar to Klatzky's [8] description of STM as a 'mental workbench'.

5.2 Symbolic Whilst Learning

It seems a feature of learning that, when presented with learning a new task or skill, the processing tends to be symbolic and procedural in the initial phases, less so in the latter phases. For, example consider learning to type, to drive a car or learning to read. Again we see similarities with the symbiotic design. Consider the clustering problem discussed above. In the initial phases the identification of new features is performed entirely within the symbolic layer. Once identified the apparatus necessary to *perceive* the feature is incorporated within the nodes and links of the connectionist layer, so this is where the processing now takes place.

5.3 Speed and Implementation

The architectural implementation of the modern computer is in most cases a serial, symbolic, Von Neumann process. Whereas there is no doubt that the brain is made up from nerves which can operate in parallel.

People seem to have two modes of operation, to quote Norman [9] 'one rapid, efficient, subconscious, the other slow, serial and conscious'. The proposed architecture also seem to perform in two modes: a fast serial symbolic process, a slow connectionist process. A distinction between processes is preserved, although the performance ratios disagree with one another. It seems possible, however, that the juxtaposition of performance ratios is attributable to the differing implementations. Certainly the connectionist network discussed above is in reality a serial emulation of a connectionist process and so it would be reasonable to expect a performance drop.

The issue of emulating a serial process within a connectionist architecture is more complex. Clark [4] has speculated for some time that 'the human mind might effectively simulate a serial, symbol processing Von Neumann architecture.' And it is interesting to note that in his book [5] he notes two shortfalls of connectionist networks in explaining human capacity:

1. to be able to perform serial reasoning in which the ordering of operations is vital.

2. to be able to utilise a control structure in order specify salient micro features for inductive generalisation.

As this is precisely the type of functionality being satisfied here by the symbolic layer.

Rumelhart and Smolensky [11] have also pondered on the human capacity to engage in conscious, symbolic reasoning, and Touretsky [12] has contributed to the debate with his proof that neural nets can be used as Turing machines. All we can do here is to leave the open ended question: would a neural implementation of a symbolic process have a limited capacity and be relatively slow to process? If so, this would fall in line with the arguments presented above.

6 Conclusion

In the above we outline a symbiotic architecture encapsulating both a symbolic and connectionist process. The architecture was conceived initially as the solution to a clustering problem which could not easily be solved using solely connectionist techniques. However the combined model has several interesting architectural properties some of which seem to reflect observations of the brain's own problem solving behaviour.

Within the model the symbolic layer provides a form of working memory for the network as it learns from new data. The connectionist layer in return provides the perceptual framework for the symbolic layer: supplying the symbols upon which the symbolic process is to operate. The two are symbiotic in that they modify each other's data. The model as a whole is providing a learning schema where data is initially analysed symbolically, and the trace of this data later imprints itself onto the connectionist network.

7 Acknowledgements

We would like to acknowledge the original flash of insight from Nick Porter, which lead to the analysis presented in this paper.

References

[1] N. Allott, P. Fazackerley, and P. Halstead. *A Clustering Algorithm for Producing Context Rich Networks*, pages 220–229. John Wiley, Chichester, 1995.

[2] R. C. Atkinson and R. M. Shiffrin. *Human Memory: A Proposed System and its Control Processes.* Academic Press, New York, 1977.

[3] P. Brown, H. Lee, and J. Spohrer. Bayesian adaptation in speech recognition. In *Proceedings of the ICCASP*, pages 761–764. Boston.

[4] A. Clark. *Connectionism and Cognitive Science*, pages 3–15. John Wiley, Chichester, 1987.

[5] A. Clark. *Microcognition.* MIT Press, Cambridge, MA, 1990.

[6] F. Jelinek, R. Mercer, and L. R. Bahl. Continuous speech recognition: Statistical methods. *IEEE Transactions on Pattern Analysis and Machine Intelligence*, PAM-5(2):179–180, 1983.

[7] F. Keenan. *Large Vocabulary Syntactic Analysis for Text Recognition.* PhD thesis, The Nottingham Trent University, 1992.

[8] R. L. Klatzky. *Human Memory: Structure and Processes.* Freeman, San Francisco, 1980.

[9] D. Norman. *Reflections on Cognition and Parallel Distributed Processing*, volume 2, pages 110–146. MIT Press, Cambridge, MA, 1986.

[10] E. Riseman and A. Hanson. A contextual postprocessing system for error correction using binary N-grams. *IEEE Transactions on Computers*, C-23:490–493, 1983.

[11] D. Rumelhart, P. Smolensky, J. McClelland, and Hinton G. *Schemata and Sequential Thought Processes in PDP Models*, volume 2, pages 110–146. MIT Press, Cambridge, MA, 1986.

[12] D. S. Tourtsky. BoltzCONS: Dynamic symbol structures in a connectionist network. *Artificial Intelligence*, 46:5–46, 1990.

[13] W. A. Wicklgren. The long and short of memory. *Psychological Bulletin*, 80:425–438, 1991.

Genetic Design of Robust PID Controllers

A. H. Jones and P. B. de Moura Oliveira
Intelligent Machinery Division, Research Institute for Design, Manufacturing, and Marketing
University of Salford, Salford. M5 4WT UK.
Email: J.P.B.MouraOliveira@Aeromech.salford.ac.uk

Abstract

Genetic algorithms are proposed as a new and novel technique to solve the problem of designing a robust PID controller for a plant with model uncertainties. The evolutionary scheme used, involves generating two separate populations, one representing the controller and the other the plant. The controller population is then co-evolved against a fixed population of plants covering the plant uncertainty search space, such that the controller can control all the plants effectively. A time domain cost function subjected to a frequency domain vector margin stability constraint, is then deployed in order to obtain a robust controller design. This evolutionary approach is illustrated by evolving a PID controller for a linear plant which has a set of prescribed model uncertainties.

1 Introduction

The majority of control system designs are based on models of the plant. However this model is often only an approximation to the real plant. The difference between the real plant and the corresponding model is referred to as model uncertainty originated by parameter variations and unmodelled dynamics.

Historically, the robustness characteristics of controllers has been accessed by looking to the gain and phase margins of the control system. These measures of robustness stem from the original work from Nyquist [12]. Since then a great deal of research effort has been directed to developing new robust design techniques like the well-known LQG optimal approach [2], and the graphical Nyquist-type technique introduced by Horowitz [7]. One more recent and already well established technique which does address robustness is H_2/H_∞ [3, 10] originate from the influential work from Zames [18]. However, in general, these techniques give rise to high order compensators which then have to be simplified by model reduction [15], in the frequency domain, prior to implementation, which makes them difficult to implement in practical industrial applications. The H_∞ design also involves the appropriate selection of uncertainty weighting functions to achieve robust stability and robust performance, which is itself not an easy task, since this selection procedure is highly problem dependent, and a research subject by itself (Lundström [11]).

One successful methodology for control systems design based on the technique of Genetic Algorithms (GAs) (Goldberg [4], Holland [6]) was introduced by Porter and Jones [14] and extended by Jones and Oliveira [9]. The same technology can be extended to the design of robust control systems which can cope with parameter uncertainty. One powerful and novel way to extend the GA to design robust control systems is by incorporating both a process of co-evolution and the vector margin concept [16].

An alternative approach to the design of PID controllers has been proposed by Jones and Oliveira [8], using the technique of co-evolution. Artificial co-evolution as a optimisation procedure was introduced by Hillis [5, 17]. The co-evolution is analogous to the biological evolution of predator and prey. Such co-evolution can be a generator of genetic diversity, which results in emergent robust solutions. Indeed the advantages of competitive environments to evolve better solutions to complex tasks has been reported by Angeline and Pollack in [1]. The co-evolutionary scheme introduced in this paper involves two separate populations, one representing the controller and the other the plant. The controller population is subject to the normal genetic operations however the plant population is not a proper genetic population in the sense that the set of plants are fixed in a similar way as the one used by Paredis [13]. The population of controllers is then evolved against the population of plants using a minimisation of a maximum technique. This results in a controller which is optimised about the lowest performance plant, whilst ensuring that the time-domain performance for all the other plants is better. Moreover, the technique also ensures the frequency domain stability constraint is met for all plants in such manner that a robust controller emerges. The controller performance is obtained by defining the cost function in the time-domain such as the *Integral of the Square Error* (ISE), and the stability constraint is defined as the frequency domain vector margin. The resulting paradigm thus provides an alternative technique for the design of robust control systems to deal with plant uncertainties.

2 Genetic Design of Robust PID Controllers

In order to use co-evolution to design robust control systems a population of controllers is co-evolved against a fixed population of plants which covers the model uncertainty space. The controller cost is defined as a time domain cost function, subject to a frequency domain stability constraint. In this case the closed loop time domain performance criteria used is the ISE, and the frequency domain robustness constraint is defined as being to the right of a circle of radius α_c centred at point -1, where α_c, has been defined by Smith [16] as the vector margin. This is illustrated in Figure 1.
Hence, in the case of the controller, the aim of the genetic optimisation is to minimise the ISE of the lowest performance plant subjected to the frequency domain stability constraint being satisfied for all the plants.

$$controller\ cost = ISE \Leftarrow \alpha > \alpha_c \quad (1)$$

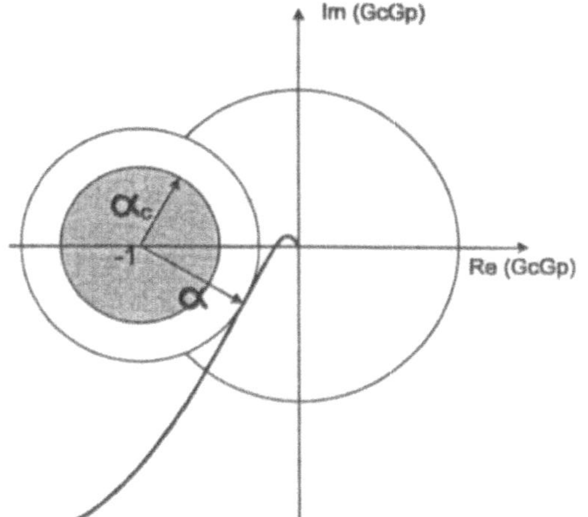

Figure 1: Frequency domain constraint function definition.

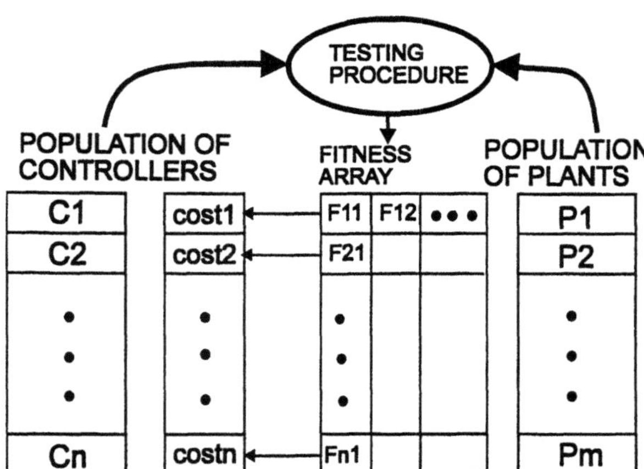

Figure 2: Fitness test procedure.

where α is the vector norm of the closed loop system, and α_c is a constraint defining the minimum acceptable vector norm.

In order to use the evolutionary methods to design robust time-domain control systems, a population of controllers is evolved against the fixed population of plants within the uncertainty space. The problem space for the controller thus consists of a population of plants representing the plant uncertainty. This testing procedure is illustrated in Figure 2.

Each individual controller C_i from the controller population is tested against all the individual plants P_j. In this way the fitness associated with a given pair is stored in the appropriate row of a bidirectional fitness array. The cost for each controller C_i is the maximum of all the fitness costs in the ith row, provided that each controller-plant pair satisfies the robustness constraint. The maximum cost is chosen because by using the GA to minimise the maximum cost, a controller emerges which can deal with all the other plants with lower cost, whilst ensuring the robustness constraint is satisfied. This procedure effectively searches the plant population for the lowest performance plant and then tunes this plant up, whilst simultaneously checking that this tuning up procedure has not forced some of the other plants to either yield a worst controller-plant pair performance than the currently worse controller-plant pair, or violate the stability constraint. The evolutionary scheme can reach one of two equilibria. The first equilibrium occurs when a low performance controller-plant pair is tuned up it will eventually force one of the other plants on to the stability constraint. The constraint then provides evolutionary pressure to detune the controller. The second equilibrium occurs when low performance controller-plant pairs compete with each other for the worst case position. This is seen when improving the ISE of one degrades the ISE of the other and vice versa. The existence of one of these two equilibrium is the key to the robustness characteristic of the evolutionary algorithm, because this process forces the controller to continuously improve through an evolutionary process against the population of plants. Once the difficult plants are found in the search space the controller population will self-adapt to these plants in such a way as to force an emergent robust solution to the PID design problem.

3 Illustrative Example

Many industrial process can be modelled using a linear simple first order model with a transfer function of the form:

$$G_p(s) = \frac{\beta K}{s+\beta} e^{-sL} \tag{2}$$

where K is the dc. gain, $\beta = 1/T$ where T is the time constant and L is the dead time. In this case the dead-time has been fixed at 1 second and the prescribed large model plant uncertainties have been defined to lie in the following range: $0.3 < K < 1.0$ and $0.2 < \beta < 0.8$.

The aim of the design is to tune up the global performance of the system, such that the performance in terms of ISE at the worst point is optimised, whilst simultaneously ensuring the performance every where else is better, and that the stability constraints are maintained. In solving this problem two populations are co-evolved. The first population represents the controller parameters $\{K_p, K_i, K_d\}$, and the second population represents the fixed plant parameters $\{\beta, K\}$. In this case both the variation in K and β were used to define a uniform grid, resulting in a plant pop size $m = 16$. The controller parameters are encoded in accordance with a system of concatenated, multiparameter mapped, fixed-point coding [4]. The sets of PID gains are then represented as strings of binary digits. Then, following random initial choice, entire generations of such strings can be processed in accordance with a basic genetic cycle. The selection procedure ensures

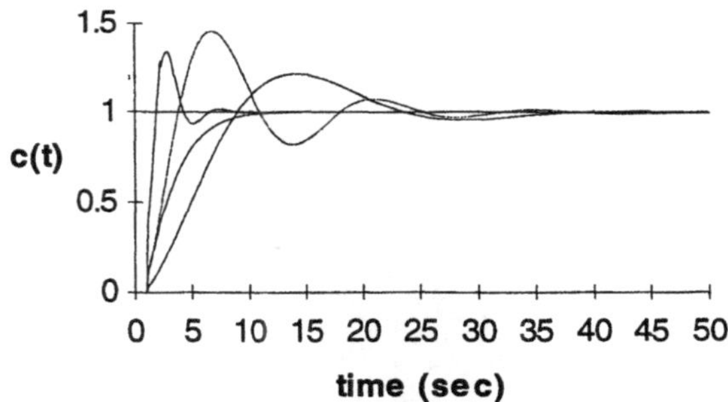

Figure 3: Step response for the worst four plants.

that the successive generations of digital PID controllers produced by the GA exhibit progressively improving behaviour in respect to the fitness measure of the ISE.

The results of solving this problem by means of co- evolving GAs, with a controller population, of size of $n = 100$, and a plant population of size $m = 16$, a crossover probability $p_c = 0.65$, and a mutation probability of $p_m = 0.005$ is shown in Figure 3 and 4. In this case the controller converged to: $K_p = 1.31$, $K_i = 0.94$ and $K_d = 0.33$, with a corresponding best ISE of 42.

Figure 3 shows that the resulting time-domain performance of the worst plant, corresponds to $K = 1$ and $\beta = 0.2$, together with three other cases corresponding to $K = 1$ and $\beta = 0.8$; $K = 0.3$ and $\beta = 0.2$ and $K = 0.3$ and $\beta = 0.8$. Figure 4 shows that two plants reached a stability constraints α_c of 0.5 and all the others being within the constraint. When the co-evolutionary process converges, an optimised controller emerges for the lowest performance plant, such that if this controller is used anywhere else in the operating envelope the time domain performance will be better in terms of ISE, and the frequency robustness criteria will be satisfied.

4 Conclusions

Genetic algorithms have been proposed as a new and novel technique to solve the problem of designing a robust PID controller for a plant with model uncertainties. The evolutionary scheme used, involves generating two separate populations, one representing the controller and the other the plant. The controller population is then co-evolved against a fixed population of plants covering the plant uncertainty search space, such that the controller can

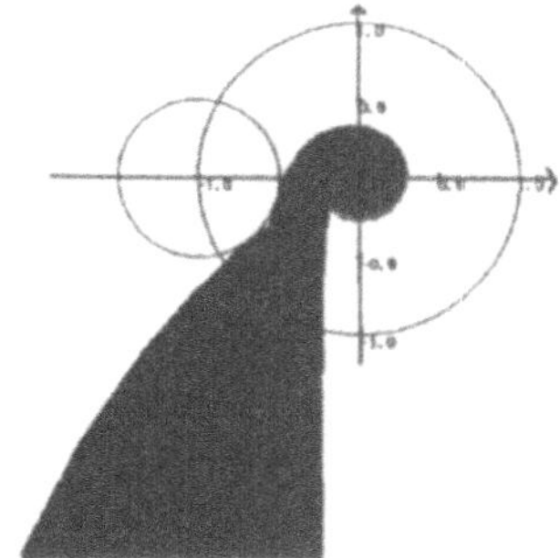

Figure 4: Frequency domain uncertainty envelope.

control all the plants effectively. A time-domain cost function subjected to a frequency domain vector margin stability constraint, was then deployed in order to obtain a robust controller design. This evolutionary approach was illustrated by evolving a PID controller for a linear plant which has a set of prescribed model uncertainties.

References

[1] P.J. Angeline and J.B. Pollack. Competitive viroments evolve better solutions for complex tasks. In *Fifth International Conference on Genetic Algorithms*, pages 264–270, 1993.

[2] M. Athans and P. L. Falb. *Optimal Control.* McGraw-Hill, 1966.

[3] L. Ching-Hsiang C. Bor-Sen, C. Yu-Ming. A genetic approach to mixed H_2/H_∞ optimal PID control. *IEEE Control Systems Magazine*, pages 51–60, October 1995.

[4] E.D. Goldberg. *Genetic Algorithms in Search, Optimization and Machine Learning.* Addison Wesley, 1989.

[5] W.D. Hillis. *Co-evolving Parasites Improve Simulated Evolution as an Optimization Procedure*, pages 313–323. Addison-Wesley, 1991.

[6] J.H. Holland. Adaptations in natural and artificial systems. *The University of Michigan Press*, 1975.

[7] I. M. Horowitz and M. Sidi. Synthesis of feedback systems with large plant ignorance for prescribed time-domain tolerances. *Int. Journal of Control*, 16(2):287–309, 1972.

[8] A. H. Jones and P. B. De Moura Oliveira. Genetic design of dual mode controllers through a process of co-evolution. In *Fourth IEEE Mediterranean Symp. on New Directions in Control and Automation*, pages 794–798, Chania, Greece, 1996.

[9] A.H. Jones and P.B. De Moura Oliveira. Genetic auto-tuning of PID controllers. In *First IEEE Conference on Genetic Algorithms in Engineering Systems: Innovations and Applications*, number 414, pages 141–145, Sheffield, September 1995.

[10] H. Kwarkernaak. Robust control and H_∞- optimization tutorial paper. *Automatica*, 29(2):255–273, 1993.

[11] P. Lundström, S. Skogestad, and Z Wang. Weight selection for h-infinity and mu-control methods-insights and examples from process control. In *Robust Control System Design and Related Methods Proc.*, pages 139–157, Cambridge, UK, 1991.

[12] H. Nyquist. Regeneration theory. *Bell System Technical Journal*, 11:126, 1932.

[13] J. Paredis. Co-evolutionary constraint satisfaction. In Y. Davidon et al, editor, *Third Conference Proceedings. Lecture Notes in Computer Parallel Problem Solving From Nature*, pages 866, 46–48. Springer-Verlag, 1994.

[14] B. Porter and A.H. Jones. Genetic tuning of digital PID controllers. *Electronic Letters*, 28:843–844, 1992.

[15] D.E. Rivera and M. Morari. Low-order siso controller tuning methods for H_2, H_∞ and μ objective functions. *Automatica*, 26(2):361–369, 1990.

[16] O. J. M. Smith. *Feedback Control Systems.* McGraw-Hill, New York, 1958.

[17] W.D. Wood. Co-evolving parasites improve simulated evolution as an optimization proceedure. In *Emergent Computation: Self-Organizing, Collective, and Cooperative Computing Networks.* MIT Press, 1990.

[18] G. Zames. Feedback and optimal sensitvity: Model referencing transformations, multiplicative seminorms and approximate inverses. *IEEE Trans. Aut. Control*, AC-26:301–320, 1981.

Coevolutionary Process Control

J. Paredis
RIKS / MATRIKS, Universiteit Maastricht
Postbus 463, NL-6200 AL Maastricht, The Netherlands
Email: jan@riks.nl

Abstract

This text describes the use of a coevolutionary genetic algorithm (CGA) for process control. A CGA combines two artificial life techniques - life-time fitness evaluation (LTFE) and coevolution - to improve the genetic search for a neural network (NN) controlling a given process.

Here, the approach is illustrated and tested on a well-known bioreactor control problem which involves issues of delay, nonlinearity and instability.

1 Introduction

Escaping from predators is clearly an essential ability which increases an entity's survival chances and hence also its reproduction chances. In general, *predator-prey interactions* often result in a selection pressure which enforces evolution's drive towards the creation of highly complex adaptations. Predator-prey coevolution is the main motor behind the coevolutionary genetic algorithm (CGA) presented here.

Earlier research studied the use of a CGA for two — completely unrelated — tasks: the search for good classification neural networks [6] and the search for solutions of constraint satisfaction problems [5]. Both applications have demonstrated the power of the CGA. In addition to this, a symbiotic variant of the CGA algorithm has been used to search for good genetic representations for a given problem [7, 10]. According to the definition of Narendra and Parthasarathy [3], process control involves the analysis and synthesis of dynamical systems in which one or more variables are kept within prescribed bounds. The current paper demonstrates how the properties of a CGA can be exploited for process control.

2 The Bioreactor

Figure 1 depicts a bioreactor. Such a reactor is a tank containing water, nutrients and biological cells. Nutrients and cells are introduced in the tank at a rate called r. The state of the process is characterized by the number of cells and the amount of nutrients. The outflow rate is equal to the inflow rate. As a consequence, the volume in the tank remains constant. The objective of the bioreactor control problem is to bring and maintain the amount of cells in the tank at an *a priori* given desired level. Anderson and Miller [1] and Ungar [11] describe the relevance, and complexity, of this problem.

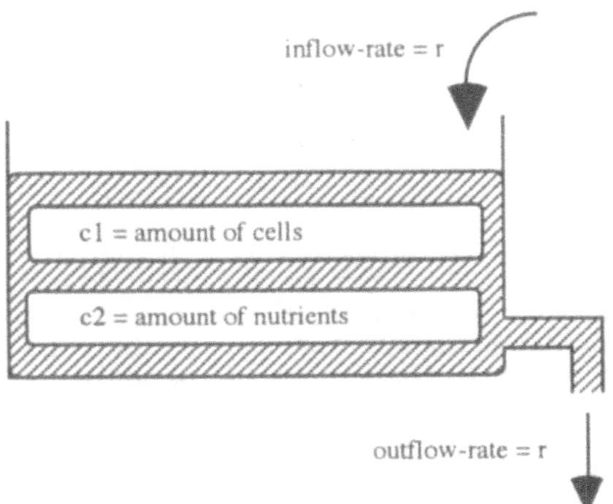

Figure 1: Schematic representation of a bioreactor.

Mathematically, the state of the tank is described by c_1 and c_2, the amount of cells and nutrients respectively. The control parameter is r. The variables c_1 and c_2 are constrained to the interval $[0, 1]$, r is in the interval $[0, 2]$. The equations of motion describing the system dynamics are given below. Here, the functions $c_1[t]$, $c_2[t]$, and $r[t]$ represent the values of c_1, c_2, and r at time t. The state of the entire system is described by the triple $(c_1[t], c_2[t], r[t])$. The initial values, $c_1[0]$, $c_2[0]$, and $r[0]$, are given. In the equations, the growth parameter, β is equal to 0.02. The nutrient inhibition parameter, g, equals 0.48. These parameters determine the rate of cell growth and nutrient consumption, respectively. A sampling interval, Δ, of 0.01 seconds is used. Hence, each application of the equations describes the state of the reactor 0.01 second later.

Equations of Motion:

$$\begin{aligned} c_1[t+1] &= c_1[t] + \\ \Delta.(-c_1[t].r[t] &+ c_1[t].(1-c_2[t]).e^{c_2[t]/\gamma}) \end{aligned} \quad (1)$$

$$\begin{aligned} c_2[t+1] &= c_2[t] + \\ \Delta.(-c_2[t].r[t] &+ c_1[t].(1-c_2[t]).e^{c_2[t]\gamma}.\frac{1+\beta}{1+\beta-c_2[t]}) \end{aligned} \quad (2)$$

The goal of this problem is to bring and keep the amount of cells at a desired level during 50 seconds. This involves 5000 applications of the equations of motion given above. During this period of 50 seconds, the control parameter r can be adjusted every half a second. Hence, the input to the controller is $c_1[t]$ and $c_2[t]$ for $t = 50, 100, \ldots, 5000$. The output is $r[t]$, again for $t = 50, 100, \ldots, 5000$. For other values of t, $r[t]$ is equal to $r[t-1]$. The overall objective is to minimize the cumulative measure:

$$\sum_{t\in\{50,100,\ldots,5000\}} (c_1[t] - c_1^*[t])^2. \quad (3)$$

Here, $c_1^*[t]$ is a constant function giving the desired level of c_1. We concentrate on the control of the bioreactor around an unstable desired state $(c_1^*, c_2^*, r^*) = (0.2107, 0.7226, 1.25)$. The initial values of $c_1[0]$, $c_2[0]$ and $r[0]$ are within 10% of their target values, e.g. $c_1[0]$ is a random number uniformly drawn from the interval $[0.9c_1^*, 1.1c_1^*]$. The controller has to steer the system such that c_1 rapidly approximates c_1^*.

3 The CGA

The NN architecture used here to control the bioreactor is a standard multilayer perceptron with two input nodes, one output node, and one hidden layer consisting of 12 hidden nodes. This NN maps the amount of cells and nutrients in the tank to the flow-rate. Hence, the input nodes are filled with the values of $c_1[t]$ and $c_2[t]$, respectively. The maximum activation value of a node is 1. The flow rate, $r[t]$, should, however, cover the interval $[0, 2]$. Hence, the activation value of the output node is doubled in order to obtain $r[t]$.

Our CGA operates on two populations which interact through their fitness. The first population contains NNs. We use a simple genetic representation of a NN: a linear string of its weights, with weights belonging to links feeding into the same node located next to each other. In accordance with [4] and [13], the weights are encoded directly as real numbers. Furthermore, the weights of the NNs in the initial population are in the range $[-1, 1]$. This in order to keep weight contributions in a range to which the (sigmoidal) transfer function is sensitive [14].

The second population consists of 200 starting states around the goal state. They take the form of triples $(c_1[0], c_2[0], r[0])$. These three variables are randomly generated but deviate maximally 10% from their 'optimal' values: c_1^*, c_2^* and r^* respectively. In the strictest sense, these states do not form a real population because they are never replaced (this in contrast with some of the other CGA applications [7, 10]. Hence, the population consists all the time of the same 200 starting states.

In the CGA, the interaction between the populations occurs during encounters between a NN and a starting state. During such an encounter, the NN is tested on the starting state. This is done in the following way: starting from the $c_1[0]$, $c_2[0]$, and $r[0]$ specified by the starting state, the equations of motion are applied 50 times in order to simulate the bioreactor for half a second. This determines the values of $c_1[50]$ and $c_2[50]$ which are then clamped on the corresponding input nodes of the NN. Next, the activation value of the output node is calculated using standard forward propagation. This value is then doubled and assigned to $r[50]$. This cycle of simulation followed by the calculation of r is repeated 100 times. The error-measure, as defined by Equation (3), is used to calculate the reward for the two objects involved in the encounter. For the state, this *pay-off* is equal to the error. The NN, on the other hand, receives a pay-off equal to the negative of the error. This is analogous to the negative fitness interaction in predator-prey systems in which success on one side entails failure for the other side and vice versa. Each encounter results in updating the fitness of the involved NN and starting state. The *fitness of an object*, NN and search state, is defined as the average pay-off received over the last 20 encounters it was involved in. For this purpose, each object has an associated history which contains the 20 pay-offs it received most recently.

Now, the complete algorithm can be described. First, the two initial populations are created. Next, the following cycle is repeated (the pseudo-code below describes this basic cycle). First, 20 encounters are executed. This involves the SELECTion of a state and a NN. This selection is biased towards highly ranked individuals in the population which is sorted on fitness. The error resulting from the ENCOUNTER is then used to push the pay-off on the history of the NN and state. At the same time the oldest pay-off is removed from both histories. Next, the pay-offs in the histories are averaged in order to UPDATE the fitness of the individuals. After the execution of 20 such encounters, two NNs are SELECTed in order to reproduce. Here, standard two-point crossover and adaptive mutation [12] is used. The fitness of the new NN-offspring is then calculated. This is done through 20 encounters of the NN with SELECTed states. This fills up the history of the NN with negative errors which are then averaged in order to obtain the initial fitness (this same procedure is used to compute the fitness of the NNs in the initial population). Finally, the new NN is INSERTed into the appropriate rank location in the NN population. At the same time, the individual with the minimum fitness is removed. In this way, the population remains sorted on fitness. The last 4 lines of the code below are identical to the basic cycle of GENITOR [12], a well-known steady-state GA.

```
DO 20 TIMES
      nn:= SELECT(nn-pop)
      state:= SELECT(state-pop)
      error:= ENCOUNTER(nn,state)
      UPDATE-FITNESS(nn,-error)
      UPDATE-FITNESS(state, error)
nn1:= SELECT(nn-pop) ; parent1
nn2:= SELECT(nn-pop) ; parent2
child:= MUTATE-CROSSOVER(nn1,nn2)
f:= FITNESS(child)
INSERT(child,f,nn-pop)
```

It is important to remark that a 'maximization GA' is used here. Hence, highly fit NNs receive small (negative) pay-offs, i.e. they keep the error small. Highly fit states, on the other hand, get relatively high (positive) pay-offs. Or, in other words, a state is considered fit when the NNs cannot achieve good control from it. We use the term *life-time fitness evaluation* (LTFE) for the continuous fitness feedback resulting from the encounters. As soon as the NNs are becoming successful on a certain set of states then these states get a lower fitness and other, more difficult states will be selected more often. This process forces the CGA to concentrate now on these states too (because of fitness proportional SELECTion of the pairs of individuals involved in an encounter). As Hillis [2] argues, this has two advantages over the traditional approach. Firstly, it helps prevent large portions of the NN population from becoming stuck in local optima. Secondly, fitness testing becomes much more efficient: one mainly focuses on not yet solved states. The brief description of the results in the next section illustrates these advantages.

4 Empirical Results

All experiments used the following parameter values: population size of the NN population = 100; population size of the population of starting states = 200; the hidden layer of the NN contains 12 nodes. Due to space restrictions only the general results of the experiment will be described here. More information can be found in a more detailed version of this paper.

A couple of general observations can be made from the experiments. Firstly, after about 1500 cycles (i.e. 1500 new NNs have been generated) good (i.e. highly-fit) NNs oscillate with an amplitude of 0.7 around the optimal concentration of $c_1(0.2107)$. After 3000 cycles, however, this amplitude is reduced to 0.07. In addition to this, the NN now also reduces the amplitude of the deviation during the control period of 50 seconds.

The experiments also showed the power of LTFE: control starting from the fit states is less good than the control starting from the less fit states. On average, the fit starting states were located the furthest away from the desired state. This clearly shows that LTFE ranks the states and helps the CGA to focus on the 'not-yet-solved' problems.

5 Conclusions

Hillis [2] provided the basic inspiration for the work on CGAs. He coevolved (using predator-prey interactions) sorting network architectures and sets of lists of numbers on which the sorting networks are tested. The partial and continuous nature of LTFE is ideally suited to deal with coupled fitness landscapes. Coevolutionary interacting species typically give rise to such coupled landscapes. The incorporation of LTFE is an important difference with Hillis' work. As a matter of fact, LTFE is much closer to reality than the traditional 'all at once' fitness calculation. Over the last two years, CGAs have been used in various applications [5, 6] . Moreover, the CGA can easily be extended to incorporate symbiotic interactions [7, 10]. All of this earlier work showed several advantages of the combined use of coevolution and LTFE, such as: increased performance (in terms of solution quality as well as computation demand) and noise tolerance.

With respect to process control, an important question remains to be answered: How large is the class of control problems which can be solved with a CGA? Narendra and Parthasarathy's definition of process control given in the introduction of this paper sheds some light on this issue. According to this definition, the main task of process control is to 'keep one or more variables within prescribed bounds'. This immediately proposes the fitness criterion to be used: the amount of deviation outside the prescribed boundaries (as, for example, in Equation (3)). Specific control problems typically specialize the general definition above. In the bioreactor example described here, control was learned from different starting states. Other problems, for example, involve various noise patterns the controller should be able to deal with. In this case, the CGA would use a population of noise patterns. In general, the second population consists of tests for the controller. These can take different forms: starting states, noise patterns, etc.

Furthermore, additional (life-time) learning algorithms (such as reinforcement learning) can be used during an encounter to tune the weights of a NN. This way, genetic and life-time learning can be combined. This point certainly warrants further research. Initial experimental results of this combination are described in [9].

References

[1] C. W. Anderson and W. T. Miller. Challenging control problems. In T. Miller W, R. S. Sutton, and P. J. Werbos, editors, *Neural Networks for Control.* MIT Press/Bradford Books, 1991.

[2] W. D. Hillis. Co-evolving parasites improve simulated evolution as an optimization procedure. In C. G. Langton, C. Taylor, J. D. Farmer, and S. Rasmussen, editors, *Artifical Life II.* Addison-Wesley, California, 1992.

[3] K. S. Narendra and K. Parthasarathy. Identification and control of dynamical systems using neural networks. *IEEE Transaction of Neural Networks*, 1(1), 1990.

[4] J. Paredis. The evolution of behaviour: Some experiments. In Meyer and Wilson, editors, *From Animals to Animats.* MIT Press/Bradford Books, 1991.

[5] J. Paredis. Coevolutionary constraint satisfaction. In Y. Davidor, H-P. Schwefel, and R. Manner, editors, *Parallel Problem Solving from Nature III, Lecture Notes in Computer Science*, volume 866. Springer-Verlag, 1994.

[6] J. Paredis. Steps towards coevolutionary classification neural networks. In R. Brooks and P. Maes, editors, *Proc. Artificial Life IV.* MIT Press/Bradford Books, 1994.

[7] J. Paredis. The symbiotic evolution of solutions and their representations. In L. Eshelman, editor, *Proceedings of the Sixth International Conference on Genetic Algorithms.* Morgan Kaufmann Publishers, 1995.

[8] J. Paredis. Coevolutionary computation. *Artificial Life Journal*, 2(4), 1996.

[9] J. Paredis. Coevolutionary life-time learning. In H-M. Voigt, M. Ebeling, I. Rechenberg, and H-P. Schwefel, editors, *Parallel Problem Solving from Nature IV, Lecture Notes in Computer Science*, volume 1141. Springer-Verlag, Heidelberg, 1996.

[10] J. Paredis. Symbiotic coevolution for epistatic problems. In *Proceedings of the European Conference on Artificial Intelligence*, Chichester, 1996. John Wiley and Sons.

[11] L.H. Ungar. A bioreactor benchmark for adaptive network-based process control. In T. Miller W, R. S. Sutton, and P. J. Werbos, editors, *Neural Networks for Control.* MIT Press/Bradford Books, 1991.

[12] D. Whitley. Optimizing neural networks using faster, more accurate genetic search. In *Proc. Third Int. Conf. on Genetic Algorithms.* Morgan Kaufmann, 1989.

[13] D. Whitley. Genetic reinforcement learning for neurocontrol problems. *Machine Learning*, 13:259–284, 1993.

[14] A. P. Wieland. Evolving controls for unstable systems. In D. Touretzky, J. L. Elman, T. J. Sejnowski, and G. E. Hinton, editors, *Proc. of the 1990 Summer School.* Morgan Kaufmann, 1991.

Cooperative Coevolution in Inventory Control Optimisation

R. Eriksson and B. Olsson
Department of Computer Science, University of Skövde,
Box 408, 541 28, Sweden
roger@ida.his.se, bjorne@ida.his.se

Abstract

This paper introduces an extension to Potter and De Jong's [5] work on cooperative coevolutionary GAs (CCGA). We discuss the problem of inventory control and discuss the potential advantages of applying an evolutionary method to this problem. We describe our approach and experimental design for solving the inventory control problem using a CCGA. We also show how the CCGA can be extended and modified in order to handle larger inventory control optimisation problems.

1 Previous Work

A number of different forms of coevolutionary relationships can be found in nature — including cooperative, competitive, predator-prey, and host-parasite relationships. There has recently been increasing attention in the evolutionary computation community to the improvements which can be achieved by exploiting such coevolutionary relationships in evolutionary algorithms. Examples include Hillis' work on host-parasite coevolution [4], Angeline and Pollack's work on competitive learning [1], and Potter and De Jong's work on cooperative coevolution [5].

In the case of cooperative coevolution, Potter and DeJong's motivation for extending a traditional genetic algorithm with coevolution is to take advantage of explicit notions of modularity. When applying GAs to complex problems, we want to allow cooperating subpolutations to evolve components of a solution, which can be assembled to form a complete solution. This raises many problems regarding the decomposition of the problem: Do we have to rely on the experimenter to find a suitable decomposition, so that a fixed number of subpopulations can be formed? If so, how can the experimenter find the most suitable decomposition for a given problem? If we want to allow the system more autonomy, by evolving a decomposition — then how can this be achieved? (This problem was addressed in [2]). The cooperative approach also faces problems regarding fitness allocation: Given that we form a complete solution by assembling components evolved by subpopulations, how do we best assign credit for a complete solution to the individuals which participated in this solution?

In [5] Potter and De Jong studied the use of a cooperative coevolutionary GA (CCGA) on traditional function optimisation problems. Their object was to study the behaviour of the CCGA on a well-defined problem, and use the results as a foundation for extending the CCGA to harder optimisation problems. Potter and De Jong started by defining an initial CCGA which they named CCGA-1 and which they tested on four functions. They found that CCGA-1 outperformed a traditional GA on all four functions.

1.1 CCGA-1

CCGA-1 uses a separate subpopulation for each function parameter. For an N-dimensional function, it uses N subpopulations where the individuals of each subpopulation evolves competing values for one of the parameters. When evaluating an individual, its chromosome is decoded and mapped to a value which is combined with values from selected individuals from the other subpopulations. This gives a vector of parameter values for which the function will return a value, and this value is used for computing the fitness of the individual in the usual GA manner.

During evolution, each subpopulation is evaluated and reproduced in turn, using the evaluation procedure for each individual. When evaluating an individual from a given subpopulation, a choice has to be made as to which individuals from the other populations it should be combined with. In CCGA-1, an individual is always combined with the fittest individual from each of the other subpopulations.

When comparing CCGA-1's results on four multimodal standard functions (the Rastrigin, Schwefel, Griewangk and Ackley functions), significant improvements over a standard GA were found. However, it was noticed that the improvement was reduced for functions with a stronger dependency among the parameters. When a function was added with very strong dependencies between its parameters, the CCGA-1 was outperformed by the standard GA. This suggests that the method for credit assignment used in CCGA-1 is inappropriate for such functions. Therefore, Potter and De Jong defined CCGA-2, with a more refined credit assignment algorithm.

1.2 CCGA-2

In CCGA-2, an individual undergoing evaluation is used to assemble two vectors of function parameters, rather than just one. The best of the two function values is used when calculating the fitness of the individual. The first vector is assembled as in CCGA-1, while the second vector is assembled by combining the individual with a random choice of individuals from the other subpopulations. Potter and De Jong found that this procedure made CCGA-2 perform as well as the standard GA on the function with strong parameter dependencies, while still outperforming the standard GA on functions with weaker dependencies.

Potter and De Jong's results suggest that on problems with a natural decomposition, where components can be optimised independently, a CCGA approach can result in significant improvements. In addition, if care is taken in defining the credit assignment algorithm, such a cooperative approach can perform as well or better across a set of problems with a wide range of strength in dependency between subcomponents.

2 Inventory Control using GAs

An inventory is a stock of goods serving as a buffer between successive stages of production and distribution. The investment in inventory is often a significant use of capital in a business. The problem of optimising inventory management is complex, since it involves resorting to a tradeoff between conflicting goals. For example, it is possible to increase customer service by building larger buffer stocks, but this results in increased investments in inventory. It is also possible to reduce inventory investment by ordering items more often and in smaller lots, but this increases the ordering costs.

Traditionally, the inventory control optimisation problem is reduced to the problem of deciding the values of two parameters for each item: when to order (order point R), and how much to order (order quantity Q). The traditional standard method involves the use of the economic order quantity (EOQ) formula. The object of the EOQ formula is to determine Q, under the assumption that the demand rate is known, so that R can be deduced from Q. EOQ however, relies on a number of unrealistic assumptions, which severely limits its use in practice. For the work presented in this paper, the following assumptions relied on in EOQ are the ones of highest relevance:
i) The demand rate r is constant and continuous.
ii) Per item cost does not depend on order quantity.
iii) No benefits for joint replenishment.
EOQ is applied for each item and time period to compute the continuous order quantity $q = \sqrt{2 * k * r / h}$, where r is demand during the period, k ordering cost, and h the carrying cost per unit during the period. The discrete order quantity Q is computed from q, using an algorithm for rounding to the optimal discrete value from a set of permissible values.

The problem faced in the following experiments is how to optimize inventory control parameters in cases where the assumptions of EOQ are violated. In the initial experiment, EOQ will be applicable, so that the results of the various evolutionary algorithms can be compared to those of EOQ. In further experiments, EOQ's assumptions will be violated in order to show that evolutionary approaches can be applied in cases where EOQ can not. Thus, there are two potential advantages of using evolutionary algorithms in general on this problem: It may be possible to find better results than those of EOQ in cases where EOQ is applicable, and it may be possible to optimise inventory control in cases where EOQ can not be applied. Also, in both of these cases, a CCGA may produce better results than a standard GA.

The task for the evolutionary algorithms we will test will be to evolve a set of parameter values for inventory control given a number of items. Two parameters are used for each item, and these specify R and Q for the item. The value of R tells us at which stock level a replenishment order should be placed for the item, whereas Q tells us how many units of the item to order. As input to the algorithm will be given a set of forecast data, showing the customer orders expected during a specific time period.

The customer orders may have been received at a constant or varying rate during the period.

To evaluate solutions a simulation will be run. At the start of the simulation, the different inventory costs will be initialised, and these will be incremented during the simulation, and the final total cost accumulated for each solution. The costs are:

- Lost sales costs, depending on the time customers must wait for orders to be delivered.
- Transportation costs, depending on the number of replenishments and their size.
- Order costs, depending on the number of orders.
- Storage space costs, depending on stock sizes.

The details of these simulations are described in [3]. Briefly, a simulation is initialised by collecting the order point and order quantity values for all items from the solution under evaluation; by initialising the above costs to zero; and by initialising the stock levels. During the simulation, a file containing a number of customer orders is read incrementally and each customer order is handled in turn. Deliveries are made, stock levels updated and replenishment orders placed when stock levels fall below order points. For each customer order, replenishment order or delivery, the different cost sums are updated. The final step is to sum the different costs to obtain a total inventory cost from which a fitness value can be computed.

Obviously, running a relatively complex simulation for every solution will be time-consuming, considering that a typical run of an evolutionary algorithm involves evaluating several thousand potential solutions. In our experiments, each simulation involves data for a period of 60 days, during which more than 100 customer orders are handled. Although our experiments show that such a simulation scenario is tractable for problems with few items, it will clearly not be tractable for more realistic scenarios with large numbers of items. Therefore, the decomposition of the problem into subpopulations will be focused on in the experimental design. We now describe three architectures, with different problem decompositions.

2.1 Three GA Architectures

The standard GA: Our first alternative is the standard genetic algorithm where a single population is used, and where each individual represents a complete solution. We use chromosomes in the form of binary strings, representing vectors of integers. Each integer is an order point or order quantity for an item, so that for a problem with N items, each chromosome encodes $N * 2$ integers.

Evaluation of individuals is straightforward: The values of Q and R are extracted from the chromosome and the simulation is run. Upon completion of the simulation, the summed total inventory cost is used as basis for the calculation of the individual's fitness.

CCGA-1R1 and CCGA-2R1: This architecture is based on Potter and De Jong's approach. For a problem including N items we use $N * 2$ subpopulations, where each subpopulation has the task of evolving either a single order point or a single order quantity for some item. In order to evaluate an individual, we either assemble a solution by combining it only with the best individual from all other populations (CCGA-1R1) or first with the best and then with a randomly chosen individual from each other subpopulation (CCGA-2R1).

CCGA-1R2 and CCGA-2R2: In inventory control optimisation there is one decomposition of the problem which seems natural: to let each subpopulation evolve the combined parameters for one item. In our experiments, there are only two parameters for each item, since only values of Q and R is needed for each item. Thus, it seems very natural to allow each subpopulation to evolve these two parameters for one item. This, of course, is based on the assumptions that the two parameters for to one item are strongly dependent, and that parameters for different items are either less dependent or totally unrelated. The latter assumption may or may not hold, depending on the actual case. We may, for example, find scenarios where some sets of items are typically ordered by the same customers, so that the demand for sets of items are very similar.

3 Experiments and Results

We designed a number of experiments — two of which will be discussed here — with different assumptions regarding demand rate, transportation costs and parameter dependencies. In each of the experiments, we tested all three architectures and their variants. In the first experiment, the assumptions of the traditional EOQ formula were met, so that all five evolutionary algorithms could be compared with EOQ, which serves as a baseline for acceptable results. All experiments concerned two items, and evaluation of individuals was based on 60 time steps with 103 to 120 customer orders.

3.1 Experiment 1

In this experiment demand for each item was constant. In practice this meant that during the 60 time steps of the simulation, one customer order was received for each item and for each time step. Every customer order was for 5 item units. In this experiment, the EOQ formula could be applied to calculate the economic order quantity, giving 53 units as its result (from which an order point of 25 units can be derived). The total inventory cost for EOQ's order quantity is plotted in Figure 1, together with the total costs of the best-of-generation individuals of all five evolutionary algorithms. The best individuals evolved had $Q = 50$ units and R values from 20 to 24 units.

As can be seen, all evolutionary algorithms outperform EOQ. There is also a clear difference between the standard GA and all four coevolutionary variants. The differences between the coevolutionary variants, however, are minor. This is in line with dependencies among parameters being very weak in this scenario, which means that there is no benefit of using the CCGA2 variants over the CCGA1 variants. That the parameters have very weak dependencies follows from the demand being constant.

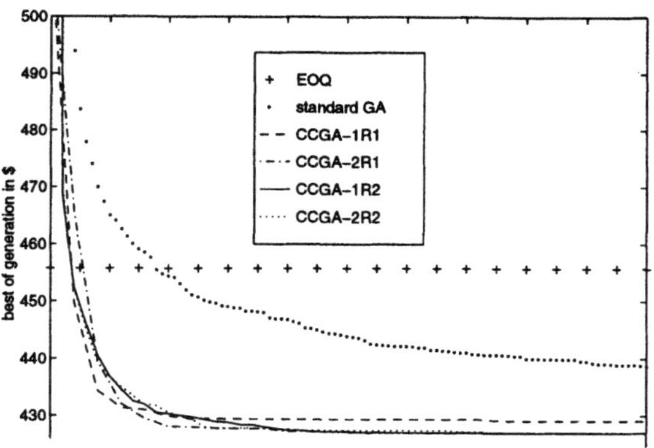

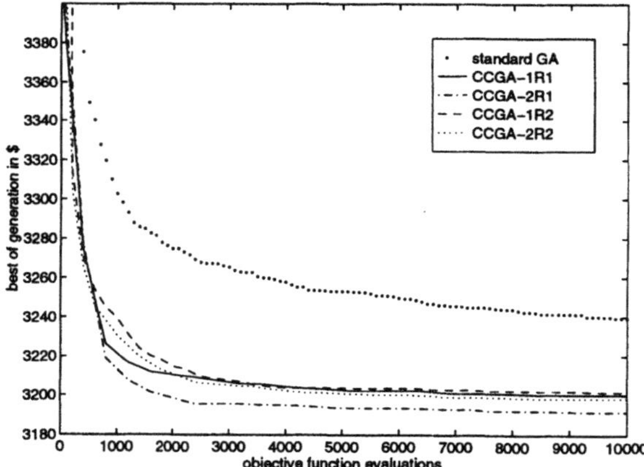

Figure 1: Experiment 1 (top) and 2 (bottom). Best-of-generation total costs, averaged over 50 runs.

3.2 Experiment 2

An element is introduced which excludes the use of EOQ: the transportation cost for a replenishment is dependent on the order quantity. This experiment also introduces varying demand. As previously, a simulation is run using customer orders from a period of 60 time units. However, instead of determining that there is a customer order of a fixed size for each item in each time period, we now draw a random integer in the range 0 to 5 for each item in each time step. The chosen integer shows the size of the customer order arriving in that time step, and in 1/6 of the time steps (when 0 is drawn) there is no customer order for that item.

The fact that customer demand varies will induce a stronger dependency between the order point and order quantity parameters of each item. If Potter and De Jong's results hold also in this domain, then we should expect CCGA2-R1 to outperform CCGA1-R1. As can be seen in Figure 1, this seems to be the case.

Another result found in Figure 1 is that CCGA1-R2 and CCGA2-R2 both perform comparably to CCGA2-R1. Our interpretation of this is that the most natural decomposition of the problem is to use one subpopulation for each item. When variable demand is introduced this leads to stronger dependencies between the pair of parameters for each item. However, the different items are still independent.

The fact that CCGA1-R2 and CCGA2-R1 perform equally in this scenario is of interest since it is a potential source of reductions in computational cost. Recall that the $R2$ variants evaluate the fitness of an individual by assembling two different solutions, rather than one. This leads to a doubling in computational cost for fitness evaluation, which is usually the most costly part of evolutionary optimisation. It is particularly costly in our case, where a simulation is made for each fitness evaluation.

4 Conclusions

What has been achieved is one further step in establishing the cooperative coevolutionary approach, which was first proposed by Potter and De Jong in [5]. We have applied the CCGA approach in inventory control optimisation, and shown that the basics of Potter and De Jong's results extend to this domain. Further, we have compared a number of alternatives regarding problem decomposition and assignment of credit and shown that a decomposition where parameters are grouped into different subpopulations according to their interdependence is beneficial in that it combines improvements in results with a reduction in computational cost.

References

[1] P.J. Angeline and J.B. Pollack. Competitive environments evolve better solutions for complex tasks. In *Proceedings of ICGA-93*, 1993.

[2] P. Darwen and X. Yao. Automatic modularization by speciation. In *Proceedings of ICEC'96*, 1996.

[3] R. Eriksson. Applying cooperative coevolution to inventory control parameter optimization. Master's thesis, University of Skövde, Sweden, 1996.

[4] W. D. Hillis. Co-evolving parasites improve simulated evolution as an optimization procedure. In *Artificial Life II*, 1992.

[5] M.A. Potter and K.A. DeJong. A cooperative coevolutionary approach to function optimization. In *Parallel Problem Solving from Nature 3*, 1994.

Dynamic Neural Nets in the State Space Utilized in Non-Linear Process Identification

R.C.L. de Oliveira[1], F.M. de Azevedo[1] and J.M. Barreto[2]
[1] GPEB-Dept. of Electrical Engineering,
[2] Dept. of Informatics and Statistics
Federal University of Santa Catarina, Brazil
Email: {limao, azevedo}@gpeb.ufsc.br, barreto@inf.ufsc.br

Abstract

This work shows the use of a novel neural model for identification of non-linear process. The neural model make use of internal dynamic with dynamical neurons. The parameters responsible for the dynamic of the neural net are adjustable, giving a high flexibility for the neural model in process identification.

1 Introduction

Static feedforward artificial neural nets (SFANN's) have been heavily used in control systems problems solution [3], where these nets are utilized as controllers [2], identifiers or predictors in adaptive control, having more focus on non-linear dynamic processes identification [7]. The use of SFANN's in order to identify and/or to control dynamic systems, normally uses the dynamics of the process, represented by its model. For a SISO discrete system, Equation (1), this model is :

$$\begin{aligned} y(k) = & f(yp(k-1), \ldots, yp(k-n-1), \\ & u(k-1), \ldots, u(k-m-1)) \end{aligned} \tag{1}$$

where $y(.)$ is the model output signal, $yp(.)$ is the process output signal, $u(.)$ is the model input signal and $f(.)$ is a non-linear function which makes the mapping $\Re^n \times \Re^m \rightarrow \Re^1$.

In those cases, the information of dynamics is determined by a signals set, delayed in time, in the neural net input. It requires the manipulation of a great number of parameters. Examples of works dealing with this case include [5, 7]. However to work in dynamical system identification is advisable to use recurrent neural nets. The reason for the use of recurrent neural nets is the use a few number of parameters, to reduce the negative contribution of the error variance in the identification [7] since it is not necessary to use the signal $yp(.)$ on the neural net input.

2 Dynamic Neural Net in the State Space

There are a vast number of specific neural networks based state-space models, which are dynamical neural nets, as shown in [4, 6, 8, 9]. In this work we present a new neural net model, that is very simple when compared with other models, containing neurons that are intrinsically dynamic. Supposing a neuron state different from the neuron output signal [1], a weighted state feedback can assign an arbitrary dynamic to the neuron, if the number of neurons is sufficient. The feedback gain is adjusted by the learning algorithm. Due to the existence of the internal dynamic in the neural model, when this kind of net is used to identify a process, it is possible to give a more succint description of the system when compared with the input-output approach. This neural model has intrinsic dynamics, and uses the dynamics representation of the process model in the state space, as shown in Equations (2) and (3), for a SISO discrete system:

$$x(k+1) = Ax(k) + Bu(k) \tag{2}$$

$$y(k) = \Phi(x(k)) \tag{3}$$

where $x(.)$ is the vector of states $n \times 1$ of the model, $u(.)$ is the model input signal and it is also the process input signal, $y(.)$ is the model output signal, A is the matrix of dynamics $n \times n$ of the model, B is the vector input $n \times 1$ of the model and $F(.)$ is the output non-linear function which makes the mapping $\Re^n \rightarrow \Re^1$. This is shown in the Figure 1.

Figure 1 shows the artificial neural net with linear dynamics in the state space (ANNLDSS) having two

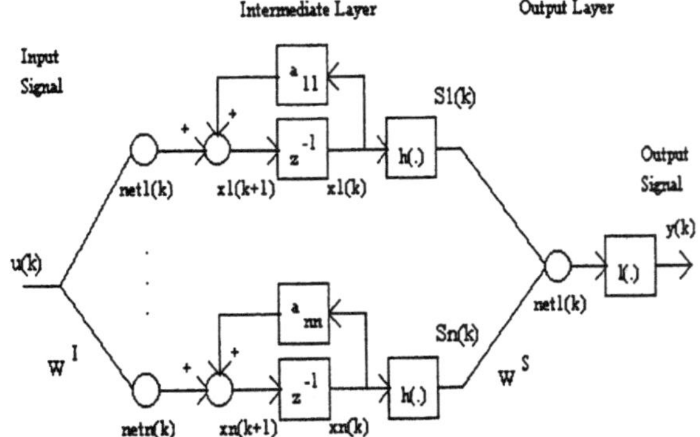

Figure 1: ANNLDSS Feedforward having one layer of neurons of linear dynamic and other layer of static neuron.

neuron layers, an intermediate dynamic layer and an output static layer.

From ANNLDSS shown in Figure 1, it can be observed that,

$$S_j(k) = h[x_j(k)] \tag{4}$$

and,

$$x_j(k+1) = a_{jj}x(k) + w_{j1}^I u(k) \tag{5}$$

From Equations (4) and (5), it follows that

$$S_j = h[a_{jj}x_j(k-1) + w_{j1}^I u(k-1)] \tag{6}$$

or,

$$S(k) = h[Ax(k-1) + W^I u(k-1)] \tag{7}$$

with,

$$A = \begin{bmatrix} a_{11} & 0 & 0 & 0 \\ 0 & a_{22} & 0 & 0 \\ \dots & \dots & \dots & \dots \\ 0 & 0 & 0 & a_{nn} \end{bmatrix}; B = \begin{bmatrix} w_{11}^I \\ w_{21}^I \\ \dots \\ w_{n1}^I \end{bmatrix} \tag{8}$$

The dynamic matrix A of the Equation (8) is a diagonal matrix, however, any other kind of matrix can be used. In this case, lateral connections between the dynamic neurons of the intermediate layer appears in Figure 1. The output signal $y(k)$ is equal to,

$$y(k) = l[W^S S(k)] \tag{9}$$

and from Equation (7), we have,

$$y(k) = l[W^S h[Ax(k-1) + W^I u(k-1)]] \tag{10}$$

or,

$$y(k) = N[W, x(k-1), u(k-1)] \tag{11}$$

that formalizes the given model for Equations (2) and (3), and the learning algorithm will adjust the elements of W in such way that $N(.) \approx \Phi(.)$. Previous knowledge about the process to be modelled, can help find the elements of A; if we know nothing about the process to be identified, the elements of A are considered as weights of the neural net and adjusted by the learning algorithms.

3 Learning Algorithm

The developed training algorithm, for ANNLDSS, is a backpropagation one, and due to the feature of neuron dynamics linearity, it presents very simple equations for the weights' updating. It happens for both situations, fixed or variable dynamic. Considering that we intend to minimize the cost function of the quadratic error $E(W)$,

$$E(W) = \frac{1}{2}[e(k)]^2 \tag{12}$$

where $e(k) = yp(k) - y(k)$ is an error signal measured according to the output signal of ANNLDSS, the weights' adjustment for the intermediate layer is given by Equation (13).

$$W_{k+1} = W_k - \eta Grad[E(W)] \tag{13}$$

The gradient $Grad[E(W)]$, for the weights of the intermediate layer is calculated by,

$$Grad[E(W)] = \frac{\vartheta E(W)}{\vartheta w_i} \tag{14}$$

where,

$$\frac{\vartheta E(W)}{\vartheta w_{ij}^I} = \delta_i^I(k)u(k) \tag{15}$$

$$\delta_i^I(k) = -[\delta_1^S w_{1i}^S]h'[x_i(k)]\left[\frac{1}{a_{ii}}\right] \tag{16}$$

$$\delta_1^I(k) = e(k)e'(k)l'[net_1(k)] \tag{17}$$

where e', l' and h' indicate the derivative of $e(.)$, $l(.)$, $h(.)$ with respect to x. The gradient $Grad[E(W)]$, for the ponderation a_{ii} of the intermediate layer is calculated by,

$$Grad[E(W)] = \frac{\vartheta E(W)}{\vartheta a_{ii}} \tag{18}$$

where,

$$\frac{\vartheta E(W)}{\vartheta a_{ii}} = \delta^I_{a_{ii}}(k) x_i(k-1) \tag{19}$$

$$\delta^I_{a_{ii}}(k) = [\delta^I_1(k) w^S_{1i}] h'[x_i(k)] \tag{20}$$

4 Identification of Non-Linear Dynamic Systems

The use of ANNLDSS is exemplified by the identification task of a non-linear, discrete second order process, in two situations: with some or none previous knowledge about the plant mathematical model. The process identification follows the procedure shown in Figure 2.

4.1 Simulation 01

For a process given by Equation (21) and the situation of knowing something about the linear part of the plant mathematical model,

$$yp(k) = 0.3yp(k-1) + 0.6yp(k-2) + f[u(k-1)] \tag{21}$$

with the input signal and the non-linear function $f(.)$ defined by equations below,

$$u(k) = \sin\left(\frac{2\pi k}{50}\right) \tag{22}$$

$$f(u) = 0.6 \sin(\pi u) \tag{23}$$

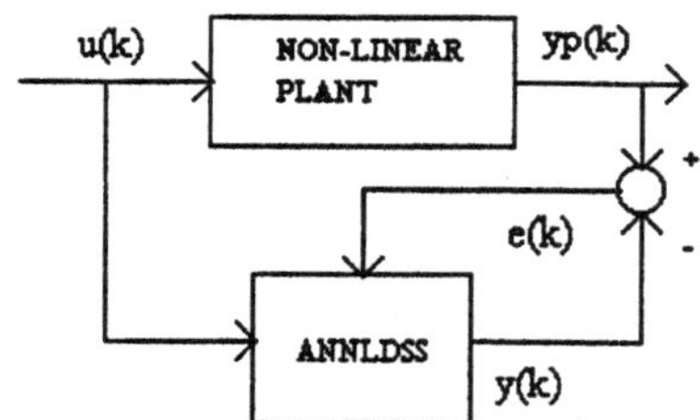

Figure 2: Identification of a non-linear plant through ANNLDSS.

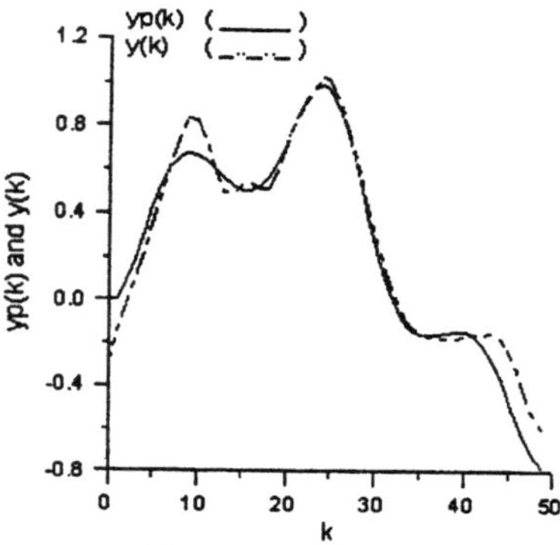

Figure 3: Signals $yp(k)$ and $y(k)$, after 6,800 iterations of learning.

the ANNLDSS has 4 neuron layers, the first two are static with function $h(.)$ as hyperbolic tangent, and 8 and 4 neurons respectively; the third layer is dynamic with two neurons of linear dynamic and linear function $h(.)$, the last layer is static, having just one neuron, with linear function $h(.)$. From the linear part of Equation (21) and using a canonical representation in the state space it is obtained the values of $a_{11} = 0.939$, $a_{22} = -0.639$ and $a_{12} = a_{21} = 0$. The result for 6,800 iterations of learning is shown in Figure 3.

4.2 Simulation 02

For a process given by Equations (24)(25) and (26) and the situation of knowing nothing about the plant mathematical model,

$$x_1(k+1) = \tanh[0.939x_1(k) + 0.6749u(k-1)] \tag{24}$$

$$x_2(k+1) = \tanh[-0.639x_2(k) + 0.9918x_1(k)u(k-1)] \tag{25}$$

$$y(k) = [0.8817x_1(k) + 0.4083x_2(k)]^2 \tag{26}$$

with the input signal defined by the Equation (22), the ANNLDSS with lateral connections that represents the matrix A given by Equation (27), where the parameters a_{ij} are adjustable,

$$A = \begin{bmatrix} a_{11} & a_{12} & 0 & 0 & 0 \\ a_{21} & a_{22} & a_{23} & 0 & 0 \\ 0 & a_{32} & a_{33} & a_{34} & 0 \\ 0 & 0 & a_{43} & a_{44} & a_{45} \\ 0 & 0 & 0 & a_{54} & a_{55} \end{bmatrix} \tag{27}$$

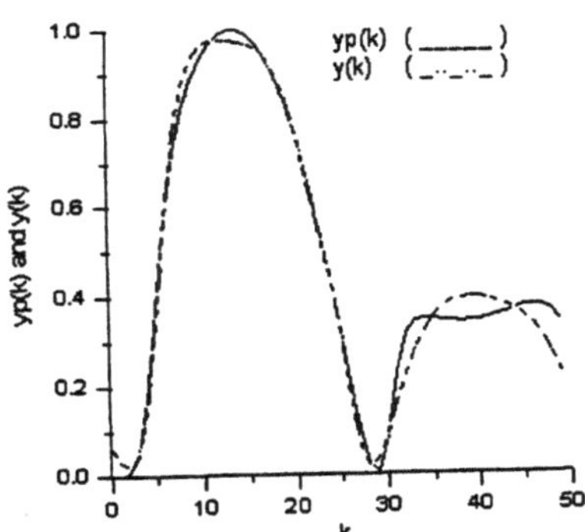

Figure 4: Signals $yp(k)$ and $y(k)$, after 26,275 iterations of learning.

have 4 neuron layers, the first is static with function $h(.)$ as hyperbolic tangent, and 10 neurons; the second layer is dynamic with 5 neurons of linear dynamic and hyperbolic tangent function $h(.)$, the third layer is static, having 10 neurons and $h(.)$ as hyperbolic tangent, and the last layer is static with just one neuron and linear function $h(.)$. The result for 26,725 iterations of learning is shown in Figure 4.

5 Discussions

For a few number of iterations, the ANNLDSS shows a good result in the task of identifying a non-linear process. The use of this novel neural model presents an original way for integrating some dynamics in the system identification, where the stability of the model is easily determined, because is necessary only to have the eigenvalues from matriz A, and the neural model uses a few number of parameters, if it compared with neural net in input-output representation.

6 Acknowledgements

The first author acknowledges the CAPES and UFPA for the material support in the development of this work.

References

[1] F.M. de Azevedo. *Contribution to the Study of Neural Networks in Dynamical Expert Systems.* PhD thesis, Institut D'Informatique, FUNDP, Namur, Belgium, 1993.

[2] R.C.L. de Oliveira, C.L. Nascimento Jr., and T. Yoneyama. A fault tolerant controller based on neural nets. In *Proceedings of the IEE International Conference on Control'91*, volume 1, pages 399–404, Endiburgh, UK, 1991.

[3] K.J. Hunt, D. Sbarbaro, R. Żbikowski, and P.J. Gawthrop. Neural networks for control systems: a survey. *Automatic*, 28(6):1083–1122, 1992.

[4] M. Jordan. *The Learning of Representations for Sequential Performance.* PhD thesis, University of California, San Diego, California, USA, 1989.

[5] C.L. Nascimento Jr. *Artificial Neural Networks in Control and Optimization.* PhD thesis, Control System Centre, Faculty of Technology, University of Manchester, Manchester, UK, 1994.

[6] L. Personnaz, I. Guyon, and G. Dreyfus. Information storage and retrieval in spin-glass like neural networks. *Journal de Physique, Lettres (Orsay, France)*, 46:L–359–L–365, 1985.

[7] J. Sjoberg. *Non-Linear System Identification with Neural Networks.* PhD thesis, Department of Electrical Engineering, Linkoping University, Linkoping, Sweden, 1995.

[8] R. Williams and D. Zipser. Experimental analysis of the real-time recurrent learning algorithm. *Connect. Sci.*, 1:179–211, 1990.

[9] R.W. Żbikowski. *Recurrent Neural Networks: Some Control Aspects.* PhD thesis, Faculty of Engineering, Glasgow University, Glasgow, UK, 1994.

Distal Learning for Inverse Modeling of Dynamical Systems

A. Toudeft[1,2] and P. Gallinari[1]
[1] Laforia, Université Paris 6
BP 169, 75252 Paris cedex 5, France
[2] Cemagref BP 509534033 Montpellier cedex 1, France
Email: {toudeft, gallinari}@laforia.ibp.fr

Abstract

This paper addresses stability issues of the learning process when the distal-in-space approach is used to learn inverse models of dynamical systems. Both direct and indirect versions of this approach are analysed for linear plants. It is shown that none of them is suitable when the plant is an unstable non-minimum phase system. When the plant is unstable, an additional problem must be solved: stability of the control system must be guaranteed at the beginning of the learning process. We do not deal with this additional problem but concentrate on the stability and the speed of the learning process. We propose solutions in the case of the direct version, applied to non-minimum phase stable plants. These solutions are compared on a linear plant control problem. Extensions to nonlinear systems are briefly discussed.

1 Introduction

One of the main problems when learning a controller for an unknown plant is the lack of information about desired network outputs. For this problem, Jordan *et al.* [2] have proposed the distal learning approach: a differentiable plant model is used to estimate the plant jacobian and to derive an estimation of the command error. The latter is obtained either directly by inverting locally the plant model or indirectly by estimating, through the model, the gradient of the plant output with respect to the controller parameters.

This paper deals with the direct and indirect versions of the distal learning approach. We show that, if the two versions are equivalent when dealing with kinematics, they can lead to completely different results in the case of dynamical plants. In this case, gradient computations involve finite difference equations and hence instability can arise. Both versions can lead to instability, depending on plant characteristics. More precisely, instability can arise either when the direct version is used for a non minimum phase plant (i.e. with unstable inverse) or when the indirect version is used for an unstable plant. This instability is due to the finite time-difference equations involved by the learning process. To solve this problem, finite time-difference equations must be avoided.

We address the stability issues of the learning process when the direct version is applied to learn a controller for non minimum phase plants. The proposed solutions will also be convenient for unstable plants if the stability of the system is ensured at the beginning of the learning process (e.g. by a classical controller).

2 The Distal in Space Learning Approach

The distal-in-space learning approach was originally proposed to extend the supervised learning paradigm to problems where desired controller network outputs are unknown and only the desired effects of network outputs on the environment are available. Jordan *et al.* [2] proposed a differentiable model of the environment to derive the network output error. This approach may be used for a variety of problems and with any differentiable parametric model.

Learning control problems are natural candidates for this approach. In this case, the environment corresponds to the plant to be controlled, and the network outputs are commands that are applied to

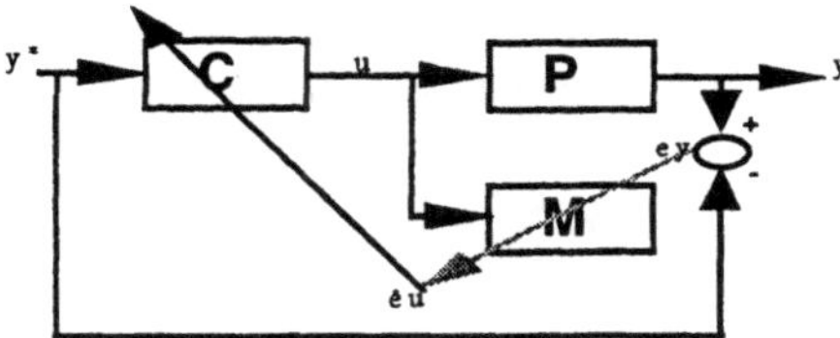

Figure 1: Distal learning approach for the control problem.

the plant (Figure 1).

Let $y*$, y_m, u and y denote the reference output, the model output and the plant input and output, respectively. Let w be any controller parameter, and e_y and e_u be the plant and the controller output errors, respectively. The two possible controller learning versions of the distal learning [7] are:

- Direct version:

$$\Delta w(t) = -\alpha.\hat{e}_u(t).\frac{\delta u(t)}{\delta w(t)} \quad (1)$$

where $\hat{e}_u(t)$ is obtained by backpropagating $e_y(t)$ through the model.

- Indirect version:

$$\Delta w(t) = -\alpha.e_y(t).\frac{\delta \hat{y}(t)}{\delta w(t)} \quad (2)$$

where the approximated gradient is also obtained by backpropagation through the model.

These versions, apparently equivalent, can lead to completely different results, especially when dealing with dynamical plants. A global comparison of these versions is presented in the following section.

3 Learning a Controller with the Distal Approach

Let the plant P to be controlled be a single input-single output linear system described by the general equation:

$$y(t) = \sum_{i=1}^{n} a_i.y(t-i) + \sum_{j=1}^{m} b_j.u(t-j-r) \quad (3)$$

Equation (3) can be written in the z-transform space:

$$y(t) = H.u(t)$$

with

$$H = \frac{\sum_{j=1}^{m} b_j.z^{-j}}{1-\sum_{i=1}^{n} a_i.z^{-i}}.z^{-r} = \frac{B(z^{-1})}{A(z^{-1})}.z^{-r}$$

The classical control performance measure is the squared error:

$$E(t) = 1/2.(y(t)-y^*(t))^2 = 1/2.e_y(t)^2.$$

Since the plant is linear, command errors e_u and plant output errors e_y are related by:

$$e_y(t) = H.e_u(t), \quad (4)$$

and the gradient of the plant output $y(t)$ with respect to the controller parameter w is given by:

$$\frac{\delta y(t)}{\delta w} = H.\frac{\delta u(t)}{\delta w}.$$

Suppose a differentiable plant model M has already been obtained by any identification technique:

$$y_m(t) = \sum_{i=1}^{n} \alpha_i.y_m(t-i) + \sum_{j=1}^{m} \beta_j.u(t-j-r),$$

i.e.

$$y_m(t) = G.u(t),$$

where G is an approximation of H. The model M is used to estimate the plant Jacobian. Hence, to apply the direct version (Equation (1)), we use:

$$\hat{e}_u(t) = G^{-1}.e_y(t),$$

which is possible only if G^{-1} is stable, i.e. when the model (and hence the plant) is in minimum phase. To apply the indirect version (Equation (2)), we use:

$$\frac{\delta \hat{y}(t)}{\delta w} = G.\frac{\delta u(t)}{\delta w}.$$

But in this case, the model (and hence the plant) has to be stable. Hence, if P is a non-minimum phase unstable plant, none of the two versions is applicable. Notice that an additional problem has to be solved when the plant is unstable: stability of the control system must be secured at the beginning of the learning process. We do not address this additional problem here but consider the application of the direct version to a non-minimum phase stable plant.

In the following section, we propose a modified version of the direct version to deal with non-minimum phase plants.

4 Controller Learning for a Non-Minimum Phase Plant

In order to avoid instability, we will not use finite time-difference equations. One way to do this is to use the 'local optimization' formulation of the distal learning approach: the plant output error $e_y(t)$ is 'statically backpropagated' through M and an error vector Δx is derived such that:

$$\Delta x = J^T.e_y(t) \tag{5}$$

where J^T is the transpose Jacobian of M : $J = (\alpha_1, \ldots, \alpha_n, \beta_1, \ldots, \beta_m)$. The $(n+1)^{th}$ component $(\beta_1.e_y(t))$ of x corresponding to $u(t-r-1)$ is considered as an estimation of $e_u(t-r-1)$ and then used as a learning error:

$$\begin{aligned} \Delta w(t) &= -\alpha.\Delta x_{n+1}.\frac{\delta u(t-r-1)}{\delta w(t)} \\ &= -\alpha.\beta_1.e_y(t).\frac{\delta u(t-r-1)}{\delta w(t)} \end{aligned} \tag{6}$$

This method has been used elsewhere [3] where the authors were motivated by the high cost of methods (1) and (2). It is clear that this method will be suitable only when the plant time-delay r is fixed and known (i.e. $\beta_1 \neq 0$), otherwise, the $(n+1)^{th}$ component of x will be near zero and learning cannot occur. We can also see, from Equation (6), that if the sign of the high frequency gain β_1 is known, backpropagation through the model can be avoided. This is well known in linear adaptive control [1]. It has been applied to linear plants [6] and to non-linear plants with adaptive learning rate [4]. We propose here to use, not only the $(n+1)^{th}$ component of x but all the components corresponding to $u(t-r-1), \ldots, u(t-r-m)$. This method takes into account all potentially required commands and will allow us to perform adequate corrections on the command. The learning rule (6) becomes:

$$\begin{aligned} \Delta w(t) &= -\alpha.\sum_{i=1}^{m} \Delta x_{n+1}.\frac{\delta u(t-r-i)}{\delta w(t)} \\ &= -\alpha.\sum_{i=1}^{m} \beta_i e_y(t).\frac{\delta u(t-r-i)}{\delta w(t)} \end{aligned} \tag{7}$$

Toudeft [5] has compared the methods described by Equations (1), (6) and (7) on a learning controller problem for a non-minimum phase system with varying time-delay.

5 Distal Learning with Partially Normalized Jacobian

Let us consider the quality of the approximation error introduced by using the transpose jacobian rather than its inverse in Equation (5). An error vector E_u at the model input involves an error $E_y = J_p.E_u$ at the plant output (where $J_p = (a_1, \ldots, a_n, b_1, \ldots, b_m)$ is the plant Jacobian). E_y is backpropagated through the model and an estimation $E'_u = N^T.E_y$ of E_u is obtained, with $N = f(J) = (n_1, \ldots, n_n, \eta_1, \ldots, \eta_m)$. In the above methods, f is the identity function. Now E'_u must be such that the amplitude of the corrected plant output error E'_y, due to the corrected model input error $(E_u - E'_u)$, is smaller than the amplitude of the error E_y. That is:

$$(E'_y)^2 < (Ey)^2 \tag{8}$$

where $E'_y = J_p.(E_u - E'_u) = J_p.E_u - J_p.N^T.E_y = (1 - J_p.N^T).E_y$. The condition (8) is equivalent to: $0 < J_p.N^T < 2$ and the error E'_y is cancelled if $J_p.N^T = 1$. Since J_p is unknown, it is replaced by J and the condition becomes: $J.N^T = 1$, which means that N is a normalized version of J and f is a normalizing function.

In the methods (6) and (7), only the $(n+1)^{th}$ component and the last m components of J are used, respectively. The condition is then changed to

$$\beta_1.\eta_1 = 1$$

and

$$\sum_{i=1}^{m} \beta_1.\eta_1 = 1$$

respectively. By replacing β_i by η_i in Equations (6) and (7), we obtain a normalized version of the methods. Note in addition that the model parameters α_i are not used in the learning process. As it can be deduced from Equation (4), when $e_u = 0$, e_y will not be instantaneously cancelled but will evolve according to

$$e_y(t) = \sum_{i=1}^{n} a_i . e_y(t-i).$$

Hence, it appears to be more realistic to reduce the filtered error:

$$(e_y(t) - \sum_{i=1}^{n} a_i . e_y(t-i))$$

instead of $e_y(t)$. Since the values of the parameters a_i are unknown, they are replaced by their estimated values α_i. The learning rule becomes:

$$\Delta w(t) = -\alpha . \sum_{i=1}^{m} \beta_i . (e_y(t) - \sum_{j=1}^{n} \alpha_j . e_y(t-j)) . \frac{\delta u(t-r-i)}{\delta w(t)}. \tag{9}$$

To summarize, we have several schemes of the distal-in-space learning approach using static backpropagation: a simple static (SS) method (Equation (6)), a multiple static (MS) method (Equation (7)), a 'recurrent' static (RS) method (Equation (9)), and their normalized versions (NSS, NMS and NRS). In the following section, these schemes are compared on a simulated linear plant.

6 Simulation Results

We present here some experimental results on a linear plant described by Equation (3), with $n = m = 2$, $a_1 = 1.1692$, $a_2 = -0.4102$, $b_1 = 0.1081$, $b_2 = 0.1318$, and $r = 8$. This plant is a non-minimum phase stable system.

The controller is a linear neural network. It is a feedforward controller with 4 inputs $[y^*(t+r), y^*(t+r+1), y^*(t+r+2), y^*(t+r+3)]$. In all experiments, the learning rate has been fixed to 0.01, and the initial weights values are the same. The reference trajectory is:

$$y^*(t) = \sin(t/10) + \sin(t/20), \text{for } t = 1, \ldots, 1100.$$

Notice that, since there is no uncertainty either in the plant model or in the environment, there is no prediction problem, i.e. the problem is the same as if $r = 0$.

We have first tried to apply the direct version (Equation (1)): instability immediately arises.

The indirect version works perfectly on this non-minimum phase plant, because it is stable. But none of the versions work when, in addition, the plant is unstable (e.g. change a_1 to 1.692).

We have tested all the methods (SS, MS, RS, NSS, NMS and NRS). Evolutions of the square sum plant output error are shown on Figure 2.

These results show that the normalized versions are much faster than the non-normalized ones. This is because partial normalization of the model jacobian allows to obtain more realistic command errors. The multiple static methods (MS and NMS) are much faster than the simple static methods (SS and NSS). This is because multiple methods use more components of Δx. Notice that superiority of the 'recurrent' methods is not evident (Figure 2). Unlike non-recurrent methods, which try to cancel the error $e_y(t)$, recurrent methods try to cancel the filtered error

$$(e_y(t) - \sum_{i=1}^{n} a_i . e_y(t-i)).$$

Adequate criteria need to be defined in order to evaluate the efficiency of recurrent methods.

7 Extension to Nonlinear Systems

The framework presented above can be applied when the plant is nonlinear. The major difference is that the model jacobian components are not constants but vary with the setpoint. These components can be obtained by backpropagation if multilayered neural models are used. An example of the application of the SS and MS methods on a nonlinear non-minimum phase system with a varying time-delay can be found in Toudeft [5].

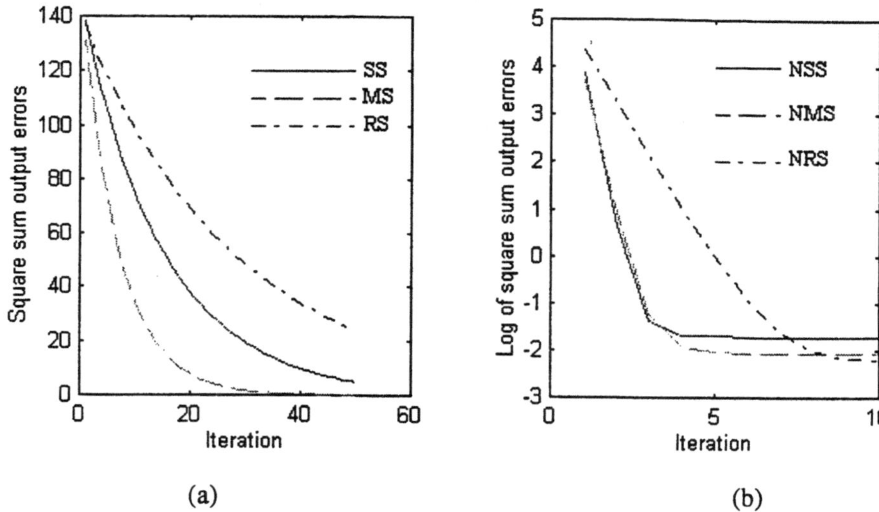

(a) (b)

Figure 2: Evolution of the square sum plant output errors when the distal learning approach with static backpropagation is used to learn a controller: (a) non-normalized schemes; (b) normalized schemes (logarithms are used to clearly display the differences).

8 Conclusion

We have argued that none of the direct and indirect versions of the distal-in-space learning approach are available when dealing with the control problem of a non-minimum phase unstable dynamical plant. To avoid instability of the learning process, the use of static backpropagation appears to be necessary but is not sufficient when the plant time-delay is variable or not exactly known. For this reason, we have proposed several modified schemes of the approach. A comparison on a linear plant has been done on the basis of the learning speed. This latter is improved by partially normalizing the model jacobian. Additional work is needed to evaluate the usefulness of the proposed methods on nonlinear systems.

References

[1] K.J. Astrom and B. Wittenmark. *Adaptive Control.* Addison-Wesley, Reading, MA, 1989.

[2] M.I. Jordan and D.E. Rumelhart. Forward models: Supervised learning with a distal teacher. *Cognitive Science*, 16:307–354, 1992.

[3] M. Saerens, J-M. Renders, and H. Bersini. Neural controllers based on backpropagation algorithm. In M. Gupta and N. Sinha, editors, *IEEE Press Book on Intelligent Control.* IEEE Press, 1993.

[4] W.H. Schiffman and H.W. Geffers. Adaptive control of dynamic systems by back propagation networks. *Neural Networks*, 6:517–524, 1993.

[5] A. Toudeft. Neural control of nonlinear non-minimum phase dynamical systems. In D.W. Pearson, N.C. Steele, and R.F. Albrecht, editors, *Artificial Neural Networks and Genetic Algorithms*, Springer-Verlag, Wien New York, 1995.

[6] A. Toudeft, P. Kosuth, and P. Gallinari. A PID neural controller for unstable delayed linear systems. In *Proc. ICANN'94*, 1994.

[7] D.A. White and D.A. Sofge. *Handbook of Intelligent Control: Neural, Fuzzy and Adaptive Approaches.* Van Nostrand Reinhold, 1992.

Genetic Algorithms in Structure Identification for NARX Models

C. K. S. Ho[1], I. G. French[2], C. S. Cox[1] and I. Fletcher[1],
[1] School of Engineering and Advanced Technology, University of Sunderland, UK.
[2] EPICC, University of Teesside, UK.

Abstract

Genetic algorithms have been recently applied to model both linear and non-linear systems. Different methods of coding the problem solutions were proposed and were claimed to have good performance. This paper presents a comparative study of three of the methods with their strengths and weaknesses highlighted.

1 Introduction

The problem of identifying a dynamic process from experimentally derived data has received much attention in the technical literature. An inspection of this literature, reveals that a primary requirement when undertaking such an identification exercise is the determination of the most appropriate model structure. For linear models the modern approach to structure selection is by exhaustive search, based on a statistical measure of model quality. In the case of a NARX model (nonlinear auto-regressive with exogenous inputs), however, the use of an exhaustive search is often infeasible, since the search space (typically 90 structures for a single-input single-output linear model) may be in the order of several hundred million structures for the NARX case.

Faced with this problem, several authors [2, 3, 4] have proposed the use of search schemes based on genetic algorithms. However, despite their common heritage, the schemes proposed differ significantly from author to author. In this paper, therefore, it is intended to present a comparison of three such methods.

2 Coding Methods

A genetic algorithm works with a population of strings or chromosomes. In general these chromosomes are constructed from a coding (typically binary, but in essence any base set of integers or characters) of the parameters which the genetic algorithm should identify.

It follows therefore, that a key feature of any successful genetic algorithm is the selection of an appropriate mechanism for coding the problem parameters (in this case the structure of the NARX model) into a usable chromosome. In the following sections it is intended to provide an outline of the three coding approaches which form the basis of the comparison study.

2.1 Coding Method 1 : Binary Coding

The binary coding method [4] is based upon the definition of a data vector with fixed structure. For example, for a single-input single output (SISO) system of second order, with time delay 2 and expanded using second order polynomials, the data vector would contain a total of 15 terms as shown below :

$$\begin{aligned} x^T = [&y(t-1)\, y(t-2)\, u(t-2)\, u(t-3) \\ &y(t-1)^2\, y(t-1)y(t-1)\, y(t-1)u(t-2) \\ &y(t-1)u(t-3)\, y(t-2)^2\, y(t-2)u(t-2) \\ &y(t-2)u(t-3)\, u(t-2)^2 \\ &u(t-2)u(t-3)\, u(t-3)^2\, d.c.] \end{aligned} \tag{1}$$

The chromosome is thus defined as a binary string containing 15 bits. In this string a '1' represents the inclusion of a term within the model and '0' represents its absence. Hence the model

$$y(t) = b_2 u(t-2) + a_1 y(t-1) + d_1 u(t-2)^2 + \text{d.c.} \tag{2}$$

may be expressed by the string

$$\text{NB}_{\text{exp}} = [1\,0\,1\,0\,0\,0\,0\,0\,0\,0\,0\,1\,0\,0\,1] \tag{3}$$

The objective of the genetic search, therefore, is to identify those terms from within the data vector needed to capture the salient dynamic features of the process; the relative proportion of each term is then determined using a least squares based algorithm.

Table 1: Assignment of integers to data vector terms.

index	Term
1	$y(t-1)$
2	$y(t-2)$
$\vdots$	$\vdots$
14	$u(t-3)^2$
15	d.c

2.2 Coding Method 2 : Integer Coding

The integer coding method [2] is based upon the assignment of a numerical value, the integer, to each of the terms in the data vector as indicated in Table 1. The chromosome is then constructed as an integer string where each bit defines a term within the model.

Thus, the model described by Equation (2) may be expressed by the string:

$$\mathrm{N_{exp}} = [1 \;\; 3 \;\; 12 \;\; 15] \tag{4}$$

It should be noted, however, that in this implementation the string length, and hence the number of terms in the model, is fixed. Once again, the approach is to search only for the terms which contribute to the model, with the relative proportion of each term being determined via a least squares based algorithm.

2.3 Coding Method 3 : Tree Structured Symbolic Coding

In a tree structured algorithm [3, 5, 6] the chromosomes, which in general are of variable length, represent expressions constructed from a base set of operators and variables. A typical base set may be:

Variables = [$y(t-1)$ $y(t-2)$ $y(t-3)$ $u(t-1)$ $u(t-2)$ $u(t-3)$]
Operators = [( ) + *]

Using this base set the terms contained in the model expressed in Equation (2) may be represented by the tree in Figure 1 or alternatively as the chromosome

$$\begin{gathered}((\text{d.c.} + (u(t-2) + y(t-1))) \\ + (u(t-2) * u(t-2)))\end{gathered} \tag{5}$$

Again, as in the previous codings, the genetic algorithm is used solely to establish the terms present within the model description.

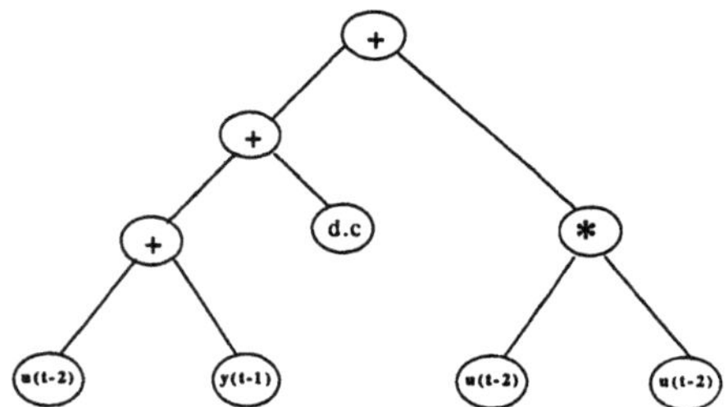

Figure 1: Tree structure representation of the terms contained in Equation (2).

3 The Fitness Function

In conventional system identification approaches, it is usual to choose the model structure based on a compromise between model accuracy and model complexity. To help with this choice it is convenient to make use of one of the standard measures.

In this study it is intended to make use of Young's Information Criterion (YIC) as defined below:

$$YIC = \ln\left(\frac{\sigma_e^2}{\sigma_y^2}\right) + \ln\left(\frac{\sigma_e^2}{n}\sum\frac{P(i,i)}{\hat{\theta}(i)^2}\right) \tag{6}$$

where σ_e^2 signifies the error variance and σ_y^2 signifies the variance of the actual process output, n is the number of parameters in the model, P is the final covariance matrix and $\hat{\theta}$ is the vector of parameter estimates. The YIC statistic is essentially a trade off between model fit and the variance of the parameter estimates. The best model will provide the smallest value of YIC. For the genetic algorithm, we define

$$Score = 50 - YIC \tag{7}$$

and the best model will maximise *Score*.

4 Termination of the Algorithm

Since, the principal aim of the genetic search is to identify those terms from within the data vector which are needed to capture the process behaviour; then it would seem sensible to terminate the search once these 'significant' terms are established.

Assuming a normal distribution, we can be 90% confident that a parameter is making a significant contribution to a model if its magnitude satisfies the relationship

$$|\hat{\theta}_i| > 1.95\sqrt{\hat{\theta}_i - Variance} \tag{8}$$

However, establishing the significance of a parameter, or indeed all parameters, within a model is not in itself sufficient for termination of the search algorithm. This is because the significance test establishes only if a term is significant within the model in which it is tested and not if it is a parameter within some form of optimum model.

One solution to the above dilemma is to first establish the model quality, in terms of the normalised residual error variance, of a candidate model. The search can then be terminated if the normalised error variance is below some predefined threshold and all terms contained within the model are significant. This threshold may be simply determined, based on a measure of the process noise variance.

$$Threshold = \frac{2\sigma^2_{e\,Steady\,State}}{\sigma^2_y} \tag{9}$$

5 Example: The Basic Algorithms

To contrast the performance of the methods we will consider a SISO process described by the expression

$$\begin{aligned} y(t) =& 0.5y(t-1) + u(t-1) + 0.8u(t-3) \\ &+ u(t-1)^2 + \epsilon(t) \end{aligned} \tag{10}$$

The input to the process, $u(t)$, is chosen as a multi level random sequence and the noise term, $\epsilon(t)$, is chosen to be uniformly distributed with an output-signal to noise ratio of 20:1. Five hundred data pairs were used during identification.

The results of the three search algorithms are presented in Table 2. In each case the data vector is defined assuming a third order system expanded using second order polynomials, i.e. 28 terms. The results presented are based on an average of 20 runs with the genetic algorithm. In the integer coding approach the string length is defined as four and in the symbolic coding the tree is assumed to have a maximum of ten branch nodes.

Inspection of the table indicates that there is little to chosen between the binary and symbolic coding approaches; the integer coding approach, on the other hand, would appear to offer a significant improvement. This result, however, is misleading since the integer coding method needs to be given the precise number of terms within the model prior to the commencement of the search, which is not always possible.

Table 2: A comparison of the number of generations required for successful identification of the test example.

Coding Method	Pop. size	Gens.	Model Evaluations
Binary	50	37	1850
Integer	50	10	500
Tree	50	33	1650

6 Improved Search Procedure

In Section 4 the parameter significance test was used in the formulation of a termination criterion. However, since the primary purpose of this test is to establish the significance of terms within a given model, then it may prove advantages if the information gathered during such a test could be used to guide the genetic search. To this end it is proposed that a local hill climber is added to the genetic search, based on the information provided via significance testing. This algorithm is outlined below:

1. Evaluate θ and confidence bounds for a candidate solution
2. Are all parameters significant - (Yes/No)?
3. Yes: Return fitness and modify solution - End
4. No: Remove insignificant terms - Goto II

However, since this hill climber is very effective at rejecting non-contributing terms, it may lead to rapid convergence of the population to non-optimal super fit individuals. To guard against this possibility, the bias of the population is monitored as an indication of the entire population development. Bias in this context is defined as the average convergence of each gene [1]. Thus for a binary coding system, the bias will approach 0.5 for a uniformly distributed population. Premature convergence can be identified by a very low/high value of the bias. Should this occur the mutation rate is increased for the affected bit location.

Table 3 illustrates the effect that the inclusion of such modifications have on the genetic search, for the example shown in section 6.

As can be seen the inclusion of the modifications has significantly reduced the number of generations required by the genetic algorithm. Basing an assessment of performance on this number alone, however, would be somewhat misleading. The reason for this is that the inclusion of the hill climber, with its

Table 3: A comparison of the number of generations required for successful identification of the test example.

Coding Method	Pop. size	Gens.	Model Evaluations
Binary	50	2	525
Integer	50	3	605
Tree	50	20	1252

obvious iterative nature, significantly increases the computational burden incurred at each generation. A far better assessment of performance, therefore, is to compare the number of model evaluations performed during each search. Again, inspection of the table indicates that, except for the integer coding, substantial improvements have also been made in this respect. Moreover, in the case of the integer coding the results presented are for a search in which the string length is set to seven. Thus the inclusion of the suggested modifications has, in this case, enabled the previous restriction, that the exact string length be known prior to the commencement of the search, to be relaxed.

7 Conclusion

The paper has presented a comparison of three coding schemes, proposed by various researchers, for the identification of the optimum structure of a NARX model, using genetic algorithms.

To form a common basis for this comparison, the paper has postulated the use of a fitness function and a termination criterion based on traditional system identification quality measures. In this way, the algorithms will naturally seek solutions which trade off model accuracy against model quality and, consequently, will tend to yield unbiased models.

The paper has also proposed the use of a local hill climbing scheme, based on significance testing, to provide guidance to the genetic search. Results to date, indicate that the inclusion of such a scheme considerably improves the efficiency of the search algorithm.

Finally, for the results presented it would appear that there is little to choose between the binary and integer coding approaches (since the problem of the fixed string length has been overcome using the hill climber). Further, it would appear that both of these methods offer a considerable efficiency gain when compared with the tree structured approach.

References

[1] J.E. Baker. *Adaptive Selection Methods for Genetic Algorithms.* In J. J. Grefenstette (editor), *Proceedings ICGA'85*, pages 101–111, Lawrence Erlbaum Associates, 1985.

[2] C.M. Fonseca, E.M. Mendes, P.J. Fleming, and S.A. Billings. Non-linear model term selection with genetic algorithms. Technical report, University of Sheffield, 1993.

[3] C.K.S. Ho. Tree structured GA in system identification. Technical report, University of Sunderland, 1995.

[4] C.J. Li and Y.C. Jeon. Genetic algorithms in identifying nonlinear auto regressive with exogenous inputs models for nonlinear systems. In *Proc. Am Control Conf.*, pages 2305–2309. IEEE Press, 1993.

[5] B. McKay, M.J. Willis, and G.W. Barton. Using a tree structured genetic algorithm to perform symbolic regression. In *GALESIA'95*, pages 487–492. IEE, 1995.

[6] M.C. South. *The Application of Genetic Algorithms to Rule Finding in Data Analysis.* PhD thesis, University of Newcastle upon Tyne, UK. 1994.

[7] P. Young. *Recursive Estimation and Time-series Analysis.* Springer-Verlag, New York, 1984.

A Model-based Neural Network Controller for a Process Trainer Laboratory Equipment

B. Ribeiro and A. Cardoso
CISUC - Centro de Informática e Sistemas da Universidade de Coimbra
Polo II, Pinhal de Marrocos, P-3030 Coimbra, Portugal
Email: {bribeiro,alberto}@dei.uc.pt

Abstract

This paper presents an application of multilayered feedforward neural networks for controlling a PT326 process trainer laboratory equipment. Firstly, the process as well as its inverse have been identified using the Levenberg-Marquardt algorithm for neural network training. Secondly an internal model control (IMC) strategy has been used for neurocontrol. Different architectures and learning methods have been investigated for model approximation. Control of the process has been implemented in real-time using the Simulink/Matlab environment. Experimental results regarding the performance of the control scheme are included in a comparative study.

1 Introduction

Recent progresses in control theory have made possible the development of advanced control systems relying on model based control strategies. Due to the capabilities of non-linear function approximation with an arbitrary degree of accuracy, neural networks are optimal tools for non-linear system modelling. Therefore, in the field of control engineering several applications using neural networks have been widely reported [3]. However, most of the published results are based on simulated work. In this study, we emphasize the experimental tests which have been carried out to achieve the desired temperature control of a laboratory scale heater. Moreover, we compare the performance of the neurocontroller in the underlying process with that one obtained with a modified PID algorithm. The process for control as well as its open-loop characteristics is first described. The synthesis of the forward model and inverse model using neural networks is next presented. On the basis of the above models, the control IMC structure is illustrated and the experimental results are finally reported.

2 Process Trainer PT326

The experimental setup "PT326 Process Trainer" as shown in Figure 1, consists of a self-contained process and control equipment whose main function is similar to a hair drier. Air, drawn from atmosphere by means of a centrifugal blower, is driven through a heater grid and forced to circulate inside the tube. The process consists of heating the air flowing inside the tube to the desired temperature level. Heat is transferred from the grid to the circulating air, the rate of heat transfer depending on the heater temperature and on the air flow velocity. A bead thermistor fitted to the end of a probe can be inserted into the air stream at any of three points along the tube, spaced by 28mm, 140 mm, and 279 mm from the heater. The control action corresponds to a signal from 0 to +10 volts which is converted in a variable power supply to the grid as shown in Figure 1. As explained in a previous work [4] open-loop tests were performed to obtain information of the process characteristics. In the experimental measurements, the air stream entrance makes an angle of 30 degrees and the temperature sensor was inserted in the third position (279mm). The turbine velocity and the ambient temperature were assumed constant.

A step change ($2 \rightarrow 8$ Volts) was considered to ensure a margin of 2 Volts to the extreme values. These are defined according to the maximum value of 10 Volts and to the lower saturation zone due to the ambient temperature. Following a step change of input the temperature rise is exponential. In a process several exponential lags are normally present, leading to a response curve with an 'S' shape and producing a time lag — the time delay — of the process. To obtain the process transfer function the Bröida method was used leading to a first order system representation with time delay

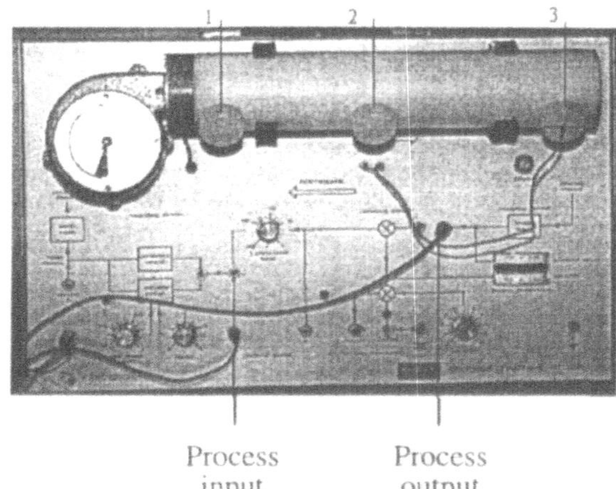

Figure 1: Layout of the front panel of the laboratory equipment.

which can be described by Equation (1):

$$G(s) = \frac{k}{1+T_1 s}e^{-sT_d} \tag{1}$$

where k is the open-loop gain, T_d is the pure time delay and T_1 is the time constant. Using the above method the values calculated for these parameters were $k = 1.0$, $T_d = 377$ ms and $T_1 = 473$ ms.

In order to obtain the discrete system representation a sample period of 100 ms was considered. It was chosen in order to guarantee that the discrete process representation was close enough to the continuous one. However, it is limited by the computation of (i) the control law, (ii) the output process visualization in the screen and (iii) the time conversion from A/D and D/A.

The discrete transfer function was obtained assuming a ZOH (zero order hold) on the inputs and can be represented in terms of the Z transform by Equation (2):

$$G(z) = (1-z^{-1})ZL^{-1}\left[\frac{G(s)}{s}\right]. \tag{2}$$

The difference equation then resulting is given by Equation (3):

$$\begin{aligned} y(k) = {} & 0.80945y(k-1) + 0.04776u(k-4) \\ & + 0.14309u(k-5). \end{aligned} \tag{3}$$

3 Process Identification

System identification methods can be divided in two parts (i) model structure identification followed by (ii) parameter estimation. Physical parameterized modelling in which all the physical insight about the plant is built into the model can be very time consuming. However, it leads to a desirable approach in which sparsity of the parameters to be identified occurs. On the other hand, in the black-box approach the model is searched in a flexible model set [2]. Instead of incorporating prior knowledge the model contains many parameters so that the unknown function can be approximated without too large bias. This approach requires less engineering time but is highly dependent on the training data. Neural networks are one possible choice within this approach. In case multilayered feedforward neural networks are selected, the use of the backpropagation (BP) algorithm (or one of its modified forms) seems to be adequate. However, the use of second order optimization techniques such as the Levenberg-Marquardt method leads to fast training and better accuracy.

3.1 Neural Network Modelling

The feedforward neural networks used had input nodes which distribute the input to the upper layer without any processing. The hidden layer(s) had symmetric logarithmoid activation function and the output nodes had a linear (identity).

The network output is given by Equation (4):

$$y = f[\mathbf{u}] = \phi\left(\sum_{i=1}^{n_{L-1}} w_i^L y_i^{L-1} + w_0^L\right) \tag{4}$$

where ϕ is the activation function, w_i^L are the connection weights between the output layer (L) and the previous layer, y_i^{L-1} are the outputs from neurons in the layer $L-1$ and w_0^L the bias. Different architectures and learning methods have been investigated for model approximation. The learning methods used were (i) the backpropagation (BP) algorithm and (ii) the Levenberg-Marquardt algorithm. The objective is to adjust the weights in order to minimize the sum of squares of residuals, calculated as the differences between the network outputs and target outputs. The training techniques have been here implemented using (i) the neural toolbox of MATLAB [1] and (ii) the neural toolbox of NNDT [5].

3.2 The Forward Model

In order to identify the forward model, a training data set (800 input-output pairs) was obtained applying to the process a square wave input signal, with period T=20 s and amplitude 4 Volts which was sufficient to cover most of the plant inputs without reaching the saturation extremes. A test data signal was obtained in the same way (800 input-output pairs), the input signal being also a square wave. In order to evaluate the network generalization capabilities, the experiments for obtaining the data test set were carried out with the temperature sensor inserted in the second position (see Figure 1). The discrete-time identification model structure used herein gives rise to the multiple input single output (MISO) model given by Equation (5):

$$y_p(k) = N\left[y_p(k-1), u(k-4), u(k-5)\right] \qquad (5)$$

where N is a neural network with three input nodes and one output node.

Multiple Input Single Output (MISO) Model

Regarding the BP learning algorithm, several configurations were tried out starting from a random weight initialization in the range of $[-1, 1]$. The activation function was the sigmoid in the range of $[-1, 1]$. The learning rate initially used was $\rho = 0.95$ and the *momentum* coefficient $\alpha = 0.5$. The root mean square error (RMS) was less than 0.1 in the tested networks. The different configurations used

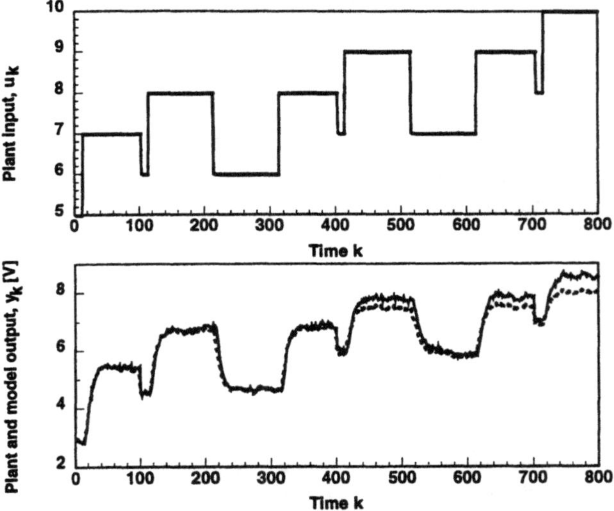

Figure 2: (a) Square wave process input, u_k; (b) Actual and predicted model response to input u_k, in the data test set (sensor position 2); Actual output (solid line), predicted model output (dashed line); network configuration $N_{3,4,1}$.

are within the general form $N_{3,a,b,1}$ which represents a neural network with three inputs, two hidden layers with a and b nodes in each one of them, and one output node. The network trained with the smallest error (0.039), had the configuration $N_{3,6,4,1}$.

In the case of the Levenberg-Marquardt algorithm it was found that no more than one hidden layer with at least 4 nodes was necessary. The training was considerable fast and the RMS error achieved with the selected network $N_{3,4,1}$ was 0.016 and less than 0.03 in all tested configurations.

3.3 Inverse Model

The inverse model identification is of relevance in the designed controller architecture, since the quality of the control depends directly on it. According to the literature, the two main learning architectures for the inverse model are: (i) the general learning mode and (ii) the specialized learning; they are also known, respectively, as direct method and indirect method.

$$u_p(k) = N\left[y_p(k), y_p(k-1), u(k-1)\right] \qquad (6)$$

In the process described above, the direct method was used and the neural network inverse model, $N_{3,4,1}$, given by Equation (6) is able to respond with the predicted process input close enough to the actual input.

4 Experimental Control Results

The internal model control (IMC) strategy was implemented by means of the forward and inverse neural network models identified in the prior phase (see Figure 3). The forward model is placed in parallel with the process, being the output error feedback to the closed-loop system. This model-based control strategy presents adequate robustness and stability for the closed-loop system. In this way, irregularities and modelling inaccuracies are taken into account. The necessary realizability of the control scheme is achieved through the use of a digital low-pass filter on the controller input (Butterword 2sd order, with cut frequency $f_c = 0.1274Hz$), preventing the large actuation effort. The filter is given by the following difference equation

$$y(k) = x(k) + 2x(k-1) + x(k-2) + 0.92387y(k-1) - 0.32671y(k-2).$$

For the sake of comparison, the results from a PID controller applied to the process using a modified algorithm are presented in Figure 4(b).

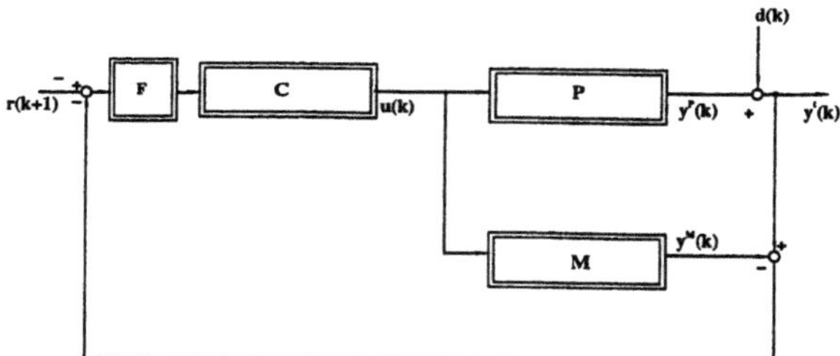

Figure 3: Internal model control strategy.

The neurocontroller ability (NN-IMC) to track the reference signal is shown in Figure 4(a) where the results have been obtained with a filter coupled to the system. Moreover, the neurocontroller ability to deal with a load disturbance (at a reference signal of 2 Volts) is depicted in Figure 4(c), where the filter influence is reducing the control action effort. The steady state response is free of off-set, as desirable, in spite of oscillations in the input.

5 Conclusions

This study investigates the applicability of multilayer feedforward neural networks for controlling a process trainer laboratory equipment,PT326, using the Simulink/MATLAB environment, the neural network toolbox NNDT, a library package that allows the real-time communication between the program and the plant.

The effectiveness of the proposed control scheme in the setting of a laboratorial heater process is illustrated by real-time results. Despite the fact that the applied technique has been used several times, the environment herein used is emphazised by real-time control results whereas the results in the literature mostly rely on numerical simulations.

The closed-loop system is stable with the output tracking error converging gradually to zero. The inclusion of a filter in the system allowed us to smooth the control signal. Despite the fact that the modified PID algorithm presents good performance, the advantages of the neural network approach, due to the lack of the physical modelling phase, offer new insights for complex systems.

Further research has to be done in order to prevent input oscillations. The neural networks were first trained off-line with noisy data and afterwards used in real-time control. In order to obtain better results the training set must cover the whole operational range of the controller.

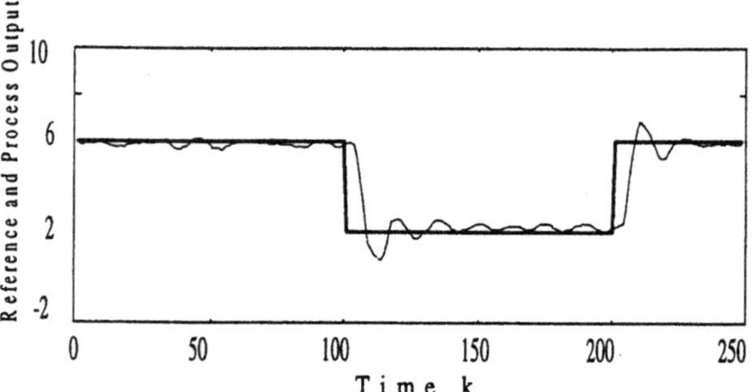

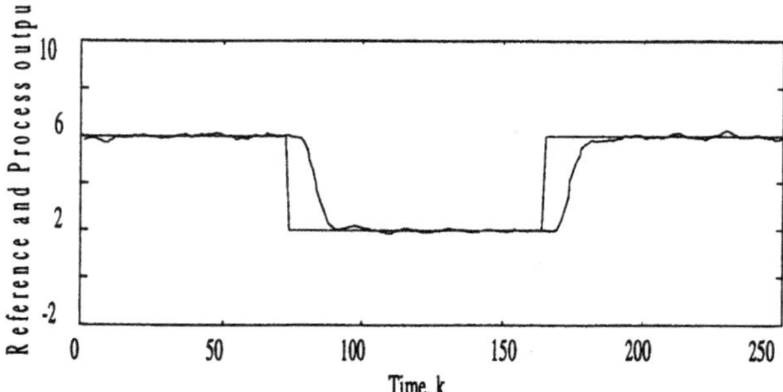

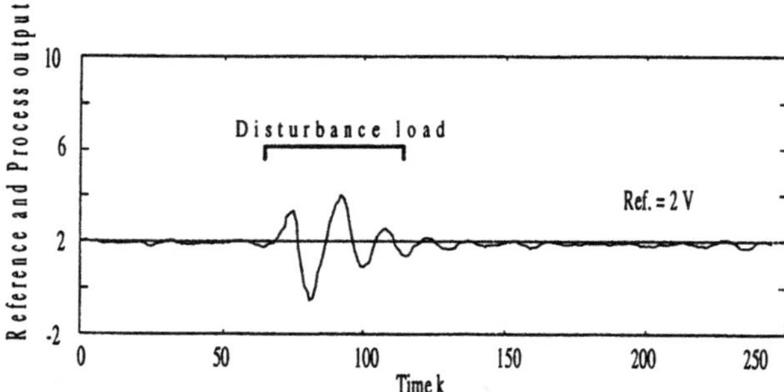

Figure 4: (a) NN-IMC + filter; (b) Setpoint tracking by PID algorithm; (c) Disturbance load rejection + filter.

References

[1] H. Demuth and M. Beale. *Neural Network Toolbox for use with MATLAB*. The MathWorks, Inc., 1994.

[2] P. Lindskog and J. Sjöberg. A comparison between semi-physical and black-box neural net modeling: a case study. In *Proceedings of the International Conference on Engineering Applications of Neural Networks (EANN'95)*, pages 235–238. Otaniemi, Finland, August 1995. A. B. Bulsari and S. Kallio, eds.

[3] K. Narendra. *Adaptive control: neural network applications*, pages 69–73. The MIT Press, London, 1995. M. A. Arbib, ed.

[4] B. Ribeiro, J. Cordeiro, I. Gomes, and A. Cardoso. Neurocontroller synthesis of a scaled heating laboratory process. In *Proceedings of the 2 Encontro Português de Controlo Automático, CONTROLO6*. Porto, September 1996.

[5] B. Saxén and H. Saxén. NNDT - a neural toolbox development tool. In D. W. Pearson, N. C. Steele, and R. F. Albrecht, (editors), *Proceedings of the International Conference on Artificial Neural Networks and Genetic Algorithms*, pages 325–328, Wien, 1995. Springer-Verlag.

MIMO Fuzzy Logic Control of a Liquid Level Process

I. Wilson[1], I.G. French[2], I. Fletcher[1], and C.S. Cox[1]
[1] Control System Centre, School of Engineering and Advanced Technology, University of Sunderland,
[2] EPICC, University of Teeside

Abstract

A large number of design strategies exist for multivariable control situations. Many of the methods require a linear time-invariant process characterisation in the form of a state space model or transfer function matrix description. Quite often this is not available and could be expensive to realise. If this latter route is pursued there needs to be considerable benefits in the quality of the resulting closed-loop performance. One alternative is to use the 'expert' approach of fuzzy logic where the plant is not modelled but the expert operator is. Application of such a controller is not so straightforward as many parameters need to be 'tuned' in order to provide precise control of a non-linear system.

1 Introduction

Multivariable systems frequently occur in industry. In many cases there is considerable interaction between the process input signals (manipulated variables) and the process output signals (controlled variables). The use of single-input single-output (SISO) design techniques in these situations should be avoided since the cross-coupling can often result in degraded or sub-standard performance.

Over the last 30 years significant progress has been made in the area of multivariable control system design. From the earliest frequency domain approaches of [7] and [5] of firstly decoupling the individual control loops so that effectively become SISO systems in the final design. To the more computationally demanding optimum approaches such as MIMO PIP (proportional-integral plus) proposed by Young *et al.* [10] and Billington [1] and the intuitive rule based methods of Heckenthaler *et al.* [3]. The TecQuipment CE105 multivariable coupled tanks apparatus has provided the vehicle to evaluate the MIMO fuzzy logic design and contrast its performance against the frequency response approach of the direct Nyquist array (DNA) [6].

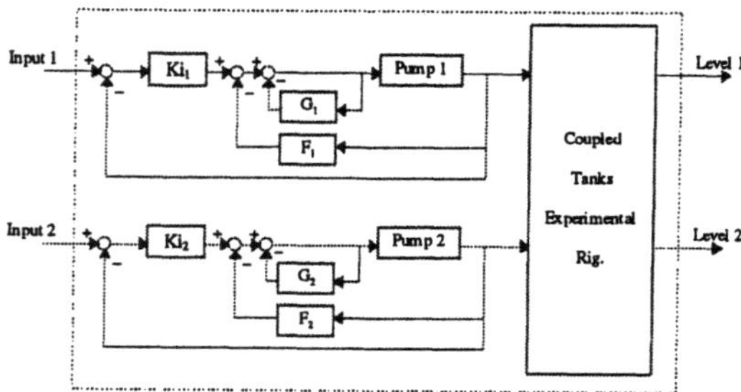

Figure 1: The PIP flow-compensated system under investigation.

2 The System

The coupled tanks, shown schematically in Figure 2, comprises two separate vertical tanks which are connected by a flow channel. Rotary valve A can be used to vary the cross-sectional area of the channel and hence change the flow characteristic between the tanks. Valves B and C provide direct discharge into the reservoir below from the left and right hand tanks respectively. Each tank is fitted with a pressure sensing liquid level transducer calibrated to provide output signals varying from 0 to +10V. An additional feature is that the output from each pump is sensed by an in-line flow transducer who's output is signal conditioned and calibrated to yield an output voltage in the range of 0 to +10V. Initial investigations of the coupled tanks rig showed how the flow was prone to fluctuate when a steady voltage was applied to the motors (see Figure 3). To counteract this two minor loop flow controllers

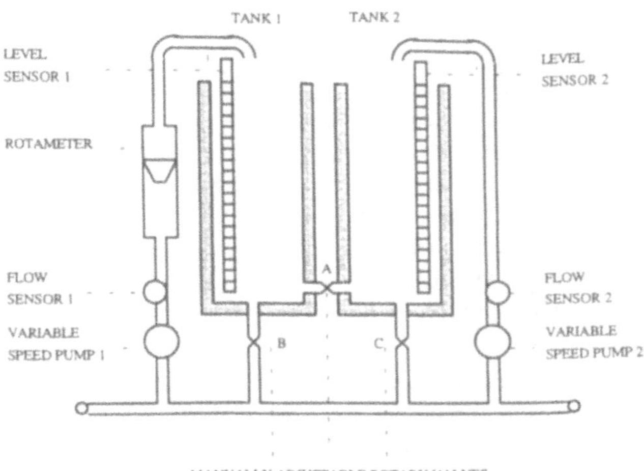

Figure 2: Coupled Tanks Experimental Rig.

were designed using the PIP approach [1] operating at a sample time of 0.2 seconds [2] (see Figure 1).

The PIP flow compensated system, Figure 3, is considered as the plant for the rest of this study. From PRBS tests the transfer function of the flow compensated system, at valve openings of $A = 60\%$, $B = 40\%$ and $C = 80\%$, was found to be:

$$\frac{\begin{bmatrix} 6451s^2 + 378.1s + 2.2723 & 166.7s + 1.7833 \\ 167.7s + 1.7944 & 5219.2s^2 + 521.8s + 4.9861 \end{bmatrix}}{265333s^3 + 18391s^2 + 259.9s + 1} \tag{1}$$

The results correspond to valve openings which resulted in tank 1 being half full and tank 2 being a quarter full when 4 volts are applied to the pumps and the system has reached steady state.

3 Fuzzy MIMO Control

A fuzzy logic controller is a control strategy that uses an expert's knowledge of a plant, and interfaces the inputs and outputs, the sensors and actuators, using fuzzy logic to this knowledge base. So the linguistic control rules of an expert can be utilised automatically, i.e.

IF *level 1 is very low* THEN *increase flow into tank 1 a lot.*

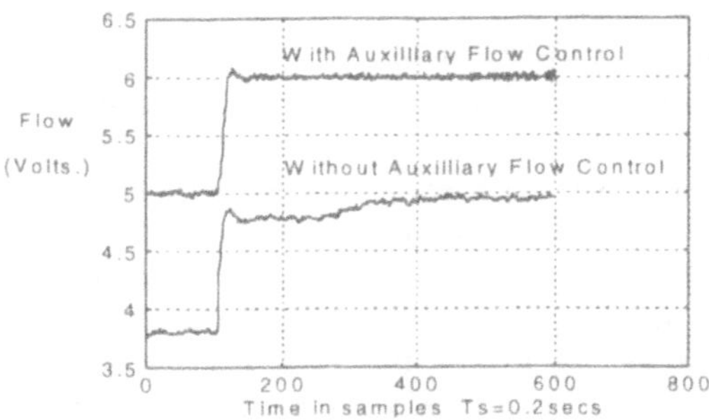

Figure 3: The improvement to the flow characteristics when under auxiliary PIP control.

IF *level 2 is slightly low* THEN *increase flow into tank 2 a little.*

...etc.

So fuzzy logic is basically a mapping tool that converts a level reading of say 3 volts to a linguistic descriptor of *slightly high*, and is then able to turn the expert's decision of *turn the flow down a little* into a signal for the pump of 2.4 volts for example. The mapping of real or crisp numbers to fuzzy sets may be done in several ways, and since one crisp value may lie in several fuzzy sets, each set must be given a percentage of truth. There are many ways of mapping crisp numbers into fuzzy sets and they are usually described by the shape of the fuzzy sets, i.e. bell shaped, exponential, triangular, trapezoidal, and the amount of overlap or width of each of the sets, i.e. thin, even, or wide. Having arrived at a fuzzy decision regarding what the output should do we then need to convert the expert decision into a control action suitable for the actuators the linguistic description of *turn the flow down a little* needs to be defuzzified. There are many methods for defuzzification including centre of gravity, mean of max, smallest of max, and largest of max.

So fuzzy control is essentially modelling of an operator as opposed to modelling the plant in order to control the system. Numerous examples of fuzzy controllers exist, especially for SISO situations, [4, 8]. Initial applications of SISO PI fuzzy control used fuzzy descriptions of error and rate of change of error to decide upon how much control action is necessary. The fuzzy PD controllers output then being integrated.

However attempts to apply this strategy directly to the MIMO plant suffered the same problems as classical SISO control methods, namely, they failed to suppress system interactions. Consequently a MIMO fuzzy control strategy was developed [3]. The approach used was to separately design the controllers and decouple in much the same way as some of the commonly used MIMO frequency response based techniques. Hence a set of rules to reduce the interactions were written to provide feedforward decoupling between the normal SISO controllers and the plant, one of the rules is given below as an example:-

IF *flow is increased a lot in level 1* THEN *decrease flow a little into tank 2*
IF *flow is decreased a lot in level 2* THEN *increase flow a little into tank 1*

In this case 28 rules to control and 10 rules to decouple the process are needed to produce the quality of result presented in Figure 4. The fuzzifier used to classify the crisp inputs from the plant into fuzzy descriptors is made up of wide triangular membership functions, to give a large overlap of sets allowing smooth transition from one state to the next. To complement this the centre of gravity method was used to defuzzify the fuzzy decisions made by the rule base, as this method also gives smooth and well weighted conversions to a single output value. The resulting controller gave impressive results, but tuning by trial and error took a full afternoon. This has led to further research and the development of a much more efficient genetic algorithm based routine that uses a cost function to penalise both excessive overshoot and long settling times [9].

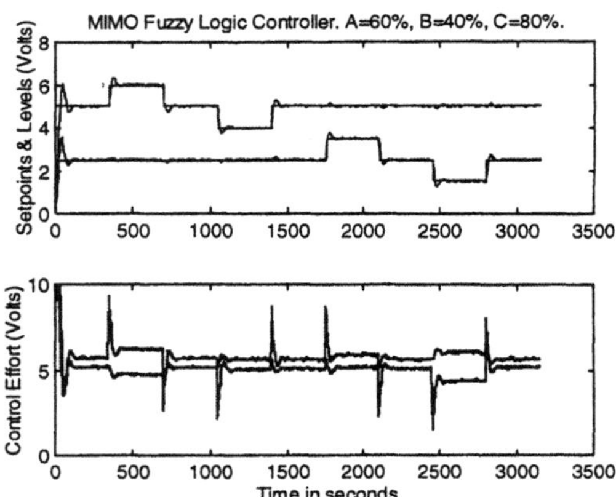

Figure 4: Performance of the fuzzy logic controller, valves at A=60%, B=40% and C=80% open.

4 Direct Nyquist Array (DNA) Compensation

One of the initial approaches to multivariable controller design [6] was simply to determine a precompensator which ensures the compensated process is diagonally dominant and then design SISO controllers for each diagonal element. For comparison purposes this exercise was initially limited to a PI structure and therefore the diagonalising compensator was essentially limited to a real gain matrix. Within MATLAB this can be achieved using the align command which identifies the best real approximation to the inverse of the frequency response at any given frequency. Obviously the accuracy of this technique is heavily dependant upon the direction of the individual elements frequency response at the selected frequency and therefore this technique only works well if these directions are similar. One method of avoiding this problem, for type zero elements, is to design the pre-compensator

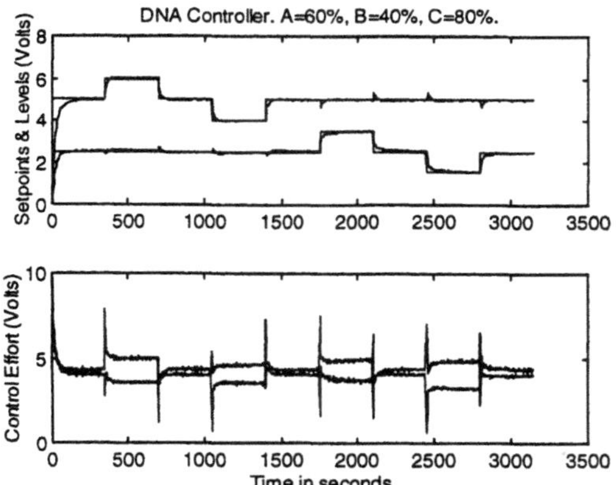

Figure 5: Performance of the direct Nyquist controller, valves at A=60%, B=40% and C=80% open.

at $\omega = 0$ which ensures that the resulting solution is real, with the added benefit that a systems steady state behaviour is easily measured accurately. This procedure was performed and achieved the required diagonal dominance. The resulting behaviour of the diagonal elements was essentially first order and therefore allowed the SISO PI controllers to be designed for specific closed loop bandwidths. Based upon the bandwidth obtained from previous SISO designs a closed loop time constant of approximately 12 seconds was chosen and Figure 5 shows the performance of this design on the experimental rig.

5 Conclusions

The major advantage of the fuzzy logic controller is that it is not designed using a process model and consequently very robust to changes in the system, this was proven by running the controller on the coupled tanks rig and altering the positions of the valves A, B and C. Table 1 provides a comparison of the designs produced for the coupled tanks using the integral of time and squared error (ITSE) performance index:

$$J = \int t.e^2 \, dt \tag{2}$$

The fuzzy logic controllers performance show a far greater consistency than that shown by the DNA design when the system is subjected to $\pm 20\%$ changes in the flow valve positions.

Table 1: The performance of the Control Strategies.

	DNA	Fuzzy Logic
Initial Plant A=60% B=20% C=80%	81.63	52.22
A=80% B=20% C=80%	330.56	75.28
A=60% B=60% C=60%	778.36	91.24

6 Acknowledgements

The authors would like to thank the University of Sunderland without whose support this work would not have been possible.

References

[1] A. J. Billington. Optimal PIP control of scalar and multivariable processes. In *IEE Int. Conf. Pus.*, volume 1 of *332*, pages 574–579, 1991.

[2] I. Fletcher, I. Wilson, and C. S. Cox. CAD supported strategies for MIMO systems: Illustrative solutions using a coupled tanks experiment. *Transactions of the Institute of Measurement and Control*, 1995. (special edition), to appear.

[3] T. Heckenthaler and S. Engell. Approximately time optimal fuzzy control of a two-tank system. *IEEE Control Systems Journal*, pages 24–30, June 1994.

[4] P. J. King and E. H. Mamdani. The application of fuzzy control systems to industrial processes. *Automatica*, 13:235–242, 1977.

[5] A. G. J. MacFarlane and J. J. Belletrutti. Characteristic locus design method. *Automatica*, 9, 1973.

[6] J. M. Maciejowski. *Multivariable feedback design.* Addison-Wesley, 1989.

[7] H. H. Rosenbrock. *Computer aided control system design.* Academic Press, London, 1974.

[8] R. M. Tong. A control engineering review of fuzzy systems. *Automatica*, 13:559–569, 1977.

[9] I. Wilson and I. G. French. Genetically tuned time optimal fuzzy control. In *Advances in Process Control IV, I Chem E.*, York, September 27-28 1995.

[10] P. C. Young, M. A. Behzadi, C. L. Wang, and W. A. Chotai. Direct digital and adaptive control by input output state variable feedback pole assignment. *Int. J. of Control*, 46(6), 1987.

A Practical Application of a Learning Classifier System in a Steel Hot Strip Mill

W. Browne[1], K. Holford[2], C. Moore[2] and J. Bullock[1]
[1] Welsh Technology Centre, British Steel Strip Products, Port Talbot, Wales, SA13 2NG,
[2] University of Wales, Cardiff, Queen's Buildings, PO Box 917, Cardiff, CF2 1XH
Email: scewb@cf.ac.uk

Abstract

The aim of this project is to improve the quality and consistency of coiling in a steel hot strip mill at British Steel Strip Products, Integrated Works. The artificial intelligence paradigm of learning classifier systems (LCS) is proposed for the processing of plant data. Improvements to a basic LCS, that allow operation on industrial data, are detailed. Initial experimental results show that the technique of LCS has the potential to become a very useful tool for processing industrial data. The stochastic computational technique will produce off- line rules to aid operator and engineering decision making. Improvements in availability, coil presentation and ultimately customer satisfaction will result in cost benefit to British Steel Plc.

1 Introduction

This paper describes the application of an artificial intelligence technique to steel strip downcoilers for improved information. Steel is produced with a wide range of properties tailored to meet modern industrial standards. A principal operation for strip products is the rolling of the cast steel (15 tonne slab) to a thin strip (typically 2 millimetres thick and up to 1.5 km long) that is coiled in order to transport the strip to future customers/processing. Coiling is critical as it should be the last process that affects the coil presentation before transportation. The routinely gathered plant data is currently only investigated in the rare event of a cobble (failure of strip to coil) as the voluminous nature limits an expert's interpretation.

Previous sources of information included the heuristics (rules of thumb) developed by plant engineers and skilled operators based on previous experience. More deterministic sources of information are the data sets taken from the downcoiler and related processes:

- the over power monitor system connected directly to the downcoilers,
- downcoiler logs from the operators describing the coil presentation,
- reworking schedules for incorrect coils,
- availability statistics for the downcoiler,
- mill central computer information on the type of steel coiled,
- the PC supervisory control and data acquisition system providing coil shape and gauge profile information.

This information is currently used only in the rare event of a serious fault occurring in order to extract information on the likely cause of the fault. A technique is needed to investigate this data for routine operational and maintenance information. The following properties of this data are important in the selection of an appropriate technique:

- incomplete knowledge of the workings of the downcoilers,
- data sets can be noisy, incomplete or voluminous,
- relationships that are complex and may contain discontinuities.

The quantity of data available indicates a computational method which, due to the data characteristics, should not rely on prior knowledge or mathematical continuity. The information produced must be understood by the operators and PCM engineers in order to facilitate changes. Quality assurance is essential in a strip mill, so the technique must be capable of being validated and verified.

The advent of cheap, available and powerful modern computers has led researchers to investigate ways in which the computational power could be used to solve problems. Classical search and optimisation techniques such as least squares regression and hill climbing have been adapted for computers. The quest for natural techniques that are not strictly based on mathematics has led to a development of stochastic methods, such as Artificial Intelligence (AI) techniques.

A rule of thumb is that if the problem field can be modelled mathematically then classic routines out perform stochastic methods for problems of less than five variables. However, by the time the number of variables has reached fifty, the computational demands of the classic techniques have grown too big for practical use [6]. The promising stochastic technique of learning classifier systems [5] (LCS) has been shown to be robust over a wide range of academic problems [2]. However, further development is needed in industrial applications due to the greater uncertainty, complexity and a critical environment.

The most industrially proven AI techniques capable of handling knowledge, expert systems (ES) and neural networks (NN), were investigated for this application. ES were rejected as the environment is not fully described and ES can not discover new rules in order to complete the domain. NN might learn these rules, but not in a transparent form. Transparency in the rule base is essential to allow operators, engineers and managers to validate and learn from the rules.

The equipment used to write the LCS is a Pentium 120MHz PC with 16 Mb ram, this is considered to be the minimum system for processing the 50+ parameters which ultimately will be analysed from the data. Instead of using an AI language, such as LISP, the computer language C/C++ was selected because of its ease of use, versatility and the speed of operation which will be critical in the later stages of this project.

Simulated data that mimicked the parameters and fault conditions of real data was generated using a bespoke C++ program. This data has many advantages over real data as the quantity, quality and volume are all controllable. Also the relationships between the parameters and the faults are known, which allows the performance of the LCS to assessed. One method is to seed the LCS with incorrect rules to test that these poor rules do gradually lose strength and hence get removed from the population. Conversely, for real data, hypothesised rules that are thought to be correct may be used to seed the population which may increase the efficiency of training depending on accuracy.

Both simulated and real data need to be encoded into a form suitable for manipulation within the LCS. The choice of alphabet was between the 'natural' approach as used by Grefenstette in his Samuel system [4] or the more common ternary system. Although the real number method greatly simplifies the industrially important encoding / decoding stage of operation it lacks definition, i.e., critical rules may contain exact parameter values that a box approach could not encode. The basic alphabet consisting of 0, 1, # (# is a 'wildcard'), is augmented with additional punctuation to aid interpretation by improving number formatting, separating parameters or marking the action statement [7].

This is necessary when real data contains signed decimal numbers as a Mantissa format is used (The format $(1.fff).2^x$ is used where only fff and x are required for encoding, e.g., 9.6 :- $1.2*2^3$:- $010;011$). The pre-processing section can also encode integer, string and binary numbers into the ternary format.

The syntax used for the rule base is common to most LCS; 'if *condition(s)* then *action(s)*'. An example of a rule as it appears in the transparent rule base and as it is used, is shown below:

Rule structure: 'if voltage > 46 then fault'

Condition: voltage has value 53, encoded as-110101
Matching Rule: 'if voltage > 46 then fault'-11####:1
Action: Coil presentation 1, decoded as FAULT.

Eventually a bespoke LCS will be written for implementation onto plant. However for familiarisation, preliminary development and evaluation, a convenient base LCS was required. A version of a simple classifier system as described by Goldberg [3] for solving the 6 bit multiplexer problem was chosen. It has been thoroughly tested over the last seven years and has been shown to be a robust system that can be used as a benchmark for performance enhancements made to LCS [1]. Significant modifications were needed to facilitate training with encoded simulated data and so are outlined below.

Achieving an exact match of the conditions of only 16 parameters, which have a length of 48 bits, even with wildcards in the rules, is extremely unlikely (~200,000:1) at normal population sizes. Initially the number of bits matched determines the rules that compete in the auction to control the systems output. As building blocks of information are discovered, the number of complete parameters matched is used. Finally complete matching of rules becomes possible. At the switch to the more exacting criterion, both techniques are used, to prevent the first exact rule dominating the population.

As the differences in functionality between the Genetic Algorithm (GA) technique and the GA used in a LCS becomes more apparent at long string lengths, the GA required updating. Convergence has to be delayed to allow building blocks to be discovered. Simple mutation, single point crossover and inversion, although good for genetic diversity, tend to destroy building blocks of information in long strings. By using a morphing operator that combines the similar information in two strings and double point circular crossover, an improvement in a slowly converging population can be achieved

As the system is primarily in stimulus-response mode the bucket brigade is not used in the apportioning of credit. The specificity is used in the bid and reward to differentiate rules in a hierarchy. To help increase generality at initial stages of training a life tax is used that penalises over specific rules. The usage of individual credit and genetic operators are adjusted automatically, by determining the current state of system performance, reproduction and rule life-span.

Initial training investigated the ability of the LCS to mine a simple relationship between one parameter and one fault within a training set of 1000 known data points. The population was seeded with a hundred random rules of the correct format from which to discover the two rules that correctly govern the known test relationship, i.e., if the parameter is

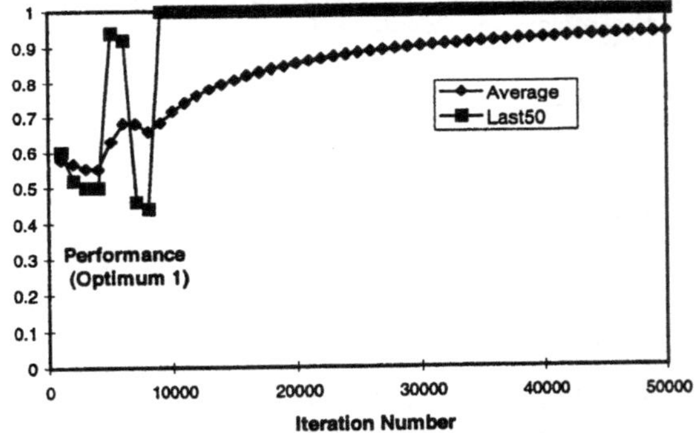

Figure 1: LCS Training Performance-Relationship of 3 Parameters with 1 Fault.

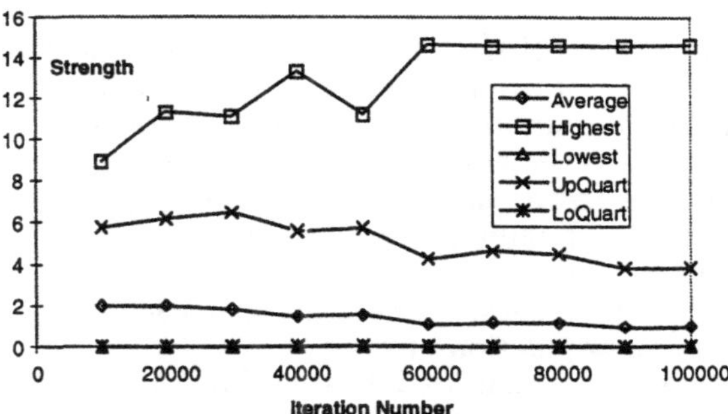

Figure 2: LCS Training Strength - Separation of High Performance Rules.

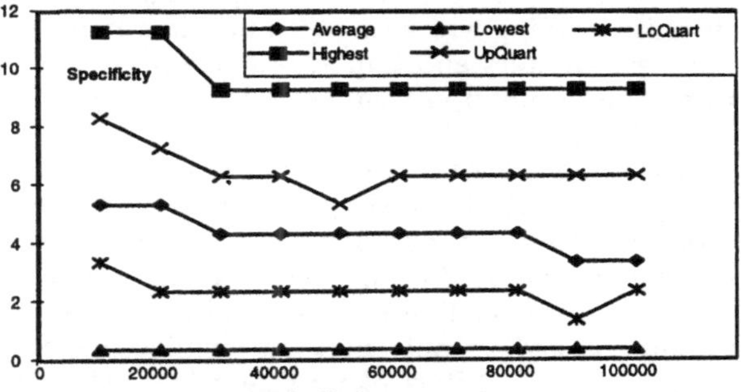

Figure 3: LCS training specificity — increase in system generality.

above a set limit then fault, whilst if below this set limit then normal operation. This was reused in each iteration block to test the hypothesised rules. Increasing the size of the data set increases the rate of convergence to a solution as well as increasing the generality of the rules discovered. Figure 1 shows that the initial performance of the LCS rules started at an average level, i.e., 60% correct. This was due to the two state fault being matched by the random rule population with at least a probably of 0.5. The spikes produced in the graph of last 50 iterations were caused by the GA searching for plausibly better rules. A near optimum relationship was identified within ten thousand iterations (3 minute computational time). Further training, if desired, does not destroy this discovered information as periodically a copy of the rule population was taken to enable high strength rules to be identified whenever they occurred.

The downward trend from 7,000 to 8,000 iterations was a partial loss of performance as the GA acted upon incorrect genetic information that could have led to an improved final solution. The ability to reject seemingly strong rules is important in preventing LCS from being trapped in local maxima. The cost is longer iteration times. The correct genetic information becomes stronger, as shown in Figure 2, as the number of tests against the known data increases as the rewards for being correct adds to the corresponding rules' strength. As the GA has a bias towards using strong rules for reproduction, the correct genetic information increasingly dominates the population of rules.

Generalities that are introduced by the wildcards are encouraged as these rules match more often, thus gaining proportionally more strength than specific rules if they are correct, e.g., Figure 3. The time for training will be significantly increased with more parameters and faults, but for most applications this is unimportant as it is accuracy of the rules that is critical.

Decoding the rule base is currently manual although it is relatively simple to automate. Operators can then set parameters to avoid faults occurring, whilst PCM engineers can determine the critical parameter limits. The results with one parameter linked to one fault on simulated data is encouraging, and it is likely that when using 50+ parameters and 10+ faults from real data, similar results could be achieved. However this is not a guarantee of success. Therefore the next stage of work will be based on simulated data that mimics the full real life situation.

Ultimately the LCS will be fed actual on-line data. The LCS will advise off-line actions that will have human validation before they are enacted. The options for reducing the number of data points needed to describe a single parameter include:

1. Reduce the number of parameters: many of the parameters are only functional at certain times in the operation of the downcoiler.
2. Increasing the time step between readings will dramatically reduce the data points.
3. Considering only the minimum and/or maximum value of a variable throughout the processing stage.
4. Variance from a mean level within a tolerance band can be used to flag parameters only when they are not operating normally.
5. Histogram, box or Pareto approaches can be used to group the values of parameters together.
6. Snap shots of the parameters at important times such as cinch, can be used to identify if the operating sequence of the coiler is correct.

2 Conclusion

The technique of LCS has the potential to become a very useful tool in the processing of data to provide information to aid industry. The desired information gained in this project is the link between input parameters and output actions. The LCS provides this in the form of 'if then' rules that are relatively transparent and connected to an autonomous strength based on correctness. Work is needed to supply the correct data to the LCS so that the useful rules can be mined in a time acceptable to industry. The basis of LCS for industrial use instead of conventional scientific studies requires more development before they can be accepted in widespread industrial use.

3 Acknowledgements

The authors would like to thank Dr B J Hewitt, Director Technical, British Steel Strip Products and Mr E F Walker, Manager, Technical Co-ordination, Welsh Technology Centre for permission to publish this paper. This work was carried out at British Steel Strip Products, Welsh Technology Centre as part of the research program of the Engineering Doctorate Centre, Wales under the auspices of the EPSRC.

References

[1] A.H. Gilbert et al. Adaptive learning of process control and profit optimization using a classifier system. *Evolutionary Computation*, 3(2):177–198, 1995.

[2] L.B. Booker et al. Classifier systems and genetic algorithms. *Artificial Intelligence*, pages 235–282, 1989.

[3] D.E. Goldberg. *Genetic Algorthims in Search Optimization and Machine Learning.* Addison Wesley, 1989.

[4] J.J. Grefenstette. The evolution of strategies for multi-agent enviroments. *Adaptive Behaviour*, 1(1):65–90, 1992.

[5] J.H. Holland. *Adaptation in Natural and Artificial.* University of Michigan Press, 1975 and 1992.

[6] A.J. Keane. *Genetic Algorthims in Design.* IEE Press, 1996.

[7] S.W. Wilson. ZCS: A zeroth level classifier system. *Evolutionary Computation*, 2(1):1–18, 1994.

Multi-Agent Classifier Systems and the Iterated Prisoner's Dilemma

K. Chalk and G. D. Smith
School of Information Systems, University of East Anglia,
Norwich, UK
Email: {kwc,gds}@sys.uea.ac.uk

Abstract

This paper describes experiments using multiple classifier system (CS) agents to play the iterated prisoner's dilemma (IPD) under various conditions. Our main interest is in how, and under what circumstances, co-operation is most likely to emerge through competition between these agents. Experiments are conducted with agents playing fixed strategies and other agents individually and in tournaments, with differing CS parameters. Performance improves when reward is stored and averaged over longer periods, and when a genetic algorithm (GA) is used more frequently. Increasing the memory of the system improves performance to a point, but long memories proved difficult to reinforce fully and performed less well.

1 Introduction

The emphasis in machine learning techniques has been on single systems learning particular tasks. Whilst there is still much work to be done on understanding and applying these systems, attention increasingly focuses on the effect and utility of interactions between multiple instances of such systems. In a set of such learning systems, each system is referred to as an *agent*. There are broadly three types of agent interaction.

- Coordination of team effort towards some shared goal, where agents explicitly co-operate to perform a certain task.
- Competition in zero sum games, where agents are in selfish competition.
- Co-operation through competition, in non-zero sum games such as the IPD.

One type of agent is the *classifier system*, which we shall use to investigate the emergence of co-operation in the IPD.

Table 1: Payoff Matrix for the IPD. C = Co-operate, D = Defect. A's payoffs are on the left of each table entry, and B's are on the right.

		Player A	
		C	D
Player B	C	30, 30	0, 50
	D	50, 0	10, 10

2 Classifier Systems

Classifier systems (CSs) are adaptive, rule-based, systems that use evolutionary computing techniques to perform reinforcement learning. Rules make a bid, proportional to their strengths, to respond to messages received by the classifier system, and the strengths of these rules are altered depending on the success of their actions. A GA is used periodically to create new rules using strong rules as parents, and these new rules replace weaker rules in the system.

3 The Iterated Prisoner's Dilemma

The prisoner's dilemma is a two player game where each player chooses one of two actions, co-operation or defection, with no knowledge of the other players choice. Reward is then given to the players depending on the combination of the moves chosen, as shown in Table 1. The dilemma lies in the fact that the best result for an individual is to defect when the other player co-operates, but if both players defect the results are poor. The Nash equilibrium point [17] is mutual defection. Mutual co-operation is the best result in terms of total payoff received by both players, but requires an element of trust. The IPD consists of repeated iterations of the dilemma, providing a history of interactions from which players may be able to learn about each other's strategies.

There is a limit to what can be learnt about the dilemma through theory; part of the nature of the problem is that it shows the inadequacy of traditional game theory to describe this type of problem [18].

Computer-based research into the IPD has mainly involved the use of finite state automata (FSA) (also known as finite state machines (FSMs)) [4, 5, 8, 10, 11, 15, 16, 19, 20], usually combined with some evolutionary strategy to improve performance. Other evolutionary work on the IPD problem has also been carried out [3, 6, 14, 21].

Previous research has concentrated on playing the IPD with agents of fixed strategies, with these strategies remaining fixed during each round of play and changing between rounds. This allows the best fixed strategy, that of mutual co-operation, to be found. The conditions conducive to this strategy emerging can then be examined.

Classifier systems allow agents to play with non-fixed strategies, allowing them to alter their strategy as they play, on two levels. Firstly, the classifier system acts as a large, stochastic FSA, which adjusts the probability of each matching rule being used depending upon its performance, round for round. Secondly, the GA is periodically invoked, to alter and improve the internal rule structure of the classifier system, changing the range of possible strategies available to the agent. Related research can be found in [7] (reduced classifier systems) and [21] (Q-learners and neural networks).

4 LEGION

Our experiments are conducted under our LEGION (Learning by Evolving Genetic Independent Operatives in a Network) system. This system enables us to define agents as individual processes, running on the same or separate computers, interacting with a separate 'environment' process which resolves the outcome of games and gives the agents their rewards. The system allows for many instances of the same problem to be run at once, and averages their results, and enables many sets of such experiments to be run, varying the parameters each time.

5 Experiments

We use the profit sharing plan (PSP), an epochal reward system [9, 12], for distributing reward. This stores up the total reward over a given number of iterations, records the rules which have been used during this period, and then evenly distributes the collected reward between the rules active during this period. A PSP period of 1 is equivalent to the bucket brigade algorithm [13], because reward is immediate.

We study the effect of varying the length of the PSP and varying the length of game memory that each classifier system has. Experiments with a payoff matrix which is not a prisoner's dilemma show a deterioration in performance as the PSP period is increased.

Classifier systems have 50 rules, use a GA with 0.02% probability, a mutation rate of 1%, and generate 20% new rules through multi-point crossover. The payoff matrix used is that shown in Table 1. Experiments take place over 300,000 iterations, in line with [21]. PSPs of length 1, 2, 4, 8, and 16 and memories with game lengths of 1, 2 and 3 are used. Results are averaged over 10 runs and all iterations. The general CS parameters given may be varied where stated in individual experiments.

5.1 Tit-For-Tat

Tit-For-Tat (TFT) is generally regarded as being the best strategy to use when playing the IPD [1, 2]. TFT simply co-operates on the first move, and afterward does what its opponent did last time. Thus, two TFT strategies playing each other would always mutually co-operate. Q-learner agents are able to learn total mutual co-operation against a TFT strategy [21].

A CS agent using PSP of length 1 should do poorly in TFT experiments due to the fact that if two rules have identical conditions but different actions, the rule with the action of 'defect' will always receive a higher payoff, become stronger and be more likely to be chosen again. PSPs negate this effect. The results can be seen in Table 2. The maximum obtainable average reward is 30, for continual mutual co-operation.

This shows a general overall increase in average

Table 2: Average Agent Payoffs Against A Tit-For-Tat Opponent Over 300,000 Iterations. The Maximum Average is 30.

	Length of Game Memory Used by Classifier Systems		
	1 Game Memory	2 Game Memory	3 Game Memory
PSP 1	15.17	16.47	15.40
PSP 2	20.54	22.67	24.02
PSP 4	21.45	23.76	21.80
PSP 8	21.48	24.35	23.54
PSP 16	21.50	27.07	24.90

Table 3: Agent vs. Agent Average Payoffs over 300,000 Iterations. The Maximum Average is 30.

	Length of Game Memory Used by Classifier Systems		
	1 Game Memory	2 Game Memory	3 Game Memory
PSP 1	15.16	15.56	16.08
PSP 2	19.02	18.17	16.32
PSP 4	20.60	19.17	16.41
PSP 8	21.48	21.42	18.18
PSP 16	22.29	23.42	17.81

reward with an increase of PSP length. Increasing the game memory length from 1 to 2 also give improvement in performance. Increasing it from 1 to 3 gives less of an improvement, because the length of time needed for a rule set to reach its full potential is proportional to the length of the memory used. A perfect score of 30 is unlikely to be achieved due to the constant exploration of the problem space by the GA.

The best rate of mutual co-operation was 67.6%, at PSP length 16 and game memory 2.

5.2 Agents Playing Agents with the Same Parameters

The general pattern of performance improvement with an increased PSP length as shown in Table 3 is as it was for the TFT experiments. The average payoffs for 2 and 3 game memories are generally lower, reflecting the increased difficulty in learning in a non-stationary environment, and the increased length of time needed to produce stable result. There is little or no improvement on payoffs over the duration of the experiment.

These experiments show the difficulty of playing an opponent with a changeable strategy, and that changes in the parameters of the experiment can lead to improved performance. Here the best level of mutual co-operation was 35.4%, again at PSP 16 in the 2 game memory experiment.

5.3 Agents Playing Agents with Different Parameters

In this set of experiments, the PSP period was fixed at 16 and game memory was fixed at 3 games. The probability of use of the GA was the factor that was altered in these experiments. Agents using GA probabilities of 0.02%, 0.1%, 1% and 5% were all tried against each other and themselves. In every contest where the GA probabilities were different, the agent with the more frequent GA probability did better than its opponent.

The smaller the GA probability was for an agent, the better it and its opponent did. 5% was the most successful frequency used, followed by the other frequencies, in order. We suggest that this is because the faster rate of change for the smaller frequencies allowed them to exploit their slower opponents and also allowed them to reach their level quickly. The results from this experiment did not compare well with the results from the previous experiments, with a best mutual co-operation rate of 24%.

5.4 Tournaments

A round robin tournament with 6 players was run, allowing the rule set of the agent to develop through interaction with several opponents. Five of the six players in the tournament were agents, fixed at PSP 4 and a game memory of 2, with different GA frequencies and number of rules in the CS. The other player used was a TFT strategy player. Each player played its 5 opponents for 60,000 iterations each, giving each player 300,000 iterations of play in total.

The TFT player came 4th in the tournament; TFT had done the best in Axelrod's tournaments [1, 2]. The player with the best average payoff was an agent with 20 rules and a GA frequency of 0.1%. However, another agent with exactly the same parameters came 5th, leading us to conclude that, due to the adaptive nature of the classifier systems, it is important which order the opponents are played in. For example, no agent lost its next game after playing the TFT strategy.

In an extended tournament with 300,000 iterations between each pair of agents, every agent managed to average 100% mutual co-operation playing against the TFT player for the last 1,000 iterations of the experiments, reinforcing the idea that with extra time, agents can improve their performance. This is consistent with [10], where FSA agents were allowed to adapt the length of their IPD encounters, resulting in a strong tendency toward the longest allowable encounters.

6 Conclusions

The IPD represents a difficult class of learning problem for CS agents. They can learn to do fairly well, even when playing other adaptive agents, but suboptimally, due to continual exploration by the GA, the nature of the problem and the way CSs work. The lack of convergence in such learning systems can also be seen in [7, 21], which contain comparative conversion results. The GA is drawn towards several local optima, and a multi-modal GA may be better suited to reinforcing such learning tasks.

Much depends on the parameters used. Performance was enhanced by increasing the period of the PSP, and by using a 2 game memory instead of 1. The 3 game memory experiments did not improve results, probably due to the increased length of time needed for them to reach their optimum performance level. Better performance was also observed when the GA probability was increased.

In a changing environment, systems with long memories to reinforce may be unable to react quickly to sudden changes. A similar effect is shown for the number of rules in a CS [7]. A balance needs to be reached between effective and efficient learning.

References

[1] R. Axelrod. Effective choice in the prisoner's dilemma. *Journal of Conflict Resolution*, 24:3–25, 1980.

[2] R. Axelrod. More effective choice in the prisoner's dilemma. *Journal of Conflict Resolution*, 24:379–403, 1980.

[3] R Axelrod. The evolution of strategies in the iterated prisoner's dilemma. In L. Davis, editor, *Genetic Algorithms and Simulated Annealing*, pages 32–41. Morgan Kaufmann, 1987.

[4] J. Bendor, R. M. Kramer, and S. Stout. When in doubt: Cooperation in a noisy prisoner's dilema. *Journal of Conflict Resolution*, 35(4), 1991.

[5] K. G. Binmore and L. Samuelson. Evolutionary stability in repeated games played by finite automata. *Journal of Economic Theory*, 57, 1992.

[6] A. Carbonaro, G. Casadei, and A. Palareti. Genetic algorithms and classifier systems in simulating co-operative behaviour. In R.F. Albrecht and C. R. Reeves, C.R. Steele, editors, *Artificial Neural Networks and Genetic Algorithms*, pages 479–483. Springer-Verlag, Wien New York, 1993.

[7] P. H. Crowley. Evolving cooperation: Strategies as hierachies of rules. *BioSystems*, 37:67–80, 1996.

[8] G. Ellison. Learning, local interaction, and coordination. *Econometra*, 61(5), 1993.

[9] A. Fairley and D. F. Yates. Improving simple classifier systems to alleviate the problems of duplication, subsumsion and equivalence of rules. In R.F. Albrecht, C.R Reeves, and N. C. Steele, editors, *Artificial Neural Nets and Genetic Algorithms*. Springer/Verlag, 1993.

[10] D. B. Fogel. On the relationship between the duration of an encounter and the evolution of cooperation in the iterated prisoner's dilemma. *Evolutionary Computation*, 3(3):349–363, 1996.

[11] J. R. Hoffmann and N. C. Waring. The localisation of learning and interaction in the repeated prisoner's dilemma. *Draft Copy, University of East Anglia*, 1996.

[12] J. H. Holland. Processing and processors for scemata. In Jacks E. L., editor, *Associative Information Processing*, pages 127–146. American Elsevier, New York, 1971.

[13] J.H. Holland. Properties of the bucket brigade algorithm. In Grefenstette J.J., editor, *Proceedings of the First International Conference on Genetic Algorithms and Applications*. Morgan Kaufmann, 1985.

[14] J. H. Holland. The effect of labels (tags) on social interactions. *Santa Fe Institute Discusion Paper 93-10-064*, 1993.

[15] O. Kirchamp. *Spatial Evolution of Automata in the Prisoner's Dilemma*. PhD thesis, 1995.

[16] Y. Mor, C. V. Goldman, and J. S. Rosenschein. Learn your opponents strategy (in polynomial time)! *In: Adaptation and Learning in Multi-Agent Systems*, (G. Weiss and S. Sen), pages 164–176, 1996.

[17] J. F. Nash. Non-cooperative games. *Annals of Mathematics*, 54:286–295, 1951.

[18] W. Poundstone. *Prisoner's Dilemma*. Oxford University Press, 1993.

[19] B. R. Routledge. *Co-evolution and Spatial Interaction*. PhD thesis, 1993.

[20] A. Rubinstein. Finite automata in the repeated prisoner's dilemma. *Journal of Economic Theory*, 39, 1986.

[21] T. W. Sandholm and R. H. Crites. On multi-agent q-learning in a semi-competetive domain. In G. Weiss and S. Sen, editors, *Adaptation and Learning in Multi-Agent Systems*, pages 191–205. 1996.

Complexity Cost and Two Types of Noise in the Repeated Prisoner's Dilemma

R. Hoffmann[1] and N. C. Waring[2]
[1] School of Management and Finance, University of Nottingham, UK.
[2] School of Information Systems, University of East Anglia, UK.
Email: Robert.Hoffmann@nottingham.ac.uk, ncw@sys.uea.ac.uk

Abstract

This study seeks to understand the effect of complexity cost and two types of transmission noise on equilibrium selection in populations of finite automata playing the repeated prisoner's dilemma. Results indicate that noise and complexity cost have a harmful effect on the types of conditionally co-operative strategies essential for the emergence of co-operative behaviour in the population. In contrast, the unconditionally defecting strategy responsible for the dominance of mutual defection is relatively unharmed under these conditions.

1 Introduction

The conditions that allow the evolution of co-operation among interacting autonomous agents have been studied extensively using the repeated prisoner's dilemma (RPD). In the game, two agents have the choice between co-operating (c) and defecting (d). The payoffs associated with the four possible outcomes obey the conditions $T > R > P > S$ and $2R > T + S$. In the finite game, the best course of action for each player irrespective of the opponent's moves is to defect in every round (see Table 1). The paradoxical character of the game arises since the corresponding equilibrium (dd) generates lower payoffs than the co-operative outcome (cc). In this notation of outcomes, the first letter represents player one's move, and the second player two's move. The game encapsulates a particular dilemma common to many important types of interaction in which the agents' individual self interests may endanger their own as well as the common welfare of the group as a whole.

Despite the compelling logic of defecting, recent computer simulations have shown conditions under which co-operation may evolve in populations of artificial adaptive agents which are matched for pairwise RPD-contests (see Axelrod [2], Miller [6], Hoffmann and Waring [5], Hoffmann [4]). In these simulations, finite automata are used to represent the game-playing strategies of agents with limited cognitive capacity. Evolutionary search techniques such as the genetic algorithm (GA) are introduced to model the agents' myopic, co-adaptive learning efforts. The automata are attributed fitnesses on the basis of their performance in interacting with other machines which are used to generate the differential propagation of alternative machines according to evolutionary selection criteria.

Hoffmann and Waring [5] as well as Hoffmann [4] report simulations of learning populations which consist of automata that generate moves on the basis of the last round of the game. In the simulations, the populations typically converge on and shift between two attractors of universal mutual defection and universal mutual co-operation respectively. Mutual defection can become established through the spread of the always defect (AD) strategy in the population on the basis of the high payoffs the exploitation of co-operators affords. Conversely, co-operation in such populations can evolve to the extent that conditionally co-operative behaviour spreads among the agents.

The results we have reported in Hoffmann and Waring [5] and Hoffmann [4] indicate the existence of two behavioural attractors in our populations of finite-state automata playing the RPD. In addition, the results show that continuous shifts between these two attractors occur regularly. These findings indicate not merely that multiple equilibria exist in the population we study, but that the individual alternatives are unstable. In the current study, we further our previous work by considering the effect of a number of alternative conditions on the selection between the two attractors of mutual defection and

Table 1: The Prisoner's Dilemma.

		Player 2	
		Cooperate	Defect
Player 1	Cooperate	R, R	S, T
	Defect	T, S	P, P

co-operation respectively. To begin with, we investigate populations with noisy strategy transmission. Players are assumed to make mistakes when playing their moves (misimplementation noise) or incorrectly observe the choices of their opponents (misperception noise). There are reasons why the addition of noise may affect the equilibrium behaviour of the population. Noise affects strategies with conditional behavioural elements more than those that generate moves independent of the opponent's behaviour. As a result, noise may lessen the chance of the evolution of co-operation since it affects primarily the strategies that are able to promote it.

The notion of complexity cost involves taxing individual players on the basis of the complexity of the strategies they employ to play the game. This notion reflects the mental effort agents expend in making decisions. The addition of complexity cost generates additional payoffs on the basis of a machine's complexity such that simpler machines are taxed less. In the populations we have studied previously, the strategies capable of spreading to fixation exhibit different complexities. As a result, extra negative payoffs determined by strategy complexity may effect the selection between them.

2 The Model

Play in the model proceeds amongst a population of finite automata. Round-robin tournaments of the RPD are played over a number generations. Following each tournament, the population is subjected to a learning phase where a proportion of the populations' strategies are updated.

Following recent research into bounded rationality (Aumann [1], Binmore and Samuelson [3], Rubinstein [7]), our agents select Moore machines to play the RPD on their behalf. The use of Moore machines enables the simulation of behaviour under bounded rationality where limits are placed on agents' cognitive ability. Strategies executed as Moore machines act as rules of thumb for playing the RPD. A Moore machine is a hypothetical computing device which generates an unconditional move in the first round of a game. Subsequently, it plays pure strategy responses to the outcome of the previous round. Moore machines can generate 26 distinct strategies for the RPD including always co-operate (AC) and AD.

Learning in the model is implemented using a modification of the canonical GA. Our version differs from the standard GA in two ways. First, we use a mating schedule developed in a previous study

1st	dd	dc	cd	cc
1	0	1	0	1

Figure 1: Encoding of a two-state Moore machine.

(Hoffmann and Waring [5]) which resembles a cellular GA. Here, of the two offspring generated per mating, only one is preserved. This is used to replace the first parent. Selection of the offspring to preserve is made arbitrarily. The first parent is then prevented from participating in further matings. Second, the fitness function here is a product of the interacting strategies employed by the population itself rather than some fixed external function as typically used. Fitness is calculated as the average of an agent's payoff gained over one generation of play against all other agents. Other aspects of the GA resemble those of a typical implementation. Single-point crossover is used with a probability of 0.6 of being used per mating. Mutation rate is set to 0.001. At each generation 50% of the individuals are mated.

Each member of the population encodes an instantiation of a two-state Moore machine. Five genes are required to represent a single encoding. The left-most is used to generate the automaton's unconditional first move of the game. The allele '0' encodes the move of defection, a '1' of co-operation. The remaining loci code responses to the four game outcomes *dd*, *dc*, *cd*, *cc* using the same encoding method as the first. At round $r + 1$, the outcome of round r is used to determine an automaton's response. Figure 1 demonstrates the format used. The strategy encoded in the diagram is the strategy Tit-For-Tat (TFT), so named because it repeats an opponent's previous move after initially co-operating.

The two forms of transmission noise are implemented as follows. Misimplementation noise is applied with some probability after the selection of an agent's current move. The value encoding the move is flipped so that '0' (defect) becomes '1' (co-operate), and vice versa. The modified value is recorded in the history of the game. Misperception noise is implemented probabilistically where on selection the value encoding the opponent's current move (using the coding described) is flipped before being recorded. Complexity cost is implemented deterministically by adding some amount to a strategy's payoff if that strategy is either AD or AC. In the current agent representation, all agents except for AC and AD have the same complexity in terms of recollecting the previous round. AD and AC, however, require no recollection. Hence, the more complex strategies are penalised by not receiving the bonus.

A simulation of the model proceeds as follows. First the population is seeded with randomly generated Moore machines which represent the strategies of the players in the population. Agents then interact in a round-robin tournament such that each agent is systematically matched to play r rounds of RPD with every other agent in the population. The GA is then invoked, and updates the population using the method described. Subsequent generations proceed in an identical manner.

The following simulation parameters were used. Populations were of size 31, and played 200 rounds of the RPD for a total of 12,000 generations. The payoffs used were $T = 1$, $R = 0.6$, $P = 0.2$, $S = 0$. Three simulations were conducted: complexity cost = 0.005, misperception noise = 5%, misimplementation noise = 5%. The results of the experiments we report are typical of simulations with alternative values of complexity cost and noise that we have observed.

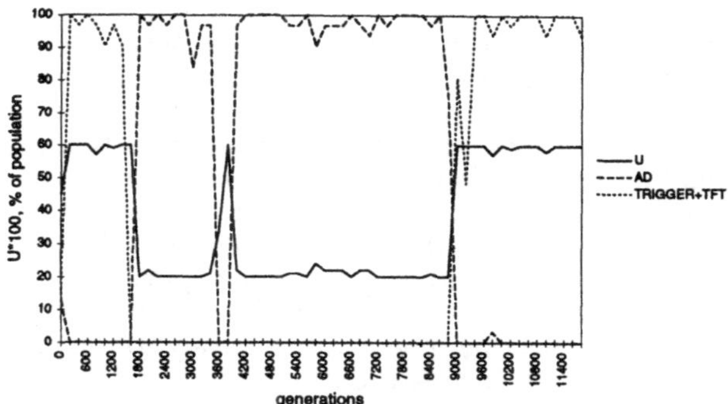

Figure 2: Benchmark simulation without transmission noise or complexity cost.

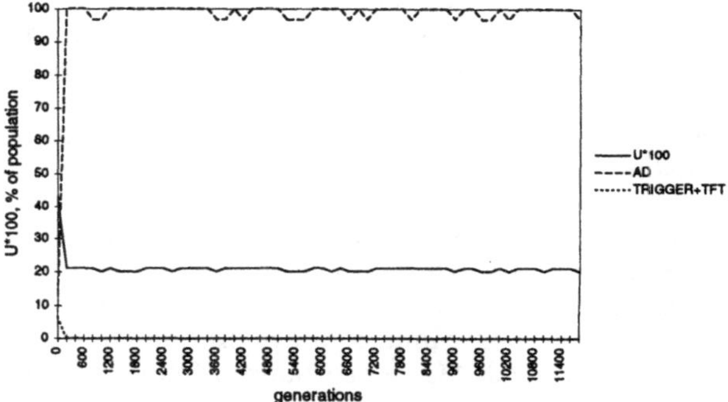

Figure 3: Simulation of complexity cost = 0.005. The relative frequencies of AD, and combined trigger and TFT are presented with U.

3 The Results

First, we simulated the current model without complexity cost or noise to provide a benchmark for the comparison of simulations under alternative conditions. The results of the simulation are presented in Figure 2. The figure depicts the spread of AD as well as of conditional co-operation as the combined representation of trigger and TFT in the population. Trigger is the strategy that co-operates until its opponent defects; thereafter it defects unconditionally. In addition, the population's average payoff U multiplied by a factor of 100 is given. A level of 60 for this latter variable indicates the convergence of the population on mutual defection, while a value of 20 implies a defecting population. The figure illustrates the periodic attractor shifts between co-operation and defection previously reported by Hoffmann and Waring [5] as well as Hoffmann [4].

3.1 Complexity Cost

The first issue concerns the effect of introducing complexity cost. Figure 3 shows the history of a simulation with a complexity cost value of 0.005. The figure indicates that complexity cost favours the evolution of defecting strategies. In the figure, AD spreads rapidly in the population to eventual fixation. Subsequently, AD remains stable but experiences the temporary appearance of other strategies caused by the GA. The stability of the evolution of defection is reflected in the plot of the population's average payoff U contained in the graph. The path of the payoff declines initially to reach a relatively stable level slightly above 0.2. This value comprises the payoff P for mutual defection with an added bonus attributed to both AD and AC due to their greater simplicity.

This result indicates that the addition of complexity cost narrows the range of possible attractors in the system to one, mutual defection. The mutual co-operative equilibrium is not visible in the simulation. The reason for this result is the following. Our previous work has shown that only conditional co-operators such as trigger and TFT as well as AD are sufficiently fit to spread to fixation in the population. The addition of complexity cost favours the selection of simpler strategies, AD and AC in the current population. Out of the two, AD alone can support the convergence of the population. As a result, the complexity cost benefits AD and the evolution of defection in general relatively to the more complex conditional strategies that can support the evolution of co-operation.

3.2 Misperception Noise

We next examine whether the addition of noise affects equilibrium selection in the current model. In this context, we investigate both misperception and misimplementation noise. Figure 4 displays the history of a simulation with 5% misperception noise. The graph indicates that the population again converges on AD only. In Figure 4 a single, unsuccessful shift from the attractor of mutual defection is in evidence. This is reflected in a single spike in the otherwise stable path of U. The population average payoff settles at a level of P. The result confirms the fact that misperception noise has no effect on the behaviour of unconditional strategies such as AD and AC. Since players following these strategies play their moves regardless of the opponents' choices, their misperception of these choices does not alter their own behaviour. Any variations in the path are caused by the intermittent arrival of mutant strategies.

Conditional strategies such as trigger and TFT may exhibit pronounced changes in behaviour as a result of the noise. For example, consider a game between two trigger-players. A single mistake by either may usher in a series of mutual defections until both players commit an error in the same round. This phenomenon undermines the chances of co-operative behaviour emerging. Hence, the attractor of mutual co-operation loses its stability under these conditions.

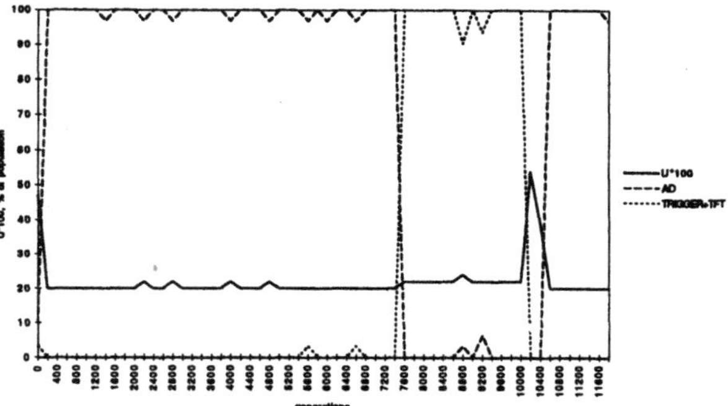

Figure 4: Simulation of misperception noise = 5%. The relative frequencies of AD, and combined trigger and TFT are presented with U.

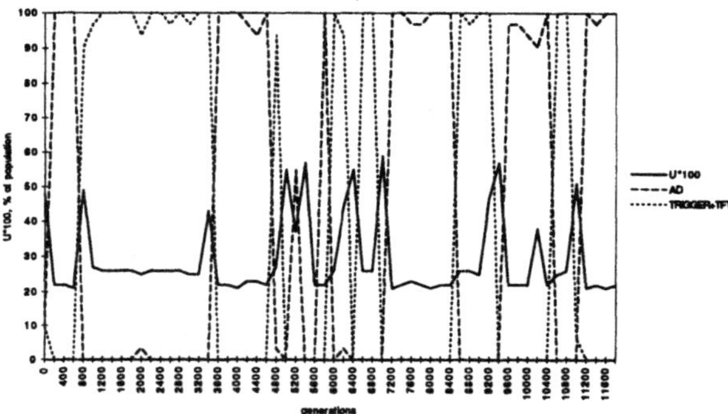

Figure 5: Simulation with misimplementation noise = 5%. The relative frequencies of AD, and combined trigger and TFT are presented with U.

3.3 Misimplementation Noise

The last issue under investigation concerns the effects of misimplementation noise. In contrast to misperception noise, the erroneous execution of moves can change the behaviour even of unconditional strategies such as AD and AC. The results of the simulation of misimplementation noise of 5% is depicted in Figure 5. The figure demonstrates that the attractor of mutual co-operation is weakened by the noise. The frequent occurrence of spikes in the population average payoff U indicates that the evolution of co-operation makes some progress in the population. However, given that the co-operative attractor is not viable under these conditions, the population quickly returns to convergence on AD. Such temporary shifts from the attractor of mutual defection indicate that to a lesser degree it too is unstable.

The magnitude of the payoffs associated with the individual attractors are influenced by the presence of misimplementation noise. For the *dd*-attractor, noise increases the magnitude of U since $T + S > 2P$. For the attractor of mutual co-operation, noise lowers U since $T + S < 2R$. Again, the conditional co-operators required for the evolution of co-operation are more affected by the noise than is AD, which is able to support the evolution of defection.

4 Conclusions

The work presented here has shed light on the effect noise and complexity cost can have on equilibrium selection in populations of learning players. The addition of noise and complexity cost to the populations studied by Hoffmann and Waring [5] as well as Hoffmann [4] reduces the number of possible but periodic attractors to a global attractor. Both noise and complexity cost have a relatively strong and harmful effect on the types of conditional co-operators that are required to establish mutually co-operative behaviour in a population. Conversely, AD, the strategy able to sustain the evolution of defection, remains relatively unaffected by both.

References

[1] R.J. Aumann. Survey of repeated games. In *Essays in Game Theory and Mathematical Economics*. Bibliographgisches Institut Mannheim, 1981.

[2] R. Axelrod. The evolution of strategies in the iterated prisoner's dilemma. In L. Davis, editor, *Genetic Algorithms and Simulated Annealing*. Pitman, 1987.

[3] K.G. Binmore and L. Samuelson. Evolutionary stability in repeated games played by finite automata. *Journal of Economic Theory*, 57, 1992.

[4] J.R. Hoffmann. The ecology of cooperation. Discussion Paper, School of Management and Finance, University of Nottingham, 1996.

[5] J.R. Hoffmann and N.C. Waring. The simulation of localised interaction and learning in artificial adaptive agents. In T.C. Fogarty, editor, *Evolutionary Computing*. Springer Verlag, 1996.

[6] J.H. Miller. The coevolution of automata in the repeated prisoner's dilemma. Working Paper 89-003, Santa Fe Institute, 1989.

[7] A. Rubinstein. Finite automata in the repeated prisoner's dilemma. *Journal of Economic Theory*, 39, 1986.

ICANNGA 97
International Conference on Artificial Neural Networks and Genetic Algorithms

Plenary Lectures

There were three plenary lectures, one to start each day of the Conference. The first was based in the field of artificial neural networks and was addressing medium term problems in autonomous robotics. The second plenary was concerned with the key issue currently besetting the GA community (and the heuristic search community at large), namely the relationship between representation and search. The final plenary, addressing issues that may arise over a much longer term, was appropriately devoted to a field that harnesses the state of the art developments in both ANNs and evolutionary computation. Abstracts follow.

Neural Networks Make Robots Intelligent: March of the Machines
Kevin Warwick
Reading University, U.K.

Artificial neural networks provide robots with a form of machine intelligence. But what can such robots achieve now, and where could this lead us? In particular, for autonomous robots, intelligence is extremely useful and can allow a range of behaviours, dependent on the robots' interaction with the outside world.

In this session, key ideas on the use of neural networks for autonomous robots will be presented. A number of intelligent autonomous robots will be demonstrated during the session, and their mode of operation will be discussed. Attendees will be able to see robots which can walk, communicate with each other, develop their own behaviours, learn to move, select a leader and teach other robots.

In his new book, 'March of the Machines', Kevin Warwick asserts that "In our lifetime there will be robots cleverer than mankind". In this session, attendees will see why he comes to such a frightening conclusion.

Representation, Search and Learning
Darrell Whitley
Colorado State University, USA

Representation plays a critical role in heuristic search and machine learning, yet constructing a good representation is still largely an art rather than a science.

The 'No Free Lunch' theorem shows that all search algorithms (and learning algorithms) are equivalent when performance is averaged over all possible discrete functions (assuming no resampling). Furthermore, all representations are the same in expectation when performance is averaged over all functions using a fixed search strategy, or over all possible search strategies when applied to a single function. However, it can be shown that relatively simple and useful ways of partitioning the space of all functions exists such that one representation is better than another over the resulting subsets of functions.

In effect, there is a free lunch.

These theoretical results shed new light on practical representation issues that have been debated for over 10 years in the genetic algorithm community.

ATR's Billion Neuron Artificial Brain Project
Hugo de Garis
ATR Human Information Processing Research Labs, Japan

I head the Brain Builder Group at ATR, a research lab in Kyoto, Japan. I expect, with the help of my group and international collaborators (from 6 countries), to build an artificial brain with a billion artificial neurons, with evolved cellular automata (CA) based neural circuits, by the year 2001. We already have 10 million neurons, and expect to achieve our target on time. By evolving neural net modules with roughly 100 neurons each, at electronic speeds (e.g. in less than a second) in special FPGA (XC6264 chip) based evolvable hardware (called a CAM-Brain Machine (CBM)), we will be able to download these CA based neural circuit modules (each with its own evolved user specified function) into user specified brain architectures embedded in a RAM based space of trillions of CA cells. The same CBM (programmable) hardware then updates the whole RAM CA space frequently enough (e.g. 30 times a second, i.e. at over 100 billion CA cells a second) for real-time operation.

By the end of 1997, our Brain Builder Group (BBG) expects to see the completion of 3 parallel tasks, namely the design and fabrication of the CBM (which was started in January 1997), the construction of a robot kitten called 'ROBOKONEKO' (in Japanese), and the creation of a 10,000 module artificial brain architecture to control the robot kitten's many behaviors. The modules for the artificial brain will be evolved in 1998 with the CBM, and put into the (life sized) kitten robot.

After 1998, the BBG hopes to work on more ambitious projects, such as household cleaner robots, and with substantially more brain builder researchers on the team. (One of my goals for Japan is to see the country create a 'J-Brain Project', which would aim to build a 10,000,000 module artificial brain with 2000 human 'EEs' (evolutionary engineers) over the time period 2000–2005). 20 years from now, brain-like computers should generate a trillion dollar industry.

Workshop Summary

UEA, Norwich, England, April 1, 1997

The Conference was preceded by a one-day Workshop in which the morning was devoted to the theory and applications of artificial neural networks and the afternoon to an introduction to heuristic search, including genetic algorithms, followed by a detailed look at some case studies. The workshop sessions and associated abstracts are as follows:

An Introduction to Artificial Neural Networks

Professor Nigel Steele,
Coventry University, UK

This session, as the opening session of the workshop, is aimed at those with limited knowledge of the field of artificial neural networks and is not primarily intended for experienced practitioners.

An introduction to the field is given, covering the basic concepts of network structure and network learning. Attention is focussed initially on the multi-layer perceptron network, with learning by error backpropagation. Ideas on improving learning performance are discussed and some practical hands-on experience incorporated.

Subsequently, the radial basis function network is introduced, in both its standard and adaptive forms. The close relationship of this type of network and fuzzy inference systems will also be discussed. Again, some practical, hands-on experience will be available.

The session will close with a discussion of an application in the field of robot navigation.

Hardware Neural Networks — Design and Application

Professor Kevin Warwick,
University of Reading, UK

Certain types of artificial neural networks are well suited for implementation in hardware. This session introduces the basic principles of such networks and describes their method of operation. Although a range of networks is considered, Kohonen and Hopfield type networks are looked at in detail and digital networks such as the n-tuple type are also presented.

Applications of the networks described are quite widespread, hence during this session, a number of actual applications will be described. Emphasis will be placed on the principles of the application, reasons for employing a neural network in each case and the positive and negative features which result.

Finally, we take a look at future implementation possibilities.

An Introduction to GAs and Other Search Paradigms

Dr Colin Reeves,
Coventry University, UK

Genetic algorithms (GAs) have become popular tools for solving difficult optimisation problems over the last decade. This part of the workshop will provide an introduction to the concept of GAs, using numerical examples. It will also explain some of our current understanding of how they work, and deal with some of the important issues involved in their implementation.

The workshop will also compare and contrast GAs with other recent general search paradigms such as simulated annealing, tabu search and perturbation methods. It will thus lead naturally into the final workshop session.

Case Studies in Genetic Algorithms and Other Search Paradigms

Mr Jason Mann,
Nortel, Harlow, UK

In this session, we look at key issues in the application of GAs and related heuristics, such as simulated annealing, tabu search and others. This is done through a number of case studies drawn from real-world applications that the UEA research group MAG has carried out. Of particular interest are representation issues and the associated choice of appropriate (genetic) operators.

These case studies, drawn particularly from applications in the telecommunications sector, are supported by laboratory sessions in which participants will have the opportunity to use some of the latest software toolkits supporting the respective search paradigms.

Subject Index

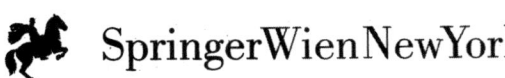

SpringerAdvances in Computing Science

Series

Advances in Computing Science

Springer-Verlag Wien New York presents the new book series "Advances in Computing Science". Its title has been chosen to emphasize its scope: "computing science" comprises all aspects of science and mathematics relating to the computing process, and has thus a much broader meaning than implied by the term "computer science". The series is expected to include contributions from a wide range of disciplines including, for example, numerical analysis, discrete mathematics and system theory, natural sciences, engineering, information science, electronics, and naturally, computer science itself.

Contributions in the form of monographs or collections of articles dealing with advances in any aspect of the computing process, or its applications are welcome. They must be concise, theoretically sound and written in English; they may have resulted, e.g., from research projects, advanced workshops and conferences. The publications (of the series) should address not only the specialist in a particular field but also a more general scientific audience. An International Advisory Board which will review papers before publication will guarantee the high standard of volumes published within this series.

Franc Solina, Walter G. Kropatsch, Reinhard Klette, Ruzena Bajcsy (eds.)

Advances in Computer Vision

1997. 96 figures. VIII, 266 pages.
Soft cover DM 69,–, öS 485,–, US $ 44.95
ISBN 3-211-83022-7

Computer vision solutions used to be very specific and difficult to adapt to different or even unforeseen situations. The current development is calling for simple to use yet robust applications that could be employed in various situations. This trend requires the reassessment of some theoretical issues in computer vision. A better general understanding of vision processes, new insights and better theories are needed. The papers selected from the conference staged in Dagstuhl in 1996 to gather scientists from the West and the former eastern-block countries address these goals and cover such fields as 2D images (scale space, morphology, segmentation, neural networks, Hough transform, texture, pyramids), recovery of 3-D structure (shape from shading, optical flow, 3-D object recognition) and how vision is integrated into a larger task-driven framework (hand-eye calibration, navigation, perception-action cycle).

Chris Brink, Wolfram Kahl, Gunther Schmidt (eds.)

Relational Methods in Computer Science

1997. 30 figures. XV, 272 pages.
Soft cover DM 69,–, öS 485,–, US $ 49.95
ISBN 3-211-82971-7

The calculus of relations turned into an important conceptual and methodological tool in computer science. The methods presented in this book include questions of relational databases, applications to program specification, resource-conscious linear logic, semantic and refinement consideration, nonclassical logics for reasoning about programs, tabular methods in software construction, algorithm development, linguistic problems, followed by a comprehensive bibliography. The reader gets an overview of the wide-ranging applicability of relational methods in computer science.

Sachsenplatz 4-6, P.O.Box 89, A-1201 Wien, Fax +43-1-330 24 26, e-mail: order@springer.at, Internet: http://www.springer.at
New York, NY 10010, 175 Fifth Avenue • D-14197 Berlin, Heidelberger Platz 3 • Tokyo 113, 3-13, Hongo 3-chome, Bunkyo-ku

Bob F. Caviness,
Jeremy R. Johnson (eds.)

Quantifier Elimination and Cylindrical Algebraic Decomposition

1998. 20 figures. XIX, 431 pages.
Soft cover DM 118,–, öS 826,–, US $ 79.95
ISBN 3-211-82794-3

George Collins' discovery of Cylindrical Algebraic Decomposition (CAD) as a method for Quantifier Elimination (QE) for the elementary theory of real closed fields brought a major breakthrough in automating mathematics with recent important applications in high-tech areas (e.g. robot motion), also stimulating fundamental research in computer algebra over the past three decades.
This volume is a state-of-the-art collection of important papers on CAD and QE and on the related area of algorithmic aspects of real geometry. It contains papers from a symposium held in Linz in 1993, reprints of seminal papers from the area including Tarski's landmark paper as well as a survey outlining the developments in CAD based QE that have taken place in the last twenty years.

Alfonso Miola,
Marco Temperini (eds.)

Advances in the Design of Symbolic Computation Systems

1997. 39 figures. X, 259 pages.
Soft cover DM 98,–, öS 682,–, US $ 79.95
ISBN 3-211-82844-3

New methodological aspects related to design and implementation of symbolic computation systems are considered in this volume aiming at integrating such aspects into a homogeneous software environment for scientific computation. The proposed methodology is based on a combination of different techniques: algebraic specification through modular approach and completion algorithms, approximated and exact algebraic computing methods, object-oriented programming paradigm, automated theorem proving through methods à la Hilbert and methods of natural deduction. In particular the proposed treatment of mathematical objects, via techniques for method abstraction, structures classification, and exact representation, the programming methodology which supports the design and implementation issues, and reasoning capabilities supported by the whole framework are described.

Norbert Kajler (ed.)

Human-Computer Interaction in Symbolic Computation

1998. 68 figures. Approx. 230 pages.
Soft cover DM 89,–, öS 625,–, US $ 59.95
ISBN 3-211-82843-5

There are many problems which current user interfaces either do not handle well or do not address at all. The contributions to this volume concentrate on three main areas: interactive books, computer-aided instruction, and visualization. They range from a description of a framework for authoring and browsing mathematical books and of a tool for the direct manipulation of equations and graphs to the presentation of new techniques, such as the use of chains of recurrences for expediting the visualization of mathematical functions.
Students, researchers, and developers involved in the design and implementation of scientific software will be able to draw upon the presented research material here to create ever-more powerful and user-friendly applications.

Springer-Verlag and the Environment

WE AT SPRINGER-VERLAG FIRMLY BELIEVE THAT AN international science publisher has a special obligation to the environment, and our corporate policies consistently reflect this conviction.

WE ALSO EXPECT OUR BUSINESS PARTNERS – PRINTERS, paper mills, packaging manufacturers, etc. – to commit themselves to using environmentally friendly materials and production processes.

THE PAPER IN THIS BOOK IS MADE FROM NO-CHLORINE pulp and is acid free, in conformance with international standards for paper permanency.

GPSR Compliance

The European Union's (EU) General Product Safety Regulation (GPSR) is a set of rules that requires consumer products to be safe and our obligations to ensure this.

If you have any concerns about our products, you can contact us on ProductSafety@springernature.com

In case Publisher is established outside the EU, the EU authorized representative is:

Springer Nature Customer Service Center GmbH
Europaplatz 3
69115 Heidelberg, Germany

Batch number: 09625682

Printed by Printforce, the Netherlands